Mobile Communications Systems Development

A Practical Introduction to System Understanding, Implementation, and Deployment

Rajib Taid

WILEY

This edition first published 2021

Registered Offices
John Wiley & Sons, Inc., 111 River Street, Hoboken, NJ 07030, USA
John Wiley & Sons Ltd, The Atrium, Southern Gate, Chichester, West Sussex, PO19 8SQ, UK

Editorial Office
The Atrium, Southern Gate, Chichester, West Sussex, PO19 8SQ, UK

For details of our global editorial offices, customer services, and more information about Wiley products, visit us at www.wiley.com.

Wiley also publishes its books in a variety of electronic formats and by print-on-demand. Some content that appears in standard print versions of this book may not be available in other formats.

Library of Congress Cataloging-in-Publication Data applied for

ISBN 9781119778684 (Hardback)

Cover Design: Wiley
Cover Image: © Metamorworks/Shutterstock, Cnythzl/iStock/Getty Images

Set in 9.5/12.5pt STIXTwoText by SPi Global, Pondicherry, India

10 9 8 7 6 5 4 3 2 1

Contents

About the Author

Rajib Taid graduated with a Bachelor of Computer Science and Engineering degree from Jorhat Engineering College, Assam, India. He began his career as an Engineer (Telemetry) at GAIL, a public-sector entity. Rajib worked there for four years in the area of design and development of application software systems for a gas pipeline. Then, he joined a Gurgaon (Haryana, India)-based communication software services major as well as product development MNC, Aricent Technologies (www.aricent.com), formerly known as Hughes Software Systems, where he started working as a software developer in the area of mobile communications. He worked there for 12 years, both as an individual contributor and in a lead role, primarily developing software in Radio Access Network, and Core Network domains. The author also visited Australia, the USA and the UAE for onsite assignments. He hails from the Jorhat district of Assam, India.

In 2013, Rajib left the mobile telecommunication software development domain and joined BCPL, a public-sector entity at Dibrugarh, close to his hometown in Assam. Currently, the author specializes in IT and enterprise business information systems management.

Rajib Taid is also a member of the Department Management Committee, as an industry expert, for the Department of Computer Science and Engineering in the Dibrugarh University Institute of Engineering and Technology, Assam, India.

Preface

Today, we are using at least one smartphone for our day-to-day voice, data communications, online gaming, and transaction services. To enable these services, we are also aware of the different mobile communications technologies that are available around us today, such as GSM, GPRS/EDGE, 3G, 4G, and 5G being the latest communication buzzword. If you are wondering and, sometimes, scouring the web regarding how a mobile communications system is designed, developed, tested, and deployed, then this book is for you! This is an introductory jump-start, foundational, and comprehensive book offering several key concepts encompassing the various practical aspects for the design and development of a mobile communications system and its various entities/elements based on the GSM, GPRS, UMTS (3G), LTE (4G), and 5G technologies. Note that this book is not about the developments of Mobile Application (apps) software that is intended and developed for a specific purpose/requirement.

The content of this book is specially tailored for the rookie computer science or electronics and communications engineering graduate engineer who has just passed out of college, or even for a lesser experienced person looking for an opportunity to work in the mobile telecommunications space through re-skilling. It starts with the various mobile communications network architectures, identification of a particular network element, the concerned 3GPP standard specifications, various protocols/stacks, as well as the development platforms such as UNIX, and multicore computing. In this book, the reader will also find the troubleshooting of any issue arising out of post-deployment and operation of a mobile communication network element. This book also introduces the "multicore processor" computing platform that is available around us and is the current buzzword in different areas of technologies, be it the desktop or mobile handset. Mobile telecommunications system development using an embedded system platform is also briefly covered.

A mobile communication network works and communicates based on the standard technical specifications related to a particular mobile communication technology such as the GSM, GPRS, UMTS, LTE, and 5G system. Also, mobile communication standard technical specifications are large in number and can be bewildering to a new learner. Reading and its implementation, through computer code, of the contents of a GSM/GPRS/UMTS/LTE/5G technical specification requires a substantial amount of effort, especially the Layer 1 and Layer 2 protocols. From a technical specification, one would come to know *what* to and *when* to transmit or receive information. But what is not available in the 3GPP technical specifications is the *how* to implement part as it is implementation dependent. This book was written keeping these facts in mind, so that students can learn the practical, real-world mobile telecommunications domain subject areas and equip themselves while in college, before starting a career in the relevant domains. To make the contents easier to understand, necessary figures, tables, and sample codes are provided to illustrate the underlying concepts. The illustrative figures and concepts are sometimes general in nature, i.e. applicable for GSM/GPRS or UMTS/LTE/5G system, or all of them, and sometimes a straight copy from the concerned 3GPP technical specification with due permissions.

This book is an overview and may not contain exhaustive descriptions or information on various individual components and protocols of a mobile communications system based on the GSM, GPRS, UMTS, LTE, and 5G system. The book attempts to provide the reader with an overall background of the various aspects of an end-to-end system development based on the available mobile communication technologies and systems. This book reflects the author's 12 years of experience with a full lifecycle of software research and development, deployment, testing, operation, and maintenance in the areas of mobile communication, Radio Access Network (RAN), and Core Network (CN) domain deployed across the available platforms, including satellite-based mobile communications systems.

Who should use this book?

Mobile Communications System Development: An Introduction to Practical Approach for Systems Understanding, Implementation, and Deployment is primarily for students who have just graduated in either computer science or electronics and communications discipline and is looking for an exciting career in the mobile communications domain. It is also appropriate for students currently studying in the above-mentioned disciplines and looking for project work assignments as a part of the academic curriculum in the mobile communication domain. An experienced person from another software domain can also go through this book for a career reboot into the mobile communication domain.

How to use this book?

Mobile communications systems protocol layers, their functions and procedures, and other related information, such as referring to figures, being presented may be brief in nature. For further details about the underlying protocols along with the materials being presented here, the concerned 3GPP technical specification(s) on its website (www.3gpp.org) [1] must be referred to while going through a chapter of this book. The concerned 3GPP technical specifications numbers are mentioned in the References section of the book. The reader is advised to refer to the mentioned 3GPP technical specification and the section number for complete information on the described protocol functions and procedures. Familiarity with the 3GPP website is also important as the reader will be required to visit it quite often to refer to its technical specifications.

Structure of this book

Overall, this book is divided into four parts, each containing several chapters. Each part begins with introductory objectives and also mentions the purposes of each chapter under it. Each chapter is followed by its summary. Also, the book starts with an introductory chapter that provides a brief description of the career opportunities offered by mobile communications systems and network ecosystems.

Part I Introduction
This part contains eight chapters containing the background and introductory aspects and areas of mobile communications systems and networks based on GSM, GPRS, UMTS, LTE, and 5G systems. The materials presented in this part are general in nature but applicable across the mobile communications systems and networks. Even if a reader is starting a career in the LTE or 5G system and network, as a developer or O&M person, one has to know the major key concepts from the legacy GSM/UMTS networks as well.

Part II Operation and Maintenance
This part contains three chapters covering various aspects and areas of the troubleshooting and resolution of mobile communications systems and network issues.

Part III Development of Mobile Communications Systems
This part contains four chapters covering various aspects and areas of the development of mobile communications systems protocol stack and layers based on the 3GPP standards and their technical specifications. This part also describes hardware platforms to be used for the development of mobile communications systems network elements.

Part IV 5G System and Network
This part contains seven chapters covering various aspects and areas of a 5G system and network based on its first Release 15 as standardized by the 3GPP. Also, an overview of the enhancements made into the existing features of the 3GPP Release 15 and the addition of new services or capabilities which have been added as part of the 3GPP Release 16 and Release 17 are covered in this part.

Dibrugarh, Assam, India *Rajib Taid*

Acknowledgments

I thank my dear friends and colleagues for offering encouragement and valuable comments during the preparation of this book. During my time in Hughes Software Systems (now known as Aricent, located in Gurgaon, India), I had the opportunity to work with very smart and talented people who were generous in sharing their knowledge and experience. Special thanks also go to Mr. Sumit Kasera (AVP, Technology at Aricent, Gurgaon, India) for his valuable feedback on this book.

I would also like to thank 3GPP for permitting me to reproduce a few snapshots from the concerned 3GPP technical specifications.

I would also like to thank and appreciate John Wiley & Sons Ltd., UK, and its acquisition, editorial, production, and publishing staff, for their continuous support and cooperation during the entire process of this book's production.

List of Abbreviations

Here are the glossaries of some of the terms used in this book for ready references. For a complete list of terms and their definitions, please refer to the 3GPP TR 21.905 [24].

3G/4G/5G	3rd /4th/5th Generation
3GPP	Third Generation Partnership Project
5GS	5G System
5G-GUTI	5G Globally Unique Temporary Identifier
5G-S-TMSI	5G S-Temporary Mobile Subscription Identifier
5GC	5G Core Network
A-bis	A-bis Interface
ACK	Acknowledged Mode
AKA	Authentication and Key Agreement
AMF	Access and Mobility Management Function
AMP	Asymmetric Multicore Processing
AP	Application Protocol
APN	Access Point Name
AF	Application Function
ARFCN	Absolute radio-frequency channel number
ARQ	Automatic Repeat Request
AS	Access Stratum
ASN.1	Abstract Syntax Notation One
AuC	Authentication Center
AUSF	Authentication Server Function
BCF	Base Control Function
BCH	Broadcast Channel (Transport)
BCCH	Broadcast Control Channel (Logical)
BICN	Bearer-Independent Core Network BICN
BIST	Built-in self-test (BIST)
BS	Base station
BSC	Base station controller
BSN	Block Sequence Number
BSS	Base Station Subsystem
BSSGP	Base Station System GPRS Protocol
BSSMAP	Base Station Subsystem Mobile Application Part
BSP	Board Support Package
BSR	Buffer Status Report

BTS	Base Transceiver Station
BWP	Bandwidth Part
C-RNTI	Cell Radio-Network Temporary Identifier
CBG	Code block group
CBGFI	CBG flush indicator
CC	Call Control
CCE	Control Channel Element
CCCH/DCCH	Common/Dedicated Control Channel
CC/CM	Call Control/Connection Management
CM	Connection Management
CN	Core Network
CORESET	Control Resource Set
CRB	Common Resource Block
CRC	Cyclic redundancy check
CRI	CSI-RS Resource Indicator
CSI	Channel State Information
CSI-RS	Channel State Information Reference Signal
CS-RNTI	Configured scheduling RNTI
CSI-RSRP	CSI Reference Signal Received Power
CSI-RSRQ	CSI Reference Signal Received Quality
CSI-SINR	CSI Signal-to-Noise and Interference Ratio
CSFB	Circuit-switched Fall-back
CP	Cyclic Prefix
CPS	Call per second
CQI	Channel Quality Indication
CS	Circuit-switched
CSN	Concrete Syntax Notation
DCI	Downlink control information
DL-SCH/UL-SCH	Downlink/Uplink Shared Channel
DM-RS	Demodulation reference signals
DN	Data Network
DNN	Data Network Name
DRB	Data Radio Bearer
DSP	Digital Signal Processor
DTAP	Direct Transfer Application Part
DTCH	Dedicated Traffic Channel
eMBB	Enhanced Mobile Broadband
EAP	Extensible Authentication Protocol
ECM	Evolved Packet System Connection Management
EDGE	Enhanced Data for Global Evolution
EGPRS	Enhanced General Packet Radio Service
eNodeB	Evolved NodeB
EIR	Equipment Identity Register
EMM	Evolved Packet System Mobility Management
EN-DC	E-UTRA NR Dual-Connectivity
EPC	Evolved Packet Core
EPS	Evolved Packet System

ESM	Evolved Packet System Session Management
ETSI	European Telecommunications Standards Institute
E-UTRA	Evolved-UMTS Terrestrial Radio Access
E-UTRAN	Evolved-UMTS Terrestrial Radio Access Network
FDD	Frequency Division Duplex
FR	Frame Relay
FR1	Frequency Range 1 (5G)
FR2	Frequency Range 2 (5G)
FTP	File Transfer Protocol
FW	Framework
GBR	Guaranteed Bit Rate
GERAN	GPRS Edge Radio Access Network
GGSN	Gateway GPRS Support Node
GMSC	Gateway Mobile Switching Center
gNB/gNodeB	5G Base Station
GPRS	General Packet Radio Service
GSM	Global System for Mobile Communication
GUAMI	Globally Unique AMF ID
GUMMEI	Globally Unique MME Identifier
HARQ	Hybrid Automatic Repeat Request
HLR/HSS	Home Location Register/Home Subscriber Server
HS-DSG	High Speed Downlink Shared Channel
HSDPA	High-Speed Downlink Packet Access
HSPA	High-Speed Packet Access
HSUPA	High-Speed Uplink Packet Access
IAB	Integrated Access and Backhaul
IDNNS	Intra Domain NAS Node Selector
IE	Information Element
IEI	Information Element Identifier
IMS	IP Multimedia Subsystem
IMSI	International Mobile Subscriber Identity
INT-RNTI	Interruption RNTI
IOT	Inter-operability Testing
IP	Internet Protocol
IPC	Inter Process Communication
IPH	IP Header Compression
ISI	Inter Symbol Interference
ISR	Idle State Signaling Reduction
IWF	Interworking Function
K_{AMSE}	LTE Key Access Security Management Entity
KPI	Key Performance Identifier
L1....Ln	Layer 1....n
LAI	Location Area Identification
LCID	Logical Channel Identifier
LDPC	Low Density Parity Check
LLC	Logical Link Control
LI	Layer Indicator

LSB	Least Significant Bit
LTE	Long-term Evolution
mMTC	Massive Machine Type Communications
mIoT	Massive Internet of Things
MAC	Medium Access Control
MANO	Management and Orchestration
MCC	Mobile Country Code
MCS	Modulation and coding scheme
MIB	Master Information Block
MIMO	Multiple-Input Multiple-Output
MM	Mobility Management
MME	Mobility Management Entity
MMEC	MME Code
MMS	Multimedia messaging service
MN	Master Node
MNC	Mobile Network Code
MOC	Mobile-originated voice call
MOCN	Multi-operator Core Network
MS	Mobile station
MSB	Most Significant Bit
MSC	Mobile Switching Center
MSC-S	MSC Server
MSIN	Mobile Subscriber Identification Number
MU-MIMO	Multi-User MIMO
N3IWF	Non-3GPP Inter-working Function
NACK	Negative Acknowledgment
NAS	Non-access Stratum
NEF	Network Exposure Function
NF	Network Function
NFV	Network Functions Virtualization
NGAP	Next Generation Application Protocol
NG-RAN	Next Generation Radio Access Network
Non-GBR	Non-Guaranteed Bit Rate
NNSF	NAS Node Selection Function
NRF	Network Repository Function
NSA	Non-Standalone
NR	New Radio
NRF	Network Repository Function
NSSAI	Network Slice Selection Assistance Information
NSSF	Network Slice Selection Function
NSAPI	Network Service Access Point Identifier
NSS	Network Subsystem
OFDM	Orthogonal Frequency Division Multiplexing
OFDMA	Orthogonal Frequency Division Multiplexing Access
O&M	Operation and Maintenance
PBCH	Physical Broadcast Channel
PCH	Paging Channel

PCF	Policy Control Function
PCO	Protocol Configuration Options
PCRF	Policy Charging and Restriction Function
PCU	Packet Control Unit
PCFICH	Physical Control Format Indicator Channel
PDCCH	Physical Downlink Control Channel
PDCP	Packet Data Convergence Protocol
PDSCH	Physical Downlink Shared Channel
PD	Protocol Discriminator
PDSCH/PUSCH	Physical Downlink/Uplink Shared Channel
PDN	Packet Data Network
PDU	Protocol Data Unit
PDP	Packet Data Protocol
PEI	Permanent Equipment Identifier
PER	Packet Error Rate
PLMN	Public Land Mobile Network
PFC	Packet Flow Context
PGW	Packet Data Network Gateway
PHICH	Physical HARQ Indication Channel
PMI	Precoding-Matrix Indicator
POST	Power-on self-test
PRACH	Physical Random-Access Channel
PRB	Physical Radio Block
P-RNTI	Paging RNTI
PS	Packet Switched
PSS	Primary Synchronization Signal
PSTN	Public Switched Telephone Network
PTI	Procedure Transaction Identity
PTRS	Phase-tracking Reference Signal
PUCCH	Physical Uplink Control Channel
PUSCH	Physical Uplink Shared Channel
QAM	Quadrature Amplitude Modulation
QCI	QoS Class Identifier
QFI	QoS Flow ID
QoS	Quality of Service
QPSK	Quadrature Phase Shift Keying
RAB	Radio Access Bearer
RAI	Routing Area Identification
RAC	Routeing Area Code
RAN	Radio Access Network
RA-RNTI	Random Access RNTI
RACH	Random Access Channel
RAT	Radio Access Technology
RB	Resource Block
RE	Resource Element
RBG	Resource Block Group
RF	Radio Frequency

RI	Rank Indicator
RAU	Routing Area Update
REG	Resource Element Group
RLC	Radio Link Control
RF	Radio Frequency
RIV	Resource Indication Value
RNA	RAN-based Notification Area
RNAU	RAN-based Notification Area Update
RNS	Radio Network Subsystem
RNC	Radio Network Controller
RNL	Radio Network Layer
RNTI	Radio Network Temporary Identifier
RoHC	Robust Header Compression
RR	Radio Resource
RRC	Radio Resource Control
RS	Reference Signal
RSRP	Reference Signal Received Power
RSRQ	Reference Signal Received Quality
RRM	Radio Resource Management
RTOS	Real-Time Operating System
S1-AP	S1 Application Protocol
SA	Standalone Mode
SAP	Service Access Point
SAPI	Service Access Point Identifier
SBA	Service-based Architecture
SBI	Service-based Interface
SCP	Service Communication Proxy
SCTP	Stream Control Transmission Protocol
SDAP	Service Data Application Protocol
SDCCH	Standalone Dedicated Control Channel
SDN	Software Defined Networking
SDU	Service Data Unit
SD	Slice Differentiator
SEAF	Security Anchor Functionality
SEPP	Security Edge Protection Proxy
SFN	System Frame Number
SFI-RNTI	Slot Format Indication RNTI
SGSN	Serving GPRS Support Node
SIB	System Information Block
SLA	Service-Level Agreement
SMP	Symmetric Multicore Processing
S-GW	Serving Gateway
SI	Skip Indicator/System Information
SM	Session Management
SMS	Short Messaging Service
SMF	Session Management Function
SN	RLC Layer PDU Sequence Number

SN	Secondary Node
SNDCP	Subnetwork Dependent Convergence Protocol
S-NSSAI	Single Network Slice Selection Assistance Information
SNPN	Standalone Non-Public Network
SSC	Session and Service Continuity
SST	Slice/Service Type
SPS	Semi-persistent Scheduling
SR	Scheduling Request
SRVCC	Single Radio Voice Call Continuity
SRB	Signaling radio bearers
SRS	Sounding reference signal
SSB	Synchronization Signal Block
SSS	Secondary Synchronization Signal
SS	Supplementary Services
SS/PBCH	Synchronization Signal Physical Broadcast Channel
SS-RSRP	SS Reference Signal Received Power
SS-RSRQ	SS Reference Signal Received Quality
SS-SINR	SS Signal-to-Noise and Interference Ratio
STL	Standard Template Library
SU-MIMO	Single-User MIMO
SUCI	Subscription Concealed Identifier
SUPI	Subscription Permanent Identifier
TAC	Tracking Area Code
TAU	Tracking Area Update
TCH	Traffic Channel
TCP/IP	Transmission Control Protocol/Internet Protocol
TDD	Time Division Duplex
TI	Transaction Identifier
TFT	Traffic Flow Template
TNL	Transport Network Layer
TPC	Transmit Power Control
TRX	Trans-receiver
TS	Timeslot
TTI	Transmission Time Interval
UCI	Uplink Control Information
UDM	Unified Data Management
UDP	User Datagram Protocol
UDR	Unified Data Repository
UE	User Equipment
Um	GSM Air Interface
UML	Unified Modeling Language
UMTS	Universal Mobile Telecommunication System
Uu	UMTS/LTE Air Interface
UPF	User Plane Function
UTRAN	UMTS terrestrial radio access network UMTS
URLLC	Ultra Reliable and Low Latency Communications
UUID	Universally Unique Identifier

VLR	Visitor Location Register
VoLTE	Voice over LTE
VRB	Virtual Resource Block
WCDMA	Wideband Code Division Multiple Access (UMTS)
Xn-C	Xn-Control plane
Xn-U	Xn-User plane
XnAP	Xn Application Protocol

1

Introduction

Career Opportunities in Mobile Communications Networks Space

You see, wire telegraph is a kind of a very, very long cat. You pull his tail in New York; and his head is meowing in Los Angeles. Do you understand this? And radio operates exactly the same way: you send signals here, they receive them there. The only difference is that there is no cat.

Source: Albert Einstein.

Gone are the days when mobile phones were considered a luxury; today, it is among our daily necessities. Mobile communications have evolved quickly and revolutionized the way we communicate and have brought people around the globe closer than ever before. At the same time, the mobile telecom industry has witnessed an explosive growth rate over the past many years, offering a wide range of services and products.

The latest trend is the availability of multitasking smartphones and various apps for different services and capabilities for our day-to-day needs. The combination of smartphones and Apps is enabling mobile broadband services to everyone, everywhere, and anytime on the go. This is leading to a growing demand for smartphones along with manifold increase in Internet traffic. The legacy mobile communications network standards such as Global System for Mobile Communication (GSM) (2G), General Packet Radio Service (GPRS) (2.5G), Universal Mobile Telecommunication System (UMTS) (3G), Long-Term Evolution (LTE) (4G), and the emerging and latest buzzword 5G, along with the growth in the number of mobile subscribers, have created a gap between the demand and supply of skilled mobile communications professionals.

The mobile communications industry is a key driver and offers immense employment opportunities. The advent of the 5G system and network is expected to create and offer more opportunities for various professionals and business owners. Communications solution service providers, network installers, original equipment manufacturers (OEMs), and communications software solution services provider firms are demanding skilled and experienced professionals. In the mobile communications space alone, career opportunities exist across these different verticals and areas. Some of these career opportunities are shown in Figure 1.1.

The career opportunity areas shown in Figure 1.1 are at a very high level only. Each of the areas shown spans across multiple systems and subsystems. Opportunities exist at an entry-level or experienced professional level for various job roles such as software developer and maintenance, team leader, project manager, system architect and engineering, site engineer, network engineer, network operation and maintenance, and so on.

Mobile communications systems and networks are built based on open technical standards, covering a wide spectrum of knowledge areas as illustrated in Figure 1.1. The knowledge areas are spread across the different system engineering areas of mobile communications systems and networks. This book attempts to cover some of those knowledge areas from a practical point of view, so that one can be self-equipped to start a career in the mobile communications domain.

Mobile Communications Systems Development: A Practical Introduction to System Understanding, Implementation, and Deployment, First Edition. Rajib Taid.

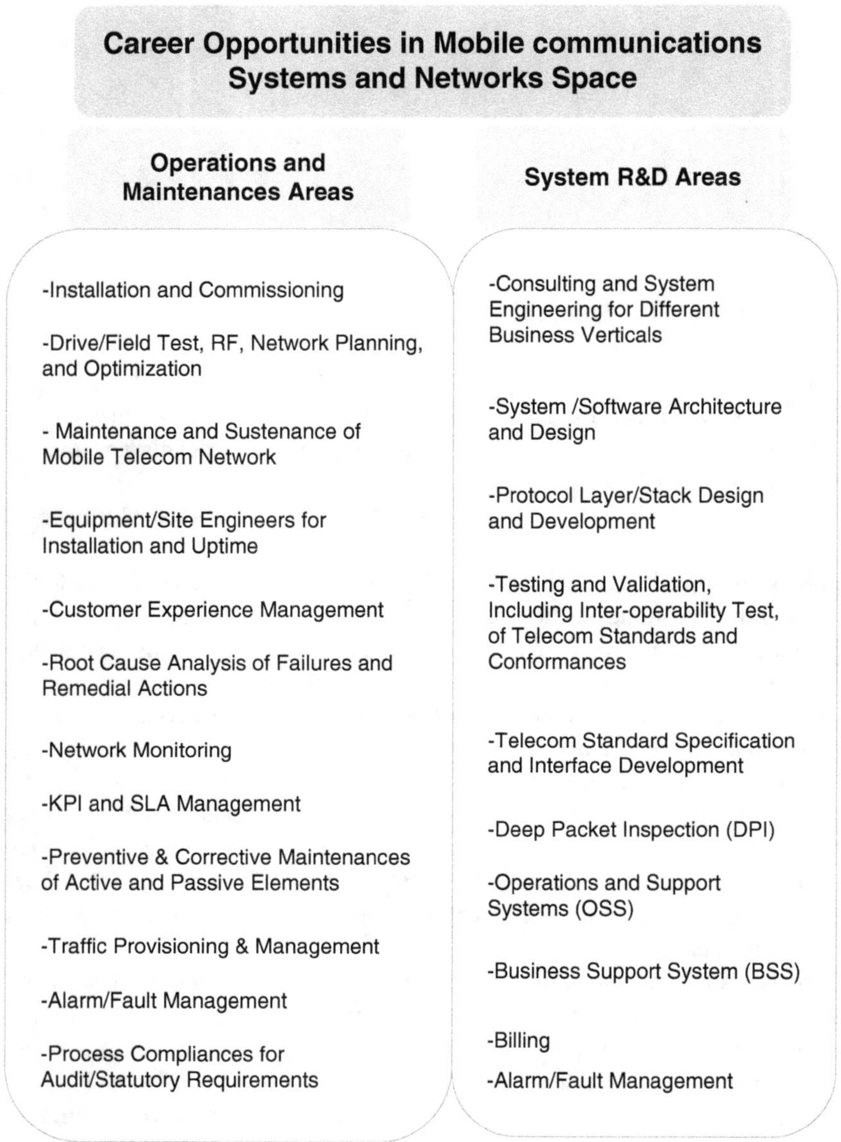

Figure 1.1 Career opportunities in the mobile communications space.

Part I

Network Architectures, Standardization, Protocols, and Functions

Mobile communications systems and networks based on the GSM/GPRS, UMTS, LTE, and 5G systems and technologies consist of telecommunications infrastructures to provide communications services to mobile users. In this first part of the book, several knowledge areas of legacy mobile communications systems and networks which are based on the GSM/GPRS, UMTS, and LTE technologies as well as 5G systems are covered. The 5G system shall be covered exclusively in Part IV of this book. Because even if a reader is starting a career in LTE, 5G, or the latest, system and network, and as a developer or O&M person, one must know the major key concepts from the legacy GSM/UMTS/LTE networks as well. The network architectures and their protocols; the 3GPP standardization processes; systems engineering; functions such as the network identities, packet encapsulations, and interworking of network aspects of the GSM/GPRS, UMTS, LTE, and 5G system are discussed. This part of the book contains eight chapters, and their purposes are as follows.

Chapter 2 describes the architectures and their domains of the GSM, GPRS, UMTS, and LTE networks. The standardization processes used by the 3GPP and evolutions of mobile communications networks based on the GSM, GPRS, UMTS, LTE, and 5G systems are also described. The *System Engineering* aspects of mobile communications networks are described briefly.

Chapter 3 describes the architecture and different types of protocols found within the mobile communications systems and networks, which are based on the legacy GSM, GPRS, UMTS, LTE system as well as the 5G system. Various interfaces for peer-to-peer protocol layers communications are also described.

Chapters 4 to 9 describes some of the essential functions that are performed by the mobile communications systems and networks, which are based on the legacy GSM, GPRS, UMTS, LTE, and the emerging 5G system to provide seamless communications services to mobile users within a particular network or across the networks.

2

Network Architectures, Standardizations Process

Introduction

This chapter provides the introductory and high-level architectures of the legacy mobile communications systems which are based on the Global System for Mobile Communication (GSM), Universal Mobile Telecommunication System (UMTS), and Long-Term Evolution (LTE) systems. However, being legacy and general, some of the contents of this chapter apply to the 5G system also. The architecture of the 5G system shall be covered in Part IV of this book. We begin with the basic network architecture of the legacy GSM system, followed by the architectures of UMTS and LTE systems as well. Following this, we present the different domains or areas of a mobile communications network along with their network elements that are interconnected together to provide communication services to users. The evolution of different mobile communications systems is presented. We also cover the system engineering aspects of a mobile communications network that span across its different domains or knowledge areas.

We then present the standardization processes used for mobile communications systems and networks that are available currently. Standardization is important for successful developments and integrations of multivendor network elements of a mobile communications system and network. Different features added, in terms of a release, during the evolution of a particular mobile communications system are also summarized.

2.1 Network Elements and Basic Networks Architectures

A mobile communications network based on the GSM (2G)/General Packet Radio Service (GPRS) (2.5G), UMTS (3G), and LTE (4G) technology comprises the interconnected network of different communicating entities/nodes, called as the *network element*. Based on a particular communications technology, a mobile communications network comprises several network elements. In the case of a GSM and GPRS network, the network elements are MS (Mobile Station), BSC (Base Station Controller), BTS (Base Transceiver Station), MSC (Mobile Switching Center), SGSN (Serving GPRS Support Node), and GGSN (Gateway GPRS Support Node). In the case of the LTE/Evolved Packet System (EPS) network, the network elements are User Equipment (UE) (mobile station), eNodeB, Mobility Management Entity (MME), Serving Gateway (S-GW), and Packet Data Network (PDN). These interconnected network elements provide end-to-end communication services like voice (Circuit Switched – CS domain in case of GSM and UMTS) and data (Packet Switched – PS domain in case of GPRS and LTE), multimedia contents to subscribers.

Note: *The abbreviated version of all the terms found in mobile communications systems and networks are being used in this book. For the complete list of glossaries of the terms, their expanded texts, and definitions, refer to the 3GPP TR 21.905* [24]. *A partial list of abbreviations is provided under the section "List of Abbreviations". More about the 3GPP, technical specifications (TS), and technical reports (TR) are described in later sections.*

Mobile Communications Systems Development: A Practical Introduction to System Understanding, Implementation, and Deployment, First Edition. Rajib Taid.

2.1.1 GSM (2G) Network Architecture

Consider the legacy GSM CS network shown in Figure 2.1. It consists of the different network subsystems and elements: MS, BSS (Base Station Subsystem), and Network switching subsystems (NSS). A GSM network provides the CS voice call services to subscribers.

As illustrated in Figure 2.1, a GSM CS communications network is broadly divided into various subsystems, which are described below.

- BSS

It consists of the network elements: BTS and BSC. A BTS is a hardware component that is installed to provide communications services in a GSM cell. A BTS transmits and receives information with mobile devices through radio frequency communications. A BSC is responsible for the allocation of radio frequency resources in one or more cells and controls one or multiple BTSs. The GSM BSS, consisting of BTS and BSC, is the interface between a mobile device and the rest of the GSM network or public switched telephone network (PSTN). A BTS is connected to a BSC through a logical interface called *A-bis interface*. An MS communicates with the BSC through the physical air interface, known as *Um*.

- Network switching subsystem (NSS)

It consists of the network elements: MSC, Home Location Register (HLR), Visitor Location Register (VLR), and so on. The MSC performs all the necessary functions to provide CS voice call services, both for the mobile originated (MO) and mobile terminated. The gateway MSC performs the routing functions on behalf of an MS that is being served by another MSC. A VLR contains information about all the MS currently being served by a particular MSC. An MSC contacts the VLR to find and retrieve the current location of an MS. The HLR is a central database that stores the permanent information of subscribers.

NSS is also known as the *Core Network (CN)* and facilitates seamless communication services to freely moving users within its coverage area or between the networks of different operators or between a mobile and fixed-line network.

The vertical dotted lines in Figure 2.1 indicates the separation of one network element from another one or an entire network from another network through a particular interface with a set of protocol layers on it. Each such interface has its name, for example, air interface (Um), A-bis interface, A-interface, and so on, as shown in Figure 2.1. More about the mobile communications network interfaces, both physical and logical, are described in Chapter 3.

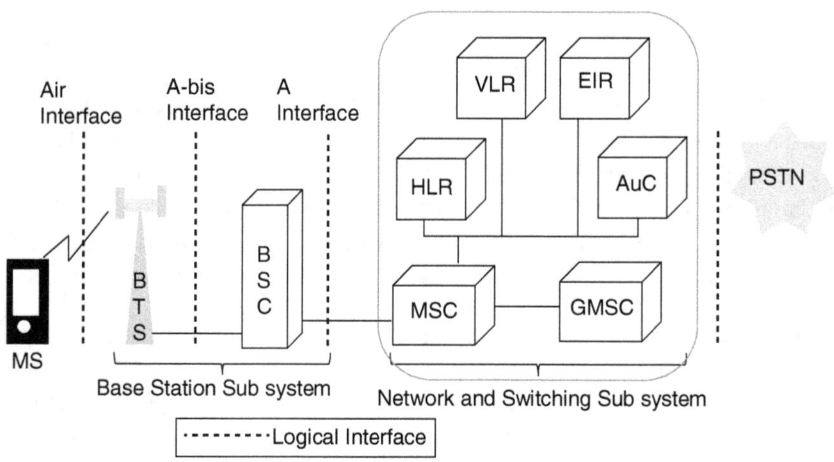

Figure 2.1 Network architecture and elements of a GSM network.

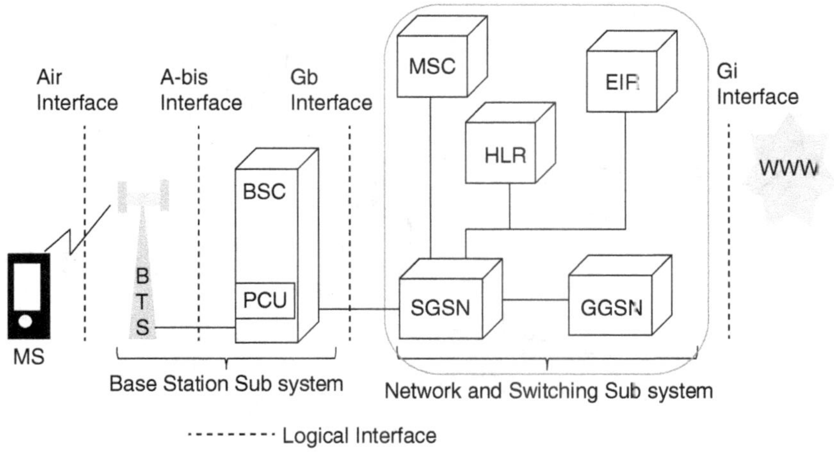

Figure 2.2 Network architecture and elements of a GPRS network.

2.1.2 General Packet Radio Service (GPRS-2.5G) Network Architecture

The architecture of a GPRS network and its elements for packet data services only is shown in Figure 2.2. As shown in this figure, a GPRS network uses the GSM BSS and NSS. However, the BSS provides the GPRS IP packet data services through a separate hardware unit as well as a software component which is known as the *Packet Control Unit (PCU)*.

A PCU performs all the packet services-related functions independently but in association with the BSC. Also, two new network elements on the NSS end are SGSN and GSSN. They communicate with each other over an IP transport network. GGSN has the interface to the external PDN over an IP transport network.

Functionality wise, an SGSN in a PS domain performs similar functions to an MSC in the CS domain core network. SGSN keeps track of the current locations of MSs and performs the PS domain control, i.e. mobility management and session management, and user data transfer functions. An SGSN can be interconnected with an MSC to deliver CS domain-related paging services to an MS in case it is engaged in a PS session. The GGSN serves as the gateway for a GPRS network to an external IP network. Apart from this, a GGSN allocates IP addresses to a registered MS. Both the SGSN and the GGSN use the IP transport network.

2.1.3 Universal Mobile Telecommunications System (3G) Network Architecture

Similarly, consider Figure 2.3 for the different network elements found in a Universal Mobile Telecommunications System (UMTS) or 3G network. A UMTS network contains network elements such as the UE (known as the MS in GSM network), NodeB (similar to a GSM BTS), Radio Network Controller-RNC (similar to a GSM BSC), SGSN, GGSN, Gateway Mobile Switching Center (GMSC), and MSC. These network elements together provide both the CS and PS data services to subscribers as illustrated in Figure 2.3.

Similar to a GSM network, the RNC together with the NodeB is called the *Radio Network Subsystem (RNS)* in a UMTS network. RNS transmits and receives information with UEs through radio frequency communications. The flow of CS voice call and PS data call is being shown with a bold solid line in the diagram Figure 2.3. The SGSN, GGSN, GMSC, and MSC network elements are reused from the GSM network. Figure 2.3 is the first version of the UMTS network architecture, also referred to as the *Release 99*. The UMTS CN elements are adapted from the *Pre-Release 99* GSM and GPRS networks. Subsequent releases of the UMTS network are known as the *Release 4*, *Release* 5, and so on, which are described later in Section 2.3.2.

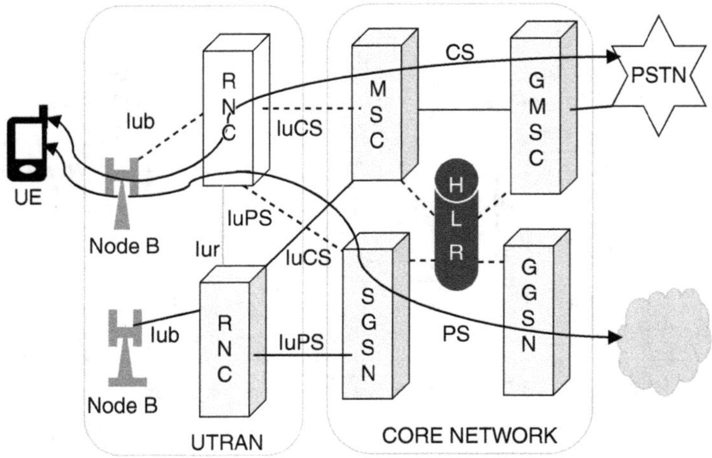

Figure 2.3 Network architecture and elements of a UMTS network.

2.1.4 LTE (4G) Network Architecture

Figure 2.4 shows the network architecture of the LTE system, refer to TS 36.300 [92], which consists of the different network elements: UE, *Evolved (e)NodeB*, and *Evolved Packet Core (EPC)* network to provide PS services only. LTE network, which provides higher data rates, was evolved from the previous UMTS system. The term called "*Evolved*", denoted by "e" or "E" can be found in any descriptions related to the LTE system. The eNodeB performs the radio communication-related functions and controls one or more cells, similar to a UMTS NodeB +RNC. Apart from this, eNodeB performs the radio resources allocation, UE scheduling, and forwarding of the user traffic/data functions to the S-GW. An eNodeB is interconnected with other eNodeBs over the X2 interface, and together, they form the Evolved-UMTS Terrestrial Radio Access Network (E-UTRAN) of an LTE/EPS network.

The LTE EPC contains the network elements such as *Mobility, Management Entity* (MME), *Serving Gateway* (S-GW), and *Home Subscriber Server* (HSS). An MME performs the signaling or controlling, e.g. mobility

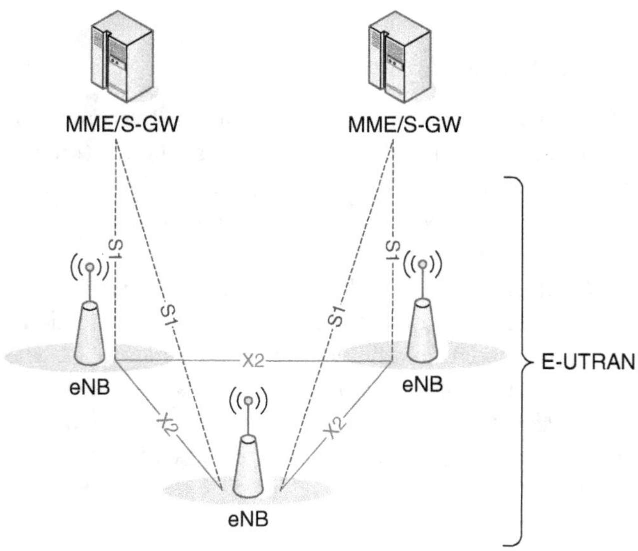

Figure 2.4 Overall network architecture and elements of an LTE network. *Source*: © 2015. 3GPP ™ TSs and TRs are the property of ARIB, ATIS, CCSA, ETSI, TSDSI, TTA and TTC who jointly own the copyright in them. © 2015, 3GPP.

management, session management, and related functions between a UE and the EPC network. The HSS performs the similar functions of an HLR found in the GSM and UMTS system. The HLR/HSS is a database that stores the subscriber's permanent information in a mobile communications network. Unlike the previous systems, i.e. GSM and UMTS, in the LTE system, the various management functions related to signaling and user data/call aspects are assigned separately to the MME and S-GW. In an LTE network, the E-UTRAN and the EPC are collectively known as the *Evolved Packet System (EPS)*. An eNodeB is connected to the MME (for signaling) and S-GW (for user traffic) over the S1 interface; refer to Figure 2.4.

The LTE system is an all IP-based network, i.e. all the network elements communicate with the existing IP transport network only. This differentiates the LTE communication network from its predecessors, GSM, GPRS, and UMTS networks, which uses other transport networks, such as ATM and Frame Relay, apart from the IP transport network.

From the comparisons of Figures 2.1–2.4, it appears that the number of network elements in an LTE/EPS network has reduced. This has led to a smaller number of protocols and interfaces between network elements compared to the GSM and UMTS system. For an overall description of the functions performed by each of the network elements of the respective mobile communications network described above, refer to the TS 23.002 [29].

2.1.5 GSM, UMTS, LTE, and 5G Network Elements: A Comparison

Based on the similar functions performed, one can compare the different network elements of a GSM, UMTS, and LTE network. A side-by-side comparison of different network elements of GSM, GPRS, UMTS, and LTE networks is shown in Table 2.1 below. The BTS and BSC of a GSM network are known as the NodeB and RNC in the UMTS system. Similarly, the functions performed by a BTS and BSC of a GSM network are performed by the eNodeB only that is found in an LTE network which was shown in Figure 2.4.

2.1.6 Circuit Switched (CS) vs Packet Switched (PS)

At the beginning of Section 2.1, the types, i.e. CS as well as PS, of services being provided by a typical mobile communications network was mentioned. In the case of a *CS call*, an end-to-end dedicated physical circuit establishment is required prior to the starting of voice call service on it. However, no such dedicated physical resources or path is required to be established for a PS call. A refresher illustration showing this basic difference between a CS and PS call is shown in Figure 2.5.

In Figure 2.5 illustration, User A wants to start a CS voice conversation (bold solid line) with User B. Prior to that, a dedicated and exclusive physical path/circuit (dotted line) for the entire duration of the call is required to be established between them through the interconnected switching or network elements. Similarly, User C wants to visit a web site, say www.abc.com, which is a typical PS data transfer scenario. In this case, the user's mobile

Table 2.1 Network elements comparisons.

	Mobile communication systems			
Network elements	GSM/GPRS	UMTS	LTE	5G
	GERAN	UTRAN	E-UTRAN	NG-RAN
	SGSN	SGSN	S-GW	UPF
	GGSN	GGSN	PDN gateway	SMF (partially)
	HLR	HLR	HSS	UDM and AUSF

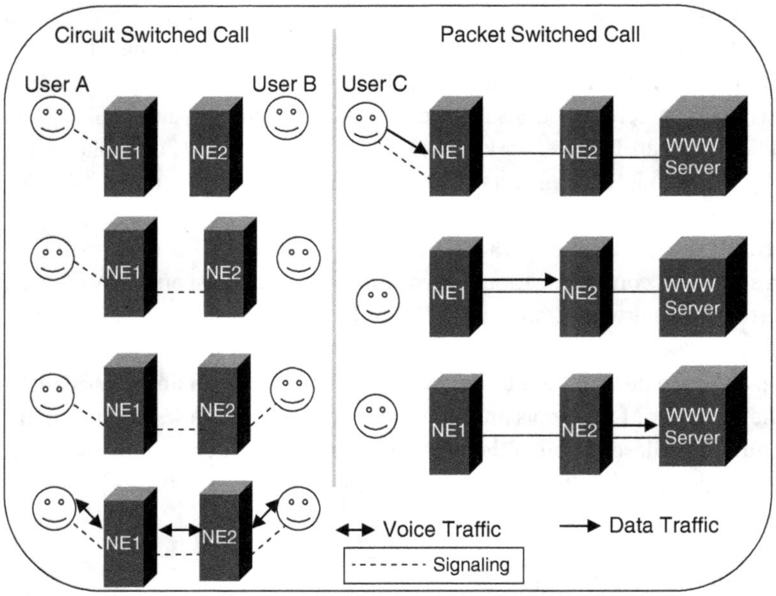

Figure 2.5 Illustration of circuit switched and packet switched data transfer.

device sends a burst of data to the concerned network element NE1. Following this, depending on the concerned timer value and response from the webserver, the mobile device may release the ongoing signaling path that was established with the network element NE1. Network element NE1, in turn, shall forward the received packets to its peer network element NE2, en route to the web site server, say www.abc.com.

2.2 Mobile Communication Network Domains

As illustrated in Figures 2.1–2.4, organizations and interconnections of different network elements constitute the *Network Architecture* for a particular mobile communications network, such as the GSM, GPRS, UMTS, and LTE networks. Interconnections are realized through different logical and physical interfaces as described later in Chapter 6. Related network elements of a mobile communications network are grouped and logically divided into three domains, as follows:

- Access Network (AN)
- Core Network (CN)
- Mobile Station or User Equipment (MS/UE)

AN and CN form the infrastructure domain of a typical mobile communications system and network.

2.2.1 AN Domain

An *AN* domain consists of equipment and systems that are responsible for radio frequency-related transmission and reception in one or more cells to or from the UE or mobile device (MS). Across the different mobile communications networks, ANs are known as follows:

- Radio Access Network (RAN) in the case of the GSM system.
- Universal Terrestrial Radio Access Network (UTRAN) in the case of the UMTS system.
- Evolved Universal Terrestrial Radio Access Network (E-UTRAN) in the case of the LTE system.
- Next-Generation Radio Access Network (NG-RAN) in the case of the 5G system.

Though the above *ANs* are responsible for radio frequency resource allocation and communication path control in general, they perform various functions differently depending on the radio access technology (RAT) being used. The GSM RAN and UMTS UTRAN perform both the CS and PS switched network functions; the LTE E-UTRAN or 5G NG-RAN performs PS network functions only.

2.2.2 Core Network (CN) Domain

A *Core Network (CN)* domain is the backbone infrastructure for a mobile communications system and network which consists of hardware and systems. A CN is also a gateway to the traditional PSTN system. CN primarily takes care of all the call control-related functions. A CN also performs various functions such as the authentication, charging, and setup of end-to-end connections required for different telecommunication services. A CN is independent of the radio connection technology for a mobile device. In the case of the LTE system, both the core network and the RAN are fully based on IP PS due to which it is also called the *Evolved Packet System* (EPS). In the case of the 5G system also, both the core network and the RAN are fully based on the IP PS network.

From Figures 2.1–2.3, one can see that the GSM/GPRS and the UMTS differ in the AN only. However, UMTS reuses the core network elements (e.g. MSC, SGSN, GGSN, and so on) from the GSM core network. In the case of the LTE system, the entire architecture is, however, different from the GSM and UMTS system. Neither has it reused the AN nor has it reused the CN elements from the previous cellular systems. The CN of an LTE network works on top of an IP transport network, whereas the GSM and UMTS CN may use mixed transport networks.

2.2.3 Network Domains and Its Elements

In the previous sections, we have described, in general, the AN and CN domains or areas of different mobile communications networks that are available today. Each network domain consists of various network elements as shown in Figure 2.6. This is a general and an introductory figure to provide the reader with an overall view of the various network domains and its elements of communication networks based on the GSM, GPRS, UMTS, and LTE systems. Note that in the LTE system, the network element HLR and Authentication Center (AuC) have been

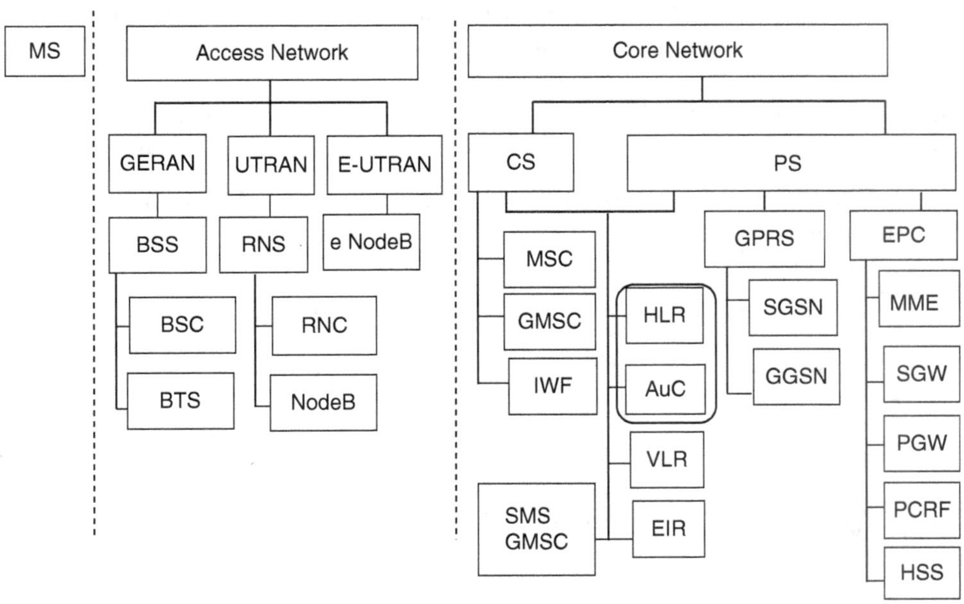

Figure 2.6 Network domains and their elements of mobile communication networks: GSM to 4G (LTE/EPS).

replaced by the HSS. For the expanded texts version of these abbreviated acronyms, refer to the abbreviation section of this book.

The vertical dotted lines shown in Figure 2.6 represent the logical interface between two network domains of a mobile communications network. More about the logical interfaces using which network elements exchange protocol information is described later in Section 3.1.2. For the description of the functions performed by each of the network elements shown in Figure 2.6, the reader is recommended to refer to the TS 23.002 [29]. Identification and other aspects of the 3GPP technical specification are described later in Section 2.5.

The network elements of the AN domain work using the respective and particular RAT of a mobile communications system.

The CN is further divided into the following domains.

- Circuit Switched (CS), which provides voice call services in case of the GSM and UMTS system.
- Packet Switched (PS), which provides data services in the case of the GPRS, UMTS, and LTE/EPS systems.
- IMS (IP multimedia subsystem), which is used to provide voice call services over an LTE/EPS network.

Note that some of the network elements are found in the CN domain only. For further details on the functions performed by the different network elements, refer to TS 23.002 [29]. In the next two sections, illustrations are presented to illustrate the end-to-end protocol information flow through the above network domains of a GSM network.

2.2.4 Example: End-to-End Mobile Network Information Flow

Figure 2.7 illustrates, in general, an end-to-end information flow starting from the MS/UE to the external network to offer communication services such as the voice, data, and multimedia contents in the case of a GSM, UMTS, LTE, or 5G network. The same figure can be used to illustrate the end-to-end information flow by replacing the RAN and CN elements of the respective communication systems, i.e. GSM, UMTS, LTE, and 5G systems.

In Figure 2.7, observe the extent of the dotted as well as the bold and solid line.

- The dotted line terminates between the MS and the RAN. This is the radio frequency signaling information flow path between the MS and the RAN, which facilitates establishing and access dedicated network resources by the MS.
- The dark solid and bold line terminates between the MS/UE and core network via the RAN. This is the actual user data/traffic information flow path between the MS and the CN, en route to an external network.

In a cellular system, radio signaling must be established prior to a user can start using various services like voice and data. So, a data/traffic path has been shown on top of the signaling path in Figure 2.7.

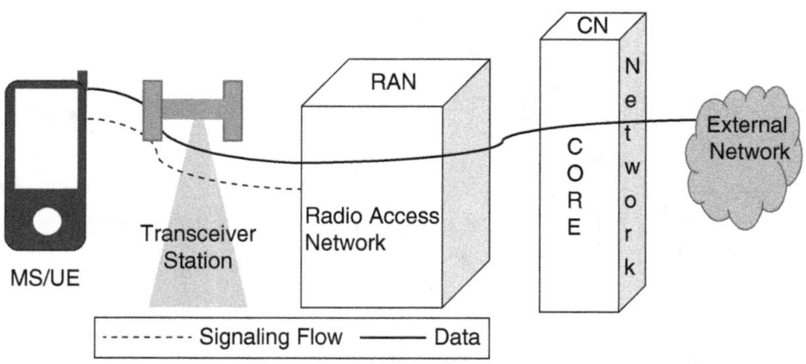

Figure 2.7 Illustration of end-to-end flow of information in a mobile communication network.

2.2.5 Example: GSM MO Call

Whenever the call button of a mobile phone is pressed, a great deal of information is exchanged in the form of a signaling message among the various network elements of a mobile communications network. In the case of a mobile-initiated call, the signaling message flow starts from the mobile station to the core network until the subscriber disconnects the voice call. This is further illustrated through a typical GSM MO call flow diagram shown in Figure 2.8. In Figure 2.8, the text inside the bracket shows the corresponding 3GPP technical specification number (described in the subsequent sections), where the definition and more details about the concerned signaling messages can be found.

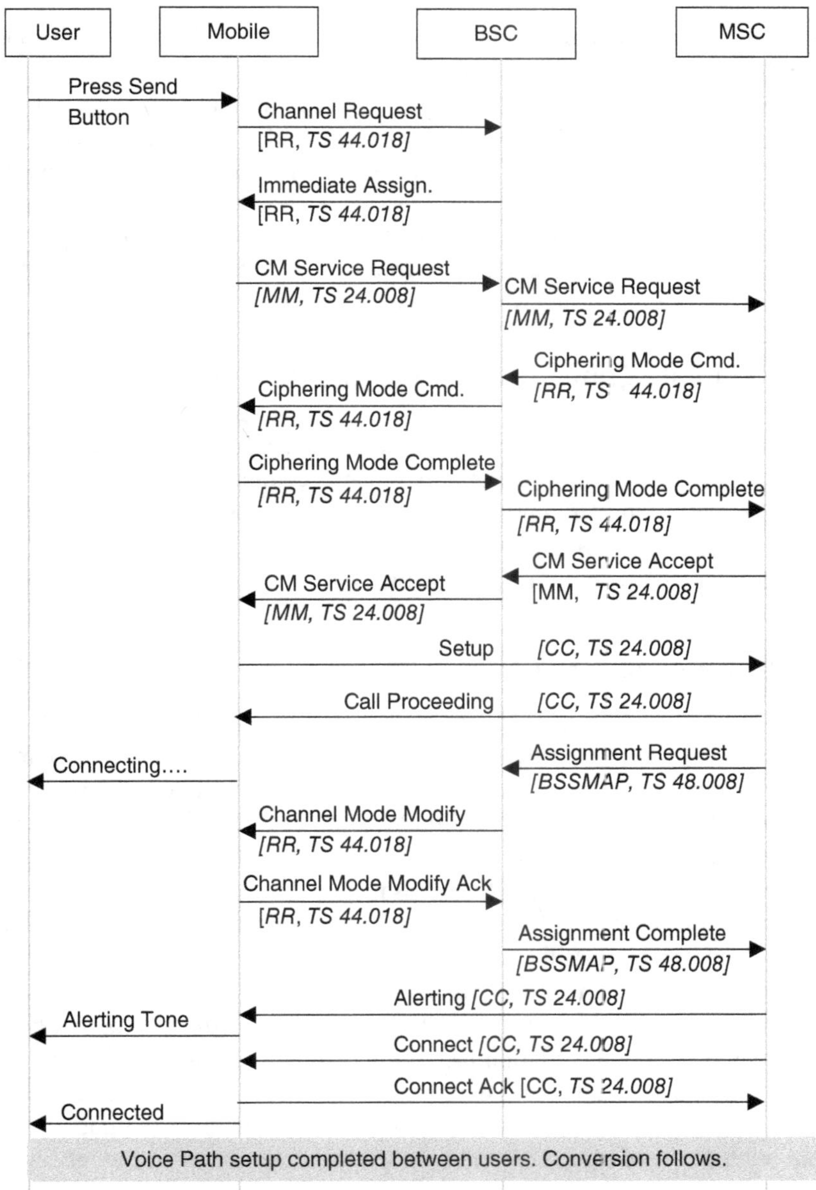

Figure 2.8 Illustration: GSM MO call flow.

In the typical GSM call flow example shown above, the following network elements are involved to provide an end-to-end MO voice call facility to the user.

- Mobile device or station
- BSS, comprising the BTS and the BSC
- MSC

Once the user initiates a voice call, it goes through several phases as summarized below. All these phases involve the exchanging of various signaling messages among the protocol layers of different network elements of the AN and CN domains.

- Establishes a Radio Resource Connection between the MS and BSC to transport the call-related information to the MSC.
- Allocates the necessary signaling connection and traffic channel by the BSC.
- MSC authenticates the mobile device/station.
- The Radio Resource Connection phase is completed and a connection between the MS and MSC is established.
- MS further sends the call setup message to the MSC. The MSC reserves the necessary resources and connects with the called party; thus, a voice/speech path has been setup between the mobile users. The voice is now in the conversation phase. Once the conversation ends and the user presses the disconnect button, the call enters into the Call Release phase where the BSS, as well as the MSC, frees the allocated radio resources. Similar steps take place whenever a mobile user tries and place a call to a PSTN landline telephone user. However, in this case, the MSC also establishes a signaling connection with the fixed PSTN.

2.3 Mobile Communications Systems Evolutions

Section 2.1 described the basic network architectures from the GSM to the LTE system. In this section, we will introduce further the evolutions of the mobile communications networks and systems, from the GSM to LTE, with respect to their air interface and network architectures. The 3GPP, which is described later in Section 2.5, is responsible for standardizing and defining the system architecture evolutions (SAEs) of the GSM to the 5G system-based mobile communications networks.

2.3.1 Evolutions of Air Interface

The air interface (known as Um in GSM: Uu in UMTS, LTE, 5G) of a mobile communications network is used to communicate between a mobile device and its RAN. In the GSM system, mobile devices use the frequency-division multiple access (FDMA)/time-division multiple access (TDMA)-based radio access methods. In UMTS, UEs use the Wideband Code Division Multiple Access (WCDMA)-based radio access method, LTE UEs use the Orthogonal Frequency Division Multiplexing Access (OFDMA) and SC-FDMA-based radio access methods, and 5G uses the OFDMA-based radio access method to communicate with the respective RAN. Figure 2.9 illustrates the evolutions of the 3GPP mobile communications systems along with the radio access method used in each evolution from the GSM to the 5G system. The arrow indicates the increased data rates in each evolution.

The air interface and its evolutions can be studied in terms of the access technologies and its modulation technique used, which is shown in Figure 2.9. One of the factors that determine the rates of data transfer or throughputs is the modulation technique used by each radio access method. Table 2.2 shows the modulation techniques, bandwidths, and data rates offered over the GSM, UMTS, LTE, and 5G air interface. From this table, it is observed that even within a particular radio access method, for example, UMTS, the data throughput offered in the downlink (DL) and uplink (UL) directions are different depending on the modulation technique used by it.

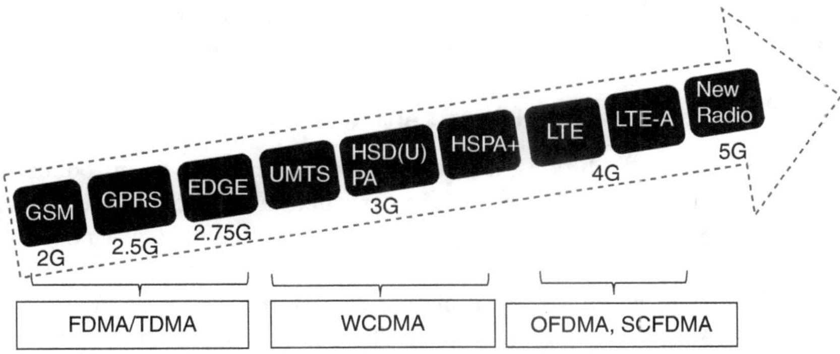

Figure 2.9 Illustration: 3GPP systems and air interface evolutions.

Table 2.2 Evolutions of 3GPP systems and their air interfaces.

System/features	Modulation techniques	Bandwidth	Throughputs
GSM (2G)	GMSK	200 KHz	40 kbps
GPRS (2.5G)	GMSK	200 KHz	171 kbps
EDGE (2.75G)	GMSK, 8-PSK	200 KHz	384 kbps
UMTS (3G)	QPSK	5 MHz	384 kbps
HSDPA (3G) Feature	DL: QPSK, 16 QAM UL: QPSK	5 MHz	DL: 14.4 Mbps UL: 384 kbps
HSUPA (3G) Feature	DL: QPSK, 64 QAM UL: QPSK, 16 QAM	5 MHz	DL: 14.4 Mbps UL: 5.8 Mbps
HSPA+(3G) Feature	DL: QPSK, 64 QAM UL: QPSK, 16 QAM	5 MHz	DL: 21–42 Mbps UL: 11 Mbps
LTE (4G)	DL: QPSK, 64 QAM	1.4, 3,5, 10, 15,20 MHz	DL: 300 Mbps UL: 75 Mbps
LTE-A (4G)	DL: QPSK, 64 QAM	1.4, 3,5, 10, 15, 20 MHz	DL: 3 Gbps UL: 1.5 Gbps
5G	QPSK, 16/64/256 QAM	5 to 400 MHz	DL: 20 Gbps UL: 10 Gbps

As shown in Table 2.2, the UMTS features and the LTE and 5G system air interface use the different modulation techniques in the UL and DL directions between the UE and RAN. Using GMSK modulation, one modulation symbol can carry 1 bit of data; one QPSK modulation symbol can carry 2 bits of data; one 16 Quadrature Amplitude Modulation (QAM) modulation symbol can carry 4 bits of data; and so on. The number of data bits transmitted per modulated symbol is called the *modulation order*. The throughput offered by different modulation techniques is shown in the fourth column.

UMTS features High-Speed Downlink Packet Access (HSDPA) and High-Speed Uplink Packet Access (HSUPA) are known as the *High-Speed Packet Access* (HSPA) method in the DL and UL directions, respectively. The HSPA+ feature is called as the evolution of the HSPA method. These access methods use the same modulation technique, i.e. QAM, but with different modulation orders. HSPA+ and HSUPA use higher (64) QAM than the HSDPA method. However, the HSPA+ method uses multiple-input multiple-output transmission techniques (MIMO) with two antennas for transmission of data from the UTRAN to UEs. More about the modulation techniques and MIMO transmissions are described later in Chapter 19.

2.3.2 Evolutions of 3GPP Networks Architectures

Along the path of evolutions, a new network element may be introduced or an existing one may be eliminated from the AN and CN domain of a particular mobile communications network and its architecture. For example, the LTE system has a simpler architecture having fewer nodes, channels, and less signaling messages than the UTRAN system. Network element such as the GPRS/UMTS SGSN also exists in an LTE/EPS network during the entire 3GPP mobile communications systems evolutions with enhanced roles being added to it. An engineer working on the GSM system will find it easy to understand the various aspects of the GPRS/Enhanced Data for Global Evolution (EDGE) system. Similarly, one will understand the LTE system better having working experience on the UMTS system. Figures 2.10–2.12 show the architectural differences among the different 3GPP systems releases, i.e. Release 99, Release 4, Release 5, and so on. Compare these architectures with the GSM network shown earlier in Figure 2.1 with respect to the network element, interfaces, and signaling systems. More about the 3GPP group and its technical specification is described later in Section 2.5.

- **Release 99: The First UMTS Release Architecture**

The 3GPP Release 99 system architecture that was shown earlier in Figure 2.3 is illustrated again in Figure 2.10 below in a slightly different way.

3GPP Release 99 evolved from the previous GSM/GPRS system in terms of new RAT only and introduces the new AN called the *UTRAN*, comprising a NodeB and an RNC. This implies that only the air interface had evolved from the TDMA/FDMA in GSM/GPRS system to the WCDMA scheme in UMTS while adapting to the CN elements from the pre-release 99 GSM/GPRS network, with necessary software enhancements. However, as shown in Figure 2.3, new interfaces are being added in the UMTS Release 99 architecture, and they are different from the GSM and GPRS networks.

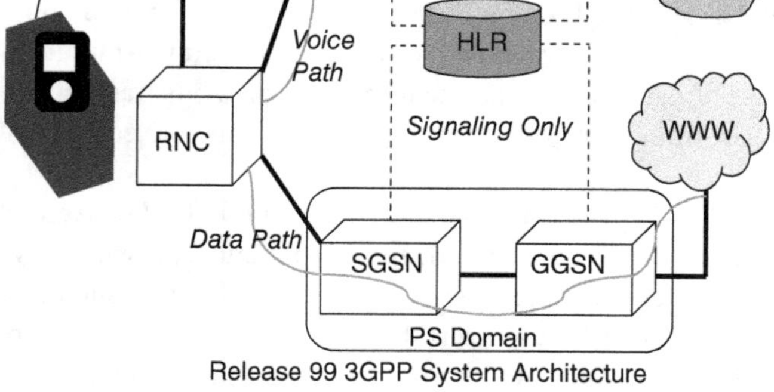

Figure 2.10 System architecture 3GPP Release 99.

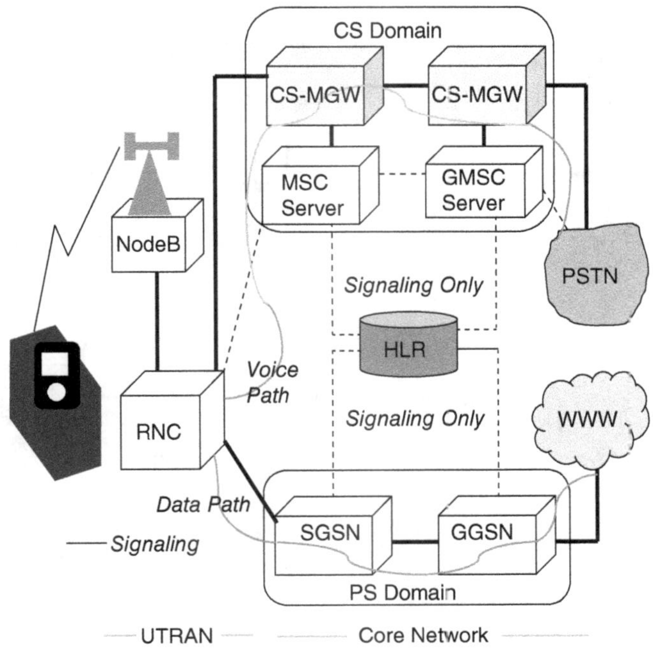

Figure 2.11 System architecture 3GPP Release 4.

- **Release 4 Architecture: Voice Call Through IP Transport Network**

  In the 3GPP Release 4 system architecture, Figure 2.11, changes were made only in the CS domain. In the case of the 3GPP Release 99 architecture, the CS domain core network elements – MSC and the GMSC – perform both the user traffic transport and the control/signaling, i.e. call control and mobility, functions. However, in the Release 4 architecture, the MSC has been separated into two entities: CS-*Media Gateway* (CS-MGW) and MSC Server. CS-MGW handles the transport of user traffic in the form of IP packets only between the RNC and the CS-MGW. The MSC/GMSC Server handles the signaling (call control and mobility management) part only as shown, the dotted line between the RNC and MSC, in Figure 2.11.

  The signaling activities last for a fraction of time only. The separation of the transport and signaling or control functions between MSC Server and the CS-MGW is achieved through a new Release 4 feature called the *Bearer-Independent Core Network (BICN)*. Because of this, in the CN, IP has been introduced as the new transport network to transport voice call traffic in the form of IP packets only instead of 64 kbits/timeslot format as it was the case during the UMTS Release 99 and pre-Release 99 core networks, i.e. GSM. In these 3GPP releases, a voice call is carried through 64 kbits/timeslot format over the A-interface between the BSC and MSC.

- **3GPP Release 5 Architecture: All IP Network**

  The 3GPP Release 5 system architecture is shown in Figure 2.12. In comparison to the previous architectures, this evolution is based on the three major changes that were introduced in the Release 5 architecture:

  - An all IP network through the usage of the IP as the only transport network right from the NodeB to the CN elements.
  - The introduction of the IMS for multimedia services, e.g. voice call and Home Subscriber Server (HSS) in place of HLR (Release 4).
  - High-Speed Downlink Packet Access (HSDPA) feature to increase the data transmission speed from the UTRAN to the UE in the DL direction.

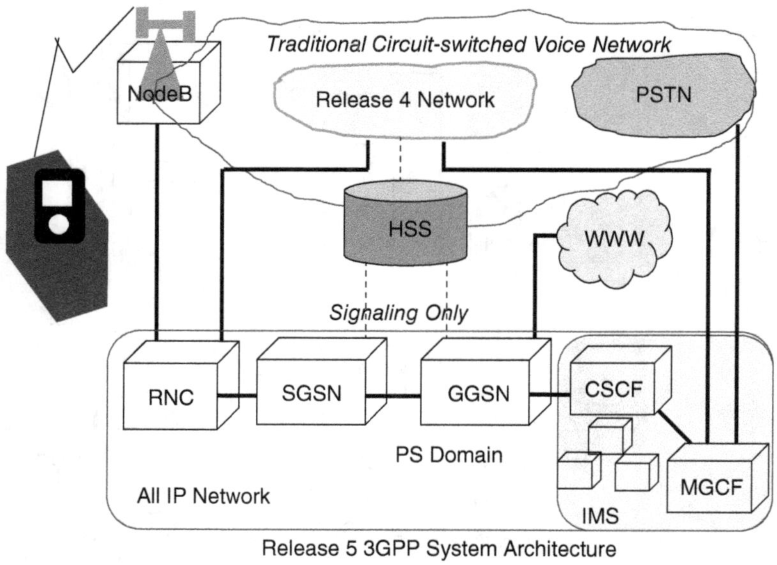

Figure 2.12 System architecture 3GPP Release 5.

In this architecture, there is no CS domain, but any CS voice call is routed through the IMS Media Gateway (MGCF) node in Release 5 to the existing Release 4 network.

- **3GPP Release 8: First Version of LTE System**

The network architecture of the Release 8 or the first version of the LTE system was shown earlier in Figure 2.4. From a comparative study of network architectures of the GSM, GPRS, UMTS, and LTE systems, the following characteristics of the Release 8, i.e. first version of the LTE network, can be summarized.

- A simpler and fully PS network based on IP transport, from E-UTRAN to *Evolved Packet Core (EPC)*. The CS domain is no longer available in the LTE and EPC networks.
- Both the AN and the CN domains of the LTE system has evolved, from the previous UMTS system, due to which it is also known as the *System Architecture Evolution* (SAE).
- LTE/EPC network has a flat architecture with fewer nodes or network elements, resulting in reduced latency and faster exchanges of information between UEs and E-UTRAN. This is because unlike the GSM and UMTS, there is no separate radio controller in the LTE system. Radio controller functionalities are integrated into the eNodeB, and it alone performs the similar functions performed by a GSM BSC and BTS or UMTS RNC and NodeB.
- New EPC network elements – MME, S-GW, and PDN gateway – have been added.

Another version of the LTE system architecture is shown in Figure 2.13. Unlike Figure 2.4, the following figure shows the interconnection of the Evolved Packet Core Network elements also, namely, the S-GW, the PDN gateway, and the HSS.

The S-GW handles and performs the user data transfer-related function, e.g. packet forwarding and routing, of the EPC network. A PDN gateway, similar to the GGSN of a GPRS network, allocates an IP address to a UE and connects the EPC network to the external IP network. For an overview of the functions by these network elements, refer to TS 23.002 [29]. The EPC network in the 3GPP Release 8 architecture support PS services only. To provide a CS voice call service for a UE registered in an LTE/EPC network, alternate features are used such as the *Circuit Switched Fall Back* (CSFB) and IMS. More about the IMS and CSFB features are described in later Sections 6.2.1.1 and 6.2.3.

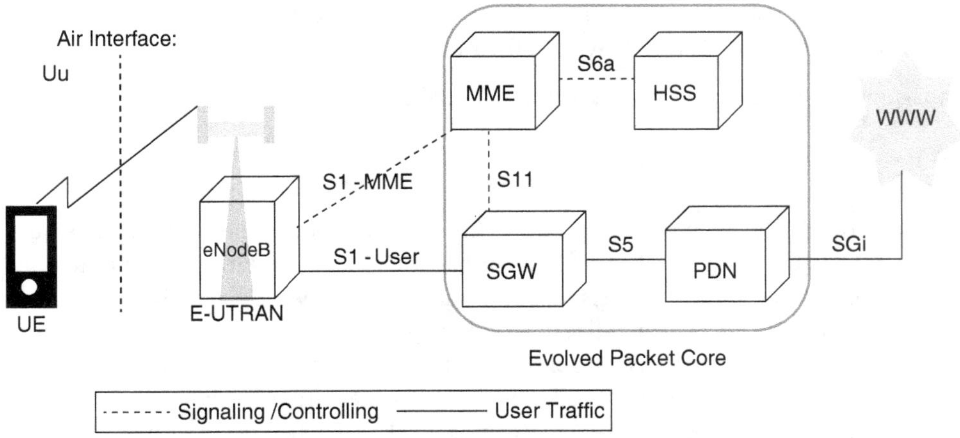

Figure 2.13 LTE system architecture with EPC nodes.

2.4 Mobile Communications Network System Engineering

In Section 2.2, the network domains of a typical mobile communications network have been introduced. Apart from these, there are several other aspects of a mobile communications network that enable network operators to run their network smoothly. Similar to any other systems engineering, a mobile communications network is also an interdisciplinary system that provides various management functions in realizing a successful network while enabling and offering various communications services to subscribers. At a high level, the essential and general systems engineering aspect of each of the mobile communications systems and networks based on the GSM, UMTS, LTE, and 5G systems can be divided into different management areas, as shown in Figure 2.14. These system engineering aspects span across the AN, the core network, and beyond.

2.4.1 Mobility Management

Mobility management aspects of a system engineering deals with the capabilities and functions performed by a mobile communications network for enabling the continuation of the current communication services being in use by a moving user. It also deals with keeping track of the current location of a mobile user so that the network could reach and alert the mobile device for an incoming call at any point in time. Other situations where mobility management functions are needed are as follows:

Figure 2.14 Mobile communications network systems engineering.

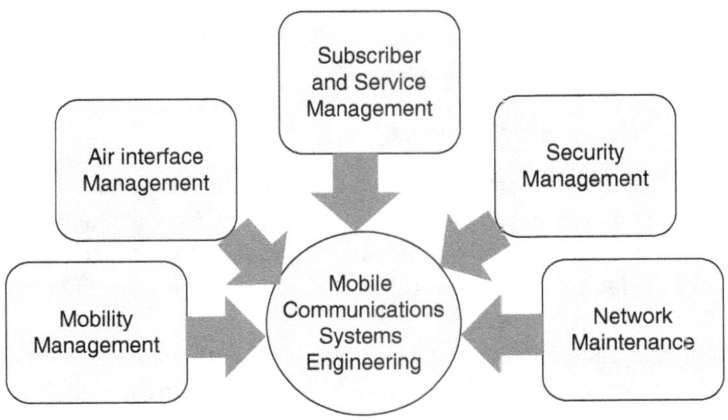

- Whenever a mobile device is switched off and on again in the same area or different service areas.
- The current state of the mobile device, i.e. idle and active and their transition.

5G system mobility management system engineering aspects are described later in Chapter 18.

2.4.2 Air Interface Management

In a mobile communications network, the air interface is used to transmit and receive data, i.e. signaling and voice traffic, over a wireless medium between a mobile device and its base trans-receiver station. The air interface uses the radio frequency transmission and forms the basis for the physical layer between the mobile device and the base trans-receiver station. Air interface management deals with the optimum allocation, re-assignment, and releasing of the allocated radio frequency resources in terms of timeslots/channels among the mobile devices. The physical properties and structure of the air interface greatly differentiate one system from another one, i.e. TDMA/FDMA in GSM, WCDMA in UMTS, and OFDMA/SCDMA in the case of the LTE system. In Section 2.3.1, we discussed how the air interface evolved from the GSM to LTE. 5G system air interface management aspects are described in Chapter 19.

2.4.3 Subscribers and Services Management

Subscribers and services management deals with various administrative tasks, as follows:

- Subscriber provisioning, i.e. establishes a new subscription, edits, or updates the existing tariff plan details, such as the supplementary services and value-added services, subscribed by a subscriber.
- Stores subscriber information in a central database.
- Generates charging, billing, and accounting for subscribers.

2.4.4 Security Management

Protecting the user's identity while engaging in a mobile communications service is a prime concern. As far as the security management functions are concerned, a mobile communications network provides the following facilities:

- Authentication, which ensures that only an authorized user/subscriber access mobile communications network services.
- Subscriber information confidentiality through ciphering/encryption method.
- Allocation of temporary identity to a mobile device to protect user identity.

Security Management aspects are described in Chapter 9.

2.4.5 Network Maintenance

Apart from network elements and their software systems, a mobile communications network consists of various active and passive infrastructures and devices. Network faults cannot be ruled out during the peak load time. Periodic and preventive maintenance minimizes the chances of failures and network downtime. Various tools are used for fault detection as well as correction of network faults. Network management aspects are described in Chapters 10–12.

Each system engineering area that is shown in Figure 2.14 is further divided into different subject areas, such as requirements, design, and signaling. The implementation details of each of the management areas shown in

Figure 2.14 shall differ as defined by the concerned 3GPP TSs, from the GSM to the 5G system. For example, in the case of the GSM system, ciphering and encryption are done for messages exchanged between the MS and the RAN only, whereas in the case of the LTE or 5G system, ciphering is also done for messages exchanged between the MS and the core network element. Also, the implementation details of a particular management area mentioned in Figure 2.14 may differ in the case of CS and PS call. For example, in the GSM system, ciphering is performed by the BTS, whereas in the GPRS system, the same is performed by the SGSN.

In this book, only the Mobility Management, Air Interface Management, Security Management, and the Network Management system engineering areas are covered. The interested reader is recommended to look for other resources for the subscribers and services management area of a particular mobile communications network.

2.5 Standardizations of Mobile Communications Networks

2.5.1 3rd Generation Partnership Project (3GPP)

In the preceding sections, we have discussed mobile communications network architectures along with their various network elements which are based on the GSM, UMTS, LTE, and 5G technologies. The interconnected network elements, either from the same or different OEM(s), provide end-to-end mobile communications and broadband services to subscribers. A network element communicates with another network element(s) through a standard protocol like the different PCs, and servers use Transmission Control Protocol/Internet Protocol (TCP/IP) and other protocols to communicate with each other over the Internet. There are international bodies that define and standardize various standard protocols such as TCP/IP, File Transfer Protocol (FTP), and Internet Control Message Protocol (ICMP). Similarly, the protocols and architectures used in a mobile communications network with interconnected network elements are defined and standardized by an organization called *3rd Generation Partnership Project*, shortly called as *3GPP*. It is a collaborative effort among different organizational partners. Those partners are ARIB, ATIS, CCSA, ETSI, TTA, TTC, and TSDSI, which are regional and national standards bodies, from Asia, Europe, and North America. See Figure 2.15.

The 3GPP and its organizational partners develop and standardize the architectures and protocols used in a mobile communications network. In a nutshell, start visiting the 3GPP site [1] right away to learn more about the 3GPP and its structures and crawl through the various information available under the different pages. The 3GPP site contains all the required information which is a key to know about the standardization processes of mobile communications systems and networks and their network elements that span across its different system engineering areas. A mobile communications network, consisting of the system engineering areas shown in Figure 2.14, that is conforming to 3GPP specifications is said to be a 3GPP compliant communications system.

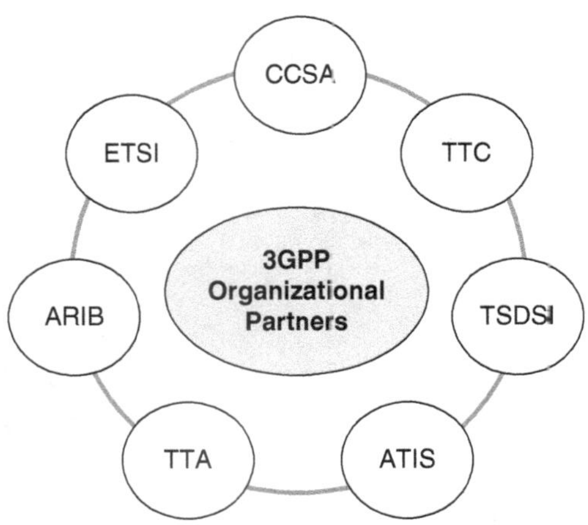

Figure 2.15 3GPP organizational partners/members.

2.5.2 3GPP Working Groups

Within the 3GPP, there are four technical specification-working groups (TSG) dealing with a particular area of work. Those groups are:

- RAN,
- Service and Systems Aspects (SA),
- Core Network and Terminals (CT), and
- GSM EDGE Radio Access Network (GERAN).

Each of the above areas/groups has further subgroups, such as RAN1, RAN2, RAN3, RAN4, RAN5, and SA1 to SA6, depending on a work area. For details, visit the specification group link on the 3GPP site [4] to know about the different working groups and their area of responsibilities. There is a table named "Project Co-ordination Group (PCG)" containing hyperlinks for various working groups which could be explored further by clicking on the same. 3GPP specifies:

- Maintenance and evolution of radio access technologies starting from GSM (2G) to 5G and beyond.
- Maintenance and evolution of core network and system architecture starting from GSM (2G) to 5G and beyond.
- Service layers such as GSM Services and IMS.

2.5.3 3GPP Technical Specification and Technical Report

In our day-to-day life, we use different electronic gadgets just in a plug and play way. Similarly, one can remove a SIM card from a phone, insert it into another phone with a different make and model, and start using it. All these are possible only when those devices are designed and works based on certain standards/protocols. The network elements of a mobile communications network are also designed and work on certain protocols as defined by the 3GPP. A particular protocol layer, and its various aspects, as defined by the 3GPP, is identified under a standard terminology called "*Technical specification*", or "*TS*" in short. There is another terminology called "*Technical Report", shortly called as "TR"*. A TR is for an informational purpose which is the result of a study phase/item on a particular area. A TR, later on, may lead to a technical specification. A technical specification is a standard specification as a result of a work item based on a TR.

A sample cover page of a 3GPP technical specification is shown in Figure 2.16. Look at the information that is available on the cover page of a 3GPP technical specification; see Figure 2.16. For example, 3GPP TS 22.060: "General Packet Radio Service (GPRS); Service Description; Stage 1" and 3GPP TS 23.060: "General Packet Radio Service (GPRS); Service Description; Stage 2"; and so on.

To decode and derive the relevant information, one must be aware of the nomenclature being used to identify a TS. Meaning and decoding of the various information available in 3GPP technical specifications are described in the subsequent sections of this book. Visit the 3GPP site [2] to get familiarity with the nomenclature. Knowing the nomenclature and picking the correct 3GPP TS is very important from the conformance point of view.

3GPP TS 22.060 V11.0.0 (2012-09)

Technical Specification

3rd Generation Partnership Project;
Technical Specification Group Services and System Aspects;
General Packet Radio Service (GPRS);
Service description; Stage 1
(Release 11)

Figure 2.16 Example of a cover page of a 3GPP technical specification. *Source*: © 2012. 3GPP ™ TSs and TRs are the property of ARIB, ATIS, CCSA, ETSI, TSDSI, TTA and TTC who jointly own the copyright in them. © 2012, 3GPP.

2.5.4 Stages of a 3GPP Technical Specification

A 3GPP technical specification may be divided into stages such as Stage1, Stage2 and Stage 3. Descriptions and purposes of each stage are shown in Figure 2.17.

In this case, the titles of the technical specifications will be the same, but they will have different specification numbers. The cover page of a sample technical specification shown in Figure 2.16 shows the "Stage1" of the technical specification.

2.5.5 Release Number of 3GPP Technical Specification

3GPP technical specifications are grouped into a particular "*Release*". A mobile communications system and the network could be developed based on the set

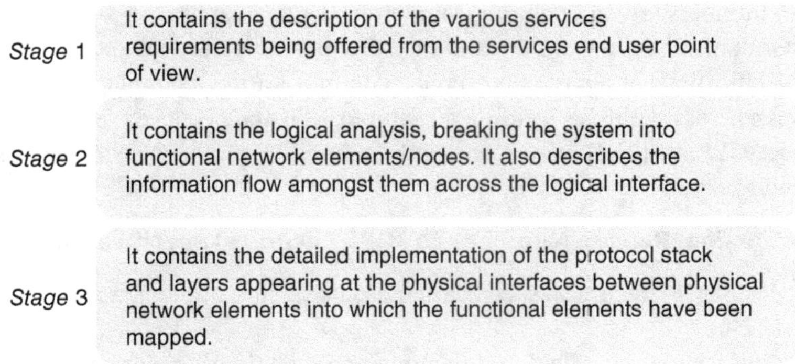

Figure 2.17 Illustration: different stages of a 3GPP TS.

of all the specifications that belong to a particular release. A mobile communications system and its node/element is said to be compliant or conforms to a particular release. A list of 3GPP releases available so far is mentioned later in Section 2.6. From the sample cover page of the TS 23.060 [31] shown in Figure 2.16, the corresponding release number is 11 that is derived from the version field: V11.0.0.

2.5.6 3GPP Technical Specification Numbering Nomenclature

- **Series Number**

Each 3GPP protocol layer has a unique technical specification number. An analogy to a technical specification is the Request for Comments (RFC) number, e.g. RFC 1234, defined by the Internet Engineering Task Force, of a particular protocol such as ICMP. The technical specification numbering system is like this: *ab.xyz or ab.xy* where the first two digits, i.e. *ab*, identifies the *series number* of the technical specification. This is further followed by either 3 digits (e.g. xyz) for series from 21 to 55 or 2 digits (e.g. xy) for series from 01 to 13. For example, consider the cover page of the sample TS 22.060, shown in Figure 2.16. The corresponding technical specification number is TS 22.060, where *22* is the *series number* here.

There are large numbers of technical specifications, or standards, for different areas of the mobile communications networks based on the GSM, GPRS, UMTS, LTE, and 5G technologies. Those technical specifications cover the different system engineering aspects of a mobile communications network as described in Section 2.4. Technical specifications of related protocols or subject areas (e.g. signaling, service aspects) of a particular mobile technology are grouped into so-called *"specification series"*, for example, 3GPP TS 44 series, 24 series specifications, and so on. Within a particular series, all the technical specifications of a related protocol or protocol stack can be found.

Visit the 3GPP site [2] for the lists of technical specifications series that are available in each of the mobile communications technologies, such as GSM, GPRS/EDGE, UMTS, LTE, and 5G (second, third, fourth column) under a particular subject area (first column). The specification is organized into the following categories:

- 3G (UMTS) and beyond/GSM (R99 and later), covering the 5G also;
- GSM only (Release-4 and later); and
- GSM only (before Release -4).

Start with and look at the standard specification(s) for a particular mobile communications technology domain (column (2) or (3) or (4)) of your interest. For the complete list of the technical specification series, the reader is advised to visit the 3GPP site [2]. Note that after Release 4 or R4, the old GSM specification numbers are increased by 40 (second column from right in [2]), whereas the UMTS standards or specification numbering is lowered by

20 numbers (second column from left in the table appearing on this web site page) corresponding to each GSM standard. Thus, the GSM standard 01.0x has now become GSM 41.00x and UMTS 21.xxx. Note that technical specifications of a particular mobile communication technology such as GSM, EDGE, UMTS, LTE, or 5G could span across multiples series. For example, LTE and 5G have the dedicated specifications series, i.e. 36 and 38; see the 3GPP site [2]. However, one can also find LTE protocol layer-related specifications/information in other series such as the TS 24.008[45].

- **Version, Release Number of a 3GPP Technical Specification**

  Each 3GPP technical specification has its version number in the form of *a.b.c*. The meanings are as follows:

  – *a*

  This is the *release* field. It is incremented each time a major new functionality is added to the concerned mobile communication technology area such as GSM, GPRS/EDGE, UMTS or 5G system.

  – *b*

  This is the *technical* field. It is incremented each time a technical change is made to the concerned technical specification. Note that the *technical* information field is reset to zero every time the release field is updated.

  – *c*

  This is the *editorial* field. It is incremented each time an editorial change is made to a specification. Note that it is reset to zero every time the technical field is updated. For example, TS 44.060 V6.0.0 means that it belongs to 3GPP Release 6. Open a technical specification and have a look at the meaning of each number. One should consider referring to a specification having the first digit of the version as 3 and above because that is the version of the specification under change control. More about the version numbering scheme could be found by visiting the 3GPP site link [5].

2.5.7 Vocabulary of 3GPP Specifications

For a list of glossaries of terms, their definition, and the abbreviations as found in various 3GPP specifications as well as in this book, refer to the 3GPP TR 21.905 [24].

2.5.8 Examples in a 3GPP Technical Specification

3GPP technical specifications are full of theories, and one will not find, except in some cases, virtually illustrating examples in any technical specifications. This may make the reader difficult to grasp a particular concept at the first go. In this case, it is highly recommended to study the contents of a particular message or a TS several times. The reader may take the help of an experienced professional/colleague in this regard.

2.5.9 Standardization of Technical Specifications by 3GPP

Standardizations of technical specifications are required and important to design interoperable systems by different vendors/OEMs. Because of standardizations, one can use various telecommunications services from different service providers using varieties of handheld devices such as Mobile and POS terminal.

2.5.10 Scope of 3GPP Technical Specification (TS)

A particular 3GPP technical specification description may cover all the systems. A TS specifies its applicability/scope to other mobile communications systems such as the GERAN, UMTS, LTE, and 5G. Pay attention to this fact. For example, click on any specification series for a particular subject area on the 3GPP site [2]. All the technical specifications shall be presented under that particular series. Now, further, click on any technical specifications. A new window shall be opened, and from there, one can find the scope and applicability, under *Radio Technology*, of the technical specification being clicked. Below the *Radio Technology*, there is a link by clicking which all the versions/releases of the particular technical specification can be downloaded.

The first section of every technical specification also describes its scope. Moreover, the first page of every technical specification displays the GSM or LTE or LTE Advanced or 5G logo depending on the scope.

2.5.11 3GPP TS for General Description of a Protocol Layer

A 3GPP protocol layer may perform several important functions and procedures. The functions and procedures performed may be split into several individual technical specifications. For such a protocol layer, it contains an introductory technical specification too, for example, LTE TS 36.201 [89] and 5G NR TS 38.201 [105], which provides an overview and a general description of the concerned protocol layer. The introductory technical specification also describes the detailed relationships among the split technical specifications. It may be also noted that for the same protocol layer, for example, in the case of the UMTS physical layer, there may be separate technical specifications depending on its duplexing modes, i.e. Frequency Division Duplex (FDD) and Time Division Duplex (TDD), of transmission which is used by the physical layer.

However, in the LTE and 5G NR system, no separate technical specification is used to describe the physical layer functions and procedures working in the TDD and FDD mode.

2.5.12 3GPP TS Drafting Rules: Deriving Requirements

To derive various technical, functional, and other requirements without any ambiguity, the contents of a 3GPP technical specification should be understood and decoded properly. A 3GPP technical specification description contains the following types of requirements.

- Normative

Such requirements must be complied with.

- Informative

Such requirements if ignored, it does not matter. Note that a technical report is always an informative one.

To identify the above requirement types, a 3GPP specification/text description may contain the following auxiliary verbs:

- Shall/Shall not – This is a normative and implies a mandatory requirement.
- May/Need not – This is a normative and optional requirement.
- Should/Should not – This is a normative and Recommendation requirement.
- Can/Cannot – This is a normative and possibility/capability requirement.

Also, a 3GPP technical description is described using an active voice rather than a passive voice. For example, "*The MS shall perform...*" rather than the "*...routing area update shall be performed by the MS*".

2.5.13 Download 3GPP Technical Specifications

1) Visit the 3GPP site [2].
2) Select a subject/area and the particular technical specification series. Select a technical specification of interest to download. Click on the *Click to see all versions of this specification* to download a specification for a particular release.

2.5.14 3GPP Change Requests

An introduction of a new feature as part of a release is documented in the form of a *Change Request*. The affected technical specifications due to a new feature are captured under a particular Change Request. Given a Change Request number, the affected technical specifications may be downloaded from the 3GPP portal https://portal.3gpp.org/ChangeRequests.aspx. One can find the relevant changes introduced in each affected technical specification due to a new 3GPP feature.

2.5.15 Learnings from 3GPP Meetings TDocs

Many technical documents (TDocs) in the form of written contributions are produced as a result of various meetings held among the 3GPP technical specifications working groups mentioned in Section 2.5.2 and other stakeholders. To produce a version of a technical specification, numerous meetings are held. Visit the 3GPP site [2] and follow the steps mentioned in the previous section. Click on the tab *Versions*. Look at the *Meetings* column under a particular *Version*. Click on any *Meeting* number to display the various written contributions (TDocs) under the "List of TDocs" link. A TDoc or a written contribution contains highly technical discussions, observations, and proposals, in a particular area/topic of the concerned technical specification. One can acquire a great deal of knowledge from such a TDoc, which can be used as supplementary information. If a particular aspect is not clear from a technical specification, a TDoc may be referred for more information on the related area.

2.6 3GPP Releases and Its Features

The 3GPP technical specifications are organized into different versions called "*releases*", such as Release 99, Release 4, Release 5, Release 6, Release 7, Release 8, Release 9, Release 10, Release 11, Release 12, Release 13, Release 14, Release 15 and beyond. Various features introduced in each 3GPP releases are summarized in Table 2.3.

Some of these 3GPP releases involve the SAE right from the GSM to the 5G system as described earlier in Section 2.3. However, some of these releases involve the introduction of additional features only to improve the existing data transmission rates. New functionality or enhancement to an existing release is also added in each subsequent release. The Release 99, shown in Figure 2.10, was the first version where UMTS came into existence which evolved from the GSM/GPRS system in terms of the new RAN, UTRAN. Subsequent releases of the UMTS

Table 2.3 Summary of 3GPP Releases and their features.

3GPP Releases	Features
Release 99 [Year: 2000]	UMTS First Release, combination of GSM and UMTS, voice call through E1 interface.
Release 4 [Year: 2001]	• UMTS core called Bearer-Independent Core Network (BICN)/Next-Generation Network, • Splitting architecture, separating control plane and data plane
Release 5 [Year: 2002]	IMS, HSDPA, all IP network
Release 6 [Year: 2005]	HSUPA
Release 7 [Year: 2007]	HSPA+
Release 8/9 [Year: 2010]	LTE E-UTRA/LTE Baseline Release
Release 10 [Year: 2011] and beyond	LTE-Advanced, Career Aggregation
Release 15 and Beyond	5G System

UTRAN introduced new features offering increased data transmission rates. Similarly, 3GPP Release 8 was the first version of the LTE system, and Release 15 was the first version of the 5G system.

Prior to Release 99, 3GPP technical specifications releases were known by the corresponding year such as Release 1997 and Release 1998. However, after the Release 99, 3GPP technical specifications releases are known by the corresponding version number only such as Release 4 Release 5 and beyond. A new 3GPP release details are described through its own and dedicated technical specifications series. However, because of a new feature or functionality, an existing technical specification from a previous release version may be also impacted. Details of such information can be derived from the 3GPP release descriptions document available on the 3GPP site [3].

Apart from the introduction of a new feature in each 3GPP release, a new message(s) or a new information element(s) or a new type of channel, either in DL or UL, to an existing protocol layer may be also added. For further details on each of the above releases, visit this 3GPP site [3].

Chapter Summary

In this chapter, we have presented the introductory architectures, the various domains or areas, and the network elements of mobile communications networks based on the GSM, UMTS, and LTE systems. The 5G system is also covered in appropriate sections of this chapter. We then presented the global 3GPP standardizations process of various technical specifications based on which mobile communications systems and networks are designed and built. The evolutions of mobile communications systems and networks in terms of 3GPP system architectures release are also discussed.

The reader is advised to get familiar with the various aspects of a 3GPP TS. To start with, select a particular mobile communications system and then use the 3GPP specification series [2] as the guiding resources to download and study the technical specifications of a particular series and the subject area of interest. Get an overview of the various protocol layers, interfaces, procedures, and functions performed by a particular network element.

This chapter was the introductory one, for the reader, whose contents are general in nature and thus applicable to the 5G system also. The various protocols, procedures, and functions performed by a mobile communications network and its elements/entities shall be described in detail in the subsequent chapters of this book. Design, development, and implementation aspects of a network element in the software code shall be also presented in the subsequent chapter.

3

Protocols, Interfaces, and Architectures

Introduction

This chapter covers the protocol architectures of network elements of mobile communications networks based on the Global System for Mobile Communication (GSM), Universal Mobile Telecommunication System (UMTS), Long-Term Evolution (LTE), and 5G systems as defined by the 3rd Generation Partnership Project (3GPP). The protocol stack, its layer, and the architecture of a mobile communications network are different from the traditional Internet Protocol (IP) networks. Nevertheless, the protocol stack and layers of a mobile communications network can be studied in parlance with the OSI-7-layer reference network architecture. Developers must have understandings of the architectural aspects of protocol stack and its layers as it is core to the design and development of network elements of mobile communications systems and networks.

We begin with the basic means of communication between two peer protocol layers of network elements irrespective of the mobile communications system and network being used. Following this, we present the concept of sublayering of a protocol layer performing different functions and procedures of each protocol sublayer. We then present how protocol layers are grouped in case of the UMTS, LTE, and 5G systems based on the peer-to-peer communication between a User Equipment (UE) and the radio access network (RAN) and between UE and core network (CN). Classification of individual protocol layers is also presented based on the nature of functions performed by each layer. We also present the working model of a protocol layer based on which it provides services to its upper layer or uses the services from a layer below it. We close this chapter with another aspect of peer-to-peer protocol layer communications, which is protocol layer termination.

3.1 Protocol Interface and Its Stack

Network elements of mobile communications networks are designed and developed based on a set of standard protocols and specifications as defined by the 3GPP. A *protocol interface* is a communication path that is established between two network elements. Each network element and its protocol layers communicate with the peer network element and its protocol layers through a particular interface. The related protocol layers or the protocol stack, supported by a network element, are organized into a protocol *interface*. Over a particular protocol interface between two network elements, they exchange various signaling messages corresponding to a particular function and procedure such as the following ones:

- Call Control Management, i.e. a call establishment, maintenance, and its releasing;
- Mobility Management and Session Management (SM);

Mobile Communications Systems Development: A Practical Introduction to System Understanding, Implementation, and Deployment, First Edition. Rajib Taid.

- Radio Resource Management, Handover, and Power Control Management; and
- Interworking, Interoperations, Roaming Management, and so on.

A protocol interface also carries user data or traffic between two network elements. A protocol interface between two network elements could be physical as well as logical as described below. A logical interface may be a point to point, i.e. direct, or may span through several network elements. A logical interface works on top of a physical interface.

3.1.1 Physical Interface

The physical interface defines the electrical characteristic of the physical transmission media being used for the actual transmission and reception of information between a sender and receiver. Consider the physical Cat5 cable or its other variants used in a Local Area Network to transmit and receive upper/lower layers information. Application data received from the upper layers are converted into Ethernet frames. The network interface card (NIC)/Ethernet chipset generates the required electrical signals to transmit the frames through the Cat5/6 physical cable/layer. The NIC/Ethernet driver also knows when to read and write binary data from the physical interface and layer.

Similarly, in the case of a mobile communications system also, various physical transmission interfaces/media are used across the different systems, i.e. GSM to 5G, to transmit either signaling or user traffic between two network elements. Being the GRPS CN, LTE/Evolved Packet Core (EPC) and 5G as an IP network-based system, the IP transport network is used to transmit and receive information among their network elements. Examples 3.1 and 3.2 below illustrate the typical physical interfaces found in mobile communications systems, from the GSM to the 5G system.

3.1.2 Logical Interface

In computer network programming, an interface is a logical point, such as a socket where two different applications/systems exchange information and communicate with each other in terms of a protocol data unit (PDU). Similarly, in the case of mobile communications networks also, two different network elements communicate, either signaling or user data, with each other using a set of predefined *messages* or PDUs. These collective *PDUs* or *messages* define the particular *logical interface* between two network elements.

A logical interface works on top of a particular physical interface. Using a logical interface, network entities exchange both the user data and signaling information in the form of a message or a PDU. For example, consider

Example 3.1 GSM E1 Physical Interface

One of the most commonly used physical interfaces is the *E1*, which is defined under the standard G.703 of ITU-T [12]. An *E1 physical interface* is not only used in the mobile communications system but also used in other systems to carry voice, data, and so on. For example, *E1 physical interface* is used between the GSM Base Station Controller (BSC) and Base Transceiver Station (BTS); GSM BSC, and Mobile Switching Center (MSC), or between the MSC and Serving GPRS Support Node (SGSN) for the Gs interface. An E1 physical interface has two pair/four wire cables for both Receive (RX) and Transmit (TX) purposes. The total bandwidth supported by the E1 interface is the 2 Mbps that is divided into 32 timeslots of 64 kbps each. Each 64 kbps timeslot is again divided into four sub-timeslots. A sub-timeslot can be allocated to a GSM voice call or General Packet Radio Service (GPRS) data communication purpose. Figure 3.1 illustrates the configuration and usages of an E1 physical interface to carry signaling as well as user traffic. Using an E1 interface, several transceivers (TRXs), along with their signaling channels, of a BTS can be configured.

Figure 3.1 Illustration: physical E1 interface configuration for GSM A-bis interface.

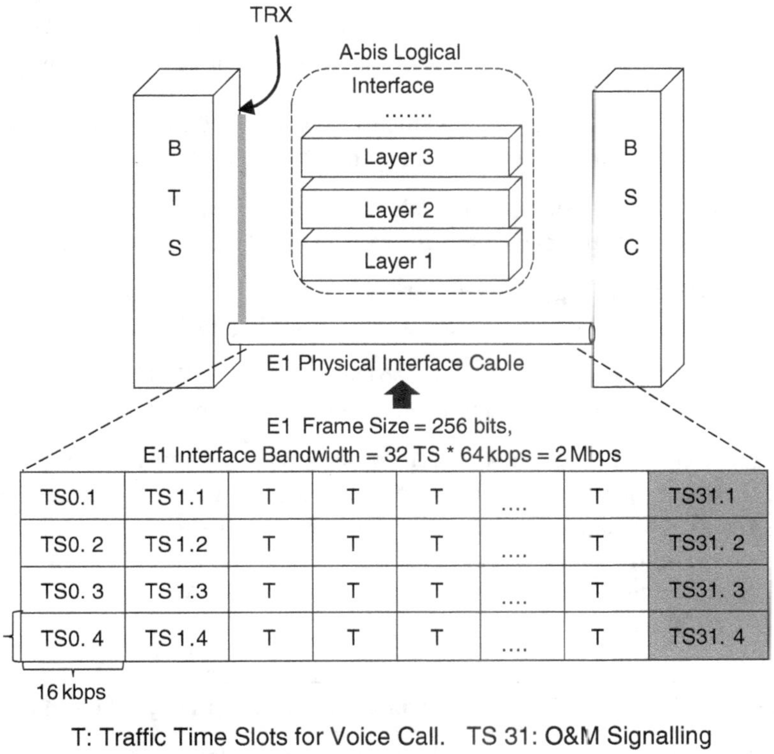

Example 3.2 Mobile Communications Networks and Their Physical Air Interface

An MS/UE communicates with the RAN of the GSM/GPRS, UMTS, LTE, and 5G systems through their respective air interface between them. The air interface is the physical interface where the actual physical link/interface/transmission media used over the air interface is the *radio wave*. A *radio wave* carries the modulated information between the UE/MS and RAN and vice versa. The air interface between the MS/UE and GSM RAN, UMTS UTRAN, LTE Evolved-UMTS Terrestrial Radio Access Network (E-UTRAN), and 5G RAN is illustrated, in general, in Figure 3.2 below.

Figure 3.2 Illustration: physical air interface for GSM, UMTS, LTE, and 5G NR systems.

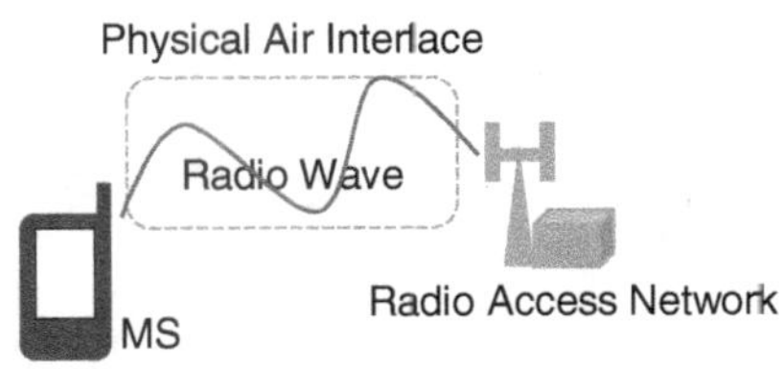

the GSM *A-bis interface* between the BTS and BSC and the *A-interface* between the BSC and the MSC. Both of these logical interfaces work on top of the physical E1 interface to transport signaling and user traffic in a GSM network. Typical signaling messages exchanged over the A-bis interface are radio resources request and allocation to an MS and its release and so on.

A logical interface may consist of a protocol stack that may contain several protocol layers in it. Protocol layers are grouped into a particular logical interface which is known by a particular logical interface name for ease of its identification. Examples 3.3 and 3.4 below illustrate the typical logical interfaces found in the LTE/EPS and the 5G system.

The S1 logical interface between the LTE eNodeB and EPC contains two types of protocol stacks that are used to carry signaling and user data; see Figure 3.3a and b:

- User data transmission protocol, also called *user plane*, S1-U.
- Signaling or control plane protocol, called *control plane*, S1-Application Protocol (AP).

Similarly, the NG logical interface between the 5G NG-RAN/gNB and UPF contains two types of protocol stacks that are used to carry signaling and user data; see Figure 3.3a and b:

- User data transmission protocol, also called *user plane*, NG-U.
- Signaling or control plane protocol, called *control plane*, NG-AP.

The S1-U or NG-U user plane protocol stack is used to transfer user traffic or data between the respective RAN (LTE eNodeB or 5G gNB) and its CN element (LTE/EPS MME or 5G UPF). The control plane protocol stack, *S1-AP(LTE/EPS) or NG-AP (5G),* is used to transfer signaling messages between the respective RAN (LTE eNodeB or 5G gNB) and its CN element (LTE/EPS MME or 5G AMF). Figure 3.3a shows side-by-side the S1-U and NG-U user plane protocol stacks, and Figure 3.3b shows the S1-AP and NG-AP signaling or control plane protocol stacks.

Example 3.3 LTE/EPC S1 Logical Interface and 5G NG Logical Interface

Consider the *S1 logical interface* between the eNodeB and the EPC (consisting of Serving Gateway (S-GW), Mobility Management Entity (MME), and packet data network gateway (P-GW)) in the case of the LTE/EPC network. Similarly, consider the *NG logical interface* between the Next Generation Radio Access Network (NG-RAN)/gNB and the *User Plane Function (*UPF*)* in the case of the 5G network. Refer to Figure 3.3a and b.

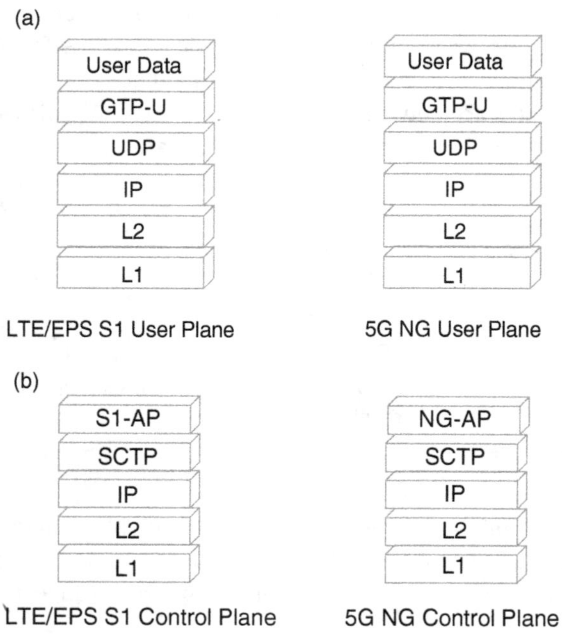

Figure 3.3 (a) LTE S1 and 5G NG logical interface: user plane. (b) LTE S1 and 5G NG logical interface: control plane.

Figure 3.3 shows that the S1-U and NG-U or the S1-AP and the NG-AP have the same protocol stack over the IP transport network. Protocol layer classifications under user plane and control plane protocols are described later in Section 3.2.

Example 3.4 LTE Logical Interfaces with the Same Protocol Stack

In the LTE/EPS mobile communications network, several logical interfaces may have the same protocol stack and its transport network. For example, consider the IP transport-based logical interfaces: S3 (MME-SGSN), S4 (S-GW-SGSN), S5 and S8 (S-GW – P-GW), S10 (MME-MME), and S11 (MME-S-GW) that are used among the several network elements of an LTE/EPS network. These logical interfaces have the same protocol stack on top of the IP transport network, i.e. UDP/IP, which is used to tunnel IP payload from one network element to its peer. Refer to the illustration shown in Figure 3.4. IP payload is tunneled through a protocol called *GPRS Tunneling Protocol* (GTP) in an encapsulated manner. GTP PDUs between two network elements are transported on top of the underlying IP transport network.

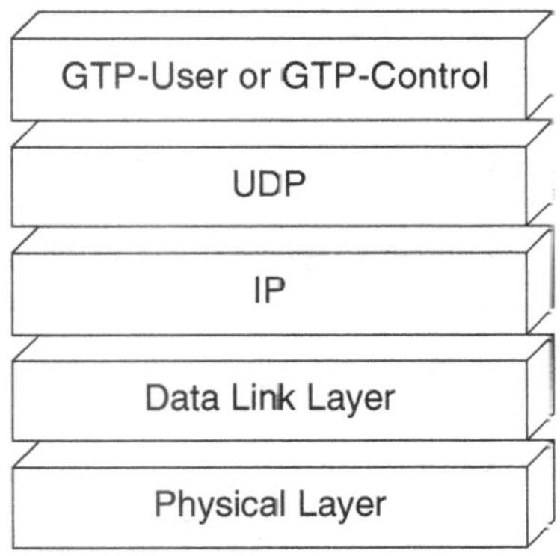

Figure 3.4 LTE/EPS logical interfaces: S3, S4, S5, S8, S10, and S11 protocol stack.

Note that in the case of the LTE/EPS system, the GTP control plane Version 2 (GTP-C v2) [70] is used in all the logical interfaces mentioned above to carry signaling information between their respective network elements. But in the case of the GPRS/UMTS system, GTP control plane Version 1 (GTP-C v1) TS 29.060 [67] is used. Similarly, the GTP-user plane, TS 29.281 [72], protocol is used to carry user traffic/data between their respective network elements.

To achieve successful interoperability between two network elements from two different vendors, it is important to understand the concerned protocol specification, encode/decode, and implement correctly each message/PDUs defined in a particular logical interface and its protocol specification. Different logical interfaces may use different physical transmission media to transport the messages of a logical interface. All the IP transport-based mobile communications protocols will use *the Stream Control Transmission Protocol* (SCTP) RFC 4960 [17] or UDP protocol suite and UTP CAT cable as physical media.

3.1.3 Logical Interfaces' Names and Their Protocol Stack

In a computer network, the TCP/IP protocol suite has a collection of protocols that are classified and grouped into four layers, namely the application layer, transport layer, network layer, and the physical layer. By referring to the TCP/IP protocol suite, one can easily recall the various layers and protocols available under it.

Similarly, a logical interface in mobile communications networks also consists of a protocol stack and its layers. To identify and refer such protocol stack and its layers, each logical interface is given an identification name that begins with a capitalized English alphabet letter. For example, in the previous section, the LTE/EPS logical interface S1 or 5G NG logical interface was discussed. Similar examples of other logical interfaces are A-interface between the BSC and MSC, Gs interface between the MSC and SGSN, A-bis interface between the BTS and BSC, Iu interface between RNC and MSC, and so on.

There are large numbers of logical interfaces connecting different network elements of GSM, GPRS, UMTS, LTE/EPS, and 5G networks. To provide a brief overview to the reader, the following Tables 3.1 to 3.4 lists the names of some of the logical interfaces, second row, along with their network elements, first row, used in the GSM, GPRS, UMTS, and LTE/EPS networks.

The reader may refer to the corresponding technical specification(s) which are available on the 3GPP site [1] for further information on the logical interfaces mentioned in Tables 3.1 to 3.4.

Table 3.5 summarizes the various logical as well as the physical interfaces being used between the respective RAN and the CN in the GSM, GPRS, UMTS, LTE, and 5G systems.

Table 3.1 GSM and GPRS system network elements and logical interfaces.

Air Interface (MS-BSC)	BTS-BSC	BSC-MSC	BSC-SGSN	SGSN-GGSN
Um	A-bis	A	Gb	Gn

Table 3.2 UMTS system network elements and logical interfaces.

Air Interface (UE-NodeB)	Node-RNC	RNC-MSC	RNC-SGSN	SGSN-GGSN	GGSN-WWW
Uu	Iub	Iu-CS	Iu-PS	Gn	Gi

Table 3.3 LTE/EPS network elements and logical interfaces.

Air Interface (UE-eNodeB)	eNodeB-MME	eNodeB-S-GW	S-GW - P-GW	P-GW - WWW
Uu	S1-AP	S1-U	S5	SGi

Table 3.4 Interworking features and their logical interfaces.

3GPP Feature: CSFB (MME-MSC)	Feature: SRVCC (MME-MSC, MSC-SGSN)
SGs	Sv

Table 3.5 Logical and physical interface between RAN and CN elements.

System	Interface Between Network Element		Logical Interface Name	Physical Interface Name
GSM	BSC	MSC	A-interface	E1
GPRS	BSC	SGSN	Gb-interface	E1, IP Layer 1
UMTS	RNC	MSC	Iu-CS	PDH, SDH, IP Layer 1
		SGSN	Iu-PS	PDH, SDH, IP Layer 1
LTE	E-UTRAN	MME	S1(S1-AP)	IP Layer 1
		S-GW	S1(S1-U)	IP Layer 1
5G	NG-RAN	AMF	NG (NG-AP)	IP Layer 1

For more information on all the 3GPP defined logical interfaces, their protocol stack/layers, and network elements of mobile communications networks, refer to the 3GPP TS 23.003 [30]. The next section illustrates the logical interfaces and their physical transmission network used in a typical GSM and GPRS network.

3.1.4 Examples of Logical Interface and Its Protocol Layers

Section 3.1.2 described the different logical interfaces used among the LTE/EPC or 5G network elements where user data or signaling information is transported over the same underlying IP transport network. However, in a GPRS and GSM network, different transport networks may be used apart from the IP network as illustrated below:

- **GPRS Gb-Interface Protocol Stack and Layers**

In the case of the GPRS system, the Gb-interface between the BSC and the SGSN is used for transferring the signaling and user data exchanged between the MS and the SGSN. The GPRS Gb interface consists of the following applications protocol layers:

- BSSSGP – *Base Station Subsystem GPRS Protocol*
- NS – *Network Service*

The Network Service (NS) layer is further divided into the following components:

- Network Service Control (NSC) Protocol
- Sub-Network Service (SNS) Protocol

The Gb interface stack with the above protocol layers follows the protocol structure shown in Figure 3.5. Below the NS layer, there are two transmission network protocols, having its physical layer and media that can be used to exchange upper-layer information between the Base Station Subsystem (BSS) and the SGSN. Those transport networks are:

- Frame Relay (FR) Protocol Transmission Network and
- IP Transmission Network.

Based on the transmission network protocol, Frame Relay, or IP, shown in Figure 3.5, the concerned physical layer shall be used accordingly. Note that the Gb-interface cannot use both the Frame Relay and the IP transmission network simultaneously but shall use only one transmission network, either Frame Relay or IP, at a time.

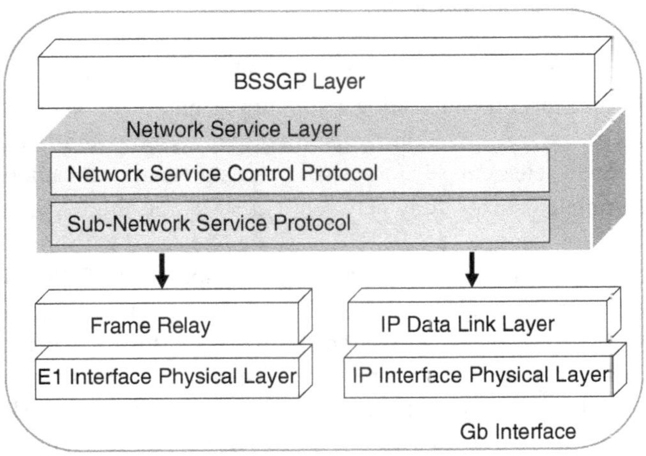

Figure 3.5 GPRS: Gb-interface protocol stack with IP and frame relay transport network.

The particular transmission network to be used is configured by the operator and maintenance personnel, and it is taken care of by the *SNS Protocol* at the NS layer level.

The *NS Control Protocol* deals with transferring of the information of the upper layer in the form of *NS Data Units* (NS SDU), whereas the *SNS Protocol* deals with the management of the underlying transmission network protocol (Frame Relay or IP) to be used at a time to transfer upper-layer information. For further details on the GPRS NS layer, refer to TS 48.016 [135].

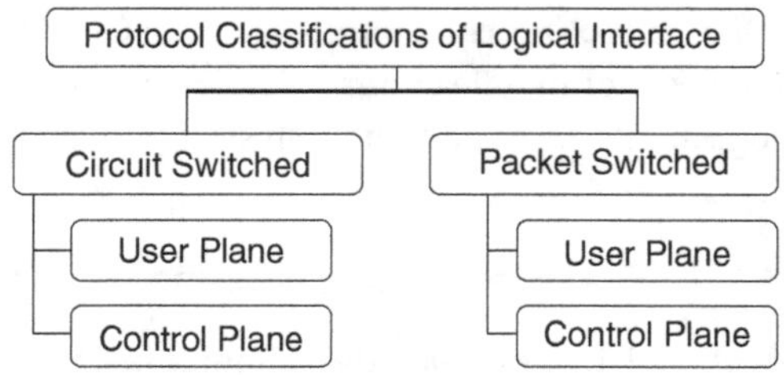

Figure 3.6 Protocol layers classifications in GSM, GPRS, UMTS, LTE, and 5G systems.

3.2 Classifications of Protocol Layers

In a mobile communications network, the circuit-switched (GSM, UMTS) or packet-switched (UMTS, LTE, 5G) domain protocol layers of a protocol stack that belongs to a particular logical interface are classified into the following categories (Figure 3.6):

- Control Plane or Signaling Plane
- User Plane or Data Plane

3.2.1 Control Plane or Signaling Protocols

- **What is a Signaling Message?**

One important aspect of a mobile communications network is the "signal" that is exchanged between the network elements. Mobile communications signaling is nothing but an exchange of information between the concerned network elements prior to providing communications services to a subscriber or another network element. Signaling messages among the network elements are exchanged to reserve resources as part of the call setup and release the resources as part of the call release procedure. In addition to this, signaling messages are exchanged to carry out various functions and procedures among the network elements of GSM/GPRS, UMTS, LTE/EPS, and 5G networks. By exchange of signaling information among the network elements, necessary logical objects, or contexts, a database of various information is created at the UE/MS, RAN, and CN end. Information such as the capabilities of a particular MS/UE, radio access, and CN capabilities and their features, information about a subscriber, and so on are also exchanged only through signaling, which is updated at the UE/MS, RAN, and CN end.

In a traditional Public Switched Telephone Network (PSTN) fixed network also, signaling functions are performed for call establishing, maintaining, and supervision and call release. Note that control/signaling plane messages last for a fraction of a second only to setup and release a call or to perform other procedures among the network elements. Before a user can start communications services, a signaling procedure or phase must be completed successfully between a UE/MS and the RAN, between the RAN and CN, and between the UE and CN. For example, an MS/UE must be registered with the CN through a successful GSM location area or GPRS routing area or LTE/EPS tracking area update or 5G Registration Request procedure. Prior to performing such signaling procedures, a UE/MS must request and be allocated the necessary radio resources by the GSM BSC, UMTS RNC, or LTE eNodeB or 5G NG-RAN through signaling only. All such signaling procedures in a particular mobile communications network are performed by its control plane protocol stack and its layers only.

If there is a signaling congestion situation in a network, an MS/UE would not get the requested resources from the network. Thus, a mobile user would not be able to make a circuit-switched (CS) or a packet-switched

(PS) call. This would further require the troubleshooting of the issue or a proper radio network planning by the operator.

- **LTE/EPS Network Control Plane Protocols: UE – eNodeB – MME**

Figure 3.7 shows the end-to-end control plane protocol layers of an LTE network, which is reproduced from TS 23.401 [39].

The LTE air interface control plane protocol layers between the UE and E-UTRAN are Radio Resource Control (RRC), Packet Data Convergence Protocol (PDCP), Radio Link Control (RLC), and Media Access Control (MAC) layer. Between the UE and the MME, the control plane protocol layer is known as the *Non-Access Stratum* (NAS) layer, consisting of the EPS Mobility Management (EMM) and ESM protocols. On the CN side, the control plane protocol is S1-AP/MME AP between eNodeB and MME. The S1-AP control messages are transported over the *SCTP* [17]. The EPS mobility and SM layer performs similar functions and procedures as described above for the GPRS control plane protocols. The RRC protocol layer is responsible for providing a reliable link between the UE and the MME in the case of the LTE/EPS network (Figure 3.7). Beyond the LTE/EPC MME, it uses the GTP control plane [11] protocol to perform tunnel management procedures with the S-GW.

The Relay as shown in Figure 3.7 is not a protocol layer but an application module that is responsible for the reconstruction or reassembly of information transmitted by the lower layer, i.e. RLC or RRC. A common module also available in GPRS, UMTS, LTE RAN, and 5G NG-RAN is the Relay module. In the GPRS/UMTS system, the Relay module forwards the reformatted data to the GPRS BSSGP or UMTS Radio Access Network Application Part (RANAP) layer in terms of PDUs for onward transmission to the SGSN over the Gb or Iu-PS interface. Similarly, in LTE/EPS, the Relay module forwards data to the MME over the S1 interface. In 5GS, the Relay module forwards data to the AMF and UPF. The relay module is required to follow the underlying protocol layer details to reassemble their data for onward transmission to the CN.

- **Functions of Control Plane Protocol: Air interface**

The control plane protocol stack performs different functions according to its logical interface. From the air interface point of view, some of the common functions and procedures, in general, performed by the control plane protocol stack and layers in the GSM, GPRS, UMTS, LTE/EPS, and 5G system are as follows. These functions are similar though their protocol layer specification and implementation aspects are different:

- Broadcasting of network system information to mobile devices,
- Radio resource allocation for CS (GSM) and PS services.
- CS (GSM) and PS call setup, supervise, and its release,

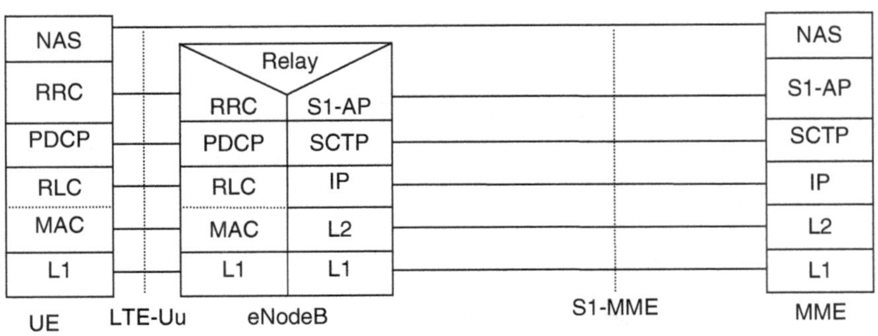

Figure 3.7 LTE network end-to-end control plane protocol layers. *Source:* © 2015. 3GPP ™ TSs and TRs are the property of ARIB, ATIS, CCSA, ETSI, TSDSI, TTA and TTC who jointly own the copyright in them. © 2015, 3GPP.

- MM functions and procedures such as the GSM location update, handover, GPRS routing area, LTE/EPS tracking area update procedure, and 5G registration area update procedure, and
- GPRS, LTE/EPS SM functions and procedures such as Packet Data Protocol Context Creation and Bearer Allocation.

However, the air interface control plane protocol stack and layers also perform different functions and procedures that are specific to a particular communications system, i.e. GSM, GPRS, UMTS, LTE, and 5G systems. For example, the following functions are performed by the UMTS, LTE, and 5G air interface control plane protocol stacks only:

- Header compression,
- Establishments of radio bearers,
- Configurations of lower layers, e.g. PDCP and RLC, by another higher-layer RRC,
- Ensuring the ciphering, integrity, and security protection of the information exchanged between UE and RAN and CN, and
- Provision for a transparent mode (TM) of user data transfer.

Using the control plane protocol layers functions and procedures, the RAN or CN controls and commands the behavior of an MS/UE. For more information on the control plane protocol functions and procedures by the individual protocol layer of a logical interface, refer to the 3GPP TSs mentioned in the References section of this book.

3.2.2 User Plane Protocols

User plane protocol layers perform the functions required to transmit and receive the user/application traffic between the source and destination network element and vice versa. As an example, Figure 3.8 shows the end-to-end LTE system, TS 23.401 [39], user plane protocol layers structure starting from the UE, eNodeB, and S-GW to P-GW.

At the UE end, the user plane protocols consist of the application layer, IP layer, PDCP, RLC, MAC, and physical layer. Note that the application and IP layers terminate at the P-GW. The IP layer contains the source and destination IP addresses using which the P-GW receives/route the user application data packets toward the external packet data network. The air interface layers (RLC, MAC, and PDCP) terminate at the eNodeB end.

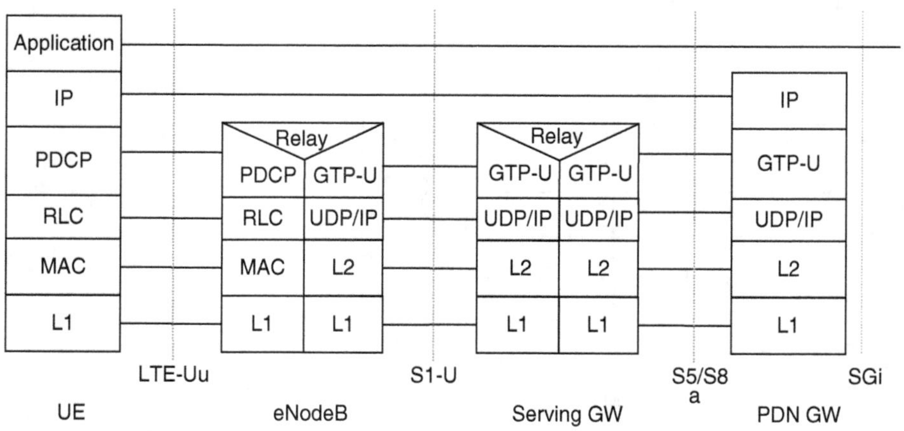

Figure 3.8 LTE UE – P-GW user plane protocol layers. *Source:* © 2015. 3GPP ™ TSs and TRs are the property of ARIB, ATIS, CCSA, ETSI, TSDSI, TTA and TTC who jointly own the copyright in them. © 2015, 3GPP.

Example 3.5 File Transfer Protocol (FTP) Protocol
Consider the File Transfer Protocol (FTP) that is used to upload and download files between two different hosts. FTP uses *port number 20* to transfer *user* data, whereas it uses *port number 21* to exchange *control* commands between the two communicating hosts. Here, one will find that the FTP also uses the so-called *user and control plane* concept to facilitate the exchange of control and data transfer between two hosts.

Example 3.6 GPRS Tunneling Protocol
The GTP has the *control* as well as the *user plane* protocols. The tunnel management is taken care of by the so-called *GPRS tunneling Control Protocol,* whereas the user IP data packets/payload is transported using the *GTP-User Plane T-PDU* (tunneled PDU) as shown later in Figure 8.1. GTP is used in a couple of interfaces in GPRS, UMTS, LTE/EPS, and 5G CN. For more information on the GTP control and user plane protocol, refer to TS 29.060 [67], 29.274 [70], and TS 29.281 [72].

At the LTE/EPC end, the user plane of the eNodeB, S-GW, and P-GW consists of the GTP-user plane [12] to tunnel user data on top of the UDP/IP transport network. Figure 3.8 also shows the respective user plane logical interfaces, i.e. S1-U and S5/S8, for carrying user data between two network elements of an LTE/EPS network.

From Figures 3.7 and 3.8, it is observed that in the LTE system, the radio interface protocol layers – PDCP, RLC, and MAC, are available in both the control plane and the user plane protocol stack. In the user plane protocol stack, the application and IP layers terminate at the CN, P-GW, which transfers user data packets to the external network. Examples 3.5 and 3.6 describe two typical protocols consisting of control plane and user plane protocol.

3.3 Grouping of UMTS, LTE, and 5G Air Interface Protocol Layers

In the previous section, we have mentioned about the classifications of protocol layers into the *control plane or signaling* and *user plane.* In the case of UMTS, LTE, and 5G systems, their air interface control plane or signaling protocol layers are further grouped into what is called the *Access Stratum* (AS)) and *NAS layer.* The word "Stratum" means a grouping of protocols that is related to one service aspect of the various services provided by one or several network elements. The LTE, 5G, and UMTS system air interface protocol layer groupings are illustrated in Figure 3.9.

3.3.1 Access Stratum (AS): UMTS UE – UTRAN; LTE UE – E-UTRAN;5G UE - NG-RAN

The *AS* contains radio air interface signaling protocol layers between UE and UTRAN; UE and E-UTRAN; and UE-5G NG-RAN only. The *AS* signaling protocol messages terminate at the UMTS UTRAN, LTE E-UTRAN, and 5G NG-RAN end. As the name implies, the AS contains all the protocols using which a UMTS or LTE or 5G UE

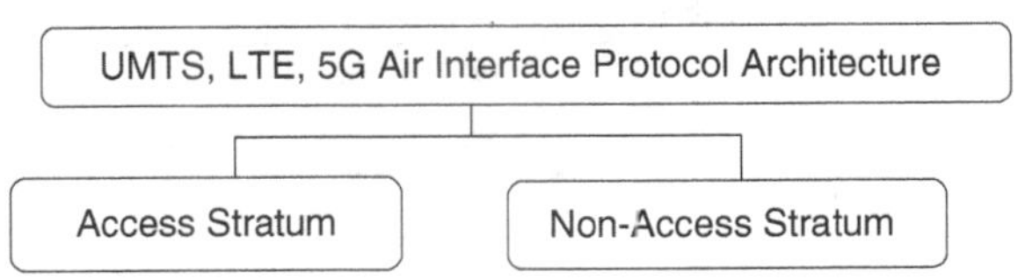

Figure 3.9 UMTS, LTE, and 5G air interface protocol layer groups.

Table 3.6 UMTS and LTE access stratum control plane protocol layers.

Protocol Category	System	
Air Interface Access Stratum Protocol Layers	UMTS/UTRAN AS Layers	LTE/E-UTRAN AS Layers
	Physical, MAC, RLC, RRC	Physical, MAC, RLC, PDCP, RRC

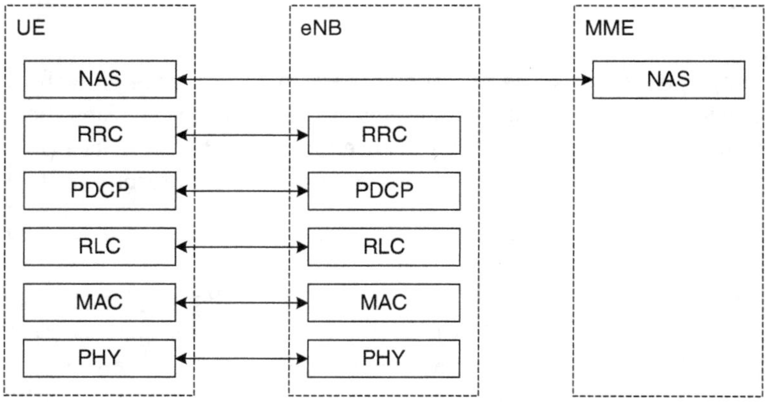

Figure 3.10 LTE E-UTRAN: access stratum and non-access stratum protocol stack. *Source:* © 2015. 3GPP ™ TSs and TRs are the property of ARIB, ATIS, CCSA, ETSI, TSDSI, TTA and TTC who jointly own the copyright in them. © 2015, 3GPP.

communicates with the UMTS UTRAN or LTE E-UTRAN or 5G NG-RAN only to transmit/receive signaling messages over their respective radio air interfaces.

The UMTS and LTE *AS* and its control or signaling plane protocol layers, over the air interface Uu, are shown in Table 3.6. The AS and NAS protocol structures for the LTE system air interface are shown in Figure 3.10, which is reproduced from TS 36.300 [92].

The typical flow of AS signaling messages, i.e. RRC layer, between a UE and the LTE/E-UTRAN is described in Example 3.7 below.

Example 3.7 LTE AS Layer: Radio Resource Connection Establishment Procedure

Let us consider the radio resource connection establishment procedure initiated by an LTE UE toward its RAN E-UTRAN. To initiate this procedure, a UE makes the radio resource requests and sends the *RRCConnectionRequest* message to the E-UTRAN, shown in Figure 3.11, which is reproduced from TS 36.331 [94]. The *RRCConnectionRequest* is an AS RRC layer message that terminates at the E-UTRAN end; refer to TS 36.331 [94]. This figure shows the RRC layer message names without showing the contents of each message.

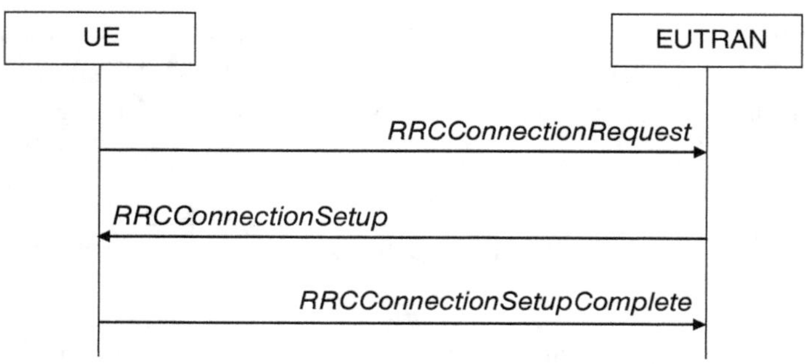

Figure 3.11 LTE access stratum RRC connection establishment procedure. *Source:* © 2018. 3GPP ™ TSs and TRs are the property of ARIB, ATIS, CCSA, ETSI, TSDSI, TTA and TTC who jointly own the copyright in them. © 2018, 3GPP.

3.3.2 Non-Access Stratum: UMTS UE – CN, LTE UE – EPC; 5G UE-Core

The NAS contains all the mobility, connection, and SM-related signaling protocols and procedures that terminate at the CN end. Using NAS signaling protocols, a UMTS UE or LTE or 5G UE communicates with the UMTS or LTE or 5G CN. Note that NAS uses the services of the AS protocols. In this case, the UMTS UTRAN or the LTE E-UTRAN or 5G NG-RAN simply forwards those transparent *NAS*-related signaling messages to their respective CN.

Figure 3.10 in the previous section shows the protocols that are grouped into AS and NAS types in the case of the LTE system only where the NAS layer terminates at the LTE/EPC MME.

NAS signaling protocols are typically used for the following services:

- CS CN procedures, such as MM and connection management (CM).
- GPRS PS core network procedures, such as mobility, SM bearer management, and security management.
- LTE/EPS core network procedures, such as the mobility, connection, and SM, bearer management, security management, EPS ciphering, and integrity protection function.
- 5G core network procedures, such as the mobility, connection, and PDU SM, security management, ciphering, and integrity protection function.

Table 3.7 shows the various NAS protocol messages exchanged between MS/UE and its concerned core network in GSM/GPRS/UMTS and LTE/EPS networks.

Example 3.8 below illustrates the typical messages flows associated with a MM layer (NAS) ATTACH procedure in the case of the LTE/EPS network.

Table 3.7 NAS protocol messages from GSM to LTE system.

System	Type of Call	CM Message	SM Message	MM Message
GSM/ UMTS	CS	Call Estb., Call Setup	–	Location Area Update
GPRS/ UMTS	PS	Call Establishments, Call Setup	Primary and Secondary PDP Context Managements	Routing Area Update, Network Attach, Detach
LTE/EPS	PS	[Fall-back] Call Establishments, Call Setup	Default and Dedicated Bearer Managements	Tracking Area Update, Network Attach, Detach

Example 3.8 LTE/EPS NAS Layer: EPS Mobility Management Layer Procedure

Let us consider the LTE UE ATTACH procedure shown in Figure 3.12 below, which is found in an LTE/EPS network. The ATTACH procedure and its messages are part of the EMM NAS protocol which is used by a UE to register with the LTE/EPS CN; see TS 24.301 [46]. This figure illustrates a successful ATTACH procedure and shows only the EMM message names without showing the contents of each message.

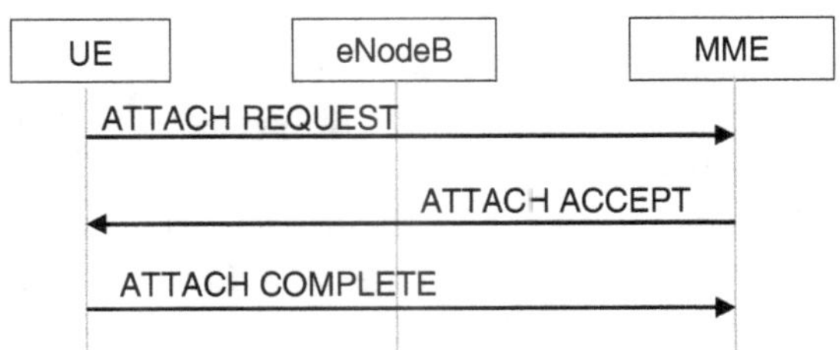

Figure 3.12 Illustration: LTE/EPS ATTACH procedure: NAS protocol messages.

Example 3.9 5G NAS Layer: 5G Session Management Layer Procedure

Let us consider the establishment of a PDU session, shown in Figure 3.13 below, a procedure that is initiated from a UE to the 5G core *Session Management Function* (SMF) network function. A PDU session is activated and is assigned to a particular network slice of a UE as part of its initial 5G UE registration procedure with a 5G CN. The establishment of a PDU session is part of the 5G Session Management (5GSM) NAS protocol. Figure 3.13 shows only the 5GSM message names without showing the contents of each message.

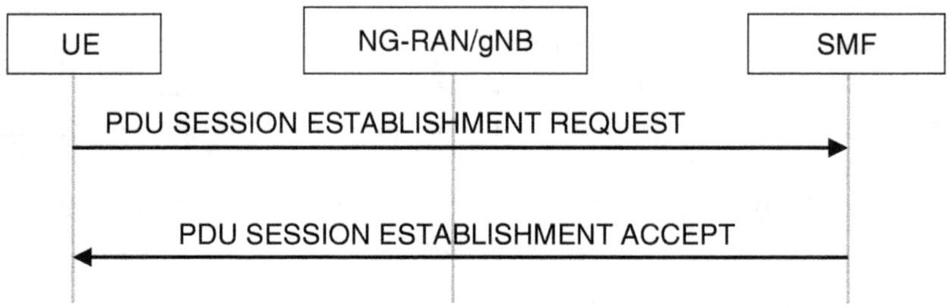

Figure 3.13 Illustration: NAS layer messages for a 5G PDU session establishment procedure.

In GPRS and UMTS networks also, an MS/UE registers with its CN using the ATTACH procedure for PS services. However, as far as the implementations are concerned, there are differences in the GPRS/UMTS and LTE/EPS ATTACH procedure. For example, unlike the GPRS system, an LTE/EPS ATTACH request message also contains a piggybacked session activation request to the CN.

Example 3.9 illustrates the typical messages flows associated with an SM layer (NAS) PDU session establishment procedure in the case of the 5G system.

3.4 Initialization of a Logical Interface

In the previous sections, several logical interfaces between the concerned network elements of mobile communications networks were described and illustrated. Different functions and procedures are performed over the respective logical interfaces. Some logical interfaces do not become ready for exchanges of information between the concerned network elements. Also, following the occurrence of an erroneous event, a logical interface may become unusable. In such scenarios, a logical interface between two network elements requires to be initialized and configured or reinitialized/reconfigured, with protocol layer-specific data, to make it ready for exchanges of information over it. The procedure for initialization and configuration with protocol layer-specific data differs from one logical interface to another one.

Example 3.10 illustrates the typical messages flows for initialization of the S1 logical interface, which is used between the LTE/eNodeB and its MME.

Similarly, the *Gb-interface* which is used in the GPRS system is also required to be initialized. The NS protocol layer of the Gb-interface in the BSC/PCU end initializes and sends the configuration data to the peer layer on the SGSN side of the Gb-interface.

Example 3.10 LTE/EPS S1-AP (eNodeB-MME) Logical Interface Initialization

In the LTE/EPS, the S1 interface is used to exchange control or signaling-related messages, between the eNodeB and the MME, which is also known as the S1-AP (Application Protocol). To make the S1 interface operational for exchanges of information and other S1-AP messages, the S1 interface is required to be setup first. For this purpose, the eNodeB sends the *S1 Setup Request* message, containing the eNodeB identity, Public Land Mobile Network (PLMN) identity, to the MME. The MME sends the *S1 Setup Response* message to the eNodeB. This is shown in Figure 3.14, which is reproduced from TS 36.413 [97]. Other examples where initialization of a logical interface is required is the 5G NG interface which is used between 5G NG-RAN and AMF network function.

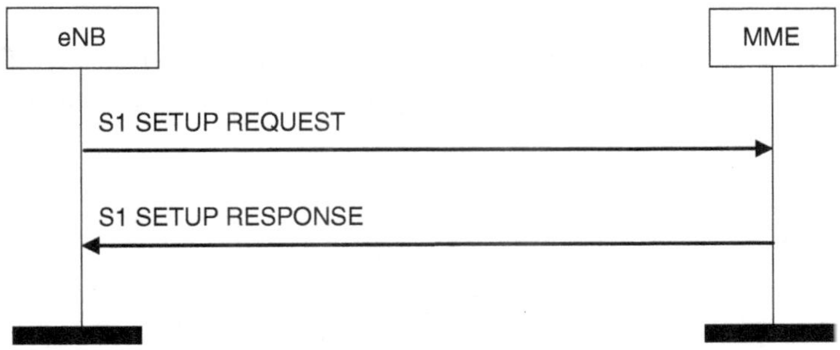

Figure 3.14 Initialization of LTE/EPS S1 interface. *Source:* © 2014. 3GPP ™ TSs and TRs are the property of ARIB, ATIS, CCSA, ETSI, TSDSI, TTA and TTC who jointly own the copyright in them. © 2014, 3GPP.

3.5 Protocol Layer Termination

The extent and working of a particular protocol layer can be further understood from the *protocol layer termination* point of view. The general working model of a protocol layer consists of functions and procedures that it requires to perform to facilitate the various services to be offered to the higher layer, as described later in Section 3.8.

Protocol layer termination refers to the making available of the various services by the concerned protocol layer to its *adjacent layers* at the same time facilitating a *peer-to-peer* communication between two network elements over a logical interface. To understand a protocol layer termination, start from the UE/MS end and proceed toward the radio access or CN. A protocol layer terminates at the destination or peer network element or domain. Find and look at the corresponding network element containing the terminated protocol layer. Next, look at its position, e.g. Layer #2, Layer #3, and so on, within the protocol layers' organization supported by the concerned network element. The protocol stack and its particular layer termination also identify the network elements that exchange various messages using the concerned layer protocol specification. For example, as mentioned in Section 3.3, the AS protocols terminate at the UMTS UTRAN or LTE E-UTRAN or 5G NG-RAN, whereas the NAS protocols terminate at the respective CN end. For more examples of protocol layer terminations, refer to TS 25.301 [49].

3.6 Protocol Sublayers

Recall the protocol layering principles of the OSI 7 protocol reference model, having a single protocol at each of the individual Layers #1–7. In the case of a mobile communications system protocol architecture, a particular protocol layer may contain sublayers also, performing its functions and procedures. For example, consider the air

Interface Layer 3 protocol stack. The GSM Layer #3, i.e. the network layer in terms of the OSI reference model, has three sublayers, as mentioned below:

- CM
- MM
- Radio Resources Control and Management (RR)

Similarly, the UMTS and LTE system air interface protocol stack, Layer #2, i.e. the data link layer in terms of the OSI reference model, has three sublayers, as mentioned below:

- Packet Data Convergence Protocol (PDCP)
- RLC
- MAC

In the case of the GPRS/Enhanced Data for Global Evolution (EDGE) system protocol stack also, Layer #2, i.e. the data link layer in terms of OSI reference model, has three sublayers, as mentioned below:

- Logical Link Control (LLC)
- RLC
- MAC

The different protocol sublayers mentioned above are illustrated in Figure 3.15.

Note that in the case of UMTS and LTE systems, sublayers of a protocol layer may spread across the *AS* as well as *NAS* groups of protocols. For example, consider the UMTS and LTE air interface Layer 3 protocol and its sublayers. Here, the RRC is the Layer 3 protocol that terminates at the UTRAN or E-UTRAN, but it is placed as part of the AS group of protocols. On the other hand, the sublayers of LTE/EPS or GPRS SM, MM, and CM are part of the NAS group of protocols that terminates at the CN, i.e. GPRS SGSN or LTE/EPC MME. Further, as illustrated, the 5G New Radio (NR) Layer 2 contains a new sublayer called the *Service Data Adaptation Protocol* (SDAP).

3.7 Protocol Conversion

A particular protocol layer communicates and exchanges information in its defined format only between two network elements. A network element could be the mobile station, BTS, BSC, MSC, SGSN, eNodeB, MME, S-GW, 5G core UPF, and so on. The communications between the protocol layers of two network elements may be direct and point to point, or it may be routed through another network element that may work on different protocol

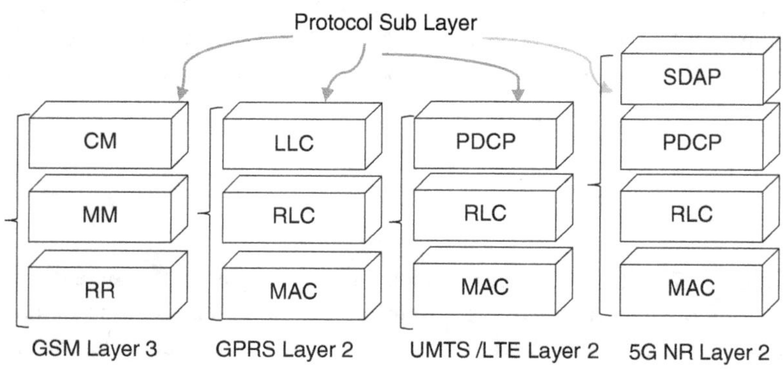

Figure 3.15 Illustration: air interface sublayers: GSM, GPRS, UMTS, LTE, and 5G.

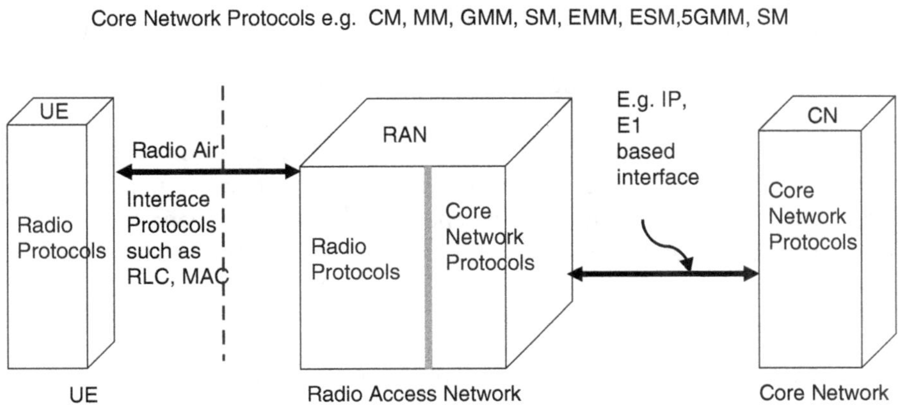

Figure 3.16 Illustration: protocol information conversions in a cellular system.

stacks. If the communication is not through a direct path, then the original message sent by the sender needs to be forwarded by an intermediate network element to the destination network element using protocol conversion. This is illustrated, in general, in Figure 3.16.

In Figure 3.16, consider that a user wants to access the Internet (e.g. www, FTP, ping, and so on) through the GPRS or LTE/EPS network. The UE will send the user's request to the RAN using RLC/MAC protocol across the air interface. The RAN will collect the RLC/MAC layer block, in the case of GPRS, or RLC layer PDU, in the case of LTE. The RAN will format the RLC/MAC layer information into an appropriate protocol layer format of the concerned CN logical interface, for example, GPRS Gb interface Frame Relay format for SGSN, or LTE/EPS S1-U format, and forward it to the SGSN or S-GW. As an analogy with a traditional IP network/Internet, in a mobile communication network also, the user data or signaling data packets pass through different protocol layers and intermediate devices between a source and destination.

3.8 Working Model of a 3GPP Protocol Layer: Services and Functions

A mobile communications network consists of several protocol stacks and layers designed to exchange information between the adjacent layers or between two peer network elements either directly or transparently through another network element. The general working model of a protocol layer consists of various *services* it offers to its higher layer or services it expects from the layer below it. The protocol layer working model also consists of collections of signaling procedures and functions. These functions are the *building blocks* of a particular protocol layer to transfer information accurately and error-free as expected by the peer layer. This working model of a protocol layer is illustrated graphically in Figure 3.17 where a protocol layer N provides services to the layer N + 1 and also uses the services from the layer N − 1.

Example 3.11 describes the LTE/E-UTRAN RLC layer with respect to its working principle.

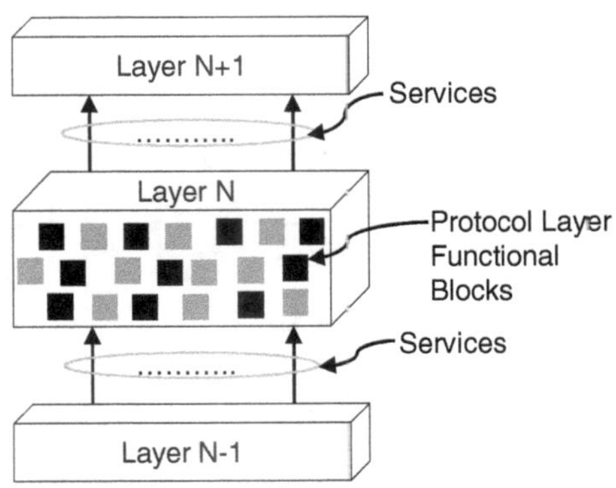

Figure 3.17 Illustration: A working model of protocol layer.

Example 3.11 Functions of LTE Air Interface RLC Layer

Consider the LTE air interface Layer 2 RLC protocol layer operating between the UE and the E-UTRAN. The RLC layer provides services for transferring higher-layer information in acknowledged (AM), unacknowledged (UM), or TM. To ensure an error-free transmission and availability of the transmitted information accurately at the destination RLC protocol layer, the transmitting RLC layer performs several important functions such as segmentation, resegmentation, retransmissions, ciphering, padding, and so on. On the other hand, the receiving RLC layer also performs functions such as reassembly and duplicate detection on the received information before it is passed to the next higher layer.

3.9 General Protocol Model Between RAN and CN (UMTS, LTE, 5G)

In Section 3.2, we have discussed the separation of protocol layers of various logical interfaces into a control plane and user plane categories which are applicable, in general, in all the mobile communications systems, i.e. GSM, GPRS, UMTS, LTE and 5G. In Section 3.3, we have also discussed the grouping of UMTS, LTE, and 5G systems only protocol layers into AS and NAS categories from the protocol layer termination at UMTS UTRAN, LTE E-UTRAN, 5G NG-RAN, and CN point of view.

The grouping of protocol layers into the AS and NAS layers in the UMTS, LTE, and 5G systems is done from their respective air interface point of view, where the air interface is the physical interface. On the other hand, the UMTS UTRAN, LTE E-UTRAN, and 5G NG-RAN communicate with their CN elements and another network element of a UTRAN, E-UTRAN, and 5G NG-RAN using the standard data transport network, e.g. ATM, IP, which is the standard protocol. It may be noted that the protocol stack and its logical interface between UMTS UTRAN and its CN; LTE E-UTRAN and its CN; and 5G NG-RAN and its CN are logically independent of the underlying data transport network used by them. Based on this, the protocol stack of a logical interface, i.e. Iu interface between UMTS UTRAN – CN; S1, X2 interface between E-UTRAN and MME or E-UTRAN; NG interface between 5G NG-RAN and 5G core, is further modeled with the following horizontal-layered structures:

- Radio Network Layer (RNL)
- Transport Network Layer (TNL)

The above protocol model of the UMTS UTRAN, LTE E-UTRAN, and 5G NG-RAN is illustrated in Figure 3.18. For more information on these layered structures, refer to UMTS TS 25.401 [51], LTE TS 36.401 [95], 5G TS 38.401 [117], and TS 38.410 [118].

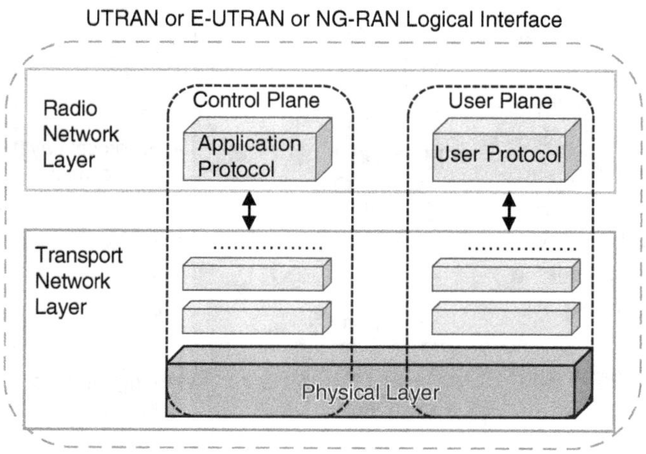

Figure 3.18 Illustration: general protocol layer model of UTRAN, E-UTRAN, and 5G NG-RAN.

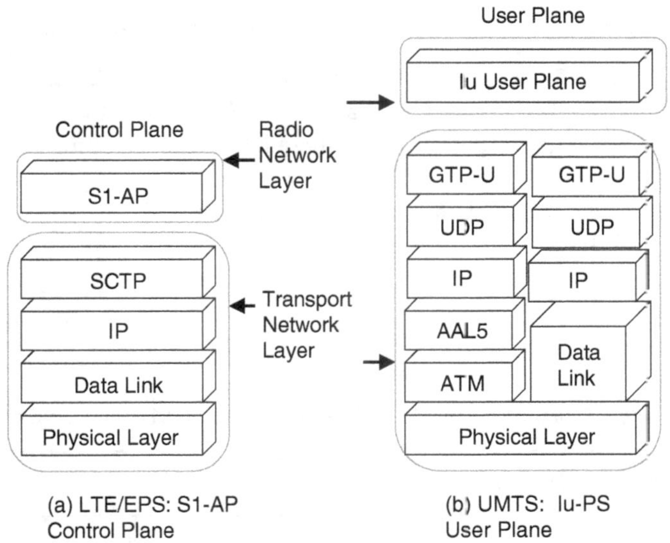

Figure 3.19 Illustration: general protocol layer model of UTRAN and E-UTRAN.

The RNL deals with the UTRAN or E-UTRAN or 5G NG-RAN-specific related various functions and procedures, for example, radio bearer management, paging, and so on, in terms of an AP. The data TNL deals with the particular transport method, e.g. ATM, IP, and so on, and possibly through the intermediate network also, to be used to transport RNL procedures messages.

RNL and TNL layers are logically independent of each other, which makes it possible to make UMTS UTRAN or LTE E-UTRAN or 5G NG-RAN protocol-related changes in the RNL without affecting the TNL. As shown in Figure 3.18, the protocol layers of the logical interfaces between the UMTS UTRAN or LTE E-UTRAN or 5G NG-RAN and their respective CN or other network element are still separated as the control plane and user plane, but they share the same physical layer. Figure 3.19 shows a side-by-side RNL and TNL of the LTE/EPS S1-AP control plane between LTE eNodeB and MME, and UMTS Iu-PS user plane protocols between RNC and SGSN.

3.10 Multiple Transport Networks, Protocols, and Physical Layer Interfaces

From the illustration shown in Figures 3.5 and 3.19, it has been observed that each of the concerned protocol stacks between the GSM BSS and its CN or the UMTS RNC and its CN has the following characteristics, i.e. multiple transport and data link networks, as described in Examples 3.12 to 3.13.

- **Provisions for multiple Data Link layers**

Example 3.12/Figure 3.20 illustrates the data link layers used for the UMTS Iu-interface. The Iu-interface transport network options are used to exchange the control plane and user place messages of both the Iu-CS and Iu-PS interfaces.

- **Provisions for Separate Transport Protocol in an IP Transport Network**

In the case of UMTS, LTE, or 5G system, if the IP transport network option is used, the IP network transport protocol layers used are listed in Table 3.8.

SCTP [17] transport network is also used for different logical interfaces in the 5G system.

Example 3.12 Multiple Transport Networks for UMTS Iu Interface

The UMTS Iu-interface is divided into two parts: Iu-CS, to support the CS domain, and Iu-PS, to support the PS domain. The Iu-interface may use the ATM or IP-based transport network as the data link layer protocol, as illustrated in Figure 3.20. For more information, refer to TS 25.412 [53].

Figure 3.20 Iu interface transport network: data link layers.

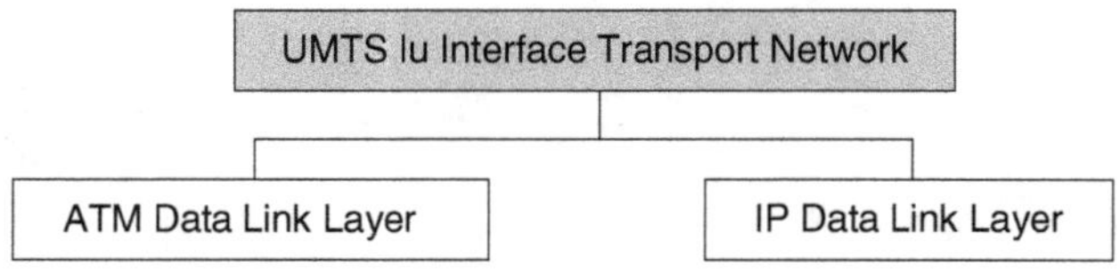

Table 3.8 Protocol layer in UMTS (Iu), LTE system (S1), and 5G NG over IP.

IP Transport Layer	Purpose	System/Interface
UDP	To transport Iu or S1 or NG user plane information over IP	UMTS (Iu), LTE (S1), 5G (NG)
SCTP [17]	To transport Iu or S1 or NG control plane information over IP	

- **Provisions for Multiple Physical Layer Interfaces**

Depending on the data link layer being used, the GSM BSS, UMTS RNC, and the core network element transport network may provide provisions for configuring multiple physical layer interfaces. This ensures a backup, in case one physical interface fails the other physical interface can be brought into service to continue communications services. The upper layers of a protocol stack are independent of the type of underlying data link layer and its physical interface being used to transport higher-layer data.

Example 3.13 GPRS Gb-Interface Multiple Physical Layer Interfaces

Consider Figure 3.5 illustrated earlier. As shown in that figure, the GSM BSC and the SGSN can exchange the Gb-interface protocol stack messages using either the *Frame Relay* network protocol or IP network. The *Sub-Network control protocol* takes care of the data link layer protocol (Frame Relay or IP) to be used by the Gb-interface. The higher-layer applications are independent of the sub-Network control protocol. In case the Frame Relay protocol and its physical E1 interface go down, the IP Transport/Ethernet interface can be brought into services immediately so that the Gb-interface is not affected and GPRS services are not down permanently.

Similarly, between the UMTS UTRAN and its CN, the Iu-interface can be configured to use either the ATM over STM or IP transport network physical layer; refer to TS 25.412 [53].

- **Reason for Multiple Data Link Layer Provisions**

Typical reasons to support multiple data links by a network element are as follows:

- Make the higher layer(s), e.g. RNL, independent of the TNL being used for the introduction of a new transport network without affecting the application or RNL;
- Create a backup transport network; and
- System and application requirements, e.g. use ATM to transport multimedia contents, whereas frame relay can transport data only.

3.11 How to Identify and Understand Protocol Architectures

3GPP defines a larger number of protocol logical interfaces, starting with the alphabet “A”, stacks, and layers which are central to the interworking of a network element with another network element of mobile communications systems and networks. In this chapter, we have covered only the following logical interfaces of mobile communications networks:

- Air interfaces, i.e. Uu, Um, between UE/MS and RAN of GSM, UMTS, LTE, and 5G systems.
- Network interfaces (A, Gb, S1, Iu, X2, Gn, NG, and so on) between the GSM, UMTS, LTE, and 5G RANs and their respective CNs.

In fact, the majority of the logical interfaces are found only in the CNs domain along with the other CN elements such as the Home Location Register/Home Subscriber Server (HLR/HSS), Visitor Location Register (VLR), and Policy Charging and Restriction Function (PCRF). Other logical interfaces are also available that can be configured to support interworking and interoperations, e.g. CS fallback, Single Radio Voice Call Continuity, and so on, between two mobile communications networks. Some of these interoperation facilities may be configured as an optional and separately licensed feature.

A developer must put the focus on a particular network element and its logical interfaces at a time. The air interface and its protocol stack is the most interesting one that consists of advanced wireless communications theories. The air interface differentiates one system from its predecessor. As a starting point, the reader is advised to go through and familiarize themselves with the list of 3GPP TSs mentioned in the Reference section of this book. There are 3GPP TSs describing the protocol layers of the respective air interface of the GSM, GPRS, UMTS, LTE, and 5G systems. The reader may, then, proceed gradually toward the other logical interfaces and their protocol stack.

3.11.1 Identifying a Logical Interface, Protocol Stack, and Its Layers

To identify and understand a particular protocol stack architecture, its layers, and their corresponding 3GPP technical specifications, the following steps may be taken:

- Choose a particular mobile communications system such as the GSM, GPRS, and UMTS, LTE, or 5G as your area of interest. Use the information available in the 3GPP site [2] (second, third, fourth column) as mentioned in Section 2.5.6.
- Next, decide the particular network element of interest such as GSM MS, BTS, BSC, and MSC; UMTS NodeB, and RNC; or LTE eNodeB, MME, and S-GW or 5G gNB and 5G core.
- Now, look at the logical interfaces supported by the chosen network element. Pick a particular logical interface and look at its protocol stack and layers. A logical interface and its protocol stack cover different subjects and specifications area. Look at the subject and specifications areas mentioned in the 3GPP site [2], first column, and pick a particular subject area. Against this chosen subject area, e.g. signaling, requirements, and so on, or a particular protocol layer, attempt to identify the corresponding technical specifications series and its specifications from the column 2, 3, or 4, 3GPP site [2].

Study the protocol layer architecture, its functions and procedures, and other details from the identified technical specification.

Example 3.14 describes the typical steps to be used to derive the 3GPP technical specification of a protocol layer and its sub-layer.

Figure 3.21 shows the protocol stack of the A-interface on the CN side.

Example 3.14 GSM Circuit-Switched BSS and Air Interface Layer 3 Technical Specifications

Suppose a developer is interested to study the GSM BSS that consists of BTSs and BSC network elements. Further, suppose that developer is interested in the air interface Layer 3 protocol, between an MS and the BSC, of a GSM system. Figure 3.21 below shows the GSM air interface Layer 3 protocols along with its sublayer protocol, as follows:

- Call Control Management (CM),
- MM, and
- Radio Resource Management (RR).

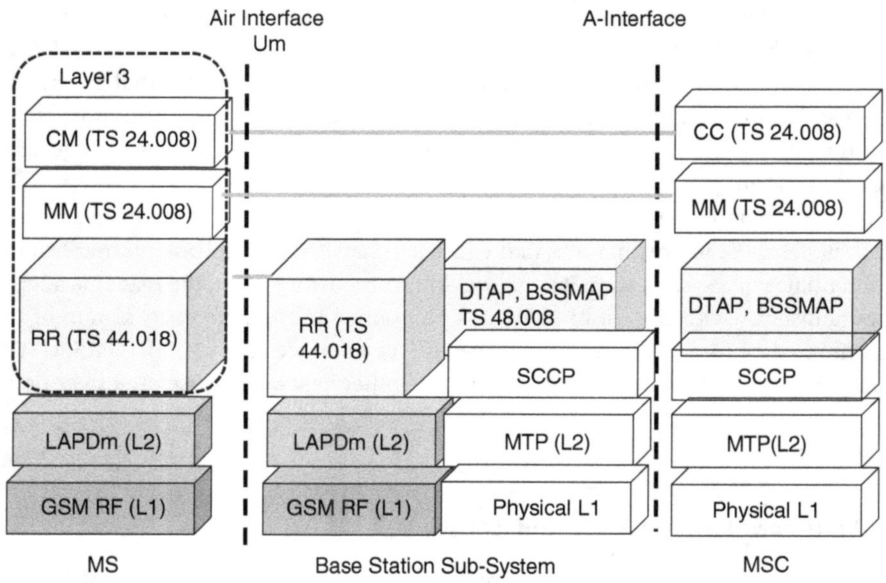

Figure 3.21 Illustration: GSM air interface Layer 3 protocol stack.

1) As shown in Figure 3.21, the GSM air interface Layer 3 consists of the CM, MM, and RR layers. The developer may be further interested in the Radio Resource Management, (RR) sublayer of the GSM Layer 3 protocol stack. The RR layer of a BSC deals with the signaling functions/messages of the GSM Layer 3 protocol stack. Now refer to the 3GPP site [2].
2) For the GSM signaling protocols/messages that are exchanged between an MS and the BSC and vice versa, the corresponding 3GPP TS series number will be either *44 Series* (After Release 4, including, GSM) or *4 Series* (Before Release 4 GSM). Assume that you have considered the 3GPP TS Series 44. Within this series, one will find all the technical specifications, listed in ascending order, related to signaling messages between MS and the BSC. Now, look for the technical specification having the title Radio Resource Management protocol. You got the desired TS. In this case, it is the 3GPP TS 44.018 [130]. For Series 4, the corresponding TS will be the 3GPP TS 04.18.

Similarly, try to find the technical specification number for the GSM CM and MM sublayer, which is TS 24.008[45]. 3GPP TS 24.008 [45] covers the entire mobile radio air interface Layer 3 specification for GSM/GPRS Edge Radio Access Network (GERAN), UMTS system. By following the same way as described above, one can find the corresponding technical specification number for the RRC protocol in the case of UMTS or LTE or 5G system. The technical specification series number for the LTE system is 36; for UMTS, it is 25 series; for 5G, it is 38 series.

In Figure 3.21, the GSM BSC and its BTSs are being shown as the combined BSS. However, a BTS and BSC of a BSS are connected by the logical *A-bis interface* that is not shown in Figure 3.21. A-bis interface is proprietary with its protocol stack. On the CN side, a BSC and the MSC are connected by the logical *A-interface* which is an open standard defined by 3GPP. Look at the protocol stack of the A-interface. At the top of the stack are the Base Station Subsystem Mobile Application Part (BSSMAP), Direct Transfer Application Part (DTAP), and Signaling Connection Control Part (SCCP) layer, which is the Layer 3 protocol. The BSSMAP and DTAP protocols are defined in the 3GP TS 48.008 [134]. The Layer 2 is the *Message Transfer Part (*MTP*).* The physical layer used for both the A-bis and A-interface is the E1 interface, as described earlier in Section 3.1.1. The SCCP, MTP is part of the standard Signaling System #7 (SS#7). For more information on the SCCP and MTP layers, refer to TS 48.006 [133].

3.11.2 Identification of Technical Requirements Using Interface Name

A 3GPP technical specification defines its scope and describes the protocols, functions, and procedures that may be applied to more than one mobile communications systems, from GSM to the 5G system. From the descriptions available in a given 3GPP TS, it is important to identify such requirements that are specific to a particular mobile communications system, i.e. GSM, GPRS, UMTS, LTE, and 5G. A 3GPP TS mentions the corresponding CN logical interface name while describing the detailed requirements of functions and procedures. Generally, a 3GPP technical specification mentions the applicability of a specific requirement/section being described using the term: *A/Gb mode* to refer to GSM/GPRS system; *Iu mode* to refer to UMTS system; *S1 mode* to refer to LTE/EPS system; and *N1 mode* to refer to 5G system. A logical interface name also identifies a particular mobile communications system that is being referred to and the requirements that belong to it.

3.12 Protocol Layer Procedures over CN Interfaces

In the preceding sections, we have discussed the CN interfaces of GSM, GPRS, UMTS, LTE/EPS, and 5G networks. The logical interfaces between the respective RAN and its CN element are as follows:

- S1, between LTE eNodeB and MME; NG between 5G NR NG-RAN and AMF
- Iu, between UMTS/UTRAN RNC and MSC; RNC and SGSN
- A, between GSM BSC and MSC
- Gb, between GSM BSC and SGSN

Over a particular logical interface, as mentioned above, the following types of signaling messages are exchanged in the forms of an *AP* between the RAN and CN for the execution of

- Interface-specific protocol functions and procedures
- Session Management, Call Management, and MM air interface Layer 3 or NAS layer messages between a UE/MS and the CN, transparently through the RAN.

The AP functions and signaling procedures executed over the respective logical interfaces are specific to a particular mobile communications system and network. There are also protocol and signaling procedures that are similar in nature, but they are implementations dependent on a particular communications system.

3.12.1 Similar Functions and Procedures over the CN Interfaces

Several similar protocol functions and procedures are performed over the LTE/EPS S1-interface, UMTS Iu-interface, GSM A-interface, GPRS Gb-interfaces, and 5G NG interface. Some of them are mentioned below. However, these high-level functions and procedures are implementation-dependent in terms of different signaling messages names and their contents but are similar in nature with respect to their applicability from GSM to the 5G system.

- **Between the MS/UE and the CN**
  - The CS or PS domain Session Management, Call Management, and MM layer signaling messages that are exchanged between an MS/UE and the CN are not processed by the GSM or UMTS or LTE or 5G RAN, but they are forwarded to the CN. Forwarding of such signaling messages by the RAN to the CN is done through the direct transfer of messages. In the GSM system, it is known as the *DTAP* as shown earlier in Figure 3.21; in the UMTS system, it is known as the *DIRECT TRANSFER*; in the LTE and 5G systems, it is known as the *NAS transport*.

A Session Management, Call Management, and MM-related message that is exchanged between an MS/UE and the CN is embedded in a DTAP (GSM) or DIRECT TRANSFER (UMTS) or NAS TRANSPORT (LTE/EPS) message. A GSM DTAP or UMTS DIRECT TRANSFER or an LTE/EPS or 5G NAS TRANSPORT message is exchanged between the radio access and its CN only. However, the initial Layer 3, i.e. CC, MM related, message from a ME/UE toward the CN is exchanged through the GSM INITIAL MS message in UMTS and the INITIAL UE message in the LTE/EPS and 5G systems.

Example 3.15 below illustrates the typical messages flow between an LTE/eNodeB and its MME through NAS transport messages in uplink and downlink direction. Such NAS transport messages between an LTE/eNodeB and its MME are used to forward signaling messages exchanged between a UE and LTE/MME and vice-versa.

Example 3.15 LTE/EPS: PS Domain NAS Transport Messages

Figure 3.12 shown earlier illustrates the LTE/EPS MM procedures and their signaling messages. This figure shows snapshots of a few end-to-end signaling messages for a PS domain LTE/EPS ATTACH procedure. In fact, between the eNodeB and the MME, the individual NAS signaling messages of a particular LTE/EPS NAS procedure will be exchanged using the uplink or downlink *NAS TRANSPORT* message as illustrated in Figure 3.22 below.

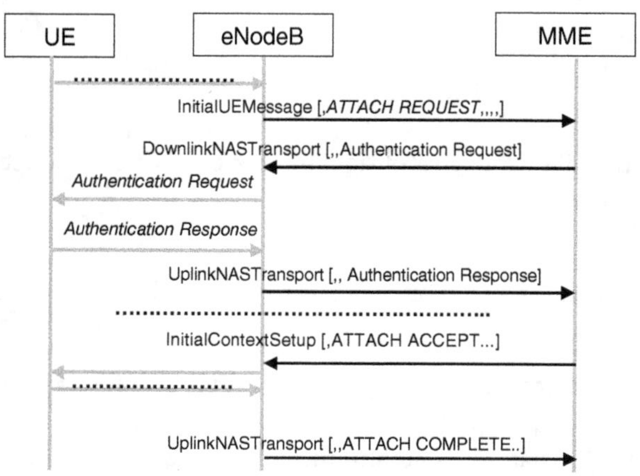

Figure 3.22 Illustration: LTE/EPS NAS transport between eNodeB and MME over S1 interface.

As illustrated in Figure 3.22, the LTE air interface initial NAS layer messages, e.g. ATTACH REQUEST, TAU, received from a UE is sent through the *InitiaUEMessage* from the eNodeB to the MME. The subsequent NAS messages between eNodeB and the MME are exchanged using the *DownlinkNASTransport* and *UplinkNASTransport* messages. The MME sends the *ATTACH ACCEPT* message to the eNodeB through the *InitialContextSetup* message. For more information on the NAS TRANSPORT messages and their contents, refer to TS 36.413 [97].

- **Between the RAN and the CN**

Several similar procedures are performed between the RAN and its CN only over their respective logical interfaces, e.g. LTE/EPS S1 and X2; 5G NG; UMTS-IuPS; and IuCS. Typical similar procedures are mentioned below:

- Handover procedure performed in case of the GSM and LTE system; relocation procedure performed in case of UMTS system. A handover or relocation procedure is executed to transfer an ongoing call from one cell to another suitable cell.
- Interface management – to setup, initialize, and release the respective interfaces, i.e. A, Gb, Iu, S1, and 5G system NG interface.
- Security and encryption.
- Paging – to notify an incoming call for an MS/UE.
- UE tracking – to track a particular UE/MS.
- Overload control from CN – which indicates the RAN to reduce the signaling load toward the CN.

Several functions are performed by a network element to complete a particular protocol layer procedure as highlighted above, which are described later in different chapters. For more information on the above functions and procedures, refer to TS 25.413 [54], TS 36.413 [97], TS 38.413 [119], and TS 48.008 [134].

3.12.2 Specific Functions and Procedures over the CN Interfaces

Some functions and procedures performed over a CN logical interface are specific to a GSM, UMTS, LTE, or 5G system. Typical examples of such GSM, UMTS, and LTE system-specific functions over their specific logical interface are mentioned below:

- BSSMAP [between GSM BSC – MSC]
  - Speech transcoding from 64 kbps, at MSC end, to 16 kbps, at BSC end
- RANAP [between UMTS/UTRAN RNC – MSC]
  - UMTS Radio Access Bearer management, i.e. establish, modify, release, create, and allocate radio access bearer to the UE
  - Transport and RNL link management functions
- S1-AP [between LTE/E-UTRAN eNodeB – MME]
  - LTE/EPS Radio Access Bearer management, i.e. setup, modify, release, to create and allocate radio access bearer to the UE
  - Transport and RNL link management functions
- NG-AP [between 5G NG-RAN – AMF]
  - 5G PDU Session Management, i.e. setup, modify, and release, a PDU session to a UE, MM, UE context management, and so on
  - Transport and RNL link management functions.

For more information on the specific functions and procedures, refer to TS 25.413 [54], TS 36.413 [97], TS 48.008 [134], and TS 38.413 [119].

Chapter Summary

This chapter has introduced the core aspects of the understanding, design, and development of logical interfaces, their protocol stack, and its layers of mobile communications systems and networks. Logical interfaces are used to communicate among network elements of a mobile communications network. Logical interfaces are also used for the interworking and interoperations of mobile communications systems and networks, which shall be described later in Chapter 6.

Different terminologies are used to describe a logical interface, its protocol stack, and layers. Protocol layers are classified into user plane and control or signaling plane distinctly based on the nature of the information that is transmitted over a particular logical interface. Within the control plane only, protocol layers are also grouped into the AS and NAS categories, especially in the UMTS, LTE, and 5G networks and systems.

We presented the protocol layer termination and sublayering of a particular protocol layer found in mobile communications systems and networks. We also presented the general working model of a protocol layer that is used to communicate with its peer layer and provide services to an upper layer. Apart from this, we presented the general protocol model and layers of the UMTS UTRAN, LTE E-UTRAN, and 5G NG-RAN and their respective CNs.

A 3GPP technical specification may cover the description of GSM, GPRS, UMTS, LTE, and 5G system protocol functions and procedures. A method to identify the functions and procedures description that applies to a particular communications system was presented. Finally, the reader is recommended to focus on a particular logical interface, protocol stack at a time, and its specifications as mentioned in the references section and then, proceed gradually toward other logical interfaces and their protocol stack.

4

Encoding and Decoding of Messages

Introduction

This chapter covers the methods for encoding and decoding of control plane or signaling messages and protocol data units (PDUs) that are exchanged between the peer protocols layers of network elements of mobile communications networks, from the Global System for Mobile Communication (GSM) to the 5G system. The method of a description of signaling messages and their encoding and decoding is part of a protocol layer specification, which differs from one logical interface to another one. Messages are exchanged between the network elements of a mobile communications network to facilitate various communication services to users. A source network element creates a protocol layer message in a predefined format using a particular encoding/packing method and sends it to the peer network element over the concerned logical interface. The peer network element decodes or unpacks a message using the same method that was used to encode it.

We begin with the description of encoding and decoding methods of air interface Layer 3 signaling messages, followed by the Layer 2 signaling messages exchanged between a Mobile station (MS)/User Equipment (UE) and the network. We also cover the encoding method used by the Radio Access Network (RAN) and core network (CN) elements. We close this chapter with the method of embedding a control plane message within another control plane message to reduce signaling overhead between two network elements, especially over the air interface.

4.1 Description and Encoding/Decoding of Air Interface Messages

One important aspect of a mobile communications network is the "*signaling message*" that is exchanged between its network elements. A signaling message is nothing but an exchange of a series of information between the network elements to establish, maintain, and release of resources allocated for communication services being provided to the service users. Information in a particular signaling message being transmitted is encoded (packed) and decoded (unpacked) differently across the mobile communications systems, i.e. from the GSM to 5G. Depending on the protocol stack and its layers supported by a network element, it may use different methods of encoding and decoding of signaling messages at each protocol layer. Also, a network element may decode the contents of a message that it receives or may not decode but transparently forward to the destination network element using another encoding method.

A protocol layer of a network element may send a signaling message to its peer layer like a long series of ordered bits where an octet alignment may or may not be required. Another protocol layer of the same network element may send a signaling message as a series of octets with octet alignments. To sum up, as far as the development of a protocol layer is concerned, two aspects of signaling messages defined in its technical specifications are required to be considered:

Mobile Communications Systems Development: A Practical Introduction to System Understanding, Implementation, and Deployment, First Edition. Rajib Taid.

- The method used to describe a signaling message
- The method used for encoding and decoding of a signaling message to transfer/receive among network elements.

It may be noted that the method of description, i.e. tabular format, and encoding/decoding of signaling messages differs from layer to layer. In the subsequent sections that follow, the following methods of descriptions, encoding, and decoding of mobile communications networks signaling messages over their respective air interfaces are discussed.

a) Encoding and Decoding of Air Interface Layer 3 Messages
   This method is used by the GSM air interface Radio Resource (RR) sublayer of Layer 3 protocol between an MS and the base station controller (BSC). This method is also used by the Call Control (CC) and Mobility Management (MM) sublayers of GSM; GPRS Mobility Management (GMM), Session Management (SM) layers of Universal Mobile Telecommunication System (UMTS); Evolved Packet System Session Management (ESM) and Evolved Packet System Mobility Management (EMM) layers of Long-Term Evolution (LTE)/Evolved Packet System (EPS) system; and 5GMM and 5GSM layers of 5G system. These protocol layers work between an MS/UE and the CN.
b) Concrete Syntax Notation.1 (CSN.1) Encoding/Decoding
   This method is used by the General Packet Radio Service (GPRS) air interface Layer 2 radio link control (RLC)/Medium Access Control (MAC) protocol between the MS and BSC.
c) Abstract Syntax Notation.1 (ASN.1) Encoding/Decoding Using Packed Encoding Rule (PER)
   This method is used by the following protocol layers over their respective logical interfaces:
   - UMTS Radio Resource Control (RRC) air interface Layer 3 between the UE and UMTS Terrestrial Radio Access Network (UTRAN)/Radio Network Controller (RNC),
   - LTE RRC air interface Layer 3 between the UE and Evolved-UMTS Terrestrial Radio Access Network (E-UTRAN)/eNodeB, and
   - 5G New Radio (NR) RRC air interface Layer 3 between the UE and Next Generation Radio Access Network (NG-RAN)/5G Base Station (gNB).

4.1.1 Encoding/Decoding: Air Interface Layer 3 Messages

To describe and encode/decode the air interface Layer 3 and its sublayers signaling messages, the *basic tabular form* of definition is used. Except for the UMTS, LTE, and 5G RRC layers, the following layers share the same tabular format to describe their air interface Layer 3 and Non-access Stratum (NAS) layers messages.

- GSM RR, MM, and Connection Management (CM) layers,
- GPRS/UMTS GMM and SM layers,
- LTE/EPS NAS layers – EMM and ESM NAS, and
- 5G System NAS layers – 5GMM and 5GSM.

An air interface Layer 3/NAS layers signaling message consists of an ordered series of octets (1 octet: 8 bits) and each message being with a protocol header. The protocol header is followed by protocol information fields. Each field is known as the *information element* (IE). An IE has certain attributes such as its unique identifier and presence requirements in the message, length, and value as described below. IEs and their attributes are described in a tabular format.

- **IE and its Identifier**

A signaling message carries various information from a sender to a receiver through a collection of *IEs*. Each IE indicates particular information of a protocol layer to a receiver and is uniquely identified by a so-called

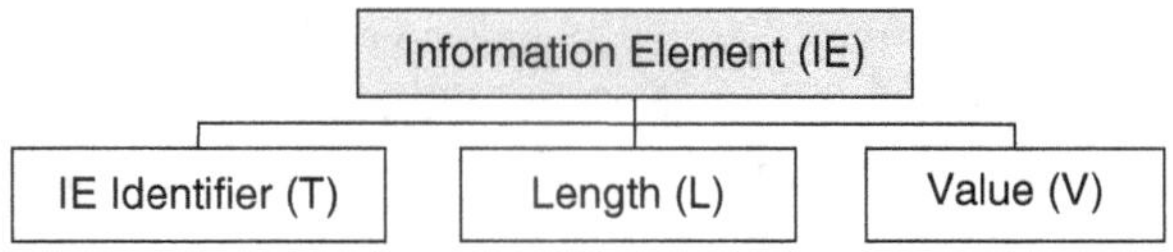

Figure 4.1 Components of an IE of a protocol message.

Information Element Identifier (IEI). An IE has a name and is represented by assigning a hexadecimal value through the IEI. IEs of a signaling message may have different lengths such as 1 octet, 2 octets, and so on. An IE of a message has the following components, also shown graphically in Figure 4.1.

- Type (T), represented by the IEI,
- Length (L), in octets, and
- Value (V), i.e. an actual value of an IE.

- **Presence Requirements of IE**

The presence of an IE in a signaling message may not be required always. Based on this, the presence of an IE is classified as shown in Figure 4.2:

- Mandatory (M) – An IE must be present always; if it is not, the receiver will consider the message as an erroneous one and reports a protocol error.
- Conditional (C) – Presence depends on the value of another IE. If a condition is met and the IE is not present, the receiver will consider the message as an erroneous one; else, it will accept the message.
- Optional (O) – The receiver will accept the received message irrespective of the presence of the IE.

- **IE Formats**

As shown in Figure 4.1, each IE of a signaling message has the type (T), represented by the IEI, along with a defined range of *values* (V), including reserved value, and its length (L). The type (T), length (L), and allowed value (V) of IEs of signaling messages of a particular protocol layer are defined by its corresponding 3rd Generation Partnership Project (3GPP) technical specification. Using the Type, Length and Value (TLV), several formats of an IE can be defined as shown in Figure 4.3; refer to TS 24.007 [44]. The formats "LV-E" and "TLV-E" indicate that these IE formats are used in the case of the LTE/EPS system only for its air interface Layer 3 MM and SM messages. GSM, GPRS, UMTS, LTE, and 5G NR air interface messages defined and encoded in TLV formats are called *Standard Messages*.

- **IE Types**

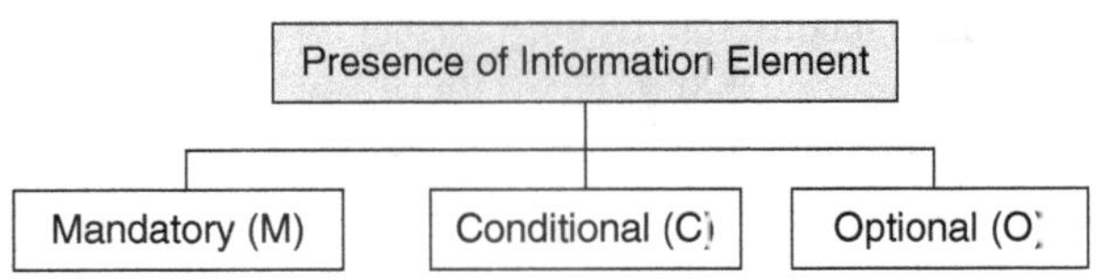

Figure 4.2 Presence requirements of an IE of a protocol message.

Table 11.1: Formats of Information Elements

Format	Meaning	IEI present	LI present	Value part present
T	Type Only	Yes	No	No
V	Value Only	No	No	Yes
TV	Type and Value	Yes	No	Yes
LV	Length and Value	No	Yes	Yes
TLV	Type, Length and Value	Yes	Yes	Yes
LV-E	Length and Value	No	Yes	Yes
TLV-E	Type, Length and Value	Yes	Yes	Yes

Figure 4.3 Standard formats of air interface layer 3 messages IEs. *Source:* © 2011. 3GPP ™ TSs and TRs are the property of ARIB, ATIS, CCSA, ETSI, TSDSI, TTA and TTC who jointly own the copyright in them. © 2011, 3GPP.

The IEs of the GSM, GPRS Layer 3 and UMTS, LTE, and 5G NR NAS layer signaling messages are octet aligned. Depending on the usage of the IEI, length (L) of an IEI, and its value (V), the IE can be represented in one of the formats as shown in Figure 4.3. The particular format used to encode an IE represents its corresponding *type*. The different types of IE are mentioned below:

- *Type 1:* IE format – V (with a half octet in length) and TV (each is half octet in length),
- *Type 2:* IE format – T, i.e. IEI only (1 octet in length),
- *Type 3:* IE format – TV (length of IEI is 1 octet; length of value: 1..n octet),
- *Type 4:* IE format – LV, TLV (length of IEI is 1 octet, length of value: 1..n octet, and length of length indicator: 1 octet), and
- *Type 6*: IE format – LV-E, TLV-E (length of IEI is one octet, length of value: 1..n octet, and length of length indicator: 2 octets).

For more information on the above IE, its format, and types, refer to the 3GPP TS 24.007 [44].

- **Encoding/Decoding of IEs**
  - The IEs of a signaling message is encoded by the sender and decoded by the receiver in order of their appearances in the tabular description. IEs are encoded as octet aligned.
  - Only the IEI/Type (T), Length (L), and Value (V) of an IE are encoded/decoded as per their tabular descriptions of the concerned message. Because of the TLV encoding, the overall message size becomes larger for transmission over the air interface.

In case an IE in a message is encoded incorrectly, a protocol error is detected and flagged by the receiver of the message which is notified to the sending network element about the detected error so that the predefined actions may be taken. For example, if an MS/UE sent an IE to the CN that is not encoded correctly, the CN element will report an error to the RAN. Further, the RAN may report about the error to the operation and maintenance personal through a predefined alarm.

Usages of IEs, their formats/lengths, i.e. TLV, and presence requirements in a NAS signaling message are illustrated through Examples 4.1 to 4.3.

From the LTE/EPS *Attach Complete* message definition which is shown in Figure 4.4, the following observations may be made:

- Tabular Description of Air Interface NAS Layer Signaling Message

All the air interface Layer 3 and NAS layer messages and their IEs are described in a tabular format with six columns. IEIs are arranged in the order they are transmitted.

- For an IE having its format type (T), the corresponding IEI is mentioned in the first column, IEI, of the table.

Figure 4.4 contains IEs with format types: V and LV.

Note that the GSM, GPRS, UMTS, air interface Layer 3, and LTE and 5G NAS layer signaling messages have protocol *header part* followed by IEs for *user information*/data part that is defined in a tabular format as shown in

Example 4.1 Illustration of LTE/EPS NAS MM Layer Message

Consider the LTE/EPS *Attach Complete* NAS layer EMM message (see Figure 4.4) reproduced from TS 24.301 [46], which is sent from the UE to the Mobility Management Entity (MME) through the eNodeB. They are described in the tabular format and also their encoding method, i.e. IE format-TLV, is the same. But there is a difference between the GSM, UMTS, GPRS and LTE/EPS, and 5G NAS layer messages. The GSM, GPRS, and UMTS air interface Mobility Management (GMM) messages are not security protected, but every LTE/EPS, 5GS NAS, MM layer message is security protected. For this purpose, the header of every LTE/EPS EMM or 5G MM message contains the *Security Header Type* IE.

IEI	Information Element	Type/Reference	Presence	Format	Length
	Protocol Discriminator	Protocol Discriminator 9.2	M	V	1/2
	Security Header Type	Security header Type 9.3.1	M	V	1/2
	Attach Complete Message Identity	Message Type 9.8	M	V	1
	ESM Message Container	ESM Message Container 9.9.3.15	M	LV-E	5-n

Figure 4.4 LTE/EPS NAS layer 3 attach complete message. *Source:* © 2014. 3GPP ™ TSs and TRs are the property of ARIB, ATIS, CCSA, ETSI, TSDSI, TTA and TTC who jointly own the copyright in them. © 2014, 3GPP.

Example 4.2 Illustration of LTE/EPS NAS SM Layer Message

Consider the LTE/EPS *ESM Information Request* NAS message (see Figure 4.5) which is sent from the MME to the UE through the eNodeB. It is described in a tabular format and their encoding method, i.e. IE format, is the same. Unlike the GSM/GPRS Layer 3 SM messages, the protocol header of every LTE/EPS NAS layer message contains an *EPS Bearer Identity* and a *Procedure Transaction Identity;* refer to TS 24.301 [46]. In the case of the 5G system NAS layer, a 5GSM message contains a *PDU Session ID* and a *Procedure Transaction Identity*.

IEI	Information Element	Type/Reference	Presence	Format	Length
	Protocol Discriminator	Protocol Discriminator 9.2	M	V	1/2
	EPS Bearer Identity	EPS Bearer Identity 9.3.2	M	V	1/2
	Procedure Transaction Identity	Procedure Transaction Identity 9.4	M	V	1
	ESM Information request Message Identity	Message Type 9.8	M	V	1

Figure 4.5 LTE/EPS NAS layer 3 ESM information request message. *Source:* © 2014. 3GPP ™ TSs and TRs are the property of ARIB, ATIS, CCSA, ETSI, TSDSI, TTA and TTC who jointly own the copyright in them. © 2014, 3GPP.

Example 4.3 Illustration of Encoding and Transmissions of Layer 3/NAS Layer Message

LTE/EPS air interface NAS layer messages and their tabular definitions/descriptions have been presented in the previous examples. It has been observed that the types, in TLV format, of the different IEs of a particular message are not the same, but they are mixed in nature, i.e. Type 1, Type 2, and so on. Figure 4.6 illustrates one such typical encoding of IEs of a Layer 3 messages header (1 octet, value only); message type (1 octet, value only); and followed by a typical IE in TLV format for transferring between an MS and the RAN over the respective air interface. The IEs of a message are encoded at the sender and decoded at the receiver in the same order as it is defined in its concerned technical specification.

Figure 4.6 Illustration: encoding and transmission of air interface message.

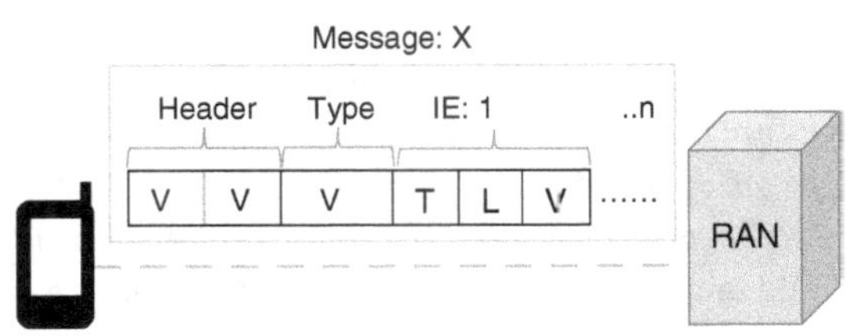

Figure 4.4. It may be noted that though the message description in tabular format is the same, the protocol headers used across the different air interfaces are different.

4.1.2 Encoding/Decoding: LTE and 5G NR Layer 2: RLC Protocol

The LTE and 5G NR air interface RLC layers provide the capability to exchange information between a UE and the LTE E-UTRAN or between a UE and the 5G NG-RAN in terms of the PDU. An RLC layer PDU facilitates transfer of higher-layer data in Transparent (TM), Unacknowledged (UNACK) or Acknowledged (ACK) mode. A PDU consists of a header part that is further followed by the data part of the PDU.

- **PDU Description**
  An RLC PDU, ACK, and UNACK mode header consist of several fields with different lengths in bits. Thus, the encoding and decoding of each field are different. Nevertheless, the protocol header and the data part of RLC PDU are octet aligned and is described in a tabular format. The TM PDU of the RLC layer does not contain the header part and is used to transfer messages such as paging and system information messages. Neither the sending RLC nor the receiving RLC layer performs any operations on a TM PDU. There is another PDU called Control PDU, which is used by the receiving RLC layer to inform the sending RLC layer on the status, i.e. lost or successfully decoded, of a PDU being received.
- **Encoding of RLC PDU**
  Though the RLC PDU is described in a tabular format, the header and data part is encoded as bit strings where the leftmost bit of the first line of the table is considered as the most significant bit and the rightmost bit of the last line of the table is considered as the least significant bit. Depending on the length of *Sequence Number* (SN) used in an RLC header, the length of the RLC header may take 1 or 2 octets at the beginning of the table and is different for the ACK mode and UNACK mode of data transfers. The 5G NR air interface RLC layer and its PDU formats are described later in Chapter 19. For more information on the RLC layer protocol header, its different parameters, and their encoding requirements, refer to TS 38.322 [114] for 5G NR.

4.1.3 Encoding/Decoding: LTE and 5G NR Layer 2: MAC Protocol

The LTE and 5G NR MAC layer facilitates the exchange of air interface Layer 2 information in terms of a PDU between a UE and LTE E-UTRAN or between UE and 5G NG-RAN. A MAC PDU consists of a MAC header and MAC SDU, which is received from the RLC layer. The method of description and encoding of the MAC layer PDU is similar to the method used by the LTE and NR RLC layers as described above. Similar to the RLC layer, the encoded bit strings of LTE or 5G NR MAC layer is an octet (8 bits) aligned.

Also, note that unlike the air interface Layer 3 message header, the MAC header and the MAC SDU are variable in size. For more information on the MAC layer protocol header, its different parameters, and their encoding requirements, refer to TS 36.321 [93] for LTE and TS 38.321 [113] for 5G NR. NR air interface MAC layer and its PDU formats are described later in Chapter 19.

4.1.4 CSN.1 Encoding/Decoding: GPRS Layer 2 Protocol (RLC/MAC)

The CSN.1 [10] notation method for describing, encoding, and decoding of air interface Layer 2 signaling messages are used by the GPRS RLC/MAC layer. It was described in Section 4.1.1 that the air interface Layer 3 messages are encoded and decoded on the octet (8 bits) level in TLV format. However, using the CSN.1 notation, only the *value, and not type or length,* of the IEs of a signaling message are encoded and decoded on a *bit* level. This results in a compact bits stream for transmission over the GPRS air interface. A CSN.1 encoded signaling message contains a *bits string* having an ordered sequence of symbols, i.e. 0 and 1. The resulting bits string may not be multiple of 8, i.e. an octet. An example CSN.1 description of a message with two IEs of length 3 bits each is shown below:

```
<Message >:: = 11 { 1 < IE1 : bit(3)> | 0 <IE2: bit(3)> }11
```

In the above CSN.1 description of the sample message, the vertical bar "|" represents a choice. If the third bit of the message is 1, then IE1 shall be encoded; if the third bit is 0, then IE2 shall be encoded. Both the IEs are illustrated with length = 3 bits. This is further followed by 11 bits, giving an overall bits stream of 8 bits.

The CSN.1 method is based on the *Backus-Naur Form (*BNF*)* that contains various rules to be followed while describing a message. For more information on such rules of encoding and decoding of RLC/MAC layer signaling message using the CSN.1 method and its other aspects, refer to Annex-B, TS 24.007 [44]. Example messages and their descriptions/structures using CSN.1 notation can be found in 3GPP TS 44.060 [131].

Reusable software modules/functions/procedures are required to be developed for CSN.1 encoding, at the sender end, and decoding, at the receiving end, of GPRS RLC/MAC signaling message. More about the implementation of encoding and decoding in the code is described in later sections.

4.1.5 ASN.1 Encoding/Decoding: UMTS, LTE, and 5G NR Layer 3 Protocol

The UMTS, LTE, and 5G NR air interface Layer 3 RRC protocol layer PDUs are described using the ASN.1 notation. Apart from these, the LTE/EPS S1-AP, X2-AP, 5G Core NG-AP, Xn-AP, and UMTS Radio Access Network Application Part (RANAP) protocol PDUs are also described using the ASN.1 notation. The ASN.1 describes the syntax of a message in an abstracted way as well as the method of encoding and decoding of message over a particular logical interface, e.g. air interface. ASN.1 notation is specified in ITU-T Rec. X.691 [11].

- **Message Description and Its Compilation**

The abstract description of a protocol layer message using the ASN.1 method is similar to a C-Language-like structure having data fields and its lengths. An example of an ASN.1 definition of a protocol message is shown below:

```
L3ProtocolMessage:: = SEQUENCE{
      Header :: =L3Header,
      Data :: =L3IEs,
      }
L3Header:: =SEQUENCE{
                  Protocol-Discriminator::= BITS STRING (SIZE(4)),
                  Skip-Indicator::= BITS STRING (SIZE(4)),
                  }
L3IEs ::= OCTET STRING(SIZE(255))
```

Consider that the C-Language is being used to develop the concerned protocol layer containing the above protocol header. From the above description of the protocol message, an ASN.1 compiler may produce the target C-Language source code/header file as illustrated below:

```
typedef struct {
      unsigned char Protocol-Discriminator:4;
      unsigned char Skip-Indicator:4;
      unsigned char L3IEs[255];
} L3ProtocolMessage
```

Further, the C-Language compiler will compile the ASN.1-generated C-header files along with the other C-source files to produce an encoding/decoding module.

- **Encoding/Decoding Method**

An ASN.1 description provides only the syntax of a message to be exchanged between two network elements. It does not define the actual method to be used for encoding and transfer of the concerned message. For encoding/decoding a signaling PDU or message and its transmission, the ASN.1 *Packed Encoding Rule (PER)* is used, which is specified in ITU-T Rec. X.691 [11]. This is also known as the *Message Transfer Syntax* in the ASN.1 paradigm. The encoded bits stream generated by the ASN.1 PER can be of two types as mentioned below:

- Octet Unaligned, e.g. RRC layer protocol PDUs.
- Octet Aligned, e.g. UMTS RANAP messages; LTE/EPS S1-AP and X2-AP; 5G NG-AP; and Xn-AP messages.

For more information on the aligned and unlighted encoding PER, refer to ITU-T Rec. X.691 [11]. There are additional protocol layer-specific encoding rules that are to be followed during the development of the concerned protocol layer. For more information on these additional rules, refer to TS 25.331 [50], TS 36.331 [94], and TS 38.331 [116].

- **How Does ASN.1 Notation Result in a Compact Encoding**

It was described in Section 4.1.1 that the air interface Layer 3 messages are encoded and decoded on the octet (8 bits) level in TLV format. Depending on the IE type, it may be required to encode the type/IEI (T), length (L), and value (V) of an IE in a message. The overall length of the encoded message becomes larger. However, in the ASN.1 method,

- The type/IEI is never encoded and transmitted, and
- Encoding the length and value is also optional if both sender and receiver already know it.

Because of these encoding rules, the size of the encoded message becomes compact in comparison to the TLV method encoding described above.

As an example, consider an IE with a value = 5 to be encoded and transmitted in TLV format as well as using PER. An illustration is shown in Figure 4.7. Type (T), Length (L), and Value (V) each have a length of 1 octet.

From Figure 4.7, it is observed that the PER encoding method results only in 3 bits of the IE's value to be transmitted, in comparison to 24 bits in its TLV encoding format. This is because in the PER method, the Type is never used, and Length may not be used during encoding of an IE.

Example 4.4 mentions typical LTE and 5G NR RRC layer signaling messages which are defined using the ASN.1 format described above.

4.1.6 Direct/Indirect Encoding Method

The encoding and decoding methods of mobile communications systems and network messages as mentioned in the preceding sections apply, in general, to all the messages defined in a particular mobile communications system such as the GSM, GPRS, UMTS, LTE and 5G NR systems. There could be other encoding and decoding methods that are applicable to a particular message defined only in a particular mobile communications domain. Two such methods are *Direct Encoding* and *Indirect Encoding*. They are used in GSM Layer 3 Radio Resource Management layer's *Immediate Assignment* message while allocating a list of hopping radio frequencies from the BSC to the Mobile station.

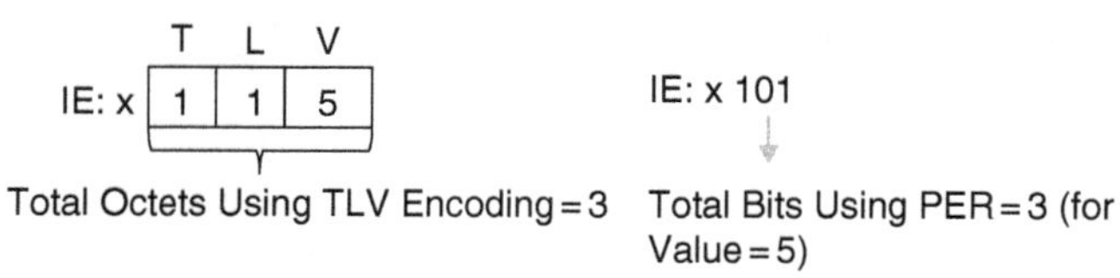

Figure 4.7 Illustration: TLV Vs PER encoding method.

Example 4.4 LTE and 5G NR Air Interface RRC Layers: RRC Connection/Setup Request ASN.1 Message

The LTE *RRCConnectionRequest* or the 5G NR *RRCSetupRequest* message is used by a UE toward the LTE eNodeB or 5G NG-RAN to request the establishment of an RRC signaling connection. For the ASN.1 definition of the LTE *RRCConnectionRequest* or NR *RRCSetupRequest* message, refer to TS 36.331 [94] and TS 38.331 [116]. The *RRCConnectionRequest or RRCSetupRequest* message carries the following information to the eNodeB:

- Initial UE identity,
- Establishment cause/purpose of the connection request, e.g. emergency call, data, signaling, voice call, and so on, and
- A randomly generated reference value.

4.1.7 Segmented Messages over the Air Interface

It was mentioned in Section 4.1.4 that the GPRS RLC/MAC layer messages are encoded/decoded at bit level using CSN.1 [10] method. Even the presence or absence of a single bit of information in a particular communications signaling message matters a lot for the receiver of the message because it may not be able to decode the rest of the received message successfully. Sometimes, if the complete information to be sent does not fit in a single signaling message, the same may be transferred in a segmented way, and this is indicated by a presence of a bit in the first message segment already sent to the receiver.

Similarly, the segmentation of the New Radio (NR) RRC layer messages i.e. *RRCReconfiguration* or *RRCResume*, in DL direction, and *UECapabilityInformation*, in UL direction, has been introduced as part of the 5G system Release 16. This is due to the limitation of the maximum PDU size of 1900 bytes as supported by the NR PDCP layer. The 5G NR air interface control plane protocol layer shall be described later in Chapter 18

4.1.8 Piggybacking a Signaling Message

To reduce the exchange of signaling messages overhead between mobile devices and the network over the air interface or other interface, sometimes a signaling message is attached to another signaling message that is being scheduled and already transmitted. Such a mechanism of attaching a signaling message as part of another signaling message is known as the *piggybacking* of a message. The piggybacked message may be used to request additional RRs from the network to start data transfer in the opposite direction or to trigger the start of a new signaling procedure between the mobile device and the network. The following Examples 4.5 to 4.7 shall clarify the concept and purpose of piggybacking a signaling message into another message.

- **Air Interface Signaling Messages**

Example 4.5 Piggybacking of GSM Air Interface Complete Layer 3 Information Using SCCP

In the case of the GSM system, the SS7 protocol stack is used to exchange messages between the GSM and the MSC over the A-interface. The *Signaling Connection Control Part (*SCCP*)* layer is a part of the SS7 protocol stack. GSM Complete Layer 3 Information is used to transport all the initial signaling connection messages, between the MS and MSC, piggyback with the SCCP protocol message. Examples of GSM system initial messages are – *CM SERVICE REQUEST*, which is used for a mobile originated call, *LOCATION UPDATE REQUEST*, which is used for location area update request from MS to the CN, etc.

Example 4.6 Piggybacking Using LTE or 5G NAS Layer Signaling Message

In the LTE or 5G system also, piggybacking of a signaling message, e.g. piggybacking an SM layer message to an MM layer message, is used. In the LTE/EPS, piggybacking of NAS messages is used for the bearer management procedure in the downlink direction. In the uplink direction, piggybacking of the NAS message is used only for transferring the initial NAS message during connection setup. For example, the LTE/EPS *ATTACH Request* message is send piggybacked with the *RRCConnectionSetupComplete* message over the LTE air interface between UE and E-UTRAN.

Similarly, in the 5G NR system, the NAS layer *RegistrationRequest* (toward the 5G Core Access and Mobility Management Function (AMF)) message is piggybacked to the *RRCSetupComplete* message from a UE to the NG-RAN.

- **CN Signaling Messages**

Example 4.7 Piggybacking GTPv2 Control Plane Messages

In the LTE/EPS and 5G systems, the GTPv2 control plane signaling messages are used in a couple of logical interfaces, for example, between the LTE/EPS MME and the S-GW and S-GW and P-GW. The fifth bit of a GTPv2 control plane header, 3GPP TS 29.274 [70], contains a field called "P" which indicates the presence of a further piggybacked GTPv2C message along with the current GTP GTPv2C messages. For example, a *Create Session Response* message from the S-GW to the MME may contain the *Create Bearer Request* message as well as the piggybacked message for the MME. In this case, the "P" flag in the header of the *Create Session Response* message shall be set to 1.

4.2 Encoding/Decoding of Signaling Messages: RAN and CN

In the previous sections, we have discussed the encoding and decoding of signaling messages between a UE/MS and their respective RAN of the GSM, UMTS, LTE, and 5G NR systems over their respective air interface. To perform certain functions and procedures between the RAN and the CN, they also exchange signaling or control plane messages over the respective logical interface. The IEs of a signaling or control plane message exchanged over the concerned logical interface between the RAN and CN are packed and unpacked using a particular encoding and decoding method, which are described below:

- **Between GSM BSC and MSC: A-Interface**

The logical A-interface is used to exchange signaling messages between the GSM BSC and the MSC. Over the A-interface, the following types of messages are exchanged between the BSC and MSC.

- Base Station Management Application Part (BSSMAP)

Signaling messages for various functions and procedures performed between the BSC and MSC are classified into BSSMAP types. BSSMAP messages, between BSC and MSC, are also encoded/decoded and are described in a tabular format similar to the air interface Layer 3 messages. However, unlike the Layer 3 messages, BSSMAP messages do not contain a message header. Each BSSAMP message starts with its message type, followed by the associated IEs.

- Direct Transfer Application Part (DTAP)

DTAP messages are exchanged between the UE and MSC only. All the air interface Layer 3 CC and MM signaling messages that are transparently forwarded by the BSC, received from the MS to the MSC without processing

Example 4.8 LTE NAS Layer: Downlink NAS Transport: MME to eNodeB

Figure 4.8 shows the definition of the downlink NAS transport message from the LTE/EPC MME to the eNodeB.

IE/Group Name	Presence	Range	IE type and Reference	Semantics Description	Criticality	Assigned Criticality
Message Type	M		9.2.1.1		YES	ignore
MME UE S1AP ID	M		9.2.3.3		YES	reject
eNB UE S1AP ID	M		9.2.3.4		YES	reject
NAS-PDU	M		9.2.3.5		YES	reject
Handover Restriction List	O		9.2.1.22		YES	ignore
Subscriber Profile ID for RAT/Frequency Priority	O		9.2.1.39		YES	ignore

Figure 4.8 LTE/EPS MME-ENodeB: S1-AP: downlink NAS transport. *Source:* © 2014. 3GPP ™ TSs and TRs are the property of ARIB, ATIS, CCSA, ETSI, TSDSI, TTA and TTC who jointly own the copyright in them. © 2014, 3GPP.

by BSC, are classified into DTAP types. DTAP messages are the air interface Layer 3 messages with a header containing *protocol discriminator* information in the header of every message that is exchanged between the MS and MSC. Because of this, the DTAP message is encoded and decoded as described in Section 4.1.1.

- **Between GSM BSC and SGSN: Gb-Interface**

The Gb-interface is used to exchange signaling messages between the GSM BSC and the SGSN. Over the Gb-interface, the following types of messages are exchanged between the BSC and SGSN of a GPRS network.

- Network Service protocol layer; refer to 3GPP TS 48.016 [135].
- BSS GPRS Protocol (BSSGP) protocol layer; refer to 3GPP TS 48.018 [136].

NS and BSSGP layer messages are also encoded/decoded and are described in a tabular format similar to the air interface Layer 3 messages. However, unlike the Layer 3 messages, NS and BSSGP layer messages do not contain a message header. Each NS and BSSGP layer message starts with their message type, followed by the associated IEs.

- **Between UMTS RNC – CN; LTE E-UTRAN and CN; 5G NG-RAN and CN**

The UMTS RANAP, over the Iu interface between RNC and CN, uses the ASN.1 notation for describing the signaling message between RNC and the CN. The LTE/EPS S1-AP, between eNodeB and MME, and X2-AP, between two eNodeBs, also uses the ASN.1 notation for encoding and decoding of signaling message over the S1 and X2 interface. Similarly, the 5G system NG-AP, between NG-RAN/gNB and AMF, and XN-AP, between two gNB, also use the ASN.1 notation for encoding and decoding of signaling message over the NG and Xn interfaces. Protocol messages specifications using ASN.1 notation was described earlier in Section 4.1.5. Note that the messages exchanged between UMTS RNC – CN; LTE E-UTRAN and CN; 5G NG-RAN and 5G core are also described using tabular notation. Example 4.8 illustrates the tabular definitions of a message described in a tabular format.

The NAS-PDU IE, which is a mandatory one, in the downlink NAS transport message contains the message to be communicated from the MME to the UE.

Chapter Summary

This chapter presented the CSN.1, ASN.1, direct, indirect, tabular format, and so on and methods of describing and encoding/decoding protocol layer messages across the different mobile communications systems from the

GSM to the 5G system. The peer protocol layers of network elements of a mobile communications network exchange signaling messages using a particular encoding/decoding method as described in this chapter. These encoding and decoding methods are used over their respective logical interfaces, e.g. air interface Layer 2, Layer 3, NAS layer, between RAN and CN, and so on, to exchange control plane information. In comparison to the other encoding methods described in this chapter, the CSN.1 encoding method produces more compact protocol layer messages that are to be transmitted over the GPRS air interface.

An encoding and decoding method represents the data structures of IEs of messages that are used during the software design and development of protocol layers supported by the network elements of a mobile communications network. An IE of a message may contain and may be encoded with several information. Unlike the traditional IP layer header and packet, information may be encoded/decoded even at a single bit level in a mobile communications message. Correctly encoding and decoding of an IE, its components, and the value of a control plane or data plane message is important for the successful operation of a mobile communications network. It is also important from the interworking and interoperation point of view. We then closed the chapter with the method of piggybacking of a signaling message over another signaling message, thus reducing signaling overhead, be it over the air interface or between CN elements.

5

Network Elements

Identities and Its Addressing

Introduction

This chapter covers the various identities that are used to identify and address different network elements or logical objects of mobile communications systems and networks, i.e. Global System for Mobile Communication (GSM), General Packet Radio Service (GPRS), Universal Mobile Telecommunication System (UMTS), and Long-Term Evolution (LTE) system. Network identities used in the 5G system are described in Chapter 16. Every network element has its own identity using which the peer network element can identify the source of data received over a particular interface.

We begin with the network identities that are used at the radio access and core network (CN) domain along with their nature. A network element identity has several other aspects such as its persistency, which may be either permanent or temporary; also, an identity may have a local or global presence. A network element can have several identities, especially if it supports multiple radio access technologies (RATs). We then present the fundamental or native and mapped identity of a network element. Mapped identities are used in case of an interworking scenario where a user and its User Equipment (UE) move from one RAT to another and vice versa.

A network identity may be used by a network element to keep track of the resources allocated to another network element. Apart from the network elements, network identities are also assigned to identify and address other logical objects such as the GSM location area, GPRS routing area, and LTE/Evolved Packet System (EPS) tracking area. Such network identities are further described in Chapter 18.

5.1 Network Elements and Their Identities

A mobile communications system and network comprise numerous physical network elements or resources as well as logical objects or resources that are used as part of the design, development, implementation, and field deployment. Network elements or resource identities are logical in nature. An identity may represent a logical connection and association between two network elements. A network element, say eNodeB, may assign an identity to another network, say UE, to uniquely identify it over a particular logical interface. The network element's identities are divided into the following categories:

- UE/Mobile Station (MS) Identities
- Radio Access Network (RAN), UMTS Terrestrial Radio Access Network (UTRAN), Evolved-UMTS Terrestrial Radio Access Network (E-UTRAN), 5G Next-Generation Radio Access Network (NG-RAN) Identities
- CN Identities

Figure 5.1 illustrates the above network identities along with their natures.

Mobile Communications Systems Development: A Practical Introduction to System Understanding, Implementation, and Deployment, First Edition. Rajib Taid.

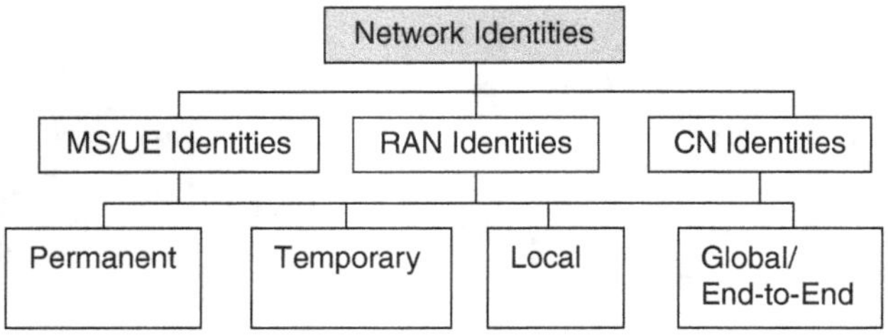

Figure 5.1 Network identities, their persistency, and presence.

A network element, or a subscriber, or a logical link between peer protocols or other logical resources are identified and addressed with their corresponding identities, which can be a permanent, temporary, static, or dynamic in nature, as shown in Figure 5.1. For example, the core network may allocate a temporary identity to a mobile device for paging purposes. The CN and the access network allocate a temporary identity to an MS/UE as a result of events such as the power off/on or changes in the current location of the mobile device. This is required to protect the real identity of a mobile device while communicating with the network over the air interface.

A permanent identity is assigned to a network element as part of its administrative provision, network planning, and configuration process. For example, an *International Mobile Subscriber Identity* (IMSI) is allocated to uniquely identify each mobile subscriber in the GSM, UMTS, LTE/EPS, and 5G systems and may be also used to page an MS/UE by the CN.

A network identity can be a *local* one, i.e. visible to a particular network element such as GSM Visitor Location Register (VLR), GPRS Serving GPRS Support Node (SGSN), and LTE/EPS Mobility Management Entity (MME), or it can be used/visible across the *end-to-end* network element, i.e. from the mobile device to base station controller (BSC), UTRAN, E-UTRAN, and 5G NG-RAN to Core Network.

A network identity can be a *global* one that is exchanged across an interface connecting two peer network elements. For example, in the LTE/EPC system, a UE and MME S1AP identity is exchanged between the eNodeB and the MME over the S1 interface to uniquely identify a UE. Note that the network identities have different lengths.

5.2 Permanent Identities

There are network identities that are permanent and unique in nature. Examples of permanent network element identities are mentioned below:

- An MS/UE is assigned an *International Mobile Equipment Identity* (IMEI) by its manufacturer that is burnt into the firmware of an MS/UE.
- An IMSI is assigned by an operator to a subscriber during its provisioning in the database of the Home Location Register (HLR). An IMSI is stored in the SIM card of an MS/UE.
- A *Public Land Mobile Network (*PLMN*)* identity of a network consists of the Mobile Country Code (MCC) and Mobile Network Code (MNC). A PLMN uniquely and globally identifies the operator of a particular network as an MCC is allocated by the International Telecommunication Union (ITU).

An operator-configured network identity also can be a permanent one as long as its configuration does not change.

5.3 Temporary Identities Assigned by CN

In an end-to-end mobile communications network, one important function of the CN is to protect the real identity of an MS/UE while exchanging information over the air interface. The confidentiality of an MS/UE identity over the air interface is accomplished by allocating a temporary identity to it by the concerned CN element. Also, the temporary network identity is provided by the concerned network element which controls a particular service area. In the case of the GSM system, a VLR controls a location area; in the case of the GPRS system, an SGSN controls a routing area; and in the case of the LTE/EPS system, an MME controls a tracking area. Some of the temporary identities assigned to a mobile device by the concerned CN elements of GSM, GPRS, UMTS, and LTE networks are described below. For more information on these temporary identities, refer to TS 23.003 [30] and TS 23.060 [31].

5.3.1 GSM System Temporary Identities

- Temporary Mobile Subscriber Identity (TMSI)

A *TMSI* is allocated by the CS domain VLR to protect an MS/mobile user's identity while communicating over the air interface. A TMSI is allocated as part of the successful CS domain location area update mobility management procedure initiated by an MS toward the CN. A TMSI is used by the MSC as the identity of an MS during the air interface Layer 3 mobility management-related procedures such as sending a paging request message to the MS to notify about an incoming call.

A TMSI has an end-to-end significance and has *32 bits* in length. TMSI with all 1's is not a valid one. A new TMSI is also allocated to a mobile device on entering a new location area under a different VLR where the MS performs a location area update procedure with the CN. The VLR maintains an association between the IMSI of the mobile device and the allocated TMSI. For multi-RATs, i.e. GPRS Edge Radio Access Network (GERAN), UTRAN, and E-UTRAN, capable MS/UE, a TMSI is allocated as a result of the combined mobility management procedures performed by a UE/MS toward the CN.

5.3.2 GPRS System Temporary Identities

- Packet TMSI (P-TMSI)

A *P-TMSI* is allocated by the GPRS PS domain SGSN to protect a mobile user's identity while communicating over the air interface. A P-TMSI is used as the identity of an MS during the GPRS air interface Layer 3 mobility management-related procedures such as the routing area update, paging request, and an attach request.

A P-TMSI has an end-to-end significance and has *32 bits* in length. The first two Most Significant Bit (MSB) bits (31, 30) of a P-TMSI are always set to 1's. A new P-TMSI is allocated to a mobile device on entering a new routing area under a different SGSN where the MS performs a routing area update procedure with the CN. The SGSN maintains an association between the IMSI of the mobile device and the allocated P-TMSI.

- Network Resource Identifier (NRI)

An *NRI* is a CN-level identifier and is used to identify a GSM MSC or SGSN or an LTE/EPS MME. Several MSCs or SGSNs or MMEs can be configured in a pool to serve mobile users in a load-balanced way. An NRI is assigned and identifies uniquely an individual CN element, i.e. MSC, SGSN, or an MME, out of all the CN elements configured in a pool.

An NRI has an end-to-end significance and has a length from *0 to 10 bits;* NRI having length 0 means the feature is not used at the CN end. More about the CN element pool and their selection are described later in Chapter 7.

- Temporary Logical Link Identity (TLLI)

A *TLLI* identifies a logical link at the Logical Link Control (LLC) layer between the MS and the SGSN of a GPRS system. At the GPRS LLC layer level, an MS is uniquely identified by a *temporary logical link identifier (TLLI)*. A TLLI is constructed from a P-TMSI that is allocated by the SGSN which has an end-to-end significance and is used across the MS, BSC, and the SGSN.

5.3.3 LTE/EPS System Temporary Identities

- MME TMSI (M-TMSI), S-TMSI

An M-TMSI is allocated by the LTE/EPS MME to a UE. An S-TMSI is constructed from an M-TMSI and the MME code, which is 8 bits in length. Using the S-TMSI, MME used to send a paging message to a UE. S-TMSI has an end-to-end significance.

For a multi-RATs capable UE/MS that supports the GSM, GPRS, UMTS, and LTE systems, a UE/MS will perform the combined registration procedures toward the respective CN. The concerned CN element will assign and allocate the three TMSIs accordingly, i.e. TMSI, P-TMSI, and S-TMSI, to a UE. For example, to notify an incoming CS call to an E-UTRAN registered with the *Circuit Switched Fall-back* (CSFB) feature, the MSC will send the paging request message to the MME with TMSI. The MME will send the paging command to the UE using the S-TMSI. The UE will send the CS paging response with TMSI allocated by the MSC/VLR.

- Network Identities for LTE/EPS Network elements

Various identities are used to identify network elements or logical objects, at the protocol layer level, of an LTE/EPS network. Such network identities, as elucidated in Example 5.1 below, may be assigned as part of a periodical

Example 5.1 Network Identities of LTE Network Elements

Figure 5.2 illustrates the fundamental identity assigned to some of the network elements of an LTE/EPS PLMN.

Several network identities also contain the PLMN Id. The purposes of the network identities shown in Figure 5.2 are described below:

- E-UTRAN Cell Identity (ECI) (28 bits in length): It is a unique cell identity and cannot be repeated in a PLMN. It is defined as the SIBType1. *CellIdentity*; refer to TS 36.331 [94]. An ECI together with the PLMN Id constitutes an E-UTRAN Cell Global Identity (ECGI).
- eNB Id (28 bits length): It identifies an eNodeB within a PLMN.
- Radio Network Temporary Identifier (RNTI) (16 bits length): It identifies a UE having a Radio Resource Control (RRC) connection and also can be scheduled for data transmission/reception.
- MMEC (8 bits length): MME Code – identifies MME with the MME group.
- MMEGI (16 bits length): MME Group Id – identifies the MME Group Id of that an MME belongs to.

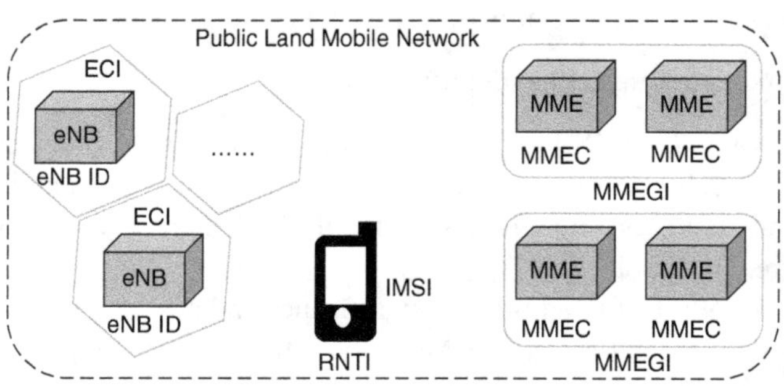

Figure 5.2 Illustration: identities for LTE network elements.

network planning or maintenance processes or can be generated and assigned by a network element dynamically to another network element as part of a protocol layer procedure.

- Globally Unique Temporary Identity (GUTI)

In the case of the LTE/EPS system, a mobile device UE is uniquely identified by a *GUTI* that is allocated by the MME. A GUTI is similar to a P-TMSI and is used as the identity of a UE during the LTE/EPS mobility management-related procedures executed between a UE and the MME. A GUTI can be a native or mapped as described later in Section 5.5. A GUTI is constructed from several fundamental network identities as illustrated in Example 5.2.

Example 5.2 LTE/EPS Globally Unique Temporary UE Identity (GUTI)

A GUTI is an LTE/EPS network-level identity that has end-to-end significance and is used to provide unambiguous identification of a UE. The construction, as well as its hierarchical structure of a GUTI, is shown in Figure 5.3.

Figure 5.3 Illustration: components and derivation of a GUTI.

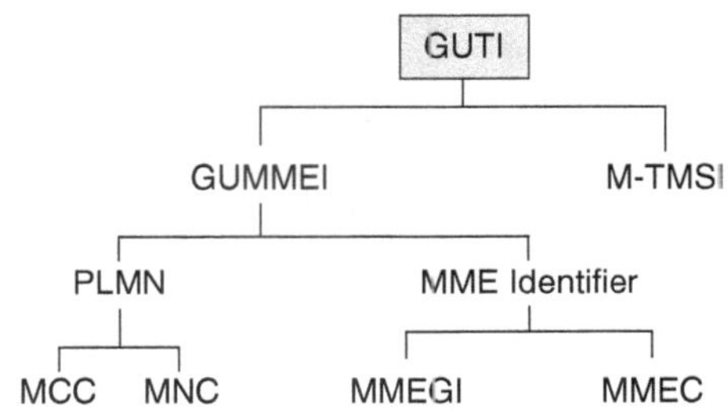

Using a GUTI, the identification of the MME and its corresponding network can be determined. It can be used by the network and the UE to establish the UE's identity during signaling between them in the LTE/EPS network.

- Physical Layer Identity of a Cell

At the physical layer level, each cell in an LTE/E-UTRAN network is identified through a *Physical Cell Identity (PCI),* which is illustrated through Example 5.3 and Figure 5.4.

As defined by the TS 36.211 [90], there are 504 unique Physical Layer Cell Identities, and they are divided into 168 (0–167) unique Physical Layer Cell Identity groups; each group has three (0, 1, 2) unique identities. A PCI has the relationships and is derived from the downlink reference signals *primary synchronization signal* (PSS) and *secondary synchronization signal* (SSS) which are transmitted from a base station. The mathematical relationship among the PCI, PSS and SSS is shown below.

```
PCI= (3* Physical Cell Identity Group) + Physical Layer Identity
       Where Physical Cell Identity Group =0 to 167;
       Physical Layer Identity=0 to 2.
```

The PSS has the link to the Physical Layer Identity, which is a Zadoff–Chu sequence, and the SSS has the link to the PCI group. A UE reads the synchronization signals during a cell search procedure in the LTE system. To avoid interference and PCI collision as well as confusion issues, the same PCI is never reused in the neighbouring cells. A PCI collision occurs when two adjacent cells have the same PCI. A PCI confusion occurs when a cell has two neighbours with the same PCI. As the number of PCIs (504) is limited, PCIs are carefully planned and repeated in other cells that are not adjacent or neighbor to each other as shown in Figure 5.4.

Example 5.3 LTE Physical Layer Cell Identity (PCI)

A cell in the LTE system can have three types of identities:

- An ECI of *28 bits* in length as mentioned in Example 5.1, which identifies a cell in a particular LTE network. It is defined and controlled by the higher layer. It is also used from the operation and maintenance point of view.
- An E-UTRAN Cell Global Identifier (ECGI), containing the PLMN Id, that identifies a cell globally, i.e. anywhere in the world,
- A *Physical Cell Identity* (PCI) is used at the LTE Physical Layer level only (Figure 5.4), and it is defined as the *PhysCellId* in the TS 36.331 [94].

Figure 5.4 illustrates the LTE Physical Layer Cell Identities (PCI) and their groups. The PCI is similar to the UMTS scrambling code concept, for which a RAN planning is required for its proper usages. A PCI helps a UE to distinguish information received from different neighboring transmitters and select a particular cell to camp-on at the physical layer level.

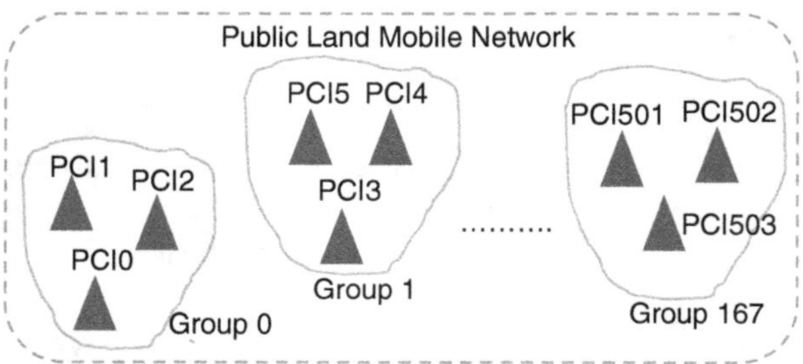

Figure 5.4 Illustration: LTE physical layer cell identities (PCI) and groups.

5.4 Temporary Identities Assigned by RAN: RNTI

RNTI stands for *Radio Network Temporary Identifier* that is allocated by the UMTS UTRAN, LTE E-UTRAN, and 5G NG-RAN to UEs. An RNTI uniquely identifies a UE and is used during the communication between UTRAN or E-UTRAN or NG-RAN and a UE over their air interface: Uu. The RNTIs used in the LTE system is described below:

- LTE E-UTRAN: Allocation of RNTIs, TS 36.300 [92]

LTE E-UTRAN also allocates several RNTI types to address a particular UE over its air interface. However, unlike the UMTS system, it is also possible to address multiple UEs in the LTE system. This is because the LTE E-UTRAN uses the physical downlink shared channel (PDSCH) to transmit control as well as user data to a UE. Thus, a mechanism is required by an LTE UE to differentiate the type of information received from the E-UTRAN. To achieve this, different types of RNTIs are used by E-UTRAN as described below. For more information on LTE RNTI values and usages, refer to TS 36.321 [93]:

- *System Information (SI)-RNTI* – It is used to broadcast system information from the E-UTRAN to the UEs/cell through the Broadcast Control Channel (BCCH). SI-RNTI is part of the cyclic redundancy check (CRC) added to the contents of the downlink control information (DCI) format 1A.
- *Paging (P)-RNTI* – It is used by the UEs for the reception of paging from the E-UTRAN through the Paging Control Channel (PCCH).
- *Cell(C)-RNTI* – C-RNTI uniquely identifies a UE having RRC signaling connection, and scheduling, with E-UTRAN in a particular cell. C-RNTI is used by the E-UTRAN to provide an uplink transmission grant and downlink assignment.

- *Temporary C-RNTI* – It is generated by the MAC layer of an eNodeB and sends it in the Random Access Response (RAR) to the UE as a response to the Random Access Preamble transmitted by the UE. Temporary C-RNTI is later changed to the C-RNTI once the UE contention is resolved at the eNodeB end.
- *SPS-CRNTI* – Semi-Persistent Scheduling RNTI is used in case the SPS activation/reactivation/retransmission is performed.
- *TPC-RNTI* – It is used to send Transmit Power Control command to a UE.
- *Random Access (RA)-RNTI* – RA-RNTI unambiguously identifies which time-frequency resource was utilized by the UE to transmit the Random Access Preamble. The eNodeB uses the RA-RNTI in the RAR from the E-UTRAN to UE. The eNodeB scrambles Physical Downlink Control Channel's (PDCCH's) CRC with RA-RNTI for transmission of PDSCH that carries RAR(s).
- *M-RNTI* – It is used to notify a UE about an MCCH information change.

5.5 Usages of Network Identities

An ongoing end-to-end call for a UE/MS involves resource allocations by the different network elements of a communication network. In this regard, each network element assigns its respective network identities, having local or end-to-end significance, which is used to keep track of the various resources allocated by them. When an ongoing call is completed, the network elements release the allocated resources as identified by the respective network identities.

At times, the end-to-end troubleshooting of an issue may be confusing using the different network identities. The corresponding identities are used during the analysis of the different call processing events, both signaling and user data, to isolate the source of the issue, i.e. network elements. Once a network element is isolated from the probable root cause, the same events may be tracked further on the peer network element using the corresponding identities that are used in that network element.

Note that the network identities are assigned by the different network elements or different protocols layers of a network element. For example, LTE RNTIs are assigned by its Layer 2 MAC layer. Knowing the network identity and its corresponding protocol layer is particularly important during the end-to-end troubleshooting of an issue using protocol analysis tools.

5.6 Native and Mapped Network Identities

Network identities assigned to a mobile device can be *native* or *mapped* ones.

- Native Identity

A network element may construct a native identity from several fundamental identities. A network element may allocate a native identity to another network element for its identification and tracking of various resources allocated to it. An example of native identity in the case of LTE/EPS is the GUTI as shown in Example 5.2; in Figure 5.3. MME assigns a GUTI to a UE and is used during the various EPS mobility management layer procedures. Similarly, in the GPRS system, the SGSN allocates a P-TMSI to an MS.

- Mapped Identity

A mapped identity is derived from a certain portion of a native identity or may assign the whole native identity. One example of mapped identity is the NRI used in a GPRS system that is derived either from a TMSI or P-TMSI. Similarly, a TLLI, which is used in the GPRS system, is derived from a P-TMSI, as shown in Example 5.4 and Figure 5.5.

- Mapped identity due to intersystem changes

An LTE E-UTRAN UE may support multiple RATs to provide communication services through legacy networks. Such a UE registers with the legacy CNs also where the respective network identities are allocated to the UE to provide seamless communication service to the user. During the intersystem changes from E-UTRAN/LTE/EPS to GERAN/UTRAN (UMTS) and vice versa due to its mobility, a UE identity shall be mapped into another identity, as shown in Example 5.5 and Figure 5.6.

Example 5.4 GPRS MS Identity: Mapping an NRI and TLLI into a P-TMSI

Consider a P-TMSI of a GPRS mobile subscriber which is of 32 bits in length. The mapping and deriving of an NRI and a TLLI from a P-TMSI is shown in Figure 5.5.

- NRI

  An NRI is mapped into the Bits 23rd to 14th of a P-TMSI.

- TLLI

  The first two MSBs of a TLLI are set to 1's. The Bits 29th to 0th of a TLLI is mapped into the corresponding bits of a P-TMSI.

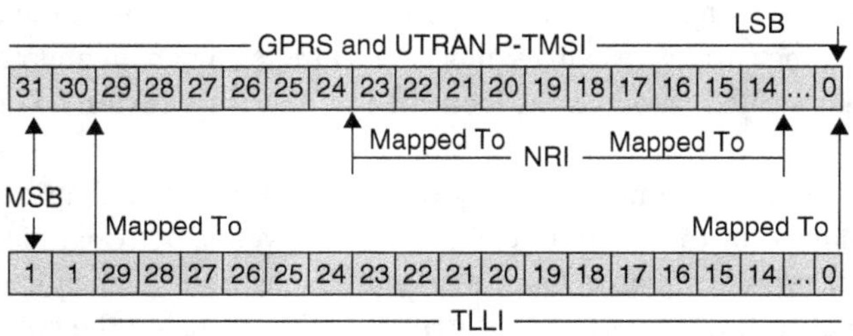

Figure 5.5 GPRS MS identity: mapping of an NRI and TLLI into a P-TMSI.

Example 5.5 UE Identity Mapping due to the Intersystem Change from GERAN/UTRAN to E-UTRAN

Due to a UE's intersystem changes from E-UTRAN/EPS to GERAN/UTRAN (UMTS) and vice versa, the new SGSN or new MME requires to retrieve the UE information from the old SGSN or old MME. Apart from this, identity mapping is also required at the UE end if it wishes to perform a mobility management procedure in the target system.

Consider that an E-UTRAN capable UE leaves a GERAN/UTRAN coverage area and enters into an E-UTRAN area and wants to perform an EPS ATTACH mobility management procedure toward the LTE/EPC network. Also, assume that the UE does not have a valid GUTI at present. In this case, the UE maps the existing GERAN/UTRAN domain identities into the E-UTRAN domain corresponding identities and sends those identities as *EPS Mobile Identity* in the Attach request message, along with the *Old GUTI Type* as "Mapped GUTI" to the MME. The GERAN/UTRAN network identities mappings to E-UTRAN network identities are shown in Table 5.1.

Table 5.1 GERAN-UTRAN to E-UTRAN-EPC: identities mapping.

GERAN-UTRAN	E-UTRAN-EPC
Mobile Country Code (MCC)	Mobile Network Code (MNC)
Mobile Network Code (MNC)	Mobile Network Code (MNC)
Location Area Code (LAC)	MME Group ID
Routing Area Code (RAC)	Bit 23rd to 16th of M-TMSI
8 Most Significant Bits of NRI	MME Code
29th to 24th bits of P-TMSI	29th to 24th bits of M-TMSI
15th to 0th bits of P-TMSI	15th to 0th bits of M-TMSI

Figure 5.6 shows graphically the mapping of LTE/EPS M-TMSI from the GERAN/UTRAN P-TMSI. In the current intersystem change scenario, the visiting and new MME performs the reverse mapping of the identities shown in Table 5.1, and based on these, the new MME is able to contact the old SGSN to retrieve the UE information.

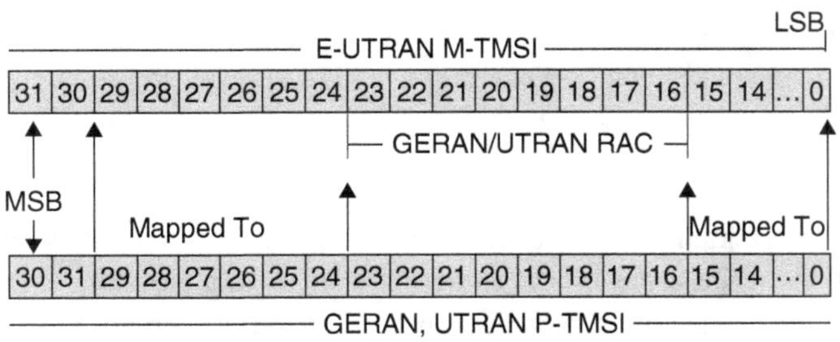

Figure 5.6 Illustration: UE identities mapping: GERAN/UTRAN to E-UTRAN.

For a UE's intersystem change from E-UTRAN to GERAN/UTRAN area, the procedure for identities mapping is almost the same. For more information, refer to TS 23.003 [30].

- LTE/EPS: Indication of Native and Mapped Identity

In the LTE/EPS, the CN elements are resolved and identified using their corresponding identity, i.e. GUTI for MME and P-TMSI/ Routing Area Identity (RAI) for the SGSN. These identities (GUTI and P-TMSI/RAI) used and assigned to a UE can be a native or mapped one. Native or mapped identity type or its nature is determined as follows:

a) Explicit indication in the Attach request and tracking area update/routing area update request message from UE to MME /SGSN.
b) Indication by setting the MSB of LAC and MME group Id.
   i) MSB of MME Group = 1 when a native GUTI is used; MSB of MME Group = 0 when the GUTI is mapped from P-TMSI and RAC.
   ii) MSB of LAC = 1 when P-TMSI/RAC is mapped from a GUTI; MSB of LAC = 0 when a native P-TMSI/RAC is used.

We have discussed and illustrated only a few network identities in the preceding sections. In fact, mobile communications systems based on GSM, GPRS, UMTS, and LTE systems use numerous network identifiers across the protocol layers and interfaces. Refer to the corresponding 3GPP specification for the different network identities and their components. Later, Section 18.4.1 further describes some of the identities associated with the mobility management procedures of GSM, GPRS, LTE/EPS, and 5G systems.

5.7 LTE UE Application Protocol Identity

In the LTE/EPC system, the eNodeB and the MME exchanges control plane protocol messages over the *S1* logical interface. Similarly, the eNodeBs exchange control plane protocol messages over the *X2* logical interface. For unique identification of a UE over the S1 interface, the eNodeB and MME assign the *eNB UE S1AP ID* and *MME UE S1AP ID*, respectively, during the initial registration request made by the UE to the CN. The eNodeB and the MME include these identities in all the application protocol (S1AP) signaling messages that are exchanged between them for the particular UE following its initial registration request message.

For example, consider the LTE/EPS *attach request* message sent by the UE to MME. The eNodeB shall assign the *eNB UE S1AP ID* to identify the particular UE within the eNB and send it in the S1AP *initialUEMessage* to the MME. For more information on the definition of the *initialUEMessage,* refer to TS 36.413 [97].

The MME will also allocate the *MME UE S1AP ID* to identify the particular UE within the MME. From this instance onward, both the eNodeB and the MME shall include the respective S1AP identities in all the application protocol S1AP messages, e.g. Uplink/Downlink NAS Transport, exchanged between them on behalf of the particular UE. Similarly, the eNodeB assigns the *eNB UE X2AP ID* to identify a UE over the X2 interface. This identity is included in the X2AP messages exchanged between two eNodeBs.

Chapter Summary

In this chapter, we have presented some of the network identities of network elements of communications networks from the GSM to the LTE system. Network identities are common as well as specific to a particular communications system. An operator may allocate network identities permanently to a network element. On the other hand, a network element can also generate network element identities and allocates temporarily to another network element at runtime. Such a network identity may address MSs/UEs over the air interface for granting, assigning, and releasing of shared radio resources by the particular RAN. This chapter discussed the usages, length, and the value range of some of the network identities.

Network identities play a very important role in troubleshooting and resolving network services-related issues. During an inter-RAT movement of a user, a network identity in the target RAT can be derived from the identity used in the source RAT. A network identity may carry and be encoded with several individual information/components. Such individual component further provides network-related information or another network element identity. For more information on the network identities described in this chapter as well as other network identities and their structures, the reader may further refer to the references mentioned at the end of this book, especially the TS 23.003 [30].

6

Interworking and Interoperations of Mobile Communications Networks

Introduction

This chapter covers one of the most important functionalities, that is, *interworking* and *interoperation*, through which mobile communications networks provide seamless communications services to subscribers. Interworking facilitates an operator to provide various communication services to its subscribers within its home network. We begin with the basic interworking among the Global System for Mobile Communication (GSM), General Packet Radio Service (GPRS), Universal Mobile Telecommunication System (UMTS), and Long-Term Evolution (LTE)/Evolved Packet System (EPS) networks and show how it can be realized within a home network using upgraded/enhanced and legacy network elements. We then present the advanced interworking features through which an operator may provide voice call services to subscribers currently registered in an LTE/EPS network. Interworking between the LTE/EPS and the 5G system is also described briefly.

Interoperations facilitate the operators to provide various communications services to their roaming subscribers when traveling outside of the home network. Interoperation works through the interworking of networks of the same or different operators. We present the interoperation cases for roaming subscribers through a visiting network that deploys either upgraded/enhanced or legacy network elements. We then close this chapter with the methods of routing roaming user data to the external network.

6.1 Requirements and Types of Interworking

Figures 2.1–2.4, in Section 2.1, provide the basic and introductory architecture of the mobile communications networks based on the GSM (2nd generation), UMTS (3rd generation), and the LTE (4th generation) systems. These architectures may be deployed individually by an operator to provide voice and data communication services to different categories of subscribers. For example, a subscriber may like to avail and subscribe to GSM service only; another subscriber may like to avail and subscribe to both the voice and data services through the UMTS network. Similarly, a subscriber may like to avail and subscribe voice call service through a UMTS network and mobile broadband services through an LTE/EPS network only. The different types of services that can be availed by a subscriber also depend on the capability of the mobile device. Within the same operator, a mobile user that is currently registered in an LTE/EPS network may move to a different part of the same city having UMTS coverage only. In this case, the User Equipment (UE) will access and register with the UMTS network and continue providing the communication services to the subscriber.

From the above examples, it appears that the mobile communications networks must have the capability and flexibility to provide communications services as per the mobilities and requirements of the subscribers are

Mobile Communications Systems Development: A Practical Introduction to System Understanding, Implementation, and Deployment, First Edition. Rajib Taid.

concerned. The UE/Mobile station (MS), access network, and circuit-switched (CS) and packet-switched (PS) domain core network elements of the GSM, GPRS, UMTS, and LTE networks provide such communications services through interworking among them.

Interworking of a mobile communications network can take place with the following types of networks:

- Traditional Public Switched Telephone Network (PSTN), as shown in Figures 2.1 and 2.3, connecting the GSM and UMTS Release 99 network with the PSTN.
- External IP data network, e.g. Internet, as shown in Figures 2.2, 2.3, and 2.13, connecting the GPRS and UMTS Release 99 network with the Internet.
- Packet domain communications networks.

Interworking among the networks is realized through a particular logical interface that connects their gateway or anchoring network element. For example, in Figures 2.2 and 2.3, GPRS and UMTS PS core network is connected with the Internet through the *Gi* interface. Similarly, in Figure 2.4, the LTE/Evolved Packet Core (EPC) network is connected with the Internet through the *SGi* interface.

In the next sections, we will discuss the interworking among the packet domain communications networks, i.e. LTE/EPS, GPRS, and UMTS networks. The interworking of these networks can be realized in the following ways:

- Interworking through enhanced network elements
- Interworking through legacy network elements

6.2 Interworking Through Enhanced Network Elements

It was mentioned in Section 2.3.2 that the legacy GPRS and UMTS PS core network element Serving GPRS Support Node (SGSN) exists in LTE/EPC network also. The SGSN is the central interworking network element for the LTE/EPC, GPRS, and UMTS PS domain core networks. Functionality wise, the LTE/EPC SGSN was upgraded to support interworking and mobility management capabilities between the LTE/EPC and GPRS/UMTS PS domain networks. This is in addition to the legacy functions that are performed by the SGSN for a GPRS and UMTS network. A scenario where the GSM, GPRS, UMTS, and LTE networks may be deployed by an operator through their interworking can be found in Figure 1(b) TS 23.002 [29].

In addition to the existing connections with the legacy GSM and UMTS network elements, the LTE/EPC SGSN also interconnects with the Mobility Management Entity (MME), Serving Gateway (S-GW) of an EPC network. Similarly, the S-GW is a centralized core network element for the interworking of user data/traffic between the LTE/EPC and the UMTS PS domain core networks.

Figure 1(b) TS 23.002 [29] contains a basic interworking configuration where the voice call services may be provided through the GSM network (left side); voice and data services through the UMTS network (middle part); only data services through the LTE/EPS network (right side) during its initial rollout. This high-level interworking configuration may appear to be complex in the first place because of the numerous network elements that are interconnected through various logical interfaces as indicated by the bold and thin solid lines. Look at this figure from the UE/MS, in the bottom-up approach, toward the core network. Identify the access network and code network followed by the CS domain and PS domain network. From Figure 1(b) TS 23.002 [29], it is observed that the interworking among the GSM, GPRS, UMTS, and LTE/EPS networks is available at the core network level only, and no interworking is available at the radio access network level.

To offer voice as well as data services through an LTE/EPS network only, the operator requires to deploy additional IP transport-based network components along with the EPC network shown in Figure 1(b) TS 23.002 [29]. On the other hand, an operator may choose to provide LTE/EPS data services only while it may continue to provide a voice call through the legacy GSM, UMTS network. Such provision requires interworking capabilities among the networks and their components deployed by an operator.

In the next sections, we will discuss the following interworking features that facilitate an LTE/Evolved-UMTS Terrestrial Radio Access Network (E-UTRAN) network registered UE and its user to make a voice call over an LTE/EPS network, in coexistence with the legacy GPRS Edge Radio Access Network (GERAN) and UMTS terrestrial radio access network (UTRAN).

- Voice-over IP Multimedia Subsystem (IMS) or voice-over LTE (VoLTE),
- Single Radio Voice Call Continuity (SRVCC), and
- Circuit-Switched Fallback (CSFB).

For these features to work, both the CS domain and PS domain core network elements, Mobile Switching Centre (MSC) and SGSN, are upgraded from its legacy networks, i.e. GSM/GPRS, UMTS. The network elements are upgraded in terms of new functions and procedures that are to be performed by the respective network element. The new functions and procedures are specified in the form of a new logical interface between two network elements. The dark dashed lines in the following illustrative figures show the existing, as well as the new logical interfaces for realizing a particular interworking feature through signaling or user traffic, flows between two network elements.

6.2.1 Interworking for Voice Call Through IMS: VoLTE

To provide both the voice and data services to subscribers over an LTE/EPS, an *IMS* may be deployed and interworked through the *SGi* interface further with the LTE EPC network. The provision for providing voice services to subscribers over the LTE system and IMS is called the VoLTE. One such network deployment scenario providing VoLTE services is illustrated in Figure 6.1.

Along with the LTE/EPS network and its IMS configuration, the handset must be also VoLTE capable to enable a user to make a voice call over an LTE network. A VoLTE-capable UE can work on different networks, i.e. GSM, UMTS, and LTE using different Radio Access Technologies (RATs).

Figure 6.1 Illustration: LTE network and IMS interworking for VoLTE call.

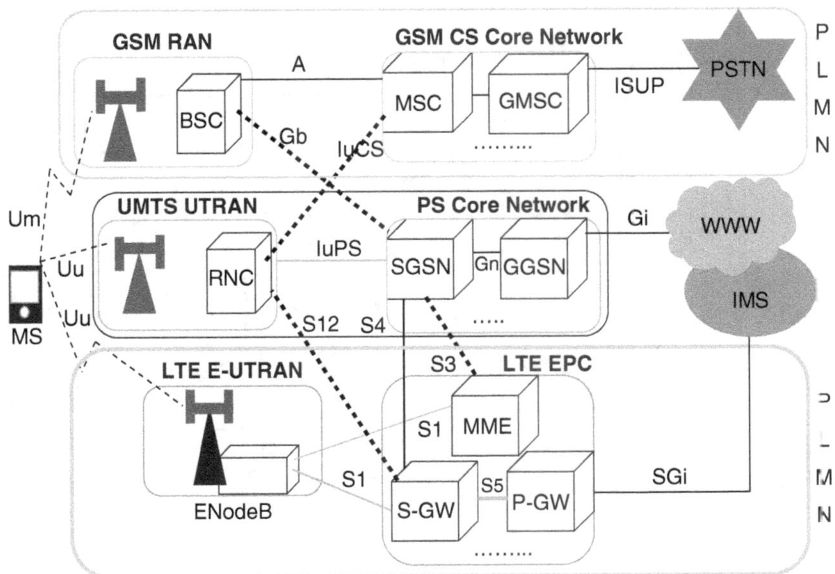

As shown in the above interworking scenario, a voice call can be facilitated to a user through the legacy GSM, UMTS network, or LTE/EPS IMS. A UE indicates its preference for a voice call through the *Voice domain preference and UE's usage setting* IE in the ATTACH Request message, TS 24.301 [46], TS 24.008 [45], that is sent to the core network during its initial registration of a UE. The same IE is also sent to the core network as part of the Routing Area Update (RAU) (GPRS, UMTS) and Tracking Area Update (TAU) (LTE/EPS) request to indicate its *voice domain preference* by a UE. An LTE UE that does not support the voice-over IMS feature will indicate the preference as the "*CS voice only*". In such a case, the UE will use the CS fallback method of providing a voice call facility to a user.

An LTE UE that support the voice-over IMS feature will indicate the preference as the *"IMS PS voice only"*. All these interworking possibilities of providing voice and data services depend on the capabilities of a UE and its usage setting, which is described in the next section. Below we briefly describe the IMS and its network elements that are interworked with an LTE/EPC to provide a VoLTE call facility to subscribers.

6.2.1.1 IP Multimedia Subsystem (IMS)

The IMS is the core network infrastructure backbone for offering common IP-based multimedia, e.g. audio, video, voice, and services to subscribers over an LTE/EPS IP transport network. The IMS contains the core network elements and interfaces, interconnecting them for routing of a VoLTE call either to a 2G/3G network or PSTN network or within an IMS network accordingly. The reference architecture, the network elements, and various interfaces of an IMS system are shown in Figure 6.2. For more information on this reference architecture, refer to TS 23.228 [36]. The IMS was standardized for mobile communications networks and was introduced as part of the 3rd Generation Partnership Project (3GPP) Release 5 architecture to provide voice calls along with other multimedia services.

The main network elements of an IMS is the *call session control function (*CSCF) and the *media gateway control function (*MGCF*).* As shown in Figure 6.2, the CSCF performs several roles: Proxy CSCF, Interrogating CSCF, and Serving CSCF.

- Proxy-CSCF (P-CSCF)

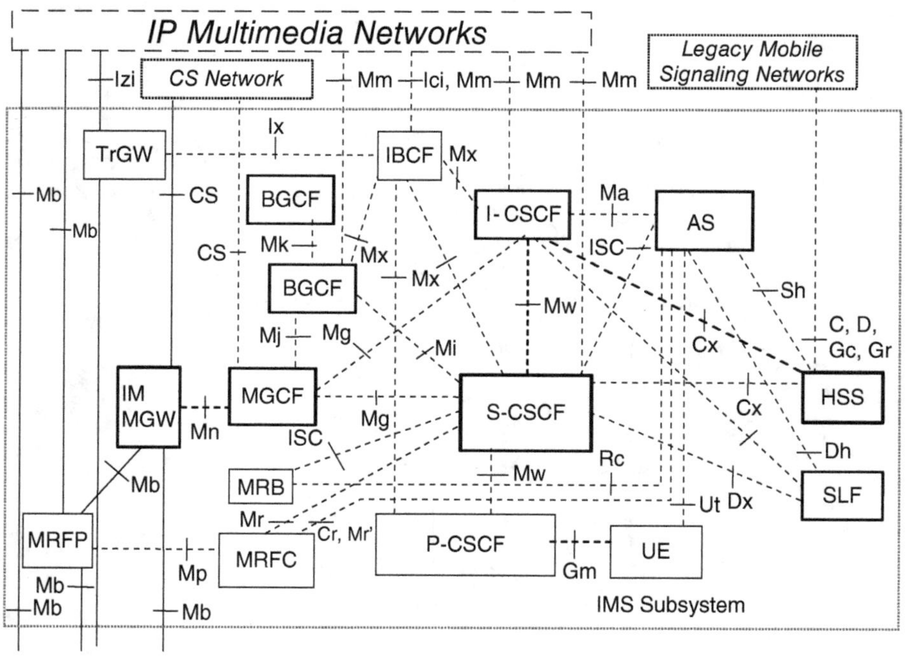

Figure 6.2 Reference architecture of an IMS. *Source:* © 2015. 3GPP ™ TSs and TRs are the property of ARIB, ATIS, CCSA, ETSI, TSDSI, TTA and TTC who jointly own the copyright in them. © 2015, 3GPP.

The information sent by a UE is received first at the P-CSCF and is forwarded to the I-CSCF.

Interrogating CSCF (I-CSCF)

I-CSCF contacts the Home Subscriber Server (HSS), and based on the response from the HSS, it forwards the call to the S-CSCF.

- Serving CSCF (S-CSCF)

The S-CSCF handles the registration of UEs, maintains sessions, and performs routing functions.

- MGCF

The MGCF interconnects the IMS with PSTN or CS domain network element. For a mobile-terminated (MT) call, the MGCF contacts the I-CSCF. For more information on the above CSCFs, refer to TS 23.228 [36].

The signaling protocol that is used to exchange information between a UE and the IMS is known as the *Session Initiation Protocol* (SIP) which uses the IP transport network. IMS uses the SIP messages to perform various functions and procedures such as registration, call establishments and release, session management, authentication, security, and charging. Though the SIP messages exchanged between a UE and the IMS are signaling/control plane information in nature with respect to the IMS, they are treated as user plane data by the E-UTRAN and its core network. The real-time voice packets are transferred through other protocols called *Real-Time Transport Protocol* (RTP) *and RTP Control Protocol* (RTCP). For more information on the IMS, the reader is recommended to refer to TS 23.228 [36].

6.2.1.2 UE Registration and Authentication

In a live network of an operator, once a UE is registered and authenticated by the LTE/EPC as part of the LTE/EPS ATTACH procedure, TS 24.301 [46], a VoLTE-capable UE is also required to register and authenticate itself with the IMS before it can start using IMS services. For these purposes, SIP signaling messages are exchanged between the UE and the IMS. Example 6.1 below describes and illustrates the flow of signaling messages, starting from registration to call establishment, conversation, and its completion, for a typical VoLTE call between two mobile devices over an IMS.

Example 6.1 Illustration: UE Registration and VoLTE Call over IMS

Let us consider a typical VoLTE call performed between two UEs: UE1 and UE2. To enable a VoLTE call, the UEs are required to be registered with both the LTE EPC and IMS for the allocation of the necessary resources, for example, EPS bearers with required Quality of Service (QoS). Several steps are involved in a VoLTE call over the IMS as described below:

- UE Registration with EPC and Allocation of Default Bearer with QoS Class Identifier (QCI) = 5

Figure 6.3 below illustrates the allocation of a default bearer to each UE as part of its ATTACH procedure requested to the core network. The allocated default bearer has the QCI = 5, Priority =1, refer to TS 23.203 [33], with nonguaranteed bit rate (NGBR) that is used to exchange IMS signaling messages between the UE and IMS. The dotted lines in Figure 6.3 indicate that some of the messages as part of the overall ATTACH procedure (which are exchanged between the UE and the EPC) are not shown in this figure. One such message is the Evolved Packet System Session Management (ESM) Information request, from EPC to UE, and its response, from UE to EPC. Using the ESM information response message, the UE provides the P-CSCF's *Access Point Name* (APN) in the case of the IMS.

- UE Registration with IMS through Authentication

The UEs register with the IMS using the default bearer with QCI = 5 that was allocated during their registrations with EPC. To register with the IMS, a UE uses the SIP signaling message that is transferred as the user plane data over the LTE/EPS toward the IMS. P-CSCF is connected with the P-GW of the LTE/EPC. For a UE, the

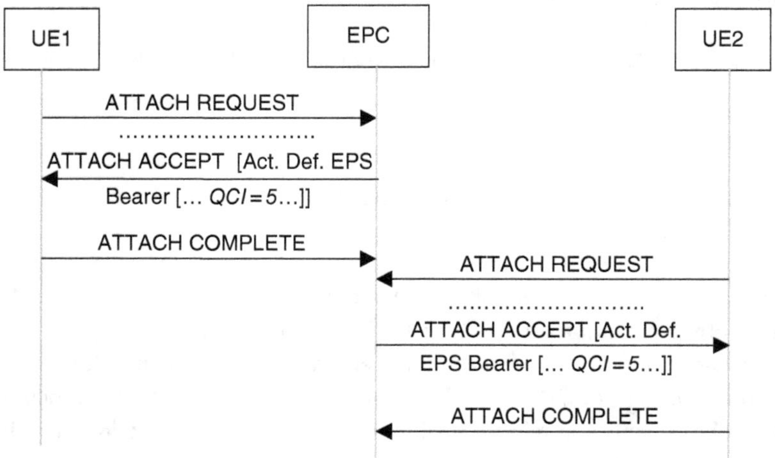

Figure 6.3 Allocation of default bearer with QCI = 5 for IMS signaling.

proxy CSCF is the entry and exit server in the IMS. A UE sends the APN of the P-CSCF server in the SIP: REGISTER message to the IMS. The UE sends its authentication response details in the REGISTER message again to the IMS following the *401: Unauthorized* request message from the IMS; see Figure 6.4.

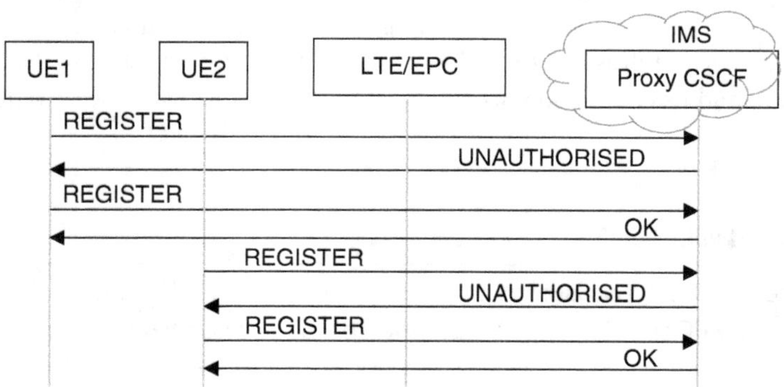

Figure 6.4 Illustration: registration of UE in an IMS.

The end-to-end message flow for a typical registration of a UE in an IMS is - UE<->eNodeB<->S-GW<->P-GW<->P-CSCF. Though not shown in Figure 6.4, once the P-CSCF receives the registration request from a UE, it is further processed by the I-CSCF, HSS, S-CSCF server and responds with the *SIP: 200 OK* messages to the registered UE.

- VoLTE Call over IMS

UEs that are EPC and IMS registered can establish a voice-over IMS to facilitate a voice call between two subscribers. The SIP signaling messages flows associated with a typical VoLTE call between two UEs – UE1 and UE2 – over IMS are illustrated in Figure 6.5. Only the P-CSFC is shown in the call flow below, though the I-CSCF and S-CSCF also take part in an IMS call.

In Figure 6.5, the E-UTRAN and EPC/NAS-related signaling messages that are exchanged with the UEs are not shown for simplicity of the flows. The INVITE method carries the calling and called party's number and is used to establish a session between the two UEs. As shown in this figure, the EPS allocates a dedicated bearer with QCI = 1, Priority = 2, refer to TS 23.203 [33], with a guaranteed bit rate (GBR) over which the voice conversation takes place between the users. The establishment of a dedicated bearer for voice conversation purposes is intimated to the called UE through SIP:UPDATE method, following which it sends the SIP:RINGING status to the caller UE.

Figure 6.5 Illustration: VoLTE call between IMS registered UEs.

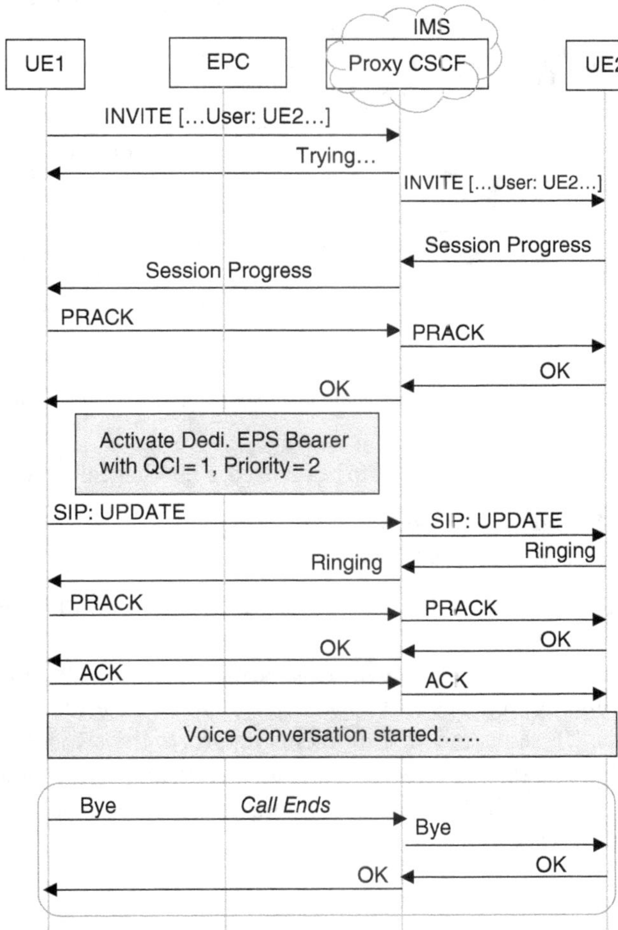

The IMS for VoLTE described above will be also the foundation for the voice services over the 5G system with essential extensions added, as part of the 3GPP Release 15, over some of the existing logical interfaces, such as the Rx, Sh.

6.2.2 Interworking for VoLTE Call Through LTE/EPS: SRVCC

The provision of a handover of an existing VoLTE call to a legacy GERAN or UTRAN system is known as the *SRVCC* (refer to TS 23.216 [35]) because of the single RAT capability of a UE at a given time only. Interworking through the SRVCC scenario is a bit similar to the one described in Section 6.2.1. However, there is a difference from the E-UTRAN UE capability point of view. SRVCC interworking scenario is suitable for an LTE UE having only one radio capability at a given time. Because of this, a VoLTE call over the IMS is required to be handed over to the legacy GERAN or UTRAN system for the continuation of the IMS voice call. The interworking of network elements, with IMS, for the SRVCC feature is illustrated in Figure 6.6.

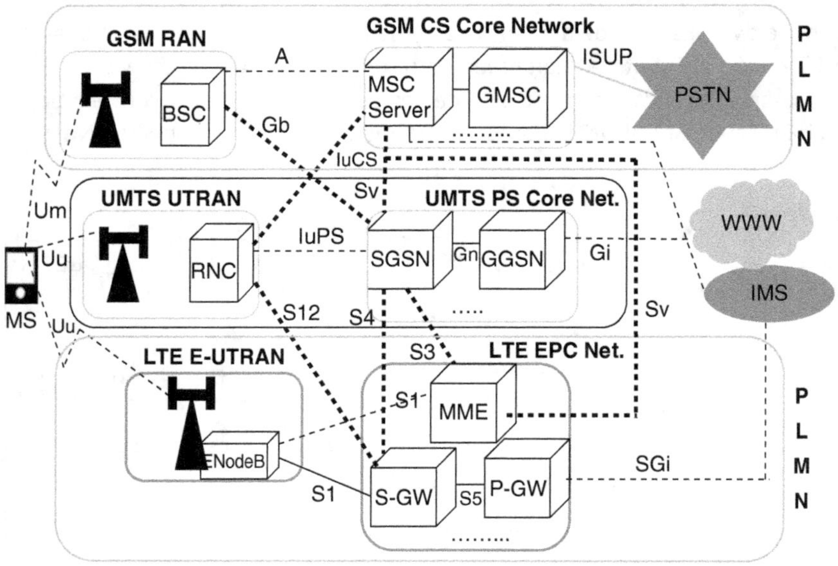

Figure 6.6 Illustration: legacy and LTE network interworking for VoLTE call HO through SRVCC.

For the SRVCC capability to be realized, a new logical interface Sv, refer to TS 29.280 [71], is defined between the

- MME and the MSC server and
- MME and SGSN, as shown above.

Figure 6.7 further illustrates the SRVCC HO of an LTE/EPS IMS voice call made by the VoLTE user A to a PSTN user B.

In Figure 6.7, user A moved to a legacy 2G/3G coverage area. As the user moved away from the LTE coverage area and approaches a 2G/3G coverage area, the eNodeB detects a better signal level from the 2G/3G networks. UE sends the measurements reports to the eNodeB which in turn sends an SRVCC HO request to the MME. In Figure 6.7, the signaling and traffic paths are shown prior to and after the SRVCC HO. All the subsequent signaling and data communication takes place through the legacy network MSC server after the SRVCC HO is completed.

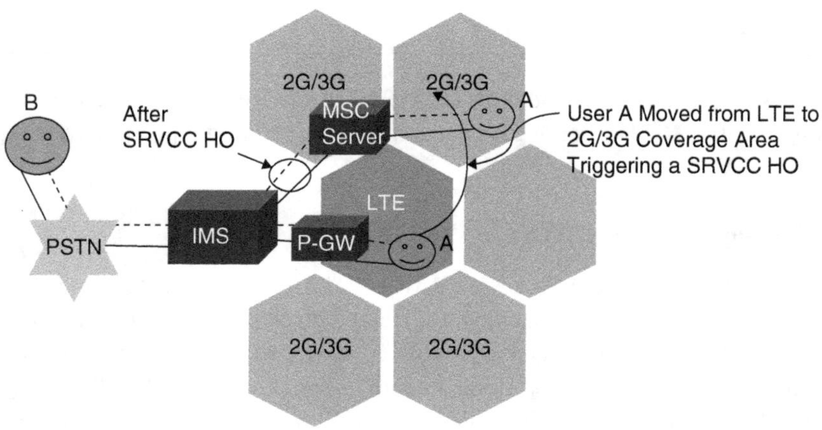

Figure 6.7 Illustration: VoLTE call HO through SRVCC feature.

- How does SRVCC work?
  - Indication of Feature Support

An LTE UE indicates its SRVCC capability information through the class mark and network capability information sent along with the *ATTACH Request* message, refer to TS 24.301 [46], to the MME. MME indicates the SRVCC capability of a UE through the *SRVCC operation possible* IE in the *INITIAL CONTEXT SETUP REQUEST MESSAGE,* refer to TS 36.413 [97], to the eNodeB. MME may also request the SRVCC capability of a UE by sending the message *UE Radio Capability Match Request* (see Section *5.3.14 TS 23.401 [39])* to the UE through the eNodeB. Any subsequent changes in the SRVCC capability of a UE are also updated to the HSS by the MME.

- Initiation of Handover of LTE IMS Voice Call (VoLTE)

SRVCC works by performing the handover of the ongoing VoLTE call as well as data session over IMS to the legacy GERAN/UTRAN network. Initiation of a handover operation works the same way as the normal handover procedure based on the signal-level measurement reports reported from the UE to the E-UTRAN. The source eNodeB sends the *HANDOVER REQUIRED* message, refer to TS 36.413 [97], to the MME with the *SRVCC HO Indication* IE for a handover of the existing VoLTE call. SRVCC HO Indication may be PS or PS and CS type. MME, in turn, initiates the SRVCC request to the MSC server. The MSC server prepares and executes the handover of the IMS voice call successfully. For further illustrations of the VoLTE call handover through the SRVCC method, refer to Figures 4.2.2-1 and 4.2.3-1, TS 23.216 [35].

- Handover of PS Call/Data Session

The handover of an ongoing PS call during an SRVCC procedure is performed as usual as per the normal inter-RAT handover procedures. For further illustrations of the inter-RAT handover procedures, refer to Section 5.2.2 TS 23.401 [39].

6.2.3 Interworking for Voice Call Through LTE/EPS: CSFB

In this interworking scenario, the provision for making a voice call for a UE registered with an LTE network is facilitated through the reuse of the legacy GERAN/UTRAN. The provision for making a voice call by an LTE/EPS registered UE through the legacy GERAN/UTRAN is known as the *CSFB* method; refer to TS 23.272 [38]. One such illustration of the interworking of network elements for CSFB arrangement is shown in Figure 6.8. A CSFB feature may be used by:

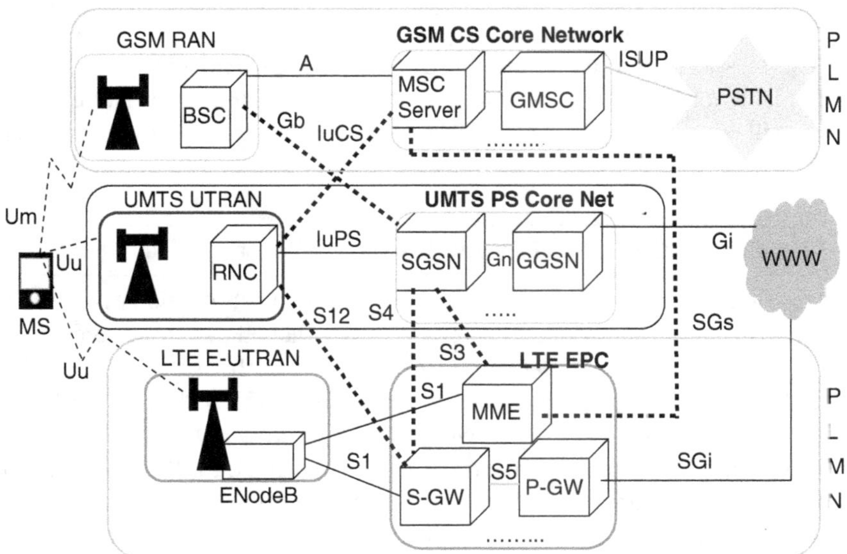

Figure 6.8 Illustration: legacy and LTE network interworking for CSFB.

- An MO or MT LTE UE that does not have the voice-over IMS or VoLTE capability
- Other scenarios such as VoLTE-capable UE that fails to register in the IMS.

Unlike the interworking scenarios for VoLTE call as mentioned in Sections 6.2.1 and 6.2.2, no IMS is deployed and interworked as part of the LTE/EPS network with the CSFB arrangement. In this interworking scenario with CSFB arrangement, the LTE/EPS network provides only data services to subscribers.

During a CSFB of a CS call, the ongoing PS session may be also transferred to the legacy network through PS HO procedure with reduced data rates. For the CSFB purpose, a new logical interface called *SGs* is configured between the MME and MSC/Visitor Location Register (VLR) to exchange signaling information only. The SGs logical interface also supports mobility management as well as MO/MT call establishment procedures. For more information on the CSFB, a protocol stack, and procedures supported over the SGs, refer to TS 23.272 [38] and TS 29.118 [68]. SGs interface uses the SCTP [17] layer and IP transport network. The SGs interface is similar to the GPRS/UMTS Gs interface that connects the SGSN and the MSC to provide CS domain services to MS engaged in a PS service.

Example 6.2 below describes the status displayed by an LTE/EPS registered UE while making a voice call through the legacy GERAN/UTRAN using the CSFB method.

- How does a CSFB work?
  - Indication of CSFB Feature Support

A UE indicates its CSFB feature support capability under the *UE Network Capability* Mandatory IE in the ATTACH Request message that is sent to the MME. The UE specifies the Attach Type as the *combined EPS/IMSI Attach Request* under the EPS attach type IE. A combined EPS/IMSI ATTACH Request from the UE enables the MME to update the UE location in the MSC/VLR end also over the *SGs* interface. Refer to TS 23.272 [38] for further details on the EMM procedures supported over the *SGs* interface. If the EPC network supports the CSFB feature, the same will be informed to the UE in the ATTACH ACCEPT, TS 24.301 [46], message through the IE *Attach Result = combined EPS/IMSI attach* and *Additional update result optional* IE. This IE will carry the value *"CS Fallback not preferred"* toward the UE, indicating that it can use the CSFB method to initiate a CS call through the legacy 2G/3G network. The same IE indicating the supporting of the CSFB feature can be also sent in a TAU ACCEPT message, TS 24.301 [46], from the MME to UE.

- Initiation of CS Call Through the CSFB Method by the UE

An LTE UE initiates a CS call through the CSFB method by sending the EMM/NAS *EXTENDED SERVICE REQUEST* message to the MME.

This message can be used to establish an MO or MT CS call through CSFB by indicating the same in the *Service type* mandatory IE; refer to Section 9.9.3.27, TS 24.301 [46]. Any existing Radio Resource Control (RRC) connection will be released by the E-UTRAN by sending the *RRCConnectionRelease,* TS36.331 [94], message to the UE with

Example 6.2 CSFB: LTE UE Fall-back to CS domain for voice call

Assume that a mobile user currently subscribes to a subscription plan consisting of LTE (4G) data along with voice call services provided by an operator. The UE will perform a combined attach, EPS as well as IMSI, procedure with the LTE/EPS network. Also, consider that the operator's LTE/EPS network does not have the IMS facility currently to provide voice-over LTE services to subscribers. The UE may display the type of network, i.e. 4G, which is currently registered with, at the top corner of the handset. Whenever the user attempts to make a voice call or an incoming call arrives for the user, the UE will fall back to either 2G or 3G network which will be shown at the top corner of the handset. Once the voice call completes, the UE will return to and access the LTE/EPS network and registers with it again, showing the 4G status at the top corner of the handset.

- *"releaseCause = cs-FallbackHighPriority"* for CS fallback to UTRAN *or*
- *"releaseCause = other" for CS fallback to GERAN.*

CSFB method works on the *redirection* concept where LTE UE is redirected to the target ARFCN of the legacy network. For redirection purposes, the *RRCConnectionRelease* message will also include the *redirectedCarrierInfo* IE containing the following information for CSFB:

- RAT/RAN, i.e. *GERAN, UTRAN, and E-UTRAN,* to be redirected.
- Corresponding Carrier Frequency, also called ARFCN, to be used by the UE.

For the various options of RATs that are allowed for CSFB purposes, refer to TS 36.331 [94]. The UE tunes to the indicated carrier frequency/ARFCN of the legacy network. To make a CS voice call, UE will start exchanging normal air interface Layer 3 signaling messages with the network as illustrated earlier in Section 2.2.5.

Example 6.3 below describes and illustrates the typical signaling messages flows during a mobile originated voice call made by an LTE/EPS registered UE through the legacy GERAN/UTRAN using the CSFB method.

Example 6.3 LTE Voice Call Through GSM Network and CSFB Feature

Figure 6.9 illustrates the MO and MT voice CS call scenario made by LTE UEs currently registered in the LTE/EPS network and shows only the important call flows from the CSFB point of view. The dotted line indicates the absence of the associated call flows, which are not shown in this figure.

Assume that the MO UE is engaged in a PS session and the MT UE is currently in the idle condition shown by its current state ECM-IDLE. Note that notification for an incoming CS call to MT UE depends on its current state. If the MT UE is currently in the

- *ECM-IDLE* state, MME notifies an incoming CS call by sending a paging message to the UE.
- *ECM-CONNECTED* state, then the MMM notifies an incoming CS call by sending a *CS SERVICE NOTIFICATION* message to it.

MSC server sends the paging request, *SGsAP-PAGING-REQUEST,* to the MME over the *SGs* interface. MME further sent the paging message toward the UE through the eNodeB. MO UE completes the MO CS call as per the normal procedures with the MSC server. MT UE sends a paging response to the MSC server and completes the MT CS call as per the normal procedures illustrated in Figure 2.8 earlier. Following this, both the UEs are connected in the legacy GSM network through CSFB arrangement and start voice/CS call conversations with each other.

In this example, illustrated in Figure 6.9, assume that the MT UE disconnected the CS call first, followed by the MO UE. In this case, the GSM BSC instructs both the UEs to release the allocated channels by sending the *Channel Release* message. This message also contains the LTE *EARFCN number,* indicating the UEs that they should return to the LTE network. Because of this, both the UEs perform their TAU procedure toward the MME to inform their current location within the LTE/EPS network.

In all of the air interface Layer 3 messages used for making a CS call in the legacy network, the flags *CSMO,* for the originating side, and *CSMT,* for the terminating side, are used and set accordingly to indicate to the receiving network element that the message is for CSFB purpose. These flags refer to TS 24.008 [45], are shown in Figure 6.9 illustration. Table 6.1 summarizes the network elements, their logical interfaces, and the related 3GPP TSs to support voice calls for LTE/EPS registered UE through the CSFB and SVRCC features.

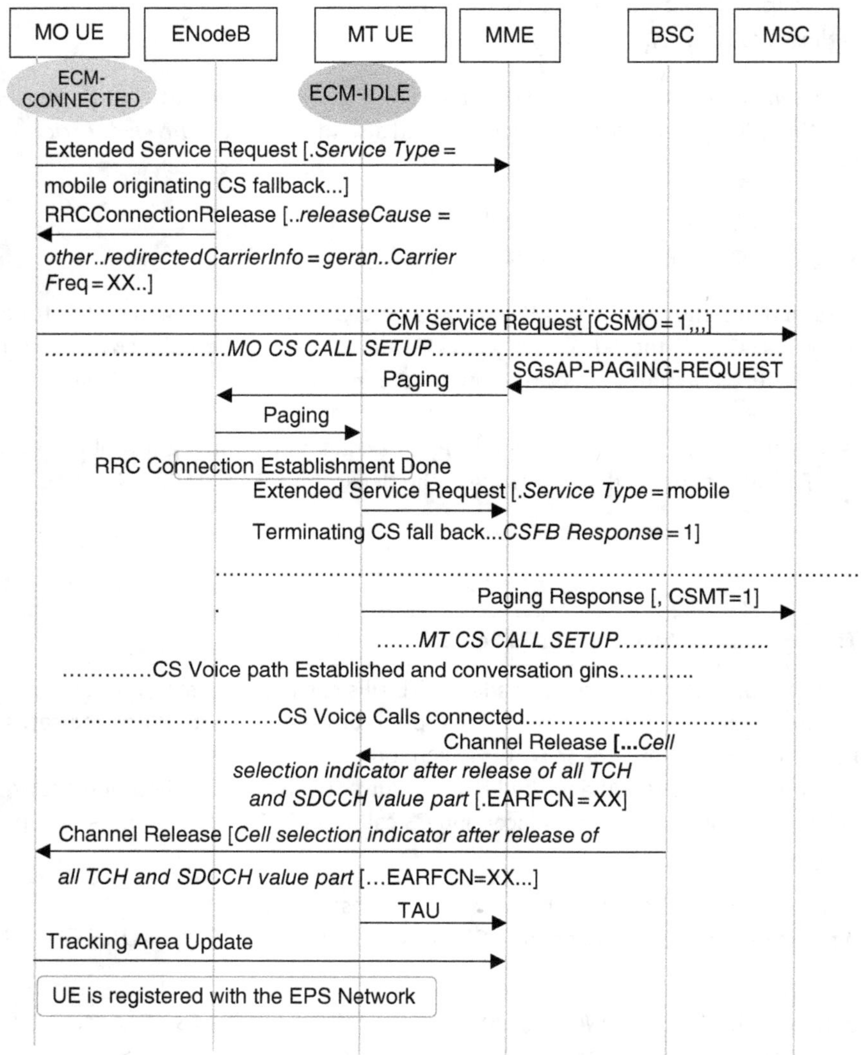

Figure 6.9 Illustration: LTE voice call: MO-MT voice call through CSFB.

Table 6.1 Interworking methods: network elements and their logical interfaces.

Interworking Methods	Network Elements	Logical Interfaces	Related 3GPP TSs
CSFB	MME, MSC Server	SGs	23.272, 29.118, 36.413, 24.301, 23.401
SRVCC	MME, MSC Server SGSN, MSC Server	Sv	29.280, 23.401, 36.413

6.3 Interworking Through Legacy Network Elements

Figure 1(b), TS 23.002 [29], presents a basic interworking of the GSM, UMTS, and LTE/EPS networks that can be deployed by an operator with entirely new and upgraded/enhanced legacy network elements, e.g. SGSN, MSC, and RNC. However, another operator may choose to interwork its LTE/EPC network with the legacy GSM, GPRS,

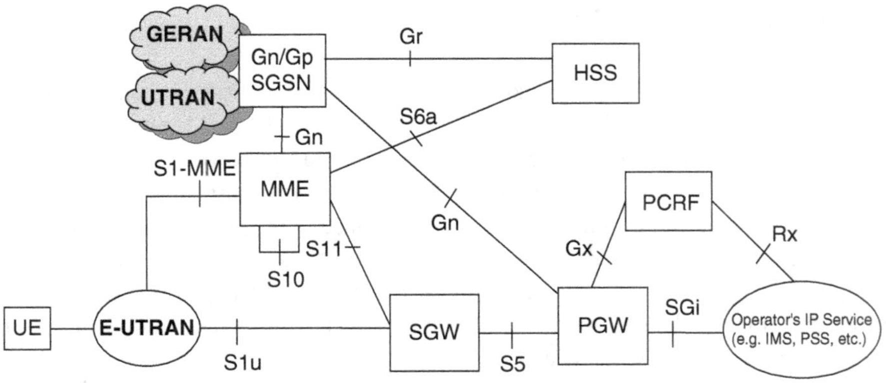

Figure 6.10 Interworking of LTE/EPS with GERAN and UTRAN through legacy network elements. *Source:* © 2015. 3GPP ™ TSs and TRs are the property of ARIB, ATIS, CCSA, ETSI, TSDSI, TTA and TTC who jointly own the copyright in them. © 2015, 3GPP.

and UMTS core network elements, e.g. SGSN and MSC, only as its initial deployment. To realize successful interworking under such a scenario with legacy network elements, the EPC network elements will be required to support the necessary interfaces/protocols also that are used by the legacy core network elements, e.g. SGSN and MSC. One such interworking scenario is shown in Figure 6.10, reproduced from TS 23.401 [39].

In Figure 6.10, only the representative figure for the GERAN and UTRAN networks along with the legacy SGSN is shown. The *Gn* interface works on top of the GTP and is used by the legacy GPRS and UMTS SGSN to communicate with the GGSN of a GPRS and UMTS network. The LTE/EPC MME communicates with the legacy SGSN over the *Gn interface*. LTE/EPC S-GW also interfaces with the legacy SGSN through the *Gn* interface. The EPC HSS communicates with the SGSN through the existing *Gr interface* that works on top of the SS7 signaling protocols. On the other hand, the HSS communicates with the MME through the *S6a* interface that works on top of IP transport.

6.4 Interworking Between LTE/EPS and 5G Systems

In the previous section, interworking among the legacy systems has been described. To support seamless mobile communications services to subscribers during the inter-system movement, the 5G system also provides interworking capabilities to operators. However, the 5G system supports interworking with the LTE/EPS only. For interworking between the 5G system and the LTE/EPS, the 3GPP defines a new inter core networks logical interface called *N26* between the Access *and Management Function (AMF)* in 5G core network and the MME in the LTE/EPC network. It may be noted that interworking between the LTE/EPS and the 5G system can also take place without the presence of the N26 interface. Further, depending on the availability of the N26 interface between the 5GC/AMF and LTE/EPS MME for interworking between them, a UE can operate in one of the following modes:

- Single Registration Mode, where the N26 logical interface is available between the 5GC/AMF and LTE/EPS MME.
- Dual Registration Mode, where the N26 logical interface is not available between the 5GC/AMF and LTE/EPS MME.

Interworking between the 5G and the LTE/EPS with or without the N26 logical interface shall be described later in Chapter 16.

6.5 Interoperations of Networks: LTE/EPS Roaming

The roaming capability of a mobile communications network makes it possible to offer seamless voice and data services by a network operator to its subscribers while traveling outside of their home network and location and enter into a network that is run by the same or different operator in a different location. Roaming services are delivered to subscribers through the interoperation of networks operated by the same or another operator. Interoperation, in turn, is achieved through the interworking of network elements and interfaces as described in the previous sections.

In the previous Sections 6.2 and 6.3, we have presented the interworking of the GSM, GPRS UMTS, and LTE/EPS networks, through the enhanced as well as legacy network elements, run by an operator. In this section, we further present the interoperations and interworking of mobile communications networks operated by different operators in different regions to provide roaming services to subscribers.

A mobile communications network operated by an operator to provide communications services in a particular area/location is known as the *Public Land Mobile Network (PLMN)*. A PLMN is uniquely identified, also see Section 5.2, by the combination of an MCC and MNC of an operator. Within a PLMN, an operator may provide both the voice (CS) and data services to subscribers. The PLMN in which an MS/UE is currently subscribed and registered in an LTE/EPS network is known as the *home PLMN* (HPLMN). The PLMN of either the same or different operator in which an MS/UE is currently roaming into and accessed the visited LTE/EPS network is known as the *visiting PLMN* (VPLMN).

6.5.1 Roaming Through Interoperations of Enhanced Networks Elements

In this interoperation and roaming scenario, the enhanced network elements for the LTE/EPS network are deployed both in the HPLMN and the VPLMN. There are two ways to transfer user data between the UE and the Internet in case of roaming between two different PLMNs with the enhanced network elements. To enable such roaming scenarios, new logical interfaces are used to interconnect among the network elements as described below:

- **Routing of User Data by the HPLMN**

In this interoperation arrangement, Figure 6.11, the roaming user data is routed from the VPLMN's S-GW back to the HPLMN packet data network (PGW) gateway. This means that a roaming user has access to the HPLMN's services and resources from the VPLMN.

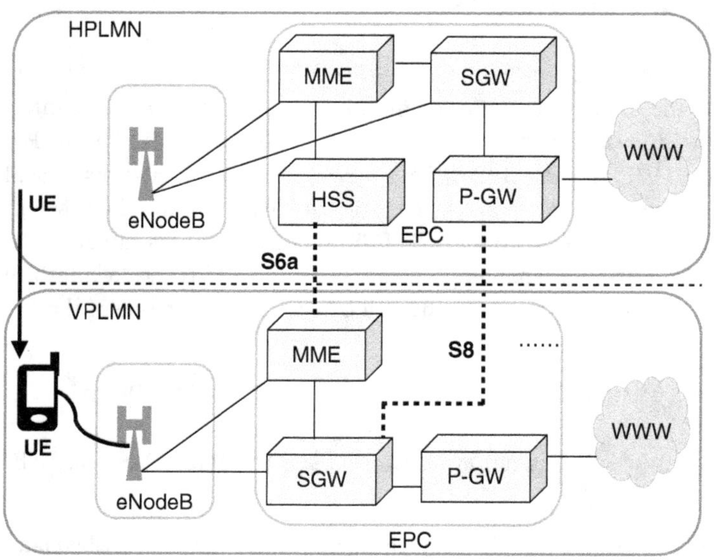

Figure 6.11 LTE/EPS Roaming: routing of user data by the HPLMN for roaming user.

To route the roaming user traffic by the HPLMN, the VPLMN S-GW and the HPLMN PGW are interconnected through the *S8* logical interface as shown in Figure 6.11. *S8 interface* uses the GTP to tunnel user data from the VPLMN S-GW to HPLMN PDN. Apart from this, the HSS of the HPLMN and the MME of the VPLMN are interconnected through the *S6a interface* through which the UE's current location and also the subscriber's information are exchanged between them.

In Figure 6.11, the thin dashed line represents the boundary that separates the HPLMN and the VPLMN. The arrow represents that the UE has left its HPLMN and registered with the VPLMN.

- **Routing of User Data by the VPLMN**

In this interoperation arrangement, Figure 6.12, the roaming user data is routed by the VPLMN PGW gateway to the external PGW.

Unlike the previous scenario, Figure 6.11, no user data is routed back from the VPLMN to the HPLMN. Because of this, it is also known as the *local breakout,* refer to TS 23.401 [39], roaming scenario. Also, the Policy and Charging Rules Function (PCRF) deployed at the HPLMN and VPLMN requires to exchange of subscriber and policy control-related information between them over the S9 interface. The PCRF is the centralized policy and charging-related control network element that is part of the LTE/EPC; refer to Figure 1(b) TS 23.002 [29]. PCRF works based on the subscription information of a user and supports both the roaming and nonroaming users. PCRF meets the charging requirement of different categories of subscribers, for example, volume-based, time-based charging. PCRF may deny the information flow of a subscriber based on its subscription profile.

For a roaming user whose traffic is routed by the VPLMN, the UE has two associations with both the home PCRF (HPCRF) and visited PCRF (VPCRF). HPCRF and VPCRF exchange policy and charging-related information for a particular subscriber and its profile. PCRF has a couple of logical interfaces toward other servers/nodes to perform various functions and requirements. Typical functions include the charging, policy enforcement, QoS control, usages monitoring, user traffic congestion detection, and mitigation at the radio access network level. For more information on PCRF, its reference architecture, and logical interfaces, refer to TS 23.203 [33].

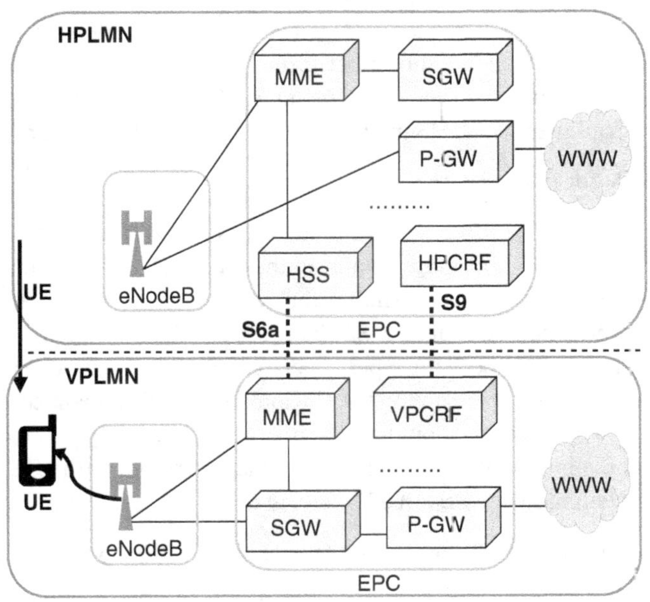

Figure 6.12 LTE/EPS Roaming: routing of user data by the VPLMN for roaming user.

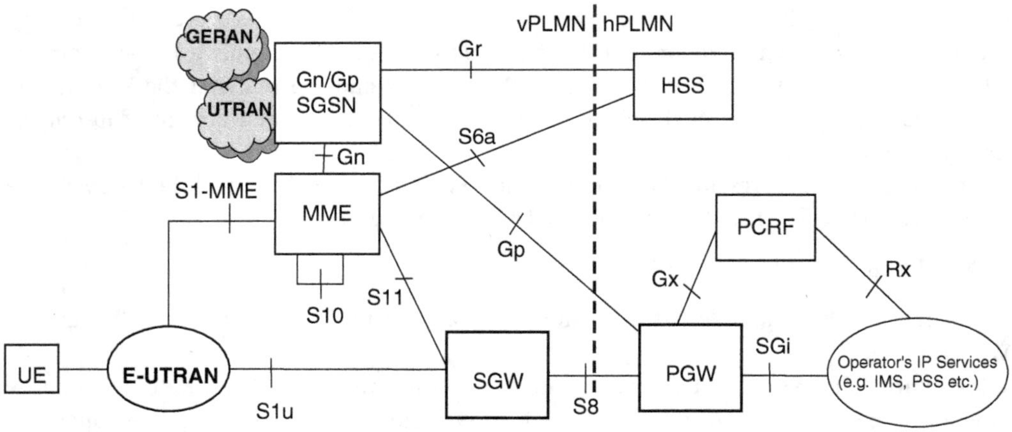

Figure 6.13 Roaming through interoperation of legacy network elements. *Source:* © 2015. 3GPP ™ TSs and TRs are the property of ARIB, ATIS, CCSA, ETSI, TSDSI, TTA and TTC who jointly own the copyright in them. © 2015, 3GPP.

6.5.2 Roaming Through Interoperations of Legacy Networks Elements

This interoperation arrangement, Figure 6.13, which is reproduced from TS 23.401 [39], is similar to the one shown in Figure 6.11, except that the legacy GPRS and UMTS SGSN with the Gn interface capability are deployed by the VPLMN operator. Also, both the VPLMN SGSN and the HPLMN PGW are connected through the *Gp* interface. The roaming user data is routed from the VPLMN S-GW back to the HPLMN PGW gateway, i.e. a roaming user has access to the HPLMNs services and resources from the VPLMN. The VPLMN S-GW and the HPLMN PGW are interconnected through the same *S8 logical interface*; refer to TS 23.401 [39].

6.6 UE Mode of Operation

In the previous sections, we have described the interworking scenarios that allow the coexistence of the legacy GSM, GPRS, UMTS with the LTE/EPS, and IMS networks. Interworking and interoperations provide communications services within a PLMN or between PLMNs for roaming users. The ability to provide a particular communications service to a subscriber through interworking and interoperations of networks depends on the capability, e.g. multiple radio access technologies (RAT), of a particular UE/MS and its network configurations. Based on its capabilities, a UE/MS may operate in a particular mode or more than one mode at a time.

- Mode of operations

Table 6.2 summarizes the mode of operations that an MS/UE can operate within the GPRS, UMTS, and LTE/EPS systems. For more details on the individual mode of operation, refer to TS 24.301 [46] and TS 23.060 [31]. A mode of operation determines the availability of types of services, i.e. CS or PS, to a UE.

The capabilities of an MS/UE are specified in terms of its ability to provide either the CS service or PS service at a time only or both the types of services at the same time to a subscriber. For these purposes, an MS/UE should be able to register itself with the CS domain and PS domain core networks simultaneously. In addition to this, for an E-UTRAN UE, it should also support multiple RATs to provide communication services through the legacy networks in case the IMS is not available/interfaced with LTE/EPC. UE mode of operations in the case of the 5G system has been described briefly in Section 6.4.

- How an MS/UE provides its mode of operation to the CN: Domain Preference

To facilitate voice or data services to a subscriber in a GPRS/UMTS and LTE/EPS network, an MS/UE provides its mode of operation preference under the *Attach Type* IE in the ATTACH Request message, TS 24.301 [46], TS

Table 6.2 GPRS, UMTS, and LTE UE mode of operation.

System	MS/UE Mode of Operation	Purposes
GPRS	Class A	MS registered with CS and PS domains, for voice and data services at the same time.
	Class B	MS registered with CS and PS domains. Either voice or data service is possible at a time.
	Class C	MS registered with PS domain only, for data service is possible.
UMTS	CS/PS Mode	Similar to GPRS Class-A mode.
	PS Mode	Similar to GPRS Class-C mode.
	CS Mode	UE registered with CS domain only, for voice service.
LTE	PS Mode 1	UE registered with the EPS network only. UE's usage is a voice-centric service only.
	PS Mode 2	UE registered with the EPS network only. UE's usage is a data-centric service only.
	CS/PS Mode 1	UE registered with EPS and CS domain legacy networks for EPS and non-EPS services. UE's usage is voice-centric service only.
	CS/PS Mode 2	UE registered with EPS and CS domain legacy networks for EPS and non-EPS services. UE's usage is data-centric service only.

24.008 [45], that is sent to the core network. In a GPRS/UMTS and LTE/EPS network, the MS/UE sets the *Attach Type* IE to *GPRS Attach* and *EPS Attach* only to enable the PS services only. Similarly, in a GPRS/UMTS and LTE/EPS network, the MS/UE sets the *Attach Type* IE to *Combined GPRS/IMSI attach* and *Combined EPS/IMSI attach* to enable both the voice and data services to a user.

A UE ATTACH Request contains an optional IE: *Voice domain preference* and *UE's usage* setting, refer to TS 24.301 [46], TS 24.008 [45], using which a UE can provide additional information on its corresponding mode of operation to the core network. Using the optional IE: *Voice domain preference and UE's usage setting*, an E-UTRAN UE indicates its CSFB (to the legacy 2G or 3G network for a voice call) capability in case the network or the UE does not support VoLTE call or the UE fails to registers with the IMS.

> **Note:** Till the TS 24.008 [45] 9.1.0 version, the CSFB capability of a UTRAN UE was indicated within the *MS Network Capability* mandatory IE in the UE ATTACH Request message to the GPRS core network. However, from TS 24.008 9.2.0 version onwards, the CSFB capability information is conveyed to the GPRS core network through the *Voice domain preference and UE's usage setting* optional IE only. This IE is also used by an E-UTRAN UE to convey its CSFB capability information in the ATTACH Request message to the LTE/EPC. CSFB capability information is no longer part of the *MS Network Capability* IE.

- E-UTRAN UE Mode of Operations and Their Transitions

In this paragraph, the mode of operation of an E-UTRAN UE, as mentioned in Table 6.2, is further described. An E-UTRAN UE registered with an LTE/EPS network for EPS only services is said to be operating in PS mode. An E-UTRAN UE registered with LTE/EPS and legacy GSM/UMTS network for EPS and non-EPS, i.e. voice, SMS, services are said to be operating in CS/PS mode. In addition to this, an E-UTRAN UE may be required to switch between the PS and CS/PS modes of operation, and vice versa, because of various reasons that take place at the network level at runtime without the user's notice. Some of the typical reasons are as follows:

- Unavailability of CSFB feature,
- Unavailability of IMS,
- Changes in preferences, network operations,

- Roaming, and
- Failure to register with an IMS.

An E-UTRAN UE may be also required to switch between the PS and CS/PS mode of operations, and vice versa, because of a user action. For example, consider that a VoLTE-capable UE is registered with the LTE/EPC and IMS. The user has turned off the VoLTE preference in the handset, making the UE unable to facilitate voice-over IMS to the user. In such a case, the UE is required to register with the CS domain core network so that it can fall back to the legacy network to facilitate voice calls to the user. As a result of the transition from the PS to CS/PS mode of operations, and vice versa, a UE requires to perform protocol-related procedures so that it can stay updated with the core network with its current location information and continue enabling seamless voice and data services to a user.

Examples 6.4 and 6.5 below highlight the information passed by a UE as part of its signaling messages to the LTE/EPC network based on the UE mode of operations described above.

Example 6.4 UE usage setting and voice domain preference during LTE/EPS Attach Procedure

During the initial rollout of an LTE/EPS network, an operator may not have the IMS to offer its VoLTE services to subscribers. In such a scenario, the operator will be required to provide voice services to subscribers through the existing legacy GSM/UMTS network. The arrangement of providing voice service to an E-UTRAN UE and its user is known as the *CSFB* method, as described earlier in Section 6.2.3, creating a UE context between the MME and the MSC over SGs interface. In this interworking scenario, the E-UTRAN UE will send the following information during its initial registration, through the ATTACH Request procedure, with the LTE/EPC.

- UE to EPC (ATTACH Request)
  - Attach Type = *Combined EPS/IMSI Attach*
  - UE Usage Setting = *Voice Centric*
  - Voice domain preference = *CS Voice Only*

The above *UE Usage Setting* and the *voice domain preference* information may also be sent by a UE if it is not VoLTE capable or fails to register with the IMS. The LTE/EPC will send the following information as part of the ATTACH ACCEPT message to the UE:

- EPC to UE (ATTACH Accept)
  - EPS Attach Result = *combined EPS/IMSI attach*
  - Additional Update Result = *CS fallback not preferred*

Example 6.5 UE usage setting and voice domain preference during LTE/EPS Attach Procedure with IMS

For a VoLTE only capable E-UTRAN UE and LTE/EPC with IMS capability, the UE will send the following information during its initial registration, through the ATTACH Request message, procedure with the EPC:

- UE to EPC (ATTACH Request)
  - UE Usage Setting = *Data-Centric*
  - Voice domain preference = *IMS Voice Only*

The LTE/EPC will send the following information as part of the ATTACH ACCEPT message to the UE:

- EPC to UE (ATTACH Accept)
  - EPS Attach Result = *EPS Only*
  - EPS network feature support = *IMS voice-over PS session in S1 mode supported*

For more details on the UE mode of operations, its usages, voice domain preference, and the tasks performed in each transition, refer to TS 24.301 [46].

6.7 Function of E-UTRAN in a VoLTE Call

The nature of VoLTE traffic is different from normal IP/data traffic. The IP packets size of a VoLTE traffic is small and is predictable, whereas the IP packets of normal data services are larger than voice packets, busty in nature, and may be downloaded or uploaded quickly. After normal IP packets transfer, radio resources allocated to the UE may be released by the eNodeB. The allocation and releasing of radio resources between UEs and the eNodeB for normal IP traffic is dynamic in nature and involves signaling overhead over the LTE air interface. Unlike normal IP traffic, the data rate of voice packets of a VoLTE call is low and remains the same throughout the conversation. Thus, the same radio resources may be allocated semi-statically for the entire duration of the session, and no additional radio resources are required to be allocated dynamically by the eNodeB. Because of this, a semi-static allocation of radio resources reduces the signaling overhead between the UE and eNodeB. In addition to this, the UE is also not required to monitor the downlink Physical Downlink Control Channel (PDCCH) for additional resources. During a voice call, only one party talks at a time while the other party listens.

To support all the above aspects of a VoLTE call over an IMS, the following functionalities are required to be made available and enabled in the E-UTRAN (UE and eNodeB) for rich user experience and an optimum allocation and usages of radio resources over the LTE air interface.

- Semi-Persistent Scheduling, for allocation of semi-static radio resources to a UE, thus reducing signaling and scheduling overhead between UE and eNodeB.
- Discontinuous Reception (DRX), to lengthen the battery life of UE that does not require to monitor the PDCCH continuously.
- Robust header Compression, for reducing the IP Header size for voice traffic.
- Transmission Time Interval (TTI) bundling, enabling a UE, located at the cell edge, to transmit in UL in four subframes in a row. TTI bundling increases the possibility of receiving of time-sensitive voice traffic packets successfully from UE to eNodeB. TTI bundling avoids the retransmission requirements that would, otherwise, involve signaling overhead and cause a delay in receiving voice packets from UE to eNodeB, resulting in poor user experience.

Support of all of the above functionalities is implemented at the LTE/E-UTRAN air interface Layer 2 Medium Access Control (MAC) protocol. For more information on these functionalities toward a VoLTE call, refer to TS 36.321 [93]. The related MAC layer parameters for a VoLTE call are configured by the RRC layer of the E-UTRAN. Apart from the MAC Layer, the Radio Link Control (RLC) layer also facilitates acknowledged mode (AM), for SIP signaling, an unacknowledged mode (UM), transfer functions for voice traffic.

Chapter Summary

This chapter presented the seamless mobile communications services provided by operators to their subscribers through the interworking and interoperations of mobile communications networks operated by them. Interoperations among the networks are important to offer communication services to roaming users. Interworking and interoperations between two network elements or different networks are established by introducing new logical interfaces that connect enhanced or upgraded network elements. Core network elements, e.g. MSC, SGSN, and MME, are upgraded to support additional interworking functions in an LTE/EPS network.

This chapter also presented the UE mode of operations that enables a user to make a voice call over a legacy as well as an LTE/EPS network. UE mode of operations is important to achieve the interworking of mobile communications networks. The interworking features, i.e. CSFB and SRVCC, may be deployed as the separately licensed feature on top of an existing network. We then presented the routing methods, either through the home network or visited network, available to route the roaming user data to an external packet network. We closed this chapter with the functions that are required to be performed from the E-UTRAN point of view to support VoLTE calls over IMS.

7

Load Balancing and Network Sharing

Introduction

One important objective and function of a mobile communications network is to ensure communication services to subscribers round the clock. One way to achieve this is by eliminating the possibility of a single point of failure of core network elements. For these purposes, the core network elements of an operator can be configured in a pool that serves the newly registered subscribers in an equally loaded manner and provides communications services in a load balancing way.

Since the beginning of the Global System for Mobile Communication (GSM), the GPRS Edge Radio Access Network (GERAN) system, the radio access network (RAN) of a mobile communications network was operated and owned by an operator in a particular area. The cost of owning and operational expenditures of dedicated networks is high for an operator; at the same time, its service coverage area is also limited. To reduce the cost of owning required network infrastructures and increase the service coverages, an operator may come into an agreement on the sharing of the RAN or core network of another network operator and provide its own communications services through the shared networks.

This chapter presents the load balancing and RAN sharing features that are described in 3rd Generation Partnership Project (3GPP) specifications. These are the optional features that an operator may choose to deploy in an existing GSM, Universal Mobile Telecommunication System (UMTS), Long-Term Evolution (LTE)/Evolved Packet System (EPS), or 5G network.

7.1 Core Network Elements Load Balancing

In a typical deployment of a mobile communications network run by a particular operator, i.e. Public Land Mobile Network (PLMN), one RAN element, i.e. a base station controller (BSC), a Radio Network Controller (RNC), or an eNodeB or a 5G Next-Generation Radio Access Network (NG-RAN), is connected to default or single-core network element, i.e. Mobile Switching Center (MSC), Serving GPRS Support Node (SGSN), Mobility Management Entity (MME)/Serving Gateway (S-GW), or 5G core Access and Mobility Management Function (AMF). Such a deployment scenario of a core network element may not ensure the availability of its services all the time and may result in service outages. To avoid a single point of failure for both the control and user plane layers of the core network point of view, a redundant and active core network element may be also configured. In such a network configuration, a RAN element is connected to more than one core network element that serves in parallel. See Figure 7.1 for an illustration, which is reproduced from TS 36.410 [96]. It shows the connection of an LTE Evolved-UMTS Terrestrial Radio Access Network (E-UTRAN) eNodeBs to respective multiple Evolved Packet Core (EPC) core network elements. In case one MME or S-GW goes down the other, one shall continue to provide mobile communications services.

Mobile Communications Systems Development: A Practical Introduction to System Understanding, Implementation, and Deployment, First Edition. Rajib Taid.

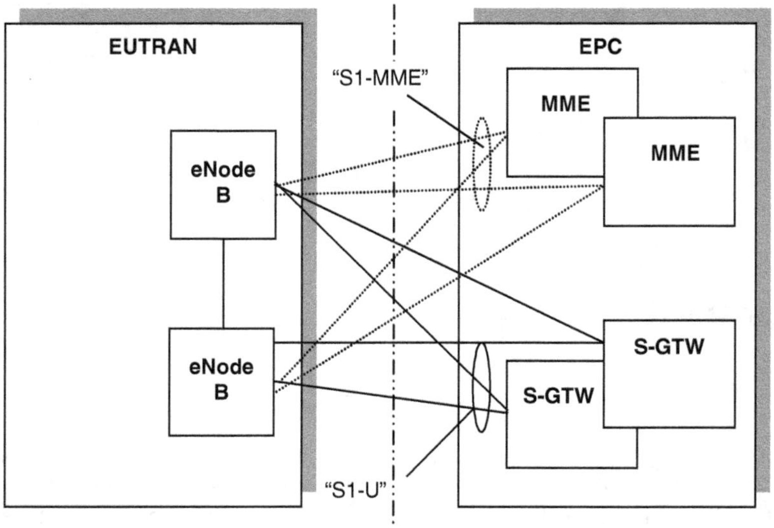

Figure 7.1 LTE/E-UTRAN eNodeB connected to multiple core network (CN) element. *Source:* © 2012. 3GPP ™ TSs and TRs are the property of ARIB, ATIS, CCSA, ETSI, TSDSI, TTA and TTC who jointly own the copyright in them. © 2012, 3GPP.

Multiple MME/S-GW may be configured to form an *MME/S-GW pool*. In case one MME/S-GW goes down the other, one will continue to provide mobile communications services to subscribers. Similarly, a GSM BSC may be connected to more than one MSC in the Circuit-switched (CS) domain or SGSN in the General Packet Radio Service (GPRS) Packet-Switched (PS) domain. The MSCs form the MSC pool area, and the SGSNs form the SGSN pool area.

In the 5G system and network also, an NG-RAN may be connected to more than one core network element, as illustrated in Figure 7.2. The *AMF* network function element performs the mobility management only-related functions, similar to the LTE/EPS MME, of a 5G network. As illustrated, a set of AMFs forms an AMF set that serves a particular service area or AMF set area. A set of AMF set forms an AMF region. On behalf of a User Equipment (UE), the NG-RAN selects an AMF set and an AMF within the set using the identity, i.e. 5G S-Temporary Mobile Subscription Identifier (5G-S-TMSI) or Globally Unique AMF ID (GUAMI), provided by the UE. These network element identities used in the 5G system shall be described later in Chapter 16.

In general, the RAN element, i.e. the BSC or RNC or the eNodeB or NG-RAN, selects and assigns an appropriate core network element, i.e. MSC or SGSN or MME or AMF during an initial mobile station (MS)/UE registration procedure with the core network. The selection procedure of a core network element is also called as the

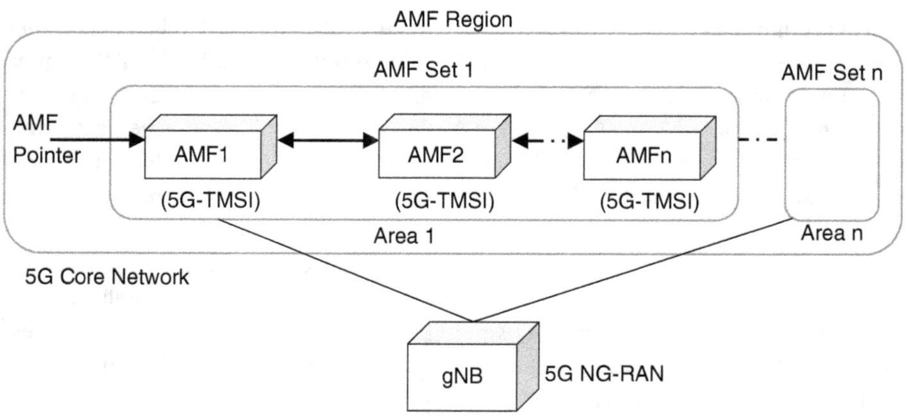

Figure 7.2 Illustration: 5G NG-RAN connected to multiple AMF sets.

Non-access Stratum (*NAS*) *Node Selection Function (*NNSF*)*. Based on the capacity or weight factor, which is an Operation and Maintenance (O&M) configurable, of the individual core network element, an NNSF selects an appropriate core network element from the available pool or set.

The NNSF procedure that is performed by the RAN element is implementation dependent. A typical NNSF is described in Section 7.1.2. Prior to that, the method of deriving the identity of a core network element, i.e. MSC, SGSN, or MME/S-GW only, is described in Section 7.1.1. Further, a RAN node connection to multiple core network elements may be an optional and separately licensed feature.

7.1.1 Identification of NAS Node: NRI and Its Source

As part of a UE/MS registration procedure with the core network, a temporary identity is assigned to the UE/MS by the concerned core network element. In the GSM system, the MSC assigns a TMSI to an MS; in the GPRS and UMTS system, the SGSN assigns a P-TMSI to an MS. In the LTE/EPS system, the MME assigns a Globally Unique Temporary Identifier (GUTI) to a UE that consists of its unique *MME code* within an MME pool. The TMSI, P-TMSI, and GUTI were described earlier in Chapter 5. In fact, a *Network Resource Identifier (NRI)* is derived from its corresponding temporary identity. Table 7.1 lists the NRI and its corresponding source in a GSM, GPRS/UMTS, and LTE/EPS network.

An NRI enables the RAN element to forward the user traffic to the correct core network element. The mapping of an NRI into a P-TMSI was illustrated earlier in Chapter 5. As shown in Table 7.1, in the case of the UMTS system, the NRI corresponds to *the Intra Domain NAS Node Selector (IDNNS),* which can be derived from the TMSI, International Mobile Subscriber Identity (IMSI), or International Mobile Equipment Identity (IMEI). The IDNNS is provided by the UMTS UE in the *INITIAL DIRECT TRANSFER* message to the UTRAN to transfer a NAS layer message to the concerned UMTS core network node. In case an E-UTRAN UE moves to a UTRAN/GERAN area, then the MME code is assigned to the 8-Most Significant Bits (MSBs) of the GERAN NRI. Similarly, in case a GERAN UE moves to an E-UTRAN area, then the 8-MSBs of an NRI is assigned to the MME code. Further, the RAN, i.e. BSC or RNC or the eNodeB, requires maintaining a database of mapping an NRI and its corresponding NAS node, i.e. MSC or SGSN or MME/S-GW to route a UE/MS uplink traffic.

Table 7.1 NRI and its source identifier.

System	CN Indicator	Source	Lengths
GSM	NRI	TMSI	10-bits
GPRS	NRI	P-TMSI	10-bits
UMTS	NRI	IDNNS	10-bits
LTE/EPS	NRI	MME code	8-bits

7.1.2 NAS Node Selection Function

As illustrated in Figures 7.1 and 7.2, a RAN element, e.g. eNodeB, can be connected to respective multiple core network elements. A RAN selects and assigns a core network element, i.e. MSC, SGSN, MME, or AMF, to a UE/MS during its registration process with the core network. For the selection and allocation of a concerned core network element by the RAN, a simple round-robin or *weighted round-robin (WRR)* procedure may be used.

The weight or capacity of a core network element, i.e. MSC, SGSN, MME, or 5G core AMF, in terms of the processing/serving capacity of MS/UE is O&M configurable. For these purposes, the concerned core network element provides its weight information to the RAN over the respective core network interface. In the case of LTE/EPS, it is the MME that provides its UE serving capacity information to the E-UTRAN. In the case of the 5G system, it is the 5G core AMF that provides its UE serving capacity information to the NG-RAN. In the case of GPRS or UMTS, it is the SGSN that provides its capacity information in terms of *signaling weight* or *data weight* to the BSC. Once a mobile device registers with a particular core network element, say the MSC or SGSN or MME or AMF, subsequent uplink traffic is routed to that particular network element only. If the RAN node, say the BSC, cannot find a core network element based on the NRI value, the RAN may drop the user traffic instead of forwarding it to the core network.

NNSF for the selection of a core network element or node from a pool of MSCs or SGSNs or MMEs or AMFs is important as it allocates and distributes MSs/UEs in a load balancing way, in proportion to the individually assigned weight factor, among the pooled core network nodes/elements. An NNSF works based on the weight/capacity information as configured by the O&M operator for each core network element. An NNSF is illustrated in Figure 7.3 in the case of an LTE/EPS network.

- NAS Node Selection Using WRR Procedure

Let us consider the selection and allocation of an MME to its UEs from an MME pool by the eNodeB. The maximum number of MMEs in an MME pool is 256; refer to TS 36.413 [97]. Each MME is assigned a unique *MME code (MMEC)*. The MME provides its processing capability, in terms of the number of serving UEs, through the *Relative MME Capacity* IE in the *S1 Setup Response* to the eNodeB; refer to TS 36.413 [97]. The value of the *Relative MME Capacity* ranges from 0 to 255. During the runtime, if the capability or weight of an MME changes due to the O&M intervention, the new capability is conveyed to the eNodeB through the *MME CONFIGURATION UPDATE* message.

Note that the capability/weight assigned by the O&M operator is a relative one with respect to the other MMEs in the pool. To select and allocate an MME from a pool in a load balancing way, the relative capability/weight of an MME may be normalized, say in the range of 0–255. This can be further considered as the granularity or scale factor for allocating several UEs proportionally to an MME. Using the granularity of 255 and the relative weights of MMEs, in terms of the actual number of UEs to be served by an MME can be calculated. NAS Node selection and allocation using the WRR method is illustrated with an Example 7.1 and Figure 7.3.

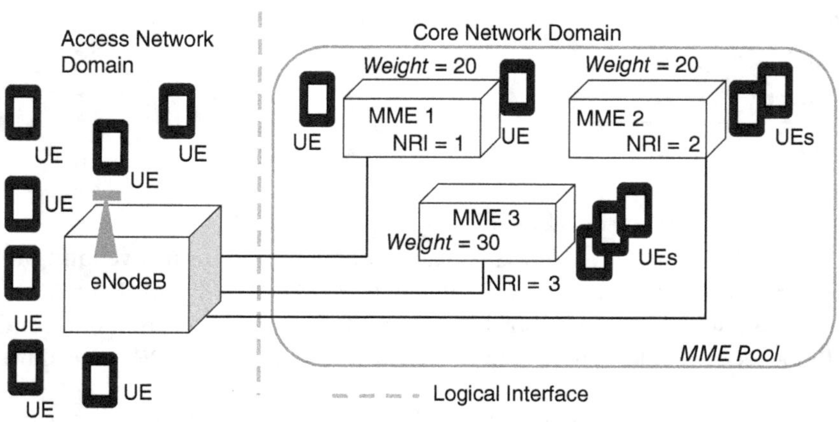

Figure 7.3 Illustration: typical MME selection and allocation to UEs using WRR method.

Example 7.1 Typical LTE/EPC MME Selection and Allocation for Load Balancing Using the WRR Method

Figure 7.3 illustrates the connection of an eNodeB to three MMEs in an LTE/EPS.

Assume that the MME 1 has NRI = 1 and relative weight = 20, MME 2 has NRI = 2 and relative weight = 20, and MME 3 has NRI = 3 and relative weight = 30, as illustrated in the MME pool shown in Figure 7.3. It is assumed that MME 3 is the high-capacity MME. The normalized weight in terms of the number of actual UEs to be assigned to the respective MME may be calculated as follows:

$$\text{Normalized Weight of MME} = \left(\textit{Relative Capacity of an MME} * \textit{Granularity}\right) \left(\textit{Sum of Weight of MME } 1 \ldots N\right)$$

$$\text{Normalized Weight of MME1} = (20 * 255) / (20 + 20 + 30) = 73$$

$$\text{Normalized Weight of MME2} = (20 * 255) / (20 + 20 + 30) = 73$$

$$\text{Normalized Weight of MME3} = (30 * 255) / (20 + 20 + 30) = 109$$

From the above typical illustration, it is observed that the MMEs are assigned with the relative weight/capability in the proportion and ratio of 20 : 20 : 30. Now, using the WRR method,

- MME 1 and MME 2 can be assigned to serve 73 UEs each in actual and
- MME 3 can be assigned to serve 109 UEs in actual.

The allocation and distribution of MMEs in the pool using the WRR method, with the granularity of 255 UEs, is illustrated in Table 7.2 below. This table shows the ideal distribution, and hence achieving load balancing, of the number of 255 UEs served by the respective MME based on its normalized weight. First of all, MME 3 is allocated to the UE1; second, MME 2 is allocated to UE2; third, MME 1 is allocated to the UE3; fourth, MME 3 is allocated UE4; and so on. In this table, the total number of UEs served by all the MMEs in the pool is 255, which is the granularity factor as mentioned above.

Similar to the LTE/EPS network, the UMTS RNC or GSM BSC may also use the WRR approach, as described above, for selecting and assigning a core network element, i.e. MSC or SGSN, to distribute UEs in a load-balanced way.

Figure 7.4 below further illustrates, in general, the core network (CN) node selection and its allocation using the WRR method.

Table 7.2 UEs distributions in MME pool using the WRR method.

System	Relative Weight	Normalized Weight	UE1	UE2	UE3	...	...	...	UE255
MME 1	20	73	0	0	1	...	...	...	73
MME 2	20	73	0	1	1	...	...	...	73
MME 3	30	109	1	1	1	...	...	...	109

- NAS Node Selection in 5G System

In the 5G system also, an appropriate AMF may be selected based on its capacity/weight factor, as configured by an O&M operator, to distribute and achieve a load balancing proportionately among the AMFs of an AMF set. The 5G core network architecture is based on the multiple instances of software-based core network functions which are deployed to provide communication services to users over different network slices. In addition to the selection of an AMF in the control plane, a particular network function shall be required to be selected in the user plane also. Different network functions and their selection in the 5G system are described later in Chapter 20.

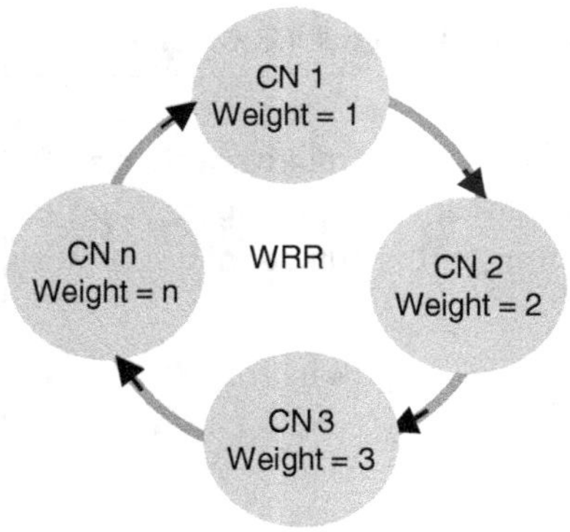

Figure 7.4 Illustration: core network (CN) node selection and allocation using the WRR method.

- Dynamic Assignments of Weights for NAS Node Selection

Different UE processing weights assigned to the core network elements for its allocation through the WRR procedure may be static or dynamic in nature. The O&M operator may reassign a new weight/capability to the concerned core network element, e.g. LTE/EPC MME or GPRS SGSN, in the pool during the runtime. This is achieved as follows:

- LTE/EPC MME can send its new capability/weight information to the eNodeB using the *MME CONFIGURATION UPDATE* message over the S1 interface; refer to TS 36.413 [97]. The eNodeB will recalculate, select, and assign an MME to UEs as per the new normalized weights.
- The SGSN in a GPRS network can send its new capability/weight information to the BSC using the *SNS-CHANGEWEIGHT* PDU over the Gb interface; refer to TS 48.016 [135]. The BSC will recalculate, select, and assign an SGSN to MSs as per the new normalized weights as the case may be.
- In the 5G system also, the AMF network function may send/update its O&M configurable capability/weight information to the NG-RAN using the *AMF CONFIGURATION UPDATE* message.

A dynamic WRR method of selection and allocation of a core network node from a pool/set offers the flexibility of performing maintenance activities on a particular core network node without service disruptions. If an O&M operator wishes to block a core network node from further allocation to UEs due to any reason, it can be achieved by assigning a weight = 0 so that the WRR algorithm does not select the concerned core network node. The blocked core network node, however, continues to serve the existing UEs.

Core network element selection using the WRR method is implementation-dependent from Original Equipment Manufacturer (OEM) to OEM. An OEM may choose to assign the capability or weight of an MME or AMF or SGSN in terms of the number of UEs/MSs to be served by it instead of using the normalized weight value of a core network element as illustrated above.

7.2 Network Sharing

Typically, a mobile communications network consisting of the RAN and Core Network Systems is owned and run by an operator. It is possible that a network operator, say A, does not have its dedicated RAN but has its core network infrastructures to provide communications services in a particular geographical location. However, another operator, say B, has the dedicated and also shared RAN infrastructures. In such a scenario, the operator A who does not have its RAN infrastructures can come into an agreement with the operator "B" which has the shared RAN infrastructures. The operator "A" provides the communication services through the shared RAN with its own identity, i.e. PLMN. This arrangement of providing communications services through RAN sharing between the operators A and B is known as the *Network Sharing* feature. As mentioned below, two approaches are available using which operators can share a RAN within a GSM/GPRS, UMTS, or LTE/EPS network.

- RAN sharing, which is known as the *Multioperator Core Network (MOCN)* feature.
- RAN as well as Core Network Sharing, which is known as the *Gateway Core Network (GWCN)* feature.

The next section describes the RAN sharing through the 3GPP MOCN feature from the GSM/GPRS, UMTS, and LTE/EPS networks point of view.

7.2.1 GSM/GPRS/LTE RAN Sharing Through MOCN Feature

In a MOCN configuration, the operators who do not have their dedicated RAN system shall provide their communications services through the shared RAN of another operator as mentioned above. In this configuration/arrangement, only the RAN of an owner operator is shared, while the other participating operators use their core network systems. An illustration is shown in Figure 7.5. The entire configuration/ arrangement is known as the *MOCN* configuration.

The sharing of RAN under the MOCN configuration shown in Figure 7.5 is similar to the network configuration described in Section 7.1 where a RAN node may be connected to a pool of core network elements of a communications network run by a particular network operator. The difference is that under the MOCN configuration, the core network elements are operated by different network operators. Some of the important aspects of the RAN sharing through the MOCN feature are described below:

- **3GPP Support of MOCN Feature**

  In the UMTS system, the RAN sharing through the MOCN configuration was made available from the 3GPP Release 6 onwards. In the LTE/EPS, the RAN sharing through the MOCN configuration was made available since its beginning which is from the 3GPP Release 8 onwards. However, in the GERAN, i.e. GSM and UTRAN, system, the RAN sharing through the MOCN configuration was made available only from the 3GPP Release 11 onwards.

- **MOCN Operator Identification – PLMN**

  Under the MOCN configuration, each core network operator is distinctly identified by the *Public Land Mobile Network* (PLMN) identity which consists of Mobile Country Code (MCC) and Mobile Network Code (MNC). The PLMN, or MCC and MNC, is also part of the IMSI of a mobile device. Using the System Information (SI) messages, a particular mobile communication network broadcasts its associated PLMN identification in a cell periodically. To extract the PLMN identity, the mobile device reads the SI messages received from the network and tries to match the received PLMN identity with the PLMN identity derived from its IMSI. If it matches, the mobile device registers with the concerned core network. The respective SI message used to broadcast the multiple PLMN identities in each network, from GSM to LTE/EPS, under the MOCN feature is shown in Table 7.3.

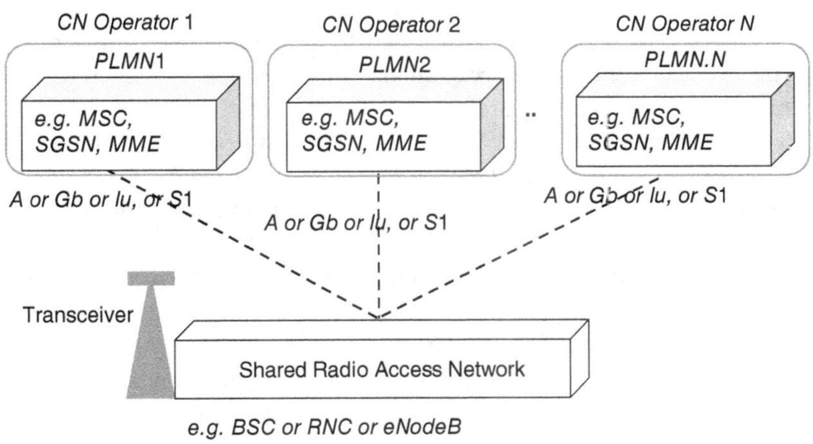

Figure 7.5 Illustration: radio access network sharing and MOCN configuration.

Table 7.3 System information (SI) messages for MOCN PLMN identities.

System	SI Message	No. of PLMN Ids
GSM/GPRS	SYSTEM INFORMATION MESSAGE Type 22 [...SI 22 Rest Octets...] TS 44.018 [130]	4
UMTS	Master Information Block [...*Multiple PLMN List....]* TS 25.331 [50]	6
LTE/EPS	SYSTEM INFORMATION BLOCK Type 1 [...PLMN Identity List...] TS 36.331 [94]	6

- **MS/UE Capability**

It has been mentioned above that under the GERAN, the MOCN feature was introduced only during 3GPP Release 11. All the pre-3GPP Release 6 and GERAN UE prior to Release 11 may not support the network sharing capability through the MOCN feature. Depending on the MS/UE and its 3GPP release version, two types of UE/MS may be found in a network, as mentioned below:

– MOCN supporting MS/UE

MOCN supporting MS/UE can decode and process all the PLMN identities that are transmitted by the shared RAN in the respective SI messages, as shown in Table 7.3. From the list of the provided PLMN identities, the UE/MS selects a particular core network operator using its PLMN identification, as specified in the 3GPP TS 23.122 [32]. The shared RAN forwards the UE data directly to the concerned core network operator and its core network element using the UE indicated PLMN identity. All the UTRAN and E-UTRAN UEs are MOCN supporting UEs.

– Non-supporting MS/UE

UMTS pre-3GPP Release 6 and GERAN UE prior to Release 11 are non-supporting MS/UE categories as these MS/UEs cannot process multiple PLMN identities of different core network operators, as broadcasted by the shared RAN. In a shared network, the non-supporting UE/MSs use the *common PLMN* identity which is also used as the identity of a normal or traditional network. A *common PLMN* identity does not identify any one of the sharing core network operators. The shared RAN broadcasts the common PLMN identity of a cell using the System Information Types 3 and 4.

For a non-supporting UE/MS, the shared RAN may select a particular core network operator in a normal round-robin manner and forward the UE initial signaling message, e.g. ATTACH request, Location Area, or Routing Area Update request, to the selected core network element, i.e. MSC or SGSN. However, the selected core network element, i.e. MSC or SGSN, may not accept the requesting UE/MS, due to the roaming agreement restriction. MSC or the SGSN determines the acceptability of the roaming agreement based on the PLMN identity (MCC + MNC) which is derived from the IMSI of the UE/MS. If the MS/UE can be admitted, the MSC or the SGSN sends an acceptance response, e.g. ATTACH accept, Location Area, or Routing Area Update Accept, to the MS/UE.

On the other hand, if the UE/MS cannot be accepted by the MSC or the SGSN as selected and forwarded by the shared RAN, the initial signaling message will be *rerouted* by the shared RAN to the core network element of another core network operator. This process is repeated until the UE/MS is not accepted by a particular core network operator or no more core network operator is left for selection by the shared RAN.

The overall time required for the completion of the initial signaling procedure and its messages sent by a non-supporting UE/MS may be slower. Because in this case the shared RAN will be required to *reroute* the MS/UE initial signaling message request to the next core network operator, in case of the preceding core network operator, node rejects the MS/UE registration. The reason for rejection may be due to nonroaming agreement between operators or MS/UE CS/PS coordination is required so that both the CS and PS services are served by the same operator.

- **Functions Performed by the Network Elements**

Several network elements are affected due to the MOCN feature and perform the typical functions as mentioned below:

- UE/MS

From a list of PLMN identities provided by the network, a supporting UTRAN and E-UTRAN UE select a particular core network operator and its PLMN identity and send it in the initial network registration message to the shared RAN. The shared RAN forward the initial message to the UE selected core network operator. In the case of GERAN, an MS/UE sends the PLMN index instead of PLMN ID to the shared RAN. The shared RAN finds the PLMN ID corresponding to the PLMN index provided by a supporting GERAN UE/MS. The shared RAN forward the initial message to the UE selected core network operator.

- UMTS RNC

The RNC extracts the selected core network operator PLMN identity from the RRC layer signaling message sent by the UE and forwards the initial message sent by the UE to the selected core network.

- LTE eNodeB

Similar to the RNC, the eNodeB extracts the selected core network operator PLMN identity, *selected PLMN-Identity,* from the RRC layer signaling message sent by the UE and forwards the initial message sent by the UE to the selected core network.

- GSM BSC

The GSM shared BSC broadcasts the list of PLMN identities through the SI message that identifies the sharing of core network operators in a particular location area. In the GSM/GPRS, a supporting MS provides the PLMN index along with the initial network registration message sent to the shared RAN/BSC. Using the PLMN index, the BSC will derive the actual PLMN identity of the core network operator and forward the initial message to the selected core network.

- MSC/SGSN

For a non-supporting MS, the MSC and SGSN will notify the shared RAN to reroute the signaling procedure originally initiated by the concerned UE to an MSC or SGSN of another operator. Once the UE is accepted by the routed operator's core network, the MSC and the SGSN will assign an NRI to the MS so that the subsequent message of the MS can be sent to the concerned MSC or SGSN through the shared RAN.

- MME

The E-UTRAN UEs are supporting types, i.e. a UE will indicate the PLMN ID of the core network operator to the shared RAN in its initial Layer 3 signaling message. The concerned MME will accept the signaling procedure and allocate a GUTI to the UE.

- CS/PS Coordination by Shared RAN for Non-supporting UE/MS

In a traditional mobile communications network, an operator provides both the CS domain and PS domain services to an MS/UE. For these purposes, a ME/UE registers with the CS and PS core network elements of a particular operator. While an MS is engaged in a PS service, an operator's core network may provide GSM CS services, such as a paging request to an MS for an incoming voice call, through the GPRS PS network. This arrangement, shown in Figure 7.6, is also known as the *CS/PS coordination* in the GSM network. To enable this, the *Gs interface,* refer to TS 29.018 [66], is configured between the MSC/VLR and the GPRS SGSN.

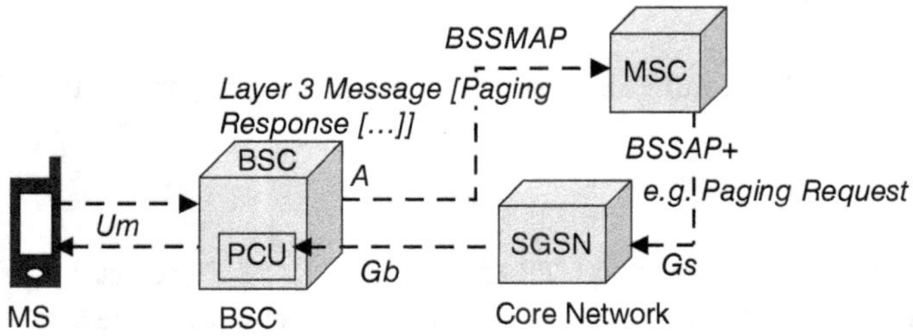

Figure 7.6 Illustration: GSM/GPRS CS/PS paging through Gs interface between MSC and SGSN.

In the above illustration, Figure 7.6, only a normal GSM CS and PS core network is being shown without a pool of MSCs or SGSNs. If the MSCs or SGSNs are configured in a pool and MSC sent the IMSI as the MS identity, then the SGSN also includes the MSC/VLR identity in the paging request message to the BSC. The MSC/VLR identity is used by the BSC to forward the paging response message from MS to the correct MSC in the pool.

Assume that the MSC wants to notify the GPRS MS about an incoming voice call. As illustrated in Figure 7.6, the MSC sent the GSM CS *paging request* messages to the SGSN through the Gs interface configured between them. The SGSN will further send the GSM CS *paging request* message to the GPRS *Packet Control Unit* of the BSC. The BSC will forward the CS *paging request* message to the MS over the air interface. In return, the MS will send the *paging response* to the BSC, which will further forward the paging response as the complete Layer 3 message over the A interface to the MSC only, as illustrated in Figure 7.6. The paging response is the Base Station Subsystem Management Application Part (BSSMAP) message between the BSC and the MSC, whereas the *paging request* through the G*s interface* is a Base Station Subsystem Application Part (BSSAP+) message.

Similarly, in a shared network also, a supporting UE will register with the same operator for the CS and PS domain services. On the other hand, for a non-supporting UE in a shared network, the network selects a core network operator and forwards the UE initial Layer 3 signaling message (e.g. ATTACH, Location Area, or Routing Area Update Request). The selected core network element may or may not accept the request made by the UE initial Layer 3 signaling message. If UE/MS is already served by the selected core network, say in the PS domain, then the initial Layer 3 signaling message received from the same UE for the CS domain will be accepted by the core network, resulting in the successful completion of the procedure initiated by the UE. However, if the initial Layer 3 signaling messages from the UE cannot be accepted, the core network will reject the UE request and send a rerouting command, as described below, to the shared RAN with an indication of CS/PS coordination requirement. For more information on the CS/PS coordination, refer to TS 48.008 [134].

From the above description, it is observed that the CS/PS coordination in a non-shared GSM/GPRS network is performed by the core network through the *Gs interface*, whereas in a shared network, the CS/PS coordination is performed by the shared RAN.

- Rerouting Function by Shared RAN for Non-supporting UE/MS

A shared network's rerouting capability makes it possible to use its services by a non-supporting UE/MS. A shared RAN, i.e. BSC or RNC, performs the rerouting functions for a nonsupporting UE/MS only for its initial Layer 3 signaling messages (e.g. ATTACH, Location Area, or Routing Area Update Request). For a non-supporting UE/MS, the shared RAN, i.e. BSC or RNC, includes the *Reroute Attempt* flag, refer to TS 48.008 [134] and TS 48.018 [136], along with the complete initial Layer 3 signaling message to the core network. A *Reroute Attempt* flag informs the core network node, MSC, to return either a *Reroute Command* or *Reroute Complete*, refer to TS 48.008 [134], message to the shared RAN. Similarly, the SGSN will return either a *Redirection Indication or Redirection Completed,* refer to TS 48.018 [136], TS 25.413 [54], information to the shared RAN. A *Reroute* or *Redirection Completed* from the MSC or SGSN indicates its acceptance and successful completion of the procedure initiated by the UE.

On the other hand, if the MSC or SGSN cannot accept the UE initial Layer 3 signaling message or procedure, a *Reroute Command*, refer to TS 48.008 [134], or *Redirection Indication,* refer to TS 48.018 [136], containing the rejection of the original initial Layer 3 signaling message, will be sent back to the shared RAN. On receiving the *Reroute Command or Redirection Indication*, the shared RAN will attempt to forward the UE initial Layer 3 signaling message again to another core network. The rerouting process is repeated by the shared RAN as long as a core network does not accept the UE initial Layer 3 signaling. In case no more sharing core network operator is left for selection, the shared RAN will inform the UE about the rejection of the original initial Layer 3 signaling message or procedure. Rejection of a UE initial Layer 3 signaling by the MSC or SGSN for a non-supporting UE/MS may be required due to the following reasons:

a) Roaming is not allowed, i.e. roaming restriction applies to the concerned UE/MS.
b) Roaming is allowed but CS/PS coordination is required for the concerned UE/MS.

- IMSI Analysis by the Shared RAN

An IMSI is included by the MSC or the SGSN as part of the *Reroute Command or Redirection Indication* message to the shared RAN in case the UE initial Layer 3 signaling procedure cannot be accepted but is rejected by the MSC or the SGSN. If the rejection is because of the CS/PS coordination requirement, then an IMSI analysis is required by the shared RAN and based on the PLMN identity derived from the IMSI, the UE initial Layer 3 signaling message will be rerouted to the identified core network.

- **How MOCN Feature Works**

The basic aspects of the MOCN feature have been described above. The following figures illustrate the MOCN feature for the supporting and nonsupporting UEs:

- Supporting UE

Figure 7.7 illustrates, in general, the signaling flow in a shared RAN through the MOCN feature for a supporting UE. As illustrated in this figure, there are three core network operators – CN-A, CN-B, and CN-C – that are identified by the PLMN 1, PLMN2, and PLMN3, respectively. Since the UE supports the shared network feature, it will decode all the PLMNs identities from the SI messages broadcasted by the shared RAN and select a particular core network operator whose PLMN identity matches with the one derived from the IMSI of the UE. Assume that the UE selected the PLMN: 3. The UE will send the initial message for a particular procedure, say the ATTACH request, to the core network operator with the PLMN: 3 through the shared RAN.

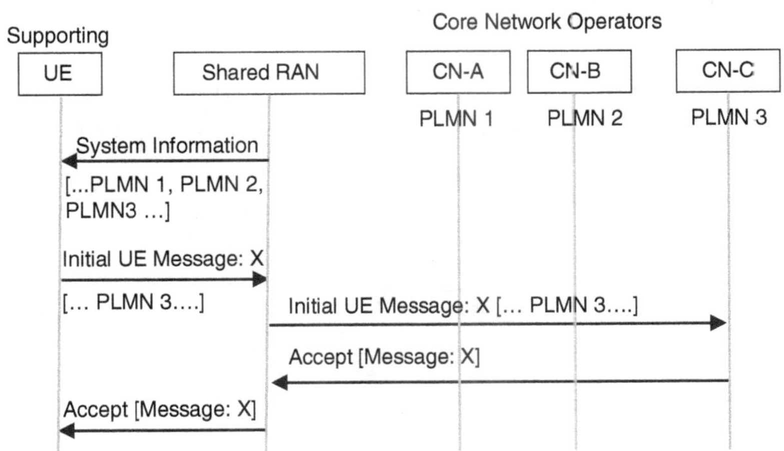

Figure 7.7 Illustration: supporting UE signaling flow for RAN sharing through MOCN.

The shared RAN extracts the PLMN identity: PLMN 3 from the signaling message received from the UE and forwards the message to the indicated core network of the operator: CN-A. The core network will accept the initial signaling procedure initiated by the UE and replies to the UE about the acceptance of its request through the shared RAN as illustrated in Figure 7.7. The accepted message from the core network contains its PLMN ID of the chosen operator.

For more details on the actual message flows for supporting UEs, refer to TS 23.251 [37].

- Non-supporting UE

Figure 7.8 illustrates, in general, the working of a RAN sharing through the MOCN feature for a non-supporting UE. As illustrated in Figure 7.8, there are three core network operators – CN-A, CN-B, and CN-C – that are identified by the PLMN 1, PLMN2, and PLMN3, respectively. Since the UE does not support the shared network feature, it cannot decode all the PLMNs identities from the SI messages broadcasted by the shared RAN. However, the UE will decode the common PLMN identity from the SI message. The UE will send the initial Layer 3 message for a particular procedure, e.g. *ATTACH, Location Area, or Routing Area Update request,* to the core network operator with the common PLMN through the shared RAN.

The shared RAN forwards the UE message mentioned above to one of the core networks of the sharing operators, say the CN-A. The core network CN-A verifies the roaming agreement for the UE from its IMSI. If a roaming agreement is permitted but the CN-A finds that a CS/PS coordination is required for the UE, as described earlier, then the request made by the UE cannot be accepted by the CN-A. If the CN-A is the MSC, it responds with a *Reroute Command*, TS 48.008 [134], to the shared RAN for *rerouting* of the UE message to another core network. The *Reroute Command* contains the rejection of the particular procedure, say *Location Update Request* reject, along with a *Reroute Indication* and the original UE initial Layer 3 message. If the CN-A is the 3G/UMTS SGSN and it cannot accept the UE request, say ATTACH or routing area update, either because of its roaming restrictions or CS/PS coordination requirement, a *Reroute NAS Request* message, TS 25.413 [54], is sent to the shared RAN for *rerouting* of the UE message to another core network identified by the P-TMSI of the UE or SGSN group identity.

If the CN-A is the 2G GPRS SGSN and it cannot accept the UE request, say ATTACH or routing area update, either because of its roaming restrictions or CS/PS coordination requirement, then a *Redirection Indication* information, TS 48.018 [136], is sent to the shared RAN along with the existing *DL-UNITDATA* PDU of the BSSGP layer of the Gb interface. The shared RAN forwards or *reroutes* the UE message to another core network.

The rerouting process of a UE initial Layer 3 message to another core network mentioned above is repeated, and finally, the core network CN-C has accepted the request made by the UE. The number of rerouting attempts to be made by the shared RAN is controlled by a timer at the RAN end. The last step in Figure 7.8 shows that the core

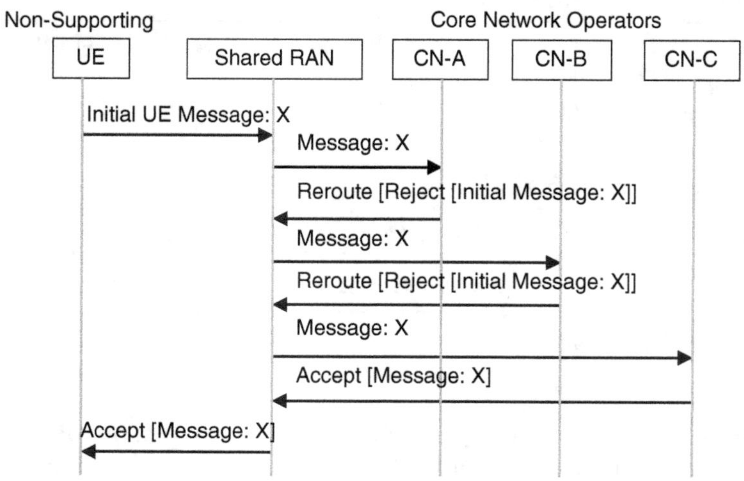

Figure 7.8 Illustration: non-supporting UE signaling flow for RAN sharing through MOCN.

network of the operator, CN-C, sent the result of the corresponding signaling procedure to the UE through the shared RAN. The final acceptance response, e.g. *ATTACH Accept or Location Area Update Accept or Routing Area Update Accept*, from the respective core network contains its common PLMN ID for the non-supporting UE. For more details on the actual message flows for non-supporting UEs, refer to TS 23.251 [37].

- Usages of NRI for Nonsupporting UE/MS

Usages of NRI were discussed in Section 7.1.1. The O&M operator configures and assigns the NRIs to the core network elements, e.g. MSC, SGSN, and MME, that are configured in a pool. Using the NRI, the MSC or SGSN generates and assigns a TMSI or P-TMSI to a UE or MS as part of its registration with the core network.

The allocation of NRIs to core network elements, e.g. MSC and SGSN, is important and must be coordinated for nonsupporting UEs in a shared network. NRIs coordination among the sharing core network operators will ensure that the UE/MS will remain CS/PS coordinated by allocating it to the NRIs in CS and PS domains of the same core network operator. Similarly, as far as the allocation of a TMSI or P-TMSI is concerned, as it uniquely identifies an ME/UE over the air interface, NRI coordination among the sharing core network operators will ensure that only the correct UE/MS responds to the paging request sent by the shared RAN.

7.2.2 5G NG-RAN Sharing Through MOCN Feature (Release 16)

The 5G system also supports the sharing of an NG-RAN of an operator with the 5G core network of other participating operators. An arrangement of an NG-RAN sharing with a core network of participating operators through the MOCN feature is illustrated in Figure 7.9. It may be noted that unlike the first 5G system Release 15, the 5G system Release 16 supports a private 5G network also through a *Stand-alone Non-Public Network (SNPN),* in addition to providing 5G services to the public through a *PLMN.* In the case of the 3GPP Release 15, the sharing of a 5G NG-RAN shall be the same as the previous generation systems, i.e. 5G system Release 15 supports NG-RAN sharing with a PLMN only. However, the support of an SNPN by the 5G system was introduced as part of the 3GPP Release 16. Thus, a 5G NG-RAN can be shared with a normal, i.e. public network, PLMN as well as the SNPN, i.e. private 5G network, as illustrated in Figure 7.9. A PLMN supports all the radio access technologies, i.e. GSM, UMTS, LTE, and 5G new radio (NR), whereas an SNPN supports the 5G NR only, i.e. N1 mode only.

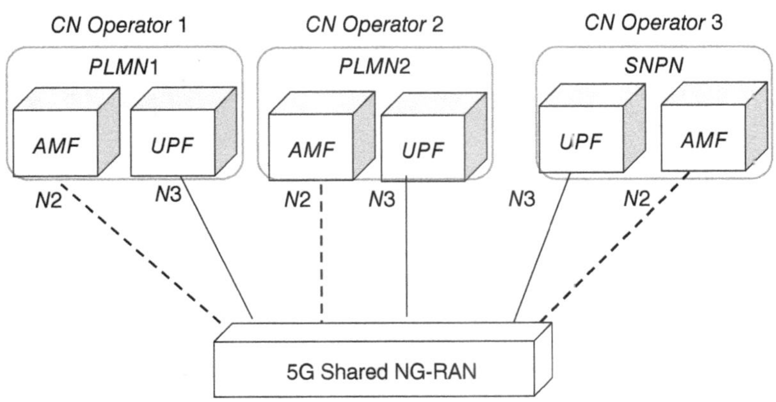

Figure 7.9 Illustration: 5G NG-RAN sharing through MOCN feature.

To transfer signaling information and user data, the shared NG-RAN is connected to the AMF and UPF network functions of the core network of a participating operator over the control plane reference point N2 and user plane reference point N3, respectively. An NG-RAN can be shared with multiple PLMNs or multiple SNPNs. Further, as illustrated in Figure 7.9, the participating operators provide their 5G services over their respective network slices.

Similar to the previous generation systems and as part of the 3GPP release 15 MOCN feature, only the list of PLMN identities of participating operators are broadcasted in a cell by the NG-RAN to indicate a shared network to UEs. However, as part of the 3GPP release 16 MOCN feature, a list of PLMN identities and *Network Identities (NID)* is broadcasted in a cell by the NG-RAN to indicate a shared network to UEs. A PLMN ID and an NID constitute an SNPN ID.

Chapter Summary

In this chapter, we presented the two important functions and features of a mobile communications network, that is, load balancing within an operator and network sharing among the participating network operators. Load balancing among the core network elements of a communications network of an operator is essential to ensure service continuity in the event of failure of a particular network element. For these purposes, core network elements are configured in a pool, and each one may be assigned either on a normal round or weighted round basis to serve UEs/MSs.

RAN and core network element sharing through the MOCN feature is of particular interest among the participating network operators as it results in a significant saving in the capital and operational costs for an operator. A participating operator can increase its services coverings by sharing with the RAN of another operator. For a RAN to be shared, cooperation and agreements among the participating operators are required. Through a RAN sharing, the participating operators also share radio resources.

8

Packets Encapsulations and Their Routing

Introduction

This chapter covers the user data packets encapsulation methods applied in a mobile communications network while transferring information between two network elements over a particular interface. In a mobile communications network, such as General Packet Radio Service (GPRS), Universal Mobile Telecommunication System (UMTS), Long-Term Evolution (LTE)/Evolved Packet System (EPS), and 5G system, packet-switched (PS) user data is not directly transported over the underlying Internet Protocol (IP) transport network; rather, the user data packets are put into another protocol whose resulting packet is further transported over the underlying IP transport network.

Data packets encapsulations may be applied while communicating over the air interface or between network elements of a core network (CN). We begin with the IP packets encapsulations applied at the CN. It is followed by packets encapsulations that are applied over the air interface while communicating between the User Equipment (UE) and the CN. We then present how the packets encapsulations are used to route user traffic between the ME/UE and the CN, irrespective of the current location of the mobile user/mobile station (MS)/UE. We close the chapter with the IP header compression and decompression that is applied to user data packets over the air interface, its methods, and usages over a low bandwidth network.

8.1 User Data Packets Encapsulations

In a mobile communications network and system, user/application data packets between the packet data network and an MS/UE are exchanged in an encapsulated and transparent manner, that is, user/application data packets are not directly put into and transported over the underlying transport network, say the IP transport. Instead of it, user data with the IP addresses of the MS/UE and an actual destination server are put and encapsulated into another protocol. The encapsulated and resulting information are put and transported using the actual underlying transport network, for example, using IP transport. User- or application-level IP data packets-only encapsulations are performed over different interfaces and network elements, as mentioned below:

- Between the CN elements of GPRS, UMTS, LTE/EPC, and 5G network.
- Between the CN elements (e.g. UMTS, LTE/EPC, 5G) and radio access network (RAN) (e.g. UTRAN, Evolved-UMTS Terrestrial Radio Access Network (E-UTRAN), 5G NG-RAN).
- Between the logical nodes, i.e. centralized unit (CU) and distributed unit (DU) of gNB of 5G NG-RAN.
- Over the air interface between the UE/MS and the CN elements of GPRS.

Mobile Communications Systems Development: A Practical Introduction to System Understanding, Implementation, and Deployment, First Edition. Rajib Taid.

Section 8.1.1 below describes the methods through which packet encapsulations are performed at the RAN and CN ends. Section 8.1.2 describes the methods through which packet encapsulations are performed over the air interface between a UE and the respective CN.

8.1.1 Packets Encapsulations at the CN and RAN

At the CN end, only user data/IP packets are encapsulated and exchanged between network elements using the *GPRS Tunneling Protocol* (GTP). The GTP is a general-purpose packet encapsulation protocol that is used across the mobile communication systems and is described in Section 8.1.1.1. Encapsulation allows the transferring of user data packets and routing transparently in a mobile communications network.

In a UMTS UTRAN, the GTP user plane is used to exchange user data packets between the Serving GPRS Support Node (SGSN) and UTRAN/Radio Network Controller (RNC) over the PS domain IuPS interface using the IP transport network. GTP user plane may be also used to exchange user data packets between the SGSN and UTRAN/RNC over the IuPS interface using the ATM transport network. For more information on the PS domain protocol stack structure of the UMTS/UTRAN IuPS interface, refer to TS 25.410 [52].

In an LTE/EPS network, the GTP user plane is used to exchange user data packets between the E-UTRAN and Serving Gateway (S-GW) over the S1 user plane interface. Apart from these, there are a couple of logical interfaces, for example, between SGSN and S-GW, S-GW, Packet Data Network Gateway (P-GW), and so on, that also use the GTP user plane to exchange user data packets between them. GTP is described in the next section.

In a 5G network, the GTP user plane is used to exchange user data packets between the 5G NG-RAN and User Plane Function (UPF) at 5G CN over the N3 user plane interface. Similarly, the GTP user plane is used to exchange user data packets between the NG-RANs over the Xn user plane interface. Apart from these, there are a couple of logical interfaces that also use the GTP user plane to exchange user data packets between them. GTP is described in the next section.

8.1.1.1 GPRS Tunneling Protocol (GTP)

In a traditional IP transport network, application/user data is directly transported as part of the payload of an IP packet that is exchanged between the source and destination hosts. But the data for a user/mobile device in a PS communications network are not directly transported over the underlying IP transport network. The mobile user's data are encapsulated using the *GTP* before transporting between the source and destination hosts over the connecting IP network. Only application or user data are transferred using the tunneling mechanism between two network elements as specified by the GTP. Each GTP tunnel has two endpoints. Each endpoint is identified through a *tunnel identifier*, an IP address, and a UDP port.

A typical user data transfer using the File Transfer Protocol (FTP) session, between a UE and Internet FTP server, over a GPRS network is illustrated in Figure 8.1 through the user plane protocol stack of the GTP (GTP-U). As shown in this figure, core network nodes, e.g. SGSN/Gateway GPRS Support Node (GGSN), encapsulates an FTP PDP PDU, called *T-PDU*, within a GTP header. Also, note that the FTP works on top of the Transmission Control Protocol/Internet Protocol (TCP/IP) protocol with the actual MS and Internet server IP address. GTP header is added to the T-PDU, giving a GTP PDU. The GTP-PDU *(G-PDU)* is put into a UDP PDU and finally put into an IP PDU of the underlying IP transport network. The underlying IP transport header contains the core network elements, or source and destination addresses of the concerned network elements in a mobile network, whereas the GTP PDU header contains the tunnel endpoint identifier that is necessary to uniquely address a PDP context.

8.1.1.2 GTP Functions

The GTP performs control and user plane function as listed below:

- Control Plane Functions

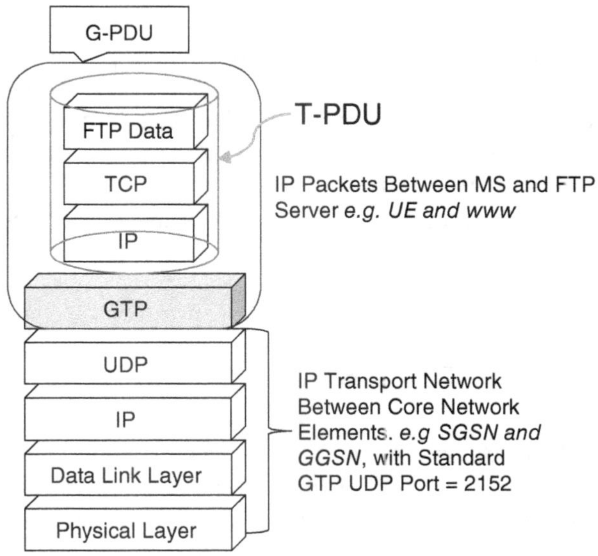

Figure 8.1 Illustration: packet encapsulation using GTP.

It is used to exchange signaling-related messages, such as the Mobility Management and Session Management between the core network elements. GTP Control Plane signaling is also used to create, use, modify, and release tunnels. The following functions are performed by the GTP control plane protocol:

- Path Management Functions, e.g. Echo-Request and Echo Response.
- Mobility Management Functions to support handover procedure, e.g. Forward Relocation Request.
- Tunnel Management Functions, e.g. Create Session Request.
- Support of circuit-switched (CS) fallback feature

For more information on the above control plane functions of GTP, refer to TS 29.274 [70].

- User Plane Function – Exchanges messages to update the status of user data packet tunnel between peers.
  - Tunnel Management Functions.
  - Path Management Functions, e.g. Echo-Request and Echo Response.

For more information on the above user plane functions of GTP, refer to TS 29.281 [72].

8.1.1.3 GTP User Plane PDU: G-PDU

The GTP user plane PDU is called as the G-PDU, which was illustrated in Figure 8.1. As illustrated in this figure, a G-PDU consists of the following components, see Figure 8.2:

- GTP Header
- T-PDU (Tunneled PDU or GTP Payload for User Data Packet).

T-PDU carries the actual user or application data packet between a UE and the external IP network. For routing the user data/IP packet, a T-PDU contains the source destination (UE or external network) and destination (UE or external network). As an example, in an LTE/EPS network, GTP user plane messages are exchanged over the S1-U, X2, S4, S5, S8, and S12 interfaces. In a 5G network, GTP user plane messages are exchanged over the N3, Xn, and so on logical interfaces. For more details on the components of the GTP header, refer to TS 29.281 [72].

Figure 8.2 Components of GTP user plane G-PDU.

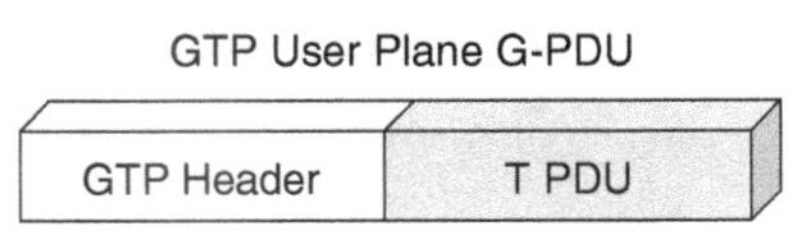

8.1.1.4 GTP Control Plane PDU

Unlike the GTP user plane protocol, the GTP control plane protocol does not use a separate encapsulated PDU to exchange information between two network elements. GTP control plane messages exchanged for various GTP procedures as mentioned in Section 8.1.1.2 are directly transported between two network elements over the underlying IP transport network. This is illustrated in Figure 8.3 between LTE/EPC Mobility Management Entity (MME) and P-GW.

GTP control plane message has its header which consists of the following typical information fields:

- GTP version
- Piggybacking information flag
- Tunnel endpoint Identifier flag
- Message type
- Message length

As an example, in an LTE/EPS network, GTP control plane signaling messages are exchanged over the S3, S4, S5, S8, S10, S11, and S16 interfaces; refer to Figure 1(b) TS 23.002 [29]. Figure 8.4 below illustrates the usages of some of the GTP control plane signaling messages that are exchanged over the LTE/EPC S3 interface between SGSN and LTE/EPC MME and the S11 interface between MME and S-GW during a typical inter-Radio Access Technology (RAT) handover from GPRS to LTE/EPS network.

To initiate an inter-RAT handover procedure from a GPRS to LTE/EPS network at the CNs level, the SGSN makes a handover request to the LTE/EPC MME, which creates a new session with the S-GW on behalf of the UE. For illustration purposes, GTP signaling messages exchanged among the CN elements are shown in italics in Figure 8.4.

For more information on the GTP control plane protocol, its header, and message details, refer to TS 29.274 [70].

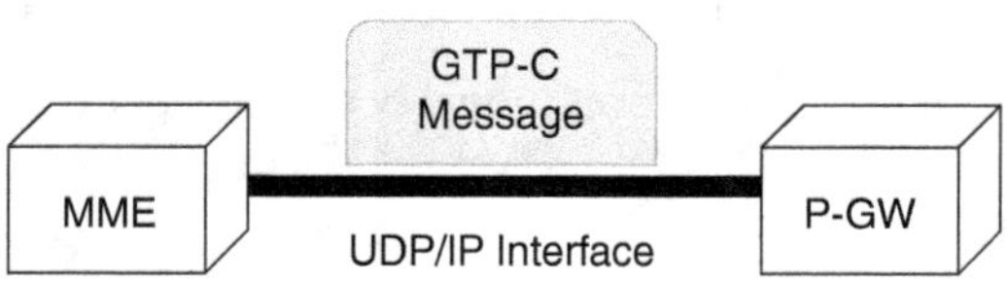

Figure 8.3 Illustration: LTE/EPC GTP-C message over UDP/IP interface.

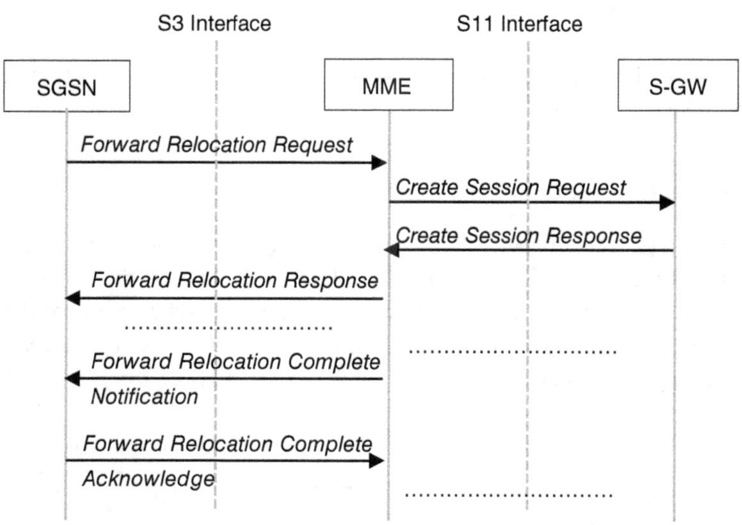

Figure 8.4 Illustration: GTP-C message exchanged during inter-RAT handover.

8.1.1.5 Example: GTP and Packet Encapsulations at LTE EPC

Figure 8.5 illustrates the transportation of UE-generated user traffic/IP packets through encapsulations using the GTP user plane protocol in an all IP LTE/EPC network. Note that the LTE UE is not aware of the GTP being used for encapsulations and transporting of its packet with the CN. As illustrated in this figure, the UE communicates with the WWW server and exchanges IP packets over the LTE/EPC network. The IP packets originated from the UE contains its IP address as the source and the WWW server address as the destination IP address. With this source and destination IP addresses remaining unchanged, UE IP packets are tunneled across the EPC network element, i.e. ENodeB, S-GW, and P-GW using the GTP.

In Figure 8.5 illustration, the S1 GTP tunnel between the eNodeB and the S-GW contains the eNodeB and S-GW IP address as the source and destination IP addresses and vice versa. The S5 GTP tunnel between the S-GW and P-GW contains the S-GW and P-GW IP address as the source and destination IP addresses and vice versa. A GTP tunnel has a tunnel endpoint identifier, IP address, and UDP port number in each node. In this figure, the tunnel is being shown as the pipe, and inside the tunnel, the flow of IP packets is shown as the bold horizontal line.

8.1.2 Packet Encapsulations over Air Interface

In Section 8.1.1, user data packets-only encapsulations at the RAN, CN, and various network elements of a mobile communications network have been described. Similar to this, user data packets are also encapsulated while transporting them over the respective air interface of different mobile communications systems as described below:

- Packets encapsulations between the CN element and MS in the GPRS network

Consider the packet data encapsulation services which are provided through the GPRS communications system. In the GRPS system user plane protocol stack, there is a protocol layer called *Subnetwork Dependent Convergence Protocol (*SNDCP*), TS 44.065* [132], terminating at the MS and SGSN end. Subnetwork refers to the GPRS network. SNDCP layer converges and multiplexes packet data, called *network PDU,* N-PDU, coming from a network/application protocol over a particular *Network Layer Service Access Point Identifier* (NSAPI*).* In the same way as the GTP, the PDU that is generated and passed by the SNDCP to the lower layer is called as the *SN-PDU.* An SN-PDU contains the user application packets between a UE and the Internet server with the corresponding source and destination addresses. The SN-PDUs are transported between a UE and the SGSN across the Global System for Mobile Communication (GSM)/GPRS air interface. The SGSN, in turn, forwards the SN-PDUs to the GGSN over the Gn interface using the GTP.

Figure 8.6 illustrates packet encapsulations in a GPRS network over its air interface between an MS and SGSN. In this illustration, the base station controller (BSC) is not being shown between the MS and the SGSN for brevity.

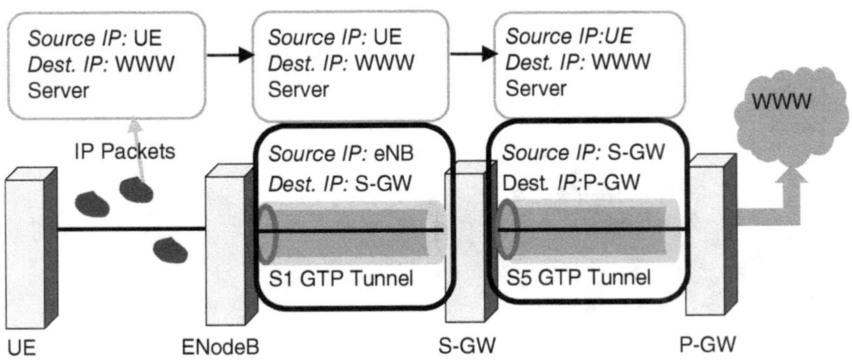

Figure 8.5 Illustration: user data/IP packets encapsulations and tunneling using GTP in LTE/EPC network.

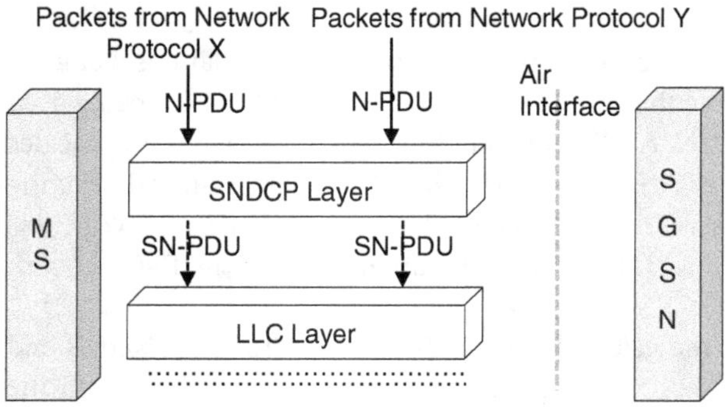

Figure 8.6 GPRS packet encapsulation, between MS and SGSN, over the air interface.

8.2 IP Packet Routing in Mobile Communications Networks

In a mobile communications system, it is interesting to see the reachability of a mobile subscriber irrespective of its current location within a particular coverage area. This is possible because of the accurate routing of the ongoing call, either CS or PS, for the concerned mobile subscriber. It becomes possible to route a call or user data correctly to the mobile user because of the mobility management functions that are performed by a UE/MS and its mobile communications network. More details on 5G mobility management functions are available later in Section 18.4.

In a PS network, a router makes packets routing decisions based on the destination IP address of a packet as well as the current routing table of the router. In the case of a traditional IP network, the location of a destination device and its IP address rarely changes. But in a mobile communications network, a subscriber can freely move at any point in time within its network coverage areas or other network areas, i.e. a roaming user. This gives rise to the requirements of a flexible as well as a correctly routing mechanism to route packets, to a UE, of a PS call by the CN.

- GPRS/UMTS CN

Consider the IP packet routing requirement in an inter SGSN user movement scenario of a GPRS/UMTS network, shown in Figure 8.7.

In the case of packets routing between the SGSN and GGSN as shown in Figure 8.7, they do not use the source (MS/UE) and actual destination (e.g. WWW server) IP address that is available in the user's IP packet; rather, the user's IP packets are routed/tunneled as an *encapsulated* payload using the GTP between the SGSN and the GGSN. The GTP over the *Gn* interface contains the IP addresses of the SGSN and GGSN as the source and destination, and vice versa, to exchange tunneled user data packets between them. This approach is

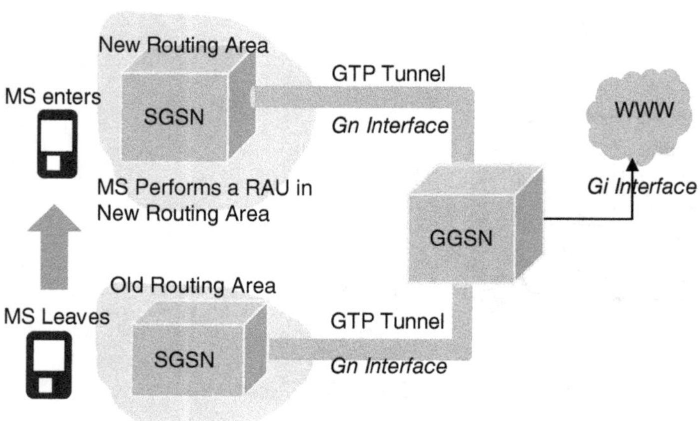

Figure 8.7 Illustration: GPRS inter SGSNs routing area update procedure.

particularly helpful in scenarios where the SGSN and the GGSN are not connected directly but connected through several intermediate routers.

Whenever a mobile subscriber enters a new coverage area controlled by another SGSN and to route the incoming packets correctly toward the mobile device, the routing tables of the intermediate routers as well as the SGSN and GGSN would have been required to be updated. But to avoid such update requirements at multiple intermediate hosts during the UE/user's movements, only the GGSN's routing table is required to be updated using the IP address of the new SGSN. Figure 8.7 illustrates an example of an inter SGSNs routing area update (RAU) scenario where the GGSN updates its routing table with the IP address of the new SGSN that is now serving the MS.

To establish and inform the current location, an RAU, as shown in Figure 8.7, procedure is initiated by the mobile device on entering and detecting a new routing area that is controlled by the new SGSN. As part of the RAU procedure and on behalf of the requesting MS, the new SGSN of the new routing area sends the *Update PDP Context Request* message, refer to TS 29.060 [67], containing the new SGSN address to the GGSN. Subsequent packets of the MS are routed from the GGSN to the new SGSN only. For more information on the inter SGSNs RAU procedure, refer to TS 23.060 [31].

- LTE/EPS CN

Similar to the GPRS/UMTS RAU, an LTE/EPS-registered UE makes the tracking area update (TAU) request to the LTE/EPC MME to update its current location. The MME will send, on behalf of the UE, the *Create Session Request* Message, refer to TS 29.274 [70], to the new S-GW if there is a change during the TAU so that subsequent packets of the UE are routed from the new S-GW to the MME. For more information on the inter S-GW TAU procedure, refer to TS 23.401 [39].

A similar procedure is used to correctly route packets to a mobile device in the 5G system also, using the GTP, independent of its location.

8.3 IP Header Compression and Decompression

- **Requirements of IP Header Compression**

All of us used to save computer storage space by compressing files through the zip utilities. Similarly, in mobile communications networks also, protocol layer *header* information exchanged may be compressed before it is sent over the air interface between the sending and receiving ends. At the receiving end, the IP layer header is decompressed again. The bandwidth resources, especially the air interface, utilization for IP data services in a mobile communications network could be reduced and utilized in an optimum way through compressing (at the sending end) of IP layer header information. This will result in an efficient transmission and quick network response, delivering an overall rich user experience. The application of the header compression procedure will be particularly useful in the case of a low bandwidth network. In a mobile communications system and network, the IP header compression/decompression procedure may be an optional one.

- **How does IP Header Compression work?**

A packet data session or stream of a user or application contains multiple IP packets having protocol header information in each packet for its proper routing across multiple hosts. Recall the components of an IPv4 header that consists of the following fields:

- Version, Internet Header Length (IHL), Differentiated Services Code Point (DSCP), Explicit Congestion Notification (ECN), Total Length
- Identification, Flags, Fragment Offset
- Time To Live, Protocol, Header Checksum
- Source IP Address
- Destination IP Address

The value of the some of the fields, e.g. source and destination IP address, UDP or TCP port, IP version, and so on, of an IPv4 header mentioned above, may not change during the entire data flow/session. IP protocol header compression algorithm will not send those fields that are determined to remain the same in each IPv4 packet of the same data flow/stream, or few bits may be used to represent those fields, thus resulting in overall smaller size IPv4 packets and saving significantly in the total number of the bits sent in each IP packet over a particular interface. See Example 8.1 and Figure 8.8 for a description and an illustration of the IP header compression method.

- **IP Header Compression/Decompression Methods**

In mobile communications networks, an IP header compression/decompression, for user data packets, related tasks are performed by the protocol layer which is immediately below the IP layer of the UE/MS protocol stack. Table 8.1 shows the protocol layers performing the header compression/decompression tasks along with their corresponding Request For Comments (RFC) number for the respective compression/decompression framework implemented by the concerned protocol layer. These layers use the well-known protocols as mentioned in the concerned RFCs defined by the IETE. Refer to the indicated RFC# for more information on the header compression/decompression procedure adopted by each protocol layer.

Example 8.1 IPv4 Header Compression for VoIP Calls

Consider a typical VoIP call made over the IPv4 network. The typical protocol stack for a VoIP call is shown below (Figure 8.8). The voice is digitized, and the coded data call is placed as the payload of the RTP. RTP packets are transported over the UDP, which is routed further by the IP transport network. The minimum size of the IPv4 header with its fields is 20 bytes. Also,

- UDP header takes 8 bytes and
- RTP header takes 12 bytes, excluding the payload part.

Thus, the total size of the uncompressed header, in this case, is 20 + 8 + 12 = 40 bytes. The header compression algorithm, at the sending end, may compress the header size of 40 bytes to fewer bytes, say 1–2 bytes, and sends it in the subsequent IPv4 packets, thus resulting in significant savings in the consumption of network bandwidth resources.

Similarly, using the IPv6 version, the size of the uncompressed IPv6 header is 40 bytes. If an application, such as FTP, uses the TCP whose transport header size is 20 bytes, the total size of the uncompressed header (IPv6 + TCP) is 60 bytes. For a large file size transfer scenario, gain from the protocol header compression may not be significant. But for data traffic, e.g. VoIP, having a small payload size less than the uncompressed header size may gain significantly in the overall bandwidth resource utilization.

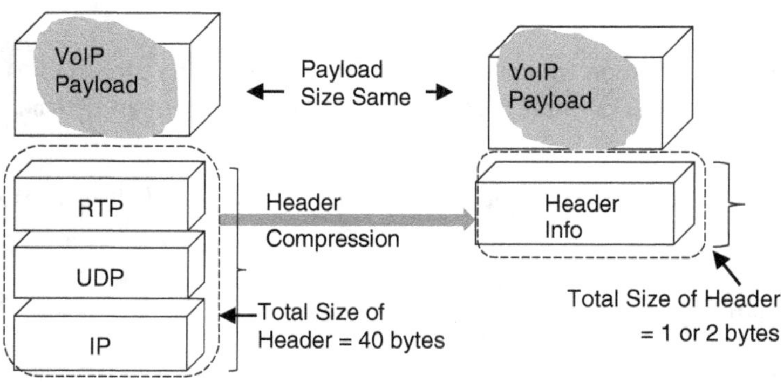

Figure 8.8 Illustration: IP header compression for a VoIP call.

Table 8.1 IP header compression/decompression methods.

System	Air Interface Protocol Layers	IP Header Compression/Decompression Method	RFC#
GPRS	SNDCP	TCP/IP Header Compression, IP Header Compression, ROHC	1144, 2507, 3095 [13]
UMTS	PDCP	IP Header Compression, ROHC	2507, 3095 [13]
LTE	PDCP	ROHC	3095, 4815, 4995
5G	PDCP	ROHC	3095, 5795, 4815 [16]

Chapter Summary

In this chapter, we have presented one of the important functions, i.e. encapsulation of user or application layer IP data traffic performed by the RAN and CN elements. We have also discussed another important function performed by the CN, which is the routing of user IP data traffic in mobile communications networks irrespective of the current location of a mobile user. In summary, the GTP is used to tunnel and route user data between CN elements in an encapsulated manner. There are a couple of logical interfaces that use the GTP to exchange information among the network elements.

Chapter Summary

9

Security Features in Mobile Communications Networks

Introduction

In mobile communications networks, the information transmitted and received by a mobile device over the air interface gets exposed and can be intercepted by unauthorized users. The protection of user privacy, as well as the information transmitted/received between a mobile device and the network, is one of the important functions that is performed by mobile communications systems and networks. Another important function of mobile communications networks is to ensure that only an authenticated and authorized user is allowed to access network services. These capabilities are realized and ensured by the security aspects of the concerned mobile communications system and network.

Security features and their mechanisms are part of the system engineering aspects of a mobile communications system, as described in Section 2.4.4. This chapter begins with the security aspects in terms of security features and mechanisms that are employed by a particular mobile communications system to authenticate its mobile users and protect their information as well as to protect signaling information being exchanged over the air interface. A mobile user's security context and its interworking requirements in the case of inter-Radio Access Technology (RAT) mobility are also discussed.

9.1 A Brief on the Security Architecture: Features and Mechanisms

Privacy protection and secure communications services to mobile users over the air interface depend on the security architecture of a particular mobile communications system, i.e. Global System for Mobile Communication (GSM), Universal Mobile Telecommunication System (UMTS), Long-Term Evolution (LTE), and 5G system. The security architecture, in general, of a mobile communications system and network, is based on the following aspects:

- Security features, which specifies the security requirements between the user and the network for a particular mobile communications system. Some of the security features, in general, used over the physical Radio Frequency (RF) path of the GSM, UMTS, LTE, and 5G system air interface are listed below:
  - Subscriber Identity Authentication,
  - Subscriber Identity Confidentiality,
  - Subscriber Information Confidentiality, and
  - Integrity Protection, in the case of UMTS, LTE, and 5G systems only.

Mobile Communications Systems Development: A Practical Introduction to System Understanding, Implementation, and Deployment, First Edition. Rajib Taid.

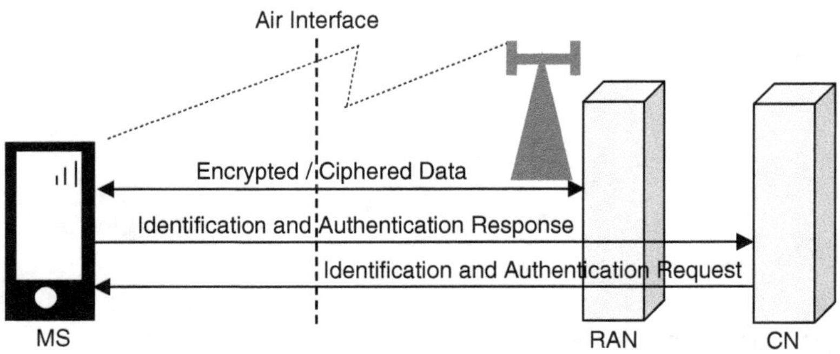

Figure 9.1 Illustration: security features for protection of user traffic in GSM, GPRS, UMTS, LTE, and 5G systems.

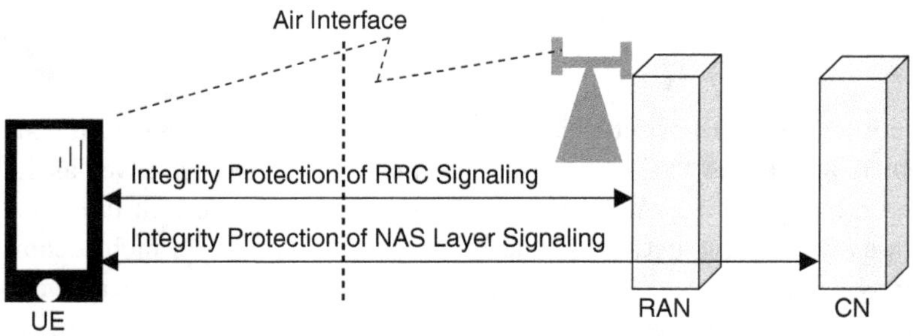

Figure 9.2 Illustration: security features for protection of signaling traffic in UMTS, LTE, and 5G systems.

- Security mechanisms, which specifies the methods to be used by a mobile device and its network to implement the above security features. For example, user data/information confidentiality is realized using the ciphering/encryption method. Similarly, user identification and authentication are performed by using security keys and algorithms that are known to the mobile device and the network only.

Figure 9.1 illustrates the security features, in general, applied between a mobile device and the radio access network (RAN) of the GSM, General Packet Radio Service (GPRS), UMTS, LTE, and 5G systems to protect user traffic over the air interface.

First of all, mobile device identification and authentication is performed by the network during a call setup phase. For example, refer to Figure 2.8, where the authentication task is performed following the GSM Connection Management (CM) *layer Service Request* message from the mobile station (MS) to the network for a GSM mobile-originated voice call (MOC). Only when the mobile device has been authenticated successfully, the core network initiates the security procedures toward the MS so that the subsequent traffic between the mobile device and the network is exchanged in a ciphered/encrypted manner over the air interface. Note that in the GSM system, ciphering is performed by the base transceiver station (BTS), whereas the same task is performed by the Serving GPRS Support Node (SGSN) in the GPRS/UMTS system. Further, Figure 9.2 illustrates the security features, in general, applied between a mobile device and the RAN and core network of the UMTS, LTE, and 5G systems only to protect the integrity of signaling traffic contents from alteration due to any reasons.

Table 9.1 summarizes the security features that are applied to protect the user as well as signaling information in GSM, GPRS, UMTS, LTE/EPS, and 5G networks.

From Table 9.1, it can be stated that the UMTS, LTE, and 5G systems have more security features and methods compared to the GSM system. Unlike the GSM system, in the UMTS, LTE, and 5G systems, the integrity of the Radio Resource Control (RRC) and Non-access Stratum (NAS) layer signaling information which is exchanged between the User Equipment (UE) and the network is also protected using different security algorithms. Also, in the UMTS system only, the air interface RRC layer messages are integrity protected. But in the LTE and 5G systems, the RRC as well as the NAS layer messages are also security and integrity protected.

Table 9.1 Summary of security features and its network elements.

Security Features	Systems	Applied Between
Ciphering (Signaling and User Traffic)	GSM, GPRS, UMTS, LTE, and 5G	MS and RAN
Identity, Authentication		MS/UE and Core Networks
Integrity Protection (Signaling Traffic)	UMTS	UE and UTRAN (AS Layer)
	LTE	UE and E-UTRAN (AS Layer), UE and EPC MME (NAS Layer)
	5G	UE and NG-RAN (AS Layer); UE and 5GC AMF (NAS Layer)

Security architectures and algorithms differ from the GSM to the 5G system. Various algorithms and keys are used to implement security features and integrity protection of the information exchanged between a mobile device and the network. For more information on the security architecture of the GSM system, refer to TS 42.009 [128]; the UMTS system, refer to TS 33.102 [85]; the LTE/EPS system, refer to TS 33.401 [86]; and the 5G system, refer to TS 33.501 [87].

As a result of the execution of a security procedure successfully, a corresponding security context containing security-related information is created at both the mobile device and the network end. A security context created in a particular RAT/system, e.g. LTE, may be also used to map and derive the security context for another RAT, e.g. 5G, during an inter-RAT handover of a mobile user.

9.2 Security Features and Its Mechanisms

- Subscriber Identity Confidentiality

To protect a subscriber/MS identity, such as its International Mobile Equipment Identity (IMEI) or Mobile Subscription Identification Number (MSIN), from eavesdropping over the air interface, the core network provides a facility to allocate a temporary identity instead of using its permanent identification such as the International Mobile Subscriber Identity (IMSI). This temporary identification is known only within the currently registered area of the mobile subscriber. Upon an MS/UE current location change, it may be allocated again to a new temporary identity by the core network. Table 9.2 shows the temporary identity assigned to an MS/UE by the respective mobile communications system to protect the subscriber's identity over its air interface.

For more details on the above temporary identities of an MS/UE, refer to Section 5.3.

- Subscriber Identity Authentication

Only an authorized subscriber is allowed to access a mobile communications network and its services. This function contains authentication procedures and algorithms to be used by the concerned core network element to verify and authenticate a mobile subscriber. Table 9.3 shows the authentication procedures used by the different mobile communications systems.

In the UMTS, LTE, and 5G systems, a subscriber and the core network mutually authenticate each other to ensure that only an authorized subscriber is allowed to access the network and also to ensure that a UE is accessing its correct network. The method used to perform such a mutual authentication task is known as the *Authentication and Key Agreement* (AKA) method where a secret key, called K_{ASME} in LTE/EPS network; K_{AMF} in 5G system; and K in UMTS, is shared between UE and core network. LTE/EPS AKA is based on the UMTS AKA method. For more information on the AKA procedure, refer to 3GPP TS 33.102 [85]; TS 33.401 [86]; and TS 33.501 [87].

Table 9.2 Temporary identities used for subscriber identity confidentially.

System	Temporary Identity Used for MS/UE	Valid Area of the Temporary Identity
GSM	Temporary Mobile Subscription Identifier (TMSI)	Location Area
GPRS	Temporary Logical Link Identifier (TLLI), P-TMSI	Routing Area
UMTS	TMSI, P-TMSI	Location/Routing Area
LTE	Globally Unique Temporary Identifier (GUTI), S-TMSI	Tracking Area
5G	5G-GUTI, 5G-S-TMSI	Registration Area

Table 9.3 Keys and algorithm used for subscriber identity authentication.

System	Keys and Algorithms
GSM	RAND, Authentication Key, SRES, A3 Algorithm, IMSI, and TMSI during LAU
GPRS	RAND, Authentication Key, SRES, A3 Algorithm, IMSI, and TLLI during RAU
UMTS	Authentication and Key Agreement (AKA)
LTE	Authentication and Key Agreement (AKA)
5G	Authentication and Key Agreement (AKA)

- Confidentially of User and Signaling Information

Information transmitted as part of the user and signaling messages between the mobile device and the network is also required to be protected so that eavesdropping on them is not possible. To achieve this, encryption of information through the *ciphering* mechanism is used. It is the common method that is employed across the communications systems, i.e. GSM, GPRS, UMTS, LTE, and 5G systems, over their respective air interface. The ciphering method is applied to the signaling as well as the user traffic information that is exchanged between the MS/UE and the network. The ciphering method works on the bit per bit addition of the user data flow and its corresponding key stream.

Table 9.4 shows the various ciphering algorithms and their keys used in the GSM, GPRS, UMTS, and LTE systems.

Ciphering is an optional security feature at the network end. For more information on the above ciphering algorithms and keys, refer to the 3GPP TS 33.102 [85], TS 33.401 [86], and TS 33.501 [87].

- Integrity Protection of Signaling Information (UMTS and LTE)

Signaling information exchanged between a mobile device and the core network may get altered because of unknown reasons. Integrity protection functionality of signaling information ensures that the message contents are not altered and their integrity is protected during transmission.

Table 9.5 summarizes the keys and algorithms used for integrity protection purposes over their respective logical interfaces of the UMTS, LTE, and 5G systems.

Table 9.6 summarizes the network elements and protocol layers that are responsible for performing the ciphering and integrity protection functions for the information exchanged between the MS/UE and network.

Table 9.4 Ciphering algorithms and keys used for information confidentiality.

System	Keys, Algorithms
GSM	Ciphering Key Kc, Key Sequence number, Algorithm: A5, A8
GPRS	Ciphering Key: GPRS-Kc, GPRS-Key Sequence number, Algorithms: GPRS-A5, A8
UMTS	UEA0 (UMTS Encryption Algorithm), UEA1 (UMTS Encryption Algorithm), UEA2 (UMTS Encryption Algorithm)
LTE	EEA0 Null ciphering algorithm 128-EEA1 SNOW 3G-based algorithm (128 bits) 128-EEA2 AES-based algorithm (128 bits) 128-EEA3 ZUC-based algorithm (128 bits)
5G	NEA0 Null ciphering algorithm 128-NEA1 SNOW 3G-based algorithm (128 bits) 128-NEA2 Advanced Encryption Standard (AES)-based algorithm (128 bits) 128-NEA3 ZUC-based algorithm (128 bits)

Table 9.5 Algorithms and keys used for integrity protection.

System	Keys and Algorithms
UMTS	UIA1 (UMTS Integrity Algorithm) UIA2 (UMTS Integrity Algorithm)
LTE	EIA0 Null Integrity Protection algorithm 128-EIA1 SNOW 3G-based algorithm (128 bits) 128-EIA2 AES-based algorithm (128 bits) 128-EIA3 ZUC-based algorithm (128 bits)
5G	NIA0 Null Integrity Protection algorithm 128-NIA1 SNOW 3G-based algorithm (128 bits) 128-NIA2 AES-based algorithm (128 bits) 128-NIA3 ZUC-based algorithm (128 bits)

Table 9.6 Protocol layers: ciphering and integrity protection.

Security Functions	GSM	GPRS	UMTS	LTE	5G
Ciphering	BTS	SGSN	Radio link control (RLC), Medium Access Control (MAC)	Packet Data Convergence Protocol (PDCP)	PDCP
Integrity Protection	—	—	RRC	PDCP, NAS	PDCP, NAS

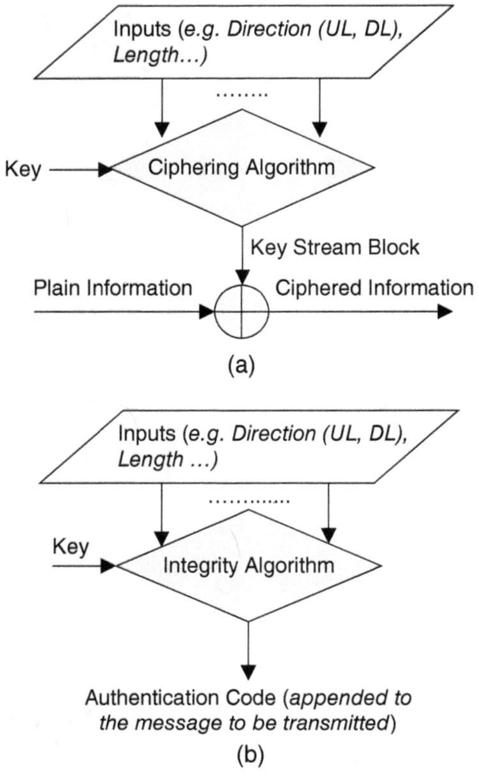

Figure 9.3 Illustration of (a) ciphering and (b) integrity protection algorithms.

Figure 9.3 illustrates the concept of working of the ciphering and integrity protection of information transmitted over the air interface.

9.3 GSM Security Procedures

The activation of a GSM security procedure, i.e. confidentiality/ciphering, for a typical mobile-originated call (MOC) is illustrated in Figure 9.4. As illustrated in this figure, authentication of a user in a GSM network is performed after receiving the initial Layer 3 message *CM Service Request* message from the MS. An MS identity is authenticated using the challenge-response method. In this method, the core network sends a 128-bits random number (RAND), as mentioned in Table 9.3, to the MS. The MS calculates a 32-bit *signed response (*SRES*)* using the algorithm A3 and also using some secret information that is stored in the SIM of an MS. The MS transmits the SRES to the core network which verifies and authenticates the MS successfully.

Once a user has been authenticated, the encryption and ciphering of signaling and voice information is started by the MSC by sending the Base Station Subsystem Mobile Application Part (BSSMAP) layer *CIPHER MODE COMMAND*, refer to TS 48.008 [134], to the BSC. The *CIPHER MODE COMMAND* message contains the list of ciphering algorithms A5/1 to A5/7. The BSC selects a particular ciphering algorithm and sends it through the RR layer *CIPHERING MODE COMMAND* to the MS; refer to TS 44.018 [130]. The enabling of the ciphering mode control process is indicated with a RR *CIPHERING MODE COMPLETE* message from the MS to the BSC, followed by a BSSMAP layer *CIPHER MODE COMPLETE* message reply from the BSC to the MSC. This also indicates the completion of RR connection establishments and the start of the Layer 3 signaling messages between the MS and the MSC through the BSC.

Figure 9.4 Illustration: activation of GSM security procedure for a MOC.

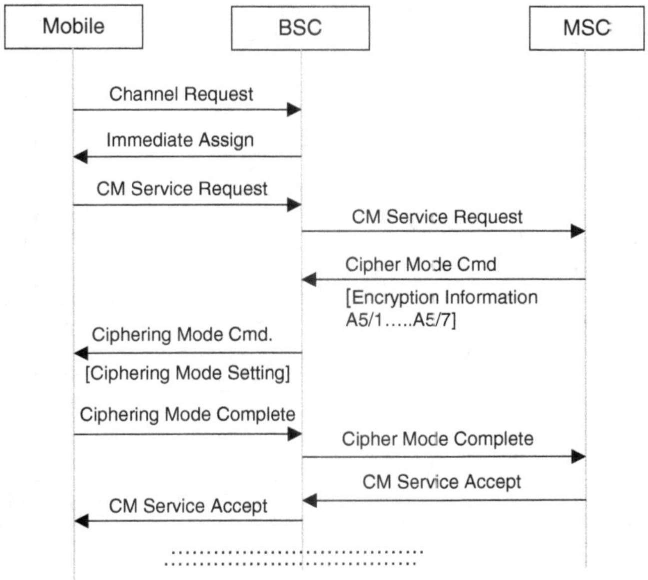

9.4 UMTS, LTE, and 5G: AS and NAS Layer Security Procedures

Unlike the GSM system, integrity protection over the air interface is applied to protect the signaling messages of the Access Stratum (AS) and NAS layers of the UMTS, LTE, and 5G systems. Information confidentiality/encryption through ciphering and also its integrity protection in the case of UMTS, LTE, and 5G systems are performed and divided into the following categories:

- Security for the AS, between
  - UE and UMTS terrestrial radio access network (UTRAN) (UMTS system),
  - UE and E-UTRAN (LTE system), and
  - UE and NG-RAN (5G system).

The protocol layers that are responsible to perform the integrity protection and ciphering functions over the respective air interface of the UMTS, LTE, and 5G NR systems were listed in Table 9.6.

- Security for the NAS Layer, between
  - UE and the Mobility Management Entity (MME) (LTE system only) and
  - UE and the Access and Mobility Management Function (AMF) (5G system only).

Figure 9.5 illustrates the integrity protection and ciphering procedure (dotted line) applied over the AS (RRC) and NAS signaling messages in the case of the LTE system. Similar security measures are also used in the case of the 5G system.

As mentioned in the previous sections, various algorithms are used by the UE, RAN, and core network elements to perform integrity protection and ciphering of information during its transmission.

- Activation of Security Features and Mechanisms

In UMTS, LTE, and 5G systems, the security features and its mechanism, i.e. integrity protection and ciphering, and their algorithms are activated and taken into account through the *SECURITY MODE COMMAND* and *COMPLETE* messages that are exchanged between the

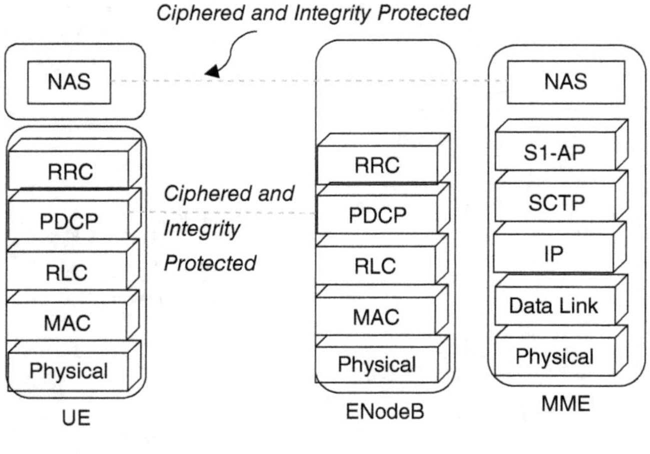

Figure 9.5 Illustration: LTE ciphering and integrity protection of AS and NAS layer signaling.

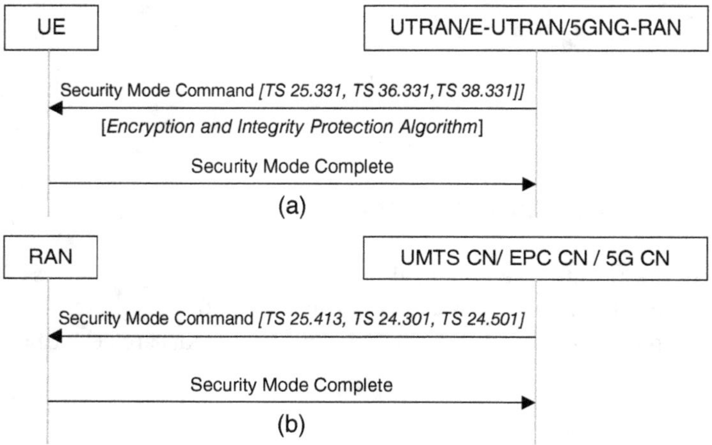

Figure 9.6 Illustration: security mode command to activate ciphering and integrity protection: (a) Access Stratum and (b) Non-access Stratum.

- UE and UTRAN (UMTS);
- UE and E-UTRAN (LTE); UE and NG-RAN (5G); and
- UE and MME, for LTE/EPS only; UE and 5G Core AMF, for 5G system only.

Figure 9.6 illustrates the activation of the security features through the *SECURITY MODE COMMAND* and *SECURITY MODE COMPLETE* messages for the security protection of AS and NAS layer signaling messages over the respective interfaces.

The UE provides its security capabilities and sends it in the initial network registration to the core network. A UE may also provide the previous generation mobile communications system's security capabilities along with its current security capabilities. Note that though the name of security command mode messages is the same between UE and UTRAN, UE and E-UTRAN, UE and the MME, as illustrated in Figure 9.6, their formats and contents are not the same. The contents of the *SECURITY MODE COMMAND and COMPLETE* messages used in the UMTS, LTE/EPS, and 5G systems can be further examined from the corresponding 3GPP TS mentioned in Figure 9.6.

The requirements of the integrity protection of NAS signaling messages are mandatory, whereas the confidentiality/ciphering of user data, Access Stratum, and NAS information through ciphering is an optional one at the network end. Also, in the UMTS, LTE, and 5G systems, there are several NAS and AS RRC layer messages where integrity protection is not applied. Refer to TS 24.301 [46], TS 25.331 [50], TS 36.331 [94], and TS 38.331 [116] for such messages where integrity protection is not applied.

Example 9.1 and Figure 9.7 describe and illustrate the AS and NAS layers security activation in the case of the LTE/EPS network.

Example 9.1 LTE/EPS Security Features Activation for AS and NAS Layers

Figure 9.7 illustrates the activation of the security features for the LTE/EPS AS and NAS layers. A UE sends its UE initial registration in an LTE/EPS network through the NAS layer ATTACH REQUEST procedure. The UE sends the supported encryption/ciphering and integrity algorithms to the core network element MME in the ATTACH REQUEST message. The security capability information from a UE is transferred under the following IEs of the ATTACH request message:

- UE network capability, which is a mandatory IE, containing the LTE/EPS and UMTS security features and its mechanisms.
- MS network capability, which is an optional IE for a UE that supports GPRS/UMTS security-related algorithms.

To activate the security features through the integrity protection and ciphering method for the NAS layer between the UE and the LTE/Evolved Packet Core (EPC), the MME resends the same security capabilities information to the UE using the message *SECURITY MODE COMMAND* under the UE security capability mandatory IE; see Figure 9.7. Once the NAS layer security features have been activated, the LTE/EPC establishes the UE initial context with E-UTRAN. To do this, the MME sends the initial context setup request message to the E-UTRAN containing the ATTACH accept message, bearer details, UE security capabilities, and security key; see Figure 9.7. A similar approach is used in the case of the 5G system also where the 5G core AMF establishes the UE initial context with NG-RAN by sending an initial context setup request message containing the registration accept message for a UE.

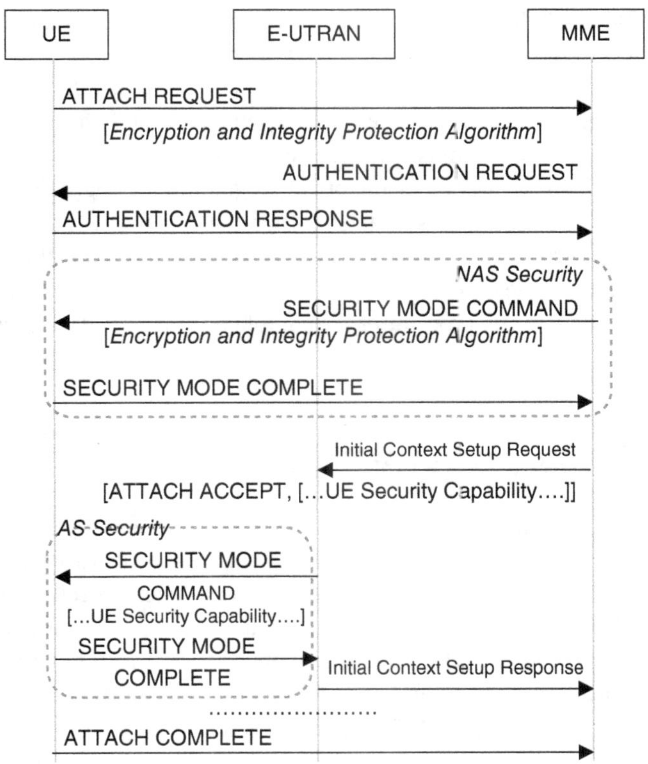

Figure 9.7 Illustration: LTE/EPS initial UE registration and its security features indication.

The UE security capabilities information contains the LTE/EPS integrity and encryption algorithms to be applied over the air interface between a UE and the E-UTRAN. Using the UE security capabilities and security key information, the E-UTRAN activates the initial security features for the AS layer between UE and the E-UTRAN by sending the SECURITY MODE COMMAND message to the UE. A SECURITY MODE COMPLETE message from the UE to the E-UTRAN ends the successful activation of the security features. For more details on the functions performed by a UE and E-UTRAN upon receipt of the initial security activation request from the MME, refer to TS 36.331 [94] and TS 24.301 [46].

Note that as shown in the above Figure 9.7, the same message name *SECURITY MODE COMMAND/COMPLETE* is used over the LTE/EPS S1 and air interface to activate the security features and their mechanism between a UE and the LTE/EPC.

9.5 Security Contexts

A security context of a UE/MS contains the important keys of the security algorithms and is created and stored at the MS as well as network. A security context is created as a result of a successful authentication procedure performed and activated by the core network during the MS/UE initial registration procedure. Refer to Figure 9.7 for a typical illustration of an ATTACH request procedure of LTE/EPS. A security context may be also created at the MS and network end as a result of an intersystem change from GSM to UMTS and vice versa; from LTE to UMTS or GSM and vice versa.

Table 9.7 shows the various security-related information that is stored as part of a security context of an MS/UE in each of the mobile communications system.

In the case of the LTE/EPS system, security context information consists of the AS and NAS layer security contexts as the security features are activated separately for them. The purpose of establishing a security context in the LTE/EPS system is to avoid the reauthentication, next time the UE reattaches, requirements at the core network end prior to processing of NAS signaling messages. For example, after a successful initial network registration, if an LTE UE was switched off and on again, the same security context can be reused while attaching to the core network. In this case, all the NAS messages from the UE shall be integrity protected. But if an LTE UE attaches to the core network for the first time, all the NAS messages from the UE will not be integrity protected until the security procedures are completed and activated successfully between UE and the core network, as illustrated in Figure 9.7.

9.6 Security Interworking

In the preceding sections, we have discussed the security features and their implementation mechanisms through various security algorithms and keys that are specific to a particular cellular system, i.e. GSM, GPRS, UMTS, LTE, and 5G. We have also discussed the cellular system-specific security context that is created as a result of the completion of a successful security procedure between a mobile device and the network.

Security interworking requirements arise as a result of the interworking and interoperations of mobile communications systems and networks, which was described earlier in Chapter 6. Two typical cases may be considered for the discussion of the security interworking requirements:

- The idle state of a mobile device and
- Inter-RAT handover.

Table 9.7 Security context contents in GSM, GPRS, UMTS, LTE and 5G systems.

Contents of GSM/GPRS Security Context	
Circuit-Switched Domain	**Packet-Switched Domain**
GSM Ciphering Key GSM Ciphering Key Sequence Number	GPRS GSM Ciphering Key GPRS Ciphering Key Sequence Number
Contents of UMTS Security Context	
Circuit-Switched Domain	**Packet-Switched Domain**
UMTS and GSM Ciphering Key UMTS Integrity Key GSM Ciphering Key Sequence Number	GPRS UMTS Ciphering Key GPRS UMTS Integrity Key GPRS GSM Ciphering Key GPRS Ciphering Key Sequence Number
Contents of LTE/EPS Security Context	
K_{ASME} with the associated key set identifier, the UE security capabilities, and the uplink and downlink NAS COUNT (NAS Sequence number) values.	
Contents of 5G Security Context	
K_{AMF} with the associated key set identifier, the UE security capabilities, and the uplink and downlink NAS COUNT (NAS Sequence number) values.	

During the idle state, a mobile device uses the security features of its home network which are operated with a particular RAT, i.e. GSM, UMTS, LTE, and 5G. However, as the user and the mobile device moves, either in idle or connected/dedicated state and enters a coverage area of a different network and RAT with good received signal strength, it has to use the security features of the target RAT. In the connected or dedicated state, the ongoing call is transferred through the inter-RAT HO procedure. In this case, the interworking of the security features and mechanism is required and is important to provide seamless communication services to users.

- Typical security interworking requirements can be described under the following scenarios:

Scenario #1 Network deployments with all the RATs, i.e. GSM/GPRS, UMTS, and LTE systems

Security interworking is required due to the following mobility states of a mobile device or user in a network that deploys all the RATs:

- Idle mode or handover from an E-UTRAN to UTRAN and vice-versa.
- Idle mode or handover from E-UTRAN to GPRS Edge Radio Access Network (GERAN) and vice versa.

Scenario #2 Network deployments with LTE/EPS and 5G systems

Security interworking is required due to the following UE modes of operation as well as the mobility states of a mobile device or user in a network that deploy the LTE/EPS and 5G systems.

- Single registration mode and dual registration mode, which shall be described later in Chapter 16.
- Idle mode or connection mode, i.e. handover, mobility between the LTE/EPS and 5G systems and vice versa.

In all of the above security interworking scenarios, the network element of the source system and RAT passes the relevant security context information such as the confidentiality key, integrity key to the network element of the target system, and RAT. Using the provided information, the target system and RAT map and derive their security keys to enforce their security features between the mobile device and the network. In other words, the source security context information is mapped and used to derive the target system security context for a successful security interworking during intersystem mobility.

For example, in a UTRAN/GERAN to LTE E-UTRAN system handover, the SGSN will derive and transfer to the MME a confidentially key and an integrity key. Based on this information and other input parameters, the MME and UE can derive K_{ASME}, a 256-bit length key. If a security interworking feature fails, the mobility of a mobile device and its use may also fail from one RAT to another RAT. For further details on K_{ASME}, refer to TS 33.401 [86], appendix A.2.

Similarly, in an LTE E-UTRAN to UTRAN/GERAN system handover, the MME derives the confidentially key and an integrity key from the key K_{ASME} and passes it to the SGSN which in turn derives its keys to be used in the target RAN.

Chapter Summary

Security is one of the most important features of the mobile communications systems and networks that are available today. Security tasks are performed by the mobility management layer of the respective communications systems, i.e. GSM, GPRS, UMTS, LTE, and 5G.

This chapter presented the various security features and their corresponding methods/algorithms that are used to provide secured communications services to subscribers by the respective mobile communications systems, i.e. GSM, GPRS, UMTS, LTE, and 5G. The security features of the 5G system shall be further described in Chapter 16.

The UMTS security architecture evolved from the GSM security system. Similarly, LTE security architecture evolved from the UMTS security system. Only user authentication and data encryption are performed in the GSM system, whereas integrity protection of signaling messages and network authentication by the mobile are also performed as part of security procedures in the UMTS and LTE systems.

Though the security features are known by the common name, they are implemented and realized using different algorithms and keys that vary from GSM to the 5G network. For more details on these security algorithms and keys, the reader is recommended to refer to the technical specifications mentioned in the reference section of this book.

This chapter also presented the network elements and protocol layers that are responsible and perform a particular security procedure. Protections of the signaling messages for the AS and NAS layers of the UMTS, LTE, and 5G systems were discussed. We closed this chapter with another important aspect, that is, the requirements of the security context and security interworking for a multi-RAT capable mobile device to support the intersystems mobility among the different communications systems and networks.

Part II

Operations and Maintenances

In this part of the book, the operations and maintenances (O&Ms) of a mobile communications network, based on the GSM to the 5G system, are discussed. O&M system makes it possible to run a communications network smoothly on a day-to-day basis. So far, the reader has learned that a mobile communications network consists of interworking network elements and systems from different vendors that work on open as well as vendor-specific technical standards. A mobile communications network consists of various physical and logical interfaces interconnecting different network elements from the same or different OEMs. One can expect the day to day field issues from a complex mobile communications network. Sometimes, the field issue could be a challenging as well as an exciting one that is hard to reproduce in a laboratory.

Day-to-day network operations issues may result from the various sources/factors which could be due to an operator's network element, other vendor's network element, inter-working, configuration, network planning issues, air interface issues, application protocols or platform software issues, and so on. For continuity of communications services, each one of these issues must be addressed to isolate and fix the actual root causes. For these purposes, a mobile communications system's network element must have the built-in capability to aid in troubleshooting day-to-day network related issues. Apart from this, a communications system must have the facility to evaluate its performance on a daily basis or periodically as required by the operator.

This part of the book contains three chapters, and their purposes are as follows.

Chapter 10 describes the *Alarm* mechanism used to notify the occurrence of an abnormal event to the O&M operator.

Chapter 11 describes the *Key Performance Indicator* mechanism through which an operator can evaluate the performance of a communications network.

Chapter 12 describes the different types of field issues that may appear in a mobile communications system and network. It also describes the available network troubleshooting tools that can be used to collect the required information from the field.

The above chapters form the network maintenance part of the system engineering aspects of a mobile communications system as described in Section 2.4.5.

Overall, the information collected from the field, the alarms, and the KPI data can be used to troubleshoot the field issue by the O&M operator or by the developer of a particular network element.

10

Alarms and Faults Managements

Introduction

In this chapter, we will discuss the *alarm* mechanism through which the runtime occurrences of various important events are reported by a mobile communications network element to the operation and maintenance (O&M) operator. Alarms are important for the day-to-day O&M as well as faults managements of a mobile communications network. An alarm draws the attention of the O&M operator, and depending on the criticality of the alarm generated, corrective actions are taken by the operator in a timely manner, to avoid probable service outages, as part of fault management steps. A 3rd Generation Partnership Project (3GPP) technical specification (TS) describes the possible runtime abnormal events for the concerned protocol layer for the various procedures performed by it. The abnormal events are converted into alarms, which is implementation specific, for the concerned protocol layer. A network element may also define its Original Equipment Manufacturer (OEM)-specific events and alarms.

This chapter begins with the definition of an alarm and its classifications. It is followed by the identifications of typical abnormal events as described in a 3GPP TS for its corresponding protocol layer and its procedures and shows how it can be converted into a possible alarm. This chapter ends with an example of definitions of typical alarms and their descriptions using which the O&M operator takes appropriate action as part of the fault management steps.

10.1 Network Elements Alarm and Its Classifications

Generally, an alarm in any system represents an important, abnormal and unexpected event being detected by the entity that had raised the alarm. Similarly, there are provisions of rising of alarms in mobile communications systems also to notify the O&M operator of the occurrences of abnormal events. Upon detection of a particular alarm, the O&M person applies the corrective/suggestive actions for each alarm. Depending on the extent of services being affected, an alarm can be classified as follows:

- Critical Alarm

A critical alarm should be addressed as soon as possible and may require a quick workaround or solution to the event or problem being identified; otherwise, it would affect the ongoing services as well as further block the services provided by the mobile communications network.

- Major Alarm

A major alarm may represent an event with a significant impact but does not disturb the ongoing operations of the network element.

- Minor Alarm

Mobile Communications Systems Development: A Practical Introduction to System Understanding, Implementation, and Deployment, First Edition. Rajib Taid.

A minor alarm does not represent an event with a significant impact.

Alarms may be raised for notification and information purposes also.

10.2 Sources of Abnormal Events and Alarms

In mobile communications networks, various abnormal events may occur during their runtime, leading to the generation of alarms of different natures as mentioned in the preceding section. Typical sources of abnormal events that are responsible for the generation of alarms are as follows:

- Software fault, for example, a process crash.
- Hardware fault, for example, base transceiver station (BTS) or trans-receiver (TRX) malfunction, leading to the unavailability of radio channels.
- A communication link breakdown between the sender and receiver, or sometimes, the receiver sends back erroneous data to the sending network element.
- Unsuccessful completion of a protocol layer signaling procedure between two peer network elements.
- Excessive retransmissions due to air interface or Transmission Control Protocol/Internet Protocol (TCP/IP) transport network issues.
- Expiry of a timer at the sender end because of a lack of response from the peer.

Generally, a hardware or software fault alarm is vendor/OEM specific and may require a proprietary solution. A 3GPP TS does not define and describe such vendor-specific hardware or software alarms and their requirements. But a 3GPP TS describes the abnormal events of a protocol layer and their associated procedures. A 3GPP TS also describes the provisions for notifications and corrective action for an alarm to the O&M operator as described below.

10.3 Identifying Sources of Alarms from 3GPP TSs

As mentioned in the previous sections, an abnormal event or an error that is detected by a network element of a mobile communications network is informed to the O&M operator in the form of an alarm. A network element supports multiple logical interfaces and protocol layers where several abnormal events may be generated at runtime. Each logical interface and its protocol layer have different sources of abnormal events for different procedures executed from time to time as described in the concerned 3GPP TS. A system or a protocol layer designer may translate each abnormal or normal event, either reported by the peer network element or generated within the same network element due to local errors, into an alarm definition.

In general, an alarm may be raised as a result of the following types of abnormal events that are generated due to the various functions and procedures of a protocol layer described in a 3GPP TS.

- Abnormal Conditions, which may be related to the signaling functions and procedures performed by a protocol layer of a particular network element with the peer. Timer-dependent protocol procedures generally result in an abnormal scenario.
- Protocol Layer Error, between two network elements over a particular logical interface. Generally, air interface-related causes result in protocol errors.
- Abnormal Conditions due to Local Errors.

10.3.1 Abnormal Conditions

Runtime abnormal conditions mostly arise from time-dependent events. In a mobile communications network, an abnormal condition may occur if a particular signaling function or procedure, e.g. a reset operation, of a protocol layer between two network elements is not completed successfully within the allowed maximum time duration and retry attempts. Failure of a signaling procedure may further impact the subsequent signaling procedures between the concerned network elements. An abnormal condition may arise due to the nonreceipt, by the sender, of a timely response from the peer (receiver) network element/protocol layer for a timer-dependent signaling procedure. The reasons could be again due to several factors such as the following ones:

- Buggy or malfunctioning software module for a particular protocol layer.
- Malfunctioning hardware devices.
- A 3GPP specification noncompliant/nonconforming network element, e.g. Mobile station (MS)/User Equipment (UE).
- Congestion in the intermediate connecting network.

Note that not all of the functions or procedures performed by a protocol layer are time dependent. Some signaling procedures may be performed without waiting for their responses from the peer.

A section called "Abnormal Conditions" is generally available in a 3GPP TS of the concerned protocol layer. Each signaling function and procedure performed by a particular protocol layer has its corresponding Abnormal Conditions section for which an alarm may be raised containing the reason for raising the particular alarm. Using the particular reason or cause, the operator may identify the fault location and take corrective actions to prevent disruptions, either ongoing or new, in services. A list of possible reasons or error causes and their meaning can be found in the concerned TS.

For example, consider the General Packet Radio Service (GPRS) Gb interface between the Global System for Mobile Communication (GSM) base station controller (BSC) and Serving GPRS Support Node (SGSN). The Gb interface protocol stack contains the Base Station System GPRS Protocol (BSSGP), refer to TS 48.016 [135], NS protocol layer, refer to TS 48.018 [136]. The concerned 3GPP TS of these layers mentions all the possible abnormal conditions that may occur between the BSC and SGSN due to the unsuccessful completion of a particular signaling procedure over the Gb interface. If such an abnormal event occurs, a necessary alarm will be raised by either of the entities (sender or receiver) for the operator's immediate attention, which is followed by a recovery action for the concerned event. Alarm generated due to abnormal events is illustrated later in Section 10.4.

10.3.2 Protocol Layer Error Handling

Protocol layer information is exchanged between two peer protocol layers in terms of various *information elements* (IEs). Sometimes, an IE may carry an erroneous and unexpected value as detected by the receiver protocol layer, due to various reasons. One of the most troublesome areas of a mobile communications network is the air interface between the MS/UE and the Radio Access Network (RAN). Because of the poor radio conditions, the information transmitted by the MS/UE may be distorted, resulting in an erroneous data at the receiving end. Other sources of errors that make a receiving network element detect erroneous information are as follows:

- Poor conditions of a physical interface between a sender and receiver.
- Buggy or malfunctioning software that is handling a particular protocol layer.
- Malfunctioning hardware devices and so on.

To handle such a protocol message's error, a 3GPP TS may also contain, in general, a section called "General Protocol Error Handling". The protocol error handling section of a 3GPP TS describes the various types of events such as the erroneous and nonerroneous along with corrective actions to be taken upon occurrences of a particular event.

For example, consider the mobile radio air interface Layer 3 signaling protocols messages, CS and PS domains, exchanged between an MS/UE and the core network through the RAN. Air interface Layer 3 protocol message consists of a header, message type, and various IEs. An incorrect value for anyone of this information may be detected at the receiver end, leading to an erroneous event. For example, if a particular air interface Layer 3 message contains a *Mandatory* (M) or *Conditional* (C) information element but the same is not present in the message being transmitted, then the receiver would flag it as an error condition. The receiver will report it to the sender with an appropriate cause, say "Missing Mandatory IE", "Missing Essential IE", or "Missing Conditional IE". A protocol error event may be also generated as a result of the receiving of a message that is not expected in the current state of the receiving protocol layer or the message is unexpected and unknown by the receiving protocol layer. All such protocol events and their error codes of various signaling procedures of a protocol layer are described by the concerned 3GPP TS.

The concerned protocol layer of a core network element has a mechanism to report, to the RAN, such an erroneous data received from the peer protocol layer of an MS/UE over the air interface. Some protocol layer provides a special protocol data unit (PDU) called "STATUS" that is used by the receiving network element, say the GSM MSC, to report the occurrence of an erroneous value, to the sender network element, say the BSC, and vice versa. The *STATUS* PDU contains the complete PDU, as received by the receiver, containing the erroneous value(s) so that the operator or developer can determine the IEs and their erroneous values. On receiving a *STATUS* PDU, an alarm is generated by the original sender network element, e.g. BSC, to its O&M operator. Alarm generated due to a protocol error event is illustrated later in Section 10.5.

10.3.3 Abnormal Conditions Due to Local Errors

A network element must also have the mechanism to alert on the occurrence of nonprotocol-related or local erroneous events so that the O&M operator can take appropriate action. Abnormal scenarios or events due to local errors are OEM and implementation specific and are not defined by the 3GPP TS. Upon detection of a local erroneous event, the concerned network element's alarm management system may raise a customized alarm to notify the operator. Examples of localized network element-specific errors due to which alarms may be raised are as follows:

- Critical process crash,
- Resource unavailability for allocation to a UE/MS,
- Channel activation failure,
- Hardware synchronization issues, and
- Disk space below a certain threshold and so on.

In summary, in the preceding text (Sections 10.3.1–10.3.3), the identification of sources of various abnormal conditions from a 3GPP TS as well as network element-specific local errors has been described. The method of reporting and handling of such abnormal conditions or protocol errors varies from layers to layers. For the specific error handling mechanism employed by a protocol layer, the reader may refer to the protocol error handling section of the concerned 3GPP TS.

In the subsequent Section 10.4, the typical design and typical components of an alarm management system of a network element with its interfaces to other protocol layers and O&M are described. In Sections 10.6 and 10.5, the identifications of typical abnormal events and protocol errors from a 3GPP TS and their corresponding sample alarms descriptions and generations are also described.

10.4 Design and Implementation of an Alarm Management System

To notify about the occurrences of abnormal events or errors in a mobile communications network, a centralized alarm management system may be designed and implemented for a particular network element only, e.g. BSC, Mobility Management Entity (MME), SGSN, and AMF. The alarm management system is implementation specific even for a particular mobile communications system. The relevant abnormal events as described in the concerned 3GPP TS may be considered and implemented in the alarm management system of a particular network element. The centralized alarm management system of a network element is responsible for the following typical tasks:

- Interface for accepting notifications of abnormal events from the different protocol layers of the concerned network element;
- Interface for the O&M operator for provisioning and configuring of alarms for different protocol layers of a network element;
- Interface for notifying the O&M operator as well as the clearing of it about the occurrences of various events; and
- Archival of the alarms generated and cleared for the entire network element.

10.4.1 Design and Components of an Alarm

The alarm management system may define and describe an alarm in the form of an alarm manual for each of the identified abnormal events as well as the informational notification to be generated from time to time in a predefined format. An alarm may be designed to have the following typical components/attributes:

- Unique identification for an alarm
- Alarm name or title
- Date and time
- Alarm type, i.e. critical, major, or minor
- A short description of the alarm
- Alarm parameters from *1. . .n*
- Instructions/corrective actions to be taken by the O&M operator upon the occurrence of the concerned alarm
- Default action, i.e. action to be taken by the alarm management system if the operator does not take any action.

Every alarm is given a unique identifier in a particular communications system. The data included in the alarm being generated contains the critical information that helps in the identification of the source of the alarm, the reason for the alarm, and so on. The reason for an alarm being generated may be specified as mentioned in the concerned protocol layer and its 3GPP TS. The alarm data may also include the related identifiers of logical objects such as MS identity (e.g. Temporary Logical Link Identifier [TLLI], Globally Unique Temporary Identifier [GUTI]), BTS identity, Temporary Mobile Subscription Identifier (TMSI), a GPRS Tunneling Protocol (GTP) tunnel identification, eNodeB UE ID, MME UE ID, and AMF set ID. Note that some of the alarm data may be designed to be mandatory, while some are optional in nature. The date and time of the occurrence of an alarm are important as they help to correlate with the time of occurrence of the actual issue.

10.4.2 Alarm Application Programming Interfaces (APIs)

The centralized alarm management system of a network element may provide generic application programming interface (API) facilities for configuring alarms and raising and clearing of alarms by different protocol layers of the concerned network element. Such generic APIs facilitate the request for raising and clearing of alarms from the different protocol layers with various alarm data in predefined formats only.

10.4.3 Alarm Database

The alarm management system may provide facilities for archival of alarms into a database for its future references and retrieval. The alarm database may contain both active and historical alarms. A typical alarm management system that may be implemented in a network element with all the typical components and interfaces is illustrated in Figure 10.1.

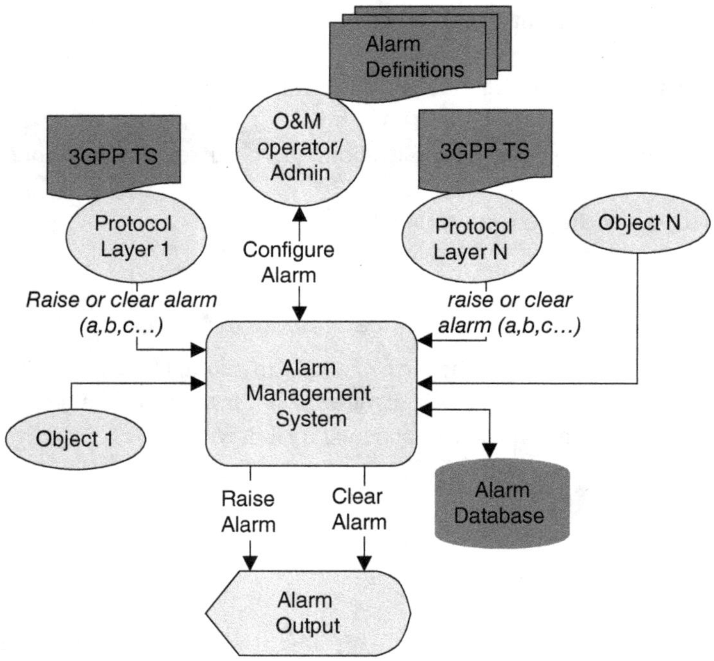

Figure 10.1 Illustration: a typical alarm management system.

As illustrated in Figure 10.1, the alarm management system may be implemented as a separate module with interfaces to various protocol layers and other modules of a network element. The O&M operator or administrator configures the alarms into the alarm management system. Logical objects, 1..N (1 to many), representing various physical hardware, for example, a BTS, BSC, eNodeB, and MME, are also configured by the alarm management system against whom an alarm is generated. Protocol layer-specific alarm, as described by the concerned 3GPP TS, is configured into the alarm management system. Upon detection of an abnormal event, a protocol error, or local error, the concerned protocol layer sends the request for raising or clearing of existing alarms through the APIs exposed, e.g. *raise_or_clear_alarm (. . ..)* by the alarm management system. The alarm is generated and sent to the display unit for the O&M operator console.

10.5 Alarm Due to Protocol Error

In this section, the handling and reporting of typical protocol errors over the air interface of GSM/GPRS and Long-Term Evolution (LTE)/Evolved Packet System (EPS) systems are illustrated and described below:

- GPRS System Alarm Due to Protocol Error

Figure 10.2 illustrates the protocol error scenarios and generation of the corresponding alarms in the case of the GPRS system. This figure illustrates the notification of protocol errors over the air interface using the *RR STATUS* message and the *STATUS* PDU over the GPRS Gb interface between the BSC and the SGSN.

As illustrated in Figure 10.2, the BSC sent an RR layer message to the MS over the air interface. If the MS finds an RR protocol layer-related error, it will be notified to the BSC through the *RR STATUS* message containing the *RR Cause*. Refer to TS 44.018 [130] for more details on the possible RR causes. Upon receiving the *RR STATUS* message, the BSC will notify the O&M operator about the occurrences of the protocol error through an alarm with mandatory as well as optional data.

Similarly, as illustrated in Figure 10.2, the BSC sent a BSSGP layer PDUs to the SGSN over the Gb interface. If the SGSN finds a protocol error for any kind of BSSGP layer PDUs, it will be notified to the BSC through the

Figure 10.2 Illustration: protocol error reporting using STATUS PDU.

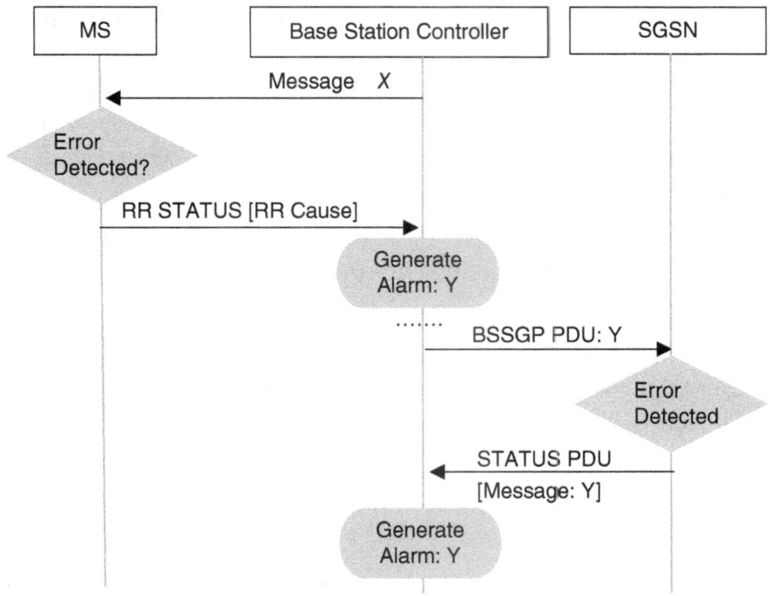

STATUS PDU containing the complete BSSGP layer erroneous PDU. Refer to TS 48.018 [136] for more details on the *STATUS* PDU of the Gb interface.

Upon receiving the *STATUS* PDU, the BSC will notify the O&M operator about the occurrences of the protocol error through an alarm with mandatory as well as optional data. The flow of the *RR STATUS* message or *STATUS* PDU is bidirectional, and it can be used by both the peer network elements for notification of a protocol error to each other.

- LTE/EPS and 5G NAS Layer Alarm Due to Protocol Error

Consider the occurrence and reporting of a protocol error that occurred between a UE and MME of an LTE/EPC network over its air interface. In an LTE/EPS network, the air interface NAS layer Evolved Packet System Mobility Management (*EMM*) *STATUS or* Evolved Packet System Session Management (*ESM*) *STATUS* message is used to notify the occurrence of a protocol error for the LTE/EPS mobility management and session management signaling procedure that is performed between a UE and the MME. The ESM STATUS or EMM STATUS message can be used by both the UE and the MME to report a protocol error to each other.

Similar protocol errors are reported by the 5G MM and SM layers also using the 5GMM STATUS and 5GSM STATUS messages, TS 24.501 [47]. A protocol error scenario between a UE and the MME of an LTE/EPC network is illustrated in Figure 10.3. In this illustration, the protocol error is notified by the UE using the *ESM STATUS* or *EMM STATUS* message.

Figure 10.3 Illustration: LTE/EPS protocol error reporting using ESM or EMM STATUS message.

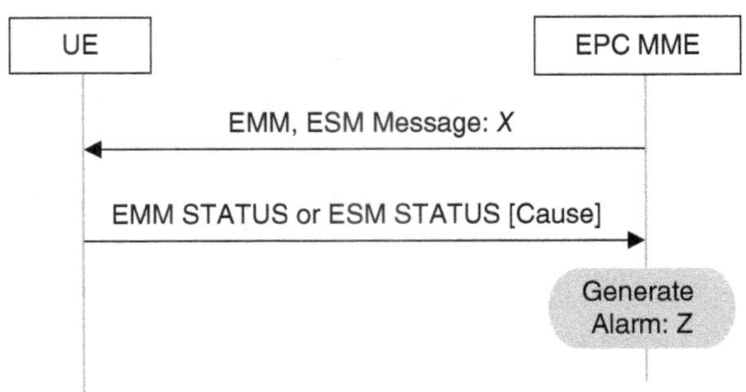

As illustrated in Figure 10.3, the *ESM/EMM STATUS* messages contain the EMM or ESM cause to report the reason for the protocol error detected by the UE. Examples of EMM causes are semantically incorrect messages and invalid mandatory information. Examples of ESM causes are *Invalid EPS bearer identity*, *Invalid PTI value,* and so on. On receiving the *ESM STATUS* or *EMM STATUS* message, the MME may generate an alarm to the O&M operator to notify the occurrence of the protocol error. For all the possible ESM and EMM protocol error causes, refer to TS 24.301 [46].

10.5.1 Sample Protocol Error Alarm Description

In the previous section, the typical protocol error reporting mechanisms for the GSM/GPRS, LTE/EPS, and 5G networks were described and illustrated. This section describes a protocol error alarm generated by the MS; BSC, UE, or MME; or 5G core AMF/SMF due to their corresponding *STATUS* message. Various alarms descriptions in detail may be found in an alarm manual of a network or network element. Such an alarm manual becomes helpful during the troubleshooting of an issue that is reported from the field.

- For GSM Air Interface Protocol Error Description, refer to Figure 10.2

```
Alarm ID: 12345
Alarm Name: GSM RR STATUS ALARM
Alarm Description: MS or BSC generates this alarm on detecting the GSM RR
layer error.
Alarm Type: Major
Parameters: MS-TEMP-ID RR_CAUSE
Instructions: .............................................................
Default Action: ...........................................................
```

- 5G System NAS Layer Protocol Error Description

```
Alarm ID: 12345
Alarm Name: 5GS 5GMM_5GSM STATUS ALARM
Alarm Description: UE or AMF/SMF generates this alarm on detecting an 5G sys-
tem NAS layer protocol error
Alarm Type: Major
Parameters: Protocol-Discriminator 5GMM_5GSM_CAUSE
Instructions: .............................................................
Default Action: ...........................................................
```

As illustrated above, a generic alarm may be designed and used to report either an LTE ESM or EMM layer or 5GSM or 5GMM NAS layer-related protocol error, though the message formats of *ESM STATUS*/*EMM STATUS* or 5G SM/5GMM *STATUS* are different. The protocol discriminator, e.g. ESM (0x2) or EMM(x7), may be included as the alarm data to identify the type of the alarm, i.e. *ESM/EMM STATUS or 5GMM/SM STATUS.*

10.6 Alarm Due to Abnormal Conditions

In this section, alarm generations due to abnormal conditions with respect to the *GPRS tunnelling protocol (GTP)* shall be described and illustrated. The GTP is used in a couple of logical interfaces in GPRS, Universal Mobile Telecommunications System (UMTS), LTE/EPS and 5G core network.

- **Alarm Generation due to GTP Path Management Failure**

Consider the GTP that is used to tunnel user data in an IP-based mobile communications system such as the GPRS, LTE/EPS, or 5G system. The GTP has the control as well as the user plane protocols for transferring control information and user data, respectively, between two network elements. One common function performed by GTP control and user plane protocol is the GTP path management between two GTP entities. For more details on the GTP, refer to TS 29.060 [67], TS 29.274 [70], and TS 29.281 [72]. A GTP path is a GTP tunnel endpoint which consists of an IP address and a User Datagram Protocol (UDP). A GTP endpoint is identified by a *tunnel identifier* (TEID).

Subsequent Sections 10.6.1 and 10.6.2 illustrate the usages of the GTPv1-U plane over the LTE/EPS S1-user plane interface, for tunneling of user data, between the E-UTRAN/eNodeB and the S-GW of LTE/EPC for normal and abnormal scenarios. There are protocol entities or software processes that implement the GTPv1-U in the eNodeB and S-GW. As mentioned above, one of the tasks performed by the GTPv1-U is the path management between the eNodeB and the S-GW. It is similar to the Keep-Alive procedure that is used in the case of the TCP layer of an IP network. The GTPv1-U path management procedure consists of the exchange of the *Echo Request* and *Response* messages between the eNodeB and S-GW. For the message format and its contents, refer to TS 29.274 [70] and TS 29.281 [72]. Note that the GTP path management procedure is performed separately and independently by each GTP entity of the network element.

10.6.1 Normal Scenario

Figure 10.4 illustrates the normal scenario where an *Echo Request* and *Response* messages are exchanged, indicating a GTP user plane tunnel is UP and running, between the LTE eNodeB and the S-GW.

As shown in Figure 10.4, the GTPv1-U entity/process at the eNodeB end sent the *Echo Request* message to the peer entity at the S-GW end and incremented the echo request counter by one. The GTPv1-U entity at the S-GW replied an *Echo Response* message to the peer at the eNodeB end, which completes the successful testing of the GTP tunnel/path between them.

10.6.2 Abnormal Scenario

Figure 10.5 illustrates an abnormal scenario where there is no *Echo Response* message, for an Echo Request sent, from the GTPv1-user plane entity at the SGW end to the LTE eNodeB.

The process is repeated at the expiry of the GTP echo request start timer with the increment of the echo request counter. As the echo request counter has exceeded the *threshold* configured by the O&M operator, the process is stopped; the GTPv1-U entity at the eNodeB end marked the tunnel or path as unavailable and raised the

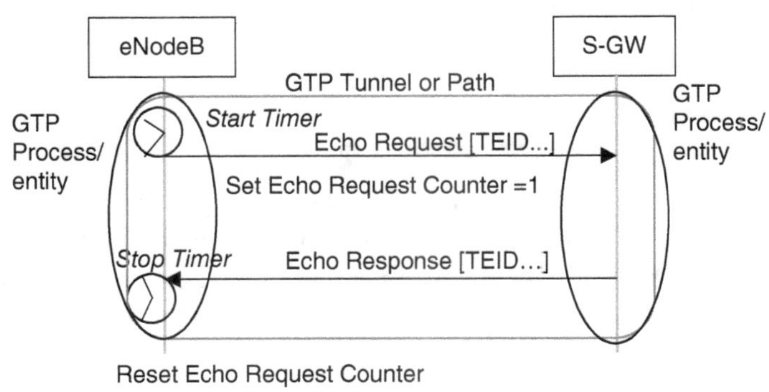

Figure 10.4 Illustration: normal GTP user plane echo request and response.

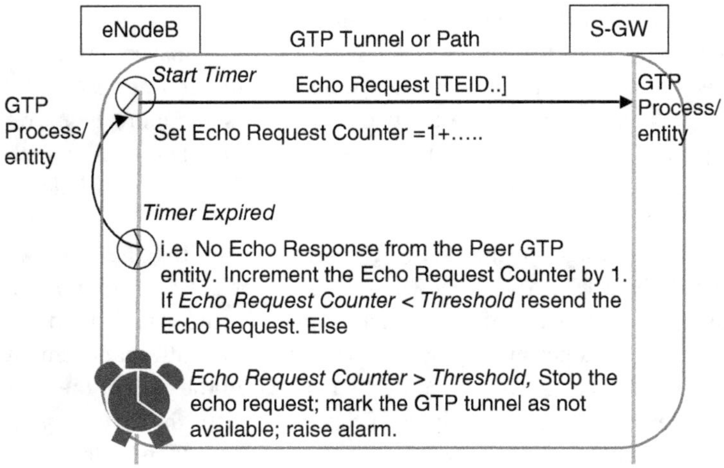

Figure 10.5 Illustration: An abnormal scenario in LTE/EPS: GTP alarm generation.

corresponding alarm as illustrated in this Figure 10.5. As a result of the GTP tunnel or path failure along with the generation of the corresponding alarm, the associated bearer context may be deleted after a certain interval. The duration of the timer and echo request counter shown in the above illustrations are maintained by the sending GTP entity. They are configurable by the O&M operator.

A similar alarm may be also generated due to the IP endpoint (IP address and UDP Port) test failure in case of the GPRS Gb interface, on top of the IP transport network, between a BSC and the SGSN.

10.6.3 Sample Alarm Description

The typical components of an alarm were described earlier in Section 10.4.1. Generation of an alarm because of the GTP user plane path failure between the eNodeB and the S-GW was illustrated in Figure 10.5. The generic description of the alarm illustrated in Figure 10.5 may be provided as follows:

```
Alarm ID: 12345
Alarm Name: GTP U-PLANE PATH TEST FAILED
Alarm Description: Generate this alarm on detecting the GTP user plane path
test failure between eNodeB and S-GW.
Alarm Type: Critical
Parameters: Tunnel-ID IP_Address UDP_Port 123 123
Instructions:
      -         Check the IP connectivity between eNodeB and S-GW
      -         Check that the GTP entities are UP and running
Default Action:
Clear the Bearer Contexts after a configurable delay.
```

As Illustrated above, the GTP user plane path test failed alarm may be designed to contain the *Tunnel identifier, IP address, and UDP Port* of the remote GTP entity as the mandatory data. These alarm data are part of the *Echo Request and Response* message that is exchanged between the eNodeB and S-GW or the concerned GTP entities. The alarm may also contain the optional data, for example, shown as 123 123, in the above illustration.

10.6.4 Sample Alarm Generation

The GTP user plane path test failed alarm described in Section 10.6.2 that may be generated on the O&M operator console may appear as illustrated below:

```
Alarm:12345 GTP U-PLANE PATH TEST FAILED DD-MM-YYYYHH:MM:SS
Parameters: 6789 123.123.123.123 123
```

As illustrated above, the Alarm ID is 12345; Tunnel ID is 6789, IP Address: 123.123.123.123; UDP Port: 123. The optional data of the alarm are not shown in the above illustration. Based on the information available in the generated alarm, its root cause analysis is carried out to isolate the source, i.e. sender, receiver, or IP transport network, of the issue. Further debugging may be carried out by a developer to rule out any software-related issue behind the generated alarm.

10.6.5 Sample Protocol Error Alarm Generation

- Alarm generated due to the GSM air interface RR layer STATUS message

Refer to Figure 10.2 for a protocol error alarm generated due to the GSM RR *STATUS* message from the MS to the BSC. A protocol error alarm generated due to the GSM RR *STATUS* message may appear on the O&M operator console as follows:

```
Alarm: 12345 RR STATUS ALARM DD-MM-YYYY HH:MM:SS
Parameters: 0x12345678 0x60
```

In the alarm illustrated above, the MS identity could be the TMSI; the RR cause of the alarm is "Invalid mandatory information" that is included in the RR *STATUS* message. Refer to the 3GPP TS 44.018 [130] for all the possible RR causes related to the RR layer protocol error.

- Alarm generated due to the LTE/EPS NAS layer ESM/EMM STATUS message

A protocol error alarm may be generated either by the UE or MME because of an ESM or EMM layer STATUS message sent/received by them over the air interface as illustrated earlier in Figure 10.3. Refer to the 3GPP TS 24.301 [46] for the definition of the LTE/EPS NAS layer EMM STATUS or ESM STATUS message. A sample LTE protocol error alarm as a result of the ESM STATUS message is illustrated below:

```
Alarm: 12345 ESM/EMM PROTOCOL ERROR DD-MM-YYYY HH:MM:SS
Parameters: 0x2b
```

In this alarm, the protocol error cause is illustrated to be the *Invalid EPS bearer identity* (0x2b).

- Alarm generated due to the 5G NAS layer 5GSM/5GMM STATUS message

Similarly, the generation of a typical protocol error alarm by the 5GMM or 5GSM layer is illustrated below due to the same error or 5G MM/SM status code: *0x2b (Invalid mandatory information)*. Refer to TS 24.501 [47] for all the probable 5GMM/5GSM cause codes. The first parameter in the alarm indicates the extended protocol discriminator, which is 0x7E for the 5GMM layer and 0x2E for the 5GSM layer.

```
Alarm: 12345 5GSM/5GMM PROTOCOL ERROR DD-MM-YYYY HH:MM:SS
Parameters: 0x7E 0x2b

Alarm: 12345 5GSM/5GMM PROTOCOL ERROR DD-MM-YYYY HH:MM:SS
Parameters: 0x2E 0x2b
```

10.7 How to Troubleshoot Protocol Error Using the Alarm Data

Supplementary information included as part of an alarm data becomes very useful while debugging the root cause of the failure of a particular abnormal event, either due to a protocol error or expiry of a signaling procedure timer. Let us consider the debugging of the protocol error issues in a 5G and LTE/EPS network from alarm data.

- LTE/EPS or 5G Network Protocol Error Issue

Consider the protocol error alarm generated due to the 5G NAS layer 5GMM/5GSM STATUS message described in Section 10.6.5. The second alarm data 0x2b is the cause of the protocol error which is the *Invalid mandatory information*; refer to Table: 9.11.3.2.1/Table: 9.11.4.2.1 in TS 24.501 [47].

Thus, the O&M operator or the RAN developer would conclude that the source of the protocol error alarm issue was the UE itself indeed because the 5G Next-Generation Radio Access Network (NG-RAN) transparently forwards UE NAS signaling messages to the 5G core AMF/SMF without processing it. Further debugging of protocol error issue may be continued by the handset research and development, followed by the RF team, if required, for any possibility of corruption of data over the LTE or 5G NR air interface.

Chapter Summary

This chapter presented the alarm mechanism which is used to report the occurrences of a runtime abnormal or informative event to the O&M person of a mobile communications network. An alarm management system was covered which is common among the network elements of mobile communications systems and is one of the important components and functions performed by a network element. An alarm management system is an integral part of a network element that provides interfaces for generation and notification on the occurrences of abnormal events to the O&M operator of a mobile communications network. Alarms generated by an alarm management system of a network element help in identifying and rectifying issues promptly on the probable causes of abnormal events, protocol errors, or other local errors.

In this chapter, we covered the various sources of abnormal events, in general. Identification of abnormal events and protocol errors from a 3GPP TS that may be used to design and implement an alarm was described. The typical description of an alarm and its generation for protocol errors and abnormal events observed over the LTE, 5G NR, and GSM air interfaces and GPRS Gb interface were illustrated. In fact, every logical interface of mobile communications networks may specify its protocol error or abnormal error handling mechanism as described by the concerned 3GPP TS, which varies from protocol to protocol. The reader may further refer to the 3GPP TS of the interested logical interface and its error handling mechanism employed over it.

We closed this chapter with the typical steps that may be used in finding the probable cause of an abnormal event or protocol error through available alarms and their data.

11

Performance Measurements and Optimizations of Mobile Communications Networks

Introduction

The network elements of mobile communications systems and networks perform various functions and procedures to provide communications services to users. As described in the previous chapter, various kinds of abnormal errors and events may occur during the runtime of a mobile communications network. Because of such abnormal errors and events, a signaling procedure between a Mobile Equipment (ME)/User Equipment (UE) and the network may end unsuccessfully, which would further block the registration of a mobile device and its user with the network. For a network operator, such failures are not acceptable from the network services quality and its performance point of view. For performance measurement purposes, mobile communications systems and networks must have the capability to report the success or failure rate of different signaling procedures and functions performed by its network elements or the usages of the service by its users.

This chapter begins with the concept of a *counter,* which is a scalar quantity that keeps track of the number of completion of particular signaling or user traffic event, either a success or failure, in a particular location or cell of a mobile communication network.

In this chapter, we also cover the method of measuring the performances and optimizations of a mobile communications network against each signaling function and procedure performed or on the various resource allocations by its network elements. A particular performance is measured through a metric. Such a performance-related metric is known as the *Key Performance Indicator (*KPI*).* A counter is used as the basic fundamental quantity in conjunction with a KPI. From a KPI value, the performance of a network element with respect to the concerned function, procedure, or services can be evaluated for its further improvements.

11.1 Counters for Performance Measurements and Optimizations

A *counter* is, generally, an identifier that can be used to record and maintain the information in a cumulative manner about the occurrences and completion, either successful or failure, of a function and procedure performed by the respective network element of mobile communications systems. Typical functions and procedures of network elements that can be tracked through their respective counters are as follows:

- Successful or unsuccessful – initial or periodic network location registration (e.g. Global System for Mobile Communication [GSM] location area, General Packet Radio Service [GPRS] routing area and Long-Term Evolution [LTE]/Evolved Packet System [EPS] tracking area update request, and 5G registration area update) signaling procedures;

Mobile Communications Systems Development: A Practical Introduction to System Understanding, Implementation, and Deployment, First Edition. Rajib Taid.

- Successful or unsuccessful resources allocation (e.g. GSM timeslots, LTE/EPS resource block, modulation coding schemes) procedures;
- Successful or unsuccessful initial network registration (e.g. GPRS ATTACH, and LTE/EPS ATTACH request, and 5G registration request) signaling procedures;
- Successful or unsuccessful handover attempts; and
- The volume of data traffic per cell and so on.

A counter is generally an unsigned integer identifier that holds the cumulative information, at a particular network element or a physical object/resource level, of a particular signaling function, procedure, or circuit-switched (CS)/packet-switched (PS) traffic volume at an equal interval, say at one-hour or two-hour interval, of a time of day. Beyond the particular interval, a counter data may be archived for a particular day and time; reset it and start updating it again at the start of the next interval. Such capability allows an operator to examine the hourly network performance for a particular function and procedure. Example 11.1 illustrates the usage of counters for a GSM CS voice call scenario.

Example 11.1 Counters to Store GSM Circuit-Switched Voice Call Processing Events

In a mobile communications network, there is a concept called "call processing". This refers to the various *stages* or *phases* that a network element, say a base station controller (BSC), becomes engaged to provide an end-to-end service right from the establishment to the successful completion of the call. Different phases of a typical GSM call processing scenario are shown in Figure 11.1 below.

As shown in the Figure 11.1, a GSM mobile-originated call processing scenario contains several phases: radio resource connection with the BSC, call setup with the mobile switching center (MSC), a voice conversation between the users, call release, and finally the releasing of the radio resource connection with the BSC. Each phase consists of exchanging signaling messages of different protocol layers between the concerned network elements. As an example, Figure 2.8 in Section 2.2.5 shows the typical GSM radio resource (RR), mobility management (MM), and call control (CC) layer signaling messages exchanged among the MS, BSC, and the MSC for a GSM mobile-originated voice call (MOC).

In each of the call processing phases shown in Figure 11.1, various tasks related to the RR layer are performed by a BSC such as the mobile admission control, selection, and allocation of a signaling and radio resources to the MS, channel activation, and so on. The result of the various tasks performed by a network element in a particular phase could be either successful or failure. For example, a BSC mail fails to allocate a traffic channel (TCH) during the radio resource allocation phase. A network element maintains and updates the respective counters for all such important and critical tasks performed by it that are associated with the

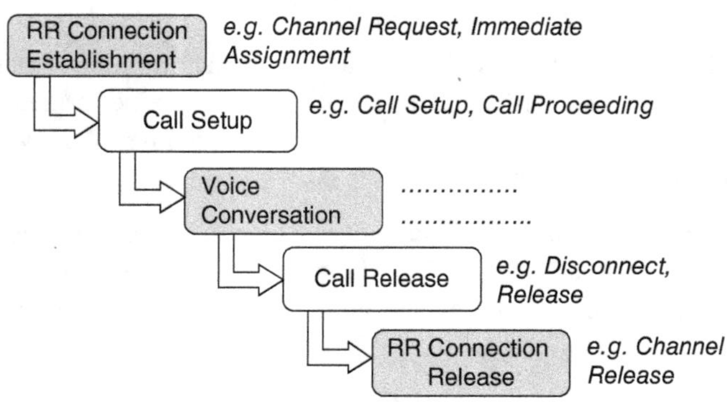

Figure 11.1 GSM CS voice call processing phases.

signaling as well as user TCHs. Those counters are updated at the predefined points of a particular call processing or signaling procedure phase as decided by the system designer from the performance and optimization requirement point of view.

11.2 Performance and Optimizations Management System

A typical centralized alarm management system of a network element was presented in Chapter 10. Similarly, a centralized performance and optimizations management system (POMS) may be also required to be designed and implemented for a particular network element of a mobile communications network. One such POMS is illustrated in Figure 11.2.

As illustrated in Figure 11.2, a POMS maintains the counters for protocol layers and their corresponding logical objects of a network element. A logical object may represent particular physical hardware such as a base transceiver station (BTS). Apart from this, a POMS maintains the counters for various signaling functions and procedures performed by different protocol layers of a network element. The Operation and Maintenance (O&M) operator configure all the counters into the POMS. Counters are updated by the concerned protocol layers per physical or logical object. Examples 11.2 to 11.4 illustrate the design and usage of typical counters for some of the scenarios/procedures of the GSM, LTE, and 5G systems.

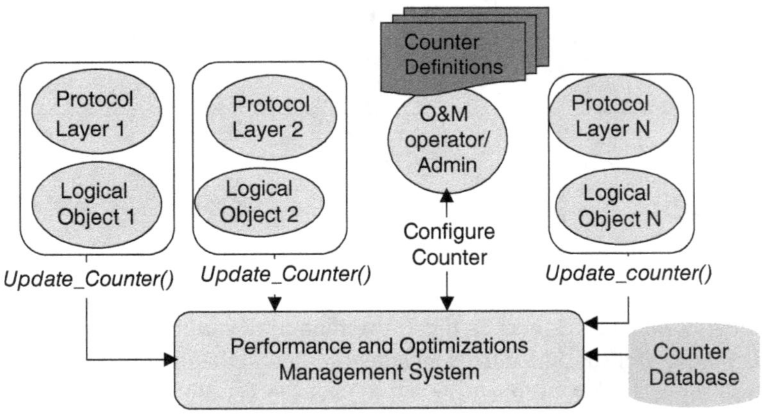

Figure 11.2 Illustration: performances and optimizations management system for mobile communications network.

Example 11.2 Resources Allocation: GSM RR Layer Counters

One of the important functions performed by the GSM RR layer is the allocation of a radio resource in terms of a timeslot to an MS from the available radio resources configured by the operator. The corresponding physical resources are the BTS and its transceivers (1: M) that transmits and receives on particular frequencies. A particular BSC serves several cells, and the RR layer may maintain separate counters for each BTS. Also, for each BTS, separate counters are maintained to record the result of the allocation of a timeslot, either a success or failure allocation. Typical counters for each BTS (XX) may be maintained by the POMS as follows:

- BTS_XX_NUMBER_OF_SUCESSFULL_ALLOCTION
- BTS_XX_NUMBER_OF_ UNSUCESSFULL _ALLOCTION

Example 11.3 Resources Allocation: LTE Air Interface Modulation and Coding Scheme Allocation Counters

One of the factors that determine the rate of data transmission in an LTE system is the allocation and usages of the Modulation and Coding Scheme (MCS) over its air interface. Each MCS is identified by an index. An MCS has its corresponding transport block, carrying upper-layer information. The size, i.e. number of bits, of a transport block is determined from the number of physical resource blocks allocated to it. As described in the 3rd Generation Partnership Project (3GPP) TS 36.213 [91], there are 32 MCS indexes. Typical counters for each MCS index may be maintained by the POMS as follows:

- NUMBER_OF_ALLOCATION_OF_MCS_INDEX0
- NUMBER_OF_ALLOCATION_OF_MCS_INDEX1
- NUMBER_OF_ALLOCATION_OF_MCS_INDEX2

Example 11.4 5G NAS Layer Procedure: Registration Request Procedure Counters

In a 5G network, a UE performs a *Registration Request* procedure during its initial registration with the 5G core network and also periodically to report its current location within a serving 5G core Access and Mobility Management Function (AMF) network function. An AMF may maintain the following typical counters for the successful, failed, and total *Registration Requests (*RA*)* received by it within its service area.

- NUMBER_OF_SUCCESSFUL_RA_REQUEST
- NUMBER_OF_FAILED_ RA _REQUEST
- NUMBER_OF_TOTAL_ RA _REQUEST

11.3 Key Performance Indicator (KPI)

11.3.1 What Is a KPI?

A KPI tells about the performance of a network element as well as the entire mobile communications system being deployed by a network operator. A KPI is carefully designed from an agreed-upon mathematical formula(s) between the network operator and Original Equipment Manufacturer (OEM) system vendor. A KPI formula could be a simple to a complex one that is derived from an aggregated or various fundamental and individual counters as collected from the individual network elements. A KPI may also contain a network parameter(s). Those measurement counters are recorded by the concerned network element during a particular event or call processing (e.g. Figures 2.8 and 11.1) scenario over a period of time as described in the previous section. KPI results can be plotted through a graph using a spreadsheet application to provide glimpses of the performance and status of the entire network and its network elements.

Examples 11.5 and 11.6 elucidate the usages of counters in designing a KPI and its associated formula for some of the protocol layer procedures of the LTE/EPS and GSM systems.

11.3.2 KPI Domains

A mobile communications network element, for example, GSM BSC or Universal Mobile Telecommunication System (UMTS) RNC, may offer both the CS (voice) and PS (data) domain services in a cell. For CS and PS domains, the concerned network element shall maintain counters separately for both the types of services that enable an operator to evaluate the performances of its network through respective KPIs results.

Example 11.5 LTE/EPS ATTACH Procedure: KPI and Measurement Counters

Consider the measurement counters and KPI for an LTE/EPS network registration ATTACH procedure performed by UEs in a particular cell X. Typical individual counters and their KPI for the ATTACH procedure can be defined as follows:

EPS_ATTACH_COUNTER_CellX = Sum of all EPS Attach Request messages between, say, 0:00 hour to 01:00 hour.

EPS_ATTACH_REJECT_COUNTER_CellX = Sum of all EPS Attach Reject messages between, say, 0:00 hour to 01:00 hour.

EPS_ATTACH_FAILURE_KPI = (*EPS*_ATTACH_REJECT_COUNTER_cellX /EPS _ATTACH_COUNTER_cellX)*100

EPS _ATTACH_SUCCESS_KPI = (1- EPS _ATTACH_FAILURE_KPI)*100

Individual counter or measurement can be collected at a predefined interval, say, 0.5 hour, 1 hour, and so on, and then put their values in the KPI formula as illustrated above. From the result, evaluate and assess the performance of the network element or the entire communications network.

A KPI result is generally expressed in *percentage (%)* term. For example, mobile-originated call success rate KPI is, say 99.99%, handover success rate KPI is 90%, network uptime and availability is 99.999%, and so on. As with other systems, the network availability KPI may indicate the uptime or downtime of the entire network on per hour, per month basis.

Example 11.6 KPI for Mobile Originating (MO) GSM Call Flow

The various call processing phases for a GSM MOC were described and illustrated in Figure 11.1. Consider again the GSM MOC flow illustrated in Figure 11.3 below from the signaling and voice traffic resource allocation point of view.

The illustration in Figure 11.3 shows the GSM call processing stages consisting of the signaling as well as user voice call TCH allocations. Look at the Figure 11.3 using the top-down approach, which shows the voice call success measurement at the top. A successful voice call results from certain activities that are shown by the horizontal arrow lines in Figure 11.3. The various tasks performed by the BSC, BTS, and MS in this regard are as follows:

- Establishment of an RR connection over the Random Access Channel (RACH), which is initiated by the MS
- Establishment and allocation of dedicated signaling channel (SDCCH) and user TCH by BSC to the MS
- Call connection and ongoing voice traffic between the users

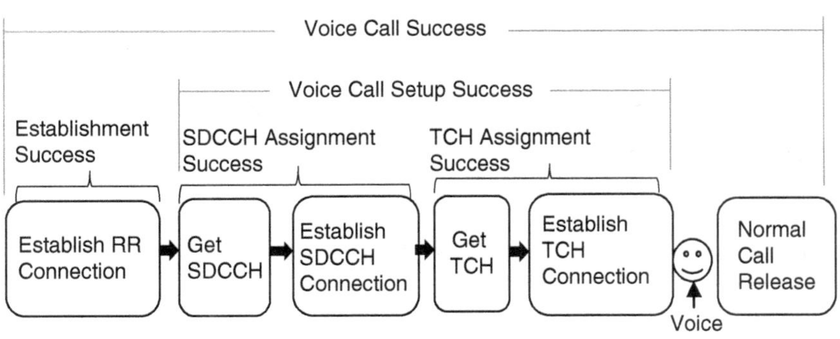

Figure 11.3 Illustration: KPI various stages of a GSM call flow.

Once an MS is granted to access the GSM network through the *IMMEDIATE ASSIGNMENT* message sent on the AGCH, refer to Figure 2.8, all the subsequent signaling messages related to a GSM voice call setup attempt are transported over the *SDCCH* only between the MS and the BTS. During a GSM voice call setup phase, a call may be released abnormally as a result of the SDCCH drop due to any reason prior to the allocation of a TCH. Similarly, an MS may not be able to establish a GSM voice call because of SDCCH blocking in a particular cell. SMSs are also transported over the SDCCH. The SDCCHs must be configured correctly by the operator. The GSM SDDCH drop rate KPI, in percentage, in a particular cell can be defined as follows:

SDDCH DROP RATE = (Total Number of SDCCH drop/Total Number of successful Voice call established on SDDCH)*100

In all of these call processing steps, the respective network element, e.g. MS or BSC, records and updates its counters. These counters are used for KPI calculation in terms of % rate, either success or failure, of various call processing tasks of a GSM mobile-originated voice call, for example, SDCCH signaling allocation rate, successful TCH allocation rate, and voice call setup rate.

In general, a call, either a voice or data, processing, or signaling procedure scenario involves the exchanging of the control or signaling plane messages among the network elements of a mobile communications network.

11.3.3 KPI for Signaling or Control Plane

In a mobile communications network, a successful establishment and completion of a signaling procedure is the prerequisite for a mobile device to use various communications services in the CS and PS domains. Some of the typical KPIs associated with the GSM, GPRS, UMTS, LTE/EPS, and 5G system air interface Layer 3/NAS layer signaling or control plane protocol layers are mentioned below:

- GSM air interface Layer 3 signaling procedures
  - Paging success rate;
  - Signaling SDCCH drop rate, e.g. user would not be able to communicate as the call establishments will fail;
  - Abnormal traffic channel TCH drop rate, e.g. user's active call will be dropped abnormally; and
  - Expiry of important network timers, counters that would lead to the completion of a signaling procedure unsuccessfully, and so on.
- GPRS/UMTS PS domain: Air interface Layer 3 signaling procedures
  - Network registration, i.e. ATTACH, success, and failure rate;
  - Session establishments, i.e. Packet Data Protocol (PDP) context activation success and failure rate and so on.
- UMTS and LTE air interface Layer 3 signaling procedures
  - RRC connection setup success rate, blocking rate, and so on.
- UMTS and LTE radio access bearer signaling procedures
  - Radio access bearer successful establishment rate;
  - Radio access bearer abnormal failure rate and so on.
- LTE/EPS air interface Layer 3 signaling procedures
  - EPS ATTACH success rate;
  - EPS ATTACH failure rate; and
  - Tracking area update success rate and so on.
- 5G system
  - Registration procedure success and failure rate;
  - Protocol data unit (PDU) session establishments success rate; control plane latency, and

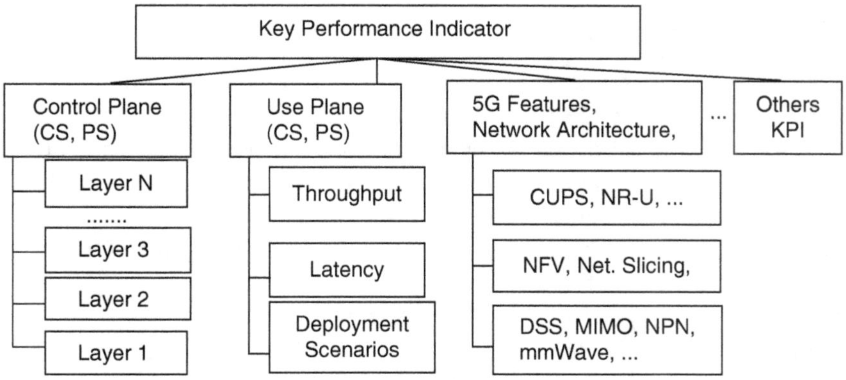

Figure 11.4 Illustration: KPI areas of a network element.

- Resources allocation success or failure rates for different network slices, i.e. Enhanced Mobile Broadband (eMBB), Ultra Reliable and Low Latency Communications (URLLC), Massive Machine Type Communications (mMTC), other technical performance and so on.

Figure 11.4 further illustrates a general overview of the different areas/domains where KPIs are designed, maintained, and updated by a network element. KPI areas /domains could be the control plane as well as the user plane protocol layers. In addition to these, there are other KPI areas such as reliability and energy efficiency, which are defined as the minimum technical performance requirements by the International Telecommunication Union—Radio communications sector (ITU-R) in their report ITU-R M.2410 in the case of the 5G system.

11.3.4 KPI for User or Data Plane

Similar to the control or signaling plane, there are also KPIs for the user plane protocols of a mobile communications network, see Figure: 11.4. Typical areas where user plane protocol KPIs are applied are as follows:

- Application latency
- Traffic volume, application or data throughput
- Different deployment scenarios such as the indoor hotspot, dense urban, rural, urban macro, user plane latency, and so on, in the case of the 5G system, refer to TR 38.913 [27].

Information about the application throughput may be maintained at the cell level. The collective throughput of all the cells shall provide the throughput for the entire network. User or application data passes through the user plane protocol stack. The user plane protocol stack information passes through some of the control plane protocol layers, e.g. radio link control (RLC) layer, which may affect the overall application throughput.

As illustrated in Figure 11.4, a network element may provide KPIs for each of its protocol layer against the different functions and procedures performed by it. The values of those KPIs are derived based on the different individual counters and network parameters maintained by the respective protocol layer.

Apart from the system-specific signaling KPIs mentioned in the preceding text, there are also KPIs in some of the common signaling areas such as the handover and paging procedures, whose success or failure rates determine the seamless mobility (either intra or inter system) experience of a mobile user. For more examples on typical KPIs, refer to 3GPP TS 32.410 [81], TS 32.450 [82], TS 32.455 [84], and TR 38.913 [27].

Example 11.7 elucidates the usage of counters as part of the root cause analysis to find the reason for a complaint of poor data services reported by a mobile user from a particular cell.

Example 11.7 LTE Air Interface Layer 2: RLC Layer and User Application Throughput Counters

Consider that a user complained about the slowness of a data session. Let us start troubleshooting the issue from the LTE air interface point of view. As far as the LTE air interface protocol stack is concerned, the user plane protocol layer data, carrying the user traffic, is transmitted by the control plane protocol layers, mainly the RLC, medium access control (MAC), and the physical layer. Similar to the GPRS, UMTS, or 5G system, the LTE air interface Layer 2 RLC layer plays an important role in delivering the expected throughput to an application. Note that the actual throughput delivered to a user application is less than the throughput of the RLC layer as the user data passes through the RLC layer up the protocol stack.

The throughput of the RLC depends on the quality of the radio air interface at present due to which the RLC layer may apply the corrective measures, if required, to deliver error-free user traffic to a receiver. To rule out the RLC layer as one of the probable causes of the low data throughput as experienced by the user and its application, the functions performed by the RLC layer is required to be considered. In the acknowledged mode of transmission, the sending LTE RLC layer retransmits the user data in case the receiving RLC layer responds with the negative Acknowledge Mode (ACK) for some of the PDUs carrying user data. During the retransmission of user data, the RLC PDU may be again resegmented. The LTE RLC layer also performs the user data discard functions. All these functions performed by the LTE RLC layer result as the overheads in terms of the additional functions performed by it, which affect the throughput delivered to a user application. Typical measurement counters as shown below may be designed and implemented at the RLC layer so that the occurrences of such overheads and excessive retransmissions may be recorded, which further helps in finding the root may cause of the low throughput issue.

- RLC_ACK_RETRANSMISSIONS_COUNT
- RLC_SDU_DISCARD_COUNT
- RLC_ACK_RE-SEGMENT_COUNT
- RLC_ ACK_POLLRETRANSMIT_TIMER_EXPIRED

In an ideal air interface condition, the value of the above RLC layer counters is typically very low. However, an abnormally high value of such counters may indicate a poor radio air interface condition, which could be also the reason for the low user data throughput.

11.3.5 KPI Categories

Regardless of the type of services, that is voice or data, and protocol layer, that is control or data plane, the KPIs provided by a mobile communications network can be divided into different categories. Some of the KPIs are common in general, whereas other KPIs are system specific. For LTE-specific KPI categories, refer to TS 32.451 [83]. For 5G system-specific KPI categories, refer to TR 38.913 [27]. Some of the common KPI categories are mentioned below:

- Accessibility

This KPI category provides how many times (successful resources granted versus the number of attempts/requested) a mobile device was granted resources i.e. channels, either traffic or signaling while trying to access the network over a given period.

*Accessibility KPI = (Successful allocation/Number of allocation requests)*100%*

- Retain ability

This KPI category provides how long the mobile user was able to continue using the services/call without an abnormal drop/disconnection. For example, if a user is not able to make a GSM call, there may be a high rate of SDDCH drop in the concerned cell.

- Integrity

This KPI category provides the throughput, say in Kbits/second, being allocated to the mobile user. An erroneous communication link may not deliver the expected throughput to the mobile user.

- Availability

This KPI category provides how long the communications service was available in a particular cell for a given period of time.

- Mobility

This KPI category provides about the success rate of various mobility management signaling procedures, CS domain or PS domain, which were attempted by a mobile device. For example, the success rate of call handover within a GSM system or between GSM and UMTS systems. Call handover types could be GSM intra BSC, inter BSC, intra/inter MSC; LTE, S1 handover, X2 handover, 5G system N2 handover, Xn handover, and so on, and inter-RAT/inter-system handover.

Apart from the above categories and also categories as defined in the concerned 3GPP TS, a network operator may also design and use operator-specific customized performance benchmarking (%) KPIs, which are agreed by the OEM or system vendor. Note that different KPIs and their formulas that are designed and used across the network elements of an OEM and operator may not be standardized by the 3GPP.

11.3.6 KPI Evaluation Steps

In a mobile communications network, KPIs data may be evaluated periodically by its network operator to assess the performance of different elements, as well as the entire network, as part of its ongoing performance management and optimizations processes. If a particular KPI category and its value drop below a threshold limit, it would affect the service-level agreement (SLA) between the OEM and its network operator. Further, a low KPI would affect the quality of network services from the end user/subscriber point of view. Thus, to maintain the desired SLA level, it is essential to evaluate the KPIs of different categories as part of the ongoing performance management and optimization processes of the mobile communication network. Figure 11.5 illustrates a typical KPI evaluation steps to be performed to meet the SLA in a mobile communication network.

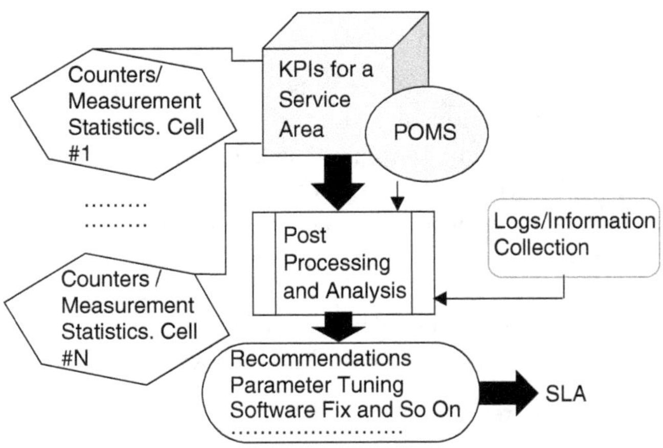

Figure 11.5 Illustration: KPI evaluation steps of a typical network element.

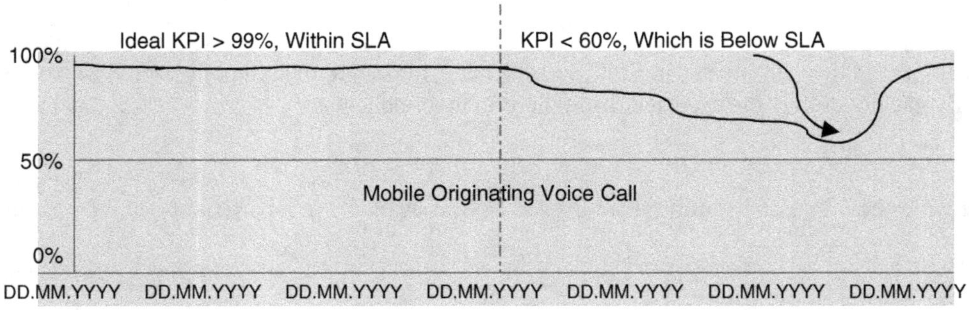

Figure 11.6 Illustration: ideal and poor SLA and KPI for GSM MOC voice call.

A mobile communication network provides services in terms of a service area, e.g. GSM service area, LTE/EPS service, 5G service area, and so on. A service area consists of several serving cells, Cell 1. . .Cell n, and they are under the control of a particular radio access network element or core network element, as shown in Figure 11.5. Measurement counters of different categories, as mentioned above, for a particular type of service are maintained on a per cell basis by the performance management and optimization (POMS) module of the network element. The raw counters and their data may be extracted from the POMS for their postprocessing and analysis using the respective KPI formula. Relevant information/logs too may be collected from the field network in case a KPI value is found to below.

The analysis of the different measurements and their counters shall provide information about the network performance in a particular cell, that is, either a problematic cell or meeting the desired SLA level. Further analysis of the measurement counters of all the cells and then putting it into the respective formulas shall provide the KPI results of different signaling procedures and categories for the entire network. From the KPI results and relevant information collected, necessary improvements and corrective measures may be taken step by step, which could be either a software fix or a fine-tuning of system parameters, etc. as described in the next section.

As an example, Figure 11.6 illustrates the KPI for a successful mobile originating GSM voice call (MOC) scenario between two particular dates (DD.MM.YYYY). Assume that the agreed SLA between the network operator and the OEM for the successful MOC case is 99% and greater. The first half (left-hand side) of Figure 11.6 illustrates the ideal KPI that is within the SLA windows, whereas the second half (right-hand side) of the figure shows an abnormal KPI where the SLA has been breached as the corresponding KPI is below the 99%.

11.3.7 Troubleshooting and Improving KPI

There are close relationships between the quality of communications services as perceived by users and the associated KPI result. A KPI result indicates the issue only and nothing about its root causes. Thus, identification and fixing of the technical root cause will improve both the quality of communications services and their associated KPI results. But the fixing of the root cause of a KPI-related issue that is complained about by a user may be a challenging one. In this case, the various factors, which may be either local or sometimes end-to-end nature, related to the KPI issue are required to be identified and isolated one by one. Past experience on the resolution of a KPI issue also matters here.

To find and isolate the actual root cause, the relevant information is required to be collected and analyzed from the concerned network elements. Some of the typical causes of KPI issues are mentioned below:

- Environmental factors (e.g. poor radio link/air interface, Interference, obstacles . . .)
- Important network parameters that might affect signaling procedures
- Incorrect network dimensioning and its fine-tuning, network coverages issues

- Signaling issues, important network functions, such as handover, and features such as fallback to CS domain issues
- Software bugs
- Hardware faults and physical interface issues
- Timing synchronization issues
- MS/UE faults
- User behavior
- Beamforming and coverage issues, in the case of the 5G NR system

Above typical field issues and their troubleshooting steps are covered in the next chapter.

11.3.8 Components of a KPI Definition

Similar to the designing and definition requirement of an alarm of a network element as described in Chapter 10, a KPI has several components/attributes that may be included as part of its design and definition. To define all such attributes, a KPI description format is used that may contain the following information:

- KPI name – an appropriate title for the current KPI
- KPI description – describes the purposes of the KPI
- KPI logical formula – formula consisting of measurement counters from the POMS and other subformulas
- KPI category, e.g. accessibility, and retainability
- KPI unit, i.e. %, time interval (seconds), Kbits/sec

For further details on the above KPI attributes in the case of the GSM/UMTS system, refer to 3GPP TS 32.410 [81].

Chapter Summary

This chapter presented the KPI method of performance measurements and optimizations of mobile communications networks in a quantified manner for various communication services in CS and PS domains. As far as a mobile communication network and its services quality aspects are concerned, the performance measurements and optimizations through KPI is important for attracting new and retaining its existing subscribers. KPIs are also one of the keys benchmarking and differentiating factors among the competing network operators in terms of their network performances and quality of services offered to subscribers.

We presented the counter concept, which is a method used by a network element to record the result of a particular event or call processing step, signaling procedure, allocation of resources, and so on that may complete either success or failure. Counters are also used to record the volume of traffic per user, per cell, and so on. Then, we presented several illustrative typical KPI formulas that are derived from individual counters. A KPI formula can be a very basic one to a complex one and is carefully designed in an agreement between an operator and its OEM of different network elements. For more KPIs formulas for different mobile communications systems, the reader may further refer to the 3GPP TS mentioned in the different sections of this chapter.

12

Troubleshooting of Mobile Communications Networks Issues

Introduction

In a mobile communications network, day-to-day issues are inevitable, which could be related to the entire network or a service area; operation and maintenances; services issues to the subscriber(s); and so on. This chapter covers the sources of some of the typical issues that may be observed in day-to-day operations of a mobile communications network. A mobile communications network operated by a particular operator consists of various network elements that are interworked together through different logical interfaces, as described earlier in Chapter 6. Also, seamless communications services to roaming users are provided through the interoperations of mobile communications networks of different operators, as described earlier in Chapter 6. Day-to-day issues from simple to complex nature also appear due to the interoperations of mobile communications networks run by different network operators.

We begin with the typical air interface-related issues and methods used to troubleshoot the same. It is followed by the Internet Protocol (IP) transport network-related issues among different network elements and methods used to troubleshooting them. We then present the typical issues that may arise due to the conformance testing, interoperability testing, and various interworking features/methods which may be deployed in a network. Finally, the typical approach to be used for debugging and resolution of a given network issue is also described.

12.1 Air Interface-Related Issues

The most troublesome part of mobile communications networks is the air interface used between an MS/UE and its radio access network (RAN). Here, the information transmitted by the, say, MS or RAN note is subjected to go under environmental disturbances, such as interference, varying weather conditions, and physical geography (city, atop a hill), leading to the corruption of bits of information. This is true regardless of the type of mobile technology, such as Global System for Mobile Communication (GSM), General Packet Radio Service (GPRS), Universal Mobile Telecommunication System (UMTS), Long-Term Evolution (LTE), or 5G network, being deployed. A network element may have been tested thoroughly, and the result may be found to be as expected. But that is at a lab condition, which is under a controlled environment where there are no external disturbances to the radio wave propagation between the sender and receiver, say, an MS and base transceiver station (BTS).

To combat errors resulting from air interface issues, each mobile communication technology uses different error correction and recovery mechanisms to protect the user data transmitted by an MS or its RAN node. Other network elements, such as MS, base station controller (BSC), mobile switching center (MSC), and Serving GPRS

Mobile Communications Systems Development: A Practical Introduction to System Understanding, Implementation, and Deployment, First Edition. Rajib Taid.

Support Node (SGSN), may also introduce errors. For example, an operator has a BSC from vendor X and an MSC or SGSN from another vendor, say Y. A problem may crop up when the operator tries to integrate both the BSC and MSC since they work on open 3rd Generation Partnership Project (3GPP) standards and it is up to the vendor regarding the compliance and implementation of their particular network element.

One could expect any kind of field issues from a mobile communications network. Whenever a customer makes a complaint about an inability to avail a service successfully, the root cause of the same may not be clear at the first instance itself. In that case, one can use intuition as well as past experiences as the starting point to proceed further in resolving the issue. One may also have detailed plans to address and resolve customer issues. Nevertheless, one can perform the following basic steps to troubleshoot a field issue being reported to a developer or support team.

12.1.1 Drive Test, Data Collection, and Its Analysis

A *drive test* is performed in a particular problematic cell/site and is used to troubleshoot issues arising out of air interface-related problems. Air interface issues may affect a single customer or multiple customers or other neighbor networks too. A drive test can be also used for an in-depth examination of the serving and surrounding radio frequency (RF) conditions as well as successful functionality testing such as circuit-switched (CS) call, packet-switched (PS) call, and so on during a new cell site commissioning phase. A drive test is performed before a cell/site is released to serve actual customer traffic from the newly rolled out site. This method comprises these steps – perform the test, capture the values, analysis them, and then recommend and tune the appropriate value for corrective actions.

For a drive test, there is a special kind of mobile equipment, fitted in a vehicle, that can be used in a particular cell either stationary or moving condition to simulate the actual mobile user's behaviors. Also, the test mobile equipment or MS can be positioned nearer to the base station, a little bit far, or farther from the base station to simulate the end user's behavior. This MS can scan and verify the current RF environment conditions, including interference, received and transmitted signal level, neighbor cells, and various system parameters being transmitted by the base station which are received by the MS.

Attached to this special MS is the software that shows the various signaling messages being exchanged between the MS and base station or the core network. One can examine and decode the contents of each message and see for any erroneous value present, which is possibly causing the current problem. A cell can be located in a city, village, having obstacles in proper signal propagation, or atop a hill. Depending on the cell site, various system parameters need to be tuned accordingly. This is possible through drive test exercises, followed by a careful analysis of the recorded values and recommendations for the appropriate one. A drive test procedure is also used to perform network optimizations, which is an ongoing activity to utilize the network resources, including the air interface, optimally while offering the desired quality of services to the subscribers. This is because a communication network run by an operator may evolve over a period of time, requiring periodic monitoring and fine-tuning.

A drive test can also be used to measure data throughput offered by, say for a GPRS or LTE or 5G PS service, a typical network and rectify the possible causes if the throughput is found to be low and not a satisfactory one. Commercial software and hardware tools are available in the market to perform drive tests, collect data, and troubleshoot air interface-related issues. Look for the online resources to know more about such tools.

Several air interface-related issues that can be analyzed through the drive test method are explained later in Section 12.6.

12.2 Debugging Issues with IP-Based Logical Interface

The LTE and 5G system are an all IP transport-based mobile communications system and network. That is, the network elements of an LTE and 5G network communicate with each other on top of the existing IP transport network. Methods for debugging of IP transport network-based mobile communications issues are described in the next sections.

12.2.1 IP Protocol Analyzer

One can download and install freely available IP-based protocol analysis tool to experiment with and understand the various IP protocols even with your desktop. An IP packet analyzer tool may support protocol drawn from mobile communications (Voice over IP [VOIP], GSM, GPRS, UMTS, LTE, 5G system, and so on) or data communications protocols (Transmission Control Protocol [TCP], IP, File Transfer Protocol [FTP], Virtual Router Redundancy Protocol [VRRP], and so on).

A developer or an IP network troubleshooter can capture IP packets from different logical interfaces that work on top of the IP transport. Then, later on, decode the captured packets using the concerned communication protocol/stack for various IP-based logical interfaces used across GPRS, LTE/Evolved Packet System (EPS), or 5G networks. Even the proprietary A-bis interface has nowadays become IP-based, called Packet over A-bis. Using an IP-based packet analyses tool, one can:

- Study the behavior and become familiar with protocol/stacks in-depth that may help you in resolving live network, data, or communication issues of different complexities;
- Study the behavior and troubleshoot the performance of an application;
- Print and export data in a customized format;
- Graphical reporting capabilities of various categories, giving a visual representing and pin-pointing the probable root cause, an issue, or behavior of the captured IP session; and
- Study the suspicious behavior user activity or protocol layer.

12.2.2 Network/Application Throughput Issue

An end-to-end mobile communications service is provided through an interconnected network of network elements such as the MS/UE, RAN, and core in association with the external packet data network. The issues in such a network could be either in the basic network signaling or application throughput issue. Some examples of mobile communications network PS call signaling protocols that work on top of the IP transport network are as follows:

- GPRS Gb interface.
- LTE S1, X2 interface (Stream Control Transmission Protocol [SCTP] [17] on top of IP); 5G NG, Xn interface (on top of IP).
- UMTS Gb/Iu PS interface.
- Gn/Gi interface.

Examples of user applications where one can experience slow performance-related issues are slow internet browsing, e-mail, and video/audio streaming. For the issues of these types, chances are that there are issues with the IP transport layer. To locate and find out the root cause of the current fault/issue, start with the bottom-up approach, i.e. isolate the application layer and start troubleshooting at the IP transport layer first. Figure 12.1 shows an example of how an application layer uses the services of the IP transport layer.

12.2.3 Switch Port Mirroring

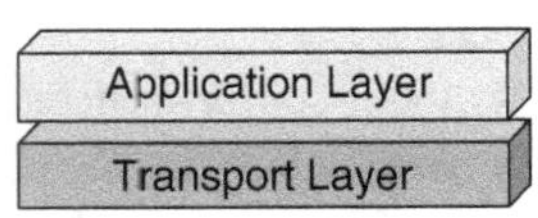

Figure 12.1 Application and transport layer with respect to IP troubleshooting.

A *Port Mirroring* is the feature that is supported by a switch, not by a hub. A mirrored port sees all the packets, without disturbing them, which are being sent/received through an actual switch port. A developer or Operation and Maintenance (O&M) operator will be required to configure a mirror port using the commands provided by a particular switch. Check the documentation of the concerned switch on how to configure a mirrored port. On the mirrored port, one can hook up the IP packet

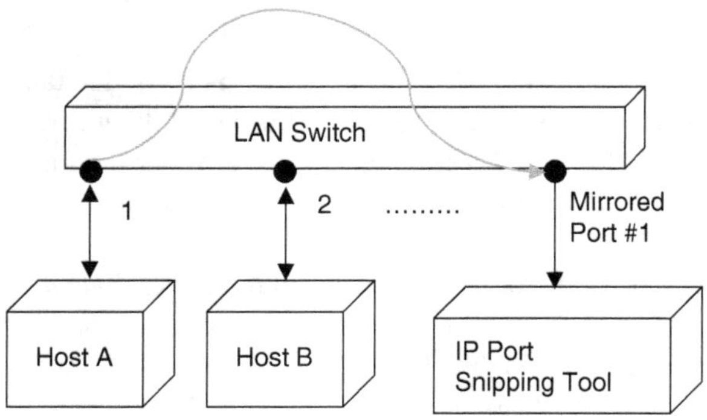

Figure 12.2 IP packets snipping through switch port mirroring.

snipper and then listen and capture the packets exchanged between two hosts. Packet snipping using port mirroring is useful to determine and isolate the root cause and origin of an issue that is arising out of the IP transport communications used between two hosts, especially from different vendors. Figure 12.2 illustrates a sample arrangement of packets snipping using the Wireshark tool through a port mirroring feature of a switch.

A thorough packet-by-packet analysis of the IP packets captured using the port mirroring mechanism shown above may lead to the source and root cause of an issue, as follows:

- Client, i.e. application at MS/UE end, can be excluded from the issue
- The network is dropping frames, say at the air interface level, leading the server to retransmit frames
- Server, i.e. application at network end can be excluded from the issue

There is the probability of dropping frames, say at the air interface level, by the communications network itself. In this case, it will be required to be troubleshot by the RF team, say by using a drive test as described in Section 12.1.1.

If the above factors do not reveal any clue to the current issue, do not be afraid to suspect the behavior of the underlying TCP or User Datagram Protocol (UDP)/IP stack itself! But before suspecting the TCP or UDP/IP stack behavior, one must be thorough with the analysis with all the supporting data.

As a developer, the task is to do the root cause analysis (RCA) of the delay experienced in the application, e.g. browsing, e-mail, chat, and so on, as reported by the communications service user. Problems can crop up anywhere in the network sometimes, for example, because of a low quality and 3GPP noncompliant MS that behaves erratically, causing troubles either for the RAN or CN. One must have an issue resolution plan in place to find its solution. A developer requires to establish the fact that the concerned module or network element is not leading to the current issue at hand. To accomplish this, one may have to collect logs from multiple interfaces across the network. This is described in Section 12.3.

12.3 Conformance Testing Issues

Protocol *conformance testing* is performed to verify the desired behavior and functionalities of a particular network element against the specified ones as mentioned in the 3GPP technical specifications, TS 34.12* series [88], TS 36.52* series [99], TS 51 series [137] for GSM/UMTS/LTE, and TS 38.50/52/53 [124] series for the 5G NR.

12.3.1 Example: Mobile Device (MS)/User Equipment (UE) Conformance Test

Consider the mobile device (MS) or user equipment (UE), which is an intelligent device having multitasking capabilities to provide various communication services to a user through different protocols layers/stacks. An MS or UE communicates with the radio access as well as the core network to transmit and receive signaling and user data. The networks process the UE request initiated for a particular signaling procedure whose result or outcome will have different possibilities.

The network processes information transmitted by the MS/UE and responds the result to the UE/MS accordingly as specified for the concerned procedure in its technical specification. The response, either success, failure, or any other, from the network for a particular procedure is dynamic in nature, and depending on it, the UE/MS requires to behave as mentioned in the concerned specification. That is the purpose of the UE/MS conformance testing. Failure to conform by a UE/MS shall result in an unexpected behavior due to which the user would not be able to use telecommunication services. Some of the typical scenarios and functional areas requiring protocol conformance testing for a UE/MS are listed below:

- An MS that supports all the available radio access technologies, i.e. GSM, UMTS, and LTE system with different duplexing modes such as (Frequency Division Duplex [FDD] and Time Division Duplex [TDD]).
- An MS that supports LTE and 5G system with different duplexing modes such as (FDD and TDD).
- Intra-radio or inter-Radio Access Technology (inter-RAT) Handover.
- Functions and procedures, e.g. radio resource control, mobility management, call control, measurements, session management, conformances requirements for CS as well as PS domain as specified in concerned 3GPP technical specifications.
- Idle mode behavior of an MS/UE in case of pure LTE, UMTS, or GSM mode, or multimode or another environment.
- Protocol conformance requirements for various layers, e.g. Layer 1, Layer 2, and Layer 3, and their mode of working, i.e. acknowledge, unacknowledged, and transparent.
- Supplementary services and short message services conformances requirements.
- MS/UE radio transmission and reception conformances requirements.

12.3.2 Example: Location Area Update Request

Consider the location area update (LAU) mobility management procedure performed by an MS in case of CS domain GSM system. Note that an LTE UE may also perform an LAU procedure as part of the combined tracking area update procedure toward LTE/Evolved Packet Core (EPC). The MS provides its identities, which is mandatory, in the *Location Updating Request* message sent to the core network. A mobile identity could be any one of the following ones:

- International Mobile Subscriber Identity (IMSI)
- Temporary Mobile Subscriber Identity (TMSI)
- International Mobile Equipment Identity (IMEI)

An *LAU request* initiated by the MS may be either accepted or rejected by the core network, and the result is conveyed to the MS through the *Location Area Update Accept or Reject* with an appropriate code. For further details on various response codes, refer to TS 24.008 [45]. Assume that the core network accepted the LAU request made by the MS as shown in the following Figure 12.3.

As shown in Figure 12.3, there are three possibilities of response that the core network may reply in the *LOCATION UPDATING ACCEPT* message to the MS. In this case, the concerned MS is required to meet all such conformance requirements. The MS is required to function normally after the completion of a *Location Updating*

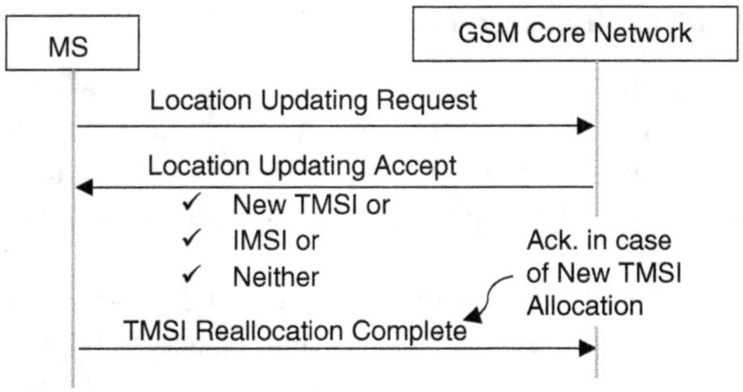

Figure 12.3 Illustration: GSM location area update procedure.

Figure 12.4 Illustration: typical UE conformance test scenarios for GSM location area update procedure.

Request procedure successfully with all the three possible responses from the network end as mentioned above. To verify the above conformances for the Location Updating Request procedure made by an MS, a series of tests shall be performed through a system simulator that behaves like a real-life core network system.

Note that 3GPP also has technical specifications that describe the conformances requirements specification, Figure 12.4, for network elements such as the MS/UE and base station. Such technical specifications span across several thousands of pages that describe the requirements of conformance for a particular procedure/function. Conformance test technical specifications also contain the execution steps in detail for a particular procedure to be tested and verified. For more information on the MS/UE conformance requirements and their specifications, refer to 3GPP TSs [88, 97, 137].

12.4 Interoperability Testing (IOT) Issues

Interoperability is the ability of a communications system and its network elements to provide and accept services successfully by exchanging information with another Original Equipment Manufacturer (OEM)'s network elements/system. Interoperability among the mobile communications networks of different operators provides seamless communication services to a roaming user while traveling outside the home network.

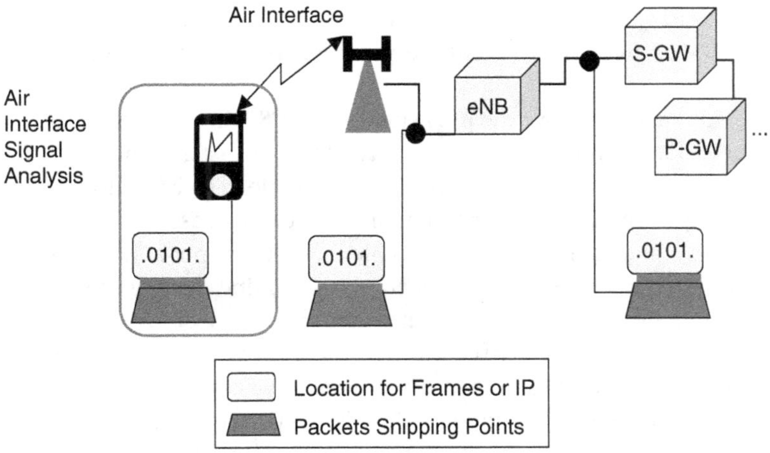

Figure 12.5 Illustration: IP packet snipping tool setup for troubleshooting IOT-related issues in an LTE/EPS network.

A mobile communications network that consists of different network elements is supplied by different OEMs. Network elements are designed and developed based on the 3GPP open standard and its technical specifications. Interoperability issues can be still expected to arise at the beginning of its deployment due to integrations of the 3GPP-compliant network elements.

- Reasons for IOT issues

Generally, an IOT issue, between two networks, may arise whenever a network element sends protocol information incorrectly to another network element. In such cases, the receiver may either discard the protocol information or respond with an error code to the sender of the message. A network element may send erroneous protocol information to a receiver as a result of ambiguity and incorrect interpretation from the concerned specification by the developer. Protocol information may also become erroneous due to poor radio conditions or other transmissions issues at different physical interfaces. Sometimes, IOT issues may also crop up because of missing mandatory protocol information as detected by its receiver. For all such cases and to find its actual root cause, necessary log/traces are collected from the respective logical interfaces that connect the concerned network elements. One such arrangement to capture log/traces from different logical interfaces of an LTE/EPS network is illustrated in Figure 12.5.

Figure 12.5 illustrates the capturing of logs/traces at UE end over the air interface and between the eNodeB and the Serving Gateway (S-GW) for user traffic throughput-related issue analysis in an LTE/EPS network. Similar approaches may be used for isolation of IOT-related issues in other communication systems, such as the 5G system.

12.5 Interworking Issues

In Chapter 6, we had presented the interworking scenarios of mobile communications networks based on the legacy GSM, GPRS, and UTMS systems along with the LTE/EPS and interworking between the LTE/EPS and 5G systems. Through interworking, an operator may provide voice, data, and other communication services to subscribers according to their needs. However, the types of services that can be used by a subscriber also depend on the capability of the MS as well as the core network. For example, an MS may provide its voice over IP Multimedia Subsystem (IMS), i.e. Voice over LTE (*VoLTE*), capability information to the LTE/EPC network. However, the LTE/EPC network may not have the IMS facility. In such a case, the MS may fall back to the legacy GSM/UMTS network to facilitate voice calls, as described earlier in Section 6.2.3.

Now, assume the scenario where the LTE/EPC does not have the provision for the CS fallback also. In such a case, a typical voice-centric UE may attempt to detach from the LTE/EPC network and register with the 2G/3G network to facilitate voice calls to the mobile user. However, a data-centric UE may be able to register successfully with the LTE/EPC network. In an interworking scenario and depending on UE mode as described earlier in Section 6.5, different UEs may behave differently, and sometimes, it is possible that a UE is not able to register with the core network at all. To address such UE network registration-related issues and find the root cause, necessary logs/traces are required to be collected from the concerned logical interfaces by repeating the test scenario. From these initial traces, the type of network registration/attach as requested by the UE, its capabilities as well as the actual network configuration-related information can be examined. This information can be further used to isolate one by one the probable causes of the UE network registration failure issue. One such arrangement to capture log/traces is shown in Figure 12.5. This figure shows a typical setup to collect the logs using an IP packet sniffer tool for troubleshooting of an IOT issue reported from an LTE/EPS network element or other communication systems using the IP transport network.

12.6 Importance of Log/Traces and Its Collections

Mobile communications network issues reported from the field can be a difficult and challenging one to resolve as the network is exposed to a variety of conditions and environments. Most of the issues reported from the field are not easily reproducible at laboratory conditions, for example, a process crash, as the field environment is very dynamic in nature. Troubleshooting of field issues becomes more challenging in case of an embedded system-based network element as it is simply a plug-in unit (PIU) with no disk-based storage facility for recording of important logs. An embedded system-based network element becomes refreshed upon its restart following an event, for example, upon a critical process crash with no memory core dump. Also, a mobile communications network and system becomes matured over a period of time with many issues being fixed, followed by delivery of the subsequent releases to the customers. If the field issue is a reproducible one, it will be fortunate enough to collect the required logs. Logs are very crucial for the troubleshooting by the development team to debug and solve the issue reported from the field.

It was mentioned in Section 3.1 in Chapter 3 that a mobile communications network consists of various network elements and physical and logical interfaces connecting them. Sometimes, an issue may be a straightforward one and requires collecting logs from the concerned network elements and logical interface only. In other cases, make an expert judgment on the possible roles of various network elements and logical interfaces that exist all along between the actual sender and receiver. Collect the required messages, at each of the logical interfaces, exchanged between each successive network elements. From the logs, verify that the contents of the messages being exchanged over the respective logical interfaces are correct among the network elements with respect to the concerned 3GPP TS. It is possible that the sender was sending a wrong value or it did not include a mandatory element as defined in the concerned TS, leading to the current issue. In that case, the defect must be fixed at the sender side only.

End-to-end, i.e. from MS/UE to core network or another external network, time-synced logs are required to be collected to correlate, troubleshoot, and fix any data throughput-related issues. From the collected logs, air interface-related issues shall be visible from the retransmission requirements of the radio link control (RLC) Layer 2 frames, blocks, or protocol data units (PDUs). Similarly, an IP packet drop or retransmission requirements shall be visible either because of air interface-related issues or network congestion. Tools that can be used to collect logs and debug various interfaces or reference points-related issues can be classified into the following types:

- *Air Interface Investigation Tool* – It can be used to collect logs for air interface-related issues at MS/UE end. Such a tool can be used to do a live analysis of the message and its contents sent and received by an MS/UE from the base station over the air interface. The air interface investigation tool can be also used to analyze the signal level and its quality for the serving as well as the neighbor cells. This information helps greatly in finding any handover or interference-related issues.
- *IP Transport Investigation Tool* – It can be used to collect logs and debug mobile communications protocols issues that work on top of the existing IP transport interface.
- *Open Interface Investigation Tool* – It can be used to collect logs and debug open protocols issues such as the SS7, ATM, and Frame Relay.
- *Proprietary Tool* – Such tools are supplied and can be used to debug an issue with an embedded system board that contains a proprietary system on chips (SOC).

12.7 Steps for Troubleshooting

In this section, we will discuss the UE/MS, radio access, and core network services-related issues and their troubleshooting approaches for a particular live communications network, i.e. GSM, GPRS, UMTS, LTE, and 5G system. The troubleshooting approaches may vary depending on several factors, deployment scenarios with various physical and logical interfaces as well as the preliminary expert judgment being applied. Below are the typical steps that can be followed to resolve network services-related issues:

Step 1. Determine the communication domain of the issue being reported, i.e. whether it is a CS domain or PS domain issue. Once the domain is identified, the concerned network elements are known toward the next step of troubleshooting.

Step 2. Gather the preliminary information on the issue reported. Some of the network services-related information to be collected are as follows, i.e. whether the issue is

- a single user related;
- a large number of users/cell outage or network outages related;
- key performance identifier (KPI) performance related;
- intra-RAT or inter-RAT related;
- 3GPP interworking and feature related, e.g. circuit-switched fallback (CSFB), Single Radio Voice Call Continuity (SRVCC), Handover procedure related issues;
- roaming related;
- voice over IMS related;
- reproducibility of issues;
- time of day the issue occurs;
- device drivers, time synchronization-related issues, and so on; and
- a particular 5G use case and its network slice-related issues and so on.

Step 3. If the issue is about the

a) Single User Call/Traffic-Related Issues

Look at it from the RAN, e.g. BSC, point of view. There could be several reasons for a call failure. For example, call failure may result because of the unavailability of radio resources/channels, signaling failure, and congestion in the network.

- Identify the target cell and look at its resource-related counters maintained by the RAN. Look at the failure-related counters (e.g. Section 11.1) maintained for the target cell that could hint at the reason for the call failure. If the call failure is due to resource unavailability, look at the allocated and available resources in the target cell at the time of serving the call. If all the available resources in the target cell are occupied by the ongoing calls, then further call failure is justified; otherwise, there may be issues in the radio resources management (RRM) program module where resources, i.e. are in blocked or in different states, are not being managed properly. For example, the RRM module may mark a timeslot in a blocked state, whereas it is working at the BTS end.

b) Signaling-Related Issues

Look at it from the RAN as well as the core network point of view.

- For RAN-related issues, look for the possibility of signaling resource unavailability at the RAN end as mentioned above.
- For CN-related issues, look for the subscriber status at the CN node, i.e. MSC, or SGSN, Home Location Register (HLR), Mobility Management Entity (MME), and 5G core Access and Mobility Management Function (AMF). Ensure that the subscriber is not barred, the MS/UE is authenticated successfully, and the subscriber subscribes to the requested services from the network. To rule out such possible issues, end-to-end logs, from each interface, from MS/UE to core network may be required.

Successful completion of a signaling procedure plays a very crucial role in all the phases of establishing and maintaining an ongoing call. Because of a signaling failure, an MS/UE would not be able to register with the core network and use its services.

c) Handover-Related Issues

Handover failure-related issues arise whenever the ongoing call for a moving subscriber could not be handed over to the target cell. The reason for handover failure could be due to the unavailability of resources at the target cell; a signaling failure; and a time synchronization issue. Scope and steps for troubleshooting of a handover-related issue shall depend on the type of handover, i.e. intra-RAT or inter-RAT/inter-system, and also network elements involved in the reported issue. For example, in the case of the LTE system, a handover type could be an X2 interface or S1 interface based. Even if an X2 interface exists between two eNodeBs, an S1 interface-based handover may be executed in case an X2 interface handover is not successful because of rejection by the target eNodeB or interface-related issues.

A handover attempt may also fail because of the physical channel failure, expiry of a timer, or handover from an LTE to UMTS terrestrial radio access network (UTRAN); LTE to 5G; or UTRAN to GSM system.

d) Air Interface-Related Issues

The air interface of a mobile communications system is exposed to varieties of environmental conditions and interferences from the neighbor cells or other Public Land Mobile Network (PLMN) cells. Air interface issue may significantly impact call, handover success rate as well as the throughput of data services. To study the air interface, its surrounding cells and their assigned frequencies, and any interferences among them as well as power levels, a drive test can be performed using an appropriate air interface protocol analysis tool. The analysis of air interface logs shall be different depending on the communications system, i.e. GSM, GPRS, UMTS, LTE, and 5G system, being used.

e) Data Throughput Issues

The quality of the RF condition/air interface plays a significant role in providing an expected data throughput to a subscriber. Based on the current state of the air interface, the RAN adjusts the resources that are allocated to a UE/MS over the air interface accordingly. This is achieved by lowering or upgrading the modulation and coding scheme (MCS) allocated to the ongoing PS session of a UE. MCS details along with the RLC Layer 2 behavior over the air

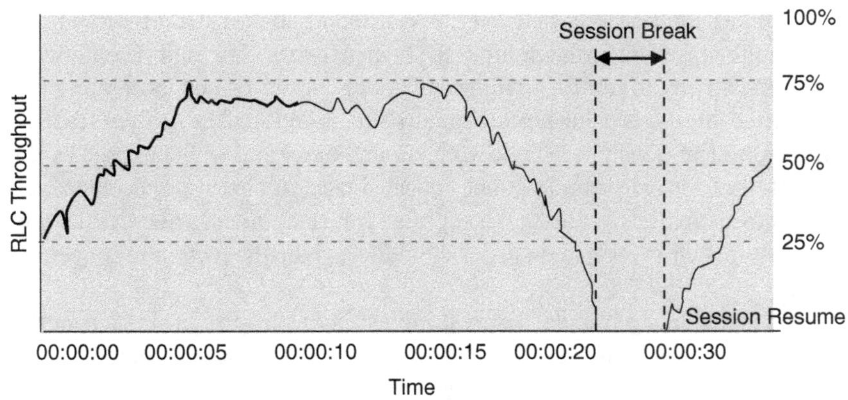

Figure 12.6 Illustration: air interface RLC layer data throughput.

interface can be studied using an appropriate air interface protocol analysis tool. Figure 12.6 illustrates a sample plot of the RLC layer throughput in the downlink direction for a typical data session in an LTE/EPS or 5G network.

As illustrated in Figure 12.6, the RLC layer behavior is a ramp-up in nature and throughput is not constant. Also, there is a break in the data session as shown by the absence of the RLC layer graph. If no issues are found from the drive test and air interface log analysis, the root cause of the throughput issue shall be attempted further by looking at the logs collected from other logical interfaces. It may be required to compare the air interface frames Receive (RX)/Transmit (TX) time along with the packets RX/TX time at other concerned interfaces. Any major RX/TX time differences shall provide hints toward the root cause of the data throughput issue. In summary, to rule out such data throughput issues, end-to-end interface logs from MS/UE to the core network may be collected and analyzed.

f) Multiple Users, Cell, or Network Outages Issues

These are severe issues affecting a large number of subscribers in the affected service area of a PLMN operator. The reason could be the hardware breakdown or interface-related issues including physical media such as the fiber optical cable cut. Services outages may also take place because critical software processes crash at the network end. For such types of outages, the issue is to be addressed in inter-domain approaches by looking at the access network and core network elements for possible causes. Available alarm data may also indicate the faulty network element. A detailed RCA is required to arrive and prevent it from a further cell or network outages.

g) Low KPI Issues

A KPI issue is observed at the network level that represents the performance degradations in terms of the signaling and services aspect of a mobile communications network. A poor KPI results call for in-depth investigation toward its root cause, which may sometimes require to look beyond its network, especially for air interface-related issues. A KPI for a particular procedure/scenario is derived from the fundamental counters that are maintained by the concerned network element and its protocol layer. The fundamental counters are updated by a network element as a result of the occurrence of a protocol layer event as well as during a call processing stage while serving both the signaling and user traffic. A dip in a KPI should be addressed by looking at the concerned scenario and its associated signaling messages where the corresponding fundamental counters are updated by the concerned protocol layer. Verify that they are being updated correctly in the code.

h) 3GPP Interworking, Features, and Roaming-Related Issues

3GPP interworking procedures and features were described earlier in Chapter 6. An interworking procedure, such as CSFB and SRVCC, involves inter-RAT communications, and it works by exchanging information through multiple physical as well as logical interfaces. 3GPP features, such as Multi-operator Core Network (MOCN) and

CS paging coordination, also involve multiples interfaces as well as PLMNs. Resolution of a 3GPP Interworking and features-related issue is more challenging as it involves multiple RATs or network elements. The network elements may be supplied by different OEMs. Thus, to resolve an internetwork or feature-related issue, appropriate logs shall be collected from the concerned interfaces using protocol analysis tools and shall be analyzed further by domain experts simultaneously. If the physical transport networks, e.g. Frame Relay and IP, being used by the concerned logical interfaces are different, then logs are required to be collected using different protocol analysis tools, which is another challenge. If a protocol analysis based on the collected information does not reveal a root cause of the interworking or feature issue, further investigation areas should be explored beyond the concerned logical interfaces.

To rule out whether the issue reported is due to a particular interworking scenario/feature or it is a generic/legacy network cause, the same can be verified by enabling and disabling the interworking scenario/feature. The outcome of such a test shall provide a new direction for further investigation of the issue reported.

i) Device Driver, Time Synchronization-Related Issues

Sometimes, the lower level platform-related issues can result in issues at the higher layer application protocols, leading to disruptions in the communication services. For example, if an embedded system-based network element is not using the IP transport network on top of Ethernet but using another interface such as Frame Relay or ATM, then the device driver for it must be a stable one so that there is no time for synchronization issues. Otherwise, upper protocol Layer 2 frames may not be transmitted and received by the device driver at the desired time, leading to out-of-synchronization frames. Such out-of-synchronized frames cannot be processed by a device driver because of cyclic redundancy check (CRC) and other errors. The out-of-sync frame shall be discarded by the device driver without forwarding it to the higher application layer. Such issues require low-level debugging skills with hardware specific tools.

In all of the above scenarios that are leading to various communications services issues, the root cause identification and its elimination require a step-by-step analysis of higher and lower layer protocols details using the available information which is collected from the field. Even within a particular network element, one should be able to analyze all the layers of a particular logical interface and look at the contents of the individual messages exchanged from the higher to the lower layer and vice versa. One should be able to identify the access network and core network protocol interfaces and their layers where performance issues are suspected. In the absence of available logs/traces from the suspected interfaces and network elements, the counters (e.g. Section 11.1) maintained at the concerned network element can be useful for offline analysis of the reported issue.

Chapter Summary

In this chapter, we presented the different types of day-to-day network issues that could prevent mobile communications networks from providing communications services smoothly to their subscribers. Typical tools and methods that can be used to debug and resolve different types of network issues were presented.

A running network issue may affect a single user, multiple users, or it may affect the entire network. Accordingly, each network issue has its challenges in terms of its root cause identification and resolution. The appearance of an issue may be a one time, random, or repeated one in nature. A one time or random issue reported from a live and matured network is especially more challenging to address and provide its resolution in terms of its non reproducibility at a lab, and also, no sufficient information, for the developer, about the issue is available from the operator end. We closed this chapter with some of the typical mobile communications network issues that may be reported from the field along with their approaches for troubleshooting and finding their solutions.

Part III

Mobile Communications Systems Development

In Part I and Part II of this book, we have presented the background information and various aspects of mobile communications systems and networks based on the available communication technologies, right from the GSM to the 5G system. In this part of the book, the actual development and realization of a 3GPP-compliant mobile communications systems network element, based on the GSM to the 5G system, shall be described and illustrated in three chapters. The purposes of the chapters are as follows:

Chapter 13 describes and illustrates the core software development fundamentals in terms of the software development life cycle which is followed during a typical software development process. This chapter also presents the other aspect of the design and development requirements of an application protocol layer or a network element.

Chapter 14 describes and illustrates the protocol layer formats and data structures used by the Air interface and core network protocol layers which are based on the concerned 3GPP technical specifications of mobile communications systems and networks. This chapter also presents the other aspect of the design and development requirements as far the 3GPP protocol layers are concerned.

Chapter 15 describes and illustrates the preparation of requirements specifications of new functions, procedures, and realization of features from the concerned 3GPP technical specifications.

The materials presented in these chapters are accompanied by sample illustrations on how the information available in a 3GPP TS for a particular protocol layer and its procedure, function, algorithm, and so on could be converted into a requirement specification, high-level design and low-level design, i.e. coding, for the realization of a particular network element.

13

Core Software Development Fundamentals

Introduction

By now, the reader must have developed application software of the following types:

- Desktop GUI applications
- Console and background applications
- Database, web applications, and so on.

The software system development requirement for a typical mobile communications system is a bit different from the traditional desktop- or server-based application developments. For the traditional, general purpose desktop and database application development environment, close interaction with the underlying hardware platform is not required. Development of a typical mobile communications application software system requires sound knowledge and skills on complex data structures, low-level programming as well as the internals and advance features of the operating system.

This chapter begins with the general and core software development fundamentals that are known as the software development life cycle (SDLC) in the software engineering paradigm. We then present the various hardware as well as the software development tools and platforms available based on which a mobile communications network element can be designed and developed. This chapter also covers some of the basic aspects of software engineering practices that can be used for the development of application protocol layers of mobile communications systems.

13.1 A Brief on Software Development Fundamentals

Let us first briefly go through the software development fundamentals, in general, that applies to any kind of application software or telecommunication software system development project as well. Generally, a software system, e.g. an application, a protocol layer, or a product, development project is divided into several tasks/phases that may be repetitive in nature depending on the scope of the project. These phases are collectively known as the *SDLC* in the software engineering paradigm, which is illustrated in Figure 13.1.

- Requirements
- Design
- Implementation
- Integration and testing
- Operation and maintenance

Mobile Communications Systems Development: A Practical Introduction to System Understanding, Implementation, and Deployment, First Edition. Rajib Taid.

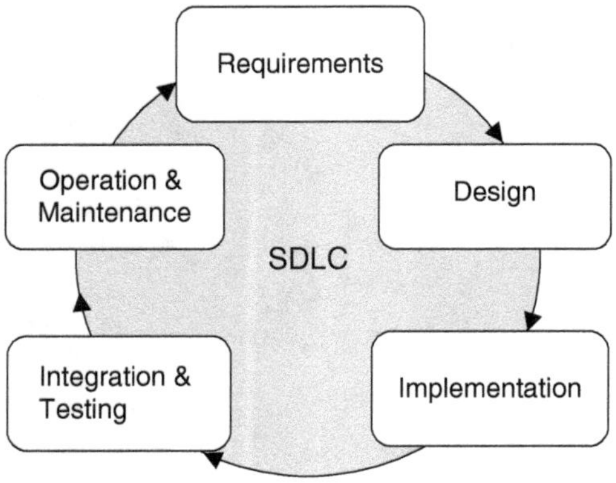

Figure 13.1 Phases of software development process and its SDLC.

It is assumed that the reader of this book is familiar with the different phases of a typical SDLC shown in Figure 13.1. Also, every project has certain deliverables that are to be produced in each phase mentioned above and is provided to the customer as part of the project plan. Typical deliverables in each SDLC phase are described below from a mobile communications system and its protocol layer development point of view.

13.1.1 Requirements Phase

- Requirements Specifications/Technical Specifications

As a developer, requirement specifications shall be derived from several sources such as the client's specifications, business objectives, and requirements derived from the standard technical/protocol specifications. For example, one can derive the function and procedure requirements of a network element of a mobile communications system from the various 3rd Generation Partnership Project (3GPP) technical specifications. A 3GPP technical specification describes a particular protocol layer or general services aspects. More about the identification of protocol layer requirements, its functions, and procedures from 3GPP technical specifications shall be described in Chapter 15.

- Requirements Traceability Matrix (RTM)

An RTM traces the requirements of the intended functions and procedures of a protocol layer against its corresponding implementation in the code or a document. Using an RTM, a client may verify that the software system implements and meets all the intended specifications or business requirements.

13.1.2 Design

- High-Level Design Document (HLD)

In an HLD, a developer will describe and document the protocol layer and its software architecture, modules, organization/integration, interfaces, various algorithms, flow diagrams, and so on, as illustrated across the chapters of this book. An HLD does not specify the low-level implementations aspect of coding.

- Low-Level Design Document (LLD)

In an LLD, a developer will describe and document the module-specific requirements at a deep level such as the kind of data structures for storing and retrieval of information; the definition of system parameters, various declaration, algorithms, procedures; and so on. The developer will mention the implementation of a particular function, algorithm, and procedure in pseudo/actual code. Chapter 21 describes and illustrates the LLD aspects in terms of software coding.

13.1.3 Implementation

In an implementation document, a developer will describe the kind of development environment that shall be used for the development of a network element. The development environment may be as follows:

- Desktop Environment: Legacy Multiuser Environment such as UNIX System.
- Embedded system including a real-time operating system (RTOS), hardware processor, and board. The processor could be either multicore or special purpose digital signal processor (DSP).
- Development languages such as the C and C++ and proprietary development tools.
- Source code control system for versions control of different releases/fixes.

13.1.4 Integration and Testing

In this phase, a developer will prepare the test plan and strategy to be used for testing a network element so that it meets the technical specifications as well as the customer's business requirements or objectives. A test plan and its strategy are shared with the client. A test plan typically includes the following aspects:

- Type of test setup environment, i.e. laboratory or field or actual or simulated network element, and network configurations.
- Release or version of the network element under test.
- Types of testing, i.e. quality assurance testing, unit testing, system integration testing, interoperability testing, performance testing, load testing, protocol layer conformance testing, release testing, drive testing, and so on.
- Preparation of test cases, test scripts, and manual or automated testing.
- Results of test cases, any exceptions.

Note that in an SDLC, a software testing activity is repetitive and ongoing in nature. Generally, a developer performs the unit and system integration testing. The remaining types of test methods are performed by the testing and verification team.

13.1.5 Operation and Maintenance

A particular network element will be shipped to the customer and deployed commercially once it has been tested and verified that it meets all the requirements, e.g. business objectives and technical requirements of the customer. During the initial phase of deployment and its operations, it is natural that hardware or software bugs/faults appear from the supplied communications software systems. The developer will fix such reported software faults during the operation and maintenance phase. In this phase, post-deployment support strategies will be prepared, such as the following ones:

- Software patches/workarounds during emergency cases such as cellular network outages
- Fixing of hardware or software bugs reported from the field operational system
- Software patches/release delivery and its management
- Development of new features/enhancement requirements based on the existing code baseline.

The subsequent sections describe the typical platforms available for the development of a network element as well as other design and development aspects of a mobile communications system.

13.2 Hardware Platforms: Embedded System, Linux Versus PC

The selection of a hardware platform, i.e. legacy versus embedded, could be a complex decision because of the complexity of the project as well as the commercially available various embedded system platforms. To select a target platform, detailed feasibility and comparative study should be performed on the available platforms that include both the technical and commercial aspects.

One should consider the various advanced features, described in Section 13.3, that support development as well as debugging tools and features provided by a particular operating system being identified. Available hardware platform could be Windows on Intel, Sun Solaris Unix on Sparc processor, and so on. The hardware platform could be also a *DSP* or an embedded version of LINUX/RTOS on an embedded processor, including multicore (more about it in a later section), based on PowerPC, ARM, and MIPS architectures.

There could be vendor-specific proprietary platforms too, both hardware and software. Choosing and selecting a target platform is the key before a particular network element could be designed and developed as its software architecture will have a close relationship with the target platform. Generally, a Unix-based or RTOS-based embedded *DSP* plug-in unit (PIU) platform is used to deploy the network elements (such as base station controller [BSC], mobile switching center [MSC], and so on) of a mobile communications network. This is because these platforms offer much more flexibility and security than other platforms such as Windows.

Apart from this, almost all of the *Unix/Linux* flavors offer a variety of useful tools (e.g. *perl, tcl*), utilities, and commands, such as *cut, awk, sed*, and so on, that can be used to perform varieties of customs tasks such as reporting, scripting, and system monitoring. A developer will be required to develop hundreds of lines of code to provide the same capabilities and functionalities for performing a similar task on a Windows platform.

13.2.1 System Development Using Embedded System Board

The reader must have already designed and developed application software using either the Windows or UNIX/Linux environment. These are disk-based development environments where the executable programs are stored on the hard disk drive. As normally does, the operating systems load the concerned program into the memory before it can be executed. But a typical runtime-embedded system or environment is a diskless system where the entire program along with the RTOS and its data is loaded into the memory and resides there till the program is executing in the memory.

On a traditional system like Windows or UNIX, one needs to just power on and log in to the system and start working on it using its development environment/tools/software development kit (SDK). However, software system design and development, for example, a network element of a mobile communications system, approach on an embedded board with an RTOS such as Linux OS is different from the traditional software developed on the Windows or UNIX platform. One such embedded system development model is illustrated in Figure 13.2.

Generally, a developer will develop a new or an embedded version of the application protocol layer module of a network element using an evaluation or prototype hardware board along with the supplied SDK. A prototype board will accompany all the necessary tools and tackles, such as a debugger kit, that are required to accomplish the development task. However, a prototype board does not include all the system support services and may not be ready to be used by an embedded operating system. A developer requires to develop all the required support services for a prototype board. One such support services package is described next.

- **Board Support Package (BSP)**

For a general purpose operating system to boot and run, there are support functions and boot processes right from pressing the power on button till the login window is presented to a user. A boot process initializes the various system services and resources such as processor, RAM, bus, clock, interrupt, NIC, and boot loader. But, in the world of embedded systems, a developer has to develop and write code to initialize the various system resources

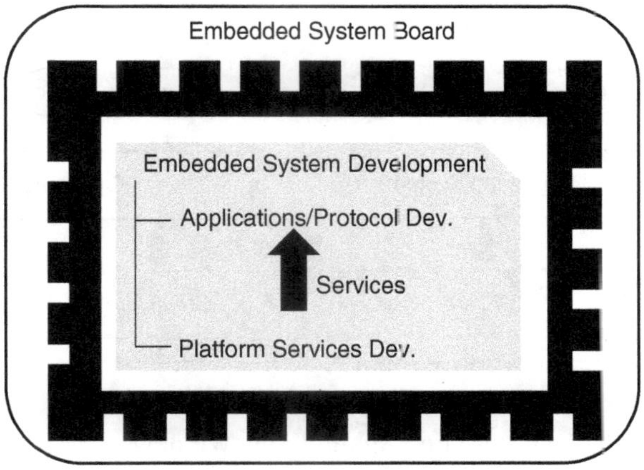

Figure 13.2 Illustration: embedded system development environment.

and devices. These are the support functions that make an RTOS as well as the other custom application software ready to run on top of the embedded system/board. Such coding, both low level and high level, and software developments that are done for different system resource areas are collectively known as the *board support package (*BSP*)*. A typical BSP consists of the following various hardware resource initialization functionalities:

- Boot loader
- Built-in self-test (BIST) or power-on self-test (POST)
- Device drivers
- Memory allocation routines
- Flash memory routines
- Interrupt controller routine
- Various register/driver initializers
- Serial port driver

If a developer chooses the Linux as the embedded operating system, it will be required to study the *GNU toolchain collection* to build a proper embedded system development environment. GNU toolchain supports the C/C++/assembly programming languages. It also contains Linker, Assembler, and an Integrated Development Environment (IDE).

Note that a BSP is dependent on the particular hardware model/type as well as the operating system selected. One needs to study thoroughly the selected hardware type and its reference manual to provide a stable platform for the applications or protocol layers to run and work properly. Apart from the development of the BSP on an embedded system using Linux, one may be required to customize the platform being used for project-specific requirements, development/modification of device drivers, interrupt controllers, and so on. More about the development of device drives can be found later in Section 14.21.

13.2.2 System Development Using Multicore Hardware Platform

Now, all of us know about a multiprocessor, multiprogramming computing platform as we have already studied them during the undergraduate course. The current buzzword is multicore, which is the new computing paradigm. The multicore platform is everywhere, be it desktop, server, smartphone, and so on.

To explore and experience a bit further on the multicore computing environment, open the desktop computer which is already running on top of general purpose commercial multicore processors. A multicore processor has multiple physical cores. Each physical core has multiple logical cores. To see the cores that are in services, open

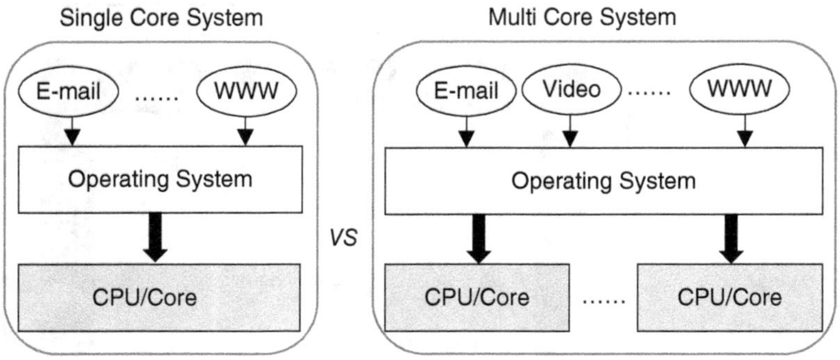

Figure 13.3 Illustration: single core and multicore processor systems.

Windows Resource Monitor (on Windows, select *Start→All Programs→Accessories→System Tools→Resource Monitor*). This shows the number of cores currently running on the desktop system.

13.2.2.1 What Is a Core?

A core is nothing but a processing engine or a unit of a processor. A multicore processor contains several cores that form a computing environment. Examples of single core and multicore platforms are shown in Figure 13.3.

13.2.2.2 Network Element Development Using Multicore Platform

An application is a good candidate for multiprocessing if it requires a large number of tasks to be completed, but the completion of one does not depend or depends only slightly on the completion of the other jobs. If the application is a single large job, but each piece of it depends on another, and also there is a high degree of required serialization, then a multicore platform may not be a good choice in this case.

But if we look at the protocol/stack and system architecture of the available mobile communications systems (Global System for Mobile Communication [GSM], General Packet Radio Service [GPRS], Universal Mobile Telecommunication System [UMTS], Long-Term Evolution [LTE], and 5G), one can choose to deploy a communications network element based on a multicore platform having 1. . .N cores.

A developer can take advantages of multiple core platform and then design and architect the software modules in the following way:

- Deploy the symmetric multicore platform (SMP) system;
- Have dedicated cores or groups of cores responsible for controlling and managing the control plan/slow path and user plan/fast path;
- Have dedicated cores or groups of cores responsible for controlling and managing the various logical interfaces such as air, Gb, Iu, and S1, and so on
- Have dedicated cores or groups of cores responsible for controlling and managing the system administration and operations and maintenance tasks; and
- Have dedicated core for handling uplink and downlink data flow.

Apart from the telecom and networking space, multicore processors and systems have their widespread use in OEM/commercial products, switches, routers, and other multiple market segments.

13.2.2.3 Runtime Choices of Multicore Processor

A multicore processor may provide multiple runtime environments or choices to run the application protocols modules:

- Simple executive environment – where a core executes and runs a stand-alone or specialized user application without the requirements of an operating system such as Linux.

- Symmetric multicore/multiprocessor environment – where an operating system could be run on multiple cores or a single operating system can control multiple cores. In this case, an application may also run in Kernel mode.

The choice of runtime environments is derived based on the application or the entire project's requirements. For example, one can run the Linux SMP operating system on certain cores, whereas the simple executive application(s) can be run on the remaining cores.

13.2.2.4 Software Programming Model for Multicore Processor

Two software programming models can be used during the design and development phase of a software module for the processing of communication services events. Those programming models are mentioned below:

- Run-to-completion programming model
- Pipelining programming model

In the *run-to-completion* programming model, a designated core would perform the processing of the entire sequences of tasks/packets for a particular communication services event. This leads to monolithic software design and can lead to unexpected issues during its maintenance.

In the *pipelining* programming model, different sequences of tasks/packets processing are divided into stages and assigned among the available cores. This leads to the modular software design and ease of maintenance in the long run. Consider the typical case of the GSM *call processing* scenario for the establishment of a mobile-originated call (Figure 2.8). In this call processing scenario, the BSC allocates the necessary radio resources to the requesting mobile device. While doing so, the GSM L3 sublayers – Call Control/Connection Management (CM), Mobility Management (MM), and Radio Resource (RR) layers – perform the required processing for the allocation of various resources for the concerned mobile device. In this case, the CM, MM, and RR protocol sublayers could be implemented separately (pipelining) and assigned to an individual core instead of a monolithic (run-to-completion) designed for processing end-to-end call processing tasks. Any bug correction or enhancements made shall affect the concerned layer only in the case of a pipelining model. The selection of a particular programming model based on a multicore processor shall depend on the application's requirements.

13.3 Selecting Software Platforms and Features

The usages of a software development platform have close links to the hardware and operating system platform that a developer has selected for a network element. Each hardware and operating system platform has its development environment in terms of the SDK.

Apart from this, development languages such as C and C++ are used to develop core elements of a mobile communications network. This is because these languages, in combination with inline assembly language, can be used to access and program hardware devices or boards in use, including processor and memory. Such capability may be required depending on the kind of system being developed as well as the operating system and hardware platform being used. Also, all the available operating systems allow the system services provided by it to be accessed through predefined sets of C/C++ application programming interfaces (APIs), for example, to perform an inter-process communication (IPC), process/thread creation/termination system calls, and so on.

As far the C/C++ languages are concerned, they expose low-level system facilities and also allow direct calls to native system libraries, so have full access to the features and performance of the platform on which the software runs can be performed using the C/C++ languages. A developer is also required to be an expert in using and handling of the various data types, operators, and advanced data structures such as the following ones while implementing protocol layers:

- Structures, Arrays, Pointers, Union, and Bit-field
- Advanced data structures – Tree, Stack, Linked List, and Queue
- Right shift(>>), left shift (<<), Bitwsie &, Bitwise | operators

Such operators/data types offered by the programming languages are used in coding/decoding of protocol information. One can also use third-party software library tools, e.g. *Standard Template Library (*STL*),* to cut down the overall development time. This will also reduce the software maintenance cost over a period of time. If one has selected the open system platform such as the Linux operating system, then the *GNU tool-chain* collection can be used to deploy that supports the C/C++/assembly programming languages. GNU tool-chain also contains a Linker, Assembler, and a rich IDE.

13.3.1 Selecting Available Data/Logical Structures

Next, while designing and developing mobile communications software, is the data structures to be used in the application protocol module. If a particular data structure is not suitable for an application protocol, the essential data structures will be required to be built by the developer. Alternatively, one may go for any third-party software library tools and integrate them into the application protocol module.

Subsequent sections describe the usages and selection of computer data structures for storing and retrieval of information of a network element.

13.3.1.1 Advanced Data Structures

A developer must have already studied the following advanced data structures as part of the graduate course:

- Linked List,
- Stack, Queue,
- Hash Table,
- Trees.

Such data structures find different applications, and a developer will be required to choose a particular, or a combination, depending on the nature of the application protocol layer, kind of data it is going to store along with its performance requirements. Figure 13.4 illustrates the storing and retrieval of a typical User Equipment (UE) context, in the 5G system, containing other structures or pointers to structures, using a hash table and linked list data structure.

Typically, there are two types of information: static, such as a mobile subscriber database with credentials, and dynamic, such as the number of mobile users that comes and goes in a radio access network. A network element contains a great deal of information of different types. In this case, one will be required to use any combinations of the above data structures for ease of storing/retrieval of a great deal of information that grows or shrinks dynamically. For example, a mobile device's index fetched from a hash table could be used to search a tree structure that stores and contains the details about the concerned mobile device.

A developer will be required to identify all types of data that are going to be stored for a particular network element and then select the best suitable data structures for storing and retrieval of each type of data.

13.3.1.2 How Data Structure Affects the Application's Performance

Another important aspect is the selection of the appropriate data structure whenever there are alternatives ways to store, retrieve, or modify information by an application protocol module. Data structures to be used by an algorithm or a function/procedure can greatly affect its performance, including the performance of the concerned

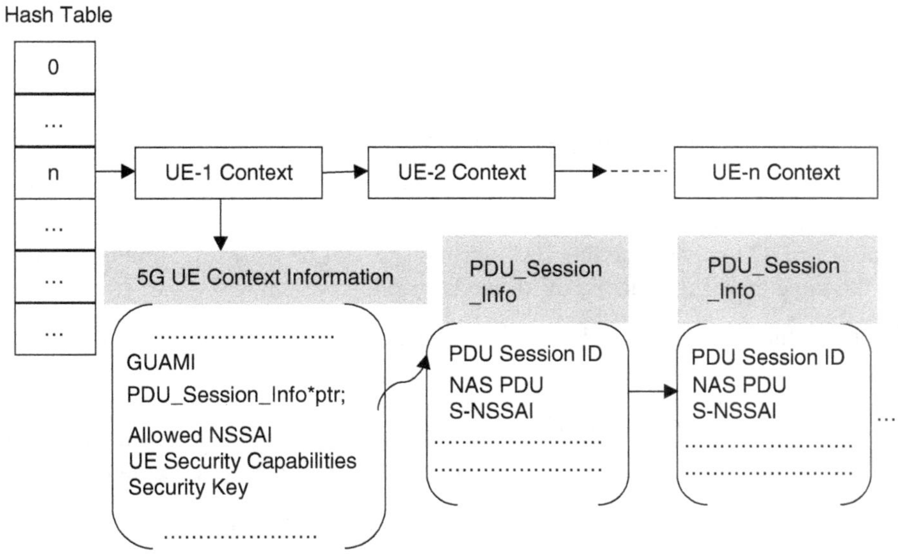

Figure 13.4 Illustration: hash table and linked list of 5G UE context structures.

application protocol module. For example, to store, retrieve, or modify a large amount of data, the best data structures to be chosen are the self-balancing binary trees such as the AVL tree, Red-Black tree, and B-tree, because they provide better searching time (Log2N) than an array (O(N)), which takes a linear searching time.

Usages of other data structures such as the array versus linked list versus hash table may be compared and adopted suitably. Hash table finds several applications for storing and retrieval of different logical objects of a mobile communications system.

13.3.2 Selecting an Operating System Services/Facilities

To design and develop mobile communications software systems, one must have the knowledge and skills on the advanced features being provided by the chosen operating system. On a UNIX platform, of particular interests are the requirements of the advanced knowledge on topics such as the threads and multithreaded programming, differences among the semaphore, mutex, condition variables,, and so on. One must also know the differences between the System-V and POSIX IPC mechanisms, including socket, its types, and usages. As a developer, one needs to make a choice that is appropriate for the application protocol module under development.

One of the most important resources being provided by an operating system platform is the *timer* facility. Every protocol layer, be it on mobile station (MS), BSC, MSC, Mobility Management Entity (MME), 5G core Access and Mobility Management Function (AMF), and so on, has its own sets of timers that are used to keep track of a particular signaling procedure being initiated toward the peer. At the expiry of the concerned timer, appropriate predefined action is taken. A sound understanding of the timer functionality being provided, either by the hardware or software platform, is very important.

13.3.2.1 Advance Features of Operating System: IPC

One of the advanced features which are of particular importance is *the Inter-Process Communication (IPC)* mechanism facility provided by an operating system. An IPC mechanism can be used for communicating and sharing of data among the co-operating processes. There are several forms of IPC mechanisms, with different properties, that are provided by the different operating systems as the programming elements. Some of the IPC mechanisms available on different platforms are as follows:

- Message Queues, Shared Memory, Pipe, and Semaphores;
- Signals (on UNIX), Events, and Remote Procedure Call (RPC) on Windows); and
- Clock and Timer.

Based on the applications as well as a project requirement, one may use several forms of IPC mechanisms. Depending on the properties of each IPC method, suitable IPC mechanisms can be used for communications among the process running on the same machine or across the different machines as follows:

- Applications/processes, representing various protocol layers, running on a local system or multiple systems that are connected by a network.
- Applications/processes, representing various protocol layers, running on different operating system platforms. For example, consider an application: X running on Windows for a particular network element whereas an application: Y running on Unix/Linux or embedded platform for another network element.

The above scenarios with different IPC mechanisms are illustrated in Figure 13.5. In this illustration, note the following:

- Processes A, B, C and D are running on a local Unix/Linux machine.
- Process E is running on a local Windows machine.
- Processes A, B, and Message Dispatcher communicate with each other using message queues. The direction of the arrow shows the direction of the message flow.
- Process Message Dispatcher is the external interface for the processes running on the Unix/Linux machine.
- Processes C and D communicate with each other using shared memory with synchronized access through a semaphore.

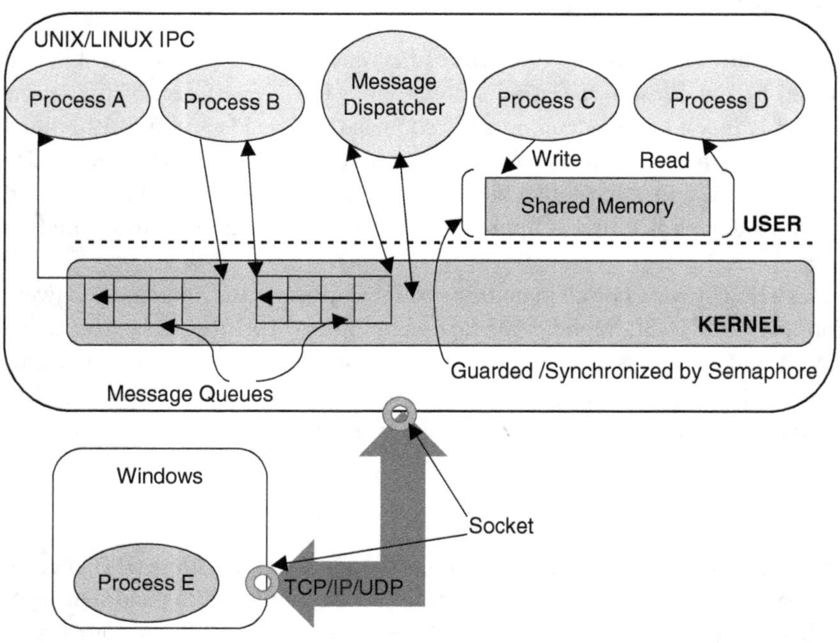

Figure 13.5 Illustration: IPC among the processes running on Windows and UNIX platforms.

- Process E on the Windows machine communicates using TCP/IP/UDP socket with the process Message Dispatcher running on the Unix machine. Message Dispatcher writes the messages intended for the Process B in its message queue.

Note that the message queues are created in the Kernel space, whereas the shared memory is created in the user space. Every message and message queue invites the Kernel's involvement making it a bit slower than the shared memory from the performance point of view.

- **Signals**

A signal is a simple but powerful way to raise an asynchronous event that is notified and delivered toward a process. One can think of it as an arrival of an interrupt either periodically or at any point in time. On Unix/Linux platform, there are around 30 system-defined signals that can be generated/raised from the different sources, i.e. the Kernel, system processes, and user processes, as illustrated in Figure 13.6.

Refer to the Unix/Linux manual pages for the description of each of the signals. Note that the Unix/Linux platform also facilitates defining and implementing a user-defined signal apart from the ones provided by the chosen operating system. Those user-defined signals are SIGUSR1 and SIGUSR2, as shown in the illustration in Figure 13.6. A user-defined signal can be used by a user process to another one to indicate the occurrence of a particular event, say the availability of a complete data frame.

Note that each of the signals has its handler function, like an interrupt handler routine, that is executed by a process upon arrival of the concerned signal. Except for SIGKILL and SIGSTOP, an operating system process has the following choices of action to be taken upon arrival of a signal. See the illustration shown in Figure 13.7.

- **Timers**

A *timer* is another important resource provided by an operating system that can be used for different purposes. Every 3GPP protocol layer contains several timers at the MS/UE and network end, which are started

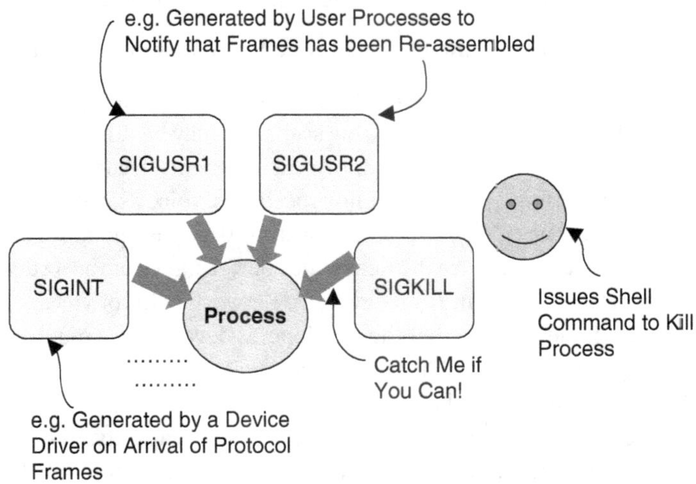

Figure 13.6 Different types of signals available on the UNIX/LINUX OS.

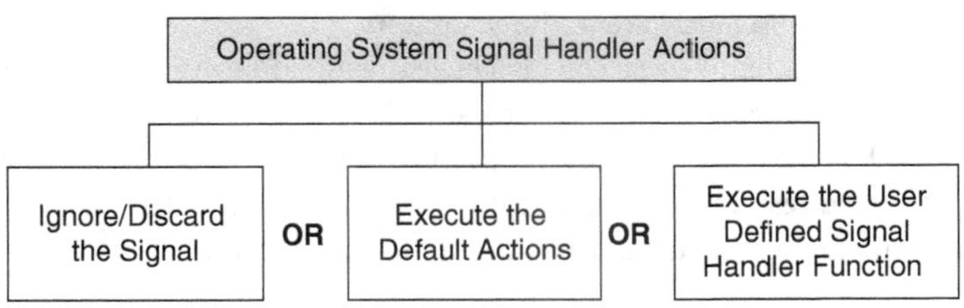

Figure 13.7 Different actions taken by a signal handler function.

independently to keep track of different activities. A developer will be required to implement various timers available in the concerned protocol layer as defined by a 3GPP technical specification. More about the implementation of timers are available in Section 14.14.

13.4 Software Simulators for a Mobile Communications Network

Now, the reader knows about a mobile communications network that consists of various network elements interconnected with each other following the open standards as defined by the 3GPP. So, during the development and testing phase of a network element, it would be required to test all the functionalities and features as defined by the concerned 3GPP technical specifications.

A network element has physical interfaces with another network element from different OEM or third-party network elements. In a laboratory, a developer may not have the required and expensive third-party network elements from different OEMs to be integrated and test it with the developer's network element. Also, sometimes, a system does not work always as expected. In such a situation, a software simulator that will behave like a real live network element/environment with the capability of creating both expected and unexpected or abnormal scenarios for various signaling and user plane functions and procedures comes handy. In the absence of an actual network element and field-like scenarios with many mobile devices, a simulator in a laboratory environment can be used for the following purposes:

- **Software Bug Fixing and Verification for Abnormal, Erroneous Scenarios**

Note that not all types of software bugs could be caught and fixed during the testing, integration phase of a network element at the laboratory. For a difficult-to-reproduce bug which is reported from a real live network at a later phase, the usage of a simulator is of great help in this case. A software simulator helps in reproducing and fixing the bug in the observed scenario by exchanging with or blocking a concerned signaling message with the actual system for both successful and failure scenarios, containing erroneous data if required.

- **System Capacity and Performance Load Test and Verification**

A network element may sometimes behave unexpectedly and fail to perform at peak load conditions in the field. Such a field-like traffic or signaling scenarios may be difficult to generate even with the actual hardware and software deployed in a typical laboratory environment. In such a condition, a software simulator can be used to generate a load test for various signaling procedures, voice as well as packet-switched (PS) call with varying input traffic conditions, time of day, and other constraints or parameters specific to a particular network. From the simulator results, one can measure the near about performance with the designed capacity of the application protocol module or network element from a real live network point of view.

For example, assume that the developer is developing a module for a customer to support a large cell site, which will require a large number of radio channels equipment such as base transceiver station (BTS) and trans-receiver (TRX). A developer's laboratory may not be equipped and have such a large number of physical resources to test the application protocol module to its full system capacity. In such a scenario, a network software simulator comes handy, where the developer needs to configure the simulator to support the required system capacity through software support only without any actual hardware and generate loads/traffic, thus saving from the requirements of actual radio hardware in the laboratory.

Using simulators, one can also measure and improve the performance of a network element in terms of its reliability, availability, and memory and CPU utilization, a number of calls being supported during a busy-hour period. A network operator or customer will verify and compare such important performance matrices from the simulated result prior to the deployment of the concerned network element. Later, in Section 14.24, various types of software testing using a simulator are also discussed.

13.5 Software Root Causes and Their Debugging

A mobile communications network service may be affected because of the unexpected behavior of a network element. Leaving aside the other external factors, such behavior could be because of poorly designed or carelessly written software. Debugging procedure of software issues could be different depending on the platform, e.g. embedded or non-embedded, being chosen to deploy the network element. In this case, a developer needs to follow the debugging procedure being identified for a particular platform with the available tools supported by it. One can use a commercially available third-party tool also to aid in debugging a software module and make it bug-free. A developer can also develop reusable test stubs or simulators to simulate a real and a difficult-to-reproduce field-like scenario in a laboratory.

13.5.1 Incorrect Usages of Software Library System Calls/APIs

Apart from the software issues that could result due to poor design aspects of a protocol layer, there are other aspects whose negligence or careless usages can result in a random and unpredictable system behavior at runtime in the field. Those are the areas that should not be overlooked while using their various facilities being provided by the particular platform/operating system/development languages. Some common facilities that are provided by every platform are memory manipulations routines, string manipulations routines, buffer manipulations routines, and so on. These facilities are provided in terms of reusable routine or C programming language library of APIs. Incorrect usage of these routines may result in the undefined behavior of an application protocol module at different points of time. Knowing the correct, intended usages and the desired result of each API is of utmost importance.

13.5.2 Incorrect Usages of System Resources

Similarly, there are quite several programming language usage rules to be followed by a developer as part of the best programming practices. It is especially true as far as the C/C++ languages are concerned. It is easy to forget or ignore such practices, but the same may throw up some surprises at a later phase after a network element has been shipped to the customer. This would add to the overall software maintenance cost. For example, it is easy to forget to release a piece of memory before returning from a function or once the piece of memory is no longer needed. This will result in memory leaks, leading to performance degradation of the concerned application protocol module. Sometimes, an application protocol module could stop responding because of a lack of required memory, leading to a failure while trying to fork a process.

13.5.3 Bad Software Programming Practices

As described in the previous section, there are development language and platform-specific facilities usages rules whose improper usages can result in a random and unpredictable system behavior at runtime at the field. Below are some of the bad programming practices using C/C++ languages that a developer should be aware of and pay close attention to during the coding and unit testing phase:

- Forget to provide a user-defined copy constructor and an assignment operator for an object, leading to undefined behavior of a program.
- Forget to release a piece of memory after it is no longer needed, leading to memory leaks and performance degradations.
- Forget to release a piece of memory before returning from an erroneous scenario, leading to memory leaks and performance degradations.

- Writing to an array beyond its boundary reading from an array beyond its boundary, either static or dynamic.
- Forget to lock a global variable using a semaphore before accessing it.
- Forget to unlock a global variable using a semaphore after accessing it.
- Lack of a virtual destructor for a base class in case memory is allocated in a derived class.

13.6 Static Code Analysis of Software

As described in Section 13.1, the last phase of an SDLC is the operation and maintenance of a particular system that has been built. As a software developer, one will spend the rest of the SDLC time in identifying and correcting software bugs once the network element or the application protocol module is designed, developed, and shipped to the customer.

An application protocol module developed without paying close attention or following the good programming practices, as mentioned in Section 13.6.3, or even if it was followed, is likely to introduce unexpected software pitfalls without the developer's knowledge. This will result in the low software quality, compliance issues, service unavailability, and requiring more maintenance of a protocol layer. Some of the typical software bugs that one could introduce are mentioned below:

- Software concurrency issues such as deadlocks, race conditions, and blocking call misuses.
- Software performance degradation problems due to memory leaks.
- Software crashes, causing errors such as a null pointer to dereference, use-after-free, double free, and improper memory allocations and mismatched array new/delete.
- Incorrect program behavior caused by dead code, uninitialized variables, invalid use of negative variables, etc.
- Improper use of APIs with STL usage errors, etc.
- Security vulnerabilities due to buffer overflows, insufficient validation, etc.

One can deploy a commercially available third-party software tool to identify and fix all such possible software bugs mentioned above and improve its quality before it is shipped to the customer. Such a software code analysis tool performs an inspection of the entire code during the compile time itself. This process is known as the *static code analysis* method which produces a pinpointing report showing all the possible bugs in the application protocol module. Such a static code analyzer tool avoids the effort-intensive manual code inspection requirements for a large project. It also increases developer productivity by finding and helping them to fix the defects at an early phase in a faster way rather than leaving it till the same is not discovered either at a laboratory or reported from the customer site.

13.7 Software Architecture and Software Organization

A fair knowledge about the software architecture and software organization will help in understanding a protocol stack and its layers. *Software architecture* refers to the different building blocks, in terms of protocol layers and associated support functions, of a particular system or a network element. As an example, Figure 13.8 shows the various management program blocks/units for a typical GSM BSC.

- **Software Organization**

Software organization refers to the arrangement of the identified software building blocks of a particular entity or system. Look at the reference architecture of the OSI 7 layers and its organization of the different layers. One may also look at the architecture and organization of the protocol stack of different protocol stacks.

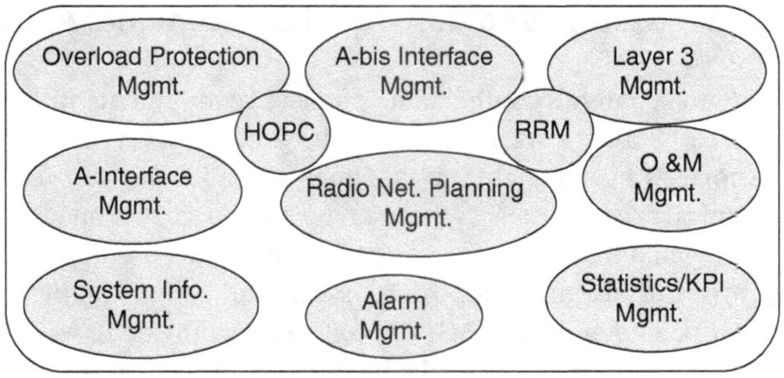

Figure 13.8 Illustration: typical software architecture for a GSM BSC.

RRM: Radio Resources Management, HOPC: Handover and Power Control

Note that the target hardware/platform (e.g. normal processor or a DSP) being chosen will have an impact on the application software architecture as well as the software organization. One may choose to run an application on a normal processor, whereas another application may be designed to run on a DSP. One may also utilize the power of a multicore processor while designing and developing a network element.

Example 13.1 illustrates the conceptual architecture and organization of the various logical objects used in a typical GSM RR allocation and management algorithm.

Example 13.1 GSM RR Allocation and Management (RRM) and Its Object Model

Consider that a developer is implementing radio resources allocation/management algorithm for a GSM system. Such an algorithm will use several logical objects, each one representing the corresponding physical radio frequency (RF) equipment. The operator will configure the actual physical RF processing hardware such as the BSC, BCF, BTS, TRX, and TSLs. The RR layer will create corresponding logical objects against each physical resource. A developer will use the logical objects in the algorithm that will store information about a particular type of physical resources. In the resource allocation and management algorithm, a developer would likely model and organize the hierarchies (see Figure 13.9) of the above logical objects, starting from the top to bottom approach in a 1 : N (many) ratios.

Figure 13.9 GSM RRM logical object model and its hierarchy.

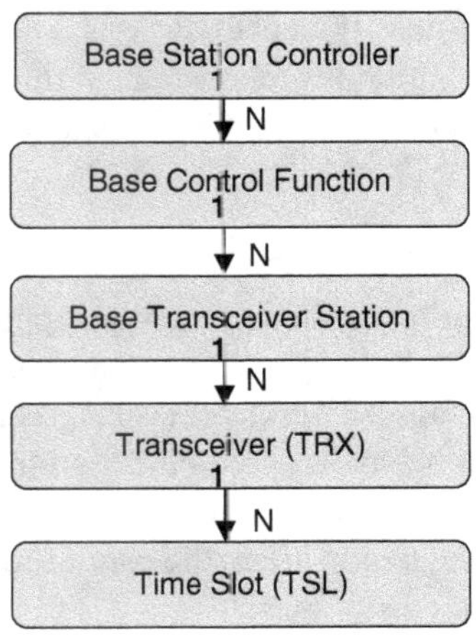

13.8 System and Software Requirements Analysis

One must carefully gather the complete requirements in terms of functionalities and procedures to be performed by a network element and its protocol layers from the concerned 3GPP technical specifications and customer's business objectives. Other dependencies and requirements shall be also derived from the related technical specifications as listed in the references section of a technical specification. Otherwise, a network element would be incomplete in terms of its intended functionalities. There would be requirement errors, project cost overheads, unexpected IOT issues, and so on, leading to an unsuccessful deployment of the concerned network element. Some 3GPP specifications provide flowcharts and algorithms in a pseudocode style. Certain functionalities and non-implied requirements are left as the implementation-dependent ones that are specified in a 3GPP TS. Such requirements should be also captured carefully during the software requirements and analysis phase.

13.9 Software Quality: Reliability, Scalability, and Availability

One must pay attention to the software quality aspects of an application protocol module during its design and development phase. Typical runtime software quality attributes are mentioned below.

13.9.1 Reliability

Reliability is the ability of a network element to continue providing its services to users in an expected way over a period of time under all the environmental conditions or circumstances. For example, the RF/antenna system, BTS, BSC, and so on of a mobile communications network must be working as expected under all the radio and environmental conditions such as terrain, atop a hill, rain, and fog. To ensure services to users, each of the mobile communications technologies (GSM, GPRS, UMTS, LTE, and 5G) provides its error control, correction, and recovery mechanism that needs to be designed and implemented as per the concerned 3GPP specifications.

13.9.2 Scalability

Based on the operator's business objectives, an application protocol module must be designed and developed in such a way that it supports network configurations from time to time. Because of such capabilities, the operator can perform a network provisioning task to achieve the scalability of a mobile communications network. For example, if an operator wants to increase its network coverage to serve more users, it should be possible to do so without software rework but through network configurations. For any major changes in the existing network element, a software release upgrade may be performed.

13.9.3 Availability

A network element and its application protocol layer modules of a mobile communications system must be available to provide communications services to users, say, 99.99% of the time. This is also defined as per the agreed Service Level Agreement (SLA) between an operator and the OEM. A network element should not behave unpredictably or crash upon receipt of an unexpected external trigger or event such as the congestions in the network due to sudden surges in the traffic. Such random unavailability of a network element would affect the communication services to users requiring a higher amount of software maintenances and costs.

Apart from the reliability, scalability, and availability attributes of a software system as described above, there are other quality attributes such as maintainability and reusability that one should pay close attention during the software design and development phase.

Chapter Summary

The software systems research and development environment and platforms used for a mobile communications system are generally different from the one used for the desktop environment or even a general-purpose server system. This chapter presented the core software development fundamentals and software development life cycle along with producing various artifacts as part of project deliverables. This chapter introduced the different choices of hardware platforms as well as the software toolchain and system, based on which a commercial and carrier-grade mobile communications system can be designed, architected, and developed.

A mobile communications system and network demands for its higher availability to provide various communications services to its users. Some of the good programming practices to be followed as well as other aspects of application software have been covered from the software engineering point of view. A good programming practice shall avoid software pitfalls, which will result in low software maintenance costs. Paying close attention to the good programming practices while designing and developing a network element also helps to realize the higher availability requirements of a communications network. We closed this chapter with another aspect of software and its quality in terms of its reliability, scalability, and availability.

14

Protocols, Protocol Stack Developments, and Testing

Introduction

So far, the reader has learned that mobile communications systems and networks consist of network elements that operate using different radio access technologies, standards, protocol stacks, layers, and interfaces among them. Even within a particular mobile communications system, different protocol stacks, and interfaces, encoding and decoding methods are used to exchange information between two network elements. Similar to the Open Systems Interconnection (OSI) reference model, mobile communications systems also apply the same protocol modeling principle where a group of protocols can be organized together to form a layered protocol stack.

In this chapter, we will cover the air interface Non-access Stratum (NAS) layer, Layer 3, Layer 2, and Radio Access Network (RAN) and core network protocol stack and layers of the available mobile communications systems, i.e. Global System for Mobile Communication (GSM), General Packet Radio Service (GPRS), Universal Mobile Telecommunication System (UMTS), Long-Term Evolution (LTE), and 5G system.

This chapter begins with the basics and typical components of a 3rd Generation Partnership Project (3GPP) technical specification (TS) based on which the requirement specifications shall be derived for a particular protocol layer and its network element. We then present the layer-to-layer as well as a peer-to-peer method of communications between protocol layers. Different message formats used by the air interface NAS layer, Layer 2, Layer 3, and core network protocols layers are also presented. Typical low-level design and its implementation requirements derived from a concerned 3GPP TS are also illustrated through sample code snippets. Several other aspects related to the design and development of the above protocol layers are also covered. We close this chapter with the protocol stack testing and validation task and its importance, given the complexities and various interfaces involved in a mobile communications network.

14.1 Components of a 3GPP Protocol TS

In Part I of this book, the reader has been introduced to the different types of 3GPP protocols being used in the GSM, GPRS, UMTS, LTE, and 5G mobile communications systems. A 3GPP protocol is described by its corresponding protocol layer TS as defined by the concerned group of 3GPP. A 3GPP TS describes the various aspects of a protocol layer. Apart from this, there are 3GPP TSs that describe the common as well as specific areas of a particular mobile communications system such as the GSM, GPRS, UMTS, LTE, and 5G. In this

Mobile Communications Systems Development: A Practical Introduction to System Understanding, Implementation, and Deployment, First Edition. Rajib Taid.

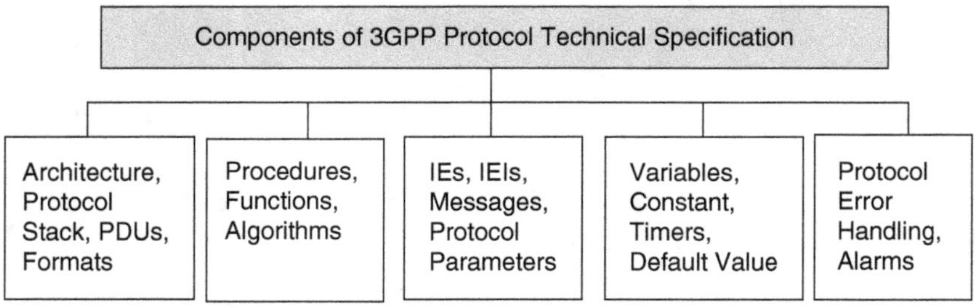

Figure 14.1 Components of a 3GPP TS for a protocol layer.

section, the components/sections of a 3GPP TS are described first. Typical components of a 3GPP TS are shown in Figure 14.1.

Each of these components of a 3GPP protocol layer specification is described below:

- **Protocol Layer Architecture**

This section contains the protocol layer architecture showing its position (layer) in the protocol stack. It also shows where does the particular protocol layer terminates, i.e. at the mobile station (MS)/User Equipment (UE), base station controller (BSC), RNC, NodeB, Mobility Management Entity (MME), or 5G Next-Generation Radio Access Network (NG-RAN) end. This section also contains the state machine architecture and the allowed functions to be performed in each state of the concerned protocol layer along with their allowed state transitions.

- **Protocol Layer Structured Procedures, Functions, Algorithms**

This section contains the detailed descriptions of each of the functions and procedures, with the accompanying flow of sequences, supported by the particular protocol layer. A TS of a protocol layer may also contain algorithms that may not be standardized and is left to the implementation-dependent.

- **Protocol Layer Protocol Data Units (PDUs), Information Elements (IEs), Formats, Definitions, and Parameters**

This section contains the PDUs, *service access point identifier* (SAPI), IEs, and other parameters. The particular format to be used, e.g. Type, Length, and Value of an IE, for the encoding of various information is also described here while communicating with an adjacent layer or peer layer on the other network element.

- **Protocol Layer Variables, Constant, Timers, and Default Value**

This section contains the concerned protocol layer's state variables, constants, and most importantly timers. Timers are used to keep track of a particular procedure being triggered toward an adjacent layer or layer on the peer network element. Upon expiry of a timer after certain retries, the appropriate action is taken by the protocol layer as specified by the concerned specification. Under this section, a protocol layer specification contains a list of system parameters, timers, and counters along with their default value to be used by the concerned protocol layer.

- **Protocol Layer Error Handling and Alarms**

Sometimes, a peer or an adjacent protocol layer may respond erroneously, giving rise to an abnormal condition or event to the initiating protocol layer entity. In such circumstances, the initiating protocol layer entity takes the appropriate actions as described in the concerned protocol layer specification and raises a corresponding alarm, described earlier in Chapter 10, to notify the system operator.

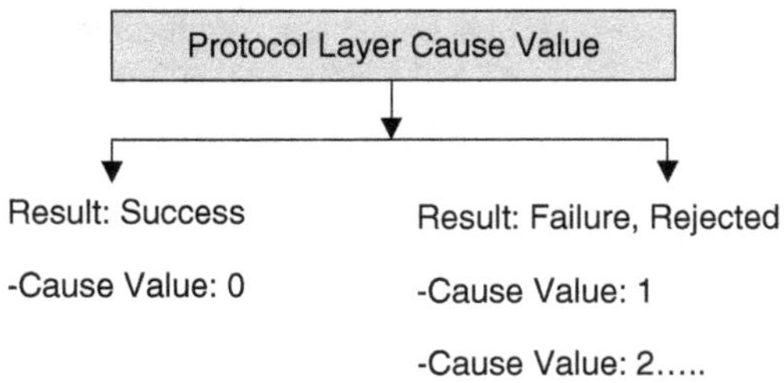

Figure 14.2 Illustration: protocol layer procedure's case value.

- **Protocol Layer Services Offered and Consumed**

In a layered protocol architecture, a particular protocol layer provides services to its upper layer, and it expects and uses services from the protocol layer below it. The services section of a 3GPP TS contains all such details of services being offered and assumed by a particular mobile communications network protocol layer.

- **Protocol Layer Procedure's Cause Value**

Every protocol layer and its sublayers perform certain procedures and their associated functions, described in Section 14.2, in response to an event or service request received from the peer protocol layer. Examples of protocol layer procedures are location area update, GPRS attach, routing area update, tracking area update, etc. The outcome of the execution of a particular procedure at the receiving end could be either a successful or a failure/abnormal one. Apart from this, a receiving protocol layer may receive invalid messages/values from the peer layer.

Sometimes, the receiving protocol layer may reject the service being requested by the peer or sending protocol layer. For all such varieties of reasons, the result is conveyed in terms of the *cause value* to the initiating protocol layer accordingly. Note that there could be several cause values for all the possibilities of outcomes from the execution of a particular procedure, either at MS/UE or network end, as illustrated in Figure 14.2. As specified in the concerned 3GPP TS, appropriate actions shall be taken by the initiating or sending protocol layer for a particular procedure cause whose value is found to be a failure.

Figure 14.3 illustrates the 5G *Registration Request* and its *Reject* procedure with the 5G mobility management (5GMM) cause code = 0x60 (*invalid mandatory information).* For more information on the action to be performed by the UE with this cause code, refer to Table 9.11.3.2.1, TS 24.501 [47].

14.2 3GPP Protocol Layer Structured Procedure Description

A 3GPP TS contains section(s) called "Procedure" for each of the functionalities or procedures supported by a particular protocol layer. A protocol layer procedure is executed in response to a particular service request received from the peer. A procedure specifies the desired functional behavior using text and typically contains a sequence of events with a *start* and *ending* of the procedure. The sequences of events are collectively known as the *structured procedure.* A procedure may also contain an accompanying diagram showing the sequences of events. The end of a procedure could be at the initiating protocol layer itself, or it could be at the peer destination layer. Figure 14.4 illustrates a normal as well as an abnormal scenario for a typical procedure: X executed by a particular protocol layer between an MS and the BSC.

Figure 14.4a shows the scenario where the BSC may either accept or reject the particular procedure: X initiated by the MS. On the other hand, it may be also possible that the procedure response message: X sent by the BSC did

Figure 14.3 Illustration: 5G registration request procedure with reject cause code = 0x60.

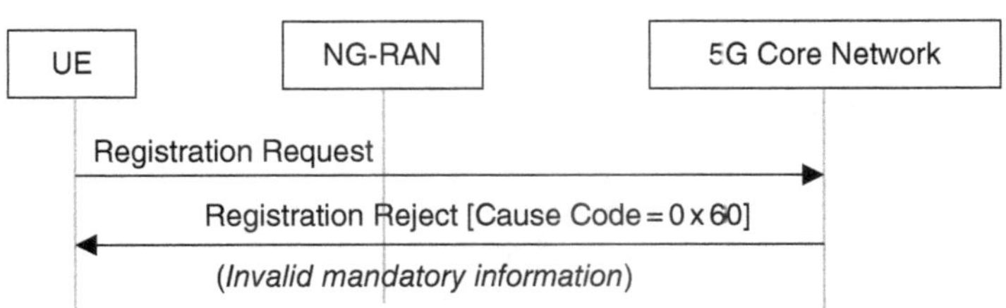

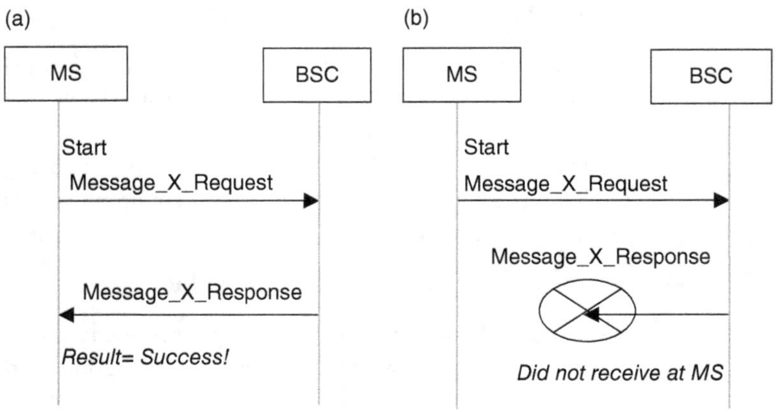

Figure 14.4 Illustration: protocol layer procedure and its result: (a) normal and (b) abnormal.

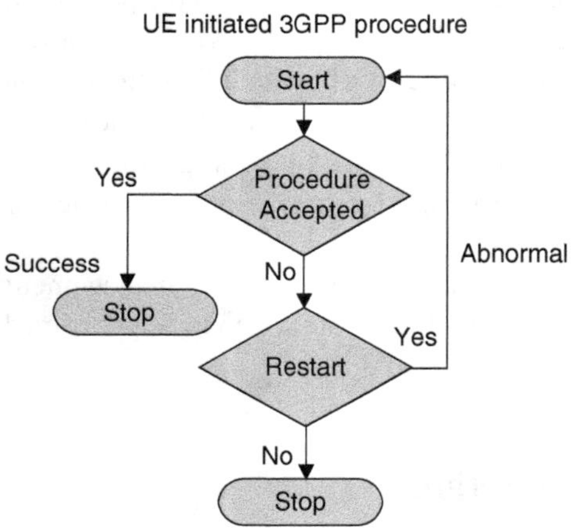

Figure 14.5 Illustration: flow of a 3GPP protocol layer procedure.

not reach the MS. This is illustrated in the same Figure 14.4b. A procedure of a 3GPP protocol layer can be completed either *successfully* or *unsuccessfully*. The overall flow of a 3GPP procedure initiated by an MS/UE toward the RAN or core network is illustrated in Figure 14.5.

The concerned 3GPP TS describes both the normal and abnormal cases that may occur at the MS as well as at the network end for a particular procedure performed by a protocol layer. The abnormal cases specify all the possible root causes of the erroneous events that may take place and notify the operation and maintenance (O&M) personal through an alarm as described earlier in Section 10.6.2. An abnormal case of a particular procedure between two network elements may affect the ongoing communications services.

As shown in Figure 14.5, a UE-initiated 3GPP procedure can be either accepted or rejected by the core network. If it is rejected by the core network due to a clear reason, the procedure comes to end. However, before the procedure comes to end, the same procedure may be restarted under the following conditions:

- Abnormal conditions at UE end.
- Abnormal conditions at network end.

Various reasons may lead to the above abnormal conditions, and action is required to be taken appropriately as described in the concerned procedure. Some typical examples of abnormal conditions causes are illustrated later in Section 14.14.

14.3 Protocol Layer Communications

As mentioned earlier, network elements of 3GPP mobile communications systems provide communications services through various protocol stacks and layers. In a layered protocol stack, two types of communication or exchange of information can take place. Those are mentioned below:

- Layer-to-layer communication between adjacent layers within a particular network element, say the LTE eNodeB or 5 NG-RAN/gNB.
- Peer-to-peer communication between two protocols layers of two different network elements, say between the UE and the eNodeB or 5G NG-RAN/gNB.

A particular logical interface supports the structure of a layered protocol that can be compared with the OSI 7 layers reference architecture. By now, the reader is aware of the various *logical interfaces*, e.g. GSM A-bis interface, LTE/EPS S1 interface, and so on, that are supported across the different cellular systems.

A logical interface facilitates the exchange of information between the peer protocol layers of two network elements. Across the different logical interfaces, the protocol used in a particular layer is different from each other. For example, in the case of the GSM system, one can find that the A-bis and air interface use the Link Access Protocol D-Channel (LAPD) and Link Access Protocol D-Channel-modified (LAPDm) protocol, respectively, as the Layer 2 protocol. Even the physical interface/layer that supports a particular logical interface is different across the cellular systems. As mentioned earlier in Section 3.6, a logical interface's protocol layer may also have sublayers as shown in Figure 14.6. For example, the GSM Layer 3 has three sublayers, namely the Call Control (CC), mobility management (MM), and Radio Resource (RR).

More details about each of the above protocol layers with respect to the air interface across the different cellular systems are described in the subsequent sections below.

14.3.1 Layer-to-Layer Communication Using Service Primitives

In a layered architecture, a protocol layer in a protocol stack uses the services from the lower layer in the stack. The logical information flow between the adjacent/sublayers is performed by the use of so-called *Service Primitives and* Service *Access Points (SAPs)*. A protocol layer of a protocol stack in a network element, say BSC, communicates with its adjacent higher layer in the form of service primitives. Service primitives work in terms of commands and their respective responses associated with the services requested from another layer. These primitives are generally called *REQ*uest, *IND*ication, *RES*ponse, and *CON*firm. Figure 14.7 illustrates the usages of these service primitives.

- *REQuest* – The REQUEST primitive type, sent by the transmitting protocol layer, is used by a higher layer to request a service from a lower layer.
- *INDication* – The INDICATION primitive type is used, on the peer end, by layer providing a service to notify its next higher layer for the activities related to the primitive type REQUEST.
- *RES*ponse – The RESPONSE primitive type is used by a layer, the service user, to acknowledge receipt from the INDICATION primitive type.
- *CON*firm – The CONFIRM primitive type is used by the lower layer providing the requested service to confirm its next higher layer that the activity has been completed.

A 3GPP TS defines all the service primitives, along with their parameters, to be supported by the concerned protocol layer. Refer to the corresponding section in the concerned TS for the service primitives to be used during the protocol layer design and implementation phase. Note that one can also define and use custom primitives with a meaningful name that indicates their purpose while designing and developing a protocol stack. For an overview of the services primitives used over the different mobile communication systems, refer to TS 24.007 [44].

14.3.2 Layer-to-Layer Communication: SAP

As mentioned in Section 14.3.1, adjacent protocol layers communicate through the *service primitives* and *SAP*. The *SAPs* are the points at which a lower layer provides services to the upper layer in the protocols stack. An SAP is identified by an *SAPI*.

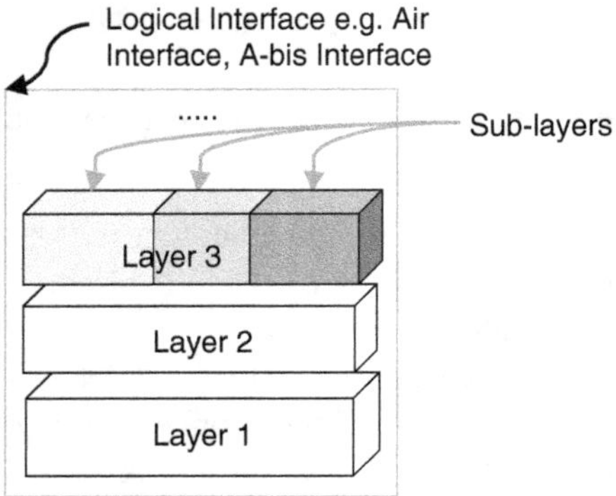

Figure 14.6 Illustration: protocol layer and sublayer architecture of a logical interface.

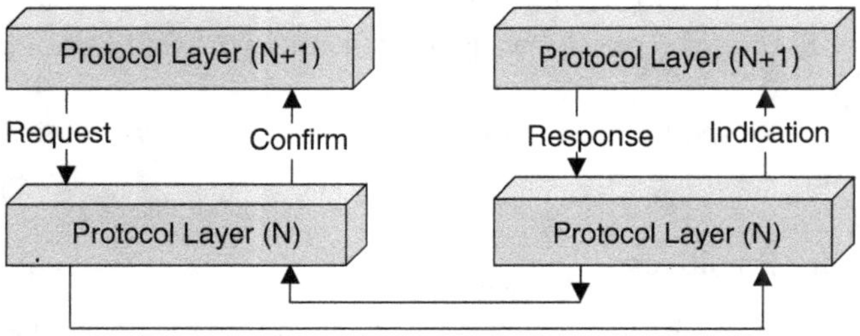

Figure 14.7 Illustration of various types of service primitives.

For example, consider the GPRS system Gb interface protocol stack between the BSC and Serving GPRS Support Node (SGSN). At the GPRS core network end, a cell is identified by an identifier called *BSSGP virtual connection identifier (BVCI)*. Each BVCI has an end-to-end significance across the Gb interface. Also, an SGSN controlling a particular routing area is identified by a so-called *Network Service Entity Identifier (NSEI)*. Now the BVCI together with the NSEI uniquely identifies a BVC within an SGSN. The BVCI and NSEI are used on the Network Service-Service Access Point (NS-SAP) for layer-to-layer communication. Usages of a Gb interface NS-SAPI are illustrated in Figure 14.8.

On the right-hand side of Figure 14.8, it is shown that different SAPIs, identifying different types of information, can be used for layer-to-layer or peer-to-peer protocol communication. It is also possible that multiple protocol layers communicate using different SAPIs to a lower layer where it performs multiplexing. For example, the LTE or 5G New Radio (NR) Medium Access Control (MAC) layer multiplexes upper layer signaling and user data, identified by their logical channel id, for transmission over its air interface.

Service Access Point Identifier(SAPI): Between Adjacent Protocol Layers

An SAPI is used as an element/field in a message/service primitive to achieve adjacent protocol layer-to-layer communication. As shown in Figure 14.8, using the Gb interface NS-SAPI=BVCI + NSEI, GPRS traffic for a particular mobile/user that is located in a particular cell and routing area is exchanged between the adjacent layers.

Service Access Point Identifier(SAPI): Between Peer Layer/Network Elements

An SAPI is also used as an element and a part of a protocol message exchanged for peer-to-peer communication between two network elements. The function and purpose of an SAPI, in this case, identify the types of protocol messages being exchanged between the two network elements. Regardless of the usages, either in layer-to-layer or peer-to-peer communication, of the SAPI field, it has a certain length and varies from protocol to protocol.

Example 14.1 illustrates the usages of SAPIs in the case of the GSM LAPD/LAPDm protocols.

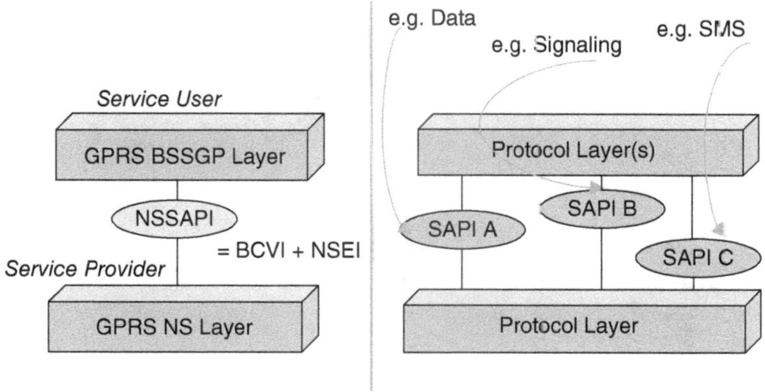

Figure 14.8 Illustration: service access point (SAP) and identifier (SAPI).

Example 14.1 SAPs Used in the GSM System

Consider the GSM LAPD protocol used on the A-bis interface between a base transceiver station (BTS) and BSC. The SAPI field is of 6 bits in length in case of GSM *LAPD* protocol message exchanged between the BSC and BTS. Similarly, the SAPI field is of 3 bits in length in case of the GSM *LAPDm* protocol message exchanged between BTS and MS over the GSM air interface. This means an SAPI can have a value ranging from 0 to 63 in LAPD protocol and 0 to 8 in the case of LAPDm. These SAPIs are used to identify different types/purposes of messages that the LAPD/LAPDm protocol can support. Types of LAPD/LAPDm messages that can be exchanged using various SAPIs are as follows:

- O&M messages, say between BSC and BTS;
- Call processing, i.e. call setup, call release, and messages.; and
- Layer 3 messages, Short Messaging Service (SMS) and Supplementary Services (SS).

Protocol layer communications using multiple SAPIs for multiple services, such as the traffic and signaling, are illustrated on the right-hand side of Figure 14.8.

14.3.3 Peer-to-Peer Layer Communication: PDU and Service Data Unit (SDU)

It has been mentioned in Section 14.3.2 that in a network element, a protocol layer provides services to the layer above it and also uses the services of the layer below it. A network layer provides and uses services by communicating through the *SAP and SAPI*. A protocol layer operating in a network element, say MS, also exchanges information, either signaling or user data, with its peer protocol layer in a network element, say SGSN. In general, the various information exchanged between the peer network elements and peer protocol layers are grouped into a unit called the *PDU*. So, a PDU is a unit of data exchanged between the *peer protocol layers* between the interconnected network elements.

On the other hand, the data that flows UP and DOWN between two layers of a protocol stack in a network element is called a *Service Data Unit (SDU)*. A PDU from a higher layer becomes the SDU for the lower layer. So, an SDU is a unit of data exchanged between the *adjacent layers* of a protocol stack in a particular network element. The flow of an SDU and a PDU, among the network elements of an LTE/EPS system, is illustrated in Figure 14.9. This is a general model of information flow, which also applies to the protocol stack of other mobile communication systems, e.g. 5G system.

As illustrated in Figure 14.9, an SDU received by a protocol layer, N, from its higher layer, N + 1, is converted into a PDU in the format of the protocol layer N and passes it in the form of an SDU to the next layer, N–1, below it, i.e. layer N. The PDU constructed by the layer N at the sending side can only be understood and decoded by its peer layer N at the receiving end. Layer N at the sending side may perform the segmentation of an SDU received from its higher layer, N + 1. The peer layer N at the receiving end performs the reassembly of the received PDUs

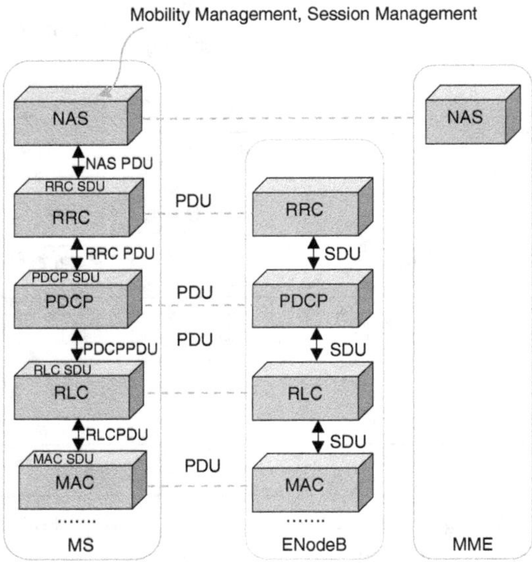

Figure 14.9 Illustration: LTE/EPS peer-to-peer communication using PDU.

before it is passed to the layer above it. Depending on the working principle of a protocol layer, a PDU of a protocol layer "N" may concatenate and carry several SDUs of protocol layer N + 1.

An SDU from a protocol layer N + 1 has a unique ID, and it may follow the header of protocol layer N. A PDU travels across a logical and physical interface and may have a defined lifetime beyond which the PDU may be discarded by the intermediate network element, say the BSC, before it reaches the destination network element, say the MS.

14.3.4 Types of PDU

A protocol layer, for example, the LTE or 5G NR air interface radio link control (RLC) and Packet Data Convergence Protocol (PDCP) layer, can have two types of PDUs. Those are as follows:

- Data PDU

  A data PDU is used to carry user data between a transmitter and a receiver.
- Control PDU

  A control PDU is used to carry signaling or control data using control channels between a transmitter and a receiver. Generally, the information and various fields being packed/encoded in a PDU are *byte aligned*.

14.3.5 Formats of PDU

A protocol layer may support several PDU formats, as defined by its 3GPP TS, using which two peer protocol layers exchange information across two network elements over the corresponding logical interface. Within a particular protocol layer, the format used for the data as well as the control PDU may be different. A protocol layer PDU format also depends on mode, i.e. acknowledged and unacknowledged, of its working. Refer to the corresponding 3GPP TS of the concerned protocol layer for the defined formats to be used for different types, i.e. control or data, of PDUs as mentioned in Section 14.3.4.

14.4 Air Interface Message Format: Signaling Layer 3

14.4.1 A Brief on the Air Interface Layer 3 Protocol Stack

Mobile communications radio air interface Layer 3 is one of the most interesting layers in all the cellular systems that are available today. Radio or air interface is used to communicate between the MS/UE and its RAN (RAN),

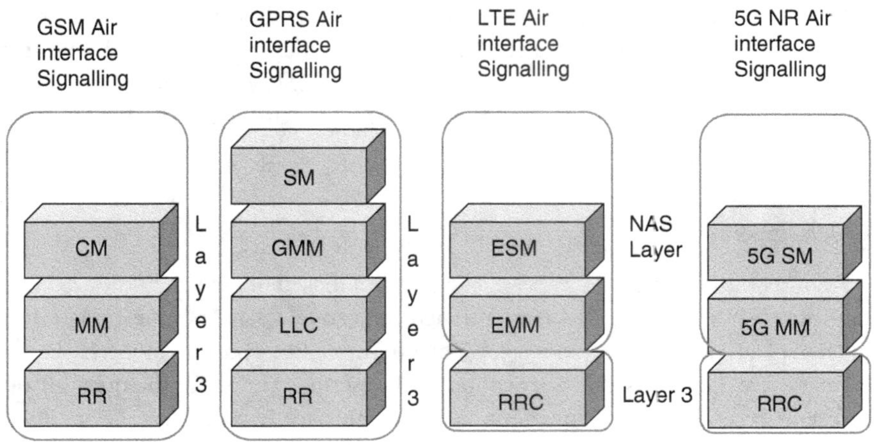

Figure 14.10 Illustration: air interface Layer 3, its sublayers, and NAS layers.

be it GSM/GPRS, UMTS, LTE, or 5G NR system. The name of the air interface (logical) in all of these cellular systems is known as the: *Um* (GSM and GPRS) and *Uu* (UMTS, LTE, and 5G NR), respectively.

As mentioned earlier, the GSM and GPRS air interface Layer 3 has three sublayers. The sublayers performs connection management, mobility management, and radio resource control management tasks. However, in the LTE and 5G NR systems, only the RRC layer is considered as the Layer 3 protocol, while the mobility and session management layers are considered as the NAS layers. Layer 3 and its sublayers and NAS layers across the different cellular systems are shown in Figure 14.10.

Only the RR/RRC layer terminates at the RAN end, while the mobility management, session management, connection management, and logical link control layers terminate at the core network end. As mentioned earlier in Section 4.1.1, the UMTS, LTE, and 5G NR NAS layer protocols also follow the air interface Layer 3 protocol message format, which is described in the next section.

14.4.2 Classification of Layer 3 Messages

Before a user can start voice conversation or packet-switched (PS) data services, the MS/UE sends the resource request messages to the respective RAN or Core Network. These messages are called *signaling messages*. As mentioned earlier, a prior successful establishment of signaling messages is required to provide various communication services to mobile users. The air interface Layer 3 and its sublayers signaling messages are divided into two categories; see Figure 14.11.

- Standard Layer 3 messages, e.g. GSM/GPRS Layer 3, LTE/EPS, and 5G NAS layer
- Nonstandard Layer 3 messages.

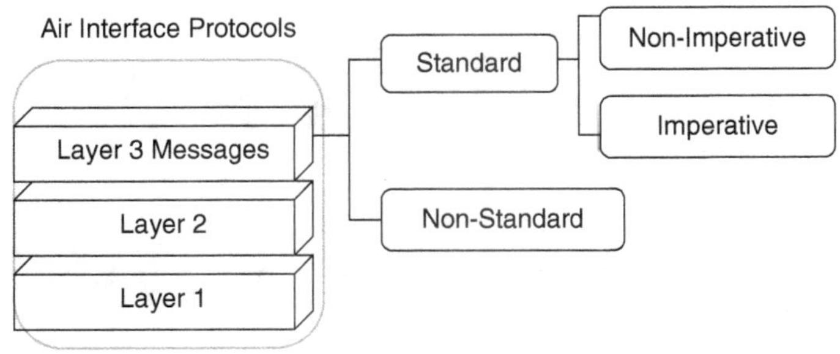

Figure 14.11 Classification of radio/air interface Layer 3 messages.

Standard Layer 3 signaling messages have a predefined format. On the other hand, some messages or protocols may not follow the standard Layer 3 format. Such messages are called *nonstandard Layer 3* messages. Typical examples of nonstandard Layer 3 messages are the GSM system information messages that are sent on the BCCH and CCCH channels. Also, as shown in Figure 14.11, the standard Layer 3 messages have been further divided into the following types:

- Imperative
- Non-imperative.

An imperative standard Layer 3 message is composed of a header, followed by mandatory standard IEs having the format T, V, LV, or LV-E, as mentioned in Section 4.1.1. The non-imperative portion follows the imperative portion of standard Layer 3 messages. The non-imperative portion of the Layer 3 message contains IEs having the format T, TV, TLV, or TLV-E. See Section 4.1.1 described earlier for the meaning of the format T, L, V, and E. These components and their organizations of the air interface Layer 3 message are illustrated in Figure 14.12.

[GSM] L2 Pseudo Length and Rest Octets IE in a Layer 3 Message

This information is part of the nonstandard air interface Layer 3 message, see Figure 14.13, as described in the preceding paragraphs.

- L2 Pseudo Length IE
  The pseudo length IE indicates the length in octets of the subsequent octet string, following this IE that can be considered and analyzed as a standard L3 message.
- Rest Octets IE
  This is the part following the standard Layer 3 message and up to the end of the complete message. For the coding format of the rest octets, refer to TS 44.008 [129], Section 10. The coding method differs from the one used in the standard Layer 3 IE.

Some examples of messages using the L2 pseudo length and rest octets IEs are GSM paging request types 1, 2, and 3 and system information messages.

14.4.3 Layer 3 Protocol Header: Signaling Message Format

Starting from the GSM to the 5G system, the mobile radio/air interface standard Layer 3/sublayers messages share the same generic message structure or format used for communicating with the access network and core network. In line with the component of a standard Layer 3 message as shown in Figure 14.12, the Layer 3 message header format is shown in Figure 14.14, which is reproduced from TS 24.008 [45]. *Bit-1* of an octet is the least significant bit (LSB), while *Bit-8* is the most significant bit (MSB).

Note that the *transaction identifier (TI)* field is used for the air interface Layer 3 GSM connection management (CM) and GPRS session management sublayer only.

Figure 14.12 Components of air interface Layer 3 message.

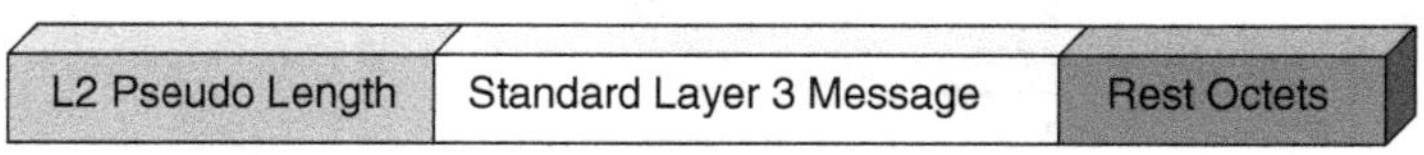

Figure 14.13 Components of a non-standard Layer 3 message.

8 7 6 5	4 3 2 1	
Transaction identifier or Skip Indicator	Protocol discriminator	octet 1
Message type		octet 2
Other information elements as required		etc...

Figure 14.14 GSM, GPRS, and UMTS radio/air interface Layer 3 message format. *Source:* © 2018. 3GPP™ TSs and TRs are the property of ARIB, ATIS, CCSA, ETSI, TSDSI, TTA and TTC who jointly own the copyright in them. © 2018, 3GPP.

8 7 6 5	4 3 2 1	
EPS bearer identity or Security header type	Protocol discriminator	octet 1
Procedure transaction identity		octet 1a*
Message type		octet 2
Other information elements as required		octet 3 ... octet n

Figure 14.15 Mobile radio/air interface NAS layer message format for the LTE/EPS system. *Source:* © 2014. 3GPP™ TSs and TRs are the property of ARIB, ATIS, CCSA, ETSI, TSDSI, TTA and TTC who jointly own the copyright in them. © 2014, 3GPP.

In the case of the LTE/EPS system, the NAS layer message (ESM and EMM) format used to exchange air interface signaling messages between the MS/UE and the core network is a little bit different from the GSM and GPRS message format shown in Figure 14.14. The LTE/EPS NAS layer message format is shown in Figure 14.15, which is reproduced from TS 24.301 [46].

As shown in Figures 14.14 and 14.15, the first octet of every air interface Layer 3 message contains the following header information:

- Protocol discriminator, 4–bits in length, in case of GSM/GPRS and LTE/EPS;
- LTE/EPS bearer identify or security header type (4 bits) and procedure transaction identity (1 octet), in case of LTE/EPS NAS layer only;
- Skip indicator (SI) – 4 bits in length, in case of GSM and GPRS systems; and
- Transaction identity – 4 bits in length in case of GSM and GPRS systems.

Further, in the case of the 5G NR air interface, its NAS layer mobility and session management layer messages also use the same air interface message format shown in the above figures. However, the header contents of the 5G NR air interface NAS layer are a little bit different from the LTE/EPS NAS layer message header contents shown in Figure 14.15. In the case of the NR system, the first octet of every NAS layer message contains the *Extended Protocol Discriminator (EPD)*, which is of 1 octet in length, unlike the 1 octet length protocol discriminator in the LTE/EPS NAS layer shown in Figure 14.15. The second octet of an NR NAS layer message contains the PDU session identity only, for the 5GSM layer, or security header type (4 bits) for the 5GMM message. Refer to TS 24.501 [47] for the header contents of the 5G NR NAS layer messages.

14.4.4 Layer 3 Protocol Header: Protocol Discriminator

It has been mentioned earlier that the mobile air interface protocol Layer 3 contains sublayers having the same message structure. So it requires a method to identify and differentiate a protocol sublayer that a particular message belongs to which will be processed by the concerned peer layer on the destination network element. For this purpose, the *protocol discriminator* field, which is of *4 bits* in length, is used in the header (first octet) portion of Layer 3 message as shown in Figures 14.14 and 14.15. The protocol discriminator field can have the values shown in Figure 14.16, which identifies the different Layer 3 air interface subprotocols. For the complete list of protocols, refer to TS 24.007 [44].

- Protocol Discriminator for 5G NAS Layer

In the case of the 5G system, the protocol discriminator is called *Extended Protocol Discriminator (EPD)*, consisting of the values: 0×2E and 0×7E for the 5G session management and mobility management messages, respectively. Refer to TS 24.007 [44].

14.4.5 Layer 3 Protocol Header: GSM, GPRS Skip Indicator

The SI following the protocol discriminator in a GSM, GPRS Layer 3 message header indicates to the receiver of the message whether it should process the message or not. For the RR and MM layers, the value of SI is always coded as *0000,* i.e. the receiver should process the Layer 3 messages. Any value assigned other than *0000* to the SI field makes the receiver ignore the received Layer 3 message. Note that a poor radio condition may lead to an erroneous value of SI, causing the receiver to ignore the received messages, finally leading to unexpected behavior of the concerned sublayer. Table 14.1 summarizes the value of SI for different sublayers of Layer 3.

- **Skip Indicator and Shared RAN Feature**

However, if a mobile communications system supports and is using the shared RAN feature, then the usage of the SI is a bit different. As mentioned above, generally, the SI will be encoded a 0 by the MS. But in a shared RAN system, the SI represents the Public Land Mobile Network (PLMN) IDs of the core networks of different operators. So, an MS can encode the SI as 0 that represents the first PLMN id in the *PLMN id list* broadcasted by the

Table 11.2: Protocol discriminator values

bits 4 3 2 1	
0 0 0 0	group call control
0 0 0 1	broadcast call control
0 0 1 0	EPS session management messages
0 0 1 1	call control; call related SS messages
0 1 0 0	GRPS Transparent Transport Protocol (GTTP)
0 1 0 1	mobility management messages
0 1 1 0	radio resources management messages
0 1 1 1	EPS mobility management messages
1 0 0 0	GPRS mobility management messages
1 0 0 1	SMS messages
1 0 1 0	GPRS session management messages
1 0 1 1	non call related SS messages
1 0 0 0	Location services specified in 3GPP TS 44.071 [8a]
1 1 1 0	reserved for extension of the PD to one octet length
1 1 1 1	used by tests procedure described in 3GPP TS 44.014 [5a], 3GPP TS 34.109 [17a] and 3 GPP TS 36.509 [26].

Figure 14.16 Mobile air interface Layer 3 protocols identified by protocol discriminator. *Source:* © 2011. 3GPP™ TSs and TRs are the property of ARIB, ATIS, CCSA, ETSI, TSDSI, TTA and TTC who jointly own the copyright in them. © 2011, 3GPP.

Table 14.1 Value and meaning of skip indicator.

Layer 3 Sublayers	Fixed Value of Skip Indicator(SI) in the L3 Header	Action by the Receiver for SI Value Other Than 0000
CM	NA	NA
MM	0000	Ignore the erroneous message
RR	0000	Ignore the erroneous message

network to MS. In this case, the core network will not treat the message as invalid with SI = 0; RAN sharing feature was described earlier in Section 7.2.

Example 14.2 illustrates the usages of the SI and protocol discriminator fields found in a typical GPRS Detach request from an MS to the core network/SGSN.

14.4.6 Layer 3 Protocol Header: GSM, GPRS Transaction Identifier

As shown in Figure 14.14 earlier, in the case of the GSM CM sublayer and its entities, i.e. CC, SS, SMS, and the GPRS session management layer, the protocol discriminator field is followed by the *TI* field. The TI field bits 5 to 8 are of 4 bits in length. If the transactions started by the originator and the responding/terminating side has

Example 14.2 GPRS MM (GMM) Air Interface Layer 3 Message: DETACH REQUEST

Whenever a GPRS network attached mobile device is switched off, it will send a *DETACH REQUEST* message to the SGSN. The *DETACH REQUEST* is a GMM message, and its tabular definition is shown in Figure 14.17 containing the SI and protocol discriminator field in the header part.

The definition of the *DETACH REQUEST* message shown in Figure 14.17 contains IEs with types V and TLV.

Table 9.4.5.2/3GPP TS 24.008: DETACH REQUEST message content

IEI	Information Element	Type/Reference	Presence	Format	Length
	Protocol discriminator	Protocol discriminator 10.2	M	V	1/2
	Skip indicator	Skip indicator 10.3.1	M	V	1/2
	Detach request message identity	Message type 10.4	M	V	1
	Detach type	Detach type 10.5.5.5	M	V	1/2
	Spare half octet	Spare half octet 10.5.1.8	M	V	1/2
18	P-TMSI	Mobile identity 10.5.1.4	O	TLV	7
19	P-TMSI signature	P-TMSI signature 2 10.5.5.8a	O	TLV	5

Figure 14.17 GMM air interface Layer 3: DETACH REQUEST message. *Source:* © 2016. 3GPP™ TSs and TRs are the property of ARIB, ATIS, CCSA, ETSI, TSDSI, TTA and TTC who jointly own the copyright in them. © 2016, 3GPP.

used the same TIs, then the source of the transaction is identified by the *Transaction Identifier flag(0/1)*, occupying the eighth bit of the TI field. The TI can be used to track and distinguish up to $2^4 = 16$ bidirectional simultaneous transactions/connections that may take place in the various entities, i.e. SM, CC, and SS, of the CM sublayer.

14.4.7 Layer 3 Protocol Header: LTE/EPS Bearer Identity

The LTE air interface NAS layer signaling message header contains an EPS bearer identity which is of 4 bits in length as shown in Figure 14.15. An LTE UE can have multiple defaults as well as dedicated bearers. An EPS bearer identity uniquely identifies an EPS bearer for one UE accessing via Evolved-UMTS Terrestrial Radio Access Network (E-UTRAN). Note that the EPS bearer identity is allocated by the MME.

14.4.8 Layer 3 Protocol Header: 5GSM PDU Session Identity

In the 5G system, IP-based mobile broadband as well as other data services such as the ultra-reliable low-latency communication, and massive internet of things communication services are provided to a UE and its user through an end-to-end *PDU session* only over a particular network slice. A network slice is established between a UE and its core network, i.e. 5GC. A *PDU session identity* is assigned to a PDU session and is included in the second octet of every 5G session management (SM) signaling message exchanged between a UE and the 5GC. The 5G system is described in Part IV.

Example 14.3 highlights the importance of reusable generic functions or software programs that can be used for the encoding and decoding of protocol layer header information described in the preceding text, or other IEs, with varying lengths.

14.4.9 Constructing a Layer 3 Message

GSM/GPRS and UMTS Layer 3, LTE/EPS, and 5G system NAS layer messages use a similar encoding format (T, L, and V) to exchange messages among the UE/MS, base station, and core network. A Layer 3 message can be constructed by encoding all of its IEs as defined in the concerned 3GPP TS.

Consider the *ASSIGNMENT COMMAND* message type of the GSM sublayer: RR layer. The complete description of the *ASSIGNMENT COMMAND* message and its various IEs are shown in Figure 14.18, which is reproduced from TS 44.018 [130]. A sample C-function can be developed as shown in the Figure 14.18 to construct the *ASSIGNMENT COMMAND* message using the generic macros/functions such as *encode_and_fill_2nibbles, encode_and_fill_1octe* and so on.

Example 14.3 Generic C/C++ Functions for Encoding and Decoding of Protocol Information

As mentioned in Section 14.3.4, the various IEs of a Layer 3 message being packed in a PDU is *byte aligned*. For an IE field whose length is more than a byte or octet, it needs to be encoded and decoded several times depending on its length. Sometimes, an IE field may be of only half octet in length. Developing generic functions for encoding/decoding of IEs of different lengths is very useful and can be packaged into a separate library as they can be called from different protocol layers. Generic functions further lead to the ease of code maintenance also.

Message type: ASSIGNMENT COMMAND

Significance: dual

Direction: network to mobile station

Table 9.1.2.1: ASSIGNMENT COMMAND message content

IEI	Information Element	Type/Reference	Presence	Format	Length
	RR management Protocol discriminator	Protocol discriminator 10.2	M	V	½
	Skip Indicator	Skip Indicator 10.3.1	M	V	½
	Assignment command Message type	Message type 10.4	M	V	1
	Description of the First Channel, after time	Channel Description 2 10.5.2.5a	M	V	3
	Power Command	Power Command 10.5.2.28	M	V	1
05	Frequency List, after time	Frequency List 10.5.2.13	O	TLV	4–132
62	Cell Channel Description	Cell Channel Description 10.5.2.1b	O	TV	17
10	Description of the multislot configuration	Multislot Allocation 10.5.2.21b	O	TLV	3–12

Figure 14.18 Air interface Layer 3: GSM RR assignment command. *Source:* © 2012. 3GPP™ TSs and TRs are the property of ARIB, ATIS, CCSA, ETSI, TSDSI, TTA and TTC who jointly own the copyright in them. © 2012, 3GPP.

```
GSM_LAYER_RR_ASSIGNMENT_COMMAND (buffer,….)
{
encode_and_fill_2nibbles (buffer, GSM_L3_PD_RR, GSM_L3_MM_RR_SI);
encode_and_fill_1octet (buffer, msgtype);
.....................................................
}
```

14.4.10 Security Protected LTE/EPS and 5G NAS Layer MM Messages

Though the MM protocol layer carries the signaling messages between a UE and its core network across the different mobile communication systems, there are differences in terms of its security protection being provided over the respective air interface. The GSM/GPRS and UMTS system MM messages are not security protected, in terms of integrity protection, over their air interfaces. But the LTE/EPS and 5G system NAS layer mobility management layer messages may be security/integrity protected over their air interfaces and are divided into the following types:

- Plain NAS layer MM message, i.e. no integrity protection for some of the NAS layer MM messages.
- Security-protected NAS layer MM message.

In addition to integrity protection, ciphering protection may be also applied to the LTE/EPS and 5G system NAS layer MM messages. The LTE/EPS and 5G system NAS layer MM messages between the UE and core network are independently integrity protected and ciphered. The integrity protection of NAS layer MM messages (with few exceptions of few MM layer messages) is mandatory, but the message can be sent either ciphered or unciphered as configured by the core network. For a comparative study to the reader, the respective NAS layer MM messages contents are shown in Table 14.2.

Table 14.2 Layer 3 and NAS message format contents.

GSM/GPRS/UMTS L3 Message Format Contents	
-Transaction Identifier or Skip Indicator and Protocol Discriminator -Message type -Other IEs, as required	
LTE/EPS and 5GS NAS Message Format Contents	
Plain NAS Message	Security Protected NAS Message
-EPS bearer identity (LTE/EPS only)	-Protocol discriminator
-PDU session identity (5G only)	-Security header type
-Security header type	-Message authentication code
-Protocol discriminator	-Sequence number
-Procedure transaction identity	-Plain NAS message
-Message type	
-Other IEs, as required	

From Table 14.2, compare the differences in the header contents of a plain and security-protected NAS layer MM message. To transfer an MM message securely over the air interface through an integrity protection mechanism, a 24-bit counter called NAS *COUNTER* is used as input to the integrity protection algorithm being used at the UE and LTE/EPS MME or 5G core AMF end. The 8 bits LSBs of a NAS COUNT are exchanged as the *NAS sequence number* in an MM signaling message, as listed in Table 14.2, between a UE and LTE/EPS MME or 5G core AMF end. The remaining 16 MSBs of the NAS COUNTER are called as NAS overflow counter. At the receiving end, the integrity of security-protected MM signaling message is verified using the same integrity protection algorithm by providing the received NAS sequence number and estimated NAS COUNT as inputs. Note that not all of the LTE/EPS or 5G system NAS layer MM signaling messages are integrity protected and ciphered. Refer to TS 24.301 [46] and 24.501 [47] for the LTE/EPS and 5G NAS layer MM messages that are not security protected over the respective air interface.

For the purpose of security protection of NAS layer MM signaling messages, both the UE and the LTE/EPC and 5G core network establish a *security context* first during the initial network registration procedure performed by the UE. It is the LTE/EPS mobility management (EMM) layer or the 5GMM layer that initiates the *security mode control command* toward the UE for the creation of a security context.

The illustration below shows the plain LTE/EPS NAS layer message contents for the *ATTACH REQUEST message*, sent from UE to network, and a security-protected *ATTACH ACCEPT* message, sent from the network to UE. The underlined texts show the NAS layer security aspects (security header type) during the initial network registration, through an ATTACH request, and the one during the registration/ATTACH acceptance by the core network after the authentication and security procedure is completed successfully between UE and the core network. Compare the format of the ATTACH request/accept message with the Layer 3 message format shown in Figure 14.15.

```
EPS ATTACH REQUEST: Non-Access-Stratum (NAS)PDU
0000.... = Security header type: Plain NAS message, not security protected (0)
....0111 = Protocol discriminator: EPS mobility management messages (0x07)
NAS EPS Mobility Management Message Type: Attach request (0x41)
```

```
EPS ATTACH ACCEPT: Non-Access-Stratum (NAS)PDU
0010.... = Security header type: Integrity protected and ciphered (2)
.... 0111 = Protocol discriminator: EPS mobility management messages (0x07)
Message authentication code: 0x12345
Sequence number: X
0000.... = Security header type: Plain NAS message, not security protected (0)
.... 0111 = Protocol discriminator: EPS mobility management messages (0x07)
NAS EPS Mobility Management Message Type: Attach accept (0x42)
```

14.4.11 Layer 3 Protocol Layer's Message Dump

In the preceding sections, we have discussed the general structure of the air interface Layer 3 and UMTS, LTE, and 5G system NAS layer messages and their contents. Example 14.4 illustrates how typical air interface Layer 3 signaling message contents may appear in a typical air interface protocol analyzer tool. Decoding the values of IEs of a signaling message becomes helpful while troubleshooting air interface-related issues and shall be performed as per its definition in the concerned 3GPP TS.

14.5 Air Interface Message Format: Layer 2

Layer 2 Encoding/Decoding of Protocol Frame

Air interface Layer 3 and NAS layer message format used in the GSM/GPRS, UMTS, LTE/EPS, and 5G system has been described in the previous section. As described in the previous section, the minimum size of an IE, e.g. SI, TI, and Protocol Discriminator used in GSM/GPRS messages, used in the Layer 3 message format is 4 bits or half octet. But in the case of the air interface Layer 2 protocol layers, the minimum size of an IE used may be 1 bit. In this section, the message format used by the air interface Layer 2 protocol layers shall be described.

Similar to the OSI reference protocol model, the air interface Layer 2 protocol in a cellular system also transmits and receives protocol information at a *frame* level and format which consists of a string of bits with an octet aligned. Like a PDU, a frame is a unit of information transmission, generally a bit string. In some of the frame-level Layer 2 protocol layers, especially the control plane message, the header consists of control parameters/flags, whose length is given in bits, in the first octet (*8 bits*) itself. The control parameters/flags control the behavior of the concerned protocol layer. So, in a frame of a Layer 2 protocol layer, even a single bit has its importance because it can

Example 14.4 GSM Layer 3 Call Control "Connect" Message

An end-to-end call flow for a GSM mobile-originated voice call was illustrated in Figure 2.8. Consider the GSM Call Control layer "Connect" Message which is sent by the network to the calling mobile station to indicate that the call is accepted by the called user. A typical air interface protocol analyzer tool may present the contents of the *Connect* message as follows. Compare the contents of the *Connect* message with the format of the Layer 3 message format shown earlier in Figure 14.14.

```
L3Message: Connect Date: DD-MM-YYYY Time: HH:MM:SS:... ...
Transaction Identifier: 0x1 Protocol Discriminator: 0x3 (Call Control)
Message Type: 0x7
Message Dump: 0x13, 0x7
```

be used to represent two (0/1) control information or flags which require an encoding and decoding function at a bit level.

Examples of cellular system air interface protocols operating at Layer 2 are RLC/MAC protocol used in case of GPRS, UMTS, LTE, and 5G NR system; Logical Link Control (LLC) protocol found in case of GPRS system; PDCP layer in LTE and 5G NR system; and SDAP layer in 5G NR system. Note that though the UMTS, LTE, and 5G NR system *Radio Resource Control (RRC)* is a Layer 3 protocol, they use the ASN.1 encoding/decoding syntax, which is also a bit string that is exchanged between the UE and UMTS Radio Network Controller (RNC) or LTE eNodeB or 5G NG-RAN/gNB over their air interface.

As an example, a generic or reusable program/macro as illustrated below can be developed for the encoding of Layer 2 protocol layer information at a bit level. Such a reusable macro can be called from multiple protocol layers that use a similar encoding and decoding method. Note that, unlike the common message format used by Layer 3/NAS layers messages, the formats used by Layer 2 protocol layers are different.

```
#define encode_IE_name (buffer, parameter1, parameter2..)
{//Macro to encode an IE
ASN_CSN_generic_encode_N_bits (buffer, parameter1, 1)
ASN_CSN_generic_encode_N_bits (buffer, parameter2, 2)
..................
}
#define ASN_CSN_generic_N_bits(buffer, parameter, no_of_bits)
{//Genecis macro to encode N-bits of an IE
..................................
}
```

14.6 RAN – CN Signaling Messages

It has been mentioned earlier in Section 4.1.5 that the ASN.1 notation and its *Packed Encoding Rule* (PER) are also used to describe and encode signaling messages over the following logical interfaces:

- S1-AP messages over the S1 interface between the LTE eNodeB and MME.
- X2-AP messages over the X2 interface between two LTE eNodeBs.
- NG-AP messages over the NG interface between the 5G NG-RAN and AMF.
- Xn-AP messages over the Xn interface between two 5G NG-RAN/gNBs.
- Radio Access Network Application Part (RANAP) messages over the Iu Interface between the UMTS UTRAN and the core network.

In the next subsections, different but common aspects of signaling messages exchanged over the above logical interfaces between the respective RAN and its core network (CN) element shall be described.

14.6.1 Protocol Layer Elementary Procedure

A mobile communication network and its various elements, say the LTE eNodeB or the 5G NG-RAN/gNB, execute a particular procedure for a task/service requirement initiated by another network element, say the LTE/EPS MME or 5G Core AMF. To complete the procedure, the network elements exchange a number of messages in sequence as a part of the concerned procedure being executed. Each message is identified through the assigned message identifier.

Some messages may be of request/response or request only and no response in nature. Related messages exchanged over the 5GS NG, Xn interfaces; LTE S1, X2 interfaces; and UMTS Iu-RANAP interfaces are grouped into the so-called *Elementary Procedure* (EP) which is described using the ASN.1 notation as well as in tabular format.

There is no concept of assigned and common EP for messages exchanged over the air interface Layer 3 and NAS layer messages. Over these interfaces, the messages exchanged have their message type or id. On the other hand, messages grouped into an *EP* do not have their *message id;* instead of it, they are identified under a common *procedure code* assigned to the concerned EP. An *EP* is illustrated through Example 14.5.

Example 14.5 Handover Preparation EP and its Messages

Consider the various phases of a typical call handover procedure which consists of the handover preparation, execution, and completion phases. The initial phase of the entire handover procedure is the handover preparation, consisting of the following individual messages for a typical handover in case of the LTE system:

1) Handover Required
2) Handover Command
3) Handover Preparation Failure

Handover-related above messages are structured and grouped into the LTE/EPS S1-AP EP called *HandoverPreparation* with the *procedure code = 0* as defined in Section 9.3.6, TS 36.413 [97]. Unlike the air interface Layer 3 message, there will be no individual message type/id for the above messages. Figure 14.19 shows the messages exchanged and the corresponding S1-AP EPs executed between the LTE eNodeB and the MME for a typical handover procedure performed in the LTE/EPS system.

Refer to TS 36.413 [97] for the ASN.1 definition of the *Handover Preparation* and *Handover Resource Allocation* EP. An illustrative decoded version of the above EPs and their messages are shown in Figure 14.20. As shown in these decoded versions, the related messages have the same S1-AP EP code, for example, the Handover Required and Handover Command messages have the procedure code = 0.

The entire handover operation between the eNodeB and the MME involves one more S1-AP EP called *Handover Resource Allocation* as shown in Figures 14.19 and 14.20.

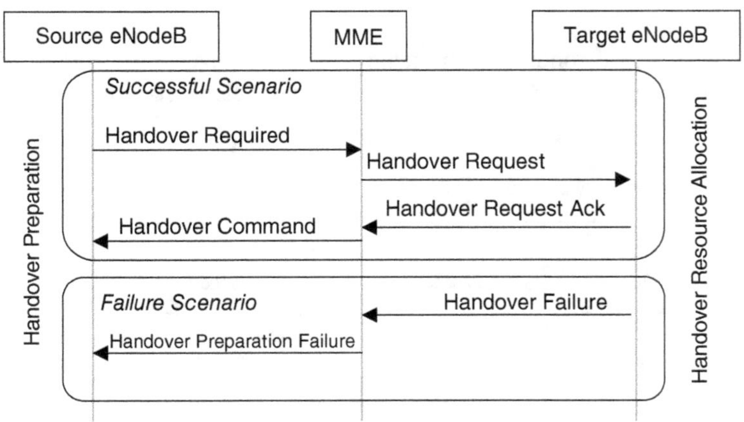

Figure 14.19 Illustration: LTE S1 handover preparation and resource allocation EP and its related messages flow.

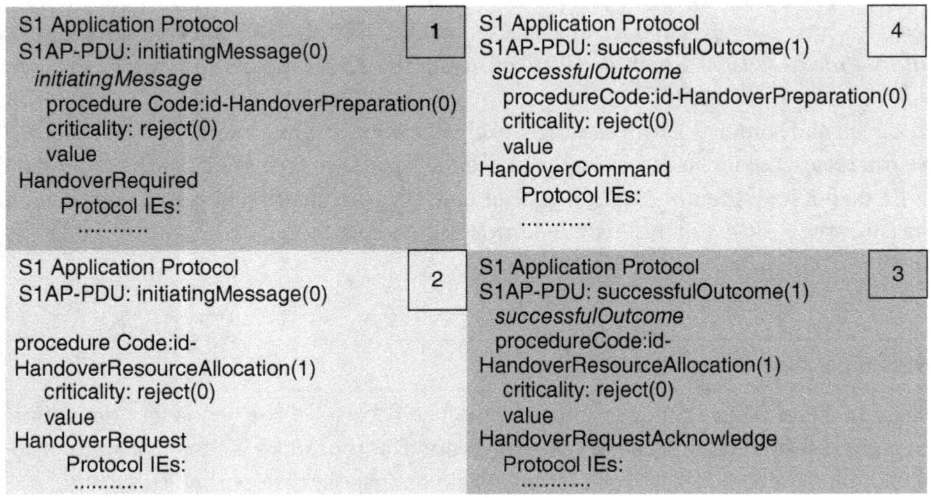

Figure 14.20 Illustration: protocol decoding of elementary procedures messages for an LTE S1-based handover procedure.

14.6.2 Types and Classes of EPs

The EPs containing the related messages of the LTE/EPS S1-AP, X2-AP; 5G NG-AP, Xn-AP; and UMTS RANAP are classified into different classes, as shown in Table 14.3.

For example, the EPs, namely *Handover Preparation* and *Handover Resource Allocation*, as described in Section 14.6.1, are of Class 1 procedures since the related messages under these procedures are *Request<>Response* (either success or failure) in nature. Similarly, the paging message sent by the MME toward the eNodeB is a Class 2 procedure as there is no response for this message from the eNodeB to the MME. For the list of Class 1, Class 2, and Class 3 EPs and their related messages, refer to TS 36.413 [97], TS 38.413 [119], TS 25.413 [54], TS 36.331 [94], TS 25.331 [50], and TS 36.423 [98]. Note that all the EPs consisting of NAS layer messages between the UE and core network are of Class 2 only. In the LTE and 5G systems, the procedures are known as the *DOWNLINK/ UPLINK NAS Transport,* and in the UMTS system, the procedure is known as the *DIRECT TRANSFER*.

14.6.3 EPs Code

As mentioned earlier, a protocol's EPs are exchanged, containing the related messages, over the LTE/EPS S1, X2 interfaces; 5GC NG, Xn interfaces; and UMTS Iu interface between the peer network elements. Unlike the air

Table 14.3 S1-AP, X2-AP, NG-AP, Xn-AP, and RANAP elementary procedure classes.

System	Elementary Procedure Classes	Class 1 Procedure	Class 2 Procedure	Class 3 Procedure
UMTS	Class 1, Class 2, Class 3	Request, Response[a] Type	Request Only Type[b]	Request, Response[c] Type
LTE, 5G	Class 1, Class 2			

[a] Successful, Unsuccessful, or Both;
[b] Always Successful;
[c] Multiple Responses.

interface Layer 3 messages, the related messages of the protocol's EP do not have their message type/id. Over the S1, X2(LTE/EPS), NG, Xn (5GS), and Iu interfaces, to identify an EP, a corresponding procedure code is assigned and shared by all the related messages under it. For example, the RANAP Iu-Release procedure has code = 1, the LTE/EPS S1-AP Paging procedure has code = 10, the LTE X2 Release procedure has code = 16, and so on. For an illustration, refer to Figure 14.20 showing the decoded version of S1-APl handover EPs and their procedure codes. For the complete list of EPs and their codes, refer to TS 36.413 [97], TS 38.413 [119], TS 25.413 [54], and TS 36.423 [98].

14.6.4 Criticality of IE

An EP exchanged between two peer network elements over the LTE/EPS S1 and X2; 5G NG and XnP; and UMTS Iu interfaces may contain a number of IEs. An IE may play a significant and critical role in successfully executing the concerned EP. The criticality information of an IE is an indication to the receiver to take the appropriate actions whenever the concerned IE is not understood by the receiver. A procedure code or an IE can have three values of the criticality information as mentioned below. See Figure 14.20 that shows the criticality of IEs, for example.

- Reject IE

  If an IE is received with critically = Reject, the receiver shall reject the EP with an appropriate error indication to the sender.

- Ignore IE and Notify Sender

  If an IE is received with critically = Ignore and Notify Sender, the receiver shall ignore the EP with an appropriate error indication to the sender.

- Ignore IE

  If an IE is received with critically = Ignore, the receiver shall ignore the EP.

14.6.5 Types of Protocol Errors and Its Handling

Similar to the reporting of a protocol error in case of receiving an air interface Layer 3/NAS layer message erroneously, a protocol error is also reported by a receiver if an ASN.1/PER encoded message contains an error. Under such circumstances, the receiver may notify the occurrences of such error to the sender of the message and take further course of actions as specified in the concerned TS. The following types of error may occur for an ASN.1 encoded message exchanged over the respective interfaces, i.e. LTE S1, X2; 5GS NG, Xn, and so on:

- Transfer Syntax Error

  This error occurs whenever the receiver is unable to decode the received message due to a missing mandatory IE, more than the defined number of IEs, received IEs in the wrong order, or violation of the ASN.1 encoding rule.

- Abstract Syntax Error

  This error occurs whenever the receiver receives IE with unknown information element identifier (IEI), duplicate IEs, wrong order of IEs, or IEs received with false conditional presence requirement. A syntax error may also result because of the wrong criticality information of an IE, as mentioned in Section 14.6.4.

- Logical Error

  This kind of error arises whenever the receiver receives a message whose contents are not valid, i.e. semantics error, or the received message is an unexpected one considering the current state of the receiver.

If an error of the above type is detected by the receiver, it will be notified to the sender using the *ERROR INDICATION* message, between the eNodeB<->MME; eNodeB<->eNodeB; 5G NG-RAN<->AMF and NG-RAN/gNB<->gNB; and UMTS RNC<->Core network. Based on the *ERROR INDICATION* and its contents, the sender of the message shall take the appropriate actions.

14.6.6 Choices of Triggering Message

It has been mentioned earlier that the EPs of Class 1 type contain the related messages that are exchanged between two peer network elements over the LTE/EPS S1, X2 interfaces; 5GS NG, Xn interfaces; and UMTS Iu interface. See the usages of the EPs illustrated earlier in Figure 14.19 for an S1-based handover procedure in the LTE system. Also, the messages of an EP of Class 1 type are of *Request<->Response* in nature. The *Response* may be either a *successful* or *unsuccessful* one. Based on this, the messages of an EP of Class 1 type are categorized into the following triggering messages choices:

- Initiating Message (Request Message)
- Successful Outcome (Response Message)
- Unsuccessful Outcome (Response Message)

The corresponding triggering message choice, mentioned above, is also encoded in the message being transmitted to the receiver, as illustrated in the protocol decoded version (Figure 14.20). For example, Handover Required and Handover Request are *initiating message* (0) triggers, whereas the *Handover Command* and Handover Request acknowledge are successful outcome (1) trigger message choices. They have the procedure codes = 0 (Handover Preparation) and 1 (Handover Resource Allocation).

14.6.7 Message Type

In the preceding sections, we have discussed the EP's procedure code as well as the choice of related messages (Initiating Message, Successful Outcome, and Unsuccessful Outcome) that are included under a particular EP. Based on this, the LTE/EPS S1-AP, X2-AP; 5G NG-AP, Xn-AP; and UMTS RANAP *message types* exchanged over their respective interfaces are encoded and have the following components:

- A unique procedure code
- Choice of a message (initiating message, successful outcome, and unsuccessful outcome)

The decoding of the above components of an EP was illustrated earlier in Figure 14.20. For example, the *HandoverPreparation* EP has the procedure code = 0 and the *HandoverResourceAllocation* EP has the procedure code = 1.

14.6.8 Message Description

The 3GPP TSs for the LTE/EPS S1-AP and X2-AP; 5G NG-AP, Xn-AP; and UMTS RANAP application protocols (APs) describe the messages exchanged over the S1, X2; NG, Xn; and Iu interface, respectively, in both the tabular and the ASN.1 notation format. The tabular format, Figure 4.8, contains the following information/columns for the description of a particular message. Compare the above components of an ASN.1 message with the IEs of an air interface Layer 3 protocol message description:

- IE/Group Name
- Presence Requirement
- Range
- IE Type and Reference

- Semantics Descriptions
- Criticality; Criticality Assigned

Some typical procedures where AP messages are used are the UE mobility management, handover procedure, paging procedure, and so on. In the next section, a typical procedure performed using AP messages over the LTE/EPS S1 interface shall be discussed.

14.6.9 Example: LTE/EPS S1 Interface: S1 Setup Procedure

Consider the LTE/EPS *S1 Setup* EP which is performed over the S1 interface. In an LTE/EPS system, the S1 logical interface is used to exchange information between the eNodeB and the core network. The S1 Setup message is used by the eNodeB to bring itself into service by providing important information such as its global unique id, tracking area code, and PLMN identity. A snapshot is shown in Figure 14.21, which is reproduced from TS 36.413 [97]. The *S1 Setup* message is divided into the following parts:

- Initiating Message, i.e. *S1 Setup Request Message* from eNodeB to MME;
- Successful Outcome, i.e. *S1 Setup Response* from the MME to the eNodeB;
- Unsuccessful Outcome, i.e. *S1 Setup Failure* from the MME to the eNodeB;
- Message Type or *Procedure Code*; and
- *Critically* of the Message.

Figure 14.21 shows the complete flow of the S1 Setup procedures.

Figure 14.22 shows the typical decoding of the complete S1 protocol stack layers using the IP transport network for the S1 Setup Request and Response command exchanged between the eNodeB and the MME shown in Figure 14.21.

The "+/−" shown in Figure 14.22 indicates the collapsed/expanded version of the particular protocol, i.e. more/fewer details can be found. The absence of +/− indicates the IEI of the concerned protocol.

14.7 Modes of Operation of a Protocol Layer

Data communication protocols such as the TCP and UDP work in the connection-oriented and connectionless modes, i.e. Acknowledged and Unacknowledged. Similarly, a protocol layer in a cellular system can also work in different modes. The possible modes of working for a protocol layer are described below:

- Acknowledged Mode (AM)

In this mode, the peer protocol layer (receiver) sends the status of the incorrectly received PDUs/blocks to the sender. Selective retransmission of incorrectly received, by the receiver, PDUs is performed by the sender, making

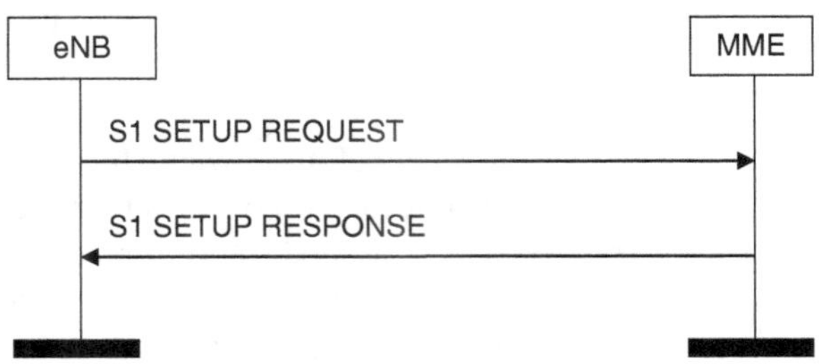

Figure 14.21 Illustration: LTE/EPS S1 Setup message over S1 interface. *Source:* © 2014. 3GPP™ TSs and TRs are the property of ARIB, ATIS, CCSA, ETSI, TSDSI, TTA and TTC who jointly own the copyright in them. © 2014, 3GPP.

```
+    Frame X: XX Bytes
+    Ethernet Src: XX,    Dst: XX
+    Internet Protocol Version 4  Src: XX    Dst:XX
+    Stream Control Transmission Protocol, Src Port: XX,Dst Port: XX
+    S1 Application Protocol
-        S1AP-PDU: initiatingMessage (0)
-            initiatingMessage
                 procedureCode: id-S1Setup (17)
                 criticality: reject (0)
-                value
-    S1SetupRequest
-                        protocolIEs: 4 items
+                        Item 0: id-Global-ENB-ID
+                        Item 1: id-eNBname
+                        Item 2: id-SupportedTAs
+                        Item 3: id-DefaultPagingDRX
+    Frame X: XX Bytes
+    Ethernet Src: XX,    Dst: XX
+    Internet Protocol Version 4  Src: XX    Dst:XX
+    Stream Control Transmission Protocol, Src Port: XX,Dst Port: XX
+    S1 Application Protocol
-        S1AP-PDU: successfulOutcome (1)
-            successfulOutcome
                 procedureCode: id-S1Setup (17)
                 criticality: reject (0)
-                value
-    S1SetupResponse
-                        protocolIEs: 3 items
+                            Item 0: id-MMEname
+                            Item 1: id-ServedGUMMEIs
+                            Item 2: id-RelativeMMECapacity
```

Figure 14.22 Illustration: decoding of S1 Setup Request Message protocol.

the AM a reliable method of transmission. Generally, PDUs belonging to the control or signaling message is transmitted using AM.

Examples of protocol layers are RLC, LLC, LAPDm, and so on. AM mode of operation of a protocol layer could be compared with data communication using a *TCP* port.

- Unacknowledged Mode (UM)

In this case, no selective retransmissions are performed by the sender even if the PDUs are received incorrectly by the receiver. Generally, data/packet services PDUs such as the Video/Audio streaming can be transmitted using the UM. Examples of protocol layers are RLC, LLC LAPDm, and so on. The UM of operation of a protocol layer could be also compared with data communication using a UDP port.

- Transparent Mode (TM)

This is a pass-through kind of mode where the concerned protocol layer does not add, unlike the AM and UM mentioned above, any protocol layer overhead such as its header, segmentation, and so on. No retransmission of PDU/blocks takes place in this mode. An example of a protocol layer is the RLC layer.

Refer to the concerned 3GPP TS of a protocol layer for more detailed information on the tasks performed by the transmitting and receiving peer layers in each mode. A network element, e.g. MS/UE, indicates the mode of

Example 14.6 Illustrative C-Pre-processor Definition: RLC Layer Modes of Operation

```
#define RLC_AM 0
#define RLC_UM 1
#define RLC_TM 2
```

operation of a protocol layer to be used for the requested service. However, the RAN system can terminate the current service and re-establish the same service on another mode depending on various factors such as the poor ongoing radio condition.

The air interface RLC layer supports and can work in all of the above modes. An MS establishes a data session in the AM, whereas another MS may establish a data session in the UM in the same direction.

Example 14.6 illustrates the typical software definitions of the modes of operation of the LTE/E-UTRAN or 5G NR RLC layer.

14.8 Example of a Protocol Primitive and PDU Definition

In the preceding sections, the reader has been introduced to the protocol layer's primitives, SDU, PDUs, SAP, and their identifiers (SAPI). An illustrative definition of a protocol layer primitive, its components and PDU using a C-language structure is given in Example 14.7.

The PRIMITIVE NAME could be replaced directly using the primitive name as mentioned in the concerned 3GPP TS such as the UNITDATA_IND_REQ. The illustrative declaration, see Example 14.7, of a protocol layer primitive contains several components, as mentioned below:

- SERVICE USER LAYER
  This is the originating layer requesting the required services from the layer below it, i.e. SERVICE PROVIDER LAYER.
- SERVICE PROVIDER LAYER
  This is the layer transmitting, to its peer protocol layer, the services being requested by the adjacent layer, i.e. SERVICE USER LAYER, above it.
- PARAMETER 1 . . .. . .PARAMETER N
  These are the primitive parameters as mentioned in the concerned 3GPP TS and the primitive under declaration.

Example 14.7 Illustrative C-Structure Definition of a Primitive

```
struct SERVICE USER LAYER_SERVICE PROVIDER LAYER_PRIMITIVE NAME_REQ
{
TYPE1 primitive_type; /*Primitive Type: REQ, IND, RES,CNF*/
TYPE1 PARAMETER 1; /* Primitive Parameter*/
................................................
TPYE2 PARAMETER N; /* Primitive Parameter*/
PDUTYPE PDU1; /* PDU to be communicated */
  };
```

- PDU
  This is the actual PDU, i.e. SDU received from the upper or SERVICE USER LAYER, to be delivered by the SERVICE PROVIDER LAYER to its peer protocol layer through this service primitive being declared.

Note that the parameters(s), 1. . .N, declared in a primitive could be part of the SAPI used between the adjacent communicating layers.

14.9 Example of a Protocol Layer Frame Header Definition

In Section 14.5, the reader has been introduced to the encoding and decoding of a frame-level Layer 2 protocol. A sample and an illustrative definition of a protocol layer frame header using a C-language structure are given in Example 14.8.

14.10 Examples of System Parameters

As mentioned in Section 14.1 overview, a protocol layer contains various system parameters such as the timers, counters, variables, and constant as defined by the concerned 3GPP TS. The value of a particular system parameter may be an implementation dependent.

- **Timer** – Indicates how long to wait for a particular procedure to complete, either successfully or unsuccessfully, initiated by a protocol layer toward its peer layer.

Example 14.8 Illustrative C-Structure Definition of a Header of Protocol Layer Message

Recall the OSI TCP IP protocol suite. A protocol layer message consists of two parts: its header and the payload/data received from the upper layer. See the illustration in Figure 14.23.

Figure 14.23 Illustration of the components of a protocol layer message.

Further, the data part as shown in Figure 14.23 may be followed by a Frame Check Sequence (FCS) or Block Check Sequence. For the exact protocol layer format to be used, refer to the corresponding Layer 2 3GPP TS.

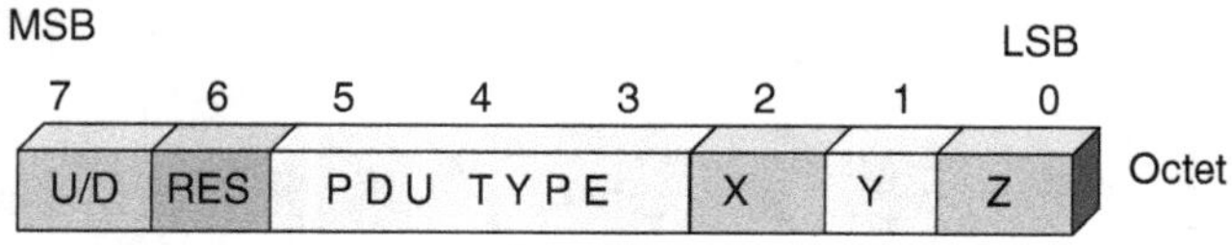

Figure 14.24 Illustration of a protocol layer header part with flags.

A sample protocol layer header consisting of different control parameters/flags and its definition is shown below:

C-Structure Definition: All the contents of a protocol layer header as illustrated in Figure 14.24 can be packed into a C-language *structure* definition. Inside the C-structure, one can also use a *C-Union* definition. One of the goals of a protocol layer design is the efficient usage of the air interface. To achieve this, a 3GPP protocol layer tries to encode information even at the bit-level that can carry two information, e.g. ON or OFF. In the code, the same requirements can be implemented using the C-language bit-level feature, as illustrated in the following definition:

```
struct{/*C Structure definition of a protocol header*/
TYPE1 U_D: 1; /* Parameter/flag U(pilnk)/D(ownlink) is 1 bit in length */
TYPE1 RESERVED: 1; /* Parameter2 is reserved for future use */
TYPE2 PDU_TYPE: 3; /* PDU_TYPE is 3 bits in length */
TYPE3 X: 1; /* Parameter/flag X is 1 bit in length */
TYPE3 Y: 1; /* Parameter/flag Y is 1 bit in length */
TYPE3 Z: 1; /* Parameter/flag Z is 1 bit in length */
} header;
```

- **Counter** – Indicates how many times to repeat for a procedure initiated by a protocol layer toward its peer layer. The counter is incremented each time the particular event is repeated. This repetition is permitted until the concerned timer is running and has not expired.
- **Constant** – It is used to compare against a value whose value is calculated during runtime protocol layer procedure or event.

System parameters for a particular protocol layer are defined separately at the UE/MS and network end, i.e. RAN and Core Network, which is used to keep track of the particular procedure initiated by the concerned network element. System parameters are also defined separately both for the CS domain and PS domain procedures. The concerned 3GPP TS also specifies the value of its range, the appropriate actions to be taken upon normal completion as well as at the first and subsequent expiring/exceeding of the particular system parameter. Example 14.9 shows the typical C-pre-processor definition of various system parameters found in a 3GPP protocol specification.

14.11 Examples of Protocol Information Elements and Its Identifier

A control plane PDU or a signaling message is used to exchange information between two peer protocol layers. A PDU or a signaling message contains various information to be communicated to the peer layer where the information is encoded in the form of IEs. For example, consider the contents of the GSM RR Layer 3 *ASSIGNMENT*

Example 14.9 Illustrative C-pre-processor Definitions of Timer, Counter, and System Variables

```
C-pre-processor Definitions
/* 3GPP TS: 44.018 [130] Timer on Network Side */
#define GSM_RR_T3101 value /* value: Network dependent */
/* 3GPP TS: 44.018 [130] Timer on MS Side */
#define GSM_RR_T3164 5 /* T3164: Duration: 5 Sec. */
/* 3GPP TS: 38.331 [116] Counter on UE Side */
#define N311 0 /*Assigning a default 0 */
#define N310 X /*Assigning a default X */
```

Example 14.10 Illustrative C-Pre-processor Definitions: IEs And IEIs

```
/* GSM Layer 3: RR Sublayer IEs and IEIs: 3GPP TS 44.018 [130] */
#define GSM_RR_CHAN_DESC_IEI 0x64
#define GSM_RR_START_TIME_IEI 0x64
#define GSM_RR_FREQ_LIST_IEI 0x05
#define GSM_RR_CCD_IEI 0x62
#define GSM_RR_MULTISLOT_IEI 0x10
```

COMMAND shown in Table 9.1.2.1, TS 44.018 [130]. Some of the IEs of the RR layer *ASSIGNMENT COMMAND* message is mentioned below:

- Protocol Discriminator, Skip Indicator, Message Type;
- Channel Description, Power Command, Frequency List; and
- Cell Channel Description, Multislot Allocation.

The above IEs and their corresponding identifiers (IEI) can be declared using C-pre-processor as shown in Example 14.10.

Look at any Layer 3 protocol message definition available in a tabular format. One will observe that the first column in the tabular definition of Layer 3 messages contains the IEI value of an IE. However, the IEI value is mentioned only for an IE whose format is T, TV, or TLV. A 3GPP TS defines all the IEI of a particular protocol layer under a separate section called "information element identifier".

14.12 3GPP Release Specific Changes Implementation

The 3GPP introduces new features and capabilities in each of its releases, such as the Release 99, Release 4, and Release 5, and so on. From a 3GPP release implementation point of view, each release will introduce several aspects such as new requirements, logical objects, structures, messages, information elements, even for an existing particular protocol layer.

Consider that the developer has developed and implemented a protocol layer as per the specification for Release 99. Now the developer wants to develop the same protocol layer, say, for Release 4. The developer has identified and wants to reuse a certain portion of the existing code being developed for the existing Release 99. A developer can incorporate new code for the next release, say Release 4, possibly in the same source file without disturbing the existing functionalities. This can be accomplished by declaring various logical objects and identifiers using C-pre-processor as shown below.

To produce the release specific protocol layer functionalities the developer will build the protocol layer program module by using a compile-time flag for the required release, say passing 3GPP_RELEASE_99 = 1, or 3GPP_RELEASE_REL4 = 1 flag to the compiler.

```
#ifdef 3GPP_RELEASE_99
#define INFORMATION_ELEMENT_IDENTIFIER_X 0x123
#endif
#ifdef 3GPP_RELEASE_4
struct Object X{ int a; int b; int c;… ….}
#endif
```

Example 14.11 Illustrative C-Pre-processor Definition of 3GPP GSM RR Message Types

```
/* GSM Layer 3: RR Sublayer Message types: 3GPP TS 44.018 [130] */
/*Channel establishment messages*/
#define GSM_RR_IMMEDIATE_ASSIGNMENT_MSG 0x3f
........................
/* Handover messages */
#define GSM_RR_ASSIGNMENT_COMMAND 0x2e
........................
/* Paging Messages */
#define GSM_RR_PAGING_REQUEST_TYPE_1 0x21
```

14.13 Examples of Protocol Messages Types

A cellular system works by exchanging various messages that carry either signaling or user traffic information between the communicating network elements. Every protocol layer/sublayer has its predefined set of messages to be exchanged between the peer layers to provide different communication services to subscribers. Within a particular protocol layer, messages being supported have a unique *identity* or *type*. For example, recall the general structure of a Layer 3 signaling message (Section 14.4.3). The second octet of a Layer 3 message header contains the message type of the message to be communicated to the peer layer. Message type definitions for some of the GSM RR layer messages through sample C-pre-processor are illustrated in Example 14.11.

14.14 Protocol Layer Timer Handling

As mentioned in the earlier sections, every protocol layer contains several timers that are maintained independently at the MS/UE and network side. A timer is used to keep track of a particular procedure or a service requested by a protocol layer that is initiated toward its peer layer. As defined by a 3GPP TS, a protocol layer may contain several timers for each of the procedures that it performs toward its peer layer. A timer contains four events and states that are associated with a particular procedure or service. Those events and states of a timer are as follows:

- Start, Stop, Expired, and Restart

The above states and their transitions are illustrated in Figure 14.25.

The typical tasks to be performed by a protocol layer timer in each of the states mentioned above are described by the concerned 3GPP TS. Apart from the specific tasks as defined by a 3GPP TS in each state of a timer, a protocol layer may also perform implementation-dependent tasks such as the following ones:

- Cleanup, at the expiry, of the internal resources being allocated at the time of start;
- Update internal statistics of the protocol layer; and
- Abort the procedure or service requested, raise alarm, and notify the O&M operator to take corrective action.

Typical definitions, implementation, and usages of timers found in the GSM CM layer are illustrated through pseudo code in Example 14.12.

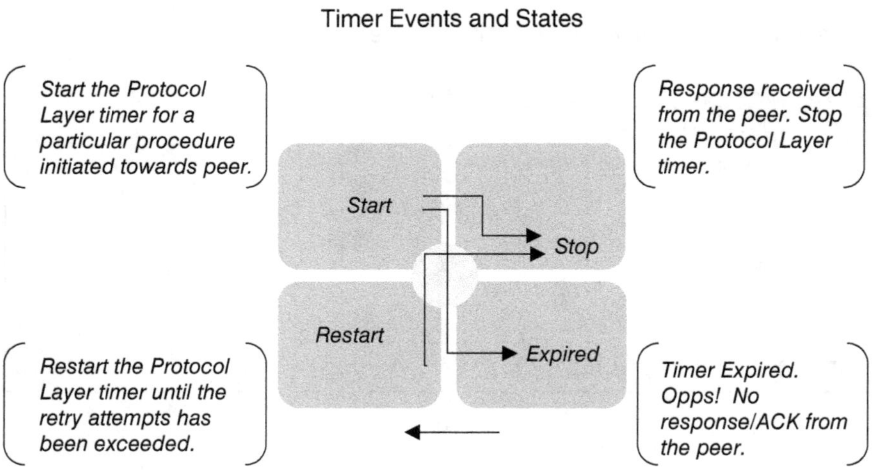

Figure 14.25 Illustration: protocol layer timer and its various states.

Example 14.12 Air Interface GSM CM Layer 3 System Timer

Consider the GSM CM sublayer of air interface Layer 3 protocol stack. There are separate timers for the CM sublayer at the MS/UE and network side. The timers available at the CM sublayer at the MS end can be declared as follows using C-pre-processor:

- Timers Definition/* *Source:* Table 11.3/3GPP TS 24.008 [45]: CM sublayer timers definition at the MS,UE side*/

```
#define CC_TIMER_T303 30 /* default 30 sec */
#define CC_TIMER_T308 30 /* default 30 sec */
#define CC_TIMER_T322 X /* implementation dependent*/
#define CC_TIMER_T323 30 /* default 30 sec */
#define CC_TIMER_T335 30 /* default 30 sec */
```

- Common function to start CM sublayer timers
  A common function can be developed, as illustrated below, to start all the timers of the GSM CM sublayers. The timer type to be started is passed to this function.

```
start_CC_Layer_Timer(timer_type...........)
{
switch (timer_type)
{
case CC_TIMER_T303:
call start_timer(CC_TIMER_T303,cc_timer_handler_function, *additional_ _
info_for_the_handler_function,.............);
....................
Break;
Case CC_TIMER_TXX:....................
....................
}
}
```

- Common timer callback/handler function

Similarly, a common timer callback function may be developed as follows:

```
cc_timer_handler_function(timer_type X,...)
{
Switch (timer_type X)
{
Case CC_TIMER_T303: /* Call timer specific handler function */
Call CC_TIMER_T303_handler (...);
Break;
....................
}
}
```

- Timer: abc-specific handler function
  A timer-specific handler function may be developed as follows:

```
CC_TIMER_Tabc_handler(...)
{
Do the necessary tasks as mentioned in the 3GPP TS for the concerned timer
Update state machine.
Update necessary statistics.
.............
}
```

- Specific timer expiry handler function
  A timer expiry handler function may be developed as follows:

```
CC_TIMER_abc_Expiry_handler(...)
{
-Do the necessary tasks as mentioned in the 3GPP TS for the concerned timer
-Do the necessary cleanup of temporary resources/data structures and other
objects
....................
}
```

- Timer-specific restart handler function
  A protocol layer procedure may be one time or repetitive in nature. In the former case, the concerned timer shall be started only once. On the other hand, if it is repetitive in nature, the concerned timer may be restarted again until the retry threshold is exceeded. For example, consider the LTE/EPS UE ATTACH request procedure which is controlled by the timer T3410, 15-second duration. At the expiry of this timer, a UE can resend the ATTACH request till the request counter is less than 5. A timer-specific restart handler function may be developed as follows:

```
CC_TIMER_Tabc_Restart_handler(time_type,...)
{
If (current retry counter <= retry Threshold as per 3GPP TS)
Call start_CC_Layer_Timer(timer_....);
/* Restart the timer and wait for the expected response from the peer */
....................
}
```

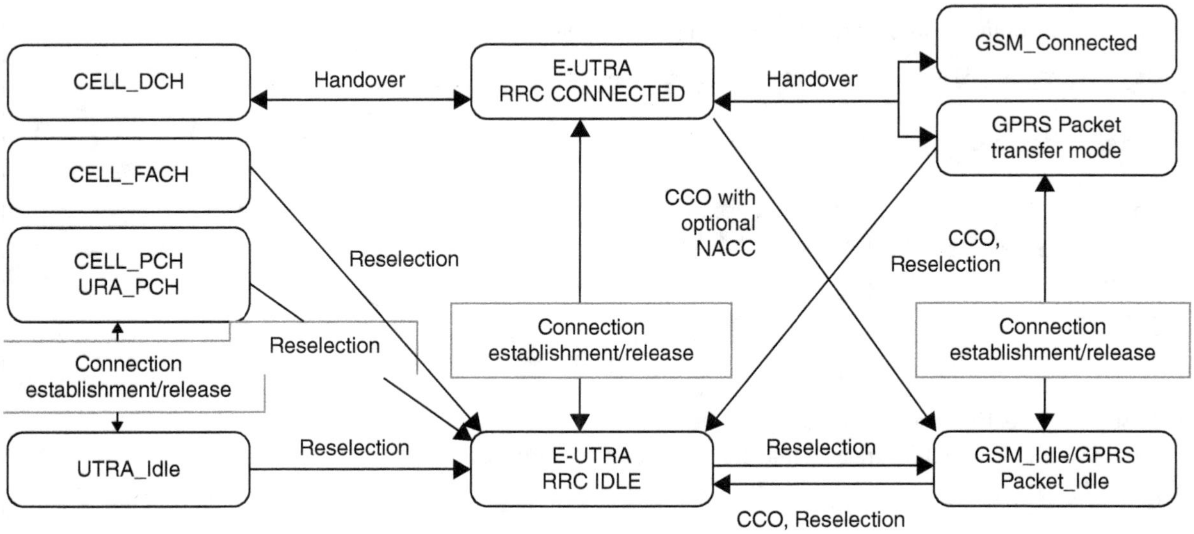

Figure 14.26 RRC layer protocol states and its transitions: GSM/GPRS to the LTE system. *Source:* © 2018. 3GPP™ TSs and TRs are the property of ARIB, ATIS, CCSA, ETSI, TSDSI, TTA and TTC who jointly own the copyright in them. © 2018, 3GPP.

14.15 Protocol Layer Development Using State Machine

A model for a protocol layer may also contain state machine architecture for its various tasks to be performed in its different states. Recall the different states of TCP. If a 3GPP protocol layer operates based on a state machine architecture, the concerned TS describes the desired behavior and the functions to be performed by the protocol layer in each state. Sometimes, a protocol layer's state machine may also contain substates. As an example, Figure 14.26 shows the different states, and their transition, of the RRC protocol layer, maintained at the MS/UE end in case of intra- and inter-RAT, i.e. LTE, GPRS and GSM network operations or procedures. A wrong implementation of the state machine of a particular protocol layer will result in unexpected behavior as well as communication services outages.

Figure 14.26, which is reproduced from TS 36.331 [94], shows the different states of the RRC protocol layer of the GSM, GPRS, UMTS, and LTE systems. An MS/UE that is capable of supporting all the radio access technologies, i.e. GSM, GPRS, UMTS, and LTE, shall have to maintain the above states and its transitions correctly at the RRC layer. This will ensure the service availability in case a user turns off and on or leaves and enters into a new coverage area with different radio access methods, for example, circuit-switched fall back (CSFB) from LTE to UMTS or GSM system.

Based on the current state of the RRC layer, the MS/UE is said to be in the corresponding state too, i.e. MS/UE is in an IDLE state or CONNECTED or DEDICATED state. The RRC protocol layer specification of GSM/GPRS (*TS 44.018 [130]*), UMTS *(TS 25.331 [50])*, and LTE (*TS 36.331 [94]*) system contains the descriptions of the various tasks that an MS/UE can perform in a particular state. The various tasks to be performed in each state shall form the functional requirements specification for the RRC layer.

The functional requirement specifications can be implemented through an event generation in the code, as illustrated below in the case of the LTE system.

```
LTE_RRC_MACHINE [] [] =
{
{/* RRC IDLE STATE MACHINE */
```

```
      /*PAGING_INDICATION_EVENT */
      {rrc_process_paging_idle_state(...)},
      /*PROCESS_SI_EVENT *
 {rrc_process_si_indication(...)},
 ..................
}
{/* RRC CONNECTED STATE HANDLING */
 /*DATA_INDICATION_EVENT */
      { rrc_process_data(...)},
      /*TIMER:x EXPIRY Event */
      {rrc_process_timerX_expiry(...)}
 ..............
}
};
```

To implement a protocol layer state machine, one can also use the C/+C++ *switch –case* statement or *function pointer tables* as illustrated in the sample pseudo code below.

```
switch (msg_type X)
{
 case X: Generate Event X;
 Call and pass the event X to LTE_RRC_STATE_MACHINE........
}
```

It may be noted that several protocol layers, e.g. EPS mobility management layer and session management layers, and their main states have substates also. The behavior of a UE or the concerned network element in each substate is required to be implemented properly. A state machine-based system architecture and its working model may be also applied to other network elements or components of a mobile communications system.

Consider the typical software architecture of BTS of a GSM communications system. A BTS can have several states as illustrated in Figure 14.27. A BTS can be in working as well as in several administrative states, e.g. LOCK, UNLOCK, and RESET states. The accompanying table on the right side shows the state's machine descriptions, i.e. actions to be taken upon receiving primitive (LOCK, UNLOCK, RESET...) during a particular state (LOCKED, UNLOCKED).

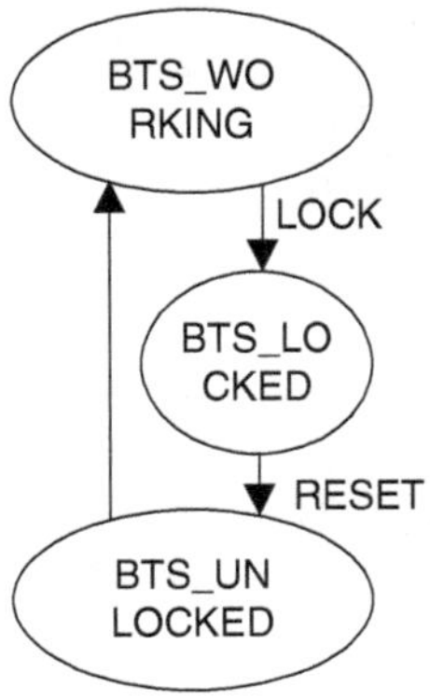

BTS STATE	Command	Actions
ANYSTATE I IDLE	LOCK	-Release BLOCKED Calls –Change STATE

BTS STATE	Command	Actions
WORKING I IDLE	RESET	-Release ACTIVE Calls –Change STATE

Figure 14.27 Illustration: states of a base transceiver (BTS) station.

14.16 Protocol Layer Development Using Message Passing

The discussion so far on the various air interface Layer 3 signaling messages belongs to a higher application/protocol layers of a particular cellular system. The application protocol layer messages are required to be transported and exchanged between network elements, i.e. intra- or inter-machine-to-machine, or even among the related processes, i.e. within a machine, at the operating system platform level. To transport and exchange information between machines or among the related processes representing the various protocol layers, various facilities and services provided by the chosen operating system should be used. The best way to exchange and share information among the protocol layers processes is the *inter-process communication (IPC)* mechanisms as discussed in Section 13.3.2.1. A suitable IPC mechanism provided by a particular operating system platform should be selected depending on the application's or protocol layer requirements.

Example 14.13 describes a method which can be used for communications among the protocol layers of a network element.

Example 14.13 Protocol Layer Communication Using IPC

Consider the illustration of an IPC shown earlier in Figure 13.5. This illustration shows the following IPCs scenarios:

a) Processes, representing protocol layers, running on a UNIX platform are using Message Queue and shared memory to exchange information among them. Note that the Message Queue and shared memory forms of powerful IPC methods are available only on Linux/Unix platform. Using Message Queues, one can develop a message router facility for routing messages among processes.
b) Protocol layer processes(s) running on a Windows platform and is communicating with the processes running on the UNIX platform using TCP/UDP socket via the message dispatcher process running on the UNIX system.

Sample C-pseudo code for wrapper function, illustrated below can be developed to realize the above IPC scenario's requirements.

- Definition of a common header for IPC

```
typedef struct
{
Type1 ipc_source_application id;
Type1 ipc_destination_application id;
Type2 ipc_protocolX_layer_message_type;
Type3 ipc_protocolX_layer_message_length;
Type4 *ipc_message_buffer;
}sample_ipc_protocolX_message_header;
```

- Wrapper function for sending an IPC message from one protocol layer to another

```
ipc_send_protocolX_message( ipc_message_header *msg,..........................)
  {
  .........
  }
```

A separate wrapper function may be also developed and used to read IPC messages belonging to the control plane and user plane. The development of such wrapper functions is illustrated below.

- Wrapper function to read an IPC message(Control Plane protocol) through the TCP socket from the peer layer

```
read_tcp_socket_data(buffer,....)
  {
  .........
  }
```

- Wrapper function to read IPC message (User/Data Plane protocol) through the UDP socket from the peer layer

```
Read_udp_socket_data(buffer,....)
  {
  .........
  }
```

A separate message dispatcher function to read IPC messages belonging to the control plane and user plane may be developed as follows:

- Message dispatcher function to read an IPC message(Control Plane protocol) through the TCP socket from the peer layer

```
message_dispatcher_ipc_tcp_socket_handler(...)
{
read_tcp_socket_data(...);
find the destination application id;
construct an IPC message and send the buffer;
ipc_send_protocolX_message(...);
................
}
```

- Message dispatcher function to read IPC message (User/Data Plane protocol) through the UDP socket from the peer layer

```
message_dispatcher_ipc_udp_socket_handler(...)
{
read_udp_socket_data(...);
find the destination application id;
construct an IPC message and send the buffer;
ipc_send_protocolX_message(...;);
...............
  }
```

14.17 Protocol Layer Data and its Types

Every 3GPP protocol layer has its protocol configuration data that is required for its smooth functioning to provide services to its upper layer. Some of the data a protocol layer requires are *static*, while some data are *dynamic* or *runtime* in nature. Dynamic or runtime data to protocol layers are provided as part of ongoing network operations

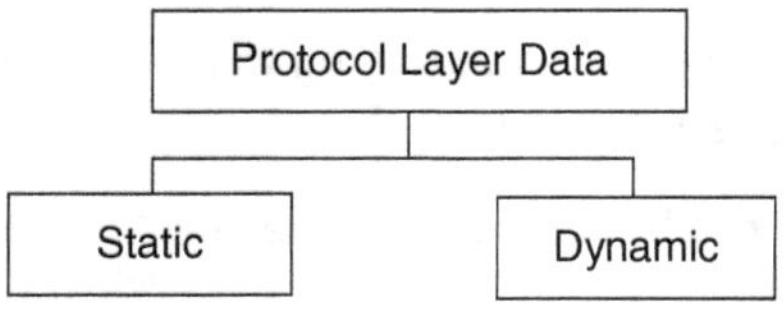

Figure 14.28 Illustration: types of protocol layer data.

and maintenance, planning, tuning, and optimizations process. Dynamic data for a protocol layer are also generated by the external environment such as the radio air interface.

For example, the mobile devices monitor and measure the downlink signal quality and its strength of the serving cell and its neighbors and report it to the base station/RRC layer from time to time. Based on the dynamically reported data, the radio resource management layer takes the appropriate actions such as the triggering of a handover requirement of a CS call or downgrading/upgrading of the current coding scheme used by a PS call. Figure 14.28 illustrates the types of data used by a protocol layer of a mobile communications network for its proper functioning.

Static data definitions through C-pre-processor were illustrated at several places in the preceding sections. Protocol layer static data may be also available in the form of a table or just as enumerated values in a 3GPP TS of a concerned protocol layer. For example, in the LTE E-UTRAN, each modulation and coding scheme is associated with a transport block size index from which the actual transport block size can be derived using a 34x100 size table (representing 1.4–20 MHz bandwidth); refer to TS 36.213 [91] Table 7.1.7.2.1-1. Such static data is maintained by the UE and E-UTRAN physical layer and are illustrated through the Examples 14.14 and 14.15.

14.18 Protocol Layer Control and Configuration

A protocol layer not only communicates with its adjacent layers but also sometimes controls and configures the runtime functions and behavior of a protocol layer below it. In this case, an upper layer, e.g. RRC layer, at the RAN end will have the interface with O&M console. The O&M operator may update and modify the various network parameters from time to time as per the ongoing radio network planning and optimization process. The peer protocol layer, e.g. RRC, at the MS/UE end will receive the network information through the system information from the RAN which further distributes and updates the network parameters to the concerned protocol layers below it at the MS/UE end. Such periodic updates of a protocol layer parameter at the UE/MS end also affect its behavior at runtime. Protocol layer configurations by another protocol layer are illustrated in Example 14.16.

Example 14.14 Modulation Index and Transport Block Index Definition for PDSCH

```
/*First Index: Modulation Index, 2nd Index: Modulation Order, 3rd Index:
Transport Block Index: TS 36.213 [91]*/
short int mcs_idx_tbs_idx[][][] ={{0,2,0}, {1,2,1},...............{28,6,26}};
```

Example 14.15 Transport Block Index and Size Definition for 1.4 MHz System Bandwidth with Six Physical Resource Blocks (PRBs)

```
/*First Index: Transport block Index from previous example, 2nd Index: #of
PRB, 3rd Index: Transport Block Size: TS 36.213 [91] Table7.1.7.2.1-1*/
int tbs_idx_tbs_size [37][110] ={{16,32,56....}, {...},{32,72,144,...}....{
840,1736, 2600,......}};
```

Example 14.16 Protocol Layer Configurations in LTE E-UTRAN and 5G NG-RAN

Consider the LTE and 5G NR air interface user plane protocol stacks illustrated in Figure 14.29 below. The RRC layer at LTE E-UTRAN or 5G NG-RAN sends the system information as well as lower layer configuration information through RRC signaling messages to the RRC layer at the UE end.

As shown in Figure 14.29, the LTE or 5G NR air interface RRC layer at UE end is in charge of configuring the user plane protocol layers, i.e. SDAP(5G NR), PDCP, RLC, MAC, and physical layer, at UE end. For more information on the parameters and their values, e.g. *RLC-Config, PDCP-Config, PDSCH-Config, PUSCH-Config*, and so on, which are configured by the LTE E-UTRAN or 5G NG-RAN RRC layer to each of these user plane protocol layers at UE end, the reader may refer to TS 36.331 [94] and TS 38.331 [116].

Figure 14.29 Illustration: LTE and 5G NR air interface protocol layer configurations by RRC layer.

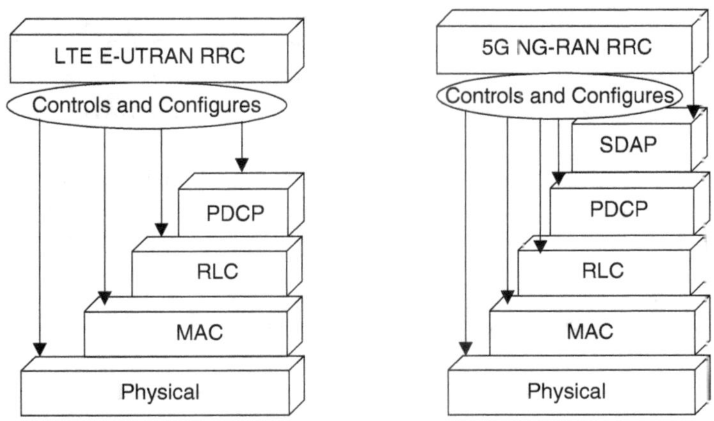

14.19 Protocol Context Information

One commonly found terminology in a 3GPP TS is the *context* information. A context is a logical repository or data structure that may store a protocol layer information or about a network element along with some implementation-dependent information. Typical context Information can be of the following types:

- Network element context information containing information of an MS/UE, SGSN, MME, 5G core AMF, and so on;
- Security context information;
- Bearer context information in case of UMTS or LTE/EPS network;
- PDP context information, in case of GPRS/UMTS; and
- Context information for protocol layers such as mobility management, RRC layer, and session management.

A protocol context contains all the related information for proper communication between peer protocol layers and may be represented by a C-structure or a C++ class, which may further contain a pointer to other context information. Context information can be created, modified, updated, or released dynamically based on the runtime protocol layer procedures that take place from time to time or upon occurrences of protocol events between the MS and the network or based on the current state of the concerned protocol layer. Some typical examples and scenarios/protocol layer events where a protocol context may be created are mentioned in Example 14.17.

Illustration of context information was shown earlier in Figure 13.4. Context information to be created and maintained may be also found in the 3GPP TS such as TS 23.060 [31] and TS 24.301 [46] under the *Information*

Example 14.17 Protocol Events for Creation and Usages of Context Information

- Upon a successful network attach or registration procedure, a *mobility management context* is established at the core network end.
- A *security context* is established and is stored in the MS/UE and network following the successful authentication procedure. The security context information of CS and PS domain is different. Refer to Table 9.7 for security context information.
- A *packet flow context* or traffic flow template may be created upon receiving a request to transmit uplink or downlink data.
- During an inter-RAT UE movement scenario, context information is exchanged between source and destination network elements. For example, a UE registered with an MME may select a UTRAN and GERAN cell serviced by an SGSN and perform an RAU procedure. In this case, the new SGSN requests the context request from the old SGSN, containing the mobility management (MM) and PDP context of the MS/UE.
- Context information is also exchanged and updated as a result of the inter-RAT handover from the 5G system to the LTE network; from the E-UTRAN to GERAN/UTRAN and vice versa; or LTE/EPS to 5G system and vice versa.
- As a result of the ISR feature activations at the LTE UE and the core network ends, UE MM contexts are created at the UE, MME, and SGSN ends.

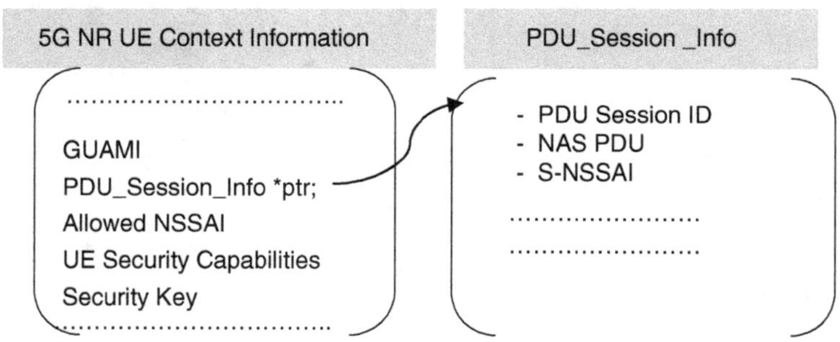

Figure 14.30 Illustration: various information contained in a logical context.

Storage section. One such example of protocol context creation for a PDU session context definition in the 5G system is illustrated in Figure 14.30.

In the 5G system, a PDU session information shall be a part of a UE context information that is stored at the UE and 5G core network, i.e. AMF network function end. Such a UE context information is exchanged between the NG-RAN and 5G core network. A UE context information is also exchanged between a source and target system during a handover procedure.

User-defined and protocol layer implementation-dependent information may be also included as part of a protocol context definition in the code apart from the relevant information as defined by the particular 3GPP TS.

14.20 Protocol Layer Message Padding

Sometimes, the information transmitted by a protocol layer may not fill up a PDU size completely. In such a scenario, all the unused space in a PDU shall be located at the end of the PDU and shall be padded with a value that is ignored by both the transmitting and receiving protocol layers. The process of adding extra information/value, usually 0 filled, toward the end of PDU is referred to as *padding*. Padding shall have a length such that the PDU as a whole has one of the predefined total lengths to make the PDU either word (32 bits) or byte (8 bits) aligned. Figure 14.31 shows the padding of 4 bits (bits 4–7) with value 0 at the end of a PDU to make it octet-aligned protocol information.

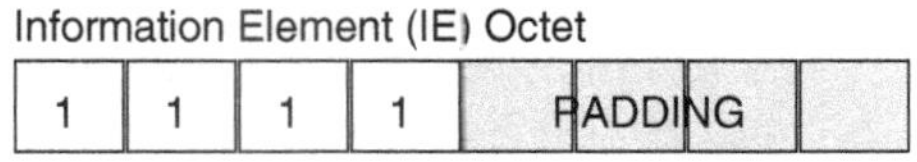

Figure 14.31 Illustration: padding in an IE of a message/PDU.

Note that padding can be also applied to an information element whose position is in the midst of a particular message. As another example, refer to the padding field of the TCP/IP protocol message format.

14.21 Device Driver Development

A *device driver* is a special kind of program written for a specific purpose and particular hardware. Generally, a device driver is a part of the system software layer that accomplishes a particular task. It can be architecture-specific, i.e. Intel x86, PowerPC MIPS, and so on. A device driver program works closely for controlling, reading, or writing into a piece of hardware or a chip. The reader is already familiar with device driver programs at the operating system level for various peripherals such as the printer, scanner, CD-ROM, and USB drive. Similarly, if a developer is developing a network element that is based on an embedded system using a particular hardware board or processor as mentioned earlier in Section 13.2, there are possibilities of design and development requirement of device driver programs. This is especially true if the developer is working at the air interface/physical layer area.

Example 14.18 describes and illustrates the development requirement of a device driver for the processing of GSM E1 Physical interface frames.

Example 14.18 Device Driver for Processing of GSM E1 Physical Interface Frames

Consider the development of a device driver to transmit and receive frames over the GSM/GPRS physical E1 interface between a BTS and BSC, as illustrated in Figure 14.32. As shown in Figure 14.32, the physical media/interface being used for the A-bis interface is called the *E1 interface* with 2 Mbps bandwidth divided into 64 timeslots of 32 kbps each. GSM E1 interface was described earlier in Section 3.1.1. The GPRS system uses a 52-radio frame structure and uses the GPRS RLC layer blocks to transfer control and user data. It may be noted that GPRS RLC blocks are used to transmit and receive information at 20 ms scheduling or transmission timing interval over the E1 interface. So, in this case, a developer may design and develop a device driver program that wakes up at every 20 ms interval. This is illustrated in Figure 14.32. At 20 ms wake-up cycle, the device driver will either:

- Transmit (write), Tx, the frames over the air interface through the BTS or
- Receive (read), Rx, the RLC block from the air interface through the BTS.

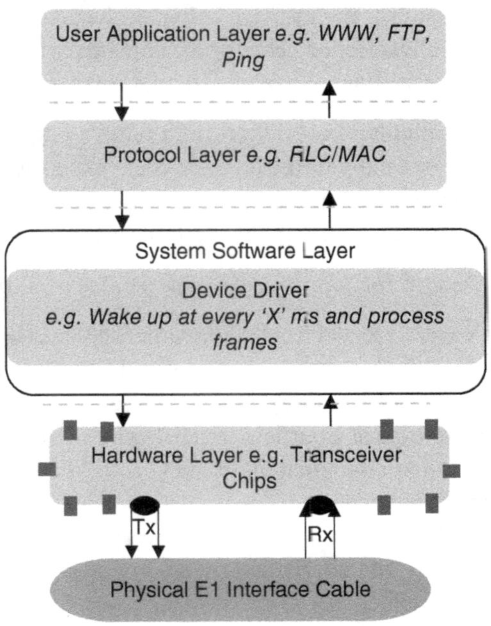

Figure 14.32 Illustration: device driver development model.

The wake-up mechanism could be either timer based or interrupt based. On receiving the air interface frames from the E1 interface, the device driver program groups the frames and sends them to the RLC layer, i.e. the application layer, for reassembling of RLC blocks. The RLC layer would further send the reassembled blocks to the intended application working at the IP layer (e.g. WWW, ping, FTP) or pass the information from the MS to the SGSN through the BSC.

To develop a device driver program, one needs a different skill set with low-level programming knowledge in assembly language and requires to fully understand the internals and hardware architecture details. A hardware chip contains registers both for general and special purposes, and as a device driver developer, the developer will be required to work closely, such as initializing and reading/writing into, with such registers. Similarly, for an IP-based system such as the LTE, one can develop a device driver to process frames at every 1 ms. This is the frame scheduling time in the case of the LTE system. The scheduling time of 1 ms is also called as the *Transmission Time Interval (TTI)*.

14.22 Guidelines for Protocol Stack/Layer Development

Here are the typical guidelines that a developer should keep in mind when designing and developing a protocol layer or protocol stack.

- **Protocol Layer Information Logging Facility**

This facility provides logging capabilities either to a file, on the console, or through the serial port containing information of various details that can be turned off and on using special-purpose flags during runtime without requiring to stop and restart a protocol layer module. For example, a protocol layer may log information such as the following messages exchanged in the uplink, downlink, or both the direction:

- Name of the message being received/send,
- Fully decoded octets of a complete particular message, and
- Binary contents of a particular octet of a message.

An inbuilt protocol layer information logging facility plays an important role during the troubleshooting of an issue especially in the absence of an expensive protocol layer message capturing and analysis tool for a hard to reproduce field issue. In the case of embedded and multicore platforms, a facility to dump the entire memory contents for offline analysis purposes is useful in case a system crash happens.

- **Design for Easy to Add New Features/Functionality from Release to Release**

Now the reader knows that a particular 3GPP system, e.g. UMTS, LTE, or 5G, has multiple releases along its path of evolutions. Such a new and subsequent release or feature may be developed using the existing release as the baseline. An existing software module for a particular protocol layer may be possible to be reused in the next release of a particular 3GPP system. To reuse an existing module, a parallel branch of the existing protocol layer may be created, which will allow multiple teams to work on the existing and new releases simultaneously.

To develop a new release, a developer will be required to gather, first of all, the requirements of the new release or feature and implement the same through the use of pre-processor directives such as the *#ifdef* and *#ifndef* features provided by the C/C++ languages. Further, to enable a selective, i.e. new release specific, code compilation, necessary compiler directives are passed during the program compiling time.

14.23 Software Profiling, Tools and Performance Improvement

One important aspect of a mobile communications network and system is its performance. Overall, the performance could be in terms of the *call per second* (CPS) that a communication network/system can process. For example, consider a 70 CPS that the RRC protocol module can process. A call-processing event may involve executions of several algorithms to assign the necessary resources either success or failure in allocation by the radio resource management. If a resource allocation was not successful, then the desired CPS would be less than the benchmark value.

To improve the performance, say, the CPS, of a mobile communications network and system, one can start working on the concerned call-processing module. To assist in improving the performance of the call-processing module, one can perform a *software profiling* of the concerned module. In this case, one can take the help of the tools provided by the chosen platform. For example, on the UNIX platform, one can use the tool called *gprof* to perform profiling of the call-processing module. See the UNIX man pages for the usages of *gprof*.

A *software profiling* allows one to learn where a program spent most of the time and which functions were called by which other functions during its course of runtime execution. This vital information can show the developer which pieces of the call-processing module are slower than expected and might be the candidates for rewriting the code (for example, by including an efficient loop) to make the call-processing module execute faster, to improve the CPS performance. A profiling data can also tell the developer which are the functions that are being called more or less often than the one expected. Another useful tool available on the UNIX platform is the *truss* to trace various system calls called by a particular module/process during runtime. Look at the UNIX manual page to learn more details on it.

14.24 Protocol Stack Testing and Validation

Now, the reader has learned that mobile communications systems and networks based on the 3GPP architectures contain more than dozens of logical interfaces, starting from the alphabet "A". Logical interfaces with protocols and protocol stacks facilitate intra- and inter-RAT communications among the various network entities of the Access Network (AN), Core Network (CN), and also the handset, i.e. MS/UE, of mobile communications systems. Prior to the releasing of a network element to field deployment, its protocols functions and procedures must be thoroughly tested and verified. But the testing and verification of field-like scenarios are not always possible within a developer's laboratory environment. To test such scenarios, customized software laboratory test stubs and simulators become useful and may be used to test exceptional cases/scenarios which are otherwise difficult to reproduce at fields.

As a starting point of protocol stacks testing and its validation requirements, identify the focus area of protocol testing, i.e. Access Network, Core Network, and MS/UE.

a) Find the various logical and physical interface(s) connecting Access Network and Core Network, or MS.
b) Study about the various messages/PDUs (either control plane or user plane) and their contents exchanged over a protocol interface.
c) Identify the tools and their usages to be used for protocol layer testing and its validation. Study the usages, and the interpretation of their report, of various testing tools available. For example, use professional tools to capture:
   - Air interface messages exchanged between the MS and AN or MS and CN;
   - Protocol messages between the different network elements and the access network, supporting the IP transport network interface;
   - To capture any IP interface-based protocol messages, use the IP packets snipping tools.
d) Drive test – Use a drive test method to observe and rectify various issues arising out of intra- or inter-RAT air interface.

e) Interoperability testing (IOT) – Use an IOT to test the network entity/system with other vendor's network entities/systems).
f) Develop customized software test stubs or simulators to:
   - Simulate the behavior of a peer protocol/network entity that is not under developer control;
   - Test every possible input to every IEs of a 3GPP message;
   - Test every possible negative, race conditions, or abnormal scenarios such as timer expiry, exceeding of max retries, i.e. counter, and so on;
   - Test and verify every hardware, software configuration, the maximum limit, or system dimension as defined in the concerned 3GPP protocol TS or as defined by the network operator;
   - Test and verify the state machine of a protocol layer.
g) Test and verify the performance, in terms of KPI, of the concerned module that implements a protocol or protocol stack.
h) Create, review, and manage test plans to coordinate with several test engineers and also test automation experience through scripts.
i) There are several categories of software testing such as the following ones. Study about the scope of each type of testing for your protocol layer:
   - Unit Test
   - Subsystem Integration
   - System Integration
   - Regression Test
   - Load Test

Example 14.19 describes and illustrates the usages of a software simulator for the performance testing of network element(s).

Example 14.19 Network Elements Performance Testing and Validation Through Software Simulator

Figure 14.33 illustrates the typical test setup environment to perform different kinds of testing and verifications as mentioned above for a typical RAN element that is under test.

In Figure 14.33, consider that the project team has developed a RAN element, say the 5G NG-RAN, whose actual performance (hardware as well as software) is required to be tested. In this setup, software simulators are being used for the core network elements (e.g. MSC or SGSN and 5G Core AMF), mobile devices, and traffic load generator. These simulators may run on legacy platforms such as Windows and UNIX/LINUX.

The software simulators for the core network elements exhibit as if they are the actual core network elements that are providing end-to-end signaling connection requirements. Note that a successful signaling establishment between the network elements is required prior to a user call, either voice or data, can be started.

Assume that the development team would like to do a performance testing in terms of the number of GSM voice or GPRS or LTE/EPS or 5G IP data call that the actual RAN can process in a given time period, say per second. This can be realized through the traffic load generator software simulator as illustrated in Figure 14.33. Using a traffic load generator, the performance of a network element can also be verified under different traffic scenarios, characteristics, or 5G network slice-specific traffic profiles. For example, a traffic load generator may be used to simulate the performance of the actual NG-RAN of a 5G system for different traffic types with

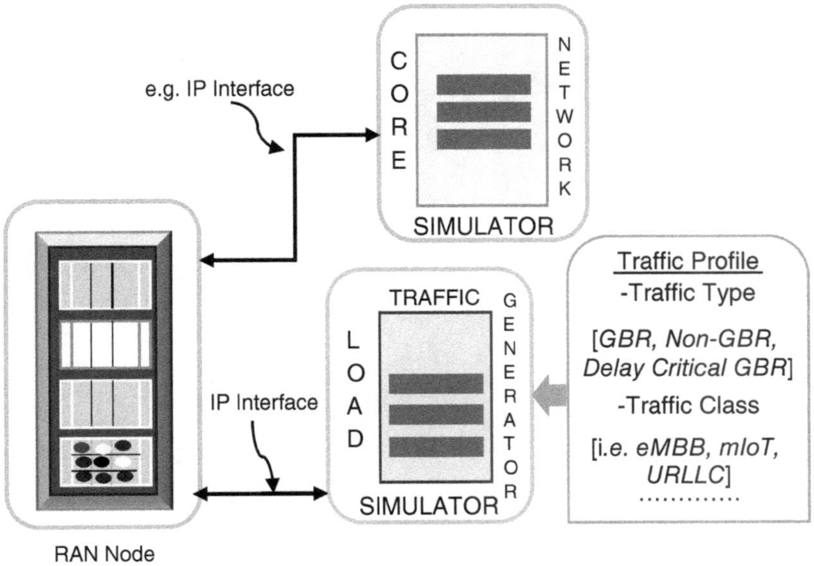

Figure 14.33 Illustration: usages of software simulator test setup for different kinds of testing.

the following characteristics as defined by the International Telecommunication Union–Radio communications sector (ITU-R) in their report ITU-R M.2410:

- High data rates, low signaling traffic and area traffic capacity for the *enhanced mobile broadband use* case;
- Low data rates, high signaling, and connection density for the *machine-type communication use case*; and
- Low latency and high reliability for the *ultrareliable low-latency communication* use case.

In the above illustration, the IP transport network and its physical interfaces are being used to connect the access network element with a simulator for data call load generation purposes. The software simulator was also described in an earlier Section 13.5.

Chapter Summary

This chapter presented some of the core aspects of the software design and development of protocol stacks and its layers of mobile communications networks which are based on the GSM, GPRS, UMTS, LTE, and 5G systems TSs. Mobile communications systems and networks consist of a large number of signaling messages, protocols layers, protocol stacks, and logical and physical interfaces through which different network elements communicate with each other. We presented the methods and formats used for the layer-to-layer and peer-to-peer communications across the different cellular systems. Signaling message formats along with their encoding and decoding methods used over the GSM, GPRS, UMTS, LTE, and 5G system air interfaces and RAN-Core Network interfaces were presented.

Sample software code snippets were presented to illustrate the design and development of a protocol layer based on the information available in the concerned 3GPP TS. Several other important aspects of the design and development of a protocol stack and its layers were also presented. We closed this chapter with the method of protocol stack testing and its validation tasks that can be performed through the usage of software simulators in the absence of an actual network element or hardware.

15

Deriving Requirements Specifications from a TS

Introduction

It has been mentioned earlier in Section 3.1.3 that the 3rd Generation Partnership Project (3GPP) standard defines a larger number of protocol interfaces and their technical specifications (TS), stacks, and layers of mobile communications networks. A 3GPP technical specification may be specific and describes the protocol aspects of a network element of a particular mobile communications system only. However, there are technical specifications that describe the logical interfaces, their protocol layers, and other general aspects of the network elements of mobile communications systems, from the Global System for Mobile Communication (GSM) to the 5G. It is important to derive the requirement specifications from all the concerned technical specifications of a specific network element of a particular communications system, which consists of various functions, procedures performed by it, protocols, and services to be provided to its upper layers.

This chapter covers how to derive and prepare the requirements from the concerned 3GPP technical specifications for typical functions and procedures for the development of a protocol layer or a new feature for a network element. Once the requirement specifications from the concerned 3GPP TSs for a protocol layer or a network element have been gathered, the formal software development life cycle can be started. As part of a Software Development Life Cycle (SDLC), detailed designs, development, and testing shall be carried out by the development team.

15.1 3GPP Protocol Layer Procedures

It has been described earlier that in a mobile communications system and network, a protocol layer of a network element performs various functions and procedures to provide communications services to users. A protocol layer procedure may be an event-driven, e.g. timer or external factor, or periodic one and is initiated either by a mobile station (MS)/user equipment (UE) or the network without user knowledge. The functions and procedures that are to be initiated also depend on the current state of a particular protocol layer or a network element. Identification and preparation of requirement specifications for typical protocol layer functions and procedures performed by a UE in an LTE/EPS network are illustrated below.

Mobile Communications Systems Development: A Practical Introduction to System Understanding, Implementation, and Deployment, First Edition. Rajib Taid.

15.1.1 LTE UE Mode of Operation Requirements

As described in Section 6.6 earlier, a Long-Term Evolution (LTE) UE has four modes of operations as mentioned below:

- Packet-switched (PS) Mode 1
- PS Mode 2
- Circuit-switched (CS)/PS Mode 1
- CS/PS Mode 2

The functions, procedures, and types of communication services provided in each UE mode of operation mentioned above are different. A snapshot of requirement specifications derived from the concerned 3GPP technical specification is illustrated below. It may be noted that the requirement specification may be required to be derived from multiple 3GPP technical specifications.

- **Requirement #1:** PS Mode 1 of operation
  - **Detailed Description:** In this mode, the UE registers only to Evolved Packet System (EPS) voice services <. . .provide further descriptions. . ..>.
  - **Requirement Type:** Mandatory
  - **Source:** TS 24.301 [46], Section 4.3

- **Requirement #2:** CS/PS Mode 1 of operation
  - **Detailed Description:** In this mode, the UE registers to both EPS and non-EPS services for voice service <. . .*provide further descriptions. . ..*>.
  - **Requirement Type:** Mandatory
  - **Source:** TS 24.301 [46], Section 4.3

Similarly, requirement specifications may be prepared for UE operating in single and dual registration modes for interworking between LTE/EPS and 5G systems, and vice versa, only with or without the N26 logical interface. A 5G UE operating in single and dual registration modes shall be described later in Chapter 16.

15.1.2 LTE UE ATTACH Procedure Requirements

- **Requirement #1:** A UE shall support ATTACH procedure for PS mode EPS services only;
  - **Detailed Description:** In this ATTACH procedure <. . .*provide further descriptions. . ..*>.
  - **Requirement Type:** Mandatory
  - **Source:** TS 24.301 [46], Section 5.5.1.2
  - **Result:**
    Success: UE shall support the ATTACH procedure accepted by the network.
    Failure: UE shall support the ATTACH procedure not accepted by the network.

- **Requirement # 2:** A UE shall support the ATTACH procedure for CS/PS Mode 1 or CS/PS Mode 2 for both EPS and non-EPS services.
  - **Detailed Description:** In CS/PS Mode 1 or Mode 2, the UE performs a combined ATTACH procedure <. . .*provide further descriptions. . ..*>.
  - **Requirement Type:** Mandatory
  - **Source:** TS 24.301 [46], Section 5.5.1.3.4/5
  - **Result:**
    Success: UE shall support a combined ATTACH procedure accepted by the network.
    Failure: UE shall support a combined ATTACH procedure not accepted by the network.

15.1.3 LTE UE DETACH Procedure Requirements

The DETACH procedure is used to disconnect and deregister a UE from the LTE core network. There are several types of DETACH procedures, and the appropriate detach type as mentioned shall be included by the UE in the detach type IE to the core network:

- The UE is switched off.
- The UMTS Subscriber Identify Module (USIM) card is removed from the UE.
- The UE wishes to detach for EPS services.
- The UE wishes to detach for non-EPS services.
- The network directs a UE to DETACH and re-ATTACH with detaching type = re-attach required.

- **UE Initiated**
  - **Requirement # 1:** A UE shall support UE-initiated DETACH procedure for EPS services only.
  - **Detailed Description:** In this DETACH procedure <. . .*provide further descriptions. . ..*>.
  - **Requirement Type:** Mandatory
  - **Source:** TS 24.301 [46], Section 5.5.2.2
  - **Result:**
    Success: UE shall support the DETACH procedure accepted by the network.
    Failure:
    a) UE shall support abnormal cases detected at the UE end.
    b) UE shall support abnormal cases detected at the core network end.

- **Network Initiated**
  - **Requirement #1:** A UE shall support network-initiated DETACH procedure for EPS and non-EPS services.
  - **Detailed Description:** In this DETACH procedure. . .. . .<. . .*provide further descriptions. . ..*>.
  - **Requirement Type:** Mandatory
  - **Source:** TS 24.301 [46], Section 5.5.2.3
  - **Result**:
    Success: UE shall support DETACH procedure initiated by the network.
    Failure:
    a) UE shall support abnormal cases detected at the UE end.
    b) UE shall support abnormal cases detected at the core network end.

15.1.4 LTE UE Tracking Area Update Procedure Requirements

A tracking area update procedure is always triggered by a UE to inform its current location to the Mobility Management Entity (MME). A tracking area update can be classified into the following types:

- Normal tracking area update;
- Periodic tracking area update upon timer T3412 expiry; and
- Combined tracking area update.

- **Requirement #1:** A UE shall support normal and periodic tracking area update procedure.
  - **Detailed Description:** A UE shall perform the normal and periodic tracking area update procedure for the following scenarios:
  - <Refer to TS 24.301 [46], Section 5.5.3.2.2 for the tracking area update scenarios>
  - **Requirement Type:** Mandatory
  - **Source:** TS 24.301 [46], Section 5.5.3.2.2
  - **Result:**
    Success: UE shall support the normal and periodic tracking area update procedure accepted by the network.
    Failure: UE shall support the normal and periodic tracking area update procedure not accepted by the network.

15.2 3GPP System Feature Development Requirements

A new 3GPP system feature on top of an existing mobile communications system is deployed to add value to the existing network in terms of:

- completely new functionality, offering new communication services/capability;
- improved operational efficiency and expenses for the operator; and
- improved system capacity, covering more service areas.

It was mentioned in Section 15.1 that it is important to derive the requirement specifications from all the concerned technical specifications, for a particular network element, that describe the various functions, procedures, protocols, and services to be supported by it. However, the preparation of requirement specifications for a new feature to be developed and integrated into an existing network/system is a bit different as the scope of system changes is wider than normal changes, including backward compatibility issues. A detailed study of the existing system functionalities, capacity, features, and any limitation shall be carried out and ensured that they are not impacted. A proof of concept may be also required to be carried out to determine the upgrade requirements toward the system's capacity and additional hardware resources. Based on the requirement specifications provided by the customer or internal source, typical areas and important tasks/studies to be performed for the development of a new feature as part of an existing or new 3GPP release are described below.

15.2.1 Identification of System/Network Elements Interfaces Changes

A new feature could be introduced in an existing 3GPP release, or it could be an entirely new release, for example, the LTE-Advanced Carrier Aggregation feature, which is the purpose of the 3GPP Release 10. A new feature may affect multiple network elements requiring changes in several logical interfaces in terms of the following:

- Addition of a new message between two network elements;
- Addition of a new IE in an existing message; and
- Addition of a new cause code in an existing IE. For this purpose, a value already marked as "Reserved" in an existing IE may be used for a new cause code value of the new feature.

Once the requirement specifications have been prepared, the formal software development life cycle can be started where detailed high-level and low-level designs, development, and testing shall be carried out.

15.2.2 Identifications of Impacts on Performance

From an implementation point of view, a new feature involves incorporation of numerous logical objects as far as its scope is concerned. This will further consume more system resources in terms of memory utilization, on top of the existing resources utilization. A high rate of memory utilization shall impact the overall performances of the entire communications system. This requires a proper system simulation and its study for performance behavior of the feature with respect to the expected resources utilization.

Consider the 3GPP system network sharing feature which works on the concept of *Redirection*. If an MS/UE cannot be served by a core network, in that case, it will reject the requested procedure and inform the shared radio access network with further redirection requirement indication. Upon receiving this indication, the shared radio access network shall select another core network and the original message sent by the MS/UE. Such network *Redirection* attempt operation shall consume additional time which will lead to delay in the overall completion of a signaling procedure.

15.2.3 Identifications of Impacts on Feature Management

A new 3GPP system feature may be an optional and licensed one. The operator should be able to activate or enable, deactivate or disable, and configure the feature dynamically as per the agreed operation and maintenance requirements with other core network operators considering the various technical, functional, and regulatory aspects. For these purposes, necessary modifications should be incorporated into the existing feature configurations and management facilities that are available with an operator.

15.2.4 Identification of Interworking Requirements with Existing Features

Inter-feature interworking and information exchange requirement analysis with any existing 3GPP system feature in a live mobile communications network should be carried out. For example, new Key Performance Identifier (KPI) counters may be added to assess the performance of the shared radio access network feature along with the existing features. Similarly, a new feature may require to exchange of a new message or may involve changes in the contents of an existing message, with the existing feature of a communications network. A new feature should be possible to be enabled or disabled without impacting the existing feature.

15.2.5 Charging and Accounting Aspects

Inter-core network operator accounting and charging requirements should be identified, and information as per the Charging Data Record (CDR) types as specified in TS 32.250/1 [80] should be exchanged between the concerned network elements of different operators.

15.3 Example Feature: Radio Access Network Sharing

Let us consider the development requirement of the radio access network sharing feature under the *Multi Operator Core Network (MOCN)* configuration. This feature can be deployed into an existing mobile communication network to share a GSM/Universal Mobile Telecommunication System (UMTS)/LTE radio access network, as described earlier in Section 7.2.1. It is an optional feature that allows different core network operators to connect to the same shared radio access network of another operator. Each operator is identified by a Public Land Mobile Network (PLMN) id. For this purpose, the network broadcasts the PLMN identities of the PLMNs sharing the cell system information. For example, in an LTE system, the PLMN ids are broadcasted in the *System Information Block Type 1*. A mobile device supporting network sharing feature uses this system information to execute a PLMN (re)selection process and indicates the selected PLMN to the base station subsystem (BSS).

Radio access network sharing feature deployed by several core network operators has impacts across the different Radio Access Technologies (RATs), i.e. GSM, General Packet Radio Service (GPRS), UMTS, LTE, and 5G system. Identifications of such impacted areas are described below in the case of the GSM, GPRS, UMTS, and LTE systems only. For further details, the reader is recommended to refer to the concerned specifications and their sections mentioned against each impacted area identified below.

15.3.1 Effects on Network Elements

The following network elements are affected because of the radio access network sharing feature:

- UE/MS, base station controller (BSC), radio network controller (RNC), eNodeB, mobile switching centre (MSC), serving GPRS support node (SGSN), and MME.

Note that there are two types of MS/UE – Supporting and Nonsupporting. A supporting MS shall specify the PLMN id of the core network to the shared radio access network, whereas the nonsupporting MS shall use the common PLMN id broadcasted in the system information. The reader is advised to refer to TS 23.251 [37] for the various functions performed by the above network elements to support the network sharing feature.

15.3.2 Effects on Logical Interfaces

The following logical interfaces and their messages are affected because of the radio access network sharing feature:

- GSM A-interface

The following complete Layer 3 information messages are affected. These messages are exchanged over the A-interface (TS 48.008 [134], Section 3.2.1.32).

- CM SERVICE REQUEST
- PAGING RESPONSE
- LOCATION UPDATING REQUEST
- CM REESTABLISHMENT REQUEST
- International Mobile Subscriber Identity (IMSI) DETACH
- IMMEDIATE SETUP

The following new information elements (IEs) are included in the messages mentioned above.

- Redirect Attempt Flag, Length: 1 Octet
- Send Sequence Number, Length: 2 Octets
- IMSI, Length: 3–10 Octets

Details on the usages of the above new IEIs are spread and available across TS 48.008 [134].

- Addition of New Messages

The following new messages/commands from the MSC to the shared radio access network have been added over the A-interface. Details on the message formats and their IEIs are described in TS 48.008 [134], Section 3.2.1.89/90.

- REROUTE COMMAND
- REROUTE COMPLETE

Further, the details of the rerouting indication and complete procedure over the A-interface are available in TS 48.008 [134], Section 3.1.32.

- GPRS Gb- Interface TS 48.018 [136]

New IEs are included in the existing DL-UNITDATA protocol data unit (PDU) that is used for downlink data transmission from SGSN to BSC. Refer to TS 48.018 [136], Section 10.2.1, for their usages, length, and format.

- Redirection indication
- Redirection completed
- Initial LLC-PDU
- Unconfirmed send state variable
- Old routing area identification
- Attach indicator

New IEs are included in the UL-UNITDATA PDU that is used for uplink data transmission from BSC to SGSN. Refer to TS 48.018 [136], Section 10.2.2, for their usages, length, and format.

- Redirect attempt flag.
- IMSI.
- Unconfirmed send state variable.
- Extended feature bitmap for MOCN is added.

Details of the GPRS rerouting indication and its complete procedures are available in TS 48.018 [136], Section 6.6, rerouting procedure in case of MOCN configuration for network sharing.

15.3.3 Selection of Core Network Operator: PLMN Id

In a shared network configuration, several core network operators share a radio access network. In such a configuration, a PLMN id, which consists of Mobile Country Code (MCC) + Mobile Network Code (MNC), is used to select and identify a particular core network operator. The methods used to select and identify a core network operator based on the PLMN id are described below.

- GSM/GPRS System: A/Gb Interface Protocol: Skip Indicator

As mentioned in Section 14.4.3/Figure 14.14, Bits-5 to –8 of the first octet of every GSM mobility management message and GPRS mobility management message protocol contain the IE id: *skip indicator*. The MS and the network use the *skip indicator* as follows:

- Action by a supporting MS: Generally, the skip indicator is specified as 0000 by the MS in the protocol header of the initial mobility management messages, e.g. LOCATION UPDATING REQUEST, CM SERVICE REQUEST, IMSI DETACH INDICATION, or CM RE-ESTABLISHMENT REQUEST message. But if a radio access network sharing is configured and the MS also supports it, then the MS will encode the skip indicator IE with the chosen core network using its corresponding PLMN identity. As mentioned earlier, the network broadcasts the PLMN identities in the system information to the MS.
- Action by the network: The shared radio access network will form the complete Layer 3 messages based on the received initial message from the MS and passes it to the selected core network identified by its PLMN id. If the skip indicator is other than 0000, the chosen core network will process the received message.

- GPRS System: Air Interface

The RLC/MAC data block contains the Temporary Logical Link Identifier (TLLI) information. A network sharing supporting MS shall provide a foreign TLLI in the first RLC data block to indicate the chosen PLMN id to the shared radio access network for onwards transmission of uplink data to the indicated core network.

- UMTS System: Iu Interface

In a shared UMTS terrestrial radio access network (UTRAN), multiple PLMN id list "Multiple PLMN List" is broadcasted by the network as part of the system information in the downlink direction toward a serving cell. A UE reads the PLMN id list and indicates the chosen PLMN identity to the UTRAN in the Radio Resource Control (*RRC*) *INITIAL DIRECT TRANSFER* message. For further details, the reader is recommended to refer to TS 25.331 [50], Section 8.1.8.

- LTE System: S1 Interface

It is mandatory to support the Evolved-UMTS Terrestrial Radio Access Network (E-UTRAN) sharing feature in the LTE system. If an E-UTRAN is shared by multiple core network operators, the PLMN id of each of the operator, in a PLMN Id list, is broadcasted in the system information message to UEs. The network can broadcast up to six PLMN ids and one tracking area code (TAC) for all the PLMNs. A UE will select one PLMN id and indicate it to the E-UTRAN during the initial network ATTACH procedure. The E-UTRAN will select the concerned MME based on the PLMN id as provided by the UE. Once a UE is attached to an MME, the same MME shall be indicated in the subsequent procedures from the UE. The LTE/EPS system supports the network sharing function over the S1 interface.

15.4 Example: Interworking/Interoperations

To provide seamless communication services to the users, one important functionality and feature among the cellular systems and networks is the *interworking*. For these purposes, the interworking and interoperation functionalities among the cellular systems were discussed earlier in Section 6.2. To realize interworking, Circuit-Switched Fall back (CSFB) and the Single Radio Voice Call Continuity (SRVCC) features were discussed. To implement them, the concerned core network elements are required to be upgraded mainly to support the required additional functionalities as defined by the concerned 3GPP technical specifications. Note that the changes to the radio access network elements of GPRS Edge Radio Access Network (GERAN)/UTRAN/E-UTRAN to support these features are minimal. In the subsequent sections below, the affected core network elements due to the CSFB and SRVCC features are identified for interworking.

15.4.1 Circuit-Switched Fall Back (CSFB)

- Effect on the Network Elements

  The following network elements have been enhanced and additional functionalities have been added to them to support the CSFB feature:

  - UE, MME, MSC server, Visitor Location Register (VLR), E-UTRAN, GSM BSS, UMTS RNS, and SGSN

- Effects on Logical Interfaces (Control Plane only)

  - A new logical interface *SGs* is introduced between the LTE MME and the GSM MSC server to support the *control plane* operation of the CSFB feature. The SGs interface is based on the Base Station Subsystem Application Part (BSSAP)+ protocol and works on top of the Stream Control Transmission Protocol (SCTP) layer for the transport of *SGs* protocol messages. SCTP [RFC2960] [17] is used to transport call controlling-related signaling messages over the IP network. The protocol stack of the *SGs* interface is shown in Figure 15.1, which is reproduced from TS 23.272 [38].

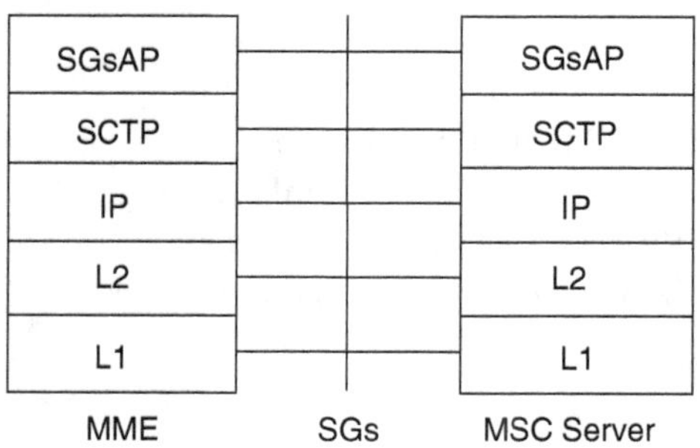

Figure 15.1 Protocol stack for SGs logical interface between MME and MSC server. *Source:* © 2014. 3GPP™ TSs and TRs are the property of ARIB, ATIS, CCSA, ETSI, TSDSI, TTA and TTC who jointly own the copyright in them. © 2014, 3GPP.

Over the *SGs* logical interface, signaling messages are exchanged to perform the following procedures between the MME and the MSC server:

- ✔ Mobility management, i.e. registration, location update, of UE
- ✔ Provision of MO and MT CS Call and SMS services to UE

For further details on the above procedures between the MME and the MSC server/VLR over the SGs interface, refer to TS 29.118 [68].

- Support for CSFB has been also added in the existing logical Interfaces: Iu interface (for Radio Access Network Application Part [RANAP]), S1 interface, Layer 3 air interface, and A-interface.

- Effects on Existing Messages

Some of the existing messages, across different cellular systems, where the support of the CSFB feature has been added are listed below.

- RRCConnectionRelease (LTE RRC layer over its air interface)

An *RRCConnectionRelease* because of CSFB redirection also triggers the release of the EPS signaling connection.

- Channel Release (GSM RRC layer over its air interface)

In this message, under the IE "Cell selection indicator after release of all TCH and SDCCH value part", the new structure "<E-UTRAN Description struct >::=" was added in TS 44.018 [130], Release 8, to specify the LTE ARFCN so that a UE can return to an LTE network after completion of the CSFB CS calls in GSM network.

- PAGING RESPONSE (GSM RRC layer over its air interface)

For Paging Response message due to the CSFB, a UE will include and set the CSMT flag (TS 24.008 [45], Section 10.5.3.14) under the *Additional Update Parameters* optional IE to the MSC server.

- *CLEAR COMMAND* (GSM A-interface: TS 48.008 [134])

CSFB indication optional IE has been added in this message that is exchanged between the MSC and BSC.

- Extended Service Request
- NAS ESM layer UE context management
- NAS EMM layer messages ATTACH REQUEST and ATTACH ACCEPT NAS layer messages

- Effects on Existing Mobility Management Procedures

Some of the existing MM procedures and their messages having the support of the CSFB feature are Attach Request, Detach Request, and Location Update procedure.

To derive all the required changes introduced due to the CSFB feature in the existing logical interfaces, procedures, and messages, refer to the related 3GPP TSs mentioned in the reference section of TS 23.272 [38].

15.4.2 Single Radio Voice Call Continuity (SRVCC)

- Effect on the Network Elements

The following network elements are affected due to the SRVCC interworking feature:

- UE, MME, GSM MSC server, LTE E-UTRAN, Home Subscriber Server (HSS), UMTS RNS, and SGSN

These network elements have been enhanced, and additional functionalities have been added to them to support the SRVCC feature.

- Effects on Logical Interfaces
  - A new logical interface *Sv* is introduced between the LTE MME and the GSM MSC servers to support the control plane operation of the SRVCC feature.
  - Support for SRVCC has been added in the existing logical interfaces: UMTS Iu (RANAP) interface and LTE/EPS S1 interface.
- Effects on Existing Messages

  Some of the existing messages where the support for the SRVCC feature has been added are listed below:
  - *INITIAL CONTEXT SETUP REQUEST* (LTE/EPS S1-AP over the S1 interface)
  - *ATTACH REQUEST, ATTACH ACCEPT* (LTE/EPS S1-AP over the S1 interface)
  - *HANDOVER REQUIRED* (LTE/EPS S1-AP over the S1 interface)
  - INTER-SYSTEM TO UTRAN HANDOVER COMMAND. In this GSM RRC message, support for SRVCC was added in Release 11.
- Effects on Existing Protocol Procedures

  Some of the protocol procedures and their messages having the support of the SRVCC feature are Attach Request, Service Request, and PS handover procedure.

To derive all the required changes introduced due to the SRVCC feature in the existing logical interfaces, procedures, and messages, refer to the related 3GPP TSs mentioned in the reference Section of TS 23.216 [35].

15.5 3GPP System Feature and High-Level Design

A 3GPP feature and its high-level design document describe the implementation aspects of a new feature that is prepared based on the requirement specifications. The feature design document is required to be prepared prior to the development of a new feature. The development of a feature design document starts following the completion of the requirement specifications phase as mentioned in the preceding sections. Requirement specification exercise may be carried out by experienced persons or even by the customer and is submitted to the development team.

The inputs to a feature and its high-level design document are as follows:

- Requirement specifications;
- Design aspects of the existing system including its features; and
- Procedurals/functional requirements as described in the concerned 3GPP TS.

Based on the above inputs, a feature design document shall be prepared and consists of the following typical descriptions:

- Functional descriptions
- Performance impact
- New algorithms and data structures
- Memory requirements and impact analysis
- Effect on other interfaces
- New definitions of alarms, KPI counters, etc.

Based on the feature and high-level design document, the corresponding low-level design document shall be prepared to kick off the entire feature development tasks.

Chapter Summary

Mobile communications systems and network requirement specifications describe the functional and procedural requirements of a network element to be developed using 3GPP technical specifications. It is the first task to be performed by a development team prior to the start of design and development of a network element. A network element or a new feature is realized from the requirement specification developed by an Original Equipment Manufacturer (OEM). Based on a requirement specification, actual design and development work for the concerned network element can be started either by the OEM or its vendor. A requirement specification becomes the basis of an agreement between an OEM and developer or vendor.

A network element may perform functions and procedures related to several subject areas as listed in the 3GPP site [2]. The requirement specification of a particular network element may be derived from the several technical specifications for all of its subject areas, functions, procedures, or features. On the other hand, a feature may affect multiple network elements. Thus, from the concerned technical specifications, a requirement specification shall capture all the new and affected areas on the existing functions, procedures, algorithms, interfaces, performance, statistics, licenses management, and so on.

This chapter presented several typical procedures and features of a network element. Also described was how to derive the requirements of each procedure and feature from its concerning technical specifications. Finally, we closed the chapter with the associated tasks, in terms of preparation of a feature and high-level design document, to design and realize a particular 3GPP feature.

Part IV

5G System and Network

In this part of the book, an introductory on the several key knowledge areas of the 5th Generation (5G) mobile communications system and network shall be described based on its first Release 15 specifications as standardized by the 3GPP. The 5G system evolved further from its predecessor, i.e. LTE/EPS system. A reader with a fair knowledge of the LTE/EPS system will find it easy to acquire knowledge on the overall 5G system architecture, protocol stacks, functions, and procedures and their differences from the LTE/EPS system. Also, a reader would discover how such an architecture of the 5G system and its fundamental key differences from the legacy LTE/EPS system envisages in providing higher data rates and diverse communications services and aims to serve heterogeneous devices.

Chapter 16 describes the different use cases and their key enablers in the 5G system. It describes the logical architecture of the next-generation radio access network, initial deployment option of 5G services, and interworking between LTE/EPS and 5G systems. This chapter also describes Network Slicing, a key enabler of different use cases or services of the 5G system. The chapter ends with security services provided by a 5G system, covering the various aspects of security requirements based on its RAN and CN architecture.

Chapter 17 provides a brief on the legacy systems (GSM to LTE) air interface.

Chapters 18 and 19 describe the 5G system New Radio (NR) air interface and its control plane and user plane protocols, covering the fundamental differences with respect to the LTE air interface protocol layers. These chapters describe how those differences would enable the 5G system and its network to provide diverse services under different use cases.

Chapter 20 describes the core network architecture of the 5G system. It describes the software-based entities and services-oriented new architecture of the 5G core network rather than using proprietary hardware-based network entities used in modern mobile communication networks. An overview of virtualization of network functions is also described, which is another key enabler of different use cases or services of the 5G system.

Chapter 21 describes some of the essential topics on the low-level design or implementation aspects of NG-RAN as well as service-oriented network functions of the 5G core network. It also describes the essential software development platform as well as typical prototype software system architecture that may be considered, given the architecture of the NG-RAN and core network of the 5G system.

Chapter 22 provides an overview of the enhancements made to the existing features and the addition of new services or capabilities which have been added as part of the 3GPP Release 16 and Release 17.

16

5G Network

Use Cases and Architecture

Introduction

Along the path of evolution of previous generation mobile communication networks from 2.5G to 4G, consumers around the world have also witnessed the evolution of speed, from Kbps to Gbps, of mobile data services with faster broadband Internet experience. Previous generations mobile communication networks meant faster Internet or broadband services. But a 5G network aims to expand its mobile broadband services beyond the current mobile Internet services to host of other diverse services or use cases such as massive machine-type communications, critical and low-latency communication services, and so on.

This chapter begins with the different use cases and their key enablers of a 5G network, compared to the previous generation mobile communication networks. It is followed by the description of the system architecture of a 5G system and network as well as the logical architecture of the 5G radio access network (RAN). As part of an early deployment strategy, the non-standalone deployment solution available for a 5G network with an existing Long-Term Evolution (LTE) system and the network is also covered. Then, interworking, through a new N26 interface as defined by the 3rd Generation Partnership Project (3GPP), between an LTE system and 5G system is also discussed. This chapter also covers network slicing and security features as part of the 5G system architecture. Overall, this chapter sets the stage for the rest of the chapters, describing some of the important aspects of a 5G network.

16.1 5G System (5GS) Use Cases

The use cases of the previous generation mobile communication systems are to deliver faster broadband services through mobile devices and other terminals such as USB dongles. However, the use cases of the 5GS *New Radio* (NR) is more than just enabling the broadband services to mobile users. The use cases of the 5GS NR system, as standardized by 3GPP, consist of the *enhanced mobile broadband (eMBB)* services in urban, rural, office, home, and other special deployment locations such as large public events. *eMBB* use case has the characteristics of high data rate requirements, in multi-Gbps, and heavy traffic densities. Other use cases of the 5G system are also defined – *massive Internet of Things (mIoT)* and *ultrareliable and low-latency communication (URLLC)* with short or low data volume. The mIoT use case has the characteristics of support requirements of a large number of heterogeneous and smart devices, within a given area, such as sensors, meters, and unmanned aerial vehicles, which will be connected and served by the 5GS base station. The URLLC use case has the characteristics of support requirements of devices with very short latency <1 ms over the 5GS NR air interface and short data transmission requirements which shall be connected and served by the 5GS base station. Typical usages URLLC include

Mobile Communications Systems Development: A Practical Introduction to System Understanding, Implementation, and Deployment, First Edition. Rajib Taid.

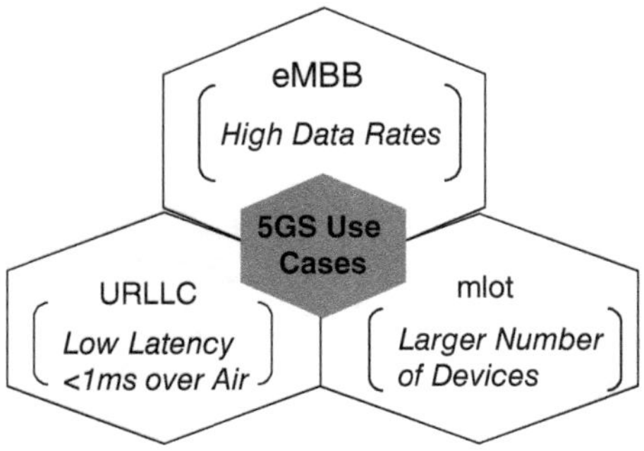

Figure 16.1 Illustration: 3GPP standardized 5G system use cases.

mission-critical applications such as industrial automation, robotics, vehicle-to-vehicle communication, and virtual/augmented reality. The genesis of the 5GS use cases, i.e. eMBB, URLLC, and mIoT, is the International Mobile Telecommunications (IMT) vision, known as "IMT-2020 and Beyond". The IMT-2020 vision is defined by the International Telecommunication Union (ITU) Radiocommunication Sector (ITU-R) of the ITU for 5G system and its services and devices. Refer to the ITU-R Recommendation M.2083 [9] for more information on these use cases and their capabilities, such as the peak data rate, latency, connection density, and so on, which are defined as part of the IMT-2020 vision. Different use cases of 5GS and their characteristics are illustrated in Figure 16.1.

In addition to the use cases and diverse services mentioned above, the 5G system is also characterized and differentiated from the previous generation mobile communication systems in terms of diverse spectrums usages as well as deployment scenarios. 5G NR system can operate in low-frequency bands (<1 GHz), medium-frequency bands (1 to–6 GHz), and high-frequency bands which are above 24 GHz. In terms of deployment, the 5G NR system can be deployed in coexistence with the LTE system in low and medium band ranges in standalone as well as non-standalone modes. For more information on the deployment scenarios of 5GS use cases, refer to TR 38.913 [27]. The next section provides an overview of some of the key enablers and principles of 5GS use cases mentioned above.

16.1.1 Enablers and Key Principles of 5GS Use Cases

As briefly mentioned above, the 5G system envisages to provide diverse services under different use cases and also aims to support a large number of heterogeneous devices apart from handheld smartphones. To enable these use cases and their services in the 5G system, new requirements and architectures have been introduced by the 3GPP which can be studied from the radio access (next-generation radio access network [NG-RAN]) as well as the core network (5GC) point of view. Such new requirements as well as features, architectures are described below at a high level to provide an overview to the reader as the key enablers for different use cases of a 5G system.

- 5GS Air Interface: Support of Diverse Spectrums

3GPP adopted the diverse new radio spectrums over its air interface. Cyclic prefix orthogonal frequency division multiplexing (OFDM) family waveforms and modulation scheme shall be used for transmissions and reception of data in a 5G system. Using OFDM family waveforms and to support different use cases, the 5G system radio frequency operating bands range from low bands to the high-frequency bands. Low-frequency bands are less than 1 GHz. Medium- or mid-frequency bands range from 1 GHz up to 6 GHz. High-frequency spectrum band range above 24 GHz, which is also called the *millimeter wave* frequency bands, differentiates a 5G system from its predecessor, i.e. LTE system. Low- and medium-frequency bands used are similar to the LTE system, thus enabling a 5G system for its coexistence with the LTE system even on the same carrier bandwidth. In a 5G system, low- and medium-frequency bands are termed as sub-6 GHz *Frequency Range 1 (FR1),* which supports both the frequency division duplex (FDD) and time division duplex (TDD) modes. FR1 supports a maximum carrier bandwidth of 100 MHz. High-frequency bands are termed as *Frequency Range 2 (FR2),* which supports the transmissions/receptions in TDD mode only. FR2 supports a maximum carrier bandwidth of 400 MHz, which is 20 times the maximum carrier bandwidth used in the LTE system.

Diverse spectrums of the 5G system will also support UEs of different radio access capabilities such as transmissions/receptions with a wideband carrier or with carrier aggregation from intra/interfrequency bands. The low-frequency band shall meet the large coverage requirements as far as the typical rural deployment areas are concerned; the high-frequency band shall meet the high-speed data service requirements as far as the typical urban deployment areas are concerned; and the medium-frequency bands shall balance the requirement as far as the coverage as well as high-speed data services are concerned.

- 5GS Air Interface: Flexible Sub Carrier Spacings, Timing, and Bandwidth Adaptation

The 5GS air interface physical layer introduces new concepts such as the flexible and multiple subcarrier spacings (15, 30, 60, 120, and 240 KHz), unlike the LTE system which supports only one subcarrier spacing of 15 KHz width. The support of multiple subcarrier spacings enables a 5G system to operate with a wide range of carrier frequencies from sub-6 GHz bands to high-frequency or millimeter wave frequency bands. Another basic and new feature called *Bandwidth Part* (BWP) has been also introduced in 5G NR. A BWP enables a bandwidth adaptation by a UE to transmit/receive over a smaller or larger part of the total carrier bandwidth, under FR1 or FR2, depending on the capability of the UE. As the subcarrier spacing increases, the duration of a timeslot becomes shorter or short, i.e. 1 ms per slot with 15 KHz subcarrier spacing down to 62.5 microseconds per slot with 240 KHz subcarrier spacing (t = 1/f). A shorter duration of a slot with a higher subcarrier spacing also meets the key characteristics in terms of low-latency requirements over the 5G air interface. Different use cases and their services have different characteristics, i.e. some of the services may be of highly delay sensitives while others services are not. Similarly, some of the services may have high bandwidth requirements while other services are not. Thus, the flexibility offered by the 5G NR air interface in terms of scalable and multiple subcarrier spacings and its varying transmission time intervals (TTIs) or timeslots would be able to meet the requirements of the different use cases and their services. Further, advance channel coding techniques – Low-Density Parity Check (LDPC) (for user data) and polar (for signaling) channel coding schemes with variable code rates – are used to support varying data rate requirements of different types of use cases of 5GS mentioned above.

As per the timing between downlink data reception by a UE and its Acknowledged Mode (ACK) to RAN is concerned, 5G NR is more flexible than the LTE system. In 5G NR, the timing between downlink data reception by a UE and its corresponding ACK to NG-RAN is not fixed, unlike in the LTE system; rather, it is configurable for FDD as well as TDD modes in 5G NR. In the case of TDD mode, a UE can provide an ACK over uplink control channel within the same slot also on which downlink data was received. Such a flexible and fast switching between downlink reception and uplink transmission results in a lower latency in 5G NR than the LTE system where transmission/reception is restricted to a slot boundary. This aspect is described later in Chapter 19.

Unlike the LTE system, the 5G NR performs code-block group-based retransmissions, beam management procedures for millimeter wave (mmWave), or very high-frequency transmissions/receptions. These aspects of the 5G NR physical layer are described in Chapter 19. On the other hand, the 5G NR system also uses the frame structure, resource grid structure, Hybrid Automatic Repeat Request (HARQ) processes, and so on, which resembles the LTE system. Also, unlike the LTE system, the 5G NR also provides a flexible air interface that allows transmission and reception in *dynamic time division (TDD)* mode. That is, depending on the instantaneous traffic load in the network, a timeslot can be used for downlink only or uplink transmission/reception by allocating and switching the same resources in the time domain. Such fast timeslot switching information is provided dynamically through downlink control information with the required timeslot format from NG-RAN to a UE, which will be described later in Chapter 19.

- 5GS Air Interface: Enhanced Protocol Layers

Protocol layers of the 5G air interface perform similar functions and procedures as the LTE system air interface protocol layers. However, compared to the LTE air interface protocol layers, some of the functions have been reorganized between air interface protocol layers of the 5G air interface. In some cases, the protocol data unit (PDU) headers have been restructured in 5G NR air interface protocol layers. Such enhancements and modifications made into the different protocol layers of the 5G NR air interface are described in Chapter 19.

In summary, the support of diverse spectrums together with the enhanced and modified air interface protocol stack of the 5G system provides flexibility in terms of multiple subcarrier spacings, slot and non-slot-based (i.e.

OFDM symbol and hence not restricted to slot boundary) scheduling, reduced latency (TTI =1 ms down to few microseconds), coverage, deployments, and so on, which will be the foundation for providing the diverse services under different use cases.

- New NG-RAN

Unlike the LTE Evolved-UMTS Terrestrial Radio Access Network (E-UTRAN)/eNodeB which had a monolithic structure, the 5G NG-RAN/gNodeB, on the other hand, has a separate logical unit handling the control plane and data plane paths of the NG-RAN air interface. NG-RAN logical architecture is the first of its kind where control plane and user plane protocol communication exists even within its logical nodes. The logical architecture of the 5G NG-RAN is described later in Section 16.4.

- New 5G Core (5GC) Network: Key Principles

At the core network level also, the architecture of 5GC is different from the previous generation's core networks. 5GC is based on *services-oriented architecture*. The previous generation core network elements are deployed on Original Equipment Manufacturer (OEM)/proprietary hardware. However, the 5GC uses software component-based architecture where the control plane (e.g. mobility management and session management) and user plane are separated/decoupled into components called *Network Functions,* which are implemented in software. All such network functions performing their functions and procedures can be run even on a general-purpose commercially available high-performance hardware server, storages, and networking resources. The basis of this approach for moving from proprietary hardware servers, for different network elements, to a general-purpose hardware server with high computing resources for deployment of different network functions is drawn from the 3GPP Release 14 feature called CUPS (Control *Plane and User Plane Separation).* CUPS feature is described later in Chapter 20. CUPS feature also enables the *Software-Defined Networking* (SDN) capability for the 5GC. Different network functions produce as well as consume different services over software-based service interfaces, which will be described later in Chapter 20.

Multiple applications/software/OSs from different OEMs/vendors can be run on a virtualized hardware server in a typical modern IT data center. Similarly, 5GC network functions, representing different network elements, from different OEMs/vendors can be also run on a high-performance hardware server with virtualized computing resources assigned to the different network functions. Thus, the network function virtualizations (NFVs) of different network functions of 5GC are achieved. Using SDN through CUPS feature and NFVs as the key foundations, different logical networks called *Network Slices* can be created, and customized, for different types of 3GPP standardized uses cases, i.e. eMBB, URLLC, and mIoT, with different features and capabilities as required by them. *NFV* and *SDN* are complement and independent of each other, but deploying them together on a consolidated hardware platform(s) would provide substantial benefits, such as a reduction in capital expenditure and recurring operational expenditures to an operator. Network slicing together with NFV is the foundation of the 5GC, which enables an operator to provide their service offerings dynamically to meet different businesses and consumer's needs with different characteristics. Network slicing is described in Section 16.9. NFV is described briefly in Chapter 20.

Figure 16.2 illustrates the 5GS use cases and their key enablers and principles, with respect to its air interface, diverse spectrums, and core network perspective for a typical 5G network.

The 5G system key enablers described above are summarized below:

- Support multiple subcarrier spacings with wider carrier bandwidth of up to 400 MHz, which is 20 times the maximum carrier bandwidth supported in the LTE system;
- Millimeter or very high-frequency wave transmission for high data rates;
- Flexible timing for uplink ACK, support of dynamic TDD, and low-latency communication for mission-critical applications/services;

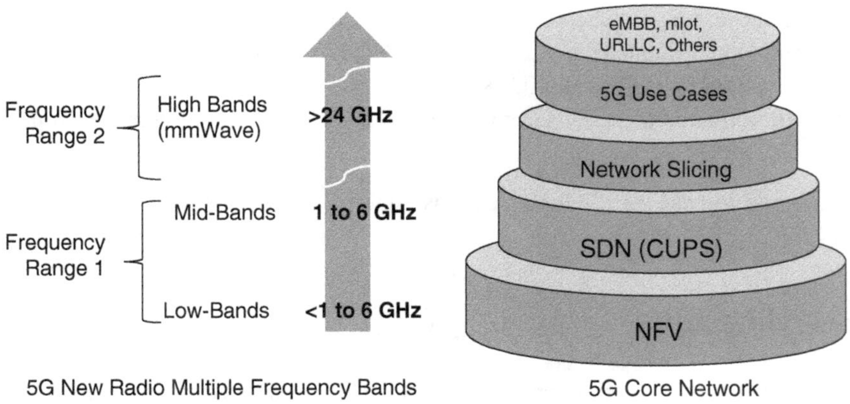

Figure 16.2 Illustration: key enablers of 5GS use cases and their services.

- New channel encoding methods used at the physical layer;
- Reorganization/restructuring of functions/procedures/protocol headers of 5G air interface protocol layers;
- Support of network slices, through SDN with diverse spectrums, for different use cases of 5GS;
- Support of new architecture for 5GC, i.e. CUPS and NFVs; and
- Support of multiradio dual connectivity (MRDC) (described in Section 16.6) for simultaneous connection with LTE/Evolved Packet System (EPS) network.

16.1.2 Other Enablers in 5G System

In addition to the key enablers described above, several other enhancements/features have been also introduced in the 5G system/network. Such enhancements/features are spread across the different protocol layers of Access Stratum (AS) and Non-Access Stratum (NAS) protocol stack. For example, the packets duplication function, described later in Section 19.3, is provided at the NR Packet Data Convergence Protocol (PDCP) layer to support a URLLC service. No such packets duplication function was performed earlier by the LTE PDCP layer. Similarly, for an energy-efficient NR base station, the NG-RAN, i.e. Radio Resource Control (RRC) layer, transmits system information, except the minimum ones, in the downlink direction on-demand request only from a UE; also, a large mobility management area, in terms of a registration area, is being introduced to reduce signaling requirements and its overhead over the air interface. Further, unlike the LTE/EPS where different IP packet flows mapped to an EPS bearer receive the same quality of service (QoS) treatment, QoS assignment in the 5G system is made on a per-packet flow basis, described later in Section 18.3.

16.2 Support of Legacy Services by 5G System

In addition to the delivery of various services under different use cases mentioned above, a 5G system will support legacy services also, including mobility, which are offered by LTE/EPS network. However, certain functionalities will not be supported by the 5G system, in its Release 15 version, at NG-RAN as well as the 5GC level as mentioned below:

- Over a GPRS Edge Radio Access Network (GERAN) or Universal Mobile Telecommunication System (UMTS) UTRAN, a UE would not be able to access a 5GC.
- Call handover between an NG-RAN and GERAN and between NG-RAN and UTRAN will not be supported.

- Fallback to GERAN or UTRAN to provide circuit-switched (CS) call continuity from 5GC will not be supported. However, this is achieved through the Single Radio Voice Call Continuity (SRVCC) feature in case a CS call fallback is required from an LTE/EPS system to a GERAN or UTRAN.
- IP address, for session continuity, will not be preserved if UE moves from a 5GS to Global System for Mobile Communication (GSM)/UMTS.

For more information on the requirements of the service for 5G use cases mentioned in Section 16.1 under different categories, refer to TS 22.261 [28].

16.3 5G System Network Architecture

5G system network architecture is described in the next subsections from the 3GPP access network and non-3GPP access network point of view.

16.3.1 3GPP Access Architecture

Similar to the previous generation mobile communication networks, the architecture of the 5G system also consists of the following network domains:

- RAN domain, which is called as *Next Generation (NG)-RAN*
- Core network domain, which is called as the *5G Core (5GC)*

Figure 16.3 shows the overall 5G system architecture which is reproduced from TS 38.300 [111]. Note that this is an MRDC architecture with a base station (ng-eNB), providing LTE air interface control plane and user plane services to a UE. The radio air interface (Uu) defined by the 3GPP for the 5G system is called the *NR*.

- NG-RAN

The 5G NG-RAN consists of two types of nodes – *gNB*, which is the NR new base station, that serves NR UEs and *ng-eNB which* serves LTE UEs. Both the gNB and ng-gNB provide the control as well as the data plane

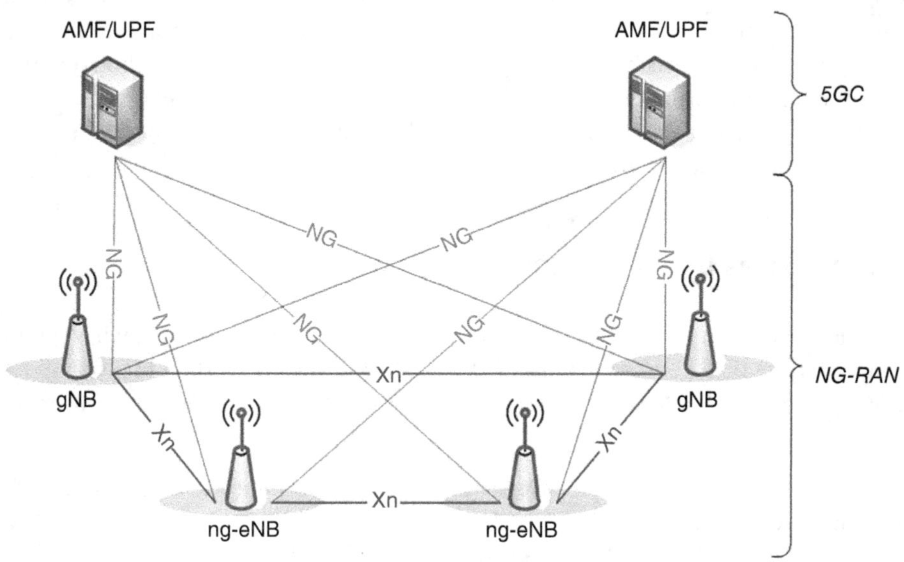

Figure 16.3 5G system network architecture. *Source:* © 2019. 3GPP™ TSs and TRs are the property of ARIB, ATIS, CCSA, ETSI, TSDSI, TTA and TTC who jointly own the copyright in them. © 2019, 3GPP.

services toward UEs over their air interface Uu. The gNBs of an NG-RAN are connected through the interface called *Xn,* which is similar to the X2 interface used in the LTE E-UTRAN for interconnecting its eNodeBs.

The NG-RAN performs the overall radio-related tasks for a 5G network in terms of radio resource allocation and management, scheduling of UEs, reliable delivery through retransmissions, dual connectivity, beam management, and so on, which shall be described in the subsequent chapters.

- 5G Core (5GC) Network

The 5GC performs tasks to provide end-to-end or complete network services such as the UE authentication, mobility management including handover and roaming, and setup of end-to-end data connections in terms of PDU sessions. To provide these communication services, 5GC contains different logical nodes/entities or building blocks in terms of network functions such as *Access and Mobility Management Function (AMF), User Plane Function (UPF), and Session Management Function (SMF).* 5GC supports virtual deployments of these network functions along with other network functions. The virtualization of 5GC network functions is the key basis for the creation of different logical networks called *network slices* which enables operators to provide varieties of services under different uses of the 5G system described earlier.

- NG Logical Interface

A gNB is connected to the 5GC through a logical interface called *NG.* It is also known as the N2 reference point. The NG interface support functions and procedures for intra-RAT handover and inter-RAT handover, transfer of NAS signaling messages between UE and AMF in terms of transfer of mobility management and session management procedures. It also supports UE context management procedures, paging procedure, and NG interface management procedures between NG-RAN and 5GC. Functions and procedures supported over the NG interface are described in TS 38.41x series. TS 38.410 [118] provides a general description of the NG interface and a roadmap on its different specifications.

NG interface also follows the general protocol stack model, i.e. *radio network layer (RNL)* on top of the *transport network layer (TNL),* which were followed by the RANs of the previous generation mobile communication systems, i.e. UTRAN and E-UTRAN, as described earlier in Section 3.9.

- NG Interface Protocol Stack: Control Plane and User Plane

NG control plane signaling messages are called *NG-Application Protocol (NG-AP)* which is transported through the Stream Control Transmission Protocol (SCTP) over the IP transport network. NG-AP protocol messages are exchanged between gNB and AMF (5GC).

The NG u(ser) plane *transfers* user data between the gNB and UPF(5GC). NG user plane data which belongs to a particular PDU session created between a UE and 5GC is transferred through a GTP user plane tunnel over the UDP/IP transport network.

- Xn Logical Interface

Xn interface interconnects two gNBs that support the exchange of signaling information between them. Xn interface also supports the forwarding of PDUs from the master node to the secondary node in case of a dual RAN connectivity setup. Xn interface supports intra-NG-RAN mobility. Functions and procedures supported over the Xn interface are described in TS 38.42x series with TS 38.420 [121] that provides a general description and specification roadmap of the Xn interface.

- Xn Interface Protocol Stack: Control Plane and User Plane

Similar to the NG interface mentioned above, the Xn interface has also a control plane as well as a user plane protocol stack. Xn control plane supports functions and procedures related to Xn interface setup/configurations UE nobilities, dual RAN connectivity, and so on. Xn user (Xn-U) plane supports functions and procedures related to data transfer functions, flow control functions, and so on.

Xn control plane signaling messages are called as *Xn-Application Protocol (Xn-AP)*, similar to the NG-AP, which is transported through the SCTP over the IP transport network. Xn-AP is the RNL, and SCTP over IP is the TNL part of the Xn interface. Xn user plane PDUs of a PDU session that carries user data are transported through the GTP-U protocol over the UDP/IP transport network.

- Access Management Function (AMF)

5G NAS layer between UE and the 5GC terminates at the *Access Management Function (AMF)* end. AMF performs similar functions and procedures of LTE/EPS Mobility Management Entity (MME), i.e. UE mobility management and connection management. However, AMF forwards UE session management-related messages to SMF. AMF also performs functions and procedures related to UE security management, creation of logical networks which is called *network slice,* and so on.

- User Plane Function (UPF)

The UPF handles the user plane aspect of a PDU session in the 5G system and performs the task related to the user packet routing and forwarding for non-roaming and roaming users, user packet filtering, and user packet inspection as per policy. UPF further acts as an anchor point for intra- or inter-RAT mobility, service data flow to QoS flow mapping, and so on. UPF enables the 5G system to support edge computing capability by deploying UPF close to a UE where the UE's traffic gets routed and delivered to it from the local network. Refer to TS 23.501 [40] for more information on functions performed by the AMF and UPF. The next section describes the non-3GPP access architecture of the NG-RAN.

16.3.2 Non-3GPP Access Architecture

Figure 16.3 from Section 16.3.1 shows the 5G system architecture with its NG-RAN and 5GC where a UE accesses the 5GC network through the NG-RAN over the NR air interface. For seamless mobility, a 5G system makes it possible for its UE to access the same 5GC through a non-3GPP RAN also. Further, a non-3GPP RAN is categorized as follows:

- A trusted non-3GPP radio access (introduced as part of the 3GPP Release 16)

Trusted non-3GPP radio access is connected to 5GC through a *Trusted Non-3GPP Gateway Function (TNGF).*

- An untrusted non-3GPP radio access

An untrusted non-3GPP access network is connected to the 5GC via a *Non-3GPP Interworking Function (N3IWF).*

Both the TNGF and N3IWF connect to the common control plane, i.e. AMF, as well as the user plane, i.e. UPF, of the 5GC using the same logical interface or reference points, i.e. N2 and N3. Reference points N2 and N3 are also used when a UE accesses the 5GC over the 3GPP access network. To register with the 5GC, a UE will establish an IPsec tunnel with the N3IWF or with the TNGF. Refer to Figure 4.2.8.2.1-1 and Figure 4.2.8.2.1-2 TS 23.501 [40] for non-3GPP access reference architecture using N3IWF and TNGF; Figure 4.2.8.2.1-2 is described in the Release 16 version of TS 23.501 [40].

16.4 5G System NG-RAN/gNB Logical Architecture

The LTE eNodeB architecture was simple in terms of the absence of logical units performing different functions and procedures over its air interface. However, the 5G NR gNB has got split logical architecture with logical nodes interfacing toward UEs as well as the 5GC. Till the LTE/EPS system, it was observed that a user plane and control plane protocol stack exist between two network elements only over a particular logical interface. However, due to

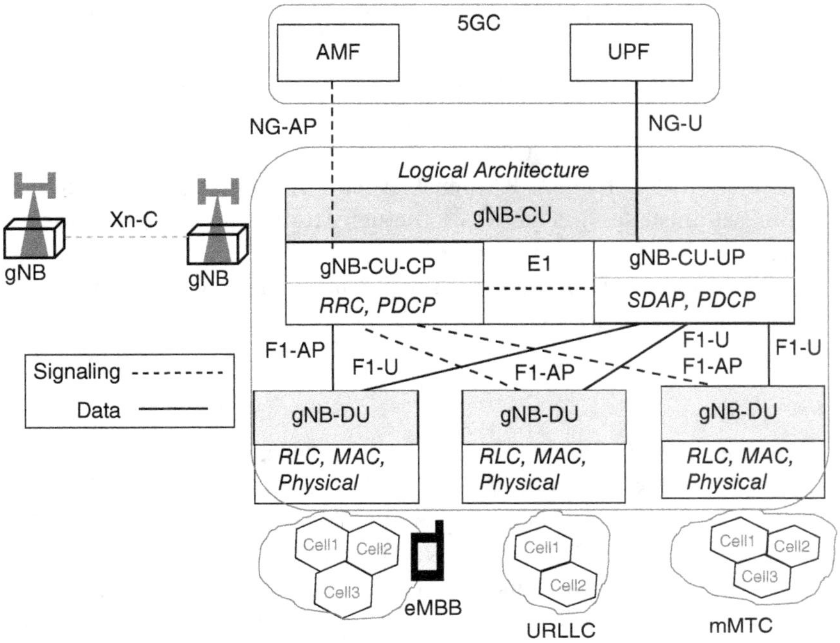

Figure 16.4 Illustration: logical architectural nodes of NG-RAN and its use cases.

the logical architecture, the user plane and control plane protocol stack exist even within the NG-RAN/gNB. As part of the NG-RAN/gNB logical architecture, 5G NR air interface protocol layers are grouped into two logical nodes called *gNB Centralized Unit (gNB-CU)* and *gNB Distributed Unit (gNB-DU)* at NG-RAN/gNB end. The gNB-CU is further divided into two nodes – *gNB-CU control plane (gNB-CU-CP)* and *gNB-CU user plane (gNB-CU-UP)*.

Figure 16.4 illustrates the logical architecture of NG-RAN/gNB with its logical nodes and interfaces as described above.

The components and functions performed by the above logical nodes of NG-RAN/gNB are as follows:

- The logical node gNB-CU-CP performs the NR air interface signaling functions and procedures, which consists of the RRC layer and control plane part of the PDCP layer. Only one gNB-CU-CP logical node can exist in an NG-RAN/gNB.
- The logical node gNB-CU-UP performs user plane functions and procedures, which consists of the Service Data Application Protocol (SDAP) layer and user plane part of the PDCP layer. Multiple gNB-CU-UP logical nodes can exist in an NG-RAN/gNB.
- The logical node gNB-DU interacts with a UE and performs functions and procedures related to scheduling, transmission/reception, and retransmissions over the NR air interface. Logical node gNB-DU consists of the Radio Link Control (RLC), Medium Access Control (MAC), and physical layers. A gNB-DU can serve one or multiple cells. Multiple gNB-DU logical nodes can exist in an NG-RAN/gNB. Refer to TS 38.401 [117] for information on the architectural description of the 5G NG-RAN.

It can be seen from the above logical architecture of NG-RAN/gNB that its logical units are separated at the PDCP layer. The choosing of split point, as defined by 3GPP, into logical gNB-CU-CP with the RRC and PDCP layers, and gNB-DU with the RLC, MAC, and physical layers, is in line with the requirement of a *split data radio bearer*. A split data radio bearer is used in case of a dual connectivity setup between LTE/E-UTRAN and 5G NR

NG-RAN to transfer user data from the LTE/E-UTRAN PDCP layer to the 5G NR NG-RAN RLC layer. The usage of a split data radio bearer will be described later in Chapter 19.

A gNB consists of one gNB-CU-CP only. It may be noted that, though not shown in Figure 16.4, one gNB-CU-CP may control multiple gNB-CU-UP nodes. Similarly, one gNB-DU node may be connected to multiple gNB-CU-UP nodes. One gNB-CU-UP is connected to only one gNB-CU-CP.

The above logical architecture with different nodes of an NG-RAN enables the 5G system to serve its different use cases and services, e.g. eMBB and mIoT, as illustrated in Figure 16.4, through a feature called *Network Slicing*, which will be described later in this chapter.

The NG-RAN/gNB logical nodes mentioned above communicate with each other using new logical interfaces, F1 and E1, as described below. F1 and E1 interface messages and their IEs are specified using Abstract Syntax Notation One (ASN.1) definition.

- F1 Interface – Control Plane and User Plane

A new logical interface called F1, supporting both the control plane and the user plane, is used to communicate between the gNB-CU and gNB-DU nodes. TS 38.47x [126] series defines the F1 interface and its functions and procedures, signaling messages, and so on. F1 control plane or signaling information is transferred as *Application Protocol (F1-AP),* TS 38.473 [127], messages between gNB-CU-CP and gNB-DU logical nodes. F1 user plane messages are transferred between gNB-CU-UP and gNB-DU logical nodes using the NR user plane protocol as defined in TS 38.425 [123]. Figure 16.5 illustrates the control plane and user plane protocol stack of the F1 interface between the gNB-CU and gNB-DU.

F1-AP between the gNB-CU and gNB-DU nodes enables the 5G system to serve its different use cases and services, e.g. eMBB, mIoT, URLLC, and so on, through a feature called *Network Slicing*, which will be described later in this chapter.

- E1 Interface

Another logical interface called E1, control plane only, is used for signaling communication between the gNB-CU-CP and gNB-CU-UP. TS 38.46x [125] series defines the E1 interface and its functions and procedures, signaling messages, and so on. Control plane or signaling information is transferred as *Application Protocol (E1-AP)* messages between gNB-CU-CP and between gNB-CU. Figure 16.5 illustrates the control plane protocol stack of the E1 interface, respectively. Note that the E1 interface between gNB-CU-CP and gNB-CU-UP does not have a user plane protocol stack. Note that the logical E1 between the gNB-CU-CP and gNB-CU-UP is not the same as the physical E1 interface that is used in the case of the GSM system.

During the setup of the E1 interface, network slice-related information is exchanged between the gNB-CU-CP and gNB-CU-UP. The NG-RAN/gNB logical nodes mentioned above enable the NG-RAN to realize a network slice by creating different logical networks over the NG-RAN part, which will be described later in Section 16.9.

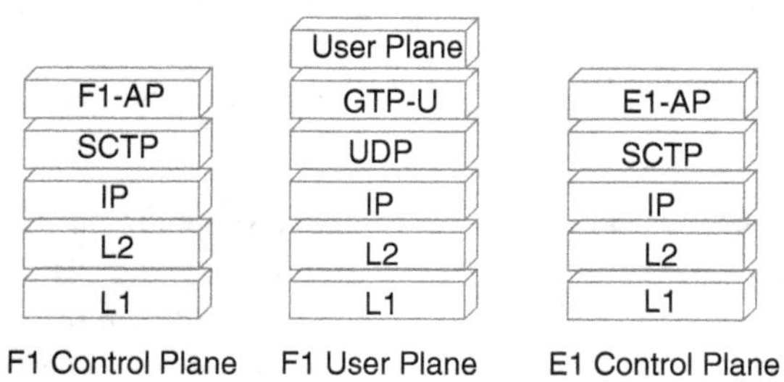

Figure 16.5 Illustration: control plane and user plane protocol stack of F1 and E1 interface of NG-RAN.

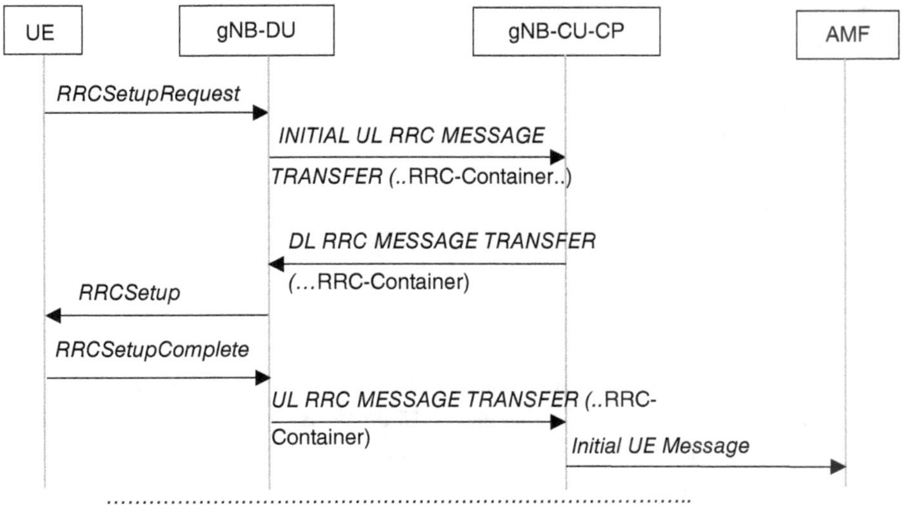

Figure 16.6 Illustration: RRC layer signaling message flow between NG-RAN logical nodes over F1 interface.

Within the NG-RAN/gNB logical node (Figure 16.4), end-to-end message flow will take place in the following way:

- End-to-end control plane or signaling message path

  [UE] – [gNB-DU] – [gNB-CU-CP] – [AMF]

- End-to-end data transfer path

  [UE] – [gNB-DU] – [gNB-CU-UP] – [UPF].

As an example, Figure 16.6 illustrates the RRC layer message flow between the logical nodes of an NG-RAN/gNB, over the F1 interface, during a UE initial registration with the 5GC. UE messages are terminated at the gNB-DU end. gNB-DU forwards the UE signaling messages as an F1-AP message to the gNB-CU-CP over the F1 interface.

It may be noted that the RRC layer messages exchanged between the UE and the gNB-CU-CP are transparently forwarded by the gNB-DU within an *RRC container*. Refer to TS 38.401 [117] for examples of typical signaling message flows within logical nodes of an NG-RAN/gNB.

16.5 5GC System Architecture Elements

One of the fundamental and architectural differences between the LTE/Evolved Packet Core (EPC) and 5GC is the separation of the control plane and user plane functions and procedures of core network elements into separate software-based entities in the 5G system. For example, LTE/EPC Packet Data Network Gateway (PGW) performs both control plane and user plane functions and procedures. In 5GC, however, such functions and procedures of a network element are performed by separate software-based entities. Table 16.1 lists and maps the LTE/EPC core network elements with their corresponding software-based entities in the 5GC.

Separation/decoupling of control plane and user plane functions and procedures of 5GC elements are described later in Chapter 20.

- Evolution of LTE System and Its Specifications

3GPP defines and specifies the entire 5G system through its technical specifications Release 15 and above. It may be noted the LTE/EPS system technical specifications have also evolved, impacting its AS, NAS, and S1

Table 16.1 LTE/EPC and 5G core network elements.

LTE EPC	5G Core
MME	AMF and SMF (Partially)
S-GW and P-GW User Plane	UPF
P-GW Control Plane	SMF (Partially)
HSS	UDM and AUSF
PDN/APN	Data Network (DN)/Data Network Name

protocols, along the path of 5G technical specifications for successful interworking between them. For example, until Release 14, the LTE RRC layer had only two states. From Release 15 onwards, the LTE RRC layer is also required to support three states, same as the NR RRC layer states, for successful interworking and mobility between LTE/EPS and 5G systems and vice versa.

16.6 5G System Deployment Solutions

3GPP defines the following deployment solutions concerning a UE RAN using which an operator can provide 5G communication services to its subscribers.

- Standalone Deployment (SA)

In a standalone 5G system deployment scenario, the entire 5G network consists of a 5G NG-RAN and its 5G packet core network, as shown earlier in Figure 16.3. Thus, in a standalone deployment solution, a 5G network would be able to provide varieties of communication services under the different use cases, i.e. eMBB, URLLC, and mIoT, through its key enablers and principles, as described in Section 16.1.

- Non-Standalone Deployment (NSA)

In a non-standalone 5G system deployment scenario, there is no 5GC. Due to this, a 5G NG-RAN is connected to an LTE/EPS user plane core network, i.e. Serving Gateway (S-GW). In the absence of a 5G system core network, an operator can deploy the non-standalone deployment solution to provide early 5G services using the 5G NG-RAN and existing LTE/EPS core network infrastructure. Due to this, only LTE/EPS core network-based communication services will be available to users/UEs. Thus, in a non-standalone deployment solution, a user would be able to use only the enhanced broadband services of a 5G system. The other features of the 5GC network, such as the network slicing, virtualization, control plane, and user plane separation, will not be available in a non-standalone network deployment solution to provide services for different types of uses cases, i.e. eMBB, URLLC, and mIoT, as described in Section 16.1.

For the interworking of a 5G system, at the NG-RAN level, with an LTE E-UTRAN and LTE/EPS core network in a non-standalone solution, new interfaces are defined among them. *X2 interface* (control as well as user planes) is defined between LTE E-UTRAN/eNodeB and 5G NG-RAN, called en-gNB. The S1 user plane only interface is defined between en-gNB and LTE/EPS core network, i.e. S-GW. No control plane signaling information is exchanged between en-gNB and LTE/EPS core network. Such an arrangement is called *E-UTRA-NR Dual Connectivity (EN-DC)*, which is based on the 3GPP Release 12 *Dual Connectivity* feature. The non-standalone deployment of the 5G system is achieved through the EN-DC feature which is a multiradio and dual connectivity with the LTE/EPC network. The LTE/E-UTRAN eNB is called as the *master or anchor node*, MeNB, and 5G NR en-gNB is called as the *secondary node*, SgNB. The EN-DC feature and its impacts are described in the next section.

16.6.1 E-UTRA-NR Dual Connectivity (EN-DC) for NSA Deployment

Under the EN-DC feature, a UE supporting the dual connectivity feature will connect to both the *master* (LTE MeNB) and *secondary* (5G NR en-gNB) RAN nodes simultaneously. It is a two-stage feature. In the first stage, a UE will register with the LTE/EPC network normally. In the second stage, the LTE/E-UTRAN will configure the UE to use 5G NR resources which are provided by an SgNB. Thus, a UE will communicate with both the LTE E-UTRAN and 5G NG-RAN base stations with their respective air interface protocol layers and be scheduled by their respective schedulers under EN-DC configuration. Because of simultaneous multiradio connections with the LTE/E-UTRAN and 5G NR base stations, EN-DC will require more UE radio capabilities resulting in its complex implementation. For an overall architecture of the EN-DC feature, i.e. connecting the 5G NG-RAN/gNB with LTE/EPC, and other dual connectivity deployment options, e.g. connecting the LTE/E-UTRAN with 5GC, refer to TS 37.340 [101]. Based on TS 37.340 [101], Figure 16.7 illustrates the non-standalone deployment of a 5GS and its NG-RAN/gNB through EN-DC configuration having multiradio and dual connectivity with the existing LTE/EPS system core network.

Compare Figure 16.3 shown earlier and Figure 16.7. In Figure 16.7, E-UTRAN eNB is called as the *master or anchor node* and NR en-gNB is called as the *secondary node*. However, in Figure 16.3, the roles are reversed, i.e. NR gNB is the master node and E-UTRAN ng-eNB is the secondary node. The 5GS architecture shown in Figure 16.3 is a multiradio and dual connectivity of the LTE E-UTRAN with 5GC.

As illustrated in Figure 16.7, in an EN-DC configuration, the master and secondary RANs nodes are interconnected through the X2 control plane (X2-C) as well as user plane (X2-U) interfaces. Secondary NG-RAN node or 5G NR en-gNB is connected with the LTE/EPC S-GW only through the S1 user plane (S1-U) interface to transfer user traffic. No control plane connection, i.e. S1-AP, exists between en-gNB and the LTE/EPC MME. Also, as illustrated in Figure 16.7, the UE is connected to the LTE as well as 5G NG-RAN over their respective air interfaces, affecting its L1, L2, and L3 protocol layers to support the EN-DC feature. For more information on such effects on the existing LTE/EPS system as well as support requirements in the 5G system due to the EN-DC

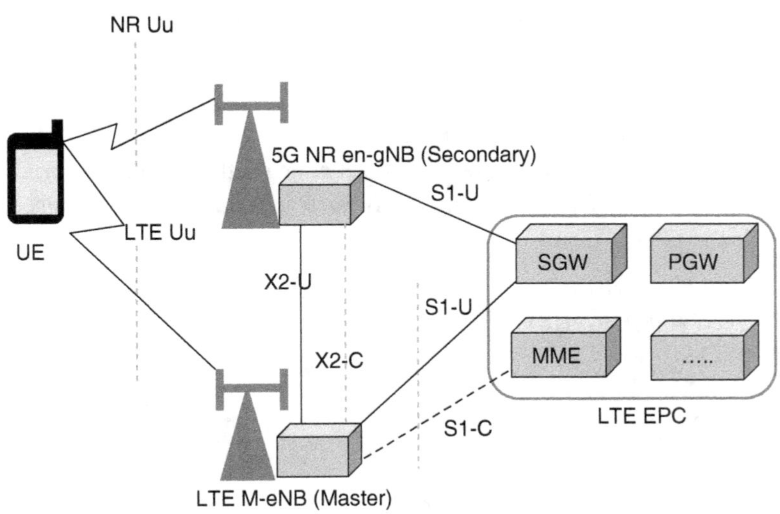

Figure 16.7 Illustration: 5GS non-standalone deployment through EN-DC feature.

feature, refer to TS 37.340 [101]. Impacts of EN-DC feature across the different network elements, interfaces, and protocol layers are described next.

- **Impacts on LTE/EPS MME: NAS Mobility Management Layer**

  – ATTACH/TRACKING AREA UPDATE Request: EN-DC Capability Indication

A UE indicates its support of dual connectivity capability to the LTE/EPS core network, i.e. MME, by specifying whether the dual connectivity with NR is supported or not during an initial UE registration or tracking area update procedure. This is specified through the Dual connectivity with NR (DCNR) bit which is a part of the IE: *UE network capability* of ATTACH request or tracking area update request procedure message sent from UE to LTE/EPS.

  – ATTACH/TRACKING AREA UPDATE Request: EN-DC Security Support Indication

A UE also indicates its 5G system security, i.e. encryption and integrity protection algorithms, support through the IE: *UE Additional Security Capability* which is sent during the ATTACH request or tracking area update procedure to the LTE/EPS, i.e. MME. If the MME supports the handling of the 5G security support as provided by UE, the same security capability is replayed back to the UE through the *SECURITY MODE COMMAND* message and its IE: "Replayed UE additional security capability".

- **Impacts on LTE/EPS MME: NAS Session Management Layer**

To support the 5G NR throughput over the LTE air interface, the following QoS-related parameters/IEs have been added as part of the EPS session management layer message, which is exchanged between LTE/EPC and UE:

  – Extended EPS QoS
  – Extended Access Point Name (APN) aggregate maximum bit rate

Refer to TS 24.301 [46] for more information on the above IEs.

- **New QoS/QoS Class Identifier (QCI) Values**

New standardized QoS and its QCIs, guaranteed bit rate (GBR) as well as non-GBR type, have been added as part of LTE/EPS QoS definitions to support different types of 5G services. For example, standardized non-GBR QCI: 80 has been added in the LTE/EPS QoS definitions for low-latency enhanced mobile broadband (eMBB) service applications. Refer to TS 23.203 [33] for more information on GBR and non-GBR QCIs.

- **Impacts on LTE/EPS S1 Control Plane Protocol, i.e. S1-AP**

  – UE Additional Security Capability

New IE: *UE Additional Security Capability*, i.e. encryption and integrity protection algorithms to be supported by a UE in 5G network, has been introduced as part of the following S1-AP messages which are exchanged between MME and eNB and vice versa.

i) INITIAL CONTEXT SETUP REQUEST (MME to eNB)
ii) UE CONTEXT MODIFICATION REQUEST (MME to eNB)
iii) HANDOVER REQUEST (MME to eNB)
iv) DOWNLINK NAS TRANSPORT (MME to eNB)
v) PATH SWITCH ACKNOWLEDGE (eNB to MME)

  – Extended QoS Values

It defines the following extended QoS values to be used for GBR type bearer over S1 and air interface:

i) Extended E-RAB-Maximum Bitrate DL
ii) Extended E-RAB-Maximum Bitrate UL

iii) Extended E-RAB-Guaranteed Bitrate DL
iv) Extended E-RAB-Guaranteed Bitrate UL
v) Extended UE Aggregate Maximum BitRate DL
vi) Extended UE Aggregate Maximum BitRate UL

- **Impacts on LTE E-UTRAN eNodeB**

In addition to the existing functions and procedures, an eNB also performs key functions to set up a dual connectivity capability for a UE. eNB uses the information provided by the MME to decide whether the dual connectivity facility should be used or not for a given UE. The eNB broadcasts whether the dual connectivity feature is supported or not to UEs in a cell. Some of the impacts on the existing interface due to the dual connectivity feature are described below:

- **Impacts on LTE X2 Interface**

In an EN-DC dual connectivity feature, a master eNB and secondary en-gNB communicate with each other over the X2 interface. New functions such as resource allocations, new procedures such as setup, the configuration of X2 interface and messages, and their IEs to perform these procedures are added to the existing X2 interface which is defined in TS 36.423 [98]. For example, to add a secondary en-gNB node, the master eNB RAN node sends the *SgNB Addition Request* message to the secondary en-gNB node. To accept the request, the secondary node replies with *SgNB Addition Request Acknowledge* message to the master node, i.e. eNB.

For an overall signaling procedure for the addition of a secondary RAN node involving the above interface, refer to TS 37.340 [101].

- **Impacts on LTE Air Interface: RRC Layer**

As part of the successful addition of a secondary en-gNB RAN node, it provides the NR radio resource information to the master LTE eNB RAN node through the message *SgNB Addition Request Acknowledge* and its IE *CG-Config*. The master eNB RAN node forwards the 5G NR radio resources, i.e. bearer related, RLC, MAC Layer resources, and so on, to the UE through the *RRCConnectionReconfiguration message*. The NR radio resources are carried by the following IEs of the *RRCConnectionReconfiguration message* from eNB to UE; refer to TS 36.331 [94]:

- nr-Config
- nr-RadioBearerConfig
- nr-SecondaryCellGroupConfig

As far as the air interface control plane protocol is concerned, a UE maintains the states of LTE/E-UTRAN RRC layer only to enable dual connectivity for a UE because the E-UTRAN RRC layer conveys all the radio resources allocated by 5G NR to a UE. Also, to enable dual connectivity for a UE, support for three types of data bearers has been added in the LTE/E-UTRAN RRC layer.

- Master cell group (MCG) bearer
- Secondary cell group (SCG) bearer
- Spilt (master cell group) bearer

In an EN-DC configuration, the air interface protocol stack to be used to route user data carried by the above bearers depends on the transmitting entity, i.e. UE or gNB or en-gNB.

An EN-DC supporting UE will contain both the LTE and the 5G NR air interface protocol stack. For example, in the case of a split bearer at UE end, its user data can be passed through either the E-UTRAN or 5G NR protocol stack. However, at the network end, if the secondary en-gNB node receives user data through the MCG bearer, the data shall be forwarded to the master node gNB over the X2 interface and further send to UE over the LTE E-UTRAN protocol stack only.

For further information on the effects on air interface Layer 2 protocols due to the EN-DC feature, refer to TS 37.340 [101].

- **Impacts on 5G NR RRC Message Container**

There is an information element (IE) in the EN-DC-related messages which are exchanged between the master (M) eNB and secondary(S) en-gNB vice versa over the X2 interface. That particular IE is called a *message container*, i.e. *SgNB to MeNB container* and *MeNB to SgNB container*. 5G NR RRC layer-related messages are passed as payload in a message container between the master eNB and secondary en-gNB and vice versa. Example 16.1 illustrates the usages of message containers for exchange of information between a master and secondary node.

- **Impacts on UE**

An LTE UE performs critical functions to provide dual connectivity services through the EN-DC feature. It performs the following functions and procedures to maintain simultaneous radio connections with LTE master gNB node and 5G NR secondary en-gNB RAN node.

- Registers with the LTE/EPS network through ATTACH procedure indicating the support of the EN-DC feature as described above.
- UE processes the 5G NR resources received from the master eNB node, which were provided by the secondary 5G NR en-gNB node during its addition procedure over the X2 interface, as illustrated in Figure 16.8.
- UE performs a random access procedure to access and become sync with the 5G NR node, i.e. en-gNB.
- UE establishes dual connectivity with LTE eNB and 5G en-gNB simultaneously.

- **Impacts on LTE/EPC Node Selection Procedures**
  - Packet Data Network (PDN) and S-GW Selection

EN-DC feature affects the selection of a core network node also. If an EN-DC feature is supported by the UE as well as the MME, a dynamic selection of a PDN GW and S-GW GW will be performed by the MME. Home Subscriber Server (HSS) provides the subscriber information to MME as the criteria of PDN GW and S-GW selection along with other information.

Various aspects of the 5G non-standalone deployment through EN-DC features have been described above. The next section describes the interworking between the standalone 5G system with an existing LTE/EPS network.

Example 16.1 Usages of Message Containers in ENDC

Refer to Figure 16.8 for illustration of a secondary en-gNB node addition to set up dual connectivity for a UE. 5G NR radio resource configuration information for EN-DC purposes is passed from the secondary(S) en-gNB to the master (M) eNB using the *CG-Config* message. *CG-Config* is put inside the *SgNB to MeNB container* of the *SgNB Addition Request Acknowledge* message which is sent from SgNB to MeNB over the X2 interface. The content of *CG*-Config is nothing but the 5G NR *RRCReconfiguration message,* which is further forwarded to the UE over the LTE/E-UTRAN air interface using its *RRCConnectionReconfiguration message* and its IE: *nr-SecondaryCellGroupConfig.* The usages of the containers are highlighted below.

- From SgNB to MeNB
  SgNB Addition Request Acknowledge {..SgNB to MeNB Container [CG-Config (...*RRCReconfiguration*)]} → *RRCConnectionReconfiguration*{.... nr-SecondaryCellGroupConfig (....*RRCReconfiguration*..)}
- From MeNB to SgNB
  RRCConnectionReconfigurationComplete {.. *RRCReconfigurationComplete*..} → *SgNB Reconfiguration Complete* {...MeNB to SgNB Container[*RRCReconfigurationComplete*]....}

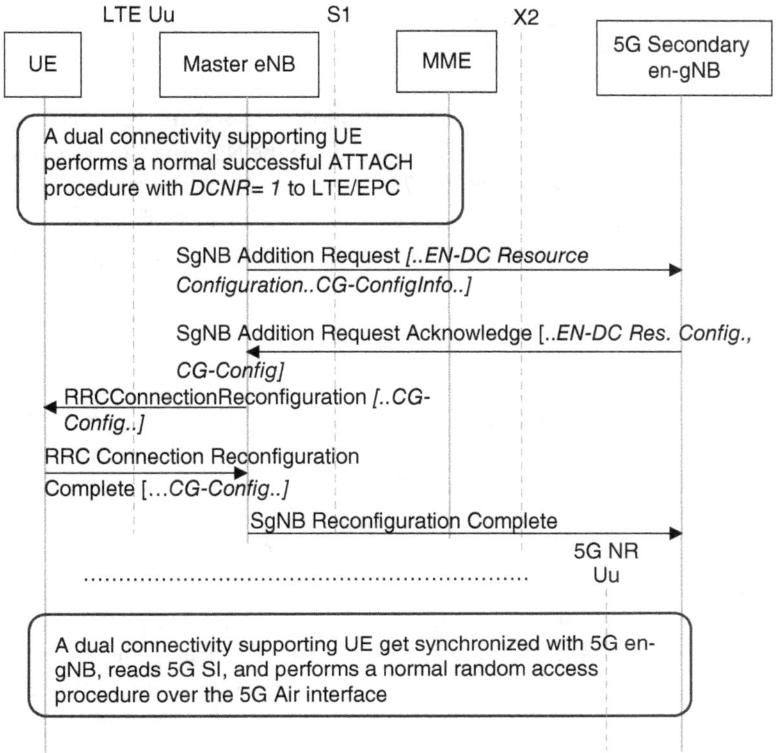

Figure 16.8 Illustration: addition of a secondary NG-RAN/gNB node as part of an EN-DC feature.

16.7 5G System and LTE/EPS Interworking

In the early phase of deployment of a 5G network with a standalone solution, having 5G NG-RAN and its 5G packet core network, an operator may not have full coverage of 5G services in all of its cells. To provide seamless 5G services to its subscribers in such a scenario, an operator can take advantage of its existing LTE network with full coverage and services. For this, a 5G network should be able to interwork with an existing LTE network.

There are two ways to perform interworking between an LTE network and a 5G network as described below. Note that the interworking of a 5G system is possible with LTE/EPS only.

16.7.1 RAN-Level Interworking

In the RAN-level interworking method, both the LTE/E-UTRAN eNodeB and the 5G NR gNB are used along with the LTE/EPC network. For this purpose, 3GPP defines a direct connection/interface, called *X2,* between the LTE/E-UTRAN eNodeB and the 5G NR gNB for the realization of interworking at RANs level.

16.7.2 Core Network (CN) Level Interworking: N26 Interface

In this method, LTE/EPS network, consisting of E-UTRAN and EPC, and 5G system, consisting of NR gNB and its core network (5GC), are used to facilitate interworking for seamless services to subscribers. However, no direct connection/interface, called X2, exists between the LTE/E-UTRAN eNodeB and the 5G NR gNB. For this purpose,

3GPP defines a new and direct inter-core network node interface, which is called *N26*, between LTE/EPC and 5GC network. N26 interface, which works on top of GTP, is defined between LTE/EPC MME and 5GC AMF.

Depending on the availability of the N26 interface and UE registration types, interworking between LTE/EPS and 5G system and vice versa is further classified into the single or dual registration modes at the core network level. A UE comes to know about the availability and support of interworking through the N26 interface from the LTE/EPC and 5GC. This information is provided through the indicator *IWKN26* of the EPS network feature and 5G network feature information element. Further, these IEs are provided as part of the acceptance of the LTE/EPS ATTACH as well as tracking area update procedure and 5G core *Registration Request* procedure. UE registration modes are described below.

16.7.2.1 Single Registration Mode with N26 Interface

In this mode, a UE can register with either LTE/EPC or 5GC network only at a time. UE and its core network maintain the mobility management contexts, i.e. LTE/EPS MM or 5GS RM, for a particular network only at a time. Due to this, the mobility management context of a UE is required to be transferred from the source to the target network for successful interworking between LTE/EPC and 5GS Core network and vice versa. This is accomplished through the N26 interface. Using the N26 interface, the target system derives/maps the UE mobility management context and identities from the source system. For example, if a UE moves from a 5G system to LTE/EPS and vice versa, the UE maps the EPS Globally Unique Temporary Identifier (GUTI) to 5G GUTI and vice versa.

As far as the session management layer is concerned in a single registration mode interworking with N26 interface, when a UE moves out of a 5G system and enters an LTE network, LTE/EPC session management/bearer contexts are created and mapped from the corresponding 5G PDU session management context information. For example, the EPS bearer identity of the default bearer in the LTE/EPS network is set to the EPS bearer identity of default QoS rule in the 5G network. A UE PDU session of the "IPv4" type in 5G becomes the PDN type of "IPv4" of a default bearer in the LTE/EPS network. UE PDU session identity from the 5G PDU session will be associated with the default bearer in the LTE/EPS network.

Similarly, when a UE moves out of an LTE/EPS system and enters a 5GS network, 5G PDU session context management information such as PDU session type, session state, QoS flows, rules, and session and service continuity mode (SSC), described later in Section 18.2.3 is created and mapped from the corresponding LTE/EPS default bearer context information. *Session continuity mode 1*, i.e. UE IP is preserved, during interworking with the N26 interface between LTE/EPS and 5G system. For example, the PDU session identity of a UE PDU session in the 5G network is set to the PDU session identity associated with the default EPS bearer of the same UE in the LTE/EPS network. LTE/EPS UE PDN type "IPv4" used in its default bearer becomes the PDU session type "IPv4" in the 5G system.

LTE/EPS and 5G network interworking using the N26 interfaces is illustrated in Figure 16.9.

- **Impacts on Mobility Management Procedure Due to Interworking with N26**
  a) In single registration mode with N26 interface, if a UE in the dedicated state, i.e. engaged in an ongoing call, moves from a 5G network to an LTE/EPS network, the UE performs a tracking area update procedure and creates default as well as dedicated bearers in the LTE/EPS after the completion of a handover procedure.
  b) In single registration mode with N26 interface, if a UE in idle state moves from a 5G network to an LTE/EPS network, the UE performs a tracking area update procedure or initial ATTACH procedure with LTE/EP network depending on the availability of a PDU session registered with 5GC.
  c) In single registration mode with N26 interface, if a UE in idle state moves from an LTE/EPS to a 5G network, the UE performs a *Registration Request* procedure toward the 5GC.

16.7.2.2 Dual Registration Mode: Without N26 Interface

In this mode, a UE can register to an LTE/EPC as well as a 5GC network independently. A UE needs to register and must be authenticated separately in the LTE/EPS and 5G networks. UE and its core network maintain their

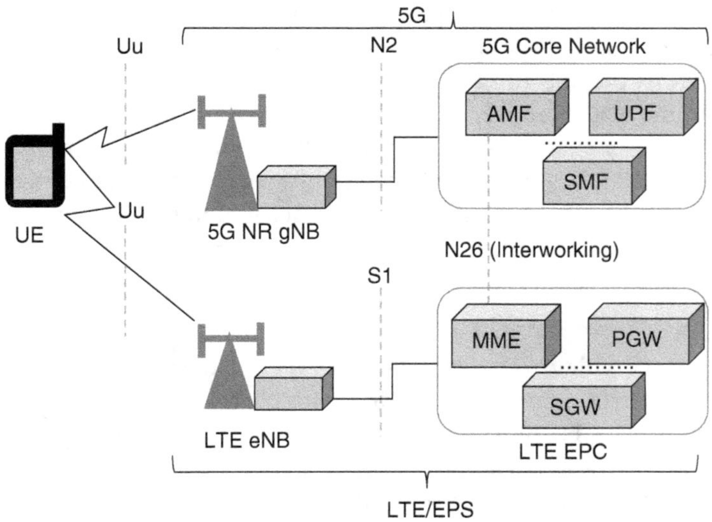

Figure 16.9 Illustration: UE single registration mode: LTE/EPS or 5G network.

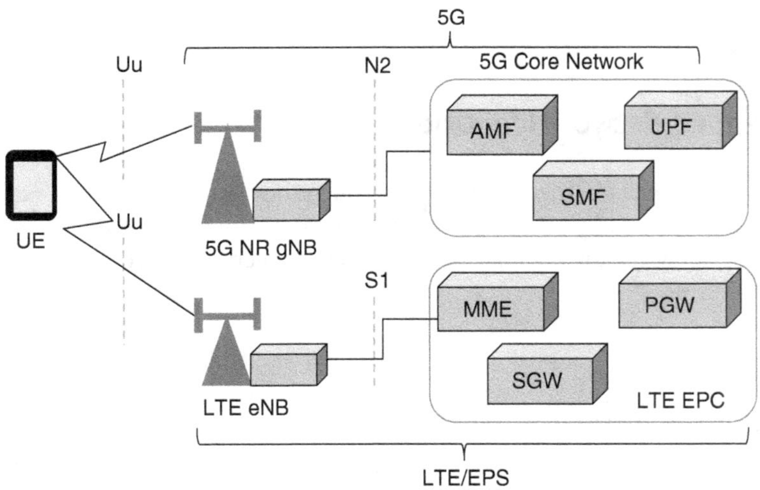

Figure 16.10 Illustration: UE dual registration mode: LTE/EPS and 5G network.

respective NAS layer mobility management contexts, i.e. LTE/EPS MM and 5GS RM, and security contexts for both the networks separately. Due to this, the N26 interface between the LTE/EPC MME and 5GS AMF is not required in the case of dual registration mode. In a dual registration mode, during interworking between the LTE/EPS and 5G system, and vice versa, a UE provides its native mobility management identity, i.e. 5G GUTI and LTE/EPS GUTI, to its corresponding core network. Also, the IP address is not preserved during interworking through dual registration mode. LTE/EPS and 5G network deployment for dual registration mode interworking without the N26 interfaces is illustrated in Figure 16.10.

- **Impacts on Mobility Management Procedure Due to Interworking Without N26**

a) In single or dual registration mode without N26 interface, if a UE moves from a 5G network to an LTE/EPS network, the UE performs an initial ATTACH procedure in the LTE/EPS network. The UE also performs a PDN connectivity procedure toward the LTE/EPC for PDU sessions that were established in the 5GC. In this case, the contents of the LTE/EPS NAS layer PDN connectivity request procedure will be derived from the contents of a UE PDU session is created in the 5G system. For example, the data network name (DNN) used in the 5G PDU session will be applied as the APN for the LTE/EPS default bearer.

b) In single or dual registration mode without the N26 interface, if a UE moves from an LTE/EPS network to a 5G network, the UE performs a registration request procedure in the 5GC. The UE also performs a PDU session establishment procedure in the 5GC for PDN connectivity established in the LTE/EPC network. For this purpose, the contents of the UE PDU session establishment procedure in the 5G network will be derived from the contents of the default bearer context information created in the LTE/EPS network. For example, the APN used in the LTE/EPS default bearer will be applied as a DNN in the 5G PDU session for the UE.

Overall, the interworking of an LTE/EPS network with a 5G network and vice versa using a single or dual mode registration UE, with or without the N26 interface, impact the NAS layer mobility management and session management functions and procedures of the LTE/EPS and 5GC. The combined entities from LTE/EPC and 5GC, i.e. HSS+ UDM, PCF + PCRF, PGW-C + SMF, and PGW-U + UPF, are used to support interworking between LTE/EPS and 5G system. Information for interworking purposes such as the PDU session identity is included through the protocol configuration options IE of NAS layer session management messages. For more information on impacts and tasks performed by the respective mobility and session management layer to support interworking between LTE/EPS and 5G network and vice versa, refer to TS 23.502 [41], 24.501 [47].

16.8 5G System Native and Mapped Network Identities

Similar to the previous generation mobile communication networks, network identities are used to address or identify different physical as well as logical objects in a 5G network. Several natives, new to the 5G system, as well as mapped network identities are used in the 5G system, which may be a temporary or permanent type as listed in Table 16.2.

16.8.1 Mobility Area Identifiers

A tracking area identity (TAI) is used in a 5GC also to identify a tracking area of a UE. It is constructed from a PLMN and the associated tracking area code (TAC). TAI and TAC are also used in the case of the LTE/EPS system.

Table 16.2 Identities mapping from 5G core to LTE/EPC.

5G Core	E-UTRAN-EPC
Mobile Country Code (MCC)	Mobile Country Code (MCC)
Mobile Network Code (MNC)	Mobile Network Code (MNC)
AMF Region ID (bits: 7-0)	MME Group ID (bits: 15-8)
AMF Set ID (bits: 9-2)	MME Group ID (bits: 7-0)
AMF Set ID (bits: 1-0)	MME Code (bits: 7-6)
AMF Pointer ID (bits: 5-0)	MME Code (bits: 5-0)
5G–TMSI	M-TMSI

16.8.2 UE/Subscriber Permanent Identifiers

- Subscriber Concealed Identity (SUCI)

  A SUCI is generated by a UE and is used as a user privacy identifier over the NR air interface which carries the *Subscription Permanent Identifier* in concealed form through a protection scheme. A UE may provide a SUCI as part of several 5GS NAS layer messages such as the *Registration Request* message to 5GC/AMF if a UE does not have a 5G GUTI. A UE may also provide a SUCI as part of an IDENTITY RESPONSE message from a UE to the 5GC/AMF. Refer to TS 23.003 [30] for the structure of a SUCI.

- Subscription Permanent Identifier (SUPI)

  For privacy protection, a mobile user/subscriber in the 5G system will be allocated and identified through a globally unique 5G *Subscription Permanent Identifier (SUPI)* which is derived either from an International Mobile Subscriber Identity *(*IMSI*)* or the network assigned identifier. A UE provides the SUPI to the core network in the concealed form under *SUCI (Subscriber Concealed Identity).* A SUPI is stored in Unified Data Management (UDM) at 5GC end. For more information on subscriber privacy protection through SUCI and SUPI, refer to TS 33.501 [87].

- Permanent Equipment Identifier

  A *Permanent Equipment Identifier (PEI),* equivalent to an IMEI, is used by a UE accessing the 5G system. The PEI uses the format of an IMEI or MAC address.

16.8.3 Core Network Identifiers

- AMF Name

  An AMF is identified by an AMF Name in the form of a fully qualified domain name (FQDN). AMF Name is globally unique.

- Globally Unique AMF ID (GUAMI)

  A GUAMI uniquely identifies an AMF in a 5G network which consists of the following identifiers:
  - MCC and MNC
  - AMF identifier which consists of AMF Region ID (8 bits), AMF Set ID (10 bits), and AMF Pointer (6 bits), where AMF Region ID identifies the region, AMF Set ID uniquely identifies the AMF Set within the AMF Region, and AMF Pointer identifies one or more AMFs within the AMF Set.

- Single Network Slice Selection Assistance Information (S-NSSAI)

  It identifies a 3GPP standardized as well as operator-specific network slice or logical network created for a particular 5G use case or service.

16.8.4 RAN Identifiers

- NR Cell Identity (NCI)/NR Cell Global Identity (NCGI)

  NCI is a 36 bits identifier assigned to an NR cell. An NCGI globally identifies an NR cell and is derived from the PLMN identity the cell belongs to and the NR cell identity (NCI) of the cell.

- gNB Identifier (gNB ID)

  It identifies a gNB within a PLMN. A gNB ID is contained within the NCI of its cells.

- Global gNB ID

  It identifies a gNB of NG-RAN/globally. The global gNB ID is constructed from a PLMN identity and the gNB ID.

16.8.5 Core Network Temporary Identities

- 5G-TMSI

  A 5G-TMSI is uniquely generated and allocated by the AMF to a UE. A 5G-TMSI have a local presence within an AMF only.

- 5G-GUTI

  In a 5G network, a *Globally Unique Temporary Identifier* (5G-GUTI) is allocated by the AMF to a UE. A 5G-GUTI identifies a UE without revealing a user's permanent identity. 5G-GUTI consists of Globally Unique AMF ID (GUAMI) and 5G-TMSI.

- 5G-S-TMSI

  It is the shortened version of a 5G-GUTI so that the associated signaling overhead is reduced while transferring over the NR air interface. A 5G-S-TMSI is used to page a UE over the 5G NR air interface. It is constructed from an AMF Set ID, AMF Pointer, and a 5G-TMSI. AMF region, pointer, and the set were illustrated earlier in Figure 7.2.

16.9 5G System Network Slicing

A *Network Slice* is an end-to-end (UE – NG RAN – 5GC) logical network, on top of shared infrastructure, with specific characteristic and capabilities which is used to serve a particular use case of a 5G network. Different use cases of a 5G system have different and specific characteristics and requirements in terms of high throughput (i.e. eMBB), reliability, low latency (i.e. URLLC), serving of a large number of devices (i.e. mIoT), and so on. The *Network Slice* feature of the 5G system enables it to serve a particular use case or scenario of a 5G network. Multiple network slices with the same characteristics and capabilities may be also deployed to serve different business segments and verticals apart from the 3GPP standardized use cases (eMBB, URLLC, and mIoT) mentioned above.

A *network slice* contains a RAN part, or slice, as well as the core network part, or slice which consists of a control plane as well as user plane network functions from a shared 5G network. A *Network Slice Instance* is a set of network resource instances allocated from the RAN as well as core network part along with their corresponding computing resources provisioned in terms of a virtualized CPU core, storages, memory, and network using which the network slice is deployed. In the previous generation mobile communication systems, network resources, i.e. RF, RAN, and core network, had to be provisioned fully. That is, one size network for all kinds of services, i.e. voice and data, irrespective of the number of subscribers in a given area and its traffic density or business model requirement. A network slice in the 5G system, on the other hand, enables an operator to configure only the slice-specific (e.g. eMBB, URLLC. . .) RF, RAN, and 5GC network resources. Similarly, various network slice-specific computing resource requirements can be also provisioned dynamically.

As far as the 5GC part or slice is concerned, its network functions (NFs) may be deployed differently depending on the requirements of the 3GPP standardized as well as other customized use cases and their network slices. Each network slice may not deploy an equal number of network functions. Also, the 5GC functions may be distributed or deployed nearer to NG-RAN or UE, enabling an edge computing in the 5GS.

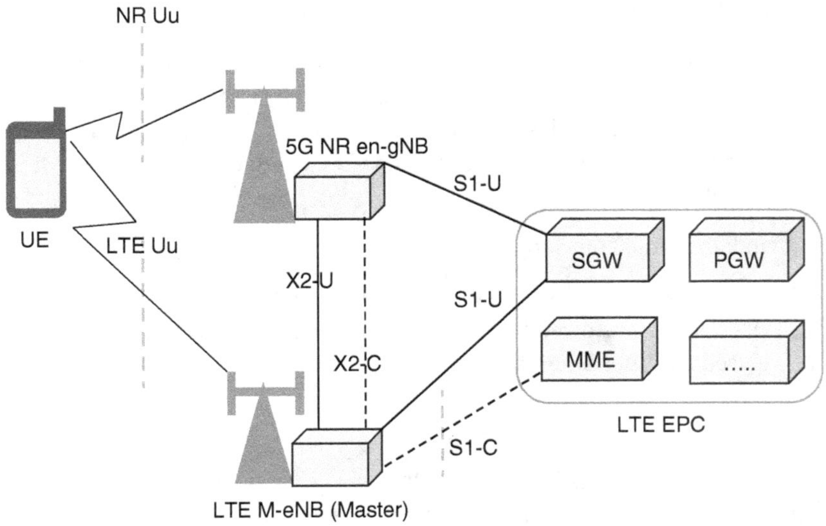

Figure 16.11 Illustration: 5G system: network slices with standardized service types.

As far as the RAN part or slice is concerned, different use cases and their network slices can be served by allocating different parts of the diverse spectrums of the 5G system as described earlier in Section 16.1.1. For example, for an eMBB use case where a delay is not the sensitive factor, a subcarrier spacing with normal timeslot duration in TDD mode may be used with large carrier bandwidth, say100 MHz, from the medium-frequency range; for URLLC use case which is a delay-sensitive one, a subcarrier spacing with short or mini-timeslot duration in TDD mode may be used with medium carrier bandwidth, say 50 MHz, from the medium-frequency range; and for mIoT use case, a channel of low-frequency bandwidth, say 25 MHz, can be allocated as network coverage is important for such services.

An abstract view of a 5G network with 3GPP standardized network slices, with RAN and core network slice part, is illustrated in Figure 16.11 within a PLMN. In this figure, the usage of Slice/Service Type (SST) is also illustrated. RAN slice or the core network slice allocated to a particular network slice is called *Network Slice Subnet (NSS)*. At NG-RAN end, a network slice is made available in one or more tracking areas of a PLMN based on the operator's policy. At 5GC end, a network slice is tracked at a registration area (RA) level.

Further, as illustrated in Figure 16.11, the network slice of each use case is connected to its DNN. A DNN in the 5G system is the same as the APN name used in the case of the LTE/EPS system. A particular network slice can be connected with and served by multiple DNNs based on the user subscription profile. For example, the eMBB network slice can have Internet (for web browsing service) and IP multimedia subsystem (for voice over IP service) as their DNNs. A UE can be served by multiple network slices also based on the user subscription profile. Each network slice is served by a separate PDU session as described in a later chapter.

16.9.1 Identities for a Network Slice

Similar to the various network identities described in Section 16.8, network slice-related identities are also used to identify 3GPP standardized as well as operator-specific network slices as illustrated in Figure 16.11. Network slice-related identities are described below.

- Network Slice Identification – S-NSSAI

Within a PLMN, a logical network slice is identified and selected by the *Single Network Slice Selection Assistance Information (S-NSSAI)* as illustrated in Figure 16.11. A logical network slice identified by an S-NSSAI has an end-to-end presence in a 5GS and is independent of other network slices. An S-NSSAI is used during a UE admission control at RAN as well as during the 5GC mobility and session management, i.e. PDU session, procedures. An S-NSSAI is made up of the following fields in order.

- *Slice/service type (SST),* which is a mandatory and 8 bits long.

As described in TS 23.003 [30], 3GPP standardized the SST with a value range from 0 to 127. SST value range of 128–255 can be used for operator-specific SST. As standardized by 3GPP, the following network slices along with their network slice/service types (SST) are used for different uses cases/services being envisioned to provide over the 5GS; refer to TS 23.501 [40].

i) eMBB services, with SST value: 1;
ii) Ultra-reliable low-latency communication (URLLC), with SST value: 2;
iii) mIoT, with SST value: 3, andVehicle to everything (V2X), with SST value: 4 (3GPP Release 16).

- *Slice differentiator (SD), which is 24 bits long.*

Though the presence of an SD in an S-NSSAI is optional, an SD can differentiate various services options available within a particular SST. For example, within a customized eMBB slice, an operator can provide data services with different plans based on subscriber categories such as enterprises, small organizations, end users, and so on. In such cases, an SD may be used to differentiate the offered services under a particular SST. Information on the network slice(s), with their SST and SD, subscribed by a subscriber is stored, as part of a profile, in the UDM network function at 5G core end.

- *Network Slice Selection Assistance Information – NSSAI*

A set of one or more S-NSSAIs subscribed by a subscriber for different network slices is called an *NSSAI*. There can be a maximum of 8 S-NSSAIs per NSSAI from a UE point of view. That is, a UE can be served by a maximum of eight network slices at a time. However, at NG-RAN and 5GC end, the maximum number of slices to be supported is 1024 per tracking area (TA). An NSSAI can be one of the following types.

- *Configured NSSAI,* which is provisioned in 5GC/Unified Data Management (UDM) for a UE. *Configured NSSAI* may be used as a default NSSAI in a PLMN. A *Configured NSSAI* can be also a *Requested NSSAI* that is used during a slice selection.
- *Allowed NSSAI,* which is provided by the 5GC in its serving PLMN to a UE as part UE Registration Accept procedure. A UE can use the S-NSSAIs provided within an NSSAI. An allowed NSSAI can be also a *Requested NSSAI* in the 5G system.
- *Rejected NSSAI,* which contains a set of S-NSSAIs rejected by 5GC as part of its Registration Accept procedure message sent from the 5GC to a UE. The set of S-NSSAIs was sent by the UE to 5GC as part of its registration request procedure. A Rejected NSSAI and its S-NSSAIs sent over the *Configuration Update Command* from 5GC to a UE also indicates that the concerned S-NSSAIs that were considered earlier as allowed are now considered as rejected by the 5GC.

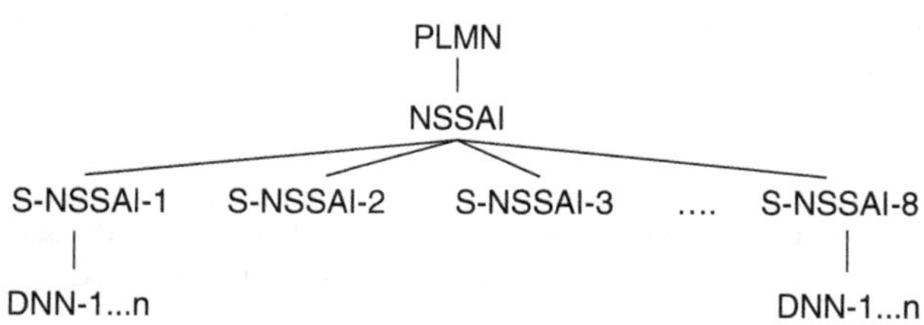

Figure 16.12 Illustration: hierarchy of PLMN and its NSSAI.

The hierarchy of a PLMN and its NSSAI is illustrated in Figure 16.12.

16.9.2 Impacts of Network Slicing Feature

Being an end-to-end feature (Figure 16.11), the realization of a network slicing impacts in several areas right from the UE to the NG-RAN as well as the 5GC. Typical impacts are described and illustrated through Figures 16.13–16.16, as part of a UE initial registration procedure with the 5GC.

- Impacts on NG-RAN Part/Slice
  - RRC Layer

During an initial registration with 5GC, a UE provides the list of S-NSSAI as part of the *RRCSetupComplete* messages toward the NG-RAN. *RRCSetupComplete* also contains the NAS layer UE *Registration Request*, containing the *Requested NSSAI* message toward the 5GC/AMF. Based on the *Requested NSSAI*, the gNB selects and forwards the UE *Registration Request* message to the concerned AMF which supports the desired network slice. Otherwise, the UE *Registration Request* message is forwarded to the default AMF. For typical signaling messages flow on AMF selection by gNB, refer to TS 38.300 [111].

If a *Registration Request* from a UE is accepted by the AMF with due authentication and authorization which is based on UE subscription information maintained in UDM, the AMF will respond with the *Registration Accept* message as part of the *Initial Context Setup Request* message, containing the PDU session to be set up, to gNB. A *Registration Accept* message is forwarded by the gNB to the UE over the *RRCReconfiguration* message. The *Registration Accept* message contains the *allowed NSSAI* which is based on UE subscription information maintained in UDM as well as the *rejected NSSAI*, if any, for the registering UE. Finally, the gNB responds with the *Initial Context Setup Response* message to the AMF upon successful provisioning of various resources for the UE-specific PDU session and its network slice.

- Impacts on Various Interfaces

An end-to-end network slice-related information is passed over the various interfaces as illustrated in Figure 16.13. At NG-RAN and UE end, network slice-related information is added to the RRC layer signaling messages exchanged over the NR air interface (Uu).

At the 5GC, network slice-related information is added to the MM as well as SM signaling procedures and their messages at the NAS layer (N1) between UE and the 5GC. Network slice-related information is also added to the signaling messages which are exchanged as part of the NG-AP between the NG-RAN/gNB and 5GC/AMF over the NG/N2 interface, e.g. *Slice Support List (Max 1024)*, and as part of the F1-AP between the gNB-CU-CP and gNB-DU logical nodes of NG-RAN over F1 interface and over the Xn interface between two gNBs.

- Slice-Specific Resource Provisioning and Configuration

NG-RAN comes to know about network slices (*S-NSSAI*) to be configured based on the *Initial Context Setup Request* message received from the 5GC as illustrated in Figure 16.13. Hence, NG-RAN should be able to configure different network slices and provision their various resources with proper demarcation so that one network slice does not interfere with another slice. For example, the URLLC network slice may have the slice-specific RAN part configured with radio resources of subcarrier spacing 120 KHz with 50 MHz transmission bandwidth. Similarly, to support a large number of diverse IoT devices or eMBB services, the mIoT or eMBB network slice may have the RAN slice configured with appropriate subcarrier spacing/numerology and transmission bandwidth.

As far as the logical architectural nodes (gNB-DU and gNB-CU-UP) of a single NG-RAN/gNB are concerned, the following RAN slice designs approach may be prototyped for a gNB to serve different network slices:

i) A single RAN slice/gNB instance with single gNB-DU and gNB-CU-UP node to serve all of its network slices. This design approach may be suitable to serve network slices of similar slice type.

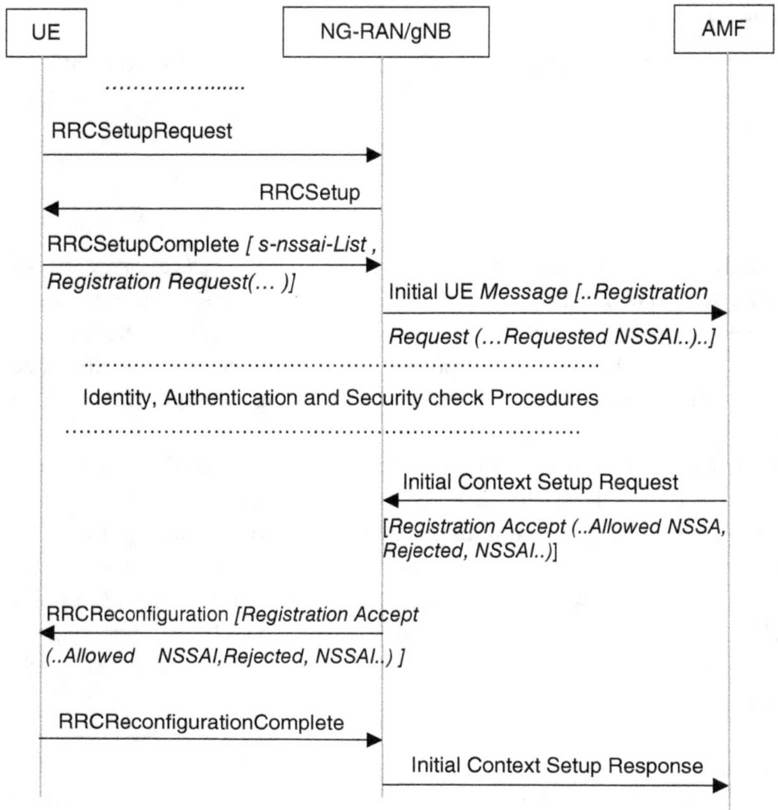

Figure 16.13 Illustration: RRC layer signaling flow for network slicing during UE initial registration with 5GC.

ii) Multiple RAN slices/gNBs (Figure 16.11) with multiple gNB-DUs and gNB-CU-UPs nodes and dedicated subcarrier spacing/numerology to serve its network slice. This design approach may be used to serve network slices of different slice types (see Figure 16.4).

- Slice-Specific Physical Radio Resource Allocation and Admission Control

Different use cases and their network slices types have different requirements from a 5G network in terms of high throughput, low latency, reliability requirements, and so on. To handle network slice with differentiated QoS requirements, NG-RAN performs slice-specific radio resources allocation tasks such as the admission control, physical radio resources allocation, i.e. timeslot with desired QoS, and its scheduling in uplink and downlink direction. Scheduling of UE, URLLC, and mIoT devices shall be required to be performed separately in uplink and downlink direction. Further, the scheduling strategy would depend on whether a single RAN slice, i.e. common scheduling, or multiple slices, i.e. slice-specific scheduling, are being deployed by the NG-RAN/gNB as mentioned above.

- Usages of Functions and Features of Air Interface Protocol Layers

Diverse services/applications of different 5G use cases have different characteristics and requirements. The functionality and features provided by the NR air interface protocol layers especially those located in gNB-DU, i.e. physical, MAC and RLC layer, and gNB-CU-UP, i.e. PDCP layer, should be utilized accordingly depending on the nature of services, e.g. eMBB, mIoT, URLLC, and so on, being attached to a network slice. For example, an RLC acknowledged mode should be used for URLLC types of services, whereas the RLC unacknowledged mode may be used for some mIoT type of services. Similarly, the usages of header compression and ciphering functions performed by the PDCP layer or usage of a hybrid automatic request by the MAC layer may be considered for the desired services only. The NR RRC layer is in control of configurations of the Layer 1 and Layer 2 protocols. Thus, an operator can activate/deactivate such desired functions and features of a protocol layer through the RRC layer for services to be offered over a particular network slice.

Similarly, as mIoT devices have limited mobility requirements, a handover procedure for such devices may be considered to be omitted which would reduce the mobility management context information to be stored at the network end.

- Impacts on 5GC

A 5GC network consists of various software-based network functions that perform similar tasks of the previous generation mobile core network elements. Due to this, the network slice feature affects the 5GC network as well, as illustrated in Figure 16.14.

5GC network is built on service-based architecture which consists of software components based on different network functions, performing the core network functions and procedures of the NAS layer as well as other functions and procedures. As illustrated in Figure 16.14, multiple instances of a network function, e.g. multiple UPFs and SMF, can be run to serve different communications services under different use cases or network slices with desired QoS. Accordingly, a network function selection function, indicated by an arrow, is required to select a particular network function based on given inputs as well its configuration in 5GC. As illustrated in Figure 16.14, during an initial UE registration, if the UE provides the *Requested NSSAI* as part of the *RRC Connection Establishment* message, the NG-RAN will select an AMF for the indicated network slice based on the S-NSSAI information and forwards the NAS layer message to the selected AMF. The AMF will also use the S-NSSAI information to select an SMF for indicated network slice and so on, for the UE.

– Impacts on NAS Layer

Network slicing impacts 5GC mobility and session management layer procedures also. Network slicing-related information is added to the mobility management layer messages including handover procedure, for example, *Registration Request and Accept* messages as mentioned above. Further, a network slice information may change dynamically due to various reasons such as a change in user subscription status, operator action, and so on. Any changes in a network slice information are updated from the 5GC to a UE over the *Configuration Update Command* sent to the UE. The service interactions among the initial AMF, NSSF, and the NRF during a UE registration process are illustrated in Figure 16.15. For more information, refer to Section 5.15.5.2.1 in TS 23.501 [40].

Network slice impacts the 5G session management layer also. At the session management layer, each network slice and its instances are associated with its PDU session. To create a PDU session, first of all, an AMF may query the *Network Slice Selection Function (NSSF)* using the S-NSSAI provided by a UE. An NSSF is a similar function performed by the NAS Node Selection Function (NNSF) procedure described earlier in Chapter 7. The NSSF returns the appropriate NRF to the AMF. Using the S-NSSAI, the AMF queries the NRF to select an SMF and data network (DN). The selected SMF creates a PDU session between the UE and the DN for the network slice identified by the S-NSSAI and its corresponding DN. AMF and SMF store a network slice instance as part of UE PDU session context information.

The service interactions among the initial AMF, NSSF, NRF, and SMF during a UE registration procedure are illustrated in Figure 16.16.

– AMF Relocation

Similar to a rerouting function which is performed as a result of a NAS Node Selection Function (NNSF), a UE *Registration Request* message may be also required to be redirected to another target AMF during a UE initial registration with the 5GC. If a UE provides a *Requested NSSAI*, the NG-RAN/gNB will forward the UE registration

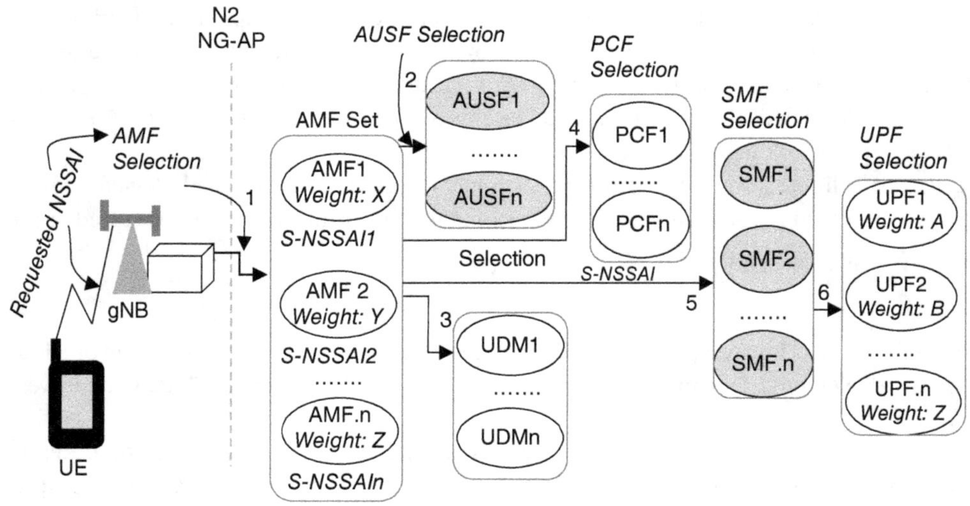

Figure 16.14 Illustration: NSSAI-based AMF selection for a network slice.

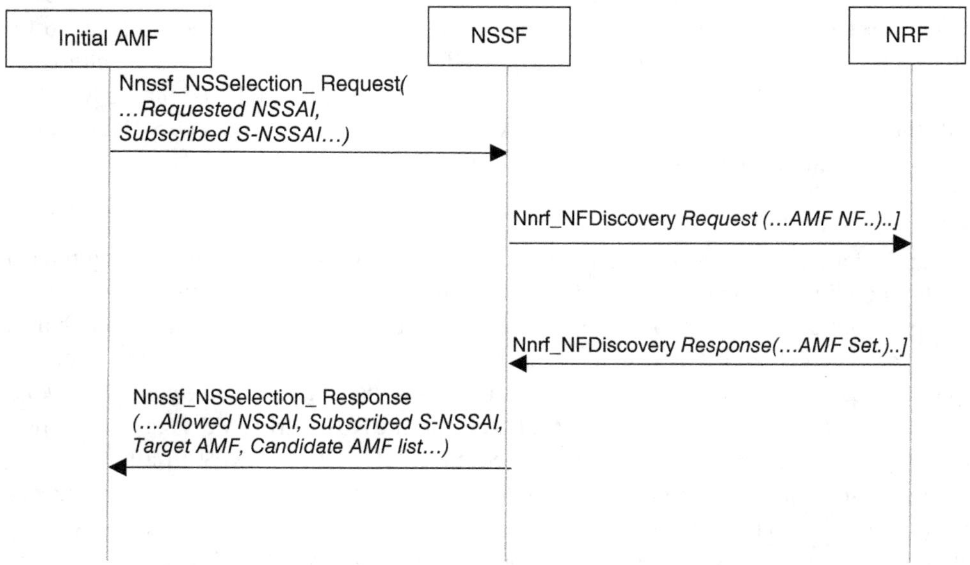

Figure 16.15 Illustration: UE registration: NF signaling flow for network slice selection.

request message to an appropriate AMF; otherwise, the UE registration message will be forwarded to a default AMF. The default AMF will perform a network slice selection function (NSSF) to select and further forward the UE registration message to an appropriate AMF.

The redirection from a source AMF to target AMF can be through direct signaling or the NG-RAN. For the selection of a new network slice, i.e. target AMF, the initial AMF sends the *Nnssf_NSSelection_Get* service operation message to the NSSF, which in turn provides the target AMF details over the *Nnssf_NSSelection_Get_Response* service operation message to the initial AMF. For more information on the impacts of NAS mobility and session management layer signaling messages due to the network slicing feature, refer to TS 24.501 [47], and for impacts on system procedures due to the network slicing feature, refer to TS 23.501 [40] and TS 23.502 [41].

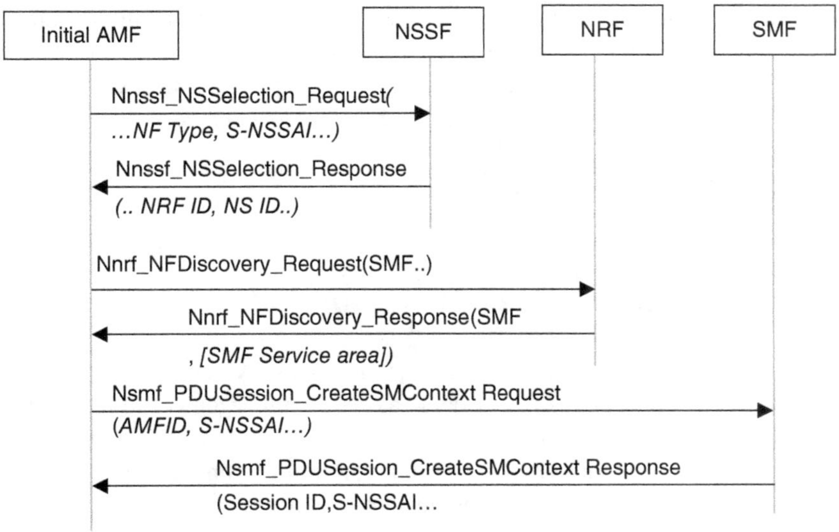

Figure 16.16 Illustration: UE registration: NF service signaling flow for PDU session creation.

- Virtualization of Network Functions

In 5GC, network slicing is realized by virtualizing its functions and procedures into different network functions (software-based) in control as well as user plane level. A network slice contains a control plane as well as user plane network functions as part of the service-based architecture of 5GC. A network function can serve multiple network slices or one slice only. For example, an AMF can be associated with multiple network slices of a particular UE. The virtualization of network function within the 5GC is a key foundation on which a network slicing can be realized in a 5GS.

- Separation of Control Plane Functions and User Plane Functions

The separation of control plane functions and User Plane functions into different network functions as part of the Release 14 CUPS feature supports the realization of network slice. The separated control plane as well as user plane network functions enable a software-defined networking capability to realize a network slice in 5GS. For an overview of the network slicing support requirements at NG-RAN and 5GC, refer to TS 38.300 [111]. CUPS is described in a later Chapter 20.

- Management and Life Cycle of a Network Slice

As described at the beginning of this section, a network slice consists of a RAN slice part as well as a 5GC network slice part. In addition to these, a network slice consists of a network slice management task, as defined by 3GPP, which deals with the managing of various tasks such as creation, activation, maintenances, and termination of network slices. A network slice management system enables an operation and management person to provision and controls its operation dynamically by managing its life cycle. The life cycle of a network slice consists of several phases as mentioned below:

- *Preparation* – In this phase, the design as well as capacity planning aspects of a new network slice is performed so that the new network slice does not interfere with the existing network slices in terms of various resources being allocated to them.
- *Commissioning* – In this phase, a network slice is created and configured with appropriate network resources, physical as well as logical database resources.

- *Operation* – In this phase, a network slice is activated or made go-live so that it can provide communication services. Performance monitoring in terms of KPI, SLA, and resource modifications and tuning is performed.
- *Decommissioning* – In this phase, a network slice is terminated, and no more communication services are provided over it. The allocated network resources, physical as well as logical will be reclaimed and reassigned.

For more information on the above management and life cycle aspects of a 5G network slice, refer to TS 28.530 [56]. The next section describes the management and orchestration (MANO) of a 5G network.

16.10 Management and Orchestration (MANO) of 5G Network

In the previous generation mobile communication networks, telecommunication network managements task is performed at a network element, e.g. UMTS NodeB or E-UTRAN eNodeB, as well as management function, e.g. Element Manager (EM), Domain Managers (DMs), and Network Managers (NMs); TS 32.101 [79]. However, as the 5G system evolved from the previous generation systems enabling various communication services under different use cases with different types and a large number of devices, the network management has also evolved.

Management and orchestration provide a management solution for system administrators who deal with the various aspects of provisioning and managing a 5G network including network slicing described in the previous section. However, unlike the previous generation network management reference model, the 5G network management and orchestration model are based on the *Management Service (MnS)* concept. Like the 5GC, MnS is also based on a service-oriented management framework with a management service provider as well as service consumer. A management service consumer uses standard service interfaces for consumer service.

MANO under the MnS reference architecture contains components which are called component A, B, and C for management of service.

- *Component Type A* consists of a group of management operations in terms of creating, reading, updating, and deleting managed object instances. Refer to Section 6 in TS 28.531 [57] for instances on Component Type A.
- *Component B* indicates the information model, also called a network resource model, which is used to represent a particular manage service.
- *Component Type C* indicates the performance or fault information for a monitored/measured entity or management service. For more information on these MnS component types used for 5G MANO, refer to TS 28.533 [58].

Overall, the description of the above components in terms of their requirements, procedures, and information model, and so on for management and orchestration of a 5G network is defined by the 3GPP technical specifications as illustrated in Figure 16.17.

Network resource model which is used to realize the management and orchestration of a 5G network is described below.

- **Network Resource Model**

In a mobile communication network, a network resource can be either a physical object, such as BTS and TRX, or a logical object representing a physical object, as described earlier in Chapter 18. A network resource can belong to a RAN and core network, and in the case of a 5G network, a resource can belong to a network slice also. A network resource is distinctly represented through an abstract concept called an *Information Object Class* (IOC), which is the same as the "class" abstract data type found in the C++ programming language. Similar to a C++ class, an IOC contains different attributes, representing the different properties or information about a network resource as well as different operations using which network as well as service management-related tasks as part of a MANO can be performed. A C++ class or an IOC can be a standalone one or it can associate with other classes or IOCs. Therefore, a *Network Resource Model* (NRM) of a 5G network contains all of its IOCs, along with

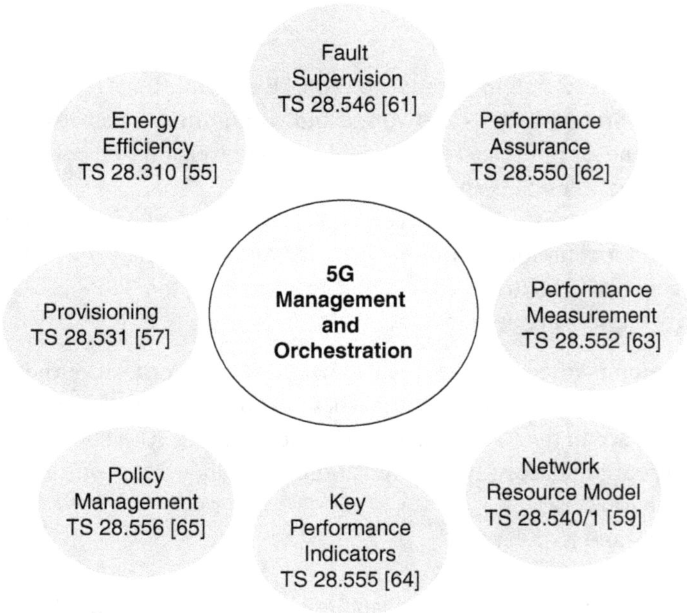

Figure 16.17 Illustration: management services for management and orchestration of 5G network.

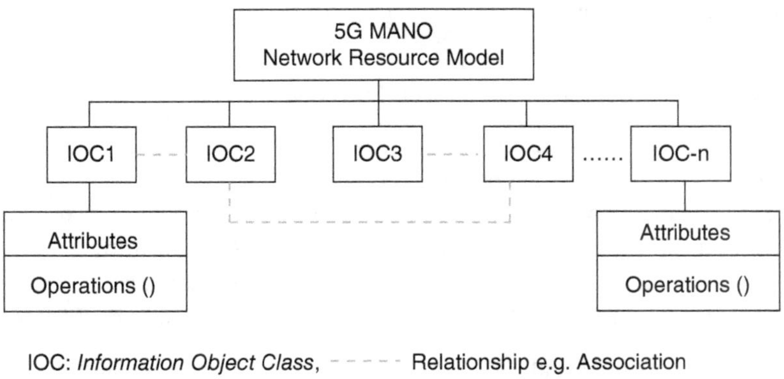

Figure 16.18 Illustration: 5G NRM, its information object class, and its associations.

their related associations, in each network domain, i.e. 5G NR or 5GC or 5G network slice, using which the overall network and services management and orchestration can be performed. A 5G NRM is illustrated in Figure 16.18, with associations, represented through solid doted thick lines, among its IOCs.

Further, JSON and XML-based network resource representation methods can be used in an NRM for data exchange purposes. For more information on all the defined IOCs under the 5G NRM for management and orchestration of network and services, refer to TS 28.540/1 [59]. 3GPP defines the 5G NRM for NG-RAN, 5GC, and so on under three different phases. Phase 1/TS 28.540 [59] defines the NRM requirements for NG-RAN, 5GC, and so on. Phase 2 and 3/TS 28.541 [60] define the IOCs and solution set, i.e. network resources representation formats, for NG-RAN, 5GC, and so on.

A 5G NRM and the relationships among the different *Information Object Classes* can be described using the *Unified Modelling Language (UML)* as part of the low-level software design phase. Usages of UML for the description of 5GC network functions and their implementation through the sample software program are described later in Chapter 21.

16.11 5G System Security

The 5G security system has evolved from the LTE security system. Similar to the LTE system, the 5G security system also consists of the features and methods to protect its UEs, networks, and communications between them. 5G system security services are provided over the air interface at AS as well as NAS level and over 5G core service-based interface and also while interworking between 5GS and LTE/EPS.

As described earlier in Section 16.3.2, a 5G system enables a UE to access a 5GC over a 3GPP as well as a non-3GPP RAN. Thus, security services are required for communications between UE and the 5GC over a 3GPP as well as a non-3GPP access network. In this section, 5G security services and features over the 3GPP access network will be discussed. Features of 5GS security services are listed below.

- Mutual authentication as the primary authentication method used between a UE and 5G network, i.e. authentication of the UE by the network and vice versa, which is known as the Authentication and Key Agreement (AKA). Further, the 5G system enhances the AKA used in the LTE/EPS, which is known as the 5G-AKA. In the case of 5G-AKA, the visited network provides proof of successful authentication, through an *Authentication Confirmation* message, of roaming UE to its home network;
- Separate security mode command procedures at AS and NAS level;
- Generation of AS and NAS security contexts;
- Confidentiality and integrity protection user plane data and control plane signaling information over the NR air interface using 128 bits ciphering and integrity algorithms. Both UE and network will support the confidentiality of user data and signaling messages through ciphering but optional to use; and
- Security interworking between 5GS to LTE/EPS only and vice versa.

However, the following security features/measures are new to the 5G security system.

- Subscriber identity confidentiality in concealed form over the NR air interface;
- Mitigation and prevention of bidding down attacks on UE and network;
- Use of *Extensible Authentication Protocol* (EAP) framework as another method of primary authentication of UE;
- New security functions due to the service-based architecture of 5GC as well as network slicing features;
- Usages of serving network anchor key (K_{SEAF}) based on which AS and NAS keys are generated; and
- Inter PLMN security services through *Security Edge Protection Proxy* (SEPP) between visited and home PLMN.

The entire UE authentication procedure is controlled and authorized by the home network, mainly the AUSF, UDM/*Authentication credential Repository and Processing Function (ARPF),* and *Subscription Identifier De-Concealing Function (SIDF)* at 5GC end as illustrated in Figure 16.19. As illustrated, the home network (i.e. AUSF in 5G core) authorizes the serving network, which consists of the AMF and SEAF of 5G core. The serving network authorizes the access network, i.e. NG-RAN/gNB, which in turn authorizes the UE.

For a detailed description of the 5G security system, refer to Figure 4-1 in TS 33.501 [87].

16.11.1 UE Authentication Frameworks and Methods

In 5G systems, two authentication frameworks are used to authenticate the UEs. The frameworks are described below with procedural differences between them.

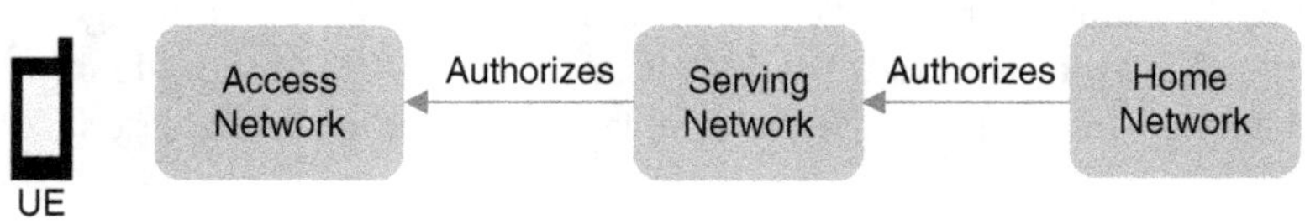

Figure 16.19 Illustration: 5G UE security controlled by home network.

- 5G UE Authentication and Key Agreement (5G-AKA)

At a high level, 5G-AKA is similar to the AKA used in the LTE system. Both the UE and the 5GC authenticate each other and generate and store different security keys based on a long-term key (K) which is available at UE USIM and 5GC, more specifically at UDM/ARPF end. However, being a service-oriented architecture of 5GC, there are significant differences between the 5G-AKA and LTE AKA.

5GS introduces the following new security entities, which are not available in the LTE/EPS, as part of its authentication framework to perform the AKA procedures between the UE and 5GC.

- *Authentication Server Function (AUSF),* in 5GC, which stores UE authentication information and provides authentication services to the AMF network function. AUSF plays the role of the *backend authentication server* under the EAP framework. Refer to Figure 4.2.2.2.2-1 in TS 23.502 [41] for more information on the tasks performed by the AMF and AUSF network functions during a UE initial registration with the 5GC.
- *Subscription Identifier De-Concealing Function (SIDF),* which de-conceals the SUCI to extract the corresponding SUPI of a UE using a protection scheme called *Elliptic Curve Integrated Encryption Scheme (ECIES),* which will be described later. The de-concealing task is performed at UDM end in 5GC using a home network private key.
- *Authentication credential Repository and Processing Function (ARPF)* which stores the *home network private key* that is used by the SIDF to de-conceal the UE/user SUCI and reconstruct the SUPI.
- *Security Anchor Function (SEAF),* which is co-located with AMF in 5GC. SEAF plays the role of *pass-through authenticator* under the EAP framework. SEAF of the serving network stores the anchor key received from the AUSF of the home network. SEAF enables the reauthentication of a UE without executing an end-to-end authentication procedure.
- *Security Edge Protection Proxy (SEPP),* which provides an inter PLMN security services by protection signaling messages exchanged between a visited and home PLMN in case of a roaming 5G system architecture.

Refer to TS 33.501 [87] for more information on the requirements and functions performed by the above network functions/security entities to provide secure communications in a 5G network. The following NAS layer authentication signaling messages are used for the 5G-AKA procedure:

- AUTHENTICATION REQUEST
- AUTHENTICATION RESPONSE

- EAP Authentication and Key Agreement (EAP-AKA)

In addition to the 5G-AKA method used for UE authentication in 5GS as mentioned above, another AKA method for UE authentication is used. This is called the EAP-AKA UE authentication method. It is based on the *Extensible Authentication Protocol* (EAP) framework which is defined in RFC 3748 [14] and the EAP-AKA is specified in RFC 5448 [19]. Under the EAP framework, three entities are involved during a UE authentication process. Those entities are *Peer* (UE), pass-through authenticator (i.e. SEAF in 5GC), and back-*end authentication server* (i.e. *AUSF in 5GC*). Thus, both the UE and serving network is required to support the EAP-AKA and 5G AKA methods of UE authentication in 5GS.

EAP-based UE authentication procedure is carried out between UE and the 5GC using *EAP message*, which is of request and response type. Similar to the 5G-AKA procedure, NAS layer *AUTHENTICATION REQUEST* and *AUTHENTICATION RESPONSE* messages are used to perform the EAP-AKA procedure between a UE and the AMF/SEAF of 5GC. However, unlike the 5G-AKA, an EAP-AKA message is carried under a separate IE: *EAP Message*, refer to TS 24.501 [47], from UE to AMF/SEAF. Also, unlike the 5G-AKA method, in the EAP-AKA method, the SEAF transparently forwards the EAP message to/from the AUSF. Further, there is a difference concerning the security derivation of the key used by the AUSF. In the case of the 5G-AKA method, the key (K_{AUSF}) to be used by the AUSF is computed by the UDM/ARPF and is returned to the AUSF. In the case of the EAP-AKA method, the key to be used by the AUSF is derived by itself, which is the most significant 256 bits derived from the *Extended Master Session Key (EMSK);* refer to TS 33.501 [87].

16.11.2 Primary Authentication and Secondary Authentication

An authentication procedure performed between UE and the 5GC is called *primary authentication,* and the authentication procedure performed between UE and the external DN is called *secondary authentication.* A primary authentication creates a network security anchor key (K_{SEAF}) from which the subsequent security keys are generated to protect information transferred between the UE and the NG-RAN as well as protect the information between UE and 5GC. That is, *a native 5G security context* for a UE is created as a result of the successful primary authentication of a UE. 5GS security key hierarchy is described in the next section.

To perform a primary authentication, 5G-AKA as well as the EAP-AKA methods are used, and to perform a secondary authentication, only the EAP-AKA method is used. The SMF network function in 5GC acts as the EAP authenticator in case of secondary authentication with an external DN. SMF communicates with an external DN over the N4 interface and is routed via UPF over the N6 interface between UPF and DN. EAP messages between a UE and the SMF are carried as part of a NAS layer message under separate IE *(EAP message).* For an end-to-end signaling flow, on a secondary authentication using the EAP method, refer to Figure 11.1.2-1 in TS 33.501 [87], and for a description, refer to TS 23.501 [40].

16.11.3 Key Hierarchy and Authentication Vector

Similar to the LTE security system, several security keys, including new ones, are used in the 5G system also for authentication and security purposes at AS as well as NAS level. Securities at AS and NAS levels are mandatory such that its UE and networks are protected while communicating over the NR air interface. Those security keys are derived and stored at different network elements/entities of the 5G system. At the root of all the different security keys of the 5G system are the *long-term key (K),* which is provisioned by the home network and is stored at the UDM/ARPF at 5GC and UE/USIM end. Based on long-term key (K), different security keys for AS and NAS security purposes are generated in a top-down approach, from left to right, as mentioned below. These AS and NAS keys are used to perform UE authentication from home as well as serving network point of view as illustrated earlier in Figure 16.19.

- (root) K → [Ciphering Key(CK), Integrity Key(IK)] → K_{AUSF} → K_{SEAF} → K_{AMF} → K_{NASint}

Purpose: Integrity protection key for UE and 5GC NAS layer signaling messages.

- (root) K → [CK, IK] → K_{AUSF} → K_{SEAF} → K_{AMF} → $_{NASenc}$

Purpose: Encryption key for UE and 5GC NAS layer signaling messages.

- (root) K → [CK, IK] → K_{AUSF} → K_{SEAF} → K_{AMF} → K_{N3IWF}

Purpose: Security key for non-3GPP interworking function.

- (root) K → [CK, IK] → K_{AUSF} → K_{SEAF} → K_{AMF} → K_{gNB} → K_{RRCint}

Purpose: Integrity protection key for UE and NG-RAN AS layer signaling messages.

- (root) K → [CK, IK] → K_{AUSF} → K_{SEAF} → K_{AMF} → K_{gNB} → K_{RRCenc}

Purpose: Encryption/confidentiality protection key for UE and NG-RAN AS layer signaling messages.

- (root) K → [CK, IK] → K_{AUSF} → K_{SEAF} → K_{AMF} → K_{gNB} → K_{UPint}

Purpose: Integrity protection key for user traffic.

- (root) K → [CK, IK] → K_{AUSF} → K_{SEAF} → K_{AMF} → K_{gNB} → K_{UPenc}.

Purpose: Encryption/confidentiality protection key for user traffic.

Table 16.3 Authentication vector for UE authentication using 5G and EAP AKA.

Authentication Method	Serving Network Authentication Vectors	Home Network Authentication Vectors
5G-AKA	RAND, AUTN, XRES* (AUSF to SEAF)	RAND, AUTN, XRES*, and KAUSF (UDM/ARPF to AUSF)
EAP-AKA	–	RAND, AUTN, XRES, CK', IK' (UDM/ARPF to AUSF)

XRES* is derived from an expected response (XRES). Further, the EAP-AKA uses the transformed version (CK', IK') of the ciphering key (CK) and integrity key (IK).

Refer to Figure 6.1 in TS 33.501 [87] for 5G system key hierarchies and their generation process for 5G-AKA and EAP-AKA of UE authentication methods. As shown in Figure 6.1 in TS 33.501 [87], the key K_{AUSF} is home network-specific, and the subsequent keys, including the anchor key K_{SEAF}, below it is the serving network specific. K_{AUSF} is derived by the UE and UDM/ARPF and provides to AUSF during the primary authentication method. On successful completion of the authentication procedure between UE and the 5GC, the key K_{SEAF} becomes the serving network-specific anchor key. The SEAF generates the K_{AMF} from the K_{SEAF} and provides it to the AMF. From the K_{AMF}, the subsequent keys are generated as mentioned above. Key K_{gNB} is derived by UE and AMF from K_{AMF}.

NAS layer integrity and encryption/confidentiality keys – K_{NASint} and K_{NASenc} – are derived by the UE and the AMF from K_{AMF}. The K_{gNB} is derived by the UE and the AMF from the K_{AMF}. Access stratum layer integrity and encryption/confidentiality keys – K_{RRCint} and $K_{RRCenc\ are}$ – are derived by the UE and the gNB from KgNB.

- Authentication Vector

As part of the UE authentication procedure, 5GC UDM/SIDF de-conceals the SUCI to obtain the UE SUPI. Based on SUPI and its subscription information, the UDM/ARPF selects the desired authentication method, i.e. either 5G-AKA or EAP-AKA. An authentication vector contains information that is generated by the UDM/ARPF and provides to the AUSF, down to the UE for its authentication purposes. 5G-AKA and EAP-AKA-based UE authentication methods have different authentication vectors as listed in Table 16.3.

- Serving Network Name (SNN)

One of the parameters used to derive keys RES*, eXpectedRES*, K_{SEAF}, and K_{AUSF} is the serving network name. In the LTE system, the serving network consists of the PLMN id (MCC and MNC) only. However, in the 5G system, the serving network consists of the concatenation of the *service code (5G)* and the serving network (SN) Id with the separation character ":" – i.e. *5G: PLMN id.* An SNN is also used during a UE authentication procedure, i.e. SNN is used to derive the serving network anchor key K_{SEAF} and mobility management procedures.

Figure 16.20 illustrates a typical UE authentication procedure using the 5G-AKA method during a UE initial registration with the 5GC. This figure illustrates the messages flow among the different security entities of the 5GC for authenticating UEs using the 5G-AKA authentication method. The reference points N12 and N13 and usages of corresponding authentication vectors mentioned in Table 16.3 along with the SNN are also indicated in Figure 16.20.

16.11.4 New Security Requirements in 5G System

In addition to the legacy security features and methods used by the 5G system as mentioned above, it also introduces the new security measures and requirements in terms of authentication and authorizations due to the following new architecture and features.

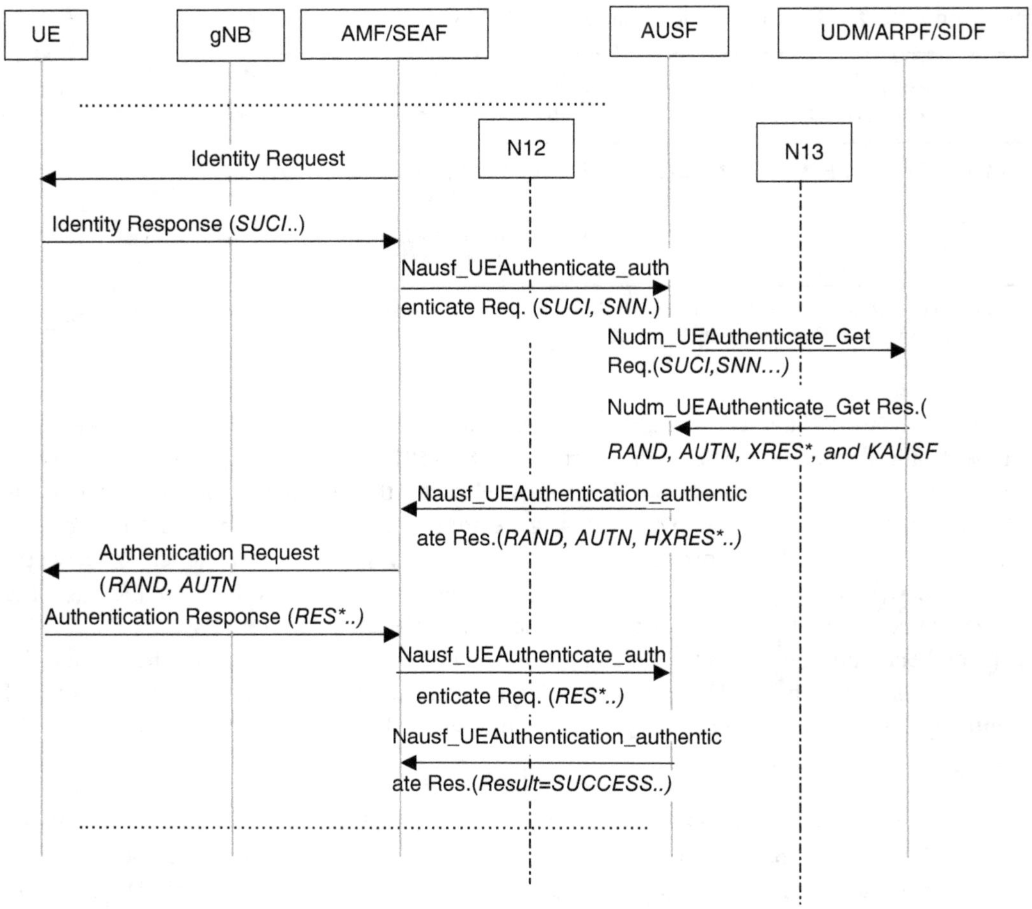

Figure 16.20 Illustration: UE authentication using 5G-AKA method through 5GC security entities.

- 5GC Network Functions Authentication Protocol

The service-oriented architecture of 5GC contains software-based network functions that perform the core network functions and procedures to provide communication services to users. The authentication method provided by the transport layer security (TLS) protocol, as specified in RFC 5246 [18], is used between two network functions of 5GC to authenticate each other. Authentication based on TLS protocol is also used between two entities or networks as follows:

- A network function and centralized authorization server called *Network Resource Function (NRF)* of a 5GC;
- A network function and *Security Edge Protection Proxy (SEPP)*; and
- Between two SEPPs, connecting two PLMNs (home and visited).

Once a network function or entity has been authenticated successfully, it must be authorized further to allow access to desired network resources or services as described below.

- 5GC Network Functions Service Access Authorization

Unlike the previous generation core network, the 5GC architecture is service-oriented and is based on various network functions, each performing a particular protocol layer functions and procedures or other functions and procedures of a 5GC. Each 5GC network function operates under the paradigm of a *service producer or a service consumer* and communicates with each other over the HTTP. However, a network function is required to be authorized first through a central authorization server before the network function can request and consume services. As part of the service-based architecture, the 5GC security feature consists of the following steps for authorization of network functions by the NRF.

- Registration of network functions;
- Discovery network functions and their services; and
- Authorization of network functions.

Within the 5GC, the first two tasks are performed by each interested network function toward a centralized authorization server called *Network Resource Function (NRF)*. The network function service access authorization method used by the NRF to authorize different consumers' as well as producers' network functions is based on the *OAuth 2.0 [20]* authorization method.

Refer to TS 33.501 [87] for more information on the functions performed by the different network functions/ security entities of 5GC to meet the 5G system security requirements.

- Security (introduced as part of the 3GPP Release 16) in Network Slicing

The management and orchestration of network slices for managing their life cycle are based on a Management Service (MnS) framework which consists of producers and consumers with the standardized interface as described in TS 28.533 [58]. Similar to the security authentication procedure used by the 5GC network functions, the MnS procedure services also use OAuth 2.0 [20] authorization method to authenticate their consumers. Further, TLS is used to provide integrity protection of information exchanged between a MnS producer and a consumer. Refer to Figure 4.2.9.2-1 in TS 23.502 [41] for more information on the tasks performed by the AMF and AUSF 5GC network functions during a network slice-specific authentication and authorization procedure using the EAP-AKA authentication framework.

- Inter PLMN Security: Security Edge Protection Proxy (SEPP)

5G system also ensures a secured communication for a roaming subscriber. In the case of a roaming architecture, the visited as well as home network/PLMNs are interconnected using the *N32* logical interface. To enable a secured communication in case of a roaming architecture, the visited as well as the home network communicates with each other through the *Security Edge Protection Proxy (SEPP)* network function. SEPPs are deployed at the perimeter or edge of the visited and home network; refer to TS 23.501 [40]. SEPPs of the visited as well as the home network communicates over the N32 logical interface as illustrated in Figure 16.21. For detailed reference architecture, refer to Figure 4.2.4-1 in TS 23.501 [40].

The SEPP provides the application layer only security by enforcing protection policies. To provide application layer security, the SEPPs use the JSON Web Encryption (JWE) as specified in RFC 7516 [22] and JSON Web Signatures (JWS) as specified in RFC 7515 [21]. Thus, using the JWE and JWS, the SEPP ensures the integrity and confidentiality of the control plane or signaling messages exchanged between the visited and the home network of two PLMNs over the N32 logical interface. That is, sending SEPP protects a signaling message before its transfer over the N32 interface. On the receiving end, the SEPP performs a security verification on the signaling message and forwards it to the desired network function of 5GC. Apart from protecting signaling messages, a SEPP also hides the network topology.

Further, if the JWE- and JWS-based protection methods are not available, TLS may be used to protect information transferred over the N32 interface. For more detailed descriptions of the functions performed by a SEPP to provide inter-PLMN security over the N32 logical interface, refer to TS 33.501 [87].

- Security During Interworking

Interworking between LTE/EPS and 5GS and vice versa was described earlier in Section 16.7. Support of a security interworking is also required whenever interworking between LTE/EPS and 5GS and vice versa takes place

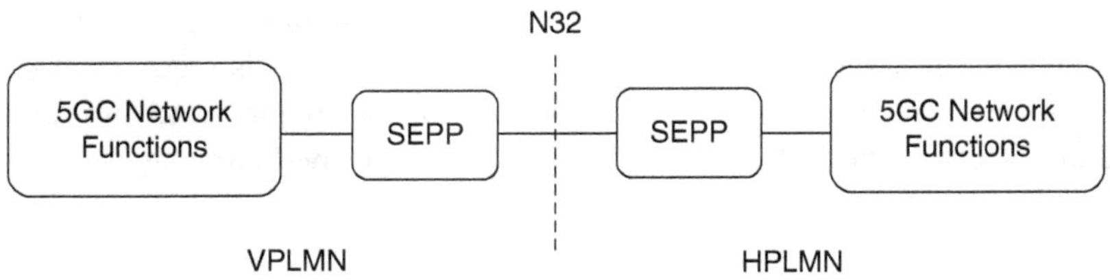

Figure 16.21 Illustration: inter PLMN secured communication through SEPP.

over the N26 interface, i.e. single registration mode. In the UE connected, i.e. RRC CONNECTED, state, interworking between LTE/EPS to 5GS and vice versa takes place through a handover procedure. In these cases, the AMF generates the mapped UE security context, i.e. LTE/EPS or 5GS, based on the corresponding security keys, i.e. LTE/EPS K_{ASME} or 5GS K_{AMF}. In the UE idle or RRC IDLE state, interworking between LTE/EPS to 5GS and vice versa takes place through a tracking area update (LTE/EPS)/ATTACH procedure or *Registration Request* (5GS) mobility procedure toward the respective core network. During such MM layer procedure also, mapped security contexts are used by the LTE/EPS MME and 5GC AMF.

In the case of dual registration mode, i.e. without the N26 interface, a UE requires registering with the LTE/EPS and 5GS network separately. To support UE mobility, LTE/EPS and 5G system-specific security functions and procedures as well as security contexts are created at UE and network end. For more information on the signaling information flow for security interworking between LTE/EPS and 5GS and vice versa, refer to TS 33.501 [87].

16.11.5 Subscriber Identities/Privacy Protection

In the LTE system, the subscriber's permanent identity, i.e. IMSI, is sent by a UE in clear text over the *IDENTITY RESPONSE* message to the LTE/EPS, exposing the privacy of a subscriber. To ensure the adequate privacy of a subscriber while communicating between a UE and the 5GC over the NR air interface, the 5G system introduces several new identities for a UE as listed below.

- Subscription permanent identifier (SUPI),
- Subscription concealed identifier (SUCI), and
- Subscription temporary identifier, i.e. 5G-GUTI and 5G-TMSI.

The above UE identities were described earlier in Section 16.8.2. As the name implies, a SUCI transfers a SUPI in an encrypted or concealed form in the 5G NAS layer *IDENTITY RESPONSE* message from a UE to the AMF. To enable subscriber privacy protection in a 5G system, two keys are used to conceal and de-conceal a subscriber's permanent identity.

- *Home Network Public Key,* which is provided by the home network and is stored at the UE/USIM end. UE uses a home network public key to conceal a SUPI into a SUCI and provides it over the *IDENTITY RESPONSE* from a UE to the AMF.
- *Home Network Private Key* which is stored at the ARPF at home network end and used by the SIDF at UDM to decrypt or de-conceal a SUPI from the SUCI provided by a UE.

In addition to the two keys mentioned above, a protection scheme, i.e. algorithm, is also used to encrypt the SUPI, i.e. IMSI. The protection scheme is specified as part of the IE: *Mobile Identity* of the *IDENTITY RESPONSE* message; refer to TS 24.502 [48]. A protection scheme is identified by an identifier which is 4 bits in length. The null scheme, operator-defined, or 3GPP defined protection schemes are allowed. 3GPP defined, TS 33.501 [87], *ECIES* is used as a protection scheme to encrypt/decrypt a SUPI into/from a SUCI at UE and 5GC end, as illustrated in Figure 16.22.

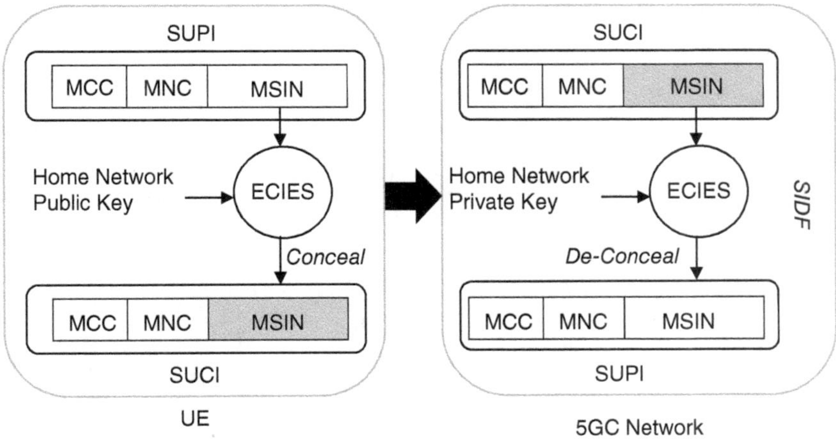

Figure 16.22 Illustration: conceal and de-concealing of a UE SUPI to SUCI and vice versa.

For more information on the 5G security architecture, features, mechanisms, and UE authentication methods mentioned above, refer to TS 33.501 [87].

Chapter Summary

This chapter presented an overall introduction to the 5G system and its key enablers, beginning with different use cases and capabilities of the 5G system along with its system architecture. The logical architecture of the NG-RAN was covered. All the 5G use cases and the diverse services being envisaged under them would require high data rates, reliability, low latency, availability, and mobility also. Key enablers, i.e. network function virtualization, network slicing, service-oriented architecture, and physical layer aspects of 5GS, were mentioned briefly and shall be described further in later chapters. A standalone and non-standalone deployment solution for a 5G system has been described. During the initial deployment phase of a 5G system, E-UTRA-NR Dual Connectivity (EN-DC) with an existing LTE/EPC network may be adopted as a non-standalone solution. Further, the interworking between a 5G system, with its NG-RAN and 5GC, and the existing LTE/EPS network, with or without the new interface N26, was described. This chapter ends with the description of the 5G security system which has evolved from the LTE system and is of paramount importance considering the increasing attacks on various interfaces of a mobile communication network, be it radio air interface or signaling messages or user data, or service-based interface and so on.

17

Introduction to GSM, UMTS, and LTE Systems Air Interface

Introduction

This chapter presents a brief and overall protocol layer architecture of the air interface used in the Global System for Mobile Communication (GSM)/General Packet Radio Service (GPRS), Universal Mobile Telecommunication System (UMTS), and Long-Term Evolution (LTE) system, along with some of the common concepts found in the 5G system (5GS) also, to provide a comparative study to the reader. The air interface and its corresponding access method evolved from GSM to the 5G New Radio (NR) system over a period of time, which also differentiates a mobile communications system and network from each other. Major differences exist at Layer 1, Layer 2, and Layer 3. However, there are similarities also at the higher layers in terms of protocol functions, procedures, and formats used to transfer messages between user equipment (UE)/mobile station (MS) and the network over the respective air interface. It will be easier for a reader to understand the various aspects of the 5G air interface protocol layers with a prior and fair knowledge of the GSM/GPRS, UMTS, and LTE system air interface protocol layers.

The air interface protocol stacks and their layers are required to be supported by both the MS/UE and its radio access network (RAN). This is also required for the interoperability of network elements of different OEMs. For interoperability, a mobile device also requires to support multiple Radio Access Technologies (RATs) and protocol stacks.

17.1 Air Interfaces Protocol Architectures

Mobile communications radio air interface protocol architectures can be studied with regard to the OSI 7 layers reference architecture. The legacy GSM/GPRS, UMTS, and LTE system air interface protocol architectures and their layers, between UE and RAN, are shown in Figure 17.1. The top part of this figure illustrates the air interface concept, in general, between a UE/MS and its base station. The logical interface name is Um for the GSM/GPRS system air interface. For the UMTS, LTE, and 5G NR systems, it is known as the Uu.

For GSM/GPRS system, the air interface protocol layers terminate at the base station controller (BSC) end; for UMTS, it terminates at the Radio Network Controller (RNC) end; for the LTE system, it terminates at the eNodeB end; and for the 5G NR system, it terminates at Next-Generation Radio Access Network (NG-RAN) end. In general, the air interface protocol layers terminate at the base station of the GSM, UMTS, LTE, and 5G networks.

Mobile Communications Systems Development: A Practical Introduction to System Understanding, Implementation, and Deployment, First Edition. Rajib Taid.

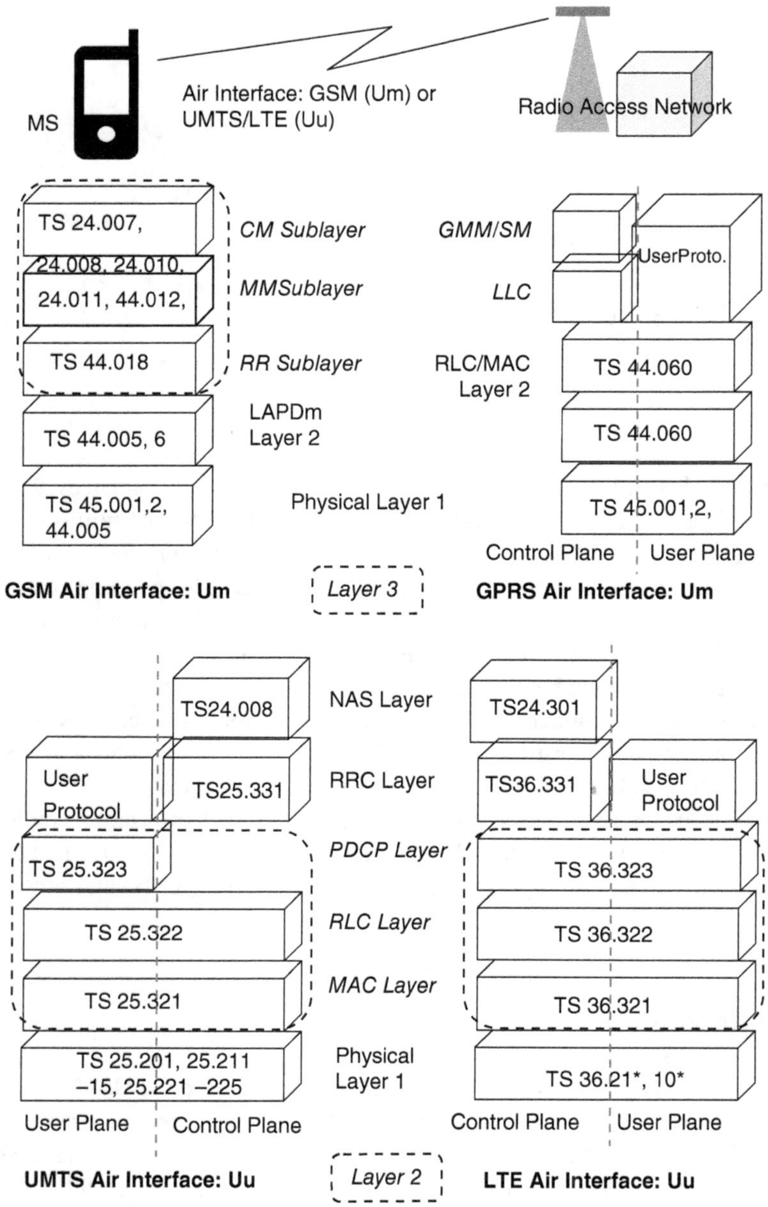

Figure 17.1 Illustration: radio air interface protocol layer architectures of GSM, GPRS, UMTS, and LTE systems.

17.2 Protocol Sublayers

A protocol layer may have sublayers. In Figure 17.1, the sublayers of a particular protocol layer are shown within the rounded and dotted rectangle. Also, against each layer and sublayer, the corresponding 3rd Generation Partnership Project (3GPP) technical specification(s) number is specified in Figure 17.1. As with OSI 7 layers, higher layer information is processed by protocol layers and sublayers before it is transmitted over the air interface.

17.3 Control Plane and User Plane Protocols

Further, the protocol layers of the corresponding air interface are grouped into the control or signaling as well as the user plane protocols which were presented earlier in Chapter 3. For illustration purposes, the control plane and user plane protocol stacks are separated with a vertical dotted line in Figure 17.1. Comparing the UMTS and LTE protocol stack from Figure 17.1, it is observed that the Packet Data Convergence Protocol (PDCP) layer is not part of the control plane protocol stack, but it belongs to the user plane protocol stack of the UMTS air interface. However, over the LTE and 5G NR air interface, the PDCP layer is part of the user as well as the control plane protocol stack. Table 17.1 summarizes the layer-wise control plane protocol layers for the respective air interface of the GSM/GPRS, UMTS, LTE, and 5G NR systems for a comparative study.

As it is clear from the Table 17.1, some of the protocol layers across the air interfaces are known by the same name, but their architecture, functions, procedures, and implementation aspects are quite different due to the different radio access methods being used at the respective physical layer.

17.4 Protocols Domains Classifications

Based on air interface Layer 3 protocol sublayers, the GSM/GPRS, UMTS, and LTE system air interface protocols can be further classified into two domains, namely the Access Network (AN) and Core Network (CN) domains, which are illustrated in Figure 17.2.

17.5 Access Stratum and Non-access Stratum

Within the protocol domain classifications as illustrated in Figure 17.2, the UMTS CN circuit-switched (CS) domain call control management (CM) and mobility management (MM); packet-switched (PS) domain MM and session management (SM); the LTE Evolved Packet System (EPS) MM and EPS SM; and also the 5G MM and

Table 17.1 Summary of air interfaces control plane protocol layers.

System	Layer 1	Layer 2 Protocols	Layer 3 Protocols
GSM(Um)	Physical Layer	LAPDm	CC, MM, RR
GPRS(Um)	Physical Layer	RLC/MAC	RRC, LLC, GMM, SM
UMTS(Uu)	Physical Layer	RLC, MAC	RRC, CC, MM, GMM, SM
LTE(Uu)	Physical Layer	RLC, MAC, PDCP	RRC, EMM, ESM,
5G NR(Uu)	Physical Layer	RLC, MAC, PDCP, SDAP	RRC, 5GMM, 5GSM,

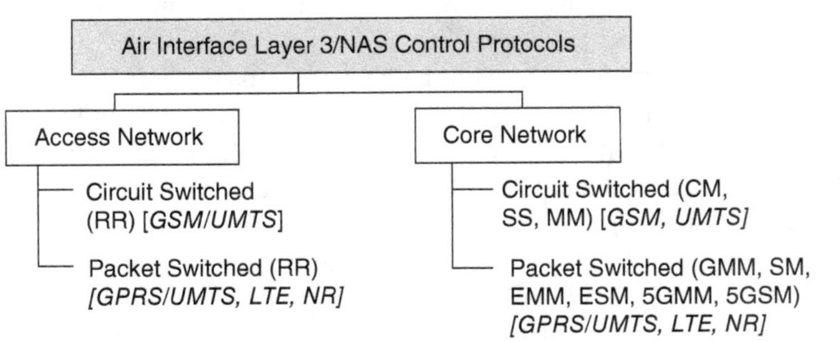

Figure 17.2 Illustration: air interface Layer 3/NAS protocols and its categories.

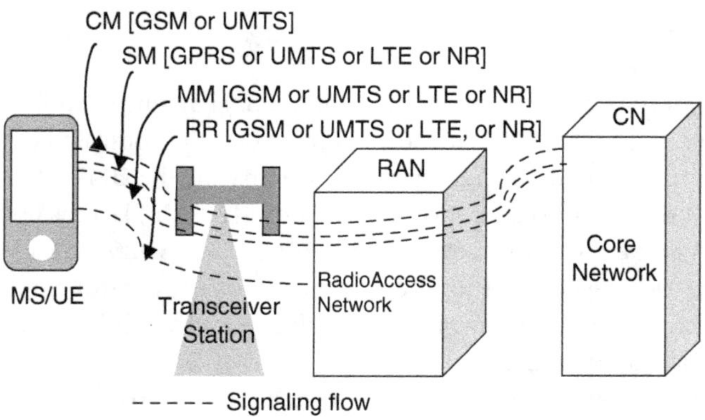

Figure 17.3 Illustration: air interface: AS and NAS layer signaling information flow between MS and network.

5GSM layers are known as the *Non-Access Stratum (NAS)* layer. Only the Radio Resource Control (RRC) layer is considered as the UMTS, LTE, and 5G NR air interface Layer 3 protocol, which is also known as the *Access Stratum* (AS) protocol. NAS layer and AS layer were described earlier in Section 3.3. Figure 17.3 further illustrates, in general, the GSM/GPRS, UMTS, LTE, and 5G NR system air interface higher layer protocol signaling connections and their terminations between the following:

- UE/MS and its RAN;
- UE/MS and the CN.

As shown in Figure 17.3, a prior establishment of a Radio Resource (RR)/RRC (Access) signaling connection between UE and RAN is required to transfer NAS layer (CM layer, MM layer, SM layer) signaling messages to the respective core network element, i.e. GSM MSC, GPRS/UMTS Serving GPRS Support Node (SGSN), LTE MME, or 5G Core AMF. If an existing RR/RRC connection to RAN does not exist, a new connection shall be established on behalf of its higher layers.

17.6 Message Formats

The GSM air interface Layer 3 protocols, i.e. RR, CC, MM, and GPRS PS domain SM, and GMM layers, use the standard Layer 3 message format to communicate with the peer network element as described earlier in Chapter 4. Similarly, though the UMTS CS and PS domain NAS layer, LTE NAS layer, and 5G NR NAS layer protocols are not part of the Layer 3 protocol, they also use the same standard Layer 3 message format to communicate with the peer network elements, i.e. UE and CN element.

Air interface signaling messages of a particular layer are identified with the help of a protocol discriminator/extender protocol discriminator (5G NR) which is part of the Layer 3 protocol header. Apart from this, an LTE or multi-RAT-capable UE performs CS domain as well as PS domain protocol functions and procedures toward the RAN and CN. Because of this, it is also required to refer GSM and UMTS CS and PS domain CM, SM, MM layer technical specifications, i.e. TS 24.007 [44] and TS 24.008 [45], along with the LTE NAS layer, comprising Evolved Packet System Mobility Management (EMM) and Evolved Packet System Session Management (ESM) layers, technical specifications, i.e. TS 24.301 [46] and TS 23.401 [39]. Similarly, one will be required to refer to TS 24.007 and TS 24.008 along with the 5G NR NAS layer, comprising 5GMM and 5GSM layers, technical specification, i.e. TS 24.501 [47].

17.7 Security Over the Air Interface

In comparison to the GSM system, the UMTS, LTE, and 5G NR systems provide more secure communication services over their respective air interface in terms of the integrity and confidentiality of signaling and user data. This is achieved through integrity protection and ciphering mechanism, which is applied on signaling, AS and NAS layers, as well as user information transmitted over the air interface. To activate such security features, dedicated security-related commands, containing its mechanisms, are exchanged between a UE and RAN; and a UE and CN element.

17.8 Piggybacking for Reduction of Signaling Overhead

One important goal of data transmissions and receptions over the air interface is to ensure its optimum utilization by reducing the signaling messages overhead between the UE/MS and the RAN element, i.e. BSC, UMTS terrestrial radio access network (UTRAN), Evolved-UMTS Terrestrial Radio Access Network (E-UTRAN), and 5G NG-RAN. To reduce the signaling overhead over the air interface, the concept of piggybacking a signaling message over another signaling message was described in Section 4.1.8. Because of this, an ongoing particular procedure of a protocol layer may further lead to other scenarios. A new scenario may be initiated over the existing signaling connection that was established for the ongoing protocol layer procedure, thus avoiding the requirement of the establishment of a new signaling message between a UE/MS and RAN and vice versa.

Depending on the current state, i.e. Idle or dedicated/connected, of the RRC protocol layer at the UE/MS end, other subscenario may also arise. Examples of such protocol layer subscenarios are described below.

17.8.1 Examples Piggybacking of Signaling Messages

1) Transmissions of Channel State Information (CSI) from a UE to LTE/E-UTRAN eNodeB or from a UE to NR NG-RAN/gNB in the case of 5G system

   A *Chanel Quality Indicator (CQI)* is the part of the downlink CSI as measured by a UE on the received signal condition from the LTE eNodeB or NR NG-RAN/gNB. A CQI is provided by a UE to the eNodeB or NG-RAN/gNB, either periodically over the physical layer PUCCH channel or aperiodically over the PUSCH channel along with the user data. A CQI over the PUCCH is a kind of signaling overhead because of its periodic nature. On the other hand, a periodic CSI may occur at the same subframe where a UE is scheduled over the PUSCH. In such a scenario, the periodic CQI is sent over the PUSCH, along with user data, only instead of the periodic PUCCH, thus reducing an additional signaling overhead between a UE and the eNodeB in the uplink direction.
2) *Current State:* An LTE or 5G NR UE has the uplink control resources allocated over the PUCCH channel.

   New Scenario: The UE Medium Access Control (MAC) layer may send a *Scheduling Request* (SR) indication to request uplink resources from the LTE eNodeB or 5G NG-RAN/gNB for a new transmission in the uplink direction.
3) *Current State:* An LTE or 5G NR UE has uplink resources allocated to it to send data over the Uplink Shared Channel (UL-SCH) transport channel.

   New Scenario: The UE MAC layer may indicate that more data has arrived to send and requires additional uplink resources. The UE MAC layer will indicate to the LTE eNodeB or NG-RAN/gNB by sending a *Buffer Status Report (*BSR) control element to the eNodeB.

In all of the scenarios described above, it is observed that a new scenario of protocol layer may be initiated in the same or opposite direction of the ongoing data transmission session which eliminates the requirements of an additional signaling establishment between a UE/MS and the RAN and vice versa. Thus, signaling overhead over the respective air interface is also reduced.

Chapter Summary

The chapter presented an introductory background to the reader on several aspects of the GSM/GPRS, UMTS, and LTE as well as some of the common concepts used over the 5G NR system air interface with its protocol stack, architectures, and its different layers. Sometimes, it is required to refer to GSM, GPRS, or UMTS or LTE air interface technical specifications also even if a developer is working on the 5G NR system air interface area. The logical name of the GSM/GPRS air interface is *Um*; for the UMTS, LTE, and 5G NR air interface, it is known as the *Uu*. The chapter ends with the method used over these air interfaces of the GSM to the 5G NR system to reduce signaling overhead while exchanging signaling messages between a UE/MS and its RAN and vice versa.

The subsequent Chapters 18 and 19 are entirely devoted to the 5G NR air interface only and describe more about the different aspects of NR air interface control plane protocol layers, their functions, and procedures. We begin with the NR air interface NAS layer first, followed by Layer 3 protocols, Layer 2 protocols, and finally the physical layer protocol.

18

5G NR Air Interface

Control Plane Protocols

Introduction

As defined by the 3rd Generation Partnership Project (3GPP), new radio (NR) shall be used in the 5G communication system. The 5G NR air interface greatly differentiates it from the previous generation mobile communication systems, namely Global System for Mobile Communication (GSM), Universal Mobile Telecommunication System (UMTS), and Long-Term Evolution (LTE) system. One distinctive feature of the NR air interface is the support of diverse operating frequency bands and greater channel bandwidth which is 20 times the maximum channel bandwidth used in the LTE system. The NR air interface protocol architecture consists of user plane and control plane protocol layers, which are grouped into a particular logical node, i.e. centralized unit (CU) and distribution unit (DU), of NG-radio access network (RAN) as illustrated in Figure 16.4. In this chapter, the control plane protocol layers of the NR air interface will be discussed, which consist of the protocol layer from the NR Access Stratum (AS), i.e. Radio Resource Control (RRC) layer, and Non-Access Stratum (NAS), i.e. Mobility Management (MM) and Session Management (SM) layers. Further, it may be noted that, unlike LTE/EPS, the 5GS architecture allows the usage of NAS protocols over 3GPP and non-3GPP access networks.

This chapter begins with the overall protocol layer architecture of the NR control plane protocol stack. Then, it starts with a top-down approach by describing the NR air interface NAS layer control plane protocol layers. NAS layer between user equipment (UE) and 5G core network (CN) introduces several new concepts to provide diverse services over the NR air interface under different uses of the 5G system. This is followed by the description of the NR AS air interface control plane protocol layer and presents how the new changes being introduced, compared to the LTE system, aim to achieve low-latency communication requirements of certain applications such as the Ultra Reliable and Low-Latency Communications (URLLC) use case of the 5G system.

18.1 NR Control Plane Protocol Layers

NR AS and NAS control plane protocol layers are illustrated in Figure 18.1, which are similar to the LTE system control protocol architecture. 5G system NAS MM and SM layers terminate at the 5G CN *Access and Mobility Management Function* (AMF) and *Session Management Function* (SMF) ends, respectively. SM-related signaling messages from UE to SMF and vice versa are forwarded by the AMF.

The 5G CN AMF and SMF are also shown for the NAS layer termination illustration purposes. However, the AMF and SMF communicate with each other through the *service-based interface* (SBI) only, over HTTP, which is based on the service-oriented architecture of the 5G CN. Further, the communication between the AMF and the SMF is identified through reference point *N11* as shown in Figure 18.1. Unlike the previous generation CN

Mobile Communications Systems Development: A Practical Introduction to System Understanding, Implementation, and Deployment, First Edition. Rajib Taid.

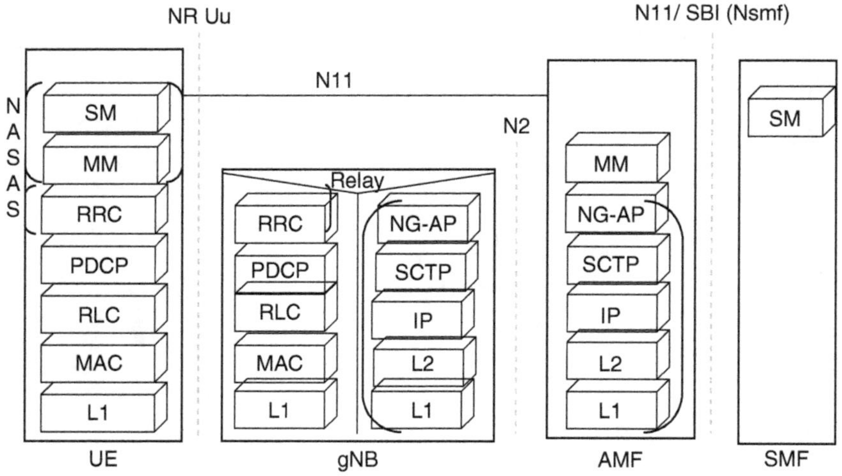

Figure 18.1 Illustration: NR control plane: AS and NAS protocol layers.

architectures, the 5G CN has the service-based architecture where different services with respect to CN protocol layers are produced and consumed by different software-based network functions. The service-based architecture of the 5G core as well as its reference points-based representation will be described later in Chapter 20.

As illustrated above, the name of the NR control plane protocol layers appears to be the same and shares similarities in terms of the similar protocol functions and procedures performed as in the LTE system. But there are significant differences in their operations as per the NR low-latency requirements as well as services delivery through the different quality of services (QoSs) are concerned. Such differences in the NR NAS and AS layers, in comparison to the LTE NAS and AS layers, shall be described in the subsequent sections.

18.2 Session Management (5G SM) Layer

In an LTE/Evolved Packet System (EPS) network, IP data services for different applications are provided through an end-to-end EPS bearer which is established between a UE and its CN. But in a 5GS network, IP data services are provided to a UE in terms of an end-to-end protocol data unit *(PDU) session* which is established between a UE and its CN, i.e. 5G Core Network (5GC) and terminates at the UPF end. Thus, the 5G SM layer works on a PDU session level and performs the PDU SM-related functions and procedures such as session establishment and its authentication, modification, release, IP address allocation, and so on. PDU session establishment task also includes a PDU session handover from LTE/EPS to 5G system; a PDU session handover between 3GPP and non-3GPP system; and a PDU session handover in case of roaming user. The SMF network function at 5GC end is primarily responsible to perform the 5G SM-related functions and procedures. The SMF also manages the user or data plane, i.e. the controlling of UPF to provide an end-to-end data transfer services through a PDU session created between a UE and its 5GC. The AMF network function acts as an interface for UEs to communicate with the SMF. Due to this, the AMF transparently forwards all the SM-related NAS layer messages that are exchanged between a UE and the SMF and vice versa. AMF and SMF communicate with each other through their respective SBIs.

It may be noted that similar to LTE/EPS MM and SM layers, the prerequisite is that in 5G system also, an MM context for a UE shall be required to be created at UE and 5GC/AMF end first before exchanging SM-related procedures between UE and the SMF via AMF. As part of a UE session establishment request, the AMF selects an appropriate SMF to handle the requested PDU session. Further, the SMF selects a user plane function (UPF) on behalf of a UE and creates the required session toward the UPF over the N4 interface. SMF also assigns a GPRS

Tunneling Protocol (GTP) tunnel, i.e. a fully qualified tunnel endpoint identifier (F-TEID), to transfer session-specific user plan data between the Next-Generation Radio Access Network (NG-RAN)/gNB and the UPF. Refer to TS 23.502 [41] for descriptions of the 5GS SM procedures by the SMF.

Note that more than one PDU session can be established between a UE and its 5GC. This is similar to the multiple packet data network (PDN) connectivity requirements that an LTE UE can establish in an LTE/EPS network. Also, as mentioned earlier in Section 16.9, each network slice has a PDU session. Some of the 5G SM procedures are described next.

18.2.1 Procedures of 5G SM Layer

- PDU Session Establishment

A UE initiates a PDU session establishment requirements procedure by sending a 5G SM *PDU Session Establishment Request* message toward the 5GC, i.e. SMF, as illustrated in Figure 18.2. This message may be also used to transfer or hand over an existing PDU session to an LTE/EPS network. A UE may specify the network slice identifier, called Single Network Slice Selection Assistance Information (S-NSSAI) in *UL NAS TRANSPORT* message between UE and AMF, during its PDU session establishment request made to the 5GC. There is a one-to-one relationship between a UE PDU session and its requested network slice. That is, a UE PDU session belongs to only one network slice instance.

A UE also specifies the type of PDU session that it wishes to establish with the 5GC through the IE: *PDU session type*. A UE may optionally specify its PDU session and service continuity mode using the *SSC Mode* IE in the message *PDU Session Establishment Request* message. Different types of PDU sessions and session continuity modes used in a 5GS network are described later in the next two sections. The 5GC (AMC and SMF) confirms the PDU

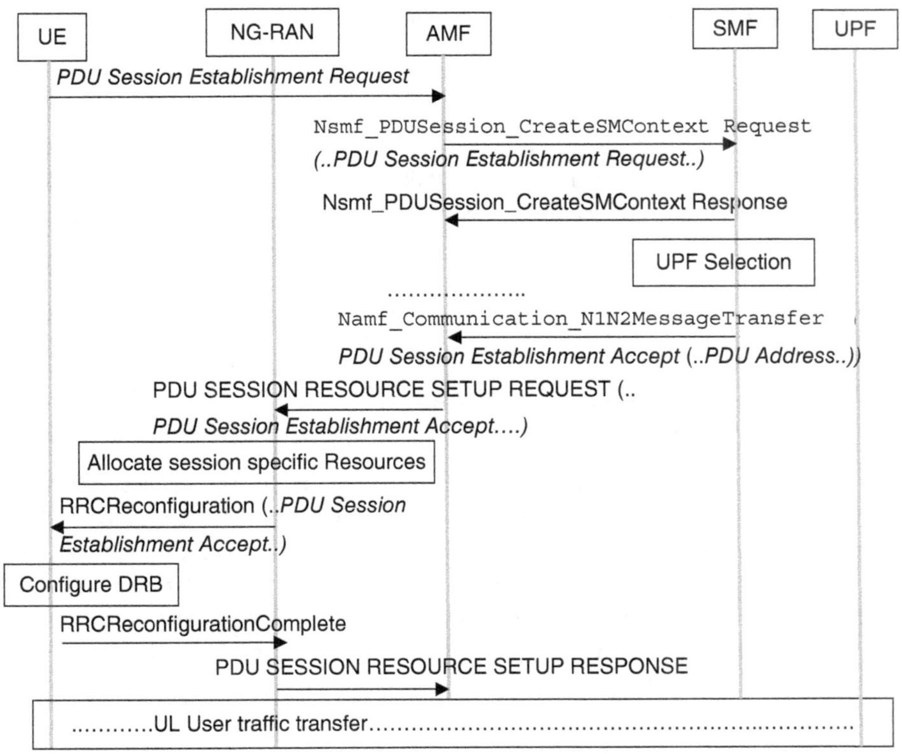

Figure 18.2 Illustration: 5G SM UE-initiated PDU session establishment procedure.

session creation by sending a *PDU Session Establishment Accept* message to the UE, containing the information from SMF to UE such as the allocated IP address, QoS profile, and session and service continuity mode.

As illustrated in Figure 18.2, NAS layer messages exchanged between UE and SMF are shown in italics, which is transferred in an encapsulated way within another message. Next-Generation Application Protocol (NG-AP) messages exchanged between gNB/NG-RAN and AMF over the N2 interface are shown in capital letters. 5GC service interface messages exchanged between AMF and SMF are shown in courier font. As part of a successful PDU session creation at 5GC end and after selection of a UPF by the SMF, the UE is allocated an IP address which is indicated by the field *PDU Address* of the *PDU Session Establishment Accept* NAS layer message that is forwarded to the UE. On the other hand, in the case of the LTE/EPS system, a UE is allocated an IP address during its initial ATTACH REQUEST procedure.

On receipt of a PDU SESSION RESOURCE SETUP REQUEST from the SMF, the NG-RAN/gNB allocates the necessary Layer 2 and Layer 1 radio resources for the corresponding data radio bearer (DRB) to be used by the UE over the NR and its requested PDU session and network slice. Such radio resources are intimated over the *RRCReconfiguration* message from NG-RAN/gNB to the UE using which the UE transfers a user/UL data to the UPF. Refer to TS 23.502 [41] for more detailed flow and descriptions of the UE-initiated PDU session establishment request procedure.

- PDU Session Modification

A PDU session may be modified because of changes in QoS parameters. Either a UE or 5GC modifies a PDU session through the NAS layer SM *PDU session modification request* message. Modification of a PDU session may be requested by a UE or commanded by the SMF to a UE.

Similarly, the 5GSM layer performs a PDU session authentication and authorization and a PDU session release procedure at a request of a UE or a command by the SMF to UE. Refer to TS 24.501 [47] for more information on the 5GS NAS layer SM procedures and their detailed signaling message flows along with roaming and non-roaming scenarios.

18.2.2 PDU Session Types

Note that an LTE/EPS network provides IPv4, IPv6, and IPv4v6 type IP data services only. However, to provide and support different types of services, devices, and applications, 5GS supports the following types of IP data services or PDU session types:

- IPv4,
- IPv6 or
- IPv4v6,
- Ethernet, to provide data services to UE over Ethernet data network, and
- Unstructured, to provide data services to a device, e.g. Internet of Things (IoT), over a non-IP data network.

In an LTE/EPS network, IP data session-related tasks are centered around an EPS bearer. However, it is the PDU session where the 5GS SM tasks are performed. Due to this, during interworking from a 5GS to an LTE/EPS network and vice versa, their respective SM layer co-ordinates with each other. For example, as a result of moving from a 5G network to an LTE/EPS network, a UE establishes a PDN type of "non-IP" for an LTE/EPS bearer corresponding to the 5GS PDU session of "Ethernet" and "Unstructured" types. Similarly, during interworking from 5GS to LTE/EPS network, the state of an LTE EPS bearer is derived from the current state of a 5GS PDU session.

As an analogy, a 5GS PDU session can be compared with a *PDP context* found in a GPRS network which is established between a mobile station (MS) and its CN, i.e. Gateway GPRS Support Node (GGSN), to transfer user plane data/traffic.

18.2.3 PDU Session and Service Continuity (SSC)

In the LTE/EPS system, the continuity of an IP session of a UE is maintained for uninterrupted services, that is, the PDN gateway and the IP address allocated to a UE remain the same. However, unlike the LTE/EPS network, the IP address which was allocated as part of a PDU session may not be preserved during a UE movement within a 5GS network. This is because in a 5GS network, multiple user plane anchor functions, i.e. UPF, may be deployed. The same UPF or a different UPF may serve, and a new IP address will be allotted to, a UE, by the new UPF, as a result of its movement. Due to this, there would be a break in the continuity of a PDU session service of a UE for a brief period. Service continuity requirements of a UE during its movement are specified in terms of a *session and services continuity (SSC)* which is provided as part of the *PDU Session Establishment Request* made to the 5GC. 5GS supports the following SSC modes for the different types of PDU sessions mentioned earlier. Depending on the service continuity requirements of the various applications being used by a UE, 5G SM defines several PDU session service continuity modes as described below. Note that the SMF network function selects an SSC mode for a UE based on its subscription information.

- SSC mode 1, i.e. there is no change in the serving UPF and, therefore, the IP address/session continuity is preserved during the movement of the UE/user.
- SSC mode 2, i.e. there is a change in the serving UPF. IP address/session continuity is not preserved during the movement of the UE/user. It is a type of *break-before-make* approach where the existing PDU session of a UE and its IP address, associated with the old UPF, is released, followed by the establishment of a PDU session/IP address with a new UPF. It is similar to a hard handover procedure found in a GSM or 4G/LTE network.
- SSC mode 3, i.e. there is a change in the serving UPF. IP address/session continuity is not preserved during the movement of the UE/user. It is a type of *make-before-break* approach where a new PDU session is established and allocates an IP address to a UE with a new UPF before releasing its existing PDU session/IP address with the old UPF. It is similar to a soft handover procedure found in a UMTS network.

An IP-based PDU session can have all the SSC modes mentioned earlier, whereas an Ethernet and Unstructured PDU session can have SSC modes 1 and 2 only. Note that the 5GC may not support the SSC mode as specified by a UE. In this case, the 5GC, i.e. SMF, selects the SSC mode and provides the same as part of *PDU Session Establishment Accept message* from the SMF to UE. Providing the session type information and SSC mode requirement by a UE is optional as part of the *PDU Session Establishment Request,* but it is mandatory to provide them by the 5GC as part of the *PDU Session Establishment Accept message.* Refer to TS 24.501 [47] for more information on these messages.

Figure 18.3 illustrates the different SSC modes – mode 1 and mode 2 – of a UE PDU session in a 5G network where two SSC modes are shown in this figure. As a result of user movement with PDU session SSC mode 1,

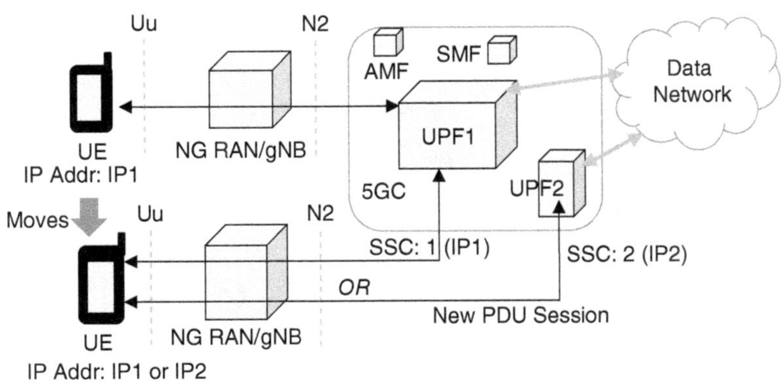

Figure 18.3 Illustration: 5GS session and service continuity (SSC) modes 1 and 2.

UE IP address IP1 will be preserved if the UE is severed by the same UPF: 1 for the same PDU session. On the other hand, in case of SSC mode 2, if the old PDU session was released and UE is served by a different UPF: 2 for a new PDU session over it, a new IP address IP2 will be assigned to the UE, as illustrated in Figure 18.3. Further, the UE is shown to be connected to the same data network in both the SSC modes.

During N26 interface-based interworking from an LTE/EPS to a 5GS network, as illustrated earlier in Figure 16.9, the UE IP address will be preserved and the type of the mapped PDU session will be set to *SSC mode 1,* based on the LTE/EPS PDN type.

18.2.4 PDU Sessions for Network Slices

In the 5G system, each network slice has its PDU session, through which a slice-specific user plan traffic is transferred between UE and its data network. A PDU session is identified through a *PDU session identifier.* A PDU session identifier is included as part of a 5G SM layer message exchanged between a UE and the 5GC SMF. Some of the 5G SM layer messages and procedures were described in Section 18.2.1. Establishing an end-to-end PDU session is the second step toward the activation of a network slice for a particular use case. Before establishing a PDU session, a network slice, i.e. AMF network function, is required to be selected, as part of a UE registration procedure, which is based on its subscription information. A UE subscription information may contain a list of subscribed network slices (S-NSSAIs), and each network slice contains a list of data network names (DNNs) and Local Area Data Networks (LADNs).

Figure 18.4 illustrates three network slices, their associated PDU sessions and associated DNN and LADN for a UE and an IoT device/service over a 3GPP access network (AN), i.e. NG-RAN. This figure also illustrates the allocation of RAN slices, in terms of different gNB-DUs, and CN slices, in terms of different network functions such as AMF, SMF, UPF, and so on for the different PDU sessions being established for different network slices on behalf of the UE and the IoT device.

As illustrated in Figure 18.4,

- A UE establishes multiple, i.e. two, PDU sessions, which are identified by unique PDU session identifiers, to different data networks. Each PDU session is associated with its network slice ID (S-NSSAI) and DNN or LADN.

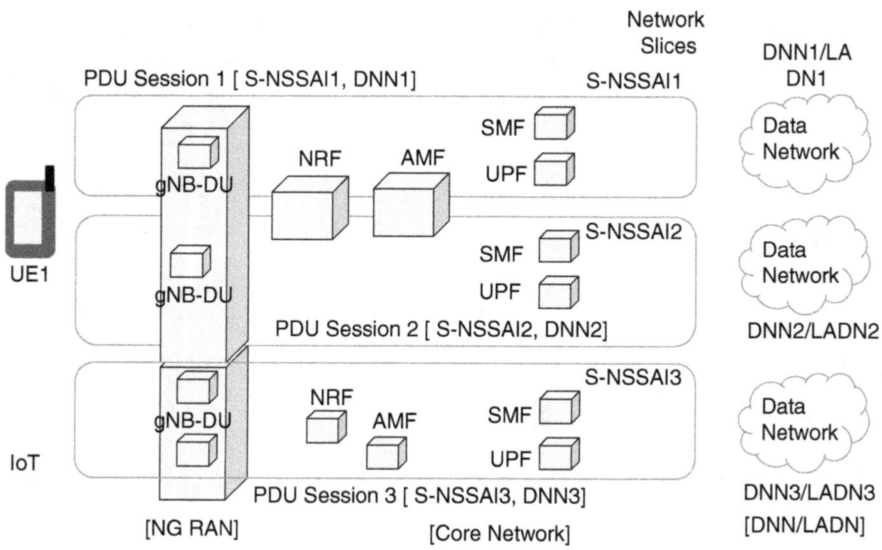

Figure 18.4 Illustration: 5G SM PDU sessions for different network slices.

- For each CN slice part and its PDU session, the SM control plane-related tasks are handled by a separate SMF.
- For each CN slice part and its PDU session, the transfer of data plane or user traffic is handled by a separate UPF. However, though the PDU sessions of a UE are being served by separate SMF and UPF, the access and mobility aspects of the UE for all of its network slices are handled by one AMF and Network Repository Function (NRF) only, as illustrated in Figure 18.4.
- Separate AMF, SMF, and UPF are allocated to the IoT device for its PDU session and network slice.
- At the NG-RAN end, each network slice is assigned a separate RAN slice in terms of gNB-DU node(s). Using the diverse spectrum of the 5G NR, each gNB-DU may perform transmission and reception with radio frequency bands which is suitable and assigned to a particular network slice. As described earlier in Section 16.4, a gNB-DU is the interface toward UEs and other communicating devices, that is, mIoT, URLLC devices, and so on. A gNB-DU may perform a slice-specific radio resource allocation, admission control, and scheduling of UEs and forwards user traffic to UPF for its routing between UE and the corresponding data network.

18.2.5 Session Management (SM) Layer States

Similar to the SM layer of the LTE/EPS system, the 5G SM layer has several states with respect to a PDU session such as the PDU SESSION ACTIVE, and PDU SESSION INACTIVE. SM layer state transition from one state to another state takes place at UE and 5GC end depending on a particular SM layer procedure, for example, a PDU session establishment request, being performed on the PDU session of a UE. For more information on 5G SM layer states and their transitions at UE and 5GC end, refer to TS 24.501 [47].

18.3 Quality of Service (5G QoS)

User plane traffic in the LTE/EPS and 5G network is delivered with different QoS characteristics as per the requirements of different types of services and applications. Though the QoS architecture in the LTE/EPS and 5G system is similar in some respects, it is fundamentally different from their architecture point of view, as described next.

18.3.1 LTE/EPS QoS Model: EPS Bearer

In the LTE/EPS network, the QoS model is based on the *EPS bearer* concept, whereas in the 5G network, the QoS model is based on the *QoS flow* concept. An *EPS bearer* in an LTE/EPS network has an end-to-end (i.e. a Radio Bearer, S1 Bearer and S5/S8 bearer) presence from UE to eNB to the EPC network. An EPS bearer is assigned to user plane traffic or IP packets which are generated from the various application(s) such as e-mail, video streaming, and so on. An EPS bearer is the finest level of granularity where a QoS is enforced. LTE/EPC network assigns a QoS to an EPS bearer in terms of a standardized QoS Class Identifier (QCI) with specific characteristics. A QoS/QCI is assigned to an EPS bearer during its activation procedure toward a UE.

18.3.2 5GS QoS Model: QoS Flow

In the 5G system, the QoS model is based on a *QoS flow* only, i.e. a QoS flow with specific QoS requirement is assigned to user data flow, or IP packets, which is generated from a particular traffic type and traffic class. In the 5G system, QoS flow is the finest level of granularity where a QoS parameters and characteristics are enforced for the transfer of user traffic which is generated from a particular application only. This is in contrast to the QoS assigned to all the packets of different applications mapped to an EPS bearer in the LTE/EPS system. A QoS flow in the 5G network also has an end-to-end presence from UE to gNB to 5G core, i.e. UPF network function.

However, unlike an end-to-end LTE/EPS bearer, in a 5G network, the concept of bearer exists between the UE and its NG-RAN/gNB only, which is called a *DRB* over the NR air interface. Tunneled user plane traffic from a QoS flow is transferred over a DRB created between NG RAN/gNB and UE. User plane traffic from more than one QoS flows can be also tunneled together and send over a DRB.

Between the NG-RAN/gNB and the UPF, a QoS flow is realized using a user plane GTP tunnel, called the N3 GTP-U tunnel. Between the NG-RAN and a UE, a QoS flow is realized through a data radio over the NR air interface. The QoS flows, DRBs, and GTP tunnel information collectively constitutes a *PDU session* for a UE in a 5G network. A UE's *PDU session* is end-to-end, i.e. UE to gNB to 5G core, in nature, which is similar in usages to an EPS bearer in an LTE/EPS network. The establishment or modification of a PDU session toward the 5G CN is initiated by a UE. This is similar to the PDN connectivity establishment procedure which is performed by a UE during its initial CN registration/ATTACH request procedure in the LTE/EPS network. A *PDU session* (UE<->gNB<->SMF) can associate with several QoS flows (UE<->gNB<->UPF) for different types of services of a user. However, there will be only one GTP-U tunnel between gNB and UPF to serve all the QoS flows of different applications and services of a user.

18.3.3 GTP-U Plane Tunnel for PDU Session

As mentioned earlier, an IP data session-related tasks are centered around an EPS bearer in an LTE/EPS network and over a PDU session in a 5GS network. Also, as far as the allocation of a GTP tunnel is concerned, there is a difference. In an LTE/EPS network, a GTP-U plane tunnel is allocated to each EPS bearer, for 1 : 1 association, to transport its user plane data. However, in a 5GS network, a GTP-U plane tunnel is allocated on a per PDU session basis only, for a 1 : M association, between NG-RAN/gNB and UPF. All the 5GS QoS flows, with different QoS, carrying user plane data of a PDU session are served and transported only over a single GTP-U tunnel. The differences in the association of a GTP-U tunnel and QoS to an EPS bearer in the LTE/EPS and 5GS PDU session/QoS flows are illustrated in Figure 18.5.

To identify the user plan data of different QoS flows over a single N3 GTP tunnel, i.e. between the 5G NG-RAN node and the UPF, a QoS flow identity (QFI) is provided as part of the GTP-U extension header type: *PDU Session Container*. Refer to TS 29.281 [72] for more information on this GTP-U extension header type and TS 38.415 [120] for the contents of the *PDU Session Container* header information. Note that security protection, i.e. integrity and confidentiality, on the information exchanged between a UE and the network is provided on a per PDU session level.

18.3.4 Service Data Flow and PCC Rule

One similarity between the LTE/EPS and 5G system with respect to QoS is that they define it at service level also for user plane traffic. QoS control for user plane traffic in terms of service level is defined by the *Policy Control and Charging* (PCC) framework of the LTE/EPS and 5G system. PCC contains several network functions/entities that allow operators to define and enforce QoS control through IP packet detection/classification and also allows charging control on user/subscriber's traffic.

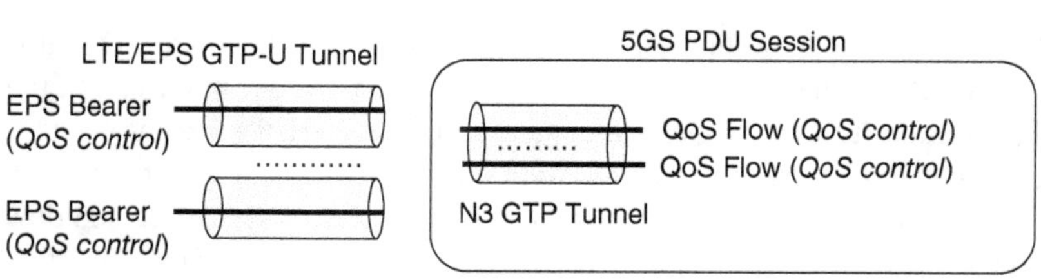

Figure 18.5 Illustration: association/control of QoS – EPS bearer versus QoS flow to 5GS.

In a PCC framework, a *service data flow* (SDF) consists of an end-to-end flow of IP packets from user applications or services which are detected using a packet filter. SDF that serves different applications or services of a subscriber may have the same or different QoS characteristics. Those QoS characteristic requirements are configured and controlled in terms of PCC rules by the *Policy and Charging Enforcement Function* (PCEF) of LTE/EPS network and SMF of a 5G system. For more information on the contents of a PCC rule used in 5GS, refer to TS 23.503 [42].

SDFs of IP packet flows are collectively called an *SDF template* in a PCC framework. An SDF template, which is similar to a traffic flow template in the LTE/EPS, consists of SDF filters. Using SDF filters, IP packet flows are classified and separated into separate SDFs. Such packet filters are configured in the form of a PCC rule by the PCC framework toward the SMF, in 5GS, or Packet Data Network Gateway (PGW), in the LTE/EPS network. A PCC rule determines whether an SDF is allowed to pass through or not by the PCEF (LTE/EPS) or SMF (5GS). If an SDF meets the condition of either policy control or charging control, or policy control and charging control both, then it will be allowed to pass through the UPF, in case of 5GS, and PGW, in the case of the LTE/EPS network.

User plane traffic/IP packets that do not meet the PCC rule are discarded. In 5GS, a PCC rule containing service data filters can be provided dynamically by the Policy Control Function (PCF) to SMF and then further from SMF to UPF in the form of a *packet detection rule* (PDR) over the N4 interface. The UPF uses the PDRs for the detection and classification of user plane traffic into different SDFs along with their mapping into different QoS flows. A PCC rule can be preconfigured also with SMF and then provide to the UPF.

18.3.5 Binding of Service Data Flow

In an LTE/EPS network, an association between an SDF and EPS bearer is performed to transfer user plane traffic over an end-to-end EPS bearer. Such an association between an SDF and an EPS bearer which is performed by the PGW is called a *binding*. Similarly, in the 5G system also, binding is performed between an SDF and a QoS flow of a PDU session by the SMF of a 5G network. To transfer IP packet flows from user applications or services with the same QoS requirements, more than one SDF could be bound to an EPS bearer in the LTE/EPS network. Similarly, in a 5G network also, more than one SDF can be bound to a QoS flow to transfer IP packet flows from user applications or services with the same QoS requirements. SDFs that are bound to an EPS bearer or QoS flow will receive the same treatment of packet forwarding in terms of their scheduling over the NR air interface.

Figure 18.6 illustrates, in general, the binding of SDFs to EPS bearers in an LTE/EPS network to ensure the QoS requirements of IP flows from different applications/services. Similarly, this figure also illustrates the binding of SDFs to QoS flows in case of a 5G system, for a typical IP session only, to ensure the QoS requirements of IP flows from different applications/services. As illustrated in Figure 18.6, SDFs that have the same QoS flow parameter/characteristic requirements are transferred as aggregated flow through an LTE/EPS bearer or 5G QoS flow.

As illustrated in Figure 18.6, IP flow1 from a user application/service is mapped to SDF1 and IP flow2 and IP flow3 are mapped SDF2. Further, the SDF1 and SDF2 are bound to an LTE/EPS bearer or a QoS flow in the case of 5GS.

18.3.6 QoS Profile and QFI

Each QoS flow of a 5GS PDU session is associated with a QoS profile in the downlink direction. A QoS profile is provided by the SMF to the NG-RAN/gNB which is used in the downlink direction to derive packet forwarding treatment and allocate DRB over the NR air interface. Each QoS flow is identified by a unique *QFI,* which is assigned by the SMF along with a QoS profile to the NG-RAN/gNB.

- **QoS Flow-Type, Parameters, and Standardized QoS Characteristics (5QI)**

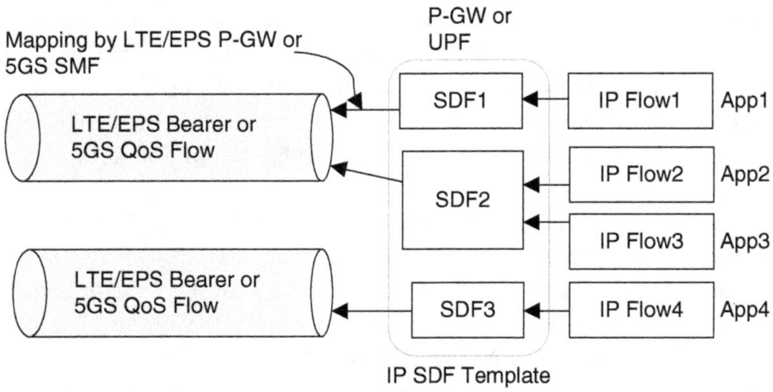

Figure 18.6 Illustration: binding of SDF and LTE/EPS bearer and SDF and 5GS QoS flows.

User plane traffic or IP flows generated from different applications/services may have different QoS requirements. To meet such QoS requirements, QoS flows are defined and assigned with specific QoS characteristics and parameters. Based on such characteristics and parameters, a QoS flow may be classified into a guaranteed bit rate (GBR), non-GBR, or delay critical GBR type.

Depending on the QoS flow type, a QoS profile may contain the following QoS parameters, per QoS flow, which are provided by the SMF to NG-RAN/gNB.

i) GBR QoS Flow Parameters
   - A 5G QoS Identifier (5QI), that points to the QoS characteristics of a QoS flow.
   - An Allocation and Retention Priority (ARP), used for admission control of UE;
   - Maximum Flow Bit Rate (MFBR) for both uplink and downlink directions;
   - Maximum Packet Loss Rate for both uplink and downlink directions;
   - Guaranteed Flow Bit Rate (GFBR) for both uplink and downlink directions;
   - Notification Control; and
   - Delay Critical Resource Type.

ii) Non-GBR QoS Flow
   - A 5QI;
   - An Allocation and Retention Priority (ARP);
   - Reflective QoS Attribute (RQA); and
   - Additional QoS Flow Information.

5QI is a scaler number that has a one-to-one association with a standardized and predefined packet forwarding characteristics/behavior to be applied on a particular QoS flow. The packet forwarding characteristics defined under a standardized 5QI are as follows:

- Priority level;
- Packet delay budget;
- Packet error rate;
- Averaging window; and
- Maximum data burst volume.

Standardized QoS parameters and characteristics mentioned above are defined in TS 23.501 [40]. Together they define a QoS profile of a user application/service. Using a QoS profile, NG-RAN/gNB maps QoS flow(s), either GBR or non-GBR, into a DRB between the gNB and UE over the NR air interface. Refer to TS 23.501 [40] Table 5.7.4-1 for more information on the QoS characteristics of different 5QIs with examples of typical services. From this table, it can be seen that some applications or services may have common QoS characteristics but with different priorities. 5G network should be able to prioritize resource allocation accordingly to meet such QoS requirements.

As an analogy, in an LTE/EPS network, the non-GBR default bearer remains active as long as a UE maintains its connection with the PDN. Similarly, in the 5G system also, a default QoS flow and QoS rule of non-GBR type are allocated to the PDU session of a UE. The default QoS flow remains established as long as the UE PDU session is active.

18.3.7 QoS Rule and QRI

Each QoS flow of a UE PDU session is also associated with a QoS rule in the uplink direction from a UE to the 5GC network. A QoS rule contains a set of packet filters that are provided by the SMF to a UE as part of a PDU session establishment or modification procedure. A QoS rule is used by a UE in the uplink direction to classify and associate uplink user traffic to QoS flows. A QoS rule is identified through an identifier called *QoS Rule Identifier* (QRI).

Some of the information provided by the 5GC network to a UE as part of a QoS rule are mentioned below:

- Type of QoS rule, i.e. a single bit indicating a default or non-default;
- QRI;
- Number of packet filters;
- IP packet filters list;
- Precedence, for evaluation of QoS rules;
- QFI along with QoS parameters; and
- EPS bearer identity, if the QoS is mapped from an EPS bearer.

• **IP Packet Filters for IP Session**

A QoS rule and PDR use IP packet filters to separate user applications/services traffic into different IP flows in the uplink direction, downlink direction, or both directions. IP packet filters contain information based on which classification of user applications/services traffic into different IP flows is performed. To perform such classifications, the header of IP traffic of user applications/services is inspected.

Typical information contained in an IP packet filter are listed below.

- Source IP address and port number;
- Destination IP address and port number;
- Protocol ID of the protocol above IP/next header type;
- Type of service: IPv4/IPv6;
- Flow label (IPv6);
- Security parameter index; and
- Packet filter direction.

18.3.8 Mapping QoS Flow to Data Radio Bearer

The Service Data Application Protocol (SDAP) sublayer of the NR air interface protocol stack, at UE and gNB end, performs the mapping and multiplexing, if required, of QoS flows into a DRB for transfer of user plane traffic over the NR air interface. A DRB is used to exchange data between the SDAP and PDCP layers of the NR Layer 2 protocol stack. The NR RRC layer at gNB end provides such mapping/multiplexing of information to the RRC layer at the UE through IE: *sdap-Config* which contains information such as the PDU session-id and QoS flows to be mapped with a DRB. All the packets of a QoS flow, of 5G core, will receive the same QoS treatment on a DRB over the NR air interface.

Figure 18.7 illustrates the partial message flows as part of a UE-initiated typical session establishment procedure with 5GS core. This illustration covers the following concepts:

a) Control Plane Signaling: UE-gNB-AMF
   - To establish an end-to-end PDU session (shown as a rounded rectangle), enabling the transfer of uplink/downlink user traffic for typical applications like www, FTP;
   - To allocate QoS flows/QFIs, say QFI: x and QFI : y, to transfer IP traffic flows.
b) User Plane: QoS flow to DRB mapping
   - Mapping of the DRBs, say DRB : x DRB : y, to the allocated QoS flows/QFIs: x and y, at the SDAP layer through RRC layer signaling message between 5G RAN/gNB and UE. Mappings between QoS flows and DRB over the NR air interface are established to transfer user plane traffic between UE and 5GC, i.e. UPF. Mapping from IP flows through SDF to QoS flow was illustrated earlier in Figure 18.6.
c) PDU Session

As illustrated in Figure 18.5, the UE PDU session consists of the QoS flows, DRBs, and the N3 GTP-U tunnel to enable user plane data transfer between UE and 5G CN.

A PDU session can be an IPv4, IPv6, IPv4v6, Ethernet, or unstructured type. PDU session-related user plane resource allocation and its setup information such as the QoS flows to be set up, session type, and QoS flow QoS parameters are provided by the SMF to the AMF. The AMF then transparently forwards that session-related information to the NG-RAN/gNB through the *PDU Session Resource Setup Request* message. The NG-RAN/gNB provides session-related information further to a UE through the RRC layer signaling message. The NG-RAN/gNB provides the list of successfully established QoS flows to the AMF through the *PDU Session Resource Setup Response* message.

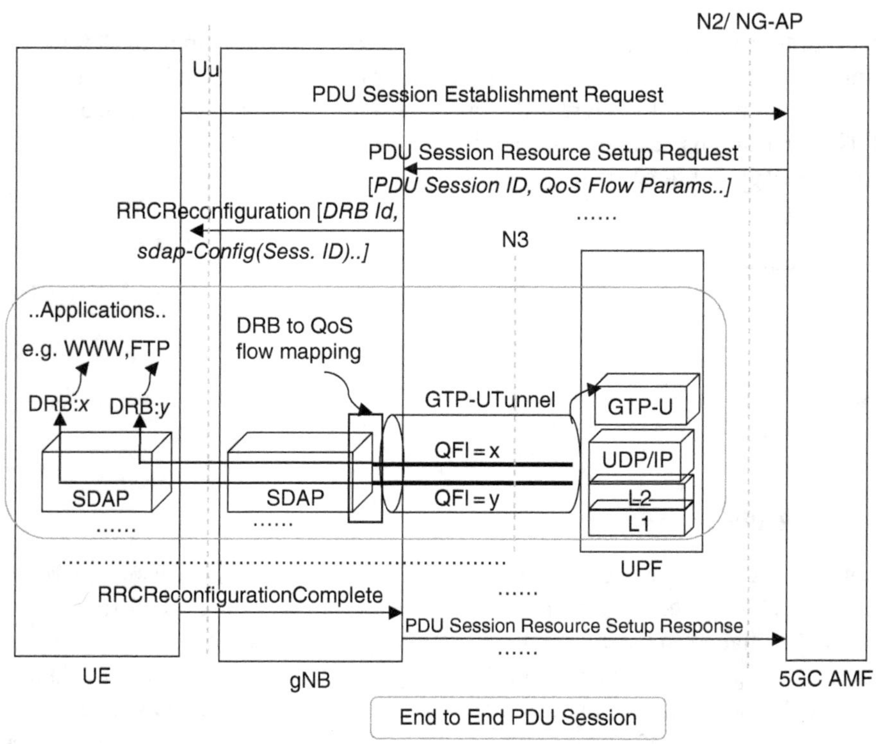

Figure 18.7 Illustration: SDAP layer: mapping between QoS flow and data radio bearer.

From the above descriptions and illustrations, it is seen that different entities such as UE, gNB, AMF, SMF, UPF, and PCF are all involved to enforce the QoS requirements in a 5G network for different types of traffics generated by user applications. 5GS QoS-related typical definitions in the software code are illustrated later in Chapter 21.

Next, a downlink data flow from 5G CN to NG-RAN for a particular PDU session through different GTP-User plane tunnels is described.

18.3.9 Downlink Data Flow Through GTP-U Plane Tunnels

Figure 18.8 illustrates the user application traffic flow in the downlink direction through GTP-user plane (GTP-U) tunnels created between the concerned network entities or logical nodes. Forwarding of the tunneled data to UE through a DRB over the NR air interface is also shown in Figure 18.8. The GTP-U plane protocol is used to transfer user application traffic over a GTP tunnel created between two network entities as mentioned below:

- Anchor UPF and forwarding UPF over the N9 reference point;
- Forwarding UPF and NG-RAN over the N3 reference point and NG-user plane interface; and
- Logical nodes: gNB-CU-UP and gNB-DU over the F1-user plane interface.

Figure 18.8 also illustrates user data flow through an anchor UPF as more than one or the intermediate UPF(I-UPF) can be deployed in a 5G CN.

Also, as mentioned earlier, in a 5G system, different IP traffic flows are classified and separated, or transferred together, into different QoS flows with their respective QoS characteristics. All such QoS flows are transferred through a GTP-U tunnel that is created between two network elements or network functions or logical nodes as mentioned and illustrated in Figure 18.8. The GTP-U tunnel created between the forwarding UPF and gNB over NG-U is associated with an end-to-end PDU session which is established between a UE and its 5G CN, i.e. SMF. Thus, there is a 1 : 1 relationship between an N3 GTP-U tunnel and an end-to-end UE PDU session. Such a

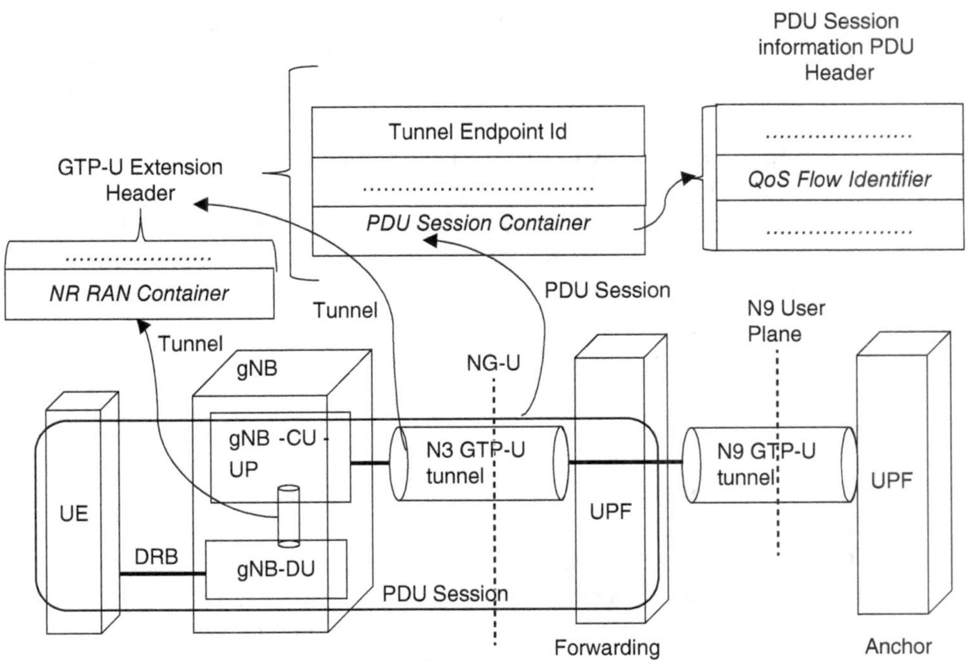

Figure 18.8 Illustration: 5GS downlink data flow through GTP-U tunnels.

relationship between a GTP-U tunnel and a PDU session is indicated by the presence of a new field, which is called a "PDU Session Container". A PDU session container field is carried as part of the GTP-U extension header, refer to TS 29.281 [72], of GTP-U protocol. A *PDU session container* is defined as part of PDU session user plane protocol over NG-U interface between NG-RAN and UPF; refer to TS 38.415 [120]. Further, to associate with and map the different QoS flows being tunneled over the N3 GTP-U tunnel, a "QoS Flow Identifier" field for each QoS flow is also carried within a *PDU Session Container->PDU Session information PDU Header*; refer to TS 38.415 [120].

At NG-RAN/gNB end, GTP-U plane data from 5G core is handled by the gNB-CP-UP logical node, more specifically, the SDAP layer. The gNB-CP-UP tunnels the user data to the gNB-DU logical node over a GTP-U tunnel created between them. However, the user plane data is passed through an "NR RAN Container" which is carried as part of the extension header of GTP-U protocol for the GTP-U tunnel created between gNB-CP-UP and gNB-DU. All such associations and mapping are illustrated in Figure 18.8 through curved arrows.

On the 5G core side, an N3 GTP-U tunnel over the NG-U interface is associated with an end-to-end PDU session. However, over the air interface, all the QoS flows tunneled by the N3 GTP-U tunnel are associated and transferred over a DRB, as illustrated in Figure 18.8. In this illustration, all the QoS flows are illustrated to be transferred over a single DRB only to a UE. QoS flows may be also transferred over multiple DRBs to a UE.

18.4 Mobility Management (5G MM) Layer

18.4.1 Mobility Area Concepts and Identifiers

In this section, the MM layer-related logical location areas (LAs) and their identifiers shall be described first as they are important from the MM layer implementation point of view. Location or mobility areas used for the MM purposes in the legacy systems, i.e. GSM, GPRS, UMTS, and LTE/EPS, are also described. A particular mobility or LA tracks the current LA of an MS/UE when moving around a network. Also, due to inter-system mobility requirements or based on the UE mode of operations, a UE may be required to perform combined, i.e. GSM, GPRS, MM, procedures toward the CN. For example, a GPRS MS may perform a combined LA and routing area (RA) update procedures to the CNs. Similarly, an LTE UE may perform a combined LA and tracking area update (TAU) procedures to update its current location to the circuit-switched (CS) and packet-switched (PS) CNs.

Each mobile communication system and the network define its CS domain and PS domain service-specific mobility areas. A GSM LA is used by the Mobile Switching Center (MSC) to notify an incoming call to an MS/UE. Similarly, a GPRS/UMTS PS domain RA is used by the SGSN to send and receive data to/from an MS/UE. Thus, different mobility areas are identified by their respective identifiers. Based on such mobility area identifiers as well as due to a user movement, an MS/UE updates its current location to its CN. Such dynamic updates of a UE mobility area can be event based, e.g. a user movement, or timer based, e.g. periodic. Common and 3GPP system-specific (mentioned within square brackets) mobility areas are described below:

- Public Land Mobile Network (PLMN)

A PLMN is a collection of interconnected network elements that provide various communication services such as voice, data, and multimedia services. The network elements could be GSM MSCs in the case of CS domain, SGSNs in the case of PS domain GPRS/UMTS system, or Mobility Management Entity (MME) in the case of the LTE system. A PLMN uniquely identifies a network operator. Within a PLMN, an operator can provide both CS and PS services. There could be several PLMNs from different telecom network operators offering various services in a particular geographical location/area. This raises the need for the unique identification of a PLMN of a particular operator. A PLMN consists of the combination of the following identities:

- Mobile Country Code (MCC) – Length: 3 octets
- Mobile Network Code (MNC) – Length: 2 or 3 octets.

PLMN	MCC	MNC

PLMN is one of the important identities as it is used to derive other network identities such as the International Mobile Subscriber Identity (IMSI), LTE/EPS Globally Unique MME Identity (GUMMEI), LTE E-UTRAN Cell Global Identity (ECGI), GSM Location Area Identity (LAI), and LTE/EPS Tracking Area Identity (TAI). An MS/UE performs a PLMN selection upon its power on and register with the PLMN. A PLMN selection could be either automatic or manual. There are two types of PLMN, as mentioned below:

- Home PLMN (HPLMN)

The PLMN of an MS/UE, which is read from its IMSI, is the same as the one currently being broadcasted by the RAN in a cell.

- Visited PLMN (VPLMN)

The PLMN of an MS, which is read from its IMSI, is not the same as the one currently broadcasted by the RAN in a cell. This MS may have entered into a network of the same operator or a different operator in another area, with an MNC which is different from the MS/UE home network.

- Cell

The cell is an area of radio coverage identified by a particular base station of a cellular system to offer various communication services to subscribers. Whenever a mobile device is in the RRC layer connected state, its current location is known to the network at the *cell level*. A cell in a mobile communications network can be in any one of the following states:

- Suitable cell

A "suitable cell" is a cell on which the MS/UE may camp on to obtain normal communication services.

- Acceptable cell

An "Acceptable cell" is a cell on which the MS/UE may camp on to obtain special communication service only, e.g. emergency call.

- Barred cell

On a *barred cell*, camping of an MS/UE is not allowed.

- Reserved cell

On a reserved cell, camping of an MS/UE is not allowed, but a selected UE may be allowed to camp on. The above states and different types of cells are illustrated in Figure 18.9.

Different types of cells described above are grouped into various mobility areas which are known by different names across the communication networks, both in the CS domain and PS domain, as described below. Also, each cell is identified by a *cell identity* (CI).

- LA (GSM and UMTS CS domain)

The LA defines a coverage area for the CS domain only in which a mobile station may move freely without subsequently updating its current location to the CN, i.e. MSC/Visitor Location Register (VLR). MSC sends a page

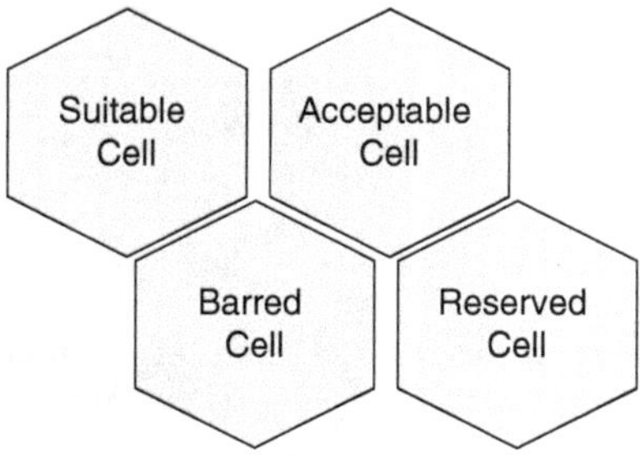

Figure 18.9 Different states of a cell in a cellular communication.

notification to MS at the LA level only. An LA is a grouping of one or several GSM cells. An LA is identified by a *Location Area Identification (LAI)* number and consists of the following information:

- MNC
- MCC
- LAC – Length: 2 octets

LAI	MCC	MNC	LAC

- RA (GPRS and UMTS PS domain)

The RA defines a coverage area in the PS domain only in which a mobile station may move freely without subsequently updating its current location to the CN, i.e. SGSN/VLR. SGSN sends a page notification to an MS at RA level only. An RA is a grouping of one or several GPRS cells. An RA is identified by a *Routing Area Identification (*RAI*)* number and consists of the following information:

- LAI

RAI	LAI	RAC

- RAC – Length: 1 octet

- Tracking Area (TA) (LTE/EPS and 5G system)

A *TA* concept in an LTE/EPS or 5G system is similar to an LA in the CS domain or RA in the PS domain only that includes one or several cells. In the case of LTE/EPS, a TA is the coverage area in which a UE that does not require a subsequent update on its current location to the MME may move freely. However, in the case of 5GS, TAs are organized into another new MM concept called *Registration Area*, which shall be described later in this section. A TA is identified by a *Tracking Area Identification* (TAI) number. A TAI consists of the following information:

- MCC and MNC
- Tracking Area Code (TAC) – Length: 2 octets.

TAI	MCC	MNC	TAC

The various mobility areas found across the different 3GPP systems as described above are updated from a UE/MS to its CN through the respective MM procedures. Such mobility areas updates by a UE/MS to its CN can take place either periodically or in an event-based manner.

- Cell Global Identity (CGI) (LTE/EPS and 5G System)

A *CGI* is derived from a PLMN identity and CI, as shown below:

CGI	MCC	MNC	CI

In the LTE/EPS network, a CGI is known as the E-UTRAN CGI or ECGI. In the 5G network, a CGI is known as the NR CGI or NCGI. A CGI can be used during an inter-radio access technology (RAT) handover procedure to identify the target network.

- Hierarchical Relationships Among Mobility Areas (GSM, GPRS, UMTS, and LTE/EPS)

From the above descriptions, it appears that a mobility identity is derived from several other mobility identities indicating hierarchical relationships among them. The fundamental mobility area identity used is the PLMN, consisting of the MCC and MNC. Based on these, the hierarchical relationships among the logical concepts and mobility areas described above are illustrated in Figure 18.10. These hierarchies of mobility areas are used to model their actual implementation in the software.

In Figure 18.10, cells are grouped into an LA or RA or a TA with a unique ID. On the right-hand side of Figure 18.10, the LA is being shown as part of LTE PLMN to support for fall back to the GSM CS domain. An LA may cover several RAs and TAs (1 : N), but an RA or TA can associate with only one LA.

- Mobility Areas Controlled by Network Elements (GSM, GPRS, UMTS, and LTE/EPS)

In the preceding paragraphs, we have described the basic mobility areas found in a GSM, GPRS, LTE/EPS, and 5G network. A TA concept is applicable for PS calls only in an LTE/EPS that is controlled by an MME. A TA in the 5G network is controlled by the AMF. Similarly, the area controlled by an LTE/EPS S-GW is known as the S-GW area. The hierarchical relationship of the different mobility areas used in GSM, GPRS, and LTE/EPS systems which are controlled by a particular CN element is illustrated in Figure 18.11. These hierarchies are used to model their actual implementation in the software:

  - GSM CS and GPRS PS domain mobility areas, refer to Figure 18.11.
  - LTE/EPS mobility areas controlled by its network elements, refer to Figure 18.11.
- MM for CS and PS Domain (GSM, GPRS, UMTS, and LTE/EPS)

Figure 18.10 illustrates the hierarchies of the different MM areas used in GSM, GPRS, UMTS, and LTE networks for the CS and PS domains. Figure 18.12 further illustrates these MM areas under the CS and PS domain service categories along with their corresponding temporary identity used by the respective CN to address a UE.

TMSI and P-TMSI identities are used in legacy GSM and GPRS networks. However, an LTE UE requires to support these legacy network identities also to enable inter-RAT mobility of a mobile user. In an LTE/EPS network, an S-TMSI is allocated to a UE by the MME and is used during an MM procedure toward the UE. Also, a

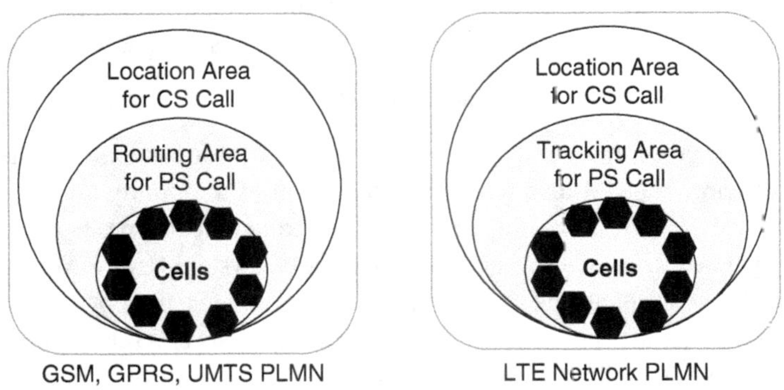

Figure 18.10 Illustration: GSM, GPRS, UMTS, LTE PLMN, and LA/RA.

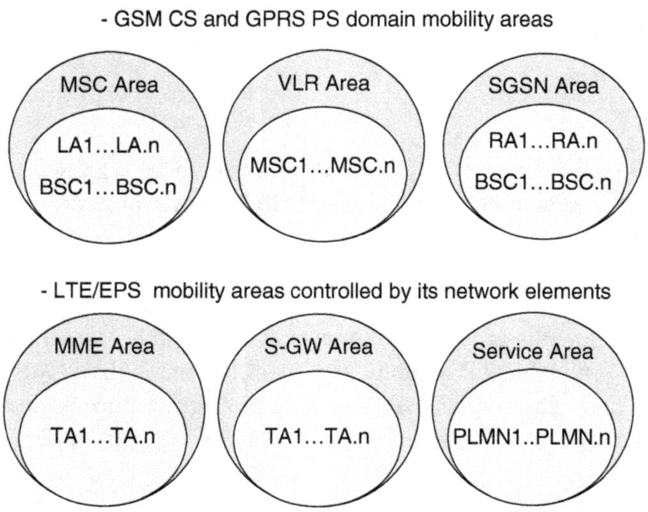

Figure 18.11 Illustration: Mobility areas controlled by core network elements.

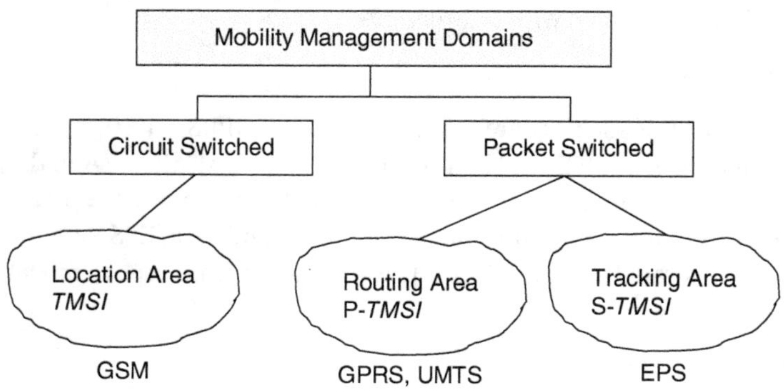

Figure 18.12 Illustration: mobility management areas for CS and PS domains.

temporary identity may be valid for a particular location or routing or TA only as long as its MM context is also valid at the CN end.

A new temporary identity is allocated again to an MS/UE as a result of new network registration with the CN.

- TA List [LTE/EPS and 5G System]

Unlike the GSM/UMTS system, the LTE/EPS introduces the concept of the TA list that may contain a list of TAs. The MME allocates the TA list to a UE whenever it registers itself with the network. Figure 18.13 illustrates the TA list concept from the LTE/EPS point of view. TAs in a TA list is served by the same MME.

The provisioning of multiple TAs in a TA list is an LTE/EPS enhancement from the previous cellular systems.

- 5GS Mobility Area: Registration Area

The preceding paragraphs described the MM areas used by the previous generation mobile communication systems. 5G MM also reuses the definition of a mobility area at the TA level, from the LTE/EPS, as described above. However, one more mobility area, called the *Registration Area (RA)*, above the TA, is defined by the 5G MM layer. An RA is a new mobility area in the 5G system NAS/MM layer and does not exist in the LTE/EPS MM layer. A TA consists of one or several cells. A TA is identified by a *Tracking Area Identifier (TAI)* which consists of a PLMN identity and TAC. Similarly, an RA consists of a set of TAs that is provided as a TAI list to a UE. The usages of these 5G mobility areas are illustrated in Figure 18.14 with three TAs.

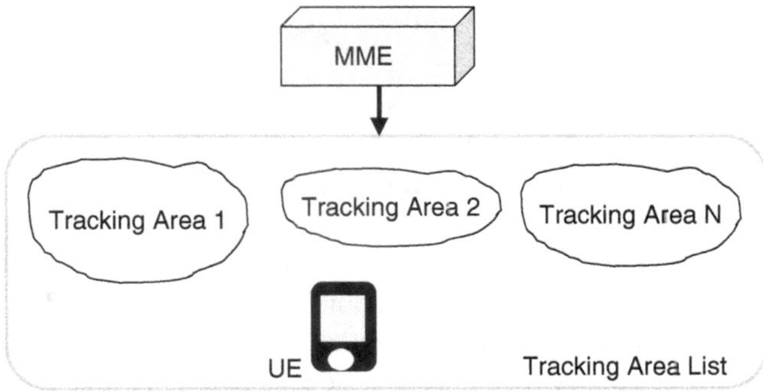

Figure 18.13 Illustration: LTE/EPS: list of TAs in the TA list.

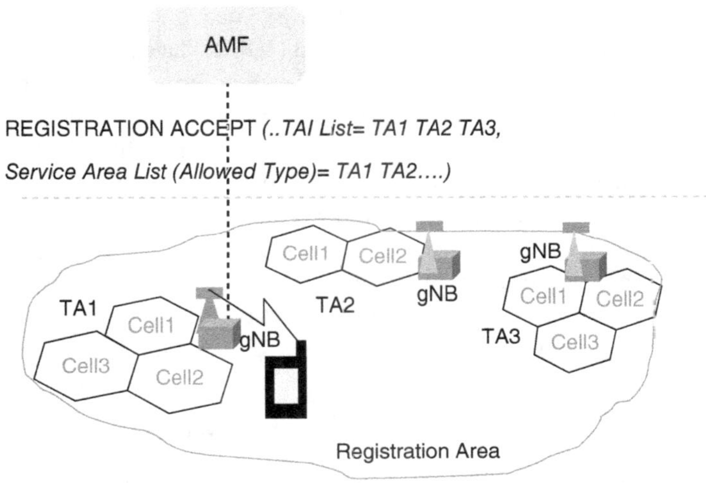

Figure 18.14 Illustration: 5G mobility areas - TA and RA.

A registration area used for MM purposes in a 5G system is similar to an LA in the GSM system, RA in GPRS/UMTS system, and a TA in the LTE/EPS. However, an MM area, i.e. an RA, in 5GC is larger in terms of coverage, as illustrated in Figure 18.14, than the MM areas of the previous generation mobile communication systems, thus reducing the NAS layer signaling overhead in the 5G system. Because of a larger registration area with a set of TAs in the TAI list, a UE does not require to perform a *Registration Area* update procedure to 5GC as long as the UE's current location is within its TAI list. The reader can compare with the TA update procedure, with respect to NAS signaling overhead, which is performed in the LTE/EPS network.

18.4.2 Requirements of Mobility Management Functions

Support of mobility features and their capability provided by the mobile communications systems and networks differentiates them from a fixed-line network. The purpose of the MM functions is to ensure communication services toward a mobile user by maintaining the current location of an MS/UE within a mobile network.

A mobile device may move freely within the logical areas, e.g. GSM LA GPRS RA, LTE/EPS TA, and 5G RA, as mentioned above. It is the responsibility of the mobile device to inform the core network about its current location, either periodically or whenever it detects a new logical area. As long as a mobile device does not cross a particular coverage area, i.e. GSM LA, GPRS RA, or LTE TA, or 5GS RA or enters a different PLMN, it is not required to inform and update the core network about its current location information. This is possible only when the CN learns dynamically and has up-to-date information about the current location of the concerned mobile device being contacted. Both the MS/UE and the CN perform the mobility management function dynamically to ensure that the network can contact a mobile device always, enabling an anytime–anywhere communication service to mobile users.

Mobility management is also required because a mobile user can roam freely with applicable mobility restrictions based on user subscription among the PLMNs or even between different cellular systems/networks or a specific Local Area Data Network (LADN), e.g. a large office. For example, an LTE UE may leave an E-UTRAN coverage area and enter a GSM/GPRS Edge Radio Access Network (GERAN) network coverage, or a 3GPP-compliant MS/UE may enter a non-3GPP communication technology domain such as CDMA2000. Mobility management also deals with the mobility restriction of a UE in terms of forbidden area, RAT restriction and core network restriction that is based on a user subscription. Another UE mobility restriction concept is the service area restriction, which is new to the 5G system. In the case of the GSM CS domain, user mobility requirements are taken care of by the MM layer; for GPRS/UMTS, it is the GPRS Mobility Management (GMM)/Packet Mobility Management (PMM) layer; in the LTE/EPS network, it is the EPS Mobility Management (EMM) layer; and in the 5GS, it is the 5G Mobility Management (5GMM) layer.

Similar to the LTE/EPS, the 5GMM layer performs the MM functions for the PS domain services only in the 5G system. The 5GMM NAS layer terminates at AMF end, at the 5GC, which performs all the 5GMM-related procedures with UEs. At the RAN level, it is the responsibility of the mobile device to select or reselect a suitable cell based on a cell selection or reselection criteria in its idle state. In the connected state of a mobile device, a handover procedure is performed to move the call into the target cell or system.

To ensure anytime–anywhere mobile communication services to users, the 5G system MM tasks perform several functions and procedures which are similar to the MM tasks performed by the LTE/EPS system. In the next section, some of the MM functions and procedures performed by the 5GC MM layer shall be described. MM functions and procedures performed by the NG-RAN/gNB, more specifically, the RRC layer, shall be described in Section 18.5.

18.4.3 Functions and Procedures of 5G MM Layer

Similar to the LTE/EPS MM layer, the 5G MM layer also performs the following MM layer procedures of different types. A 5GMM MM procedure may be initiated either by the 5GC network or the UE.

- Common procedures, such as the Primary Authentication (*Authentication Request/Response*), as described earlier in Section 16.11, UE identification procedure (*Identity Request/Response),* security mode procedure (*Security Mode Command/Complete),* and so on.
- Specific procedures, such as UE Registration (*Registration Request/Accept)* and De-Registration, which are similar to the UE ATTACH and DETACH procedures in the LTE/EPS system.
- Connection management (CM) procedures (*Service Request/Accept, Paging Request),* which are used to make service requests by the UE to 5GC to transfer either uplink or downlink data, by the 5GC to UE. This procedure is also used by a UE to respond to a paging message sent by the 5GC.

UE initial and periodic registration procedures are described below as part of the specific procedures of the 5GMM layer.

- UE Initial Registration Procedure

For normal communication services, a UE requires to register first by sending a *REGISTRATION REQUEST* message, with the IE: *5GS registration type set to* "initial registration", which is a specific 5GMM procedure, to the 5G CN; refer to TS 24.501 [47]. This is similar to the ATTACH request message used to register a UE in the LTE/EPS system. After due identification, authentication, and authorization of the registering UE, which are 5GMM common procedures, and the creation of a PDU session for it, a *REGISTRATION ACCEPT* message is returned from the 5GC/AMF to the UE. The *REGISTRATION ACCEPT* message contains, among other information, the TAI list and allowed/restricted service area list. The AMF manages the allocation of a registration area to a UE in terms of allowed TAs in a TAI list. But a TAI list may contain up to 16 TAs. Accordingly, the 5GC/AMF pages a UE in all the cells of the TAs indicated in the TAI list, which is similar to a UE paging in the LTE/EPS.

It may be noted that there is no TA update procedure in the 5G system even though a UE *Registration Area* contains TAs. Instead, in a 5G system, a *Registration Area update* procedure is performed as mentioned below.

- UE Periodic Registration Procedure

Similar to a TA update procedure in the LTE/EPS, in 5GS also, a UE performs a periodic registration area update procedure, within the same RA, at the expiry of its timer *T3512*. However, unlike the LTE/EPS, there is no separate NAS layer message for a periodic registration area update procedure in the 5GS. To perform a periodic registration area update procedure in 5GS, the UE will send the *REGISTRATION REQUEST* message, with the IE: 5GS *registration type set to* "periodic registration updating".

- UE Mobility Registration Procedure

Because of a user movement, if a UE enters a new TA which is not in its AMF-allocated TAI list, the UE will perform a Mobility Registration Procedure toward the 5GC using the *REGISTRATION REQUEST* message, with the IE: 5GS *registration type set to* "mobility registration updating".

For more detailed information on the UE registration procedure along with abnormal scenarios, refer to TS 24.501 [47].

- UE Registration Management (RM) States

A UE RM has two associated states, new in the 5G MM layer, which are based on the outcome of the registration procedure. These states are described below with respect to a UE.

- RM-DEREGISTERED: This is an initial state, i.e. a UE is not known to the AMF, and hence its current mobility area, for packet routing. This state may be also entered when the AMF rejects a UE *REGISTRATION REQUEST*.
- RM-REGISTERED: This is the state when the UE's initial *REGISTRATION REQUEST* is accepted, and hence, the AMF knows the UE's current mobility area for packet routing. In this state, a UE performs the periodic registration area update procedure at the expiry of its timer T3512 or registration area update procedure when it detects new TAs as a result of user movement.

18.4.4 Mobility Management Layer States

As a result of a successful UE initial registration with the 5GC, an MM context is created at the UE as well as the AMF end. A UE MM context contains various information such as 5G Globally Unique Temporary Identifier (5G-GUTI), allocation of the registration area, and so on that are part of the *REGISTRATION ACCEPT* message from AMF to a UE. Based on the outcome of the UE registration procedure, 5GMM also has two states, as described below, with respect to the creation of a UE MM context:

- 5GMM-DEREGISTERED, where no MM context exists at AMF end; thus, the UE is not reachable. A UE may enter into the 5GMM-DEREGISTERED if a UE initial registration request is rejected by the 5GC/AMF or deregistered by the UE itself or 5GC. The corresponding MM state in the LTE/EPS is EMM-DEREGISTERED.
- 5GMM-REGISTERED, where an MM context exists at UE and AMF ends along with one or more PDU sessions. Thus, a UE is ready for communication services, such as service request procedure, paging response, and registration update procedure, in the 5GMM-REGISTERED state. The corresponding MM state in the LTE/EPS is EMM-REGISTERED.

There will be co-ordinations between the corresponding MM states mentioned above when interworking on behalf of a UE takes place between 5GS and LTE/EPS and vice versa over the N26 interface. 5G MM layer has two more states – MM-IDLE and MM-CONNECTED – as far as the CM layer is concerned. MM-IDLE state corresponds to the idle state of CM, and the MM-CONNECTED state corresponds to the connected state of CM. CM in the 5G system is described next.

18.4.5 Connection Management (CM) and Service Request

5GC *CM* deals with the establishment and releasing of an end-to-end NAS signaling connection over the N1 interface between a UE and AMF. An end-to-end ANS signaling connection is required to exchange signaling messages between a UE and AMF. Similar to end-to-end signaling in the LTE/EPS system, an end-to-end NAS signaling connection in the 5G system over the N1 interface between a UE and AMF consists of the following connections:

- An RRC connection over the NR air interface (Uu) between a UE and the NG-RAN/gNB.
- An RRC NG-AP association/connection over the N2 interface between NG-RAN/gNB and the 5GC/AMF.

Similar to the 5G MM states, CM also has two states at the UE and the 5GC/AMF ends, depending on the availability of a NAS signaling connection between them.

- CM-IDLE, where a NAS signaling connection over the N1 reference point does not exist between UE and the 5GC/AMF, and
- CM-CONNECTED, where a NAS signaling connection over the N1 reference point exists between UE and the 5GC/AMF.

CM state transition from CM-IDLE to CM-CONNECTED takes place whenever a UE NAS MM layer makes a *SERVICE REQUEST* procedure to the 5GC/AMF, as illustrated in Figure 18.15, to send signaling information or transfer uplink data. Similarly, the 5GC may also trigger the establishment of service toward a UE, for example, a paging message to a UE in which case the UE responds to the paging message by sending a *SERVICE REQUEST* to the 5GC/AMF, or transfer downlink data to a UE.

CM state transition from CM-CONNECTED to CM-IDLE state takes place whenever the RRC layer signaling connection or NG-AP signaling connection is released. For detailed signaling flow for UE as well as 5GC-initiated service request procedure, refer to Section 4.2.3, TS 23.502 [41].

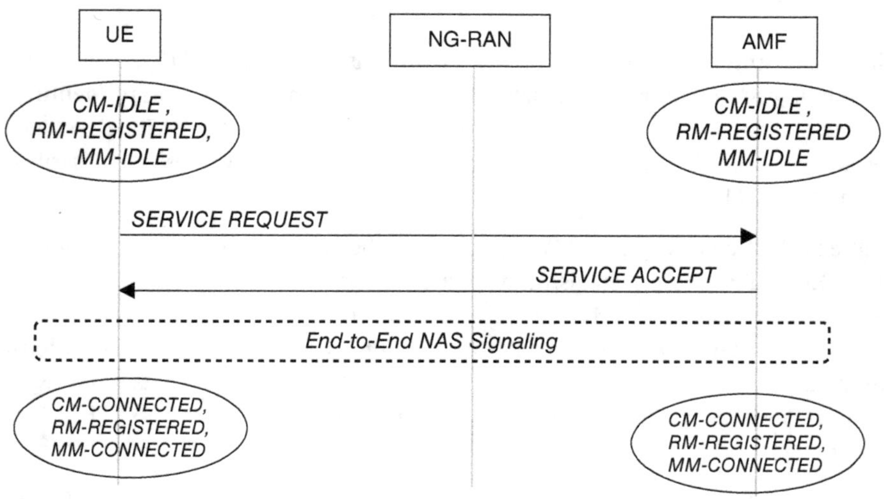

Figure 18.15 Illustration: CM and MM state transitions in a UE-initiated service request procedure.

Figure 18.15 illustrates the CM, RM, and MM state transitions before and after a UE-initiated service request procedure and its acceptance by the 5GC/AMF. Also, apart from the MM-REGISTERED and MM-DEREGISTERED states of the 5G MM layer mentioned above, it has the MM-IDLE and MM-CONNECTED states at UE and AMF ends depending on the availability of a NAS signaling connection between UE and 5GC/AMF.

Overall, the 5GS NAS layer service request procedure is similar to the LTE/EPS service request procedure.

18.4.6 Mobility Pattern of UE

It was mentioned above that a registration area reduces the NAS layer signaling overhead in the 5G system because a large registration area may consist of a set of TAs (up to 16) in the TAI list resulting in a large coverage area. Due to this, a UE does not require to perform a Registration Area Update procedure to 5GC as long as the UE's current location is within the TAI list. However, if the 5GC wishes to contact a UE, it has to be paged in all the cells of the TAs of the registration area, thus generating paging signaling messages overhead across the registration area. To reduce further such signaling overhead, registration area allocation, in terms of TAs, by the AMF is optimized. Such MM procedure optimization method is based on the *Mobility Pattern* of a UE, which is a new concept in the 5G system. For example, for fixed Internet of Things (IoT)-based devices whose mobility is less or fixed, a registration area with an appropriate number of a TA(s) and cell(s) under it may be allocated by the AMF so that network signaling overhead is reduced.

The 5GC/AMF learns about the mobility pattern of a UE which may be based on the services subscribed by a UE, its history of mobility, network local policy, etc. Refer to TS 23.501 [40] for more information on mobility patterns of UE used by 5GC/AMF.

Next, the NR RRC layer is described as part of the AS protocol between UE and NG-RAN.

18.5 RRC Layer

Similar to the previous generation mobile communication system, the NG-RAN RRC layer is the central controlling layer of NR air interface radio resources. Air interface resource allocation to a UE consists of various radio resources allocated from the physical layer, Medium Access Control (MAC) layer, and Radio Link Control (RLC) layer, which is similar to the LTE system. The NR RRC layer is in charge of these Layer 2 sublayers and configures them for various radio resources to be allocated to a UE for different use cases. In addition to this, the RRC layer performs other functions and procedures also, which resembles the LTE RRC layer, which is described next.

18.5.1 Functions and Procedures of RRC Layer

Some of the functions and procedures performed by the RRC layer are as follows:

- Minimum System Information (SI) broadcast in the downlink direction for synchronization and physical layer cell access by UEs. Additional SI is also provided by the RRC layer on-demand request from a UE;
- Establishment, maintenance, and release of RRC connection between UE and NG-RAN for standalone as well as between NG-RAN and E-UTRAN for non-standalone deployment, i.e. E-UTRA NR Dual-Connectivity (EN-DC), solution;
- Support of UE mobility function through handover procedure within the NR NG-RAN or inter-RAT, i.e. LTE E-UTRAN;
- Forwarding of NAS message between UE and the 5GC/AMF;
- Radio link failure detection and recovery;
- UE RF measurement reporting and its configuration using RRC reconfigurations;

- UE beam-level measurement reporting, which is a new NR RRC layer;
- UE paging by NG-RAN;
- Radio bearer management (signaling and data)-related tasks such as its establishment, configuration using RRC reconfigurations, maintenance, and release;
- UE security-related tasks in terms of activation of ciphering and integrity protection of control as well as user plane information between UE and NG-RAN using security mode command.

18.5.2 System Information (SI) Broadcast

Similar to the LTE system, in the NR system also, the downlink SI is transmitted in the form of information blocks carrying different types of information for UEs. But unlike the LTE system, the downlink (SI) is divided into two categories – minimum SI and other SIs. The master information block (MIB) and the system information block1 (SIB1) are part of the minimum SI, and SIBs: 2 to–9 are part of the other SIs. Similar to the LTE system, the MIB is transmitted, every 80 ms, using the BCCH logical channel, Broadcast Channel (BCH) transport channel, and PBCH physical channel. SIB1 and other SIs are transmitted over the scheduled physical downlink shared channel. The MIB carries the following minimum SI which enables a UE to decode and process the SIB1.

- System Frame Number (SFN);
- *subCarrierSpacingCommon,* which contains the subcarrier spacing or numerology to be used for transmission of SIB1, RRC Layer 3 messages for UE to make requests for on-demand SI message from NG-RAN, and UE paging by NG-RAN;
- *Pdcch-ConfigSIB1* contains a common *Control Resource Set* (CORSET 0), a common search space that indicates how and where to search for a Physical Downlink Control Channel (PDCCH) containing downlink control information for scheduling of SIB1; and
- Cell Barring Status, which contains the current status of the cell UE trying to access.

Similar to the LTE system, the NR SIB1 contains cell access and cell selection-related information. However, the NR SIB1 contains information that differentiates it from the LTE SIB1 type, as described below.

- **Periodic and On-Demand SIs: Difference Between LTE and NR**

In the LTE system, the SI is broadcasted by the E-UTRAN periodically in the downlink direction, which means systems information is always available from E-UTRAN even if there is no UE in its cell. However, in the NR, a different approach has been adopted by the 3GPP where the other SIs, excluding the MIB and SIB1, may be transmitted either periodically by the NG-RAN or on-demand requests made from a UE, thus reducing the power consumption at the base station of the NG-RAN. The other SIs may be transmitted either periodically by the NG-RAN or on demand request from a UE. The NG-RAN indicates the on-demand transmissions of other SIs to a UE through the IE: *si-BroadcastStatus* (*broadCasting*, *notBroadCasting*) included as part of the SIB1. On-demand requests as well as aperiodic transmission of SIs are illustrated in Figure 18.16.

To make an on-demand request for a list of other SIs, the UE performs the random-access procedure, if configured, toward the NG-RAN followed by the transmission of the *RRCSystemInfoRequest*, containing the required SIs. To perform a RACH request for on-demand SIs, a UE uses the RACH resources, such as the preamble index which is provided by the NG-RAN through IE: *SI-RequestConfig* as part of the SIB1.

In the NR, MIB is transmitted periodically at 80 ms interval, compared to 40 ms in the case of LTE. SIB1 is transmitted periodically at 160 ms interval, compared to 80 ms in the case of LTE. For more information on periodic and on-demand SI and the contents of the different SIBs, refer to TS 38.331 [116].

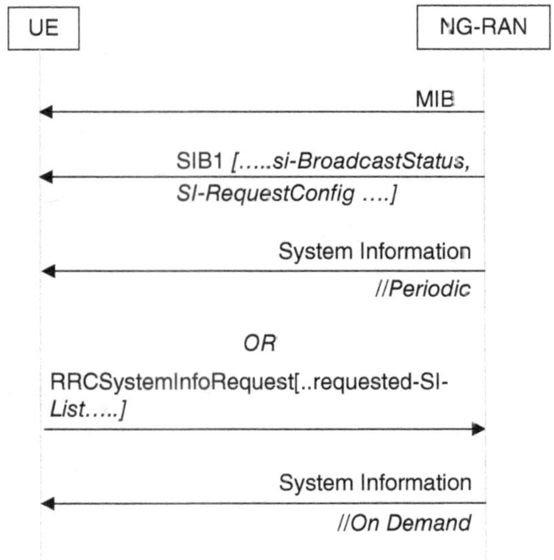

Figure 18.16 Illustration: on-demand SI request by UE.

18.5.3 RRC Layer States

As far as the LTE RRC protocol layer state machine/architecture is concerned, it has only two states – RRC_IDLE and RRC_CONNECTED – till the 3GPP Release 14. However, the 5G NR RRC layer state machine architecture introduces one more new state called RRC_INACTIVE, which is an intermediate state between the RRC_IDLE and RRC_CONNECTED states. In the RRC_IDLE and RRC_CONNECTED states, the NR RRC layer performs functions and procedures and provides services similar to the RRC layer of the LTE system. From Release 15 onwards, the LTE RRC layer also contains three states, as a result of the introduction of the RRC_INACTIVE in the 5G NR RRC layer.

The RRC layer configures UE and cell-specific parameters between the NG-RAN/gNB and UE. The RRC layer performs the PLMN selection, broadcast SI, cell reselection, 5GC-initiated paging, acquires SI, and discontinues reception (DRX) in the RRC_IDLE state. In the RRC_CONNECTED state, the RRC layer performs the network-controlled mobility, e.g. handover, within the NR or inter-RAT and uplink/downlink data transfer; acquires SI; neighbor cell signal measurements; and channel state/quality reporting to the NG-RAN.

The RRC_INACTIVE state is new to the 5G NR RRC layer which differentiates it from the LTE system RRC layer. No RRC connection exists between a UE and the E-UTRAN or NG-RAN in the IDLE state of the RRC layer. But an RRC connection also exists in the RRC_INACTIVE state of the NR RRC layer, apart from the existence of an RRC connection in the RRC_CONNECTED state.

In the RRC_INACTIVE state, the RRC layer performs the same function as in the RRC_IDLE state. In addition to these, the RRC layer manages the RAN-based notification areas (RNAs) and performs RNA-based RAN paging and periodic RNA update (RNAU) procedure toward the NG-RAN. It also performs neighbor cell measurement, acquires SI, and stores the UE AS context information.

Figure 18.17 illustrates the state transitions among the different states of the NR RRC layer, along with the particular trigger or an RRC layer procedure leading to a state transition. Along with each state of the RRC layer, the corresponding 5G MM and CM states are also shown, which was described earlier in Section 18.4.4.

Note that while a UE is in RRC INACTIVE state, it is in a CM-CONNECTED state from the NAS layer point of view. The next section describes the new RRC_INACTIVE state of the NR RRC layer.

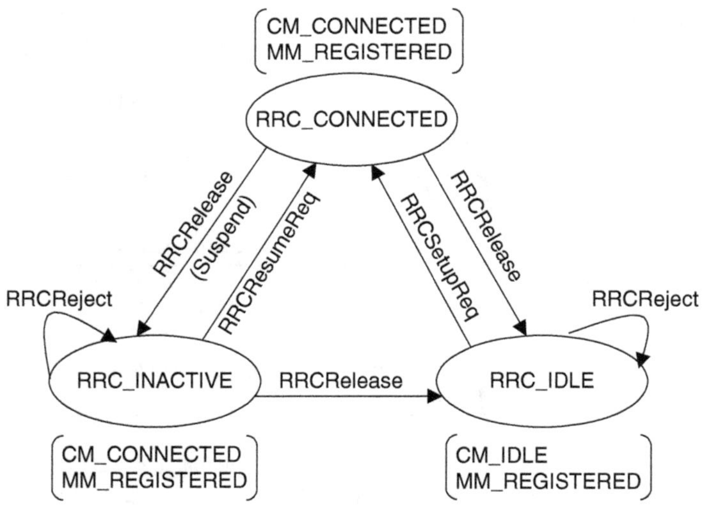

Figure 18.17 Illustration: NR RRC layer state's machine, its triggers, and transitions.

18.5.4 RRC INACTIVE State

As part of an RRC connection release procedure, either in the LTE or 5G NR system, the E-UTRAN or the NG-RAN releases the allocated radio bearers and their resources allocated at the RLC, MAC, and physical layer levels. If a UE wishes to establish an RRC connection again, it is required to exchange and process several signaling messages with the E-UTRAN. This not only leads to a signaling overhead and energy inefficiency over the E-UTRAN air interface but also increases the overall access latency for the UE while attempting to establish an RRC connection. Thus, to reduce the existing access latency in the LTE during an RRC signaling establishment requirement and also to achieve low-latency requirements of the 5G access system, 5G NR introduced the INACTIVE state of the RRC layer as part of the Release 15 specifications, both in the LTE and NR. In the NR RRC_INACTIVE state, the signaling activities of the RRC layer over the signaling radio bearer: 0 remain active while the other signaling radio bearers and DRBs are suspended. The RRC connections are restored upon arrival of a paging message toward the UE or availability of data or signaling activities in the uplink or downlink direction. During such events, fast state transition from the UE RRC INACTIVE to CONNECTED state takes place, in comparison to state transition from RRC IDLE to RRC CONNECTED, thus contributing to the overall low-latency requirements of the 5G system. For this reason, UE context information is stored at the NG-RAN end.

In the NR, the RRC_INACTIVE state introduces the following MM concepts which are used by a UE during its NR RRC_INACTIVE state.

- RAN Notification Area (RNA), which consists of a list of NR cells or RAN area IDs, within a RA, as defined by the Operation and Maintenance (O&M). RNA is configured on a per UE basis;
- Periodic RAN update timer: T380 value, as defined by the O&M; and
- Suspended UE context identification which is identified by an Inactive-Radio Network Temporary Identifier (I-RNTI).

Figure 18.18 illustrates the mobility management and reachability of a UE in its NR RRC_INACTIVE state.

An RNA consists of NR cells that belong to a registration area which is assigned to a UE by the AMF during the UE initial registration with the 5GC. During the NR RRC_INACTIVE state, a UE is known at an RNA level to its NG-RAN in contrast to cell level at the RRC_CONNECTED state. Therefore, the NG-RAN performs a paging procedure toward a UE in its NR RRC_INACTIVE state, due to which it is also called as the *RAN paging*. For paging purposes, the serving gNB sends the RAN paging message over the Xn-AP/Xn interface to all the neighbor gNBs of the particular RNA so that they can page the UE. It may be noted that a UE is paged by the 5GC in RRC_IDLE/CM_IDLE state, whereas the NG-RAN pages a UE in the NR RRC_INACTIVE/CM_CONNECTED state.

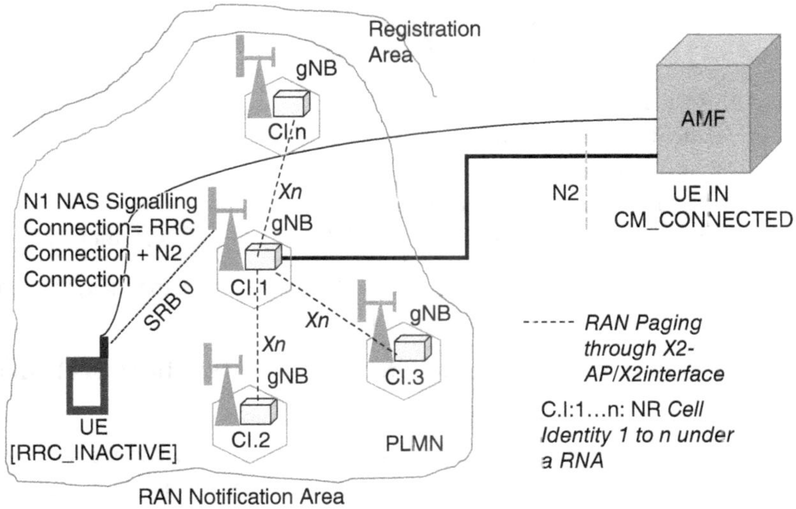

Figure 18.18 Illustration: UE RRC_INACTIVE state and RNA.

A UE periodically performs an RNAU procedure with the serving NG-RAN/gNB to inform the current RNA as part of the UE mobility task performed in the RRC INACTIVE state. A periodic RNAU procedure is controlled by the timer T380. A UE may also perform the RNAU procedure if it leaves the current RNA and enters a new RNA/cell.

To support the NR RRC_INACTIVE state, the NG-RAN maintains the AS context of a UE. Such a context is identified by an *I-RNTI* which consists of two parts; refer to TS 38.423 [122]:

- NG-RAN identity that allocated the I-RNTI;
- UE context identity is stored within the concerned NG-RAN.

I-RNTI has two versions: long (40 bits long) and short (24 bits long). NG-RAN informs UEs to use either the long or short I-RNTI through the SIB1. I-RNTI is for paging by the NG-RAN or as a UE identity during its RRC connection resume request sent to NG-RAN.

Also, as illustrated in Figure 18.18, the N1 NAS signaling connection, which is similar to LTE Evolved Packet System Connection Management (ECM) signaling connection, remains active during the RRC_INACTIVE state, i.e. the UE is CM_CONNECTED state at AMF end.

- **Method to Put into RRC INACTIVE State**

A UE is put into the RRC_INACTIVE state through the RRCRelease message. Unlike the LTE system, the 5G NR RRC layer *RRCRelease* message has dual purposes, as mentioned below:

- Put a UE into the idle state by releasing the radio bearers and their associated underlying radio resources allocated at RLC, MAC, and physical layer, or
- Put a UE into the INACTIVE state by transitioning its state from RRC_CONNECTED to RRC_INACTIVE.

A UE enters into the INACTIVE state as well as returns or resumes to RRC_CONNECTED state by suspending and resuming the allocated resources as described below.

- **Suspending of RRC Connections of a UE in RRC_INACTIVE State**

The above information is provided as part of the *SuspendConfig* optional IE. To indicate and put a UE into the RRC_INACTIVE state, the RRC layer includes the *SuspendConfig* IE as part of the RRC Release message to

UE. Further, the UE as well as the gNB stores the following UE context information which is received as part of the RRC Release message from the NG-RAN:

- AS security keys;
- UE identity, i.e. Cell Radio-Network Temporary Identifier (C-RNTI);
- Cell identity;
- Physical layer cell identity; and
- RAN notification area, consisting of cells, where a UE can move without informing the NG-RAN.

When a UE enters into the RRC_INACTIVE state, i.e. suspended, the signaling radio bearers, except SRB0, and the DRBs between the UE and the NG-RAN are suspended.

- **Resumptions of RRC Connections of a UE in RRC_INACTIVE State**

At the expiry of the periodic RNAU timer: T380, the UE triggers the periodic RNAU procedure toward the NG-RAN. As part of the periodic RNAU procedure, the UE triggers the resumption of an RRC connection with NG-NR/gNB with *resumeCause* as *rna-Update*. After the RNAU procedure, the UE is again put into the RRC-INACTIVE/suspend state by sending a *SuspendConfig* IE in the *RRCRelease* message to it. A UE will also trigger an RNAU toward the NG-RAN followed by the resumption of an RRC connection due to the following, either UE-initiated or network-initiated, events:

- Upon receiving a network-initiated paging request message from the NG-RAN over the RNA. In the RRC_INACTIVE state, a UE location is known to the NG-RAN at the RNA level. A paging message intended for a particular UE is paged on its behalf in all the NR cells configured under the concerned RNA.
- Upon availability of data, user, or signaling, in uplink or downlink direction.
- Leaving the current RNA and entering a new NR cell under a different RNA.

The response from the NG-RAN can be as follows for an RRC Resume Request message from the UE:

- Resume the UE RRC connection,
- Release the existing RRC connection,
- Set up a new RRC connection, or
- Continue suspending the same RRC connection and put the UE into RRC INACTIVE state.

- **Reduction of Signaling Messages Due to UE in RRC_INACTIVE State**

Figures 18.19 and 18.20 illustrate the RRC layer signaling messages exchanged between a UE and the NG-RAN in the following cases:

i) A UE state transition takes place from RRC_IDLE to RRC_CONNECTED state.
ii) A UE state transition takes place from RRC_INACTIVE to RRC_CONNECTED state due to the RRC resume procedure.

As illustrated in Figure 18.19, UE enters into the RRC_CONNECTED state upon receipt of the *RRCSetup* message from the gNB, carrying the IE: *RadioBearerConfig* to set up the SRB1.

By comparing the idle to connected state and inactive to connected state as illustrated in Figure 18.19 and Figure 18.20, it appears that there is a reduction in the number of RRC signaling messages or delay while transitioning from RRC_INACTIVE to RRC_CONNECTED state to restore the RRC connection and its bearers of a UE. Thus, a suspended UE in the RRC_INACTIVE state can resume and return to the RRC_CONNECTED state faster without requiring to exchange all the RRC signaling messages with the NG-RAN. Reduction of RRC signaling messages reduces the access delay between a UE and the NG-RAN during the RRC resume procedure, thus improving the overall access latency and service experience of an end-user.

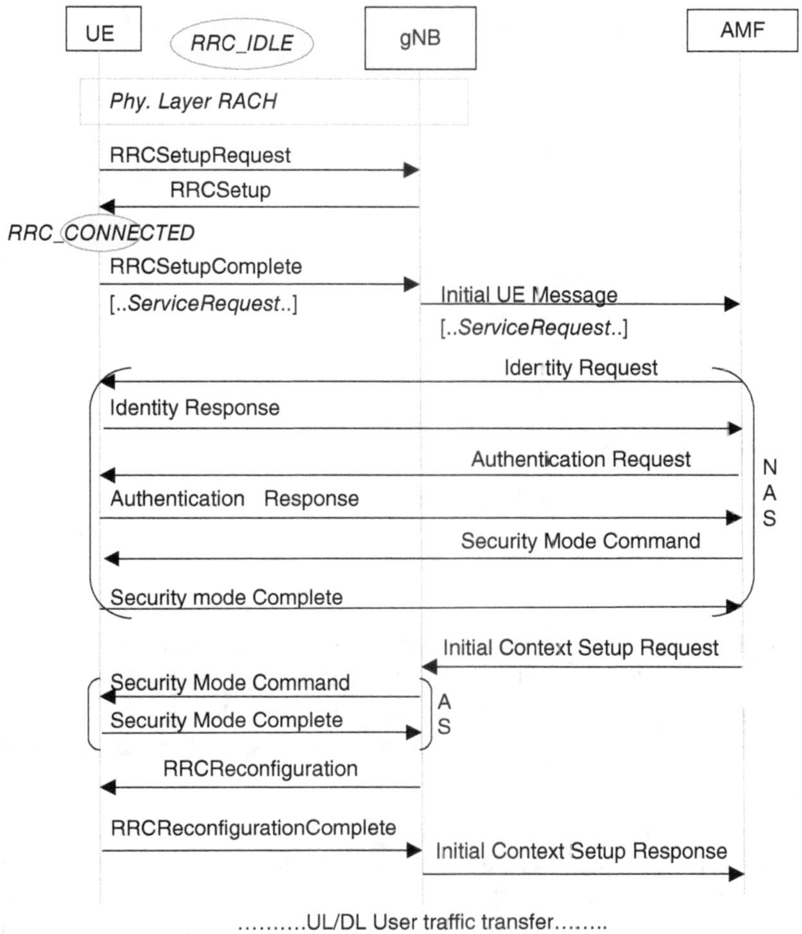

Figure 18.19 Illustration: UE triggered transition from RRC_IDLE to RRC-CONNECTED state.

It may be further noted that the CN or 5GC is not involved while transitioning back and forth between the RRC_CONNECTED and RRC_INACTIVE state. Thus, the NAS layer signaling connection between a UE and the 5GC remains active, i.e. CM_CONNECTED state.

For more information on the UE behavior, RRC connection suspend, and resume procedure, refer to TS 38.331 [116]. Further, refer to TS 38.300 [111] and TS 23.501 [40] for more information on the UE mobility and CM aspects during their RRC_INACTIVE state.

- **5GC AMF Assistance and UE Reachability in RRC_INACTIVE State**

The 5GC AMF network service function also assists the NG-RAN/gNB in deciding to put a UE into the RRC INACTIVE state. For this purpose, the AMF provides several information to the gNB during the initial UE registration request and accept the procedure. This is called as the *Core Network Assistance Information* that contains the RRC_INACTIVE state transition-related configuration information. It is provided as part of the *Initial Context Setup Request* or subsequent context modification procedure triggered from the AMF to a UE during the UE initial registration process with the 5GC. Further, the AMF may request the NG-RAN to provide the status of the UE RRC layer by including IE *RRC Inactive Transition Report Request* in the Initial Context Setup Request or subsequent context modification procedure to the NG-RAN. The NG-RAN will provide the UE RRC layer state to the AMF through the RRC INACTIVE TRANSITION REPORT message.

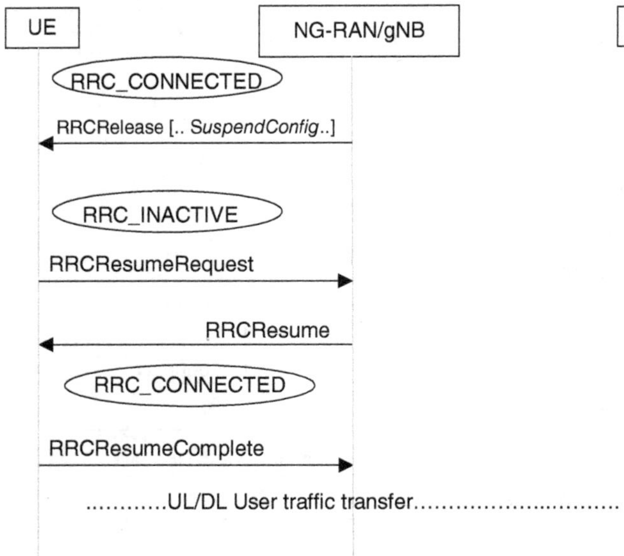

Figure 18.20 Illustration: RRC layer state transition from RRC_INACTIVE to RRC_CONNECTED state for data or signaling information transfer.

Figure 18.21 illustrates with partial messages (. . .) flows on the periodic RNAU procedure at the expiry of the T380 timer at the UE end in its RRC_INACTIVE state. In this illustration, it is assumed that the UE is within the same RAN notification area (RNA) and is also served by the same gNB.

Further, two scenarios, i.e. normal and erroneous, as illustrated in Figure 18.21 are described below:

1) Normal Scenario
   A UE performs and completes an initial registration with the 5GC with no immediate data transfer following the UE registration. As there is no data transfer in downlink or uplink direction, after some time, the UE will be put into RRC_INACTIVE state through the *RRCRelease* message from the gNB to the UE with the *SuspendConfig* information. The UE will start the periodic RNAU timer T380. At the expiry of the T380, the UE will send *RRCResumeRequest* with resume cause as *rna-Update*. This periodic RNAU procedure will be repeated as long as the UE is RRC_INACTIVE state.
2) Erroneous Scenario
   Figure 18.21 further illustrates an erroneous scenario for the recovery of resources allocated for a UE context and signaling connections in case a UE becomes unreachable in its RRC_INACTIVE state. For example, UE switched off during its RRC_INACTIVE state or a UE misbehaves due to which there is no periodic RAN notification area update, i.e. *RRCResumeRequest*, from the UE toward the NG-RAN. The UE context and other resources will remain allocated at gNB and AMF ends in its RRC_INACTIVE state. However, to recover RAN and CN resources (control and data plane) in such a case, a guard timer called *Periodic Registration Update Timer* is provided by the AMF to the gNB as part of the *Initial UE Context Setup Request* step, refer to TS 38.413 [119], performed during the registration of a UE as illustrated in Figure 18.21.

The timer T380 runs at the UE end, and the *Periodic Registration Update Guard Timer* runs at the gNB end. *Periodic Registration Update Guard Timer* runs longer than T380. The additional time added to the Periodic Registration Update Guard Timer at the gNB end is implementation-dependent. As illustrated, both the timers may be started at the same time at UE and gNB ends. If no periodic RNAU request from a UE is received at the expiry of T380, in its RRC_INACTIVE state, by the gNB, the *Periodic Registration Update Guard* Timer will eventually expire after some time. On the expiry of the *Periodic Registration Update Guard Timer,* UE signaling connections, its context, and resources allocated at NG-RAN and CN and control and user planes

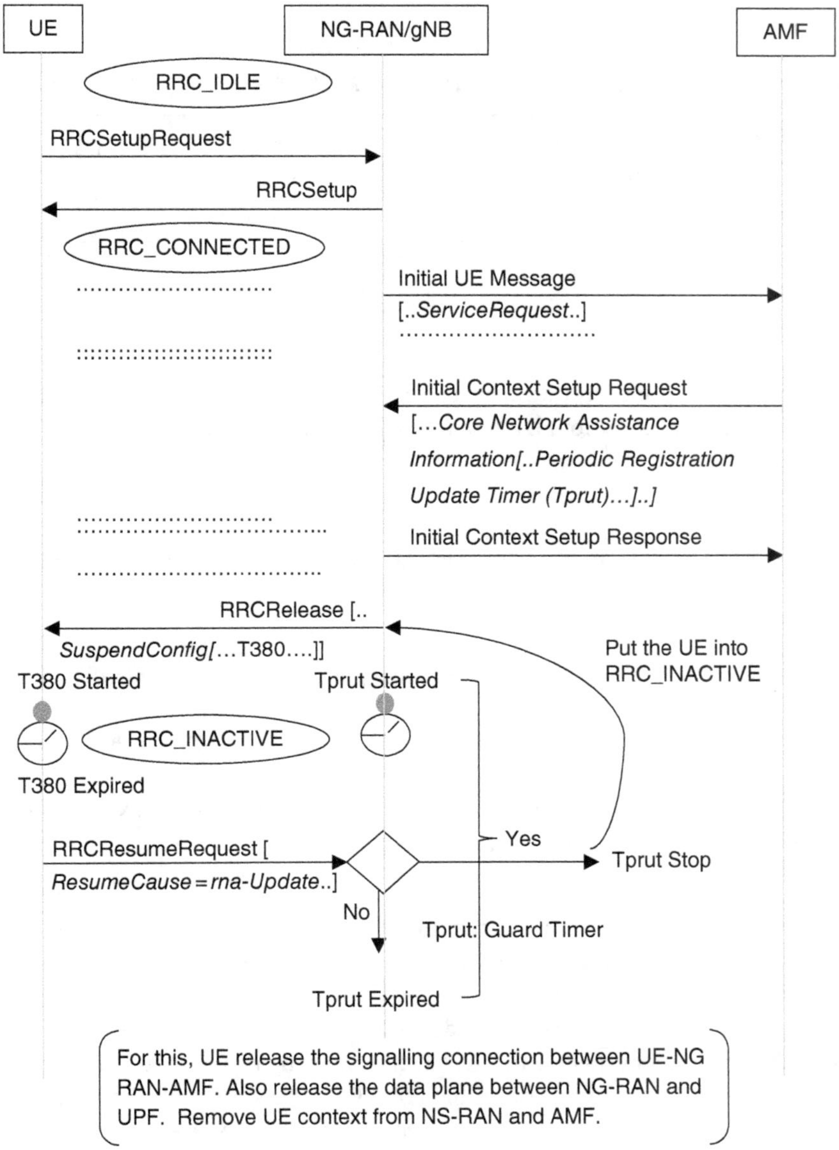

Figure 18.21 Illustration: RNA update procedure: normal and erroneous scenarios.

will be released as part of the Access Network (AN) release procedure. Refer to AN release procedure described in TS 23.50 2 [41].

For more information on the periodic RNAU procedure within or outside the current RNA area in the UE RRC_INACTIVE state, refer to TS 38.300 [111]. For more information on the UE mobilities in different states of the RRC layer, refer to TS 23.501 [40].

- **Comparisons of RRC States With Respect to UE Mobility**

Table 18.1 summarizes the mobility management functions and procedures performed by the NG-RAN and 5GC in different states of the RRC layer.

Table 18.1 RRC states vs. UE mobility.

5G MM Functions	RRC IDLE/CM_IDLE	RRC INACTIVE/ CM_CONNECTED	RRC CONNECTED/ CM_CONNECTED
Paging	Core Network/AMF	NG-RAN/gNB	NG-RAN/gNB (UE to reacquire SIs)
UE Identity for Paging	5G-S-TMSI	I-RNTI	Not Applicable
Mobility	Core Network/AMF	RAN (RNA Update)	RAN (Handover)

18.5.5 Mobility of UE

UE mobility-related functions and procedures performed by the 5G MM layer within a 5G CN area, i.e. Registration Area, have been discussed in Section 18.4. In this subsection, UE mobility-related functions and procedures performed by the RRC layer with respect to its different states shall be described below.

18.5.5.1 UE Mobility in RRC IDLE State

In the idle state, the RRC layer at the UE end performs the following tasks as far as the support of UE mobility is concerned:

- PLMN selection;
- Cell selection at NR physical layer level, which is based on Synchronization Blocks (SSB), which shall be described later in Chapter 19;
- Suitable cell selection at PLMN level, which is based on cell selection criteria so that a UE can be camped on to it;
- Cell reselection, which is based on UE NR intra- or inter-frequency or inter-RAT, i.e. LTE/E-UTRAN, measurements, and cell reselection criteria. A UE may reselect a better suitable cell; and
- 5GS to LTE/EPS interworking.

Interworking between LTE/EPS and 5G systems was described earlier in Section 16.7. Two options are available for UE mobility – through interworking with N26 or without the N26 interface. When the N26 interface is used, the LTE/EPS PGW-C + 5GC SMF acts as the anchor point of interworking for control plane messages transfer, and PGW-U + UPF acts as the anchor point of interworking for user plane data transfer to support UE mobility.

In case of interworking with N26 interface with SSC mode 1 in RRC-IDLE state, when a UE moves from a 5G to LTE/EPS system, it performs a *TAU* MM procedure or a fresh initial ATTACH procedure with the LTE/EPS network. Similarly, a UE in the E-UTRAN RRC-IDLE state performs a *Registration Request* procedure with the 5GC when the UE moves from an LTE/EPS to a 5G system.

18.5.5.2 UE Mobility in RRC INACTIVE State

RRC INACTIVE state was described in Section 18.5.4. In the RRC INACTIVE state, the RRC layer maintains a UE context information. RRC layer performs the following functions/procedures as far as the support of UE mobility is concerned in RRC INACTIVE state.

- RAN Notification Area (RNA) update procedure by triggering an *RRC Connection Resume* procedure from UE to NG-RAN. An RNAU procedure may be a periodic one or whenever a UE is moved out of the current RNA and enters into a new RNA. Refer to Figure 18.21 for an illustration of RNAU procedure.
- Paging by the same or last NG-RAN/gNB to UE. In case the UE moved to a new gNB area, the new gNB obtains the UE context from the last or old gNB and performs an admission control procedure to allocate desired radio resources. The old gNB releases the UE context information.

- Cell reselection, which is performed based on UE NR intra- or inter-frequency or inter-RAT, i.e. LTE/E-UTRAN, measurements, and cell reselection criteria, a UE may reselect a better suitable cell.
- 5GS to LTE/EPS interworking.

A UE performs a cell reselection procedure to select a suitable cell in the 5G system or LTE/EPS system when it moves from a 5G system to LTE/EPS and vice versa.

18.5.5.3 UE Mobility in RRC CONNECTED State

There are two aspects of UE mobility in the RRC CONNECTED state as described below.

a) **Beam-level mobility within a physical layer cell**
   In case mmWave frequency is used in a cell, physical layer transmission, and reception is performed using the beamforming technique. As a user moves within a cell, a new beam is selected by the UE physical layer whose *Synchronization Signal-based Reference Signal Receive Power (*SS-RSRP*)* is greater than the threshold *rsrp-ThresholdSSB* as configured by the RRC layer. Though the RRC layer configures UE on beamforming transmission and reception, no RRC layer signaling message is exchanged between UE and NG-RAN to support beam-level mobility. The beamforming technique used, at the NR physical and MAC layers, for beam-level mobility shall be described later in Chapter 19.
b) **Cell-level mobility through handover at Layer 3 (RRC)**
   While the RRC layer is in its ACTIVE state, a cell-level UE mobility is accomplished through the RRC layer signaling and a handover procedure between a source and target cell or system. A *handover* procedure is a UE/MS-assisted but network-controlled critical mobility function which is supported across the different mobile communications systems and networks. Support for handover capabilities for ongoing communication services differentiates a mobile communications network from a fixed network. The handover of an active call, as determined by the RAN, is performed whenever a moving user and its mobile device leaves its current coverage area (e.g. in terms of location/routing are in GSM/UMTS system and TA in LTE system) and enters a new coverage area. As a result of a handover procedure, a mobile device is assigned different radio resources in the new coverage area. A handover procedure could be also performed whenever the currently active call cannot be continued further with the assigned radio resources because of various reasons.

A handover procedure is performed at the command of the RAN toward the mobile device with its assistance. So a handover attempt is always performed whenever the RRC and management protocol state is at the DEDICATED (GSM system) or RRC CONNECTED (UMTS, LTE system, 5G NR) state at the mobile device as well as RAN ends. Different aspects of a handover procedure, in general, are described next.

- How a Handover Requirement is Determined

While a mobile device is engaged in a call within a serving cell, it also monitors and measures the downlink signal level and quality of its neighboring cells. For example, in the case of the GSM system, a mobile device monitors the signal levels of six neighboring cells around the serving. The measurements of the neighboring cells are reported to the serving cell periodically that is under the control of the RAN, say, BSC, once the call becomes active. The handover procedure algorithm running, at the BSC, uses the samples of measurement inputs being received from the mobile device, and that is why, a handover attempt is called a UE/MS-assisted procedure. The handover algorithm at the RAN end determines the requirements of the initiation of a handover procedure if the RF signal quality of the serving cell drops below a certain threshold limit. Upon the RF condition and its criteria being met, the RAN element, say the BSC, initiates the handover requirement of the ongoing call to a new cell. Downlink signal level and quality measurement by a UE for its serving as well as neighboring cells in the 5G system shall be described later in Chapter 19.

Figure 18.22 Different phases of a handover procedure.

- Different Phases of a Handover Procedure

In general, an end-to-end handover procedure between a source and target cell or system for an ongoing call comprises several phases, as shown in Figure 18.22. These phases of a handover procedure are supported across the different cellular systems.

- Handover Preparation

In this phase, the target cell/network element, e.g. BSC, MSC, NodeB, eNodeB, and 5G gNodeB, assign the necessary radio resources and then returns an ACK containing the NR resources to the source cell/network element.

- Handover Execution

In this phase, the MS/UE tunes and connects with the target cell with the NR resources being allocated to it.

- Handover Completion

In this phase, the old radio resources allocated to the handed over MS/UE are released at the source network element, e.g. GSM BSC and MSC.

- Types of Handover Procedure

An ongoing call within a particular network may be transferred from a source cell to a target cell through different types of handover supported in the GSM, UMTS, and LTE networks. Based on user mobility, an ongoing call may be also transferred from one network and RAT to another network with a different RAT. To support such handovers within and across networks, with different RATs, a handover procedure may be classified into the following types at a broad level:

- Intra-RAT handover – which is attempted within a particular RAT, e.g. intra-GSM/GERAN or intra-UMTS or intra-LTE network or intra-5G NR network.
- Inter-RAT or inter-system handover – which is attempted across the different RATs, e.g. LTE to UTRAN, LTE to GSM, UTRAN to LTE; GERAN to LTE network, and 5G NR to LTE network.

Handover procedures to support the above scenarios and types can be a simple, intra-RAT, to a complex, inter-RAT, one involving different network elements and logical interfaces. Though the basic purpose of a handover procedure is the same across the mobile communications systems and networks, their implementation aspects are quite different. Further, each of the above handover types has several scenarios as described in the subsequent sections below.

- Handover Scenarios Based on the RF Link Establishment

Two types of handovers are possible based on how the ongoing call is being handed over to a new cell by establishing radio link(s) with the base station(s).

- Hard Handover

A hard handover type is available in GSM, UMTS, LTE, and NR systems. In case of *hard handover*, the existing RF connection and its radio resources assigned to an MS are released in the old cell for a very brief period, in the order of a millisecond, that cannot be perceived by the mobile user. The ongoing call is handed over to the new cell that is under the control of the target base station. This is a type of *break-before-make* handover procedure having less perceived interruptions, in order of milliseconds, on the ongoing call. A hard handover takes place between cells operating at different frequencies. This means a UE can have only one active radio link with a base station at a time.

- Soft Handover

A soft handover is supported in the case of the UMTS system only. In the case of *soft handover*, the UE has the active radio link connections with more than one NodeB (source as well as destination) since they operate with the same frequency but with a different scrambling code that identifies individual NodeBs. This type of scenario arises in case of overlapping cell boundaries. It is a type of connect/*make-before-break* handover procedure where there are no interruptions on the ongoing call. Once the handover procedure is completed, UE gets disconnected from the source NodeB and remains in communication with destination NodeB in the target cell.

- Handover Scenarios Based on the Controlling Network Element

Different scenarios and handover types are also possible depending on their scope and which network elements are involved or control the handover procedure. Based on this, Table 18.2 shows the handover types along with their scenarios.

Next, the different handover types and procedures specific to the 5G NR system are described and illustrated below.

- Handover Types and Procedures Specific to 5G NR

In the 5G system, a handover procedure can take place over the *Xn* and *N2* logical interface as described below. They are similar to the handover procedure performed over the X2 and S1 logical interface in the LTE/EPS system.

- Xn Handover

A handover procedure between a source gNB and a target gNB over Xn interface is called as *Xn handover*. Inter-gNB handover procedure-related signaling messages exchanged over the direct Xn interface between a source gNB and target gNB are illustrated in Figure 18.23. Note that the AMF is not involved in an Xn-based handover procedure.

As illustrated in Figure 18.23, the logical node, i.e. gNB-CU-CP, of the source and target gNB is being used to illustrate the flow of handover signaling messages between them. As part of the preparation phase of an Xn-based handover procedure, the source gNB sends the *HANDOVER REQUEST* message to the target gNB. As part of the resource allocation phase of the Xn-based handover procedure, the target gNB allocates the necessary radio resources and responds with a *HANDOVER REQUEST ACKNOWLEDGE* message to the source gNB, thus completing the handover preparation phase. The *HANDOVER REQUEST ACKNOWLEDGE* message also contains the *HandoverCommand* message, which further contains the target gNB RRC resources to be reconfigured by the UE. *HandoverCommand* is the part of the execution phase of the Xn handover procedure.

Table 18.2 Handover types and scenarios by network element.

System	Handover Types	Handover Scenarios with Controlling Network Element
GSM	Hard	Intra-BTS, Intra-BSC, Inter-BSC, Intra-MSC, and Inter-MSC
UMTS	Hard, Soft	Intra-RNC, Intra-3G_MSC, and Inter-3G_MSC
LTE	Hard	X2 Handover (Inter-eNodeB) and S1 Handover (eNB and MME)
5G NR	Hard	Xn Handover (Intra-NG-RAN/AMF) and N2 Handover (Inter-NG-RAN/AMF)

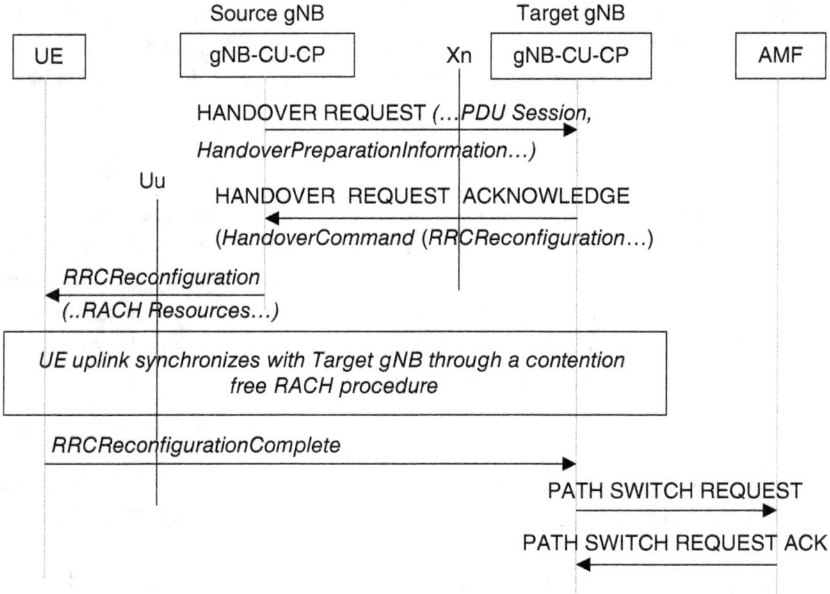

Figure 18.23 Illustration: Xn handover and its signaling messages.

The source gNB forwards the RRC resources, allocated by the target gNB, to the UE using the *RRCReconfiguration* message which contains, among others, information about the RACH parameters to be used by it to perform a contention-free random-access procedure toward the target gNB. Refer to TS 38.331 [116] for more information on the contents of the *RRCReconfiguration* message. The UE successfully processes the allocated radio resources and gets uplink synchronized with target gNB through a contention-free random-access procedure. It is followed by the *RRCReconfigurationComplete* message from UE to the target gNB, thus completing the inter-gNB handover execution procedure successfully. UE data transmission/reception will be interrupted, due to the break-before-make approach, for a short duration (in order of milliseconds) during the handover execution phase.

As part of the completion phase of the handover procedure, target gNB requests, by sending a *PATH SWITCH REQUEST* message to the AMF to switch the data transmission path from source gNB to the target gNB. Though not shown as part of Figure 18.23, the AMF further advises the SMF, and the SMF advises the UPF, to switch to the target gNB for each PDU session received in the *PATH SWITCH REQUEST* message context information in this regard.

- N2-based handover

In the absence of an Xn interface between a source NG-RAN and target NG-RAN, an N2 reference point-based handover procedure is performed with the help of the 5GC/AMF. The N2 reference point is used between the NG-RAN/gNB and the 5GC/AMF over which NG-AP control plane messages are exchanged between them. An N2-based inter-NG-RAN handover procedure in a 5G system is illustrated in Figure 18.24. As illustrated, there is no change in AMF as well as UPF.

As illustrated in Figure 18.24, the signaling messages used in the N2 handover procedure are different from the Xn-based handover. As part of the preparation phase of an N2-based handover procedure, the source NG-RAN indicates the requirement of a handover to the AMF through a *HANDOVER REQUIRED* message with handover type, PDU sessions to be handed over, availability of direct IP path between the source, and target NG-RAN. As part of the resource allocation phase of the N2-based handover procedure, AMF requests the target NG-RAN to allocate necessary radio resources by sending a *HANDOVER REQUEST* along with the required handover type.

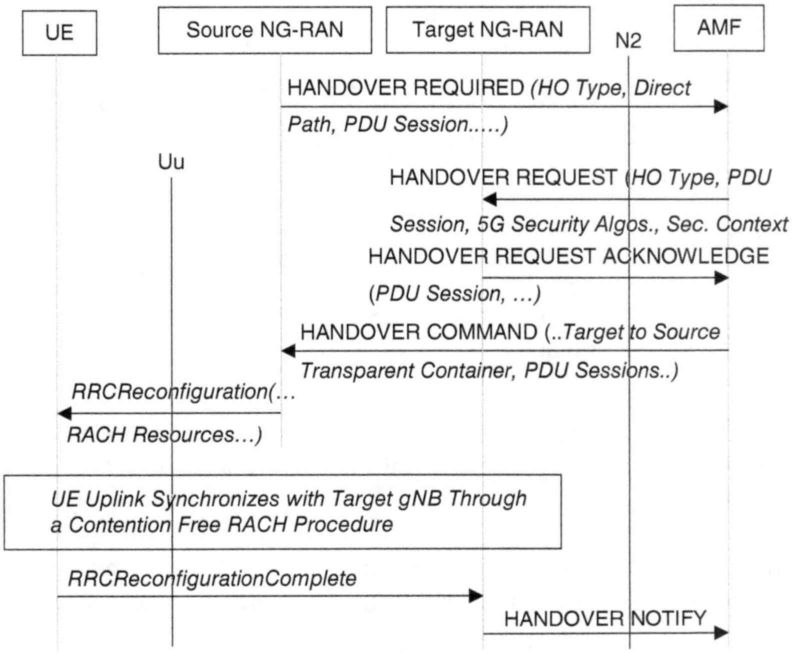

Figure 18.24 Illustration: N2-based handover and its signaling messages.

Target NG-RAN allocates radio resources and responds with *HANDOVER REQUEST ACKNOWLEDGE* message to AMF, which in turn provided to the source NG-RAN, thus completing the handover preparation phase. The *HANDOVER REQUEST ACKNOWLEDGE* message also contains the *HandoverCommand* message. A *HANDOVER COMMAND* message, which is a part of the execution phase of the N2-based handover procedure, further contains the target NG-RAN RRC resources to be reconfigured by the UE. The source NG-RAN provides the radio resources to the UE over the *RRCReconfiguration* message containing the contention-free RACH and the radio resources allocated by the target NG-RAN.

The UE successfully processes the allocated radio resource and gets the uplink synchronized with target NG-RAN through a contention-free random-access procedure. UE sends the *RRCReconfigurationComplete* message to the target NG-RAN, followed by a *HANDOVER NOTIFY* message to the AMF, which is the part of the completion phase of the N2 handover procedure. Thus, the N2 handover procedure is considered to be completed successfully.

For more information and detailed signaling message flows during Xn- and N2-based handover procedures, refer to TS 23.502 [41].

- **5GS to LTE/EPS Handover**

Interworking between LTE/EPS and the 5G system was described earlier in Section 16.7. Two options are available for UE mobility – through interworking with N26 or without the N26 interface. When the N26 interface is used, the LTE/EPS PGW-C + 5GC/SMF acts as the anchor point of interworking for control plane messages, and PGW-U + UPF acts as the anchor point of interworking for user plane data transfer to support UE mobility. In addition to this, the combined entities Home Subscriber Server (HSS)+ UDM and PCF + PCRF also support the successful completion of interworking between LTE/EPS and 5G system.

In case of interworking with N26 interface with SSC mode 1 in RRC - CONNECTED (5G/NR) or CONNECTED(LTE/E-UTRAN) state, a handover from 5GS to LTE/EPS and vice versa takes place through the exchange of RRC layer signaling messages as mentioned above.

In case of interworking with N26 interface with SSC mode 1 in RRC-IDLE state, when a UE moves from a 5G to LTE/EPS system, it performs a *TAU* MM procedure or a fresh initial ATTACH procedure with the LTE/EPS network. Similarly, a UE in the E-UTRAN RRC-IDLE state performs a *Registration Request* procedure with the 5GC when the UE moves from an LTE/EPS to a 5G system.

For more information on typical signaling messages flows during interworking with or without the N26 interface, refer to TS 23.502 [41].

- **UE Mobility Support by gNB Logical Nodes**

The logical architecture of the NG-RAN was described earlier in Section 16.4. An NG-RAN/gNB contains a single CU and multiple DUs. A DU can serve a single cell or multiple cells. Because of these, UE mobilities within the logical nodes of the NG-RAN/gNB can be classified as follows:

i) Intra-DU UE mobility
 In the case of intra-DU UE mobility, RRC layer signaling requirements do not exist. In this case, UE mobility is supported through a UE context modification request procedure, which can be initiated by either of the gNB-DUs over the F1 interface.
ii) Inter-DU UE mobility under the control of a CU
 In the case of inter-DU UE mobility, RRC layer signaling takes place between a source gNB-DU and target gNB-DU; both are under the same CU, which is similar stepwise to the Xn or N2 handover scenario described above.

Radio resource is allocated by the target gNB-DU, responds to the CU, which is further communicated to the source gNB using the CU-initiated UE context modification request message, over the F1 interface, containing the *RRCReconfiguration* message. The remaining steps till *RRCReconfigurationComplete* are the same, as described above. Refer to Figure 8.2.1.1-1: TS 38.401 [117] for more information on the signaling messages flows for an inter-gNB-DU UE mobility support.

For more information on typical signaling messages flows during inter- and intra- DU mobility over the F1 interface, refer to TS 38.401 [117].

18.5.6 Admission Control

Similar to the previous generation mobile communications systems, *Admission Control* is the first step as part of the overall radio resources allocation and management procedures executed by the RAN element on behalf of a mobile device. The admission control function is performed by the RAN for every mobile device requesting network resources and determines whether the network resources can be granted or rejected. Further, if the RAN is a shared one among different participating CN operators, identified by its PLMN ID, the admission control procedure shall be required to be performed for each of the CN operators also.

In the NR system, an admission control procedure is performed at a network slice and its QoS flow level. However, considering the diverse use cases and their services and UEs/devices, the NR admission control procedure shall be much more complex and broader than the admission control procedures used in the previous generation mobile communication systems. An admission control procedure shall be performed during the creation of a new or modification of an existing QoS flow. As part of an admission control procedure at the NG-RAN end, necessary resources are checked and reserved based on the type of radio resources and QoS requirements, which can be a GBR, non-guaranteed bit rate (NGBR), or delay critical GBR, for a particular network slice and traffic type. The outcome of an admission control procedure at NG-RAN/gNB end may be successful, i.e. admission control is accepted, or failure, i.e. admission control is rejected.

During a handover procedure also, an admission control procedure is performed at the target NG-RAN/gNB end as part of its resource allocation phase. The admission control procedure in terms of allocation radio resources for all the QoS flows of the source NG-RAN/gNB may not be successful due to the limitations of resources at the target NG-RAN/gNB. In that case, the services corresponding to the affected QoS would be affected. If admission control is rejected by the target NG-RAN, the handover procedure would be considered as a failure.

An admission control procedure requires to ensure that the ongoing services of existing users are not impacted because of new admission of a mobile device and its requested services. If the requested QoS cannot be admitted or the RAN is overloaded, the same will be rejected by the admission control procedure. In general, among other inputs, an admission control procedure shall take into account the following typical information for the allocation of radio resources to an MS/UE to establish a QoS flow. Such an admission control algorithm is said to be a slice aware/QOS flow aware procedure.

- Requested QoS profile;
- Overall availability of current radio resources as per the required GBR and NBR;
- Existing load on RAN element;
- Existing resources in use; and
- The direction of data transfer, i.e. uplink/downlink.

Such an admission control procedure used by a RAN element is OEM implementation specific and is not standardized by the 3GPP. Also, the RAN admission control procedure/algorithm(s) shall be a particular communications system, i.e. GPRS, UMTS, LTE, or NR system air interface specific.

Figure 18.25 illustrates a typical admission control procedure that may be implemented as a prototype in the NR system to serve the different use cases of the 5G system, i.e. eMBB, mIoT, and URLLC. This figure further illustrates the requirement of an admission control procedure in case the 5G NG-RAN is shared among different CN operators under the Multi-Operator Core Network (MOCN) feature as described earlier in Section 7.2. Further, the

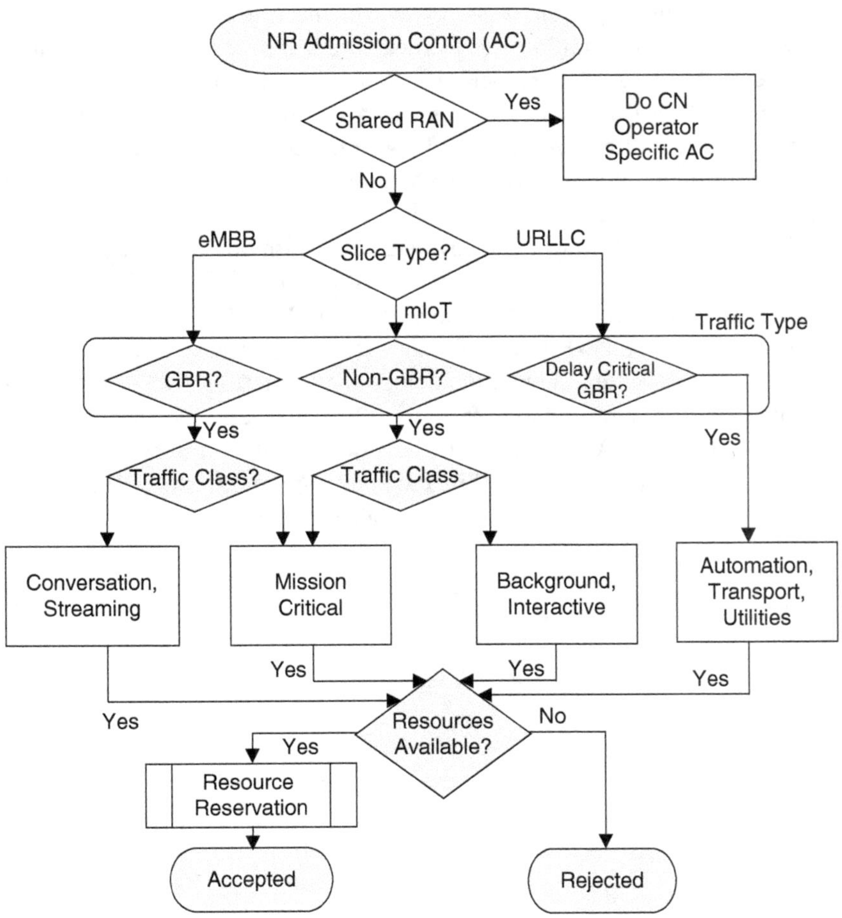

Figure 18.25 Illustration: NR admission control procedure for different network slices of 5G.

3GPP Release 16 supports the *Standalone Non-Public Network (*SNPN*)* also, see Section 7.2.2, in addition to the normal PLMN. Thus, the NR admission control procedure shall be required to accommodate the reservation of radio resources for a UE supporting an SNPN. Different traffic classes and types of a 5G system are mentioned below:

- Traffic types such as the GBR, non-guaranteed bit rate (Non-GBR), and delay critical GBR;
- Traffic classes such as voice conversation, streaming, gaming, mission critical applications, and background traffic (e.g. FTP, file download); high data rates; discrete automation; intelligent transport systems; and process automation; utilities such as electrical distributions; refer to TS 22.261 [28].

Different traffic types and their traffic classes mentioned above have different QoS requirements with different characteristics. QoS was described earlier in Section 18.3. An NR admission control procedure may accept or reject a requested service if the required QoS cannot be met.

Chapter Summary

This chapter presented an overall introduction to the AS and NAS control plane protocol layers of the 5G system over its NR air interface. Control plane protocol layers facilitate the transfer of signaling information between a UE and the Radio Access Network (RAN) as well as between a UE and the 5G core network. 3GPP introduces several new concepts as part of the AS and non-AS protocol layers of the 5G system, given the requirements to support different use cases and their services being envisaged to provide by the 5G system. At the NAS layer, 5G system communication services shall be provided through a PDU session, which has been described in Section 18.2. Also, in the 5G system, the diverse services to be provided by a particular use case will have their own QoS characteristics and requirements, which will be met by the new QoS model of the 5G system as described in Section 18.3. Interworking for UE mobility through a new logical interface and co-existence with LTE/EPS is another key requirement for the 5G system, which will be supported by the AS and NAS protocol layers of the 5G system. Further, unlike LTE/EPS, the 5GS architecture allows the usages of the NAS protocols over both the 3GPP as well as non-3GPP access networks, as described in Section 4.2.8, TS 23.501 [40].

This chapter ends with the description of the NR RRC layer and covered, along with other aspects, the most startling differences from the RRC layer in the LTE system. Differences exist in terms of reduction of some signaling overhead over the NR air interface as well as faster state transition from NR RRC INACTIVE to the CONNECTED state than from IDLE to the CONNECTED state, and hence faster resumption of communication services to a UE and its user. Another improvement added, in comparison to the LTE RRC layer, is the support of the transmission of the on-demand system information from the NR RRC layer, which enables a UE to request specific system information from the NG-RAN. This avoids the requirements of the periodic transmissions of system information from NG-RAN to UEs in a cell, thus the consumption for radio resources is reduced. The typical admission control procedure of the 5G NR system was also covered.

19

5G NR Air Interface

User Plane Protocols

Introduction

Access stratum and non-access stratum protocol layers over the new radio (NR) air interface have been described in the previous chapter. In this chapter, some of the functions and procedures of the user plane protocol layers of the NR air interface will be discussed. As described earlier in Section 16.1.1, NR user plane protocol layers also play key roles in realizing the different use cases of the 5G system in terms of providing low-latency communications, high data throughputs, and so on using the diverse spectrums of the 5G system. To enable such capabilities, the 5G NR user plane protocol layers perform functions and procedures which resemble the functions and procedures performed by the user plane protocols of the Long-Term Evolution (LTE) system. But the internal working of such functions and procedures has been enhanced and reorganized across the different user plane protocol layers of NR.

This chapter begins with the overall protocol layer architecture of the NR user plane protocol stack. Then, the new user plane protocol layer is introduced. This is followed by the description of the protocol layers in order of their appearance in the NR user plane protocol stack. An attempt has been made to cover the various new concepts being introduced in the NR user plane protocol layers with necessary illustrations. This chapter ends with the NR physical layer, covering some of the important aspects as briefed earlier in Section 16.1.1.

19.1 NR User Plane Protocol Layers

NR user plane protocol layers are illustrated in Figure 19.1, which are similar to the LTE system user plane protocol architecture. However, a new protocol layer called *Service Data Application Protocol* (SDAP) is introduced in the NR user plane, which supports the flow-based Quality of Service (QoS) model in the 5G system. SDAP together with the Packet Data Convergence Protocol (PDCP), Radio link control (RLC), Medium Access Control (MAC) layers form Layer 2 of the NR user plane. All these sublayers of Layer 2 are configured by the Radio Resource Control (RRC) layer.

Unlike the LTE Evolved-UMTS Terrestrial Radio Access Network (E-UTRAN)/eNodeB, the NR Next-Generation Radio Access Network (NG-RAN)/5G Base Station (gNB) has logical nodes, i.e. gNB-CU-CP, gNB-CU-UP, and gNB-DU, as described earlier in Section 16.4. From the NR user plane protocol stack shown in Figure 19.1, the SDAP, which is a new layer, and the PDCP user plane part perform the functions and procedures for the gNB-CU-UP logical node. The RLC, MAC, and physical layers perform the functions and procedures for the gNB-DU logical node. Subsequent sections describe the NR user plane protocol layers shown in Figure 19.1.

Mobile Communications Systems Development: A Practical Introduction to System Understanding, Implementation, and Deployment, First Edition. Rajib Taid.

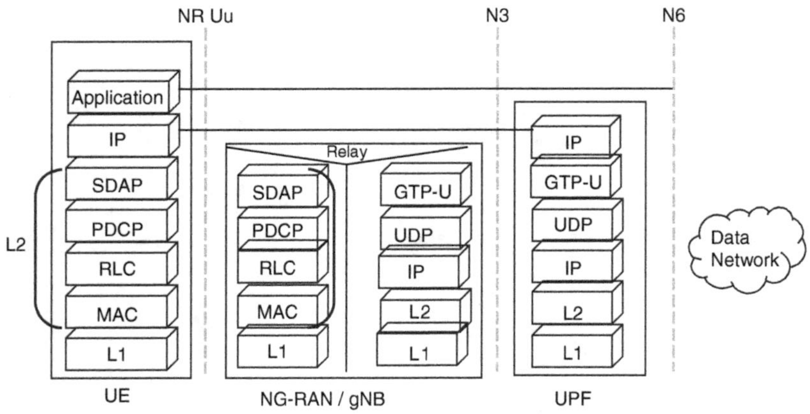

Figure 19.1 Illustration: NR air interface user plane protocol layers.

19.2 SDAP Layer

SDAP layer provides user plane data transfer services to its higher layer. In 5GS, user data transfer takes place through different QoS flows, whereas in the LTE/Evolved Packet System (EPS), user data transfer takes place through different EPS bearers. SDAP layer takes care of the NR QoS framework and its flows by performing the following functions:

- Mapping/multiplexing of different QoS flows to different data radio bearer(s) (DRBs) in the downlink (DL) and uplink (UL) directions, between the SDAP and PDCP layers, for transmission over the NR air interface. Various application IP flows are mapped into different QoS flows, which, in turn, are organized into a *Protocol Data Units* (PDU) *session* in the 5G system;
- Marking of different QoS flow IDs into the packets of a PDU session in DL and UL directions; and
- Assignment of reflective QoS flow to the UL data based on DL QoS flow. Note that UL QoS flow to a DRB mapping can be indicated either through the RRC layer signaling or reflective mapping.

Peer-to-peer SDAP layer information is transferred in terms of PDU. SDAP layer has control as well as data PDU. A data PDU used for DL data transfer may or may not have a header. If a UE is required to use a reflective QoS to DRB mapping in the UL direction which is based on the corresponding DL QoS flow, the SDAP layer at the gNB end would include an SDAP header with a *reflective QoS flow* indication field. This is indicated by the DL *PDU SESSION INFORMATION* message, and its reflective *QoS Indicator* (RQI), sent from UPF in 5GC to gNB; refer to TS 38.415 [120]. SDAP layer at the UE end always uses a header for UL data transmission. For more information on the above functions and procedures of the SDAP layer as well as its PDU formats, headers, and their contents, refer to TS 37.324 [100]. Next, the NR PDCP layer is described.

19.3 PDCP Layer

Like the LTE system, the NR PDCP layer provides data, user traffic as well as control or signaling information, and transfer services to its upper layer. Such data transfer services provided by the PDCP layer may be ciphered, integrity protected as well as the header of data can be compressed as configured by the RRC layer under the IE: *PDCP-Config*. To provide data transfer services, the PDCP layer performs several functions and procedures which are similar to the LTE PDCP layer.

- Maintaining of PDCP sequence number (SN) for packet reordering purposes at the receiver end;
- Ciphering and deciphering of user plane as well as control plane data transferred through an RRC signaling or DRB;
- Integrity protection and integrity verification of RRC control plane data transferred using a signaling radio bearer through the IE: *PDCP-Config* [*...integrityProtection...*] which is enabled by default. Integrity protection may be also applied to user plane data transferred using a DRB as configured by the RRC layer. To enable integrity protection on information exchanged between the sender and receiver PDCP layers, a MAC is attached to a PDCP header; and
- Header compression and decompression using the Robust Header Compression (RoHC) standard protocol for user plane data.

IP header compression feature decreases the IP packet header size which would result in an overall smaller packet size for transmission over the air interface. Application of such an IP header compression feature is particularly suitable for transmission of small packets, for example, Voice over Internet Protocol (VoIP) Real-time Transport Protocol (RTP) packets over UDP/IP, which would result in optimum utilization of its limited bandwidth over the NR air interface. Header compression can be also for data transmission with other transport protocols such as Transmission Control Protocol/Internet Protocol (TCP/IP).

The sending PDCP layer produces a compressed, for the header part only, packet against a PDCP Service Data Unit (SDU) received from its higher layers. The receiving PDCP layer decompresses the header part of the PDCP PDU and forwards the entire information to its higher layer. PDCP layer applies only the RoHC protocol families which are described by different Internet Engineering Task Force (IETF) Request for Comments (RFCs) for different protocols. The 3rd Generation Partnership Project (3GPP) defines the allowed IP protocols, identified through respective profile identifiers, on which a header compression can be applied with their respective RFC numbers. For example, RFC 3095 [13] and 4815 [16] can be used for header compression of RTP on top of the UDP/IP header using profile ID = 1; refer to TS 38.323 [115].

- Timer-based SDU discard;
- Duplication of PDCP data PDU across two RLC entities in case packet duplication is activated at the PDCP layer, which differentiates it from the LTE PDCP layer. Duplication of packets of a PDCP data PDU and their transmission through independent transmission paths or RLC entities is the enabler for typical 5G services that require very highly reliable and low-latency communications such as the Ultra Reliable and Low-Latency Communications (URLLC);
- Duplicate PDCP PDU discarding as a result of handover from old to new gNB, at receiving PDCP entity;
- Reordering and in-order delivery of PDCP PDU to the upper layer at the receiving PDCP entity; and
- Routing of packets for DL split bearers in case of dual connectivity setup.

For an overview of the organizations of the above functions of the PDCP layer and flow of data among these functional blocks, refer to, Figure 4.2.2-1, TS 38.323 [115].

- **Role of PDCP in Delivering High-throughput Split Bearer**

Figure 19.2 illustrates the split of the user plane protocol stack at the PDCP end of the LTE eNB. It illustrates the packet routing role of the PDCP layer PDU in a dual connectivity network setup arrangement between LTE and 5G NR systems. Such a dual connectivity setup can be used as an early deployment solution to provide some of the 5G services through an existing LTE/EPS system. This is also referred to as the *non-standalone* (NSA) *deployment* scenario since the 5G NR system coexists with the existing LTE/EPS system.

As illustrated in this dual connectivity setup, LTE eNB is called as the *master node* (MN) and the NR gNB is called as the *secondary node* (SN). Cells served by the MN are called *Master Cells Group* (MCG), and the cells served by the SN are called *Secondary Cells Group* (SCG). The UE has the simultaneous control plane and user plane connections

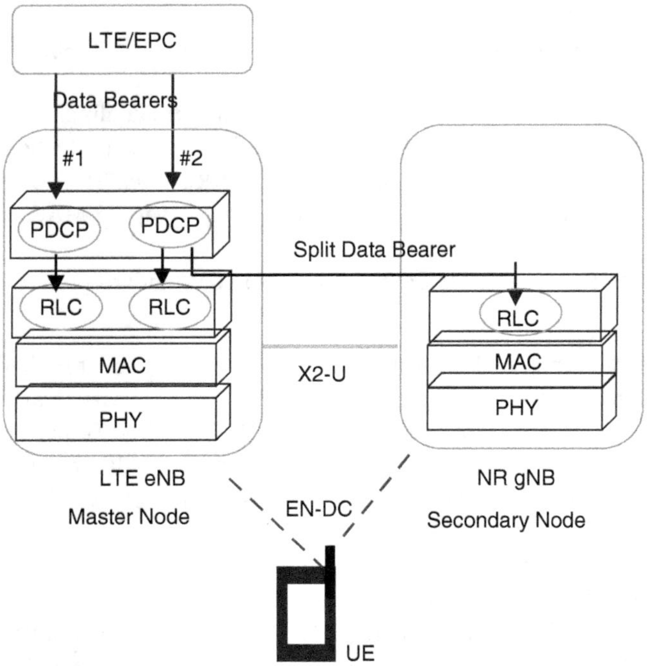

Figure 19.2 Illustration: splitting of DRB by the PDCP layer in an EN-DC setup.

with the MN and the SN. The MN/eNB has the S1 user plane as well as a control plane connection with LTE/EPC network. However, the SN/NR gNB is not connected to the LTE/EPC but is connected to the LTE eNB only through the X2 interface. Thus, a data radio bearer split takes place at the E-UTRAN/eNB end only.

User data from the LTE/EPC network arrives through two DL DRBs at the LTE eNB PDCP layer with two entities. Each PDCP entity is being represented by an oval symbol in the illustration. A DRB is generally handled either by the MCG or SCG. However, a DRB may be required to be handled by both cell groups, i.e. MCG and SCG. The PDCP layer enables this capability by splitting the concerned DRB and routing its packets to both cell groups. As illustrated, one PDCP entity splits the DRB #2 and routes the PDUs to the LTE eNB RLC and the NR gNB RLC layer. Further, a split bearer has two RLC layer entities, operating either at ACK or UNACK mode – one at LTE eNB and one NR gNB. Further, as illustrated, the non-spilt DRB #1 is handled by a separate PDCP and an RLC entity. Note that, in this illustration, Figure 19.2, the PDCP layer splits only one unidirectional DRB #2, i.e. in the DL direction, at the E-UTRAN/eNB end.

Dual connectivity and split bearer-related configuration in the NR PDCP layer is performed by the RRC layer. E-UTRA NR Dual-Connectivity (EN-DC) feature was described earlier in Section 16.6.1.

- **Role of PDCP in Delivering Reliable Communication**

Some critical applications may require highly reliable and low-latency data transfer services from a 5G network. Overall latency may increase due to the retransmission requirements of some of the PDUs. One way to provide such a highly reliable service is to duplicate the packet transmissions between 5G NR and the UE. The PDCP layer provides the packet duplication capability in the 5G NR system. Packet duplication functionality is similar to the split bearer capability provided by the PDCP layer in a dual-connected network arrangement as illustrated in Figure 19.2. In case of split bearer arrangement, the packets of a DRB are transmitted through two separate bearers, through the NN and SNs, to a UE. However, in case of packet duplication by the PDCP layer at the

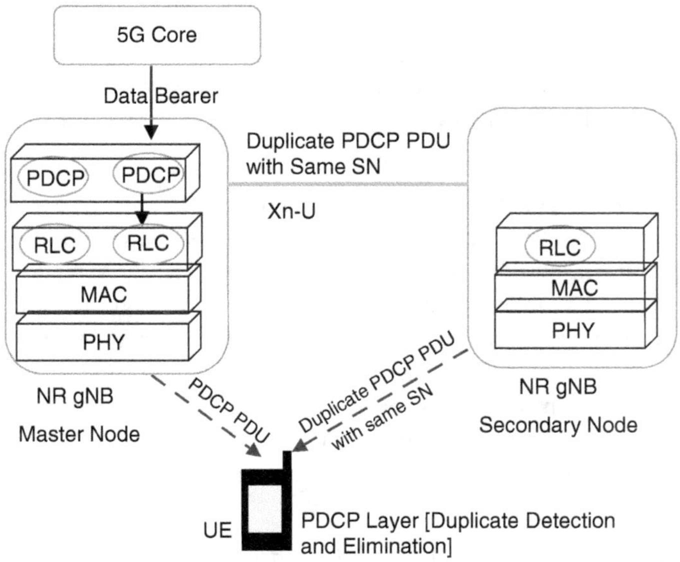

Figure 19.3 Illustration: packet duplication by the PDCP layer.

transmitting end, the same PDCP PDU, excluding the PDCP control PDU, is transmitted through two RLC/MAC entities of two separate networks, i.e. MCG and SCG, to a UE. Thus, a packet duplication avoids the necessities of retransmissions over the air interface in case erroneous packets are received by a UE from the MCG NG-RAN. The RLC entity at the MCG end is called as *Primary RLC*, and the RLC entity at the SCG end is called as the *Secondary RLC*. The receiving PDCP layer discards the duplicate PDCP PDUs, based on the same PDCP sequence number, in case the same PDUs are already received correctly. Packet duplication by the PDCP layer is illustrated in Figure 19.3.

NR user plane protocol is used to transfer DL packets related to a split bearer or packet duplication feature over the X2-U or Xn-U interface. For more details on the NR user plane protocol, refer to TS 38.425 [123]. The RRC layer activates/deactivates dynamically the packet duplication functionality over a DRB of the PDCP layer through the IE: *PDCP-config->pdcp-Duplication*. Packet duplication is activated at the bearer level only since all the applications may not require highly reliable data transfer services in a 5G network. Note that a packet duplication may be also activated for transmission of control plane information over a signaling radio bearer (SRB).

- **PDCP PDU Types and Formats**

NR PDCP layer has similar PDU types and format to the LTE PDCP layer. PDCP PDU types can be DATA PDU to transfer higher layer signaling as well as user traffic and a CONTROL PDU to transfer PDCP layer-related control information between peers. However, in the NR, the SN field used in PDCP data PDU for user traffic has a length larger than the SN field used by the LTE PDCP layer for the same PDCP data PDU. Also, as mentioned above, the NR user plane traffic/data may be also integrity protected apart from the control plane data. Therefore, in the NR, the PDCP data PDU for user plane data also contains the optional Message Authentication Code-I (MAC-I) parameter for integrity protection purposes, which is not present in the LTE system. However, the presence of the MAC-I parameter in the case of PDCP data PDU used to transfer signaling or control plane data is mandatory.

For more information on the PDCP, PDU formats, and their parameters, refer to TS 38.323 [115]. Next, the NR RLC layer is described.

19.4 RLC Layer

The RLC layer transfers the upper-layer PDUs, user data as well as signaling over the NR air interface between UE and NG-RAN. Like the LTE RLC layer, the NR RLC layer also provides the following types of data transfer services to meet the services' quality requirements of different types of applications.

- Acknowledged mode (AM), which is used for transmission of information or services of point-to-point type over the Dedicated Control Channel (DCCH) and Dedicated Traffic Channel (DTCH) logical channels. RRC layer messages and user applications traffic such as background file transfer may be transferred using the RLC AM mode.
- Unacknowledged Mode (UM), which is used for transmission of information or services of point-to-point type over the DTCH logical channel. Delay-sensitive user traffic such as voice over IP packet may be transferred using the RLC UM mode.
- Transparent Mode (TM), which is used for transmission of information or services of point-to-multipoint type over the Broadcast Control Channel (BCCH), Common Control Channel (CCCH), and Paging Control Channel (PCCH) logical channels. TM is also used to transfer the RRC layer message over the Signaling Radio Bearer #0 only.

Similar to the LTE RLC layer, the NR RLC layer is also *entity* based, see Figure 19.4, where different entities provide data transfer services to its upper layers through their respective logical channel.

As illustrated, an AM entity provides data transfer services in both the UL and DL directions (represented by up and down arrows). On the other hand, there are two entities in TM or UM mode, each entity providing a unidirectional data transfer services either in UL or DL direction only. To provide data transfer services in any one of the transmission modes mentioned above, an RLC entity performs several functions and procedures as mentioned below.

- **Functions and Procedures in AM Mode Only**
  - Retransmission of missing PDUs – In AM, the RLC layer ensures an error-free transmission through the automatic repeat request (ARQ) method. ARQ works through its retransmission feature that operates between the transmitting and receiving RLC AM entities. To enable the retransmissions of missing PDUs based on their SNs, the transmitting RLC entity requests a status report from the receiving RLC entity through the *polling* function which is indicated by a *Poll* bit in its header. Successful (ACK) as well as missing (NACK) PDUs are reported by the receiving RLC layer to the transmitting RLC layer through a *STATUS* (ACK/NACK) reporting mechanism. Based on the status report, the transmitting RLC layer retransmits the missing PDUs to the receiving RLC layer to ensure error-free delivery of information. Further, to ensure efficient usage of radio resources, the frequency of transmission of a *STATUS* report from the receiving to the transmitting RLC layer entity is controlled by the timer *t-StatusProhibit.*
  - Resegmentation of retransmitted RLC PDU at transmitting end in case the total size of the RLC PDUs to be transmitted does not fit into the scheduled transport block size. The size of a transport block, and hence the data rate, varies based on the prevailing radio conditions.
  - Duplicate detection and its removal from the reception buffer at the receiving end so that no duplicate data is delivered to the upper layer.

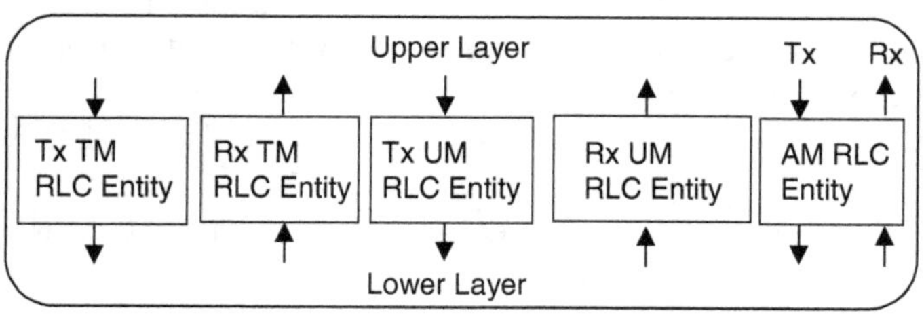

Figure 19.4 NR: RLC layer Entity Model.

- Protocol error detection upon receiving an RLC PDU with reserved value or invalid value of a field.

- **Functions and Procedures in AM and UM Modes**
  - Segmentation of SDU – RLC layer performs the segmentation, if required, of an SDU or a PDU received from the PDCP layer into RLC PDUs of appropriate size at the transmitting end. The size of an RLC PDU is not fixed but varies dynamically based on the transport block size indicated from the physical layer. However, the maximum size of the data field of an RLC PDU is 9000 bytes; refer to TS 38.322 [114], 38.323 [115]. Each segment of a segmented SDU is identified through an SN which is included as part of RLC header information.
  - Reassembly of an SDU at the receiving RLC end (in case the SDU was segmented at the transmitting RLC end) and deliver it to the receiving upper layer as soon as it is available.
  - RLC SDU discards by the transmitting end, at the direction of the PDCP layer.

Figure 19.5 illustrates side by side the *segmentation and concatenation* processes of three PDCP SDUs – 1,2,3 by the LTE RLC layer (left side) as well as only the *segmentation* process of the NR RLC layer (right side) along with the generation of a MAC PDU from the RLC PDU(s) or segment. As illustrated, the SDU1–SDU3 belong to a particular DRB. SDU3 is shown to be a segmented one in this illustration. An LTE RLC layer PDU is formed by concatenating the SDU1, SDU2, and a part/segment from the SDU3. RLC layer also adds the necessary header (H) information in front of it and submits it to the lower layer. The LTE MAC layer generates its PDU from the RLC PDU and adds its header (H) information in front of the MAC PDU.

On the other hand, an individual NR RLC PDU is generated from each SDU received from the PDCP layer. As illustrated, an SDU may be segmented by the NR RLC layer also, if required. The NR RLC layer does not concatenate the PDUs and segment of an SDU, but the NR MAC layer combines such RLC PDUs, and if required one part of the segmented SDU into a MAC PDU, as illustrated in Figure 19.5. The other part of the segmented SDU may fit completely into an RLC and a MAC PDU.The MAC PDU is transmitted as a transport block through the respective transport channel over the LTE or NR air interface.

The RLC layer entities use transmission and reception windows for its operation in AM and UM modes. PDUs with an SN that is within the transmission or reception window only is transmitted or received by the peer RLC layer entities. PDUs which are not within transmission or reception windows are discarded.

- **NR RLC Layer: Meeting Low-Latency Requirements of 5G System**
  - Unlike the LTE RLC layer, the NR RLC layer does not provide in-sequence delivery of RLC SDUs to upper layers by reordering them based on their SNs at the receiving end. Due to this, the receiving RLC layer forwards the received PDU(s) and its SDU to the PDCP layer as soon as they become available since no PDU

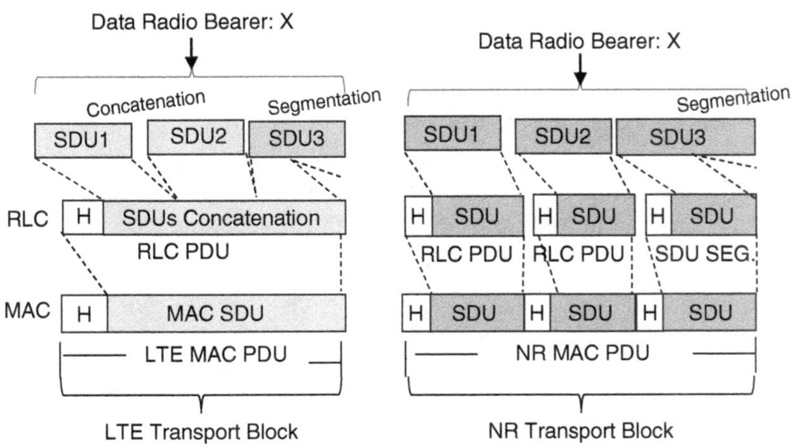

Figure 19.5 Illustration: comparisons of LTE and NR RLC layer SDU segmentation and concatenation processes.

reordering timer is running at the receiving RLC layer end, which is one of the differences between the LTE and 5G NR RLC layers. Thus, to forward an SDU to the higher layers, the receiving NR RLC layer does not have to wait for the retransmission (and hence no reordering requirement) of a PDU of any missing packet sent earlier by the transmitting RLC layer. The absence of SDUs reordering task at the receiving end saves time for the RLC layer, which reduces overall latency and meets the low-latency requirements of the 5G system. Hence, the receiving RLC layer forwards the received SDUs, after the reassembly from its PDUs in case of a segment SDU, to the upper layer, i.e. PDCP, immediately without waiting for the expected SDUs in sequence. However, the task of reordered and in-sequence delivery of data is taken care of by the receiving PDCP layer based on its SN.

- Unlike the LTE RLC layer, the NR RLC layer at the transmitting end does not perform concatenation of small-size SDUs into the *Data* field of the concerned PDU, which are received from the PDCP layer. The *Data* field of an LTE RLC PDU contains and combines RLC SDUs and RLC SDU segment, whereas the *Data* field of an NR RLC PDU contains either an RLC SDU or RLC SDU segment. Also, for SDU concatenation purposes, the size of the transport block is required to be known in advance by the RLC layer, which is not available to transmitting the RLC layer until resources are granted. Thus, the absence of the SDUs concatenation task at the NR RLC layer enables the MAC layer to assemble RLC PDUs in advance and avoid its waiting time, unlike in the LTE, for the corresponding size of a transport block to be used for transmission.

Due to the absence of the above processing tasks at the transmitting and receiving NR RLC layer, the overall latency in transmitting of upper-layer data over the NR interface and its processing is reduced, which further helps to meet the overall low-latency requirement of the 5G system.

- **RLC PDU Types and Formats**

NR RLC layer has DATA and CONTROL PDU types. A data PDU type is used to transfer user or signaling data in TM, AM, or UM mode. An AM or UM mode RLC layer DATA PDU may contain a complete SDU or a segment of an SDU. Such SDUs are received from the PDCP layer and, if required, the RLC layer performs segmentation of an SDU. TM mode is not used to transfer user traffic and is a pass-through PDU as no RLC layer overhead, in terms of a header, is added to its SDU. A CONTROL PDU, in case of AM only, is used to provide ACK information to the sender on the status of RLC PDUs received by the receiving RLC layer. Unlike the LTE RLC layer header, there is no length indicator (LI) field in the NR RLC layer header. This is because, unlike the LTE RLC layer, an NR RLC layer PDU does not contain concatenated or combined SDUs from the upper layer. For more information on the RLC layer protocol behaviour, PDU types and their header information and other implementation aspects details, refer to TS 38.322 [114]. RLC layer is configured by the RRC layer under the IE: *rlc-Config*.

19.5 MAC Layer

Like the LTE MAC layer, the NR MAC layer also provides data, i.e. user and signaling/control, transfer, and radio resource allocation services to the upper layers. To provide these services, the NR MAC layer performs several functions and procedures, in a UL or DL, which are similar to the functions performed by the LTE MAC layer. An overview of some of the NR MAC layer functions and procedures is described in the subsequent sections. For more information on the functions and procedures of the NR MAC layer, refer to TS 38.321 [113].

19.5.1 Functions and Procedures

- A mapping between transport channels and logical channels in UL and DL. A transport channel can transfer multiplexed data originated from different logical channels.

- Multiplexing of MAC SDUs, at transmitting end, into a transport block for their delivery to the physical layer through the concerned transport channel. MAC SDUs received from the RLC layer may belong to a single or different logical channel, i.e. CCCH, DTCH, and DCCH. Apart from multiplexing of different logical channels and information carried by them, a MAC PDU can also carry MAC Control Elements (CE) over the downlink shared transport channel (DL-SCH) or uplink shared (UL-SCH) transport channel. MAC CE is a kind of signaling information which is exchanged between a UE and gNB and vice versa in UL or DL direction, requesting a resource (from UE to gNB), or giving a command from gNB to a UE. Each MAC CE is associated with its corresponding logical channel identifier (LCID) which is a part of the MAC subheader; refer to Tables 6.2.1-1/6.2.1-2 in TS 38.321 [113].
- Demultiplexing of MAC SDUs, at receiving end, into a single or different logical channel from the transport block delivered by the physical layer. Demultiplexed MAC SDUs are delivered to the different RLC entities at the receiving end.
- Triggering of a *Scheduling Request* (SR) to the UE physical layer. Upon the arrival of new data in its transmission buffer, a UE MAC layer may trigger an SR to its physical layer to request UL resources grant from the NG-RAN. SR is another way, other than a random access procedure, for a UE to get assigned resources and schedule it by the NG-RAN in the UL direction.
- Logical Channel Prioritization in UL, which is used by a UE for its data transmission of a bearer in the UL direction as scheduled by the gNB. The priority of a logical channel is configured by the RRC layer through the IE: *LogicalChannelConfig*. As the priority information value increases, the priority level decreases. Signaling messages transmitted over the CCCH or MAC control element (CE) with Cell Radio-Network Temporary Identifier (C-RNTI), LCID #58, have the highest priority. On the other hand, the MAC CE for buffer status report (BSR) with LCID: 59–62 has the lowest priority. Depending on the priority, the MAC layer at UE end multiplexes the data of different logical channels for transmission in the UL direction to the gNB.
- MAC layer supports the management of beam or directional antenna-based transmissions and receptions, as compared to LTE MAC layer, using carrier frequency in millimeter wave (mmWave) range. UE MAC layer performs the beam management operation such as beam failure detection and recovery procedure by triggering its physical layer to send a random access procedure toward the NG-RAN.

The NR MAC layer performs several user plane and control plane procedures in UL and DL, as mentioned below:

- Random access procedure, which is used by a UE to make UL resource requests as well as synchronization purposes with NG-RAN;
- DL data reception over the shared channel and its physical resources which is indicated dynamically through a physical downlink control channel (PDCCH) from NG-RAN to UE;
- UL data transfer through a shared channel and its UL resource which is granted dynamically through the physical PDCCH from NG-RAN to UE. A UE may be also granted UL resources as part of a random access response and configured scheduling (CS) procedure;
- Discontinuous reception, which is configured by the RRC layer to save UE battery by allowing the UE to monitor DL PDCCH discontinuously at specified times/intervals;
- Semi-persistent scheduling (SPS);
- Bandwidth Part (BWP) operation, which is described in a later section;
- Paging Channel (PCH) Reception, which monitors a paging request intended for a UE and delivers a MAC PDU to the upper layer;
- Physical Broadcast Channel (PBCH) Reception, which monitors broadcasted system information sent by the NG-RAN and delivers a MAC PDU to the upper layer; and
- Activation and deactivation of PDCP duplication.

Some of the procedures of the MAC layer are described in the subsequent sections. For the actual procedural steps used in the above procedure, refer to TS 38.321 [113].

19.5.2 Scheduling Procedure

Similar to the LTE system, in 5G NR also, UE scheduling functions in DL and UL directions are performed by the MAC layer at the gNB end, more specifically at the *gNB-DU node*. However, in the LTE system, a UE scheduling is performed at a subframe level, whereas in the NR system, it is a timeslot-based scheduling whose scheduling interval may be less than that of an LTE subframe. Also, due to the different use cases of 5GS and the characteristics and requirements of their services, the scheduling procedures and algorithms used in the NR would be more complex and versatile than the LTE scheduling procedure. Implementation-dependent UE/device scheduling strategies and procedures would be the key success factor for the entire NG-RAN. Tasks such as the slice-specific UL and DL scheduling of UEs/devices and allocation of shared radio resources dynamically in terms of physical layer resource blocks (RB) in the frequency domain and Orthogonal Frequency Division Multiplexing (OFDM) symbols and slots in the time domain are required to be performed by the MAC layer. DL scheduling operation controls to which UE the data should be transmitted to from the gNB over the shared DL-SCH channel. UL scheduling operations grant radio resources and control which UE is allowed to transmit data to the gNB over the shared UL-SCH channel.

UE scheduling by the MAC layer in UL and DL directions may be of the following types:

- DL Scheduling

Types: Dynamic and Semi-persistent.

- UL Scheduling

Types: Dynamic and CS.

• Dynamic Scheduling

In the previous generation RANs, radio resources were allocated and scheduled to transfer data packets of burst and mostly larger packets. However, in the NR system, admission control and radio resources are required to be optimally allocated and scheduled to UE, considering the QoS requirements of the different use cases and diverse services under them. Some of the factors to be considered for the allocation of radio resources to a UE are as follows:

- Network slice type, i.e. Enhanced Mobile Broadband (eMBB), URLLC, Massive Internet of Things (mIOT), other slices;
- Type of traffic;
- Delay/latency sensitivity; and
- Traffic size, i.e. small to large size data.

DL and UL scheduler allocate radio resources accordingly, i.e. timeslot(s), physical resource block(s) (PRB), and dynamically to UEs to meet the QoS, in terms of bandwidth, reliability, latency, and requirements of different applications. The concept of working of a slice-specific dedicated scheduler algorithm at NR MAC layer for allocation of radio resources to a particular use case, i.e. eMBB, mIoT, and URLLC, of 5GS, is illustrated in Figure 19.6. As illustrated, the allocation of radio resources in the DL or UL direction by the MAC scheduler depends on several factors such as the 5G use case and its corresponding slice, amount of data available or buffer status at UE end, QoS requirements for each DRB, and current radio/channel conditions/qualities. As illustrated, the output of the scheduling algorithm in NG-RAN/gNB may be of the following types:

- A single timeslot or multislots-based scheduling for large data transmission;
- Non-timeslot-based scheduling, for short/small data, which may be called as the mini-slot or short slot for application such as the URLLC. In this type of UE scheduling, a UE is assigned resources at OFDM symbol level, which can be 2, 4, or 7 symbols in length per mini-slot; refer to Table 5.1.2.1-1 in TS 38.214 [109]. Thus, a UE with a mini-slot allocation does not have to wait for full slot duration, i.e. 1 ms, to start its scheduling, meeting the low-latency requirements of the 5G system.

UL data originated from a group of logical channels is buffered at UE, and the readiness for its transmission is intimated by sending a *BSR* MAC CE to gNB as part of a MAC PDU. However, if the current UL resources cannot be used,

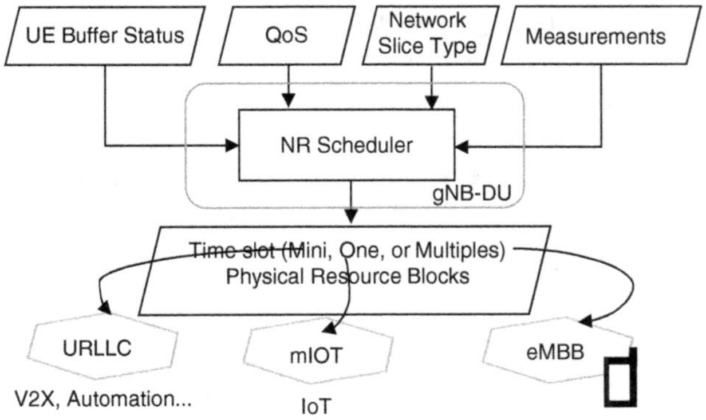

Figure 19.6 Illustration: typical inputs and outputs of a scheduler algorithm at NR MAC layer.

a UE can request the gNB to allocate UL resources for a new transmission of the buffered data by triggering an *SR* to the gNB over the PUCCH, if available. Otherwise, a UE, by its physical layer, will make a random access request procedure toward the gNB to request UL resources for transmitting the UL data buffered at UE end. Note that the UE MAC layer controls the trigger of an SR to its UE physical layer, which in turn sends the SR over PUCCH to the gNB.

Measurement information consists of current channel states as perceived and measured by a UE in the DL and UL directions based on which the gNB determines channel qualities. DL channel state information (CSI) is provided through the *CSI*, which is similar to LTE CSI, from UE to gNB. UL CSI is provided through the *Sounding Reference Signal* (SRS), which is similar to LTE SRS, from UE to gNB.

Dynamically scheduled DL radio resources, for transmission of information over the Physical Downlink Shared Channel (PDSCH), are allocated to a UE using the downlink control information (DCI) 1_0/DCI 1_1. Similarly, dynamically scheduled UL radio resources, for transmission of information over the Physical Uplink Shared Channel (PUSCH), are allocated to a UE using the DCI 0_0/DCI 0_1. UEs monitor the PDCCH, which is scrambled with its corresponding C-RNTI, to find the DL or UL radio resources assigned to them by the gNB. DCIs are described later in Section 19.6.10.5. In addition to the dynamic scheduling type as described above, a UE may be also assigned radio resources and scheduled in DL and UL directions in the following ways.

- Semi-Static Resource Allocation in DL

Radio resources allocation to user traffic which is semi-persistent, such as for VoIP packets, can be carried out semi-statically by the NG-RAN/gNB in the DL direction. Such resources are assigned to a UE and scheduled by the RRC layer through its IE: *SPS-Config* which is configured through the RRC layer signaling message. RRC layer configures resources such as the modulation and coding scheme (MCS) to be used, the number of hybrid automatic repeat request (HARQ) processes, and periodicity for an SPS. A separate identity which is called the *configured scheduling RNTI* (cs-RNTI) is assigned to a UE for SPS purpose. cyclic redundancy check (CRC) of PDCCH carrying the SPS resources allocated to a UE is scrambled with the assigned cs-RNTI. In the LTE system, it is known as the *Semi-Persistent Scheduling* C-RNTI, see TS 36.321 [93].

- CS: Flexible Resource Grant in UL for PUSCH

A UE can be also configured, through RRC layer signaling, to schedule for UL transmission only without dynamic scheduling by the MAC layer at NG-RAN/gNB end. Such a CS of a UE in UL may be of two types – Type 1 and Type 2 – which are configured/activated by the RRC layer per serving cell. In Type 1, the RRC layer activates the resource parameters such as the frequency time-domain resources, periodicity, MCS, and number of HARQ processes under the IE: *ConfiguredGrantConfig -> rrc-ConfiguredUplinkGrant*. Clearly, in Type 1, UL resource allocation does not involve the physical layer in terms of allocation of PUSCH resources through the PDCCH/DCI.

In Type 2 UL grant, however, PUSCH resources are not allocated through RRC layer signaling. Type 2 resource allocation involves the physical layer where semi-static UL radio resources are assigned over the PDCCH/DCI whose CRC is scrambled with *cs-RNTI*. CS-RNTI indicates the Type 2 UL resource allocation to a UE. In this case, the IE *ConfiguredGrantConfig* does not contain the *IE:rrc-ConfiguredUplinkGrant* information. In both the types, a UE stores the UL grants allocated to it by the gNB. Such a UL grant, i.e. without through dynamic grant, the resource may be released through its deactivation. Figure 19.7 illustrates side-by-side comparisons of transmission of UL data over the PUSCH with Type 1 and Type 2 UE scheduling.

Figure 19.7 illustrates the UL transmissions by a UE using NR allocated resources under the following scheduling types:

a) Normal dynamic UL scheduling as indicated by a DCI whose CRC is scrambled with C-RNTI;
b) UL CS – Type 1 as well as dynamic scheduling (C-RNTI); and
c) UL CS – Type 2 as indicated by a DCI whose CRC is scrambled with CS-RNTI.

Using the scheduling type shown in Figure 19.7a, a UE transmits only once over the PUSCH. Using the scheduling type shown in Figure 19.7b, a UE performs periodic PUSCH transmissions using Type 1 CS. It also illustrates the allocation of UL resources and PUSCH transmission through dynamic scheduling with UE C-RNTI. Such allocation overrides the CS. Using the scheduling type shown in Figure 19.7c, a UE performs periodic PUSCH transmissions using Type 2 CS.

19.5.3 Random Access Procedure

Similarly, as in the case of the LTE system, a UE initiates a random access channel (RACH) procedure to get UL synchronization with a 5G NG-RAN as well as to request for allocation of the UL radio resource for transmission of initial Layer 3 (Msg3) message toward the NG-RAN. A UE may initiate a RACH procedure toward the NG-RAN as a result of any one of the following events/reasons:

- Initial access from the idle state, e.g. switched on and followed by the transition to the active state;
- Upon the arrival of DL (through PDCCH order: *Random Access Preamble index* in DCI 1_0) or UL data;
- UE transition from RRC_INACITVE to ACTIVE state;
- Call handover;

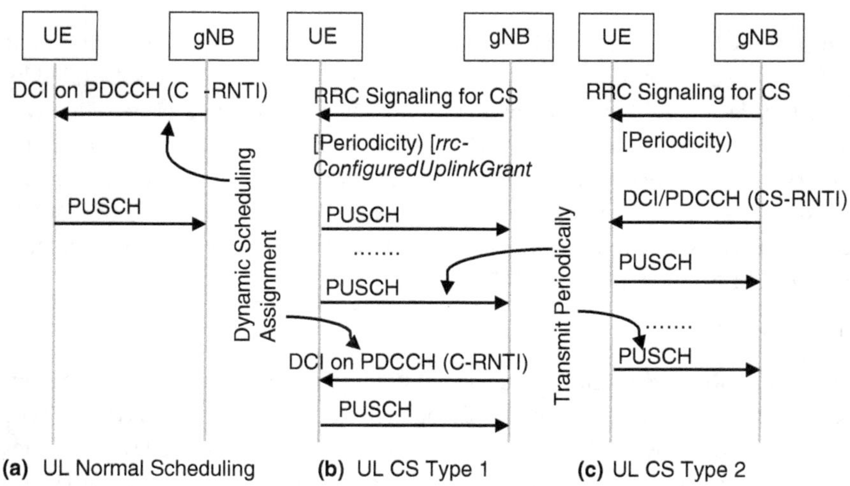

Figure 19.7 Illustration: uplink: dynamic and configured scheduling Type 1 and Type 2.

- RRC connection re-establishment procedure;
- Beam failure recovery in case of detection of a beam failure with high-frequency or mmWave transmission/reception, as indicated by the physical layer to the MAC layer. The MAC layer initiates and instructs the physical layer to start a RACH procedure to the NG-RAN;
- UE Request for other SIs.

- **Role of MAC Layer**

Similarly, as in the case of the LTE system, an NR RACH procedure also consists of the following tasks that are performed by the UE MAC layer (TS 38.321 [113]):

- Selection and transmission of a random access preamble;
- Processing of a random access response (RAR);
- Transmission of the first UL scheduled Layer 3 message;
- Contention resolution in case of a contention-based random access procedure.

It may be noted that the NR UE physical layer also performs associated tasks as part of an overall random access procedure which shall be described in a later Section 19.6.20. The NR UE MAC layer controls the RACH procedure-related tasks performed by its physical layer. That is, the UE MAC layer controls how and when to generate the RACH request information and transmit it over the UL Physical Random-Access Channel (PRACH) by its physical layer. For example, the UE MAC layer triggers a RACH procedure to its physical layer as part of a beam failure recovery.

The UE MAC layer may also trigger a RACH procedure to its physical layer at the order of NG-RAN over PDCCH DCI 1_0. In case of a contention-based random access procedure, the MAC layer at UE end starts a contention resolution timer after transmitting the Layer 3 (Msg3) message and keeps track to declare the RACH procedure either as success or failure. Further, the MAC layer at UE end processes random access response received from the NG-RAN and declares whether the random access attempt was a success or not as described above. For more information on the above tasks performed by the NR UE MAC layer as part of a RACH procedure and its PDU header/subheader structure, refer to TS 38.321 [113]. Tasks mentioned above as part of an NR random access procedure are described below.

- **Selection and Transmission of a Random Access Preamble**

Similar to the LTE system, in the NR system also, there are 64 preambles available in a cell. Some of these preambles are used as dedicated for events such as UE beam failure recovery, handover, and system information request purposes. The remaining preambles are used to make contention-based RACH procedure requests by UEs in a cell. However, the preambles available for contention-based RACH procedure requests are divided into two groups: A and B. RRC layer configures the number of RACH preambles under the Group A based on which the UE derives the number of preambles under the Group B. Appropriate RACH preamble group to be used is determined at runtime by a UE based on the size of the initial Layer 3 (Msg3) message. RACH resource selection steps are illustrated in Figure 19.8 based on the description available at TS 38.321 [113].

- **Transmission of a RACH Preamble**

A UE transmits a RACH preamble on a particular PRACH occasion which is nothing but radio resources in time and frequency domain. Further, a *Random Access* (RA)-*RNTI* is associated with each preamble occasion where a UE transmits a RACH preamble. An RA-RNTI is calculated by the UE using the first index (0–13) of the allocated OFDM symbol and timeslot (0–79) and index (0–7) of the PRACH in the frequency domain and UL carrier ID being used for transmission of the associated preamble. For more information on the formula used by a UE to calculate an RA-RNTI, refer to Section 5.1.3, TS 38.321 [113]. The UE MAC layer triggers its physical layer to transmit the RACH preamble with selected PRACH occasion, preamble index for the selected preamble, and RA-RNTI.

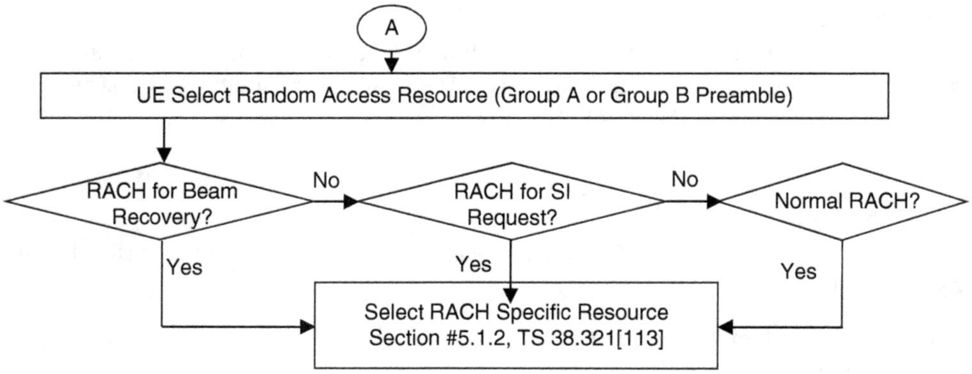

Figure 19.8 Illustration: NR RACH resource selection procedure.

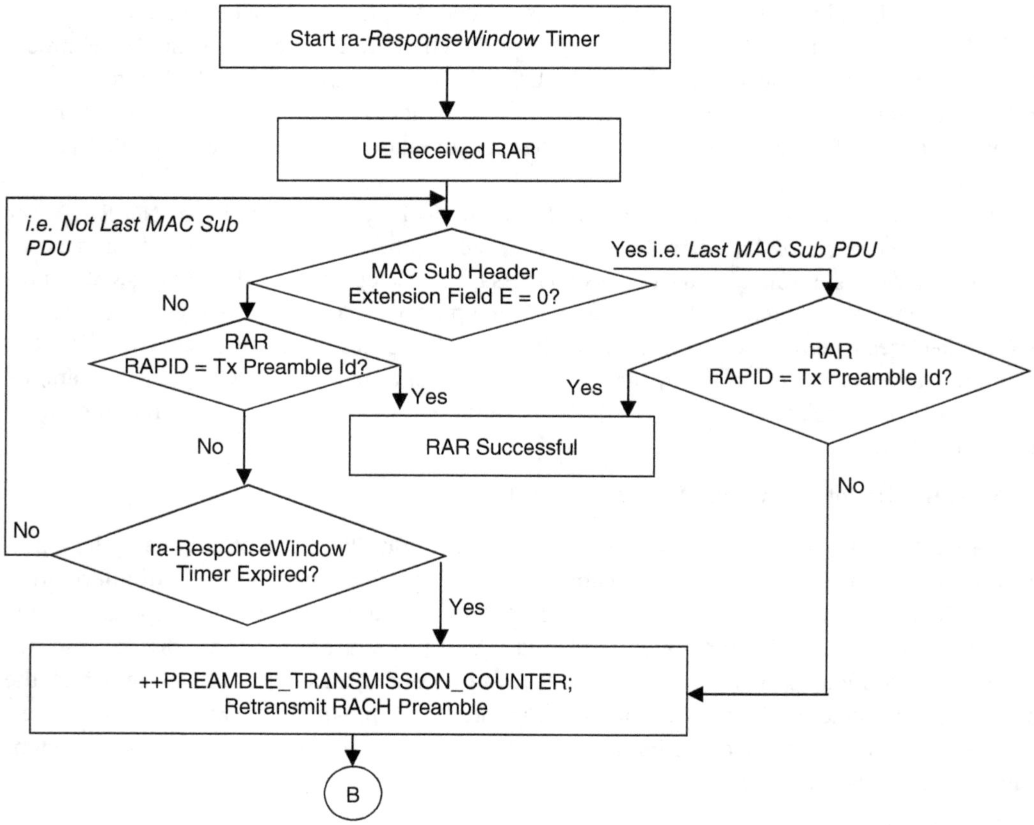

Figure 19.9 Illustration: NR UE RACH RAR process flow, TS 38.321 [113].

- **Processing of RAR**

The steps to process a RAR message, received from the NG-RAN, at the UE MAC layer end are illustrated in Figure 19.9 which is based on Section 5.1.4, TS 38.321 [113].

After transmitting a preamble, UEs are expected to receive a RACH response from the NG-RAN within the RACH response window *(ra-ResponseWindow)* time as configured by the RRC layer. To detect a RACH response, UEs look for a DL PDCCH whose CRC is scrambled with *RA-RNTI*. A PDCCH further indicates the DL transport block resource information, in time and frequency domain, to UEs. The MAC PDU in the DL transport block may contain several MAC sub-PDUs, refer to Figure 6.1.5-3, TS 38.321 [113], containing *RARs* for several UEs that might have transmitted RACH preamble in UL with the same RACH resources.

A RAR message from the gNB carries the UL resource grant (time, frequency, and modulation scheme; refer to Table 8.2-1 in TS 38.321 [113]), as payload, and a temporary C-RNTI, which are provided by the NG-RAN to attempting UEs. Also, a RAR message from gNB to UE carries the *Random Access Preamble Identifier (*RAPID) as part of the MAC layer subheader. As the RAPID is the same for all the contending UEs, they would send their initial Layer 3 message (Msg3), i.e. *RRCSetupRequest*, with the allocated temporary C-RNTI and UL resources granted over the PUSCH to the NG-RAN.

- **Types of Random Access: Contention Based vs. Contention Free**

Similar to the previous generation mobile communications systems, the RACH procedure in the NR system can be also a contention-based as well as contention-free access. Being random, multiple UEs may select the same RACH resources, in terms of the same preamble sequence and same OFDM symbol, timeslot, frequency allocation, and initiate the random access procedure toward the NG-RAN. Random access attempts by multiple UEs with the same RACH resources result in contention among the UEs. The method of resolution of contention-based as well as the method used in a contention-free random access procedure is described below, which are similar to the resolution methods used in the LTE system.

- Contention-Based Random Access Procedure

In a contention-based random access procedure, the contention resolution mechanism is based on the UE detection of the same initial Layer 3 message (Msg3: *RRCSetupRequest*) sent by them to the NG-RAN. A UE transmits a RACH preamble with the corresponding RA-RNTI. An RA-RNTI is calculated by the UE using the first index of the allocated OFDM symbol, timeslot, and index of the PRACH in the frequency domain and UL carrier ID being used for transmission of the associated preamble. For more information on it, refer to Section 5.1.3, TS 38.321 [113].

After transmitting a preamble, UEs are expected to receive a RACH response from the NG-RAN within the RACH response window *(ra-ResponseWindow)* as illustrated in Figure 19.9. A RAR message carries the UL resources grant (time and frequency), as payload, and a temporary C-RNTI, which are provided by the NG-RAN to attempting UEs. Also, a RAR message from gNB to UE carries the *RAPID* as part of the MAC layer subheader. As the RAPID is the same for all the contending UEs, they would send their initial Layer 3 message (Msg3), i.e. *RRCSetupRequest*, Figure 19.10, over the PUSCH to the NG-RAN with the allocated temporary C-RNTI and UL resources granted.

The *RRCSetupRequest* is the first scheduled message in the UL direction from the UEs, which contains an establishment cause and their respective UE identity. The UE identity is a combination of the rightmost 39 bits of 5G-S-TMSI as well as a random value of 40 bits long. A contention resolution timer is started by each attempting UEs to keep track of the reception of DL PDCCH from the NG-RAN. A UE stops the contention resolution timer upon receiving a PDCCH DCI 1_0, whose CRC is scrambled with the temporary C-RNTI.

The NG-RAN replies with the *RRCSetup(Msg4)* message to the UE. The MAC PDU, which is decoded from the PDSCH using its DL time and frequency resources information as indicated in PDCCH DCI 1_0 for Msg4, carries the *UE Contention Resolution Identity* (CRI*) MAC Control Element*, with DL-SCH LCID = 62. Refer to Section 6.1.3.3 in TS 38.321 [113]. The CRI MAC CE contains the Msg3, i.e. original *RRCSetupRequest with UE identity* sent by UE, from NG-RAN to UE. If the Msg3 sent (UE to NG-RAN) and received (NG-RAN to UE) is found to be the same through the comparison of UE identity by the UE MAC layer, UE RACH attempts contention is resolved. As the concerned UE MAC layer finds the sent-received UE identities as the same, the UE MAC layer would consider the random access procedure as successful. The successful UE responds with a HARQ-ACK as an uplink control information (UCI) over the PUCCH to the NG-RAN; see Figure 19.10. Temporary C-RNTI becomes the C-RNTI for the concerned and successful UE. Figure 19.11 further shows the steps for the contention resolution flow which is based on the description in TS 38.321 [113].

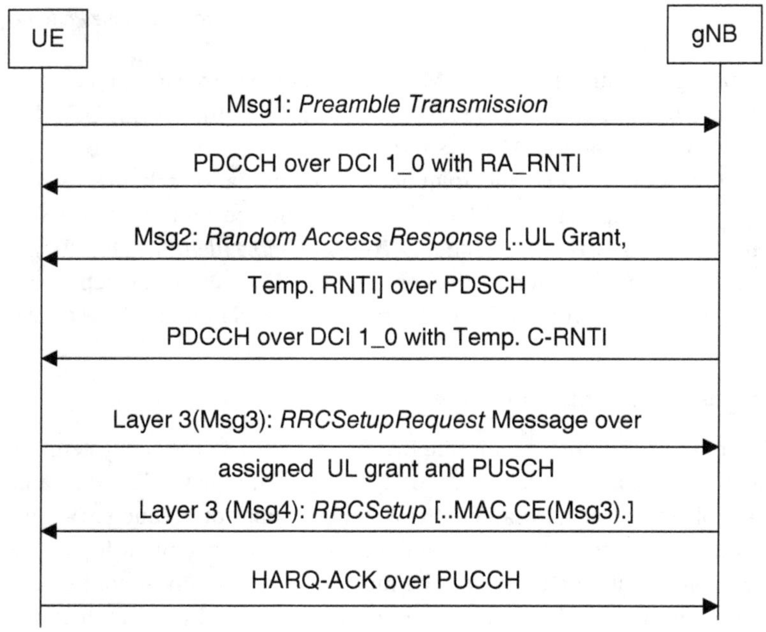

Figure 19.10 Illustration: NR UEs contention-based RACH procedure.

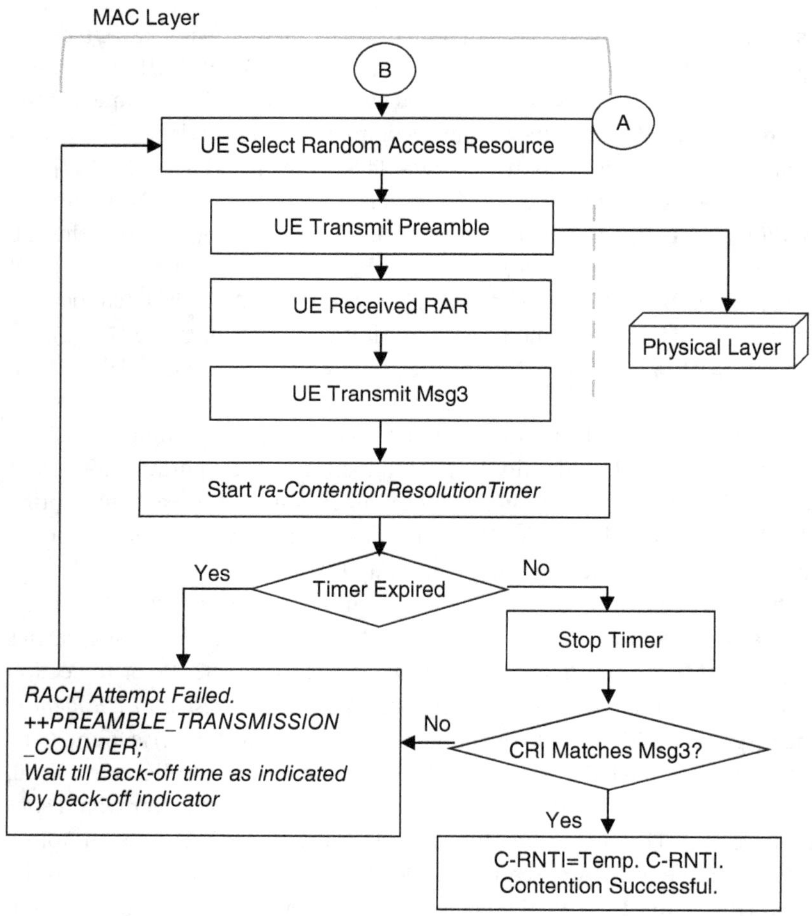

Figure 19.11 Illustration: NR UE RACH procedure contention resolution flow.

Other contending UEs will consider the contention resolution as failed and would restart the RACH procedure by transmitting a RACH preamble again with power being ramped up till the maximum attempt is not exceeded by a value given by the RRC layer IE: *preambleTransMax*.

- Contention-Free Random Access Procedure Through a Dedicated Preamble

In case of a contention-free random access procedure, whose needs arise in a typical handover or a beam failure recovery scenario, a UE uses the dedicated preamble as assigned by the NG-RAN through the RR layer IE *RACH-ConfigDedicated*. Because of this, there is no probability of collision of the concerned RACH procedure with other contending UEs. After transmitting an NG-RAN assigned RACH preamble, a UE looks for the DL PDCCH with CRC scrambled with an *RA-RNTI*. Note that if the RACH procedure attempted was part of a UE beam failure recovery step, the UE searches the PDCCH in the search space *BeamFailureRecoveryConfig -> recoverySearchSpaceId* provided by RRC layer, whose CRC is scrambled with the UE C-RNTI.

The UE decodes the RAR message sent by the NG-RAN from the PDSCH RB information as indicated in the PDCCH DCI 1_0. A RAR message over the PDSCH from gNB to UE carries the RAPID as part of the MAC layer sub-header. As the RAPID in RAR message contains the dedicated preamble transmitted by the attempting UE in UL, the UE MAC layer considers the random access procedure as successful. The payload of a RAR message also contains the UL resources grant allocated for the UE using which the UE transmits the initial Layer 3(Msg3) to the NG-RAN. A contention-free random access procedure with a dedicated preamble described above is illustrated in Figure 19.12.

19.5.4 Error Correction Through HARQ Procedure

Similar to the LTE MAC layer, the NR MAC layer also provides a HARQ-based fast retransmission method to correct MAC PDU-related transmission errors between the transmitting and receiving MAC layer. MAC layer HARQ retransmission is performed in both UL and DL directions for information transmitted over the UL-SCH or DL-SCH. HARQ function is handled by a separate HARQ entity in UL and DL directions. To perform HARQ retransmissions, HARQ entities employ multiple and parallel stop-and-wait HARQ processes, which are similar to the LTE HARQ procedure. Also, unlike the LTE HARQ, which is asynchronous in DL and synchronous type in the UL, the NR MAC layer supports asynchronous HARQ only in both the UL and DL directions. Thus, in the NR system also, an asynchronous UL HARQ procedure is performed with the help of a process ID. The maximum number of HARQ processes, as configured by the RRC layer, in the DL direction is 16, and the default number of DLs HARQ processes is 8. In the UL direction, the maximum number of HARQ processes is 8. Figure 19.13 illustrates the DL data reception over the PDSCH/DL-SCH and its HARQ ACK/NACK process.

Figure 19.13 illustrates eight parallel HARQ processes in the DL direction. An identifier, called *HARQ process number*, is associated with each HARQ process to keep track of the transmission and reception of HARQ data between the transmitting and receiving HARQ process. In the case of the LTE system, a HARQ process number is

Figure 19.12 Illustration: NR contention-free UE RACH procedure.

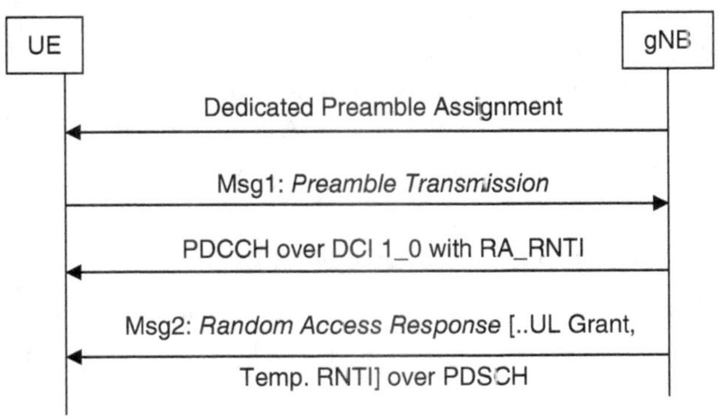

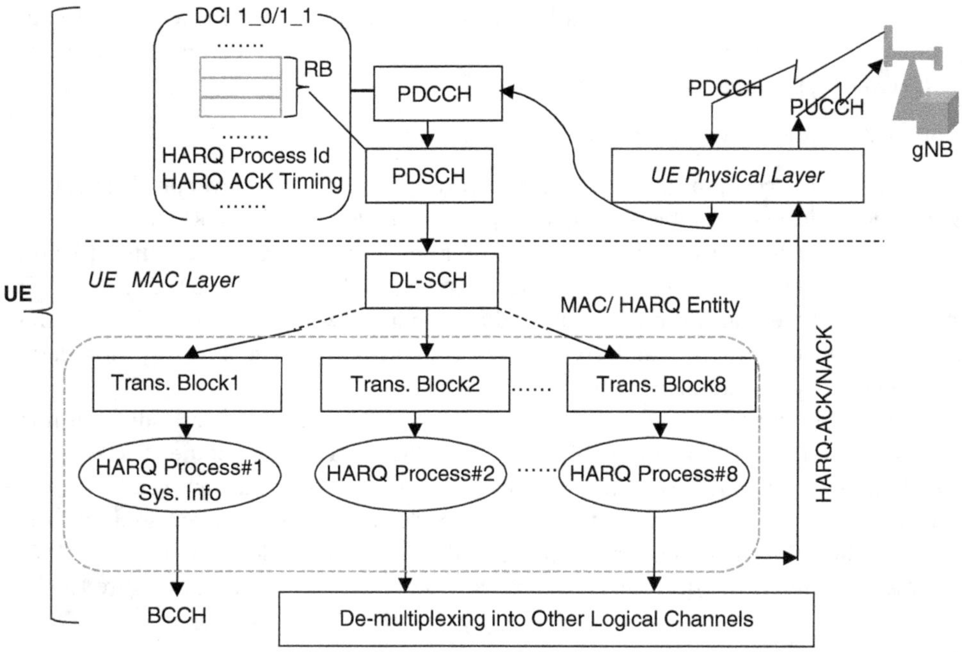

Figure 19.13 Illustration: UE downlink data reception and its HARQ ACK/NACK process.

transmitted by the eNodeB as part of the DL only scheduling information through a DCI over the PDCCH. However, in the case of the NR system, a HARQ process number is transmitted by the gNB through a DCI over the PDCCH as part of the UL as well as DL scheduling information due to its asynchronous nature.

As illustrated in Figure 19.13, a UE decodes the DCI from the PDCCH transmitted by the gNB. Among other information, a DCI contains the assigned RB information to a transport block that carries higher layer data for the UE, the corresponding asynchronous HARQ process number, HARQ ACK/NACK timing (UE to NG-RAN), and so on. From the given RB information and mapped PDSCH (physical) to downlink DL-SCH, the MAC layer attempts to decode the transport block. In Figure 19.13, the DL broadcast HARQ process #1 and its transport block are shown to be associated with the system information (other than *Master Information Block* [MIB]) as broadcasted by the NG-RAN. Thus, the system information is passed to the corresponding logical channel, i.e. BCCH. On the other hand, the transport blocks, from other HARQ processes #2–8, carrying other higher layer data are forwarded for demultiplexing purposes for separation of the decoded data into their respective logical channels, i.e. CCCH, DTCH, and DCCH.

As also illustrated in Figure 19.13, the status of decoding of a transport block is provided from the UE MAC layer to its physical layer in the form of an ACK/NACK. The UE physical layer, in turn, transmits the HARQ ACK/NACK information as a UCI over the PUCCH channel to the NG-RAN/gNB. Based on the HARQ status, the entire transport block or the erroneous code block group (CBG), as a result of segmentation of transport block, only will be retransmitted by the sender HARQ process to the receiving HARQ process. NR physical layer and its HARQ procedure will be described later in Section 19.6.11.

19.5.5 Buffer Status Reporting (BSR) Procedure

The purpose of a BSR from a UE to NG-RAN in the NR system is the same as in the LTE system. A buffer status information is reported by the UE MAC layer through its BSR CE. A BSR from a UE to the NG-RAN contains the information on the amount of UL data available for transmission in terms of buffer size for a group of logical channels, which is identified through *logical channel group ID* (LCG ID). Refer to Tables 6.1.3.1-1/6.1.3.1-2 in TS 38.321 [113]. Thus, the NG-RAN can allocate UL resources accordingly in an optimum way, if requested by the UE. In this case, a UE MAC layer may trigger an SR to its physical layer. The UE will send an SR to the NG-RAN requesting UL resources in the form of UCI over the designated PUCCH, which shall be described later in Section 19.6.10.6.

Similar to the LTE MAC layer at UE, BSR in the NR MAC layer also may contain buffer information for several logical channel groups. NR MAC layer BSR CE can carry the buffer information of up to maximum of eight logical channel groups. But LTE MAC layer BSR CE can carry the buffer information of up to maximum of four logical channel groups only. Accordingly, short and long formats of a BSR CE are used. Short BSR CE contains buffer information for one logical channel group, which is identified by the LCG ID field in its format. Long BSR CE contains buffer information for all the logical channel groups; thus, there is no LCG ID field in its format. The length of the LCG ID field is 3 bits in the NR MAC BSR CE, refer to TS 38.321 [113], whereas it is 2 bits long in the LTE MAC BSR CE. Similar to the LTE system, a BSR in the NR UE MAC layer is also an event-based trigger that can be due to the arrival of new data for a logical channel or expiry of a periodic BSR timer. UE BSR parameters are configured by the RRC layer through IE: *BSR-Config*.

19.5.6 Scheduling Request (SR) Procedure

The purposes of an SR in the NR system, from a UE to NG-RAN, are the same as in the LTE system. An SR is sent by a UE through UCI to the NG-RAN to make a UL resource grant, i.e. PUSCH transmission, for new data transmission arriving from higher layers. Note that an SR is sent by the physical layer of a UE, upon request from its MAC layer, to the NG-RAN.

An SR and its UCI are transmitted over the UL PUCCH, either in Format 0 or 1, from a UE to the NG-RAN. There can be a maximum of eight SR configurations per cell group. Each SR is associated with a PUCCH resource (PUCCH-Config (...*PUCCH-ResourceId*...), a timer, its maximum transmissions limit, periodicity of SR retransmission, and so on. All these resources for a particular SR are configured at the UE MAC layer end by the RRC layer IE: *SchedulingRequestConfig;* TS 38.331 [116]. The timer is stopped when enough UL resources are granted to the requesting UE.

On the other hand, if a UL resource grant is not received by the UE from NG-RAN, the UE will retransmit the SR at a defined periodicity, which can be either at symbol level or timeslot level, as described in TS 38.213 [108]. SR periodicity also depends on the subcarrier spacing (SCS) being used and is configured at the UE MAC layer through the RRC layer IE: *SchedulingRequestResourceConfig*. The periodicity of an SR retransmission can be equal to a one-timeslot, less than a timeslot, or greater than a timeslot. Given the periodicity and offset as configured by the RRC layer, the actual retransmission occasion of SR over PUCCH can be determined using the formula described on TS 38.213 [108]. If an SR retransmission exceeded its maximum limit *(sr-TransMax)* at UE end and to transmit the new data, the UE makes a UL resource request through a normal random access procedure toward the NG-RAN. Figure 19.14 illustrates an SR sent by the UE to NG-RAN based on the description available in TS 38.321 [113].

19.5.7 Low Latency in the NR Due to Configured Scheduling

As illustrated in Figure 19.7, periodic transmission using CS reuses the allocated UL grants. Thus, the overhead in terms of delay associated with establishing signaling connections between UE and network to request/assign UL resource is avoided. This results in a low-latency communication between a UE and the NG-RAN over the NR air interface.

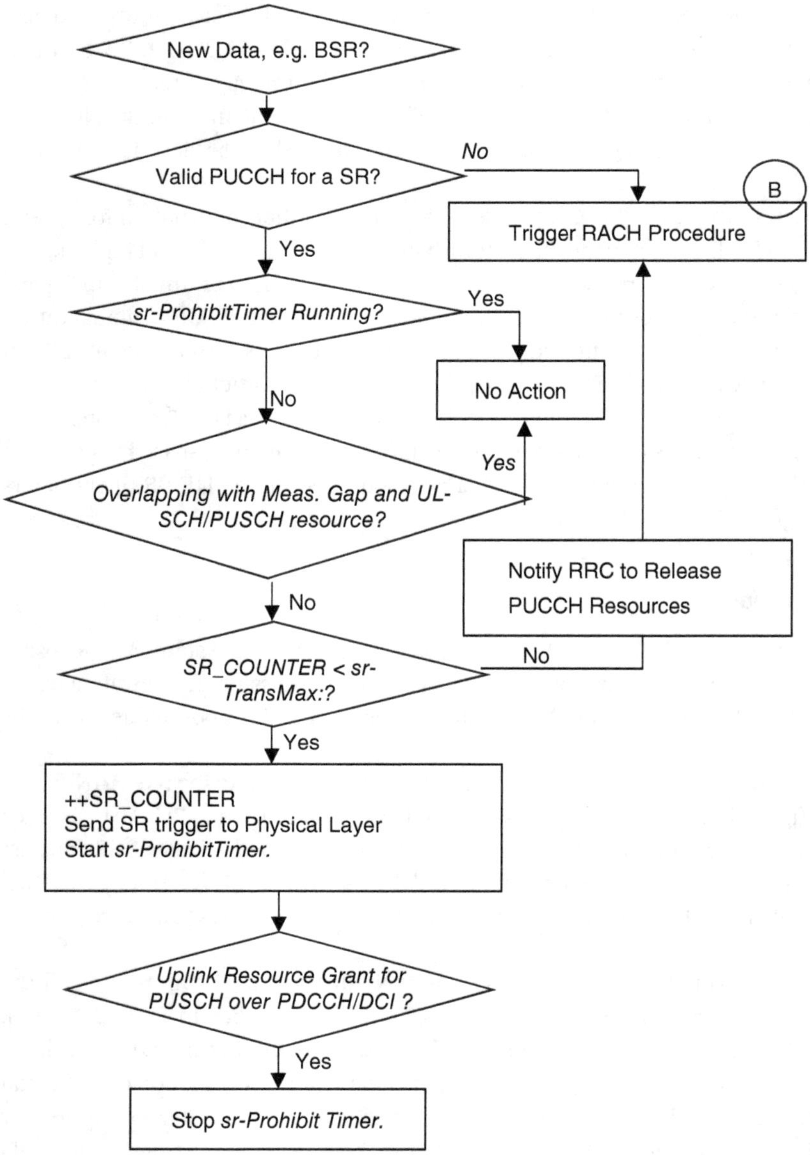

Figure 19.14 Illustration: NR UE MAC layer scheduling request to NG-RAN.

19.5.8 MAC Layer PDU and Header Structures

Both the LTE MAC and NR MAC layers transfer higher layer information in the form of a PDU over the UL and DL-SCHs. A MAC layer PDU of appropriate size, corresponding to a transport block size, is generated from the incoming RLC layer SDUs. But as far as the components of a PDU and their placements are concerned, there are differences between the LTE MAC layer and the NR MAC layer.

In the case of the LTE MAC layer, a MAC PDU consists of a MAC header, zero or more MAC SDU, zero or more MAC CEs, and optional padding. An LTE MAC PDU begins with a MAC header, which consists of subheaders. A subheader is provided for each MAC SDU received from the LTE RLC layer. All the MAC header information is placed at the beginning of an LTE MAC PDU.

In the case of the NR MAC layer, a MAC PDU consists of *sub-PDUs*. Each sub-PDU may carry the following information:

- A MAC subheader only, including padding;
- A MAC subheader and a MAC SDU;
- A MAC subheader and a MAC CE such as BSR, power headroom report (PHR), and PDCP packet duplication activation/deactivation, etc.; and
- A MAC subheader and padding.

A subheader is provided for each NR MAC SDU, CEs, or padding. Using the header information, the MAC layer at the receiving end demultiplexes a MAC PDU into different SDUs for their delivery to the RLC layer.

- **Placement of Components of MAC PDU**

Within an NR MAC layer PDU, the sub-PDUs, CEs, and padding are placed differently in the DL and UL directions.

- In the DL direction, *MAC sub-PDUs with CEs* is placed together, followed by the *sub-PDU with MAC SDU* and *sub-PDU with padding*.
- In the UL direction, the *sub-PDUs with MAC SDUs* are placed first, followed by the *sub-PDU with MAC CEs* and *sub-PDU with padding*.

In general, the NR MAC layer PDU provides a subheader with each type of information being transmitted from the transmitting end, i.e. upper-layer SDU, CEs, random access response, back-off indicator, and so on. On the other hand, in the case of the LTE MAC layer, all the subheader(s) for different SDUs are placed at the beginning of the PDU.

The assembly of two MAC SDUs, variable size MAC CE, and padding into a MAC PDU in the UL and DL directions are illustrated in Figure 19.15. A subheader "H" may contain four fields (R/F/LCID/length) or two fields (R/LCID) depending on the size of a MAC CE. The size of a MAC CE, e.g. short BSR, may be fixed, with two fields in its subheader, or variable, e.g. long BSR, with four fields in its subheader. LCID represents the logical channel from which the MAC SDU is received from the RLC. Length represents the length of the MAC SDU or CE.

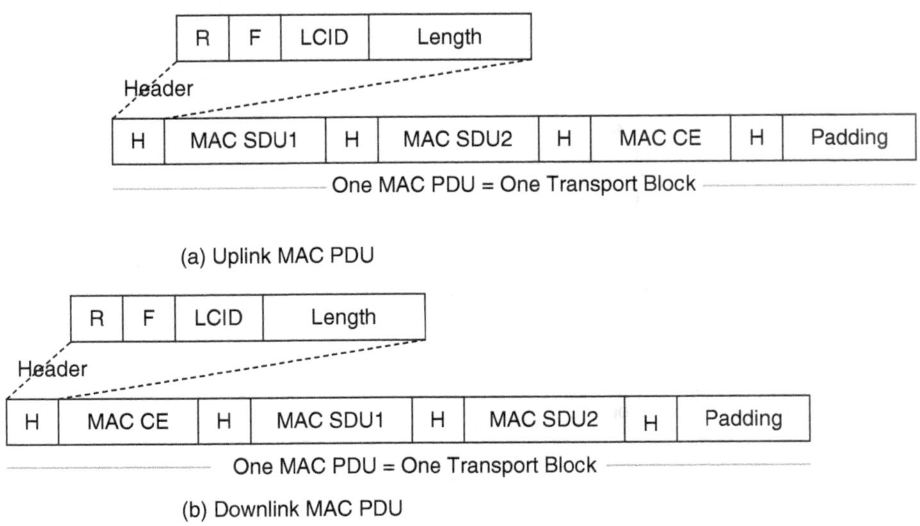

Figure 19.15 Illustration: organizations of NR MAC SDU, CE, and padding of a MAC PDU.

19.5.9 How MAC Layer Ensures Low-Latency Requirements

In the case of the LTE MAC layer, all the subheader(s) information for different SDUs is placed at a single place and beginning of an LTE MAC PDU. Due to this, the assembly of the different MAC SDUs into a MAC PDU cannot be started until the scheduling grant and its transport block size are available to the LTE MAC layer. Thus, there is a delay or latency in assembling and transmitting a MAC PDU. On the other hand, in the case of the NR MAC layer, the subheaders are distributed within a MAC PDU. Every NR MAC SDU has its subheader and is placed in front of it, see Figure 19.15. The assembly of the different MAC SDUs into a MAC PDU can be started as soon as the corresponding RLC PDU becomes available to the MAC layer without waiting for the scheduling grant and availability of its transport block size from the physical layer. This avoids the processing time delay associated with assembling MAC SDUs to generate a MAC PDU and a transport block. Padding can be added at the end of the NR MAC PDU to make it aligned with the transport block size assigned for its transmission over the NR air interface.

- **Differences in Packet Processing by LTE and NR L2 Layers: A Comparison**

Figure 19.16 illustrates user data flows in DL direction through all the Layer 2 sublayers of the air interface of the LTE and 5G NR systems. It compares the differences as far as the user plane packet processing by the RLC and MAC layers of LTE and 5G NR systems is concerned. As described earlier in Section 19.2, the NR Layer 2 contains a new sublayer called SDAP.

As illustrated in Figure 19.16, the LTE RLC layer (left side) performs the concatenations of PDCP PDUs/RLC SDUs into an RLC PDU, whereas the NR RLC layer does not perform such concatenations function. On the other hand, the NR MAC layer (right side) performs the concatenations of RLC PDUs/MAC SDUs. The LTE MAC layer generates its PDU from a single RLC PDU or MAC SDU and adds its header information in front of the MAC PDU. On the other hand, the NR MAC layer combines different RLC PDUs or MAC SDUs where each RLC PDU

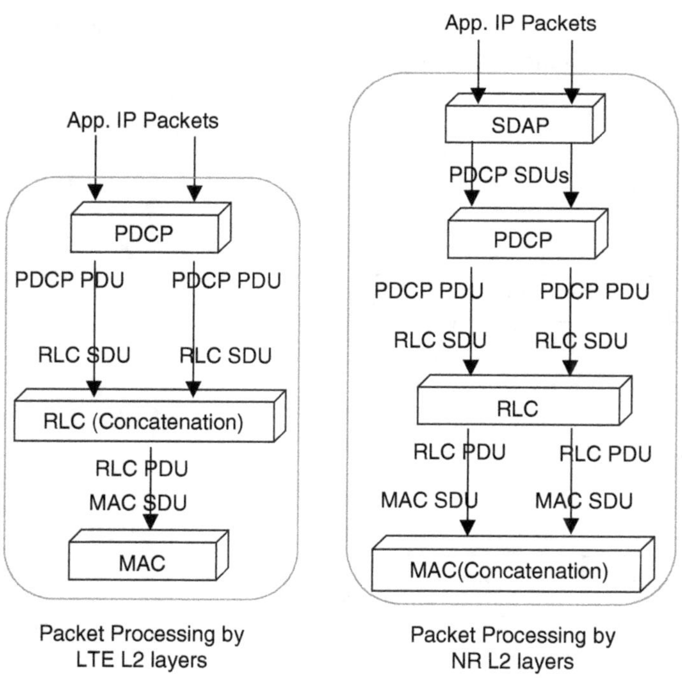

Figure 19.16 Illustration: packet processing by LTE and NR L2 layers.

has its header. The benefit gained from such an approach used in the NR RLC and MAC layers results in overall low-latency access over the 5G NR, as described earlier in Section 19.4. MAC layer is configured by the RRC layer under the IE: *mac-Config*.

19.5.10 Channel Structures in NR

- Channel Types

In a mobile communications system, a *channel* refers to a dedicated or shared transmission path/medium through which information, either signal or user traffic, is exchanged between a sender and receiver over a particular air interface. A channel could be either a *physical*, for example, a radio timeslot, or *logical* in nature. For example, the physical *E1 interface* used in the legacy Global System for Mobile Communication (GSM) system has 32 timeslots (channels) of 64 kbps each, with a total bandwidth of 2 Mbps supported by it.

A channel can be used for transmission of user traffic or control/signaling information over the NR air interface between a UE and NG-RAN and vice versa. Knowing the nature, purpose, and function of each type of channel used over a particular radio access method and air interface is core to the understanding and implementation of an entire air interface of GSM, General Packet Radio Service (GPRS), Universal Mobile Telecommunication System (UMTS), LTE, or NR system.

As a comparison, the GSM system air interface has only two types of channels – logical and physical channels. However, the UMTS, LTE, and the NR system air interface define one more type of channel, between the logical and physical channel, which is known as *the transport channel*. Naturally, the implementation of the UMTS, LTE, and NR air interface functions and procedures is more complex than the GSM air interface. The purposes of the various types of channels used over the NR air interface are described below.

- Logical Channel

A logical channel identifies the type of information, *either user or signaling data*, to be transmitted through one of the physical channels. Similar to the LTE system, a logical channel exists at Layer 2 of the NR air interface and is used to pass information, for example, from the RLC layer to the MAC layer. Logical channels are further classified into the *Control channel or Traffic Channel*.

A control channel may be either a dedicated or common in nature. The NR MAC layer provides services to the RLC layer through the logical channels. User traffic from a DRB is transmitted using the traffic logical channel. On the other hand, signaling information from the RRC layer through a signaling radio bearer is transmitted using the common control logical channel. Logical channels are implemented through logical objects in the software.

- Transport Channels

A *transport channel* specifies how the information received from a logical channel will be transferred over the air interface. It prepares the information that is received from logical channels for transmission over the air interface either on a common or dedicated channel. Similar to the LTE system, they exist between the air interface Layer 1 and Layer 2. The transport channel is used to pass information from the MAC layer to the physical layer and vice versa. The NR physical layer provides services to the MAC layer through the transport channels.

- Physical Channels

A physical channel corresponds to certain resource elements (REs) of a PRB of the NR air interface which is defined in the frequency and a time domain. The NR air interface defines several physical channels for transmission of user traffic and control information in the UL and DL directions. A PRB and its RE are described in later sections. In the NR system, physical channels are allocated from a set of subcarriers, which shall be described later on.

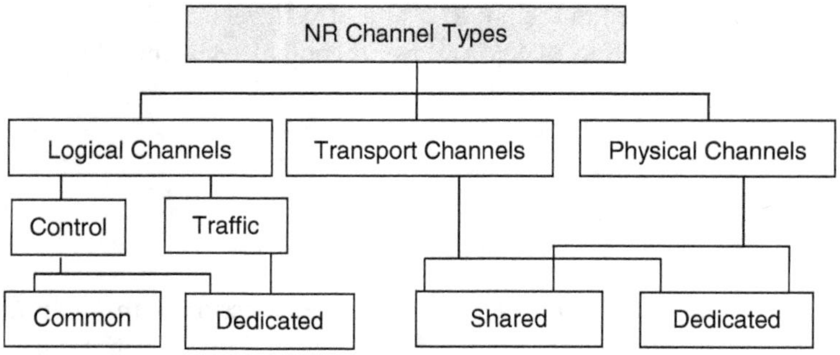

Figure 19.17 Illustration: NR air interface channel types.

Table 19.1 GSM, UMTS, LTE, and NR physical channel definitions.

Systems	Air Interface Physical Channel Description
GSM and GPRS	RF Carrier, Timeslot
UMTS	RF Carrier, Scrambling Code, Timeslot
LTE	Resource Element, Resource Block
5G NR	Resource Element, Resource Block

Channels types used over the NR air interface are shown in Figure 19.17.

For a comparative study to the reader, a description of a physical channel and its allocated resources in GSM to the NR system for transmission of information over their respective air interface is shown in Table 19.1.

From a protocol stack and its implementation in software point of view, different types of channels are realized and implemented through a logical concept called *Service Access Point* (SAP) between two adjacent layers. This is illustrated in Figure 19.18, where a small circle represents an SAP between two adjacent layers.

Similar to the LTE air interface channel structures, the 5G NR system also uses the logical channels, transport channels, and physical channels to transfer higher layer data over its air interface. These logical channels, in general, are also available in the LTE system which is used for similar purposes. The MAC layer operates on logical as well as transport channels. The MAC layer provides data transfer services to the RLC layer through logical channels.

- **Logical Channel Services to RLC Layer**

Similar to the LTE system, each NR logical channel indicates what information is being transferred by it. Different logical channels supported by the NR MAC layer for services to the RLC layer are described below.

- BCCH, using which the NG-RAN broadcasts minimum system information called *Master information Block*, in the DL direction to UEs in a cell. Using minimum system information, a UE acquires other system information to access a cell and register with the 5G network.
- PCCH, using which the NG-RAN pages UEs in the DL direction.
- CCCH, using which control information request/response is exchanged between the NG-RAN and UEs.
- DCCH, using which control information request/response is exchanged between the NG-RAN and a UE.
- *Dedicated Traffic Channel* (DTCH), using which user data is transferred between the NG-RAN and a UE.

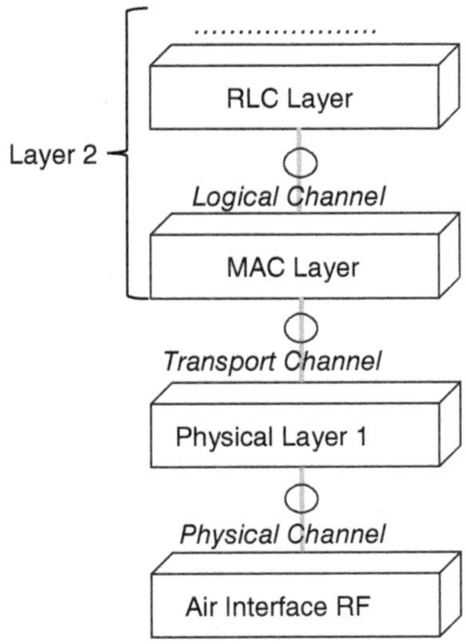

Figure 19.18 Illustration: types of NR channels and their implementation using SAP.

Table 19.2 Channel mappings at NR MAC layer.

Logical Channels	Transport Channels	Direction
BCCH	BCH	Downlink
PCCH	PCH	
BCCH, CCCH, DCCH, DTCH	DL-SCH	
CCCH, DCCH, DTCH	UL-SCH	Uplink

The NR MAC layer performs the mappings and multiplexing of different logical channels mentioned above onto their corresponding transport channels, which is similar to the channel mappings used over the LTE air interface. A unique LCID, Tables 6.2.1-1/2 in TS 38.321 [113], is used by the MAC layer in case of multiplexing of logical channels onto a transport channel. The NR logical channels to transport channels have the following mapping relationships among them; see Table 19.2.

Each of the transport channels listed above undergoes a processing chain with a series of tasks. Different tasks associated with the processing chain of an NR transport channel are described later in Section 19.6.9.

19.6 Physical Layer

The NR physical layer provides higher layer signaling information as well as user data transfer services to the MAC layer over the DL-SCH and UL-SCH. As mentioned in Chapter 16, the NR physical layer is one of the key enablers of the different use cases and their services in the 5G system. NR physical layer performs the necessary functions and procedures to provide multiplexed data as well as signaling information transfer services to the higher layers. In the subsequent sections below, some of the important aspects of the NR physical layer shall be described.

19.6.1 Principles of Transmissions and Its Directions

- Directions of Transmission

There are two directions of transmission in a mobile communications system and network between a UE/MS and the network and vice versa, as mentioned below:

  - UL direction, i.e. transmission from the MS/UE to the network;
  - DL direction, i.e. transmission from the network to MS/UE; and

- Duplex methods

Similar to any other communications systems, in a mobile communications system and network also, duplex transmission methods are used between the mobile device and the network. These duplex methods used in frequency and time domains are as follows:

  - Frequency Division Duplex (FDD)

In the FDD mode of transmission, a separate carrier frequency is used for both the UL, i.e. transmit, and DL, i.e. receive, direction, and requires a pair of frequency carriers. Separate carrier frequencies for UL and DL transmissions do not interfere with each other, and also, no switching is required between the DL and UL directions.

  - Time Division Duplex (TDD)

In this transmission method, only one carrier frequency is used for both the UL and DL directions, but the transmissions are separated in a timeslot, due to which transmissions do not interfere with each other. For example, in the case of the GSM system, UL and DL transmissions for an MS over a single carrier (SC) frequency are separated by three timeslots. The principles of duplex transmissions are illustrated in Figure 19.19.

19.6.2 Physical Layer Functions, Procedures, and Services

The NR physical layer performs the following functions and procedures at the transmitting end to provide data and signaling information transfer services to the higher layers. The physical layer at the receiving end performs the opposite of some of these functions and procedures and forwards the user data as well as signaling information to the upper layers.

- **Physical Layer Functions**
  - Frequency and time synchronizations during UE initial registration;
  - Mapping transport channels to physical channels;
  - Transport channel processing in terms of encoding with the low-density parity check (LDPC) coding method for upper-layer data and segmentation, rate matching, and interleaving so that data are received at the receiving end correctly;

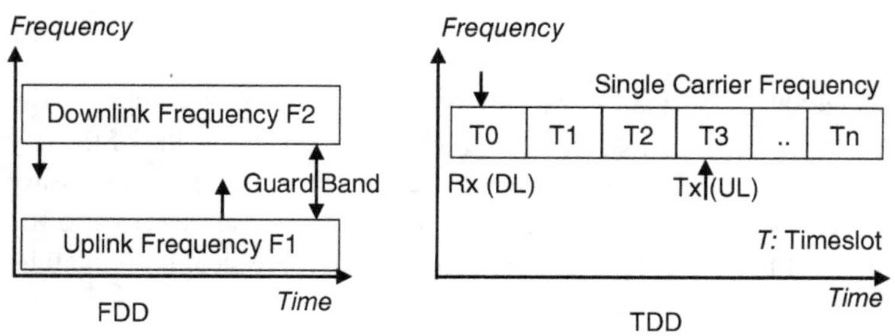

Figure 19.19 Illustration: duplex (FDD, TDD) transmissions methods.

- Physical channel processing in terms of polar coding for control information so that signaling information is received at the receiving end correctly. In addition to these, tasks such as scrambling, modulation, layers mapping, transform precoding and precoding, and mapping of coded symbols onto allocated resources are also performed by the physical layer at the transmitting end;
- Radio link measurements and reporting;
- HARQ with soft combining; and
- Multiple Input Multiple Output (MIMO) antenna processing.

• **Physical Layer Procedures**

In addition to the various functions mentioned above, the NR physical layer also performs several procedures to enable a UE to access an NR cell and register with a 5G network.

- Cell search procedure to decode the physical layer cell identity (PCI);
- Random access-related procedure;
- UL synchronization with the network, timing control, and power control;
- Beam management, channel state measurement-related procedures;
- Hybrid ARQ (HARQ)

HARQ retransmission is used to correct transmission error as well as to provide a fast retransmission mechanism compared to the retransmission mechanism provided by the upper layer, i.e. RLC. HARQ functions spread across the MAC and physical layers. HARQ is the transmission triggered by the MAC layer, but its actual soft combining is performed at the physical layer.

• **Transport Channels Services to MAC Layer**

The NR physical layer provides users as well as signaling data transfer services to the MAC and higher layers over several transport channels which resembles the LTE system. These transport channels transfer information received from multiplexed logical channels of NR. Similar to the LTE system, NR transport channels deal with how and with what characteristics, called *transport format*, data shall be transmitted over the NR air interface.

- *Broadcast Channel (*BCH*)*, which is used to transfer master information block received from the BCCH logical channel. BCH has a fixed transport format.
- *Downlink Shared Channel (*DL-SCH*)*, which is used to transfer DL user and signaling information from different logical channels in a multiplexed way. The transport format of DL-SCH is not fixed as resources are allocated dynamically as well as semi-statically by the NG-RAN. By using multiple antenna transmissions and the spatial multiplexing method, which is described later in Section 19.6.10.3, up to two transport blocks per *transmission time interval* (TTI) can be transmitted over the DL-SCH in the DL direction to a UE.
- *Paging Channel* (PCH), which is used to transfer paging information from the PCCH logical channel in the coverage area of a cell.
- *Uplink Shared Channel)* (UL-SCH), which is used to transfer UL user and signaling information from different logical channels in a multiplexed way. The transport format of UL-SCH is not fixed, as resources are allocated dynamically as well as semi-statically by the NG-RAN.
- Random Access Channel (RACH), which is used to transmit a random access preamble by a UE in the uplink direction. There is no corresponding logical channel for the RACH. Thus, a RACH does not transfer higher-layer information.

• **Channel Coding**

Similarly, as in the case of the LTE physical layer, the NR physical layer processes a transport block received from the MAC layer over a transport channel. At the NR physical layer, a transport channel processing chain involves several tasks, such as information encoding using the LDPC coding method, which is performed in UL and DL directions separately. In addition to this, the NR physical layer performs the control channel coding function in UL and DL directions, which also contains several steps as part of the entire processing chain for the transmission of control information.

- **Modulation Schemes**

  NR physical layer supports and uses different modulation schemes such as the Quadrature Phase Shift Keying (QPSK), 16 Quadrature Amplitude Modulation (QAM), 64 QAM, and 256 QAM to transfer user data and signaling information in UL and DL directions.

- **Physical Layer Measurements**

  Another function performed by the UE and NG-RAN physical layer is the measurement of the current radio conditions and taking the appropriate actions. Radio measurement results affect the dynamic UE scheduling decision also at the network end.

- **Physical Layer Access Method**

  Similar to the LTE system, the 5G NR system also uses the Orthogonal Frequency Division Multiplexing Access (OFDMA) method of communication between UE and NG-RAN over its air interface; see Table 19.3.

- **Operating Frequency Ranges (FR) and Duplexing Modes**

  Two types of frequency ranges (see Table 19.4) are used in the NR system: FR1 and FR2. FR1 uses a frequency range of 410–7125 MHz, called sub-6 GHz bands. FR2 works on the mmWave, above 20 GHz, i.e. frequency range of 24 250–52 600 MHz. Maximum channel bandwidth (CBW), i.e. NR carrier bandwidth, is up to 100 MHz in FR1 and up to 400 MHz in FR2. Within a CBW, the actual transmission bandwidth is also defined, in terms of PRBs, for transmission/reception by a UE. NR uses the FDD mode to operate in low-frequency bands. TDD mode is used to operate in high-frequency bands. In each frequency range type, FR1 and FR2, several operating bands are defined for their usages by different operators. FR1 can operate both in FDD and in TDD modes, whereas FR2 operates in TDD mode only. TDD operates both in mid and high operating frequency bands. For more information, refer to Table 5.2-1 in TS 38.104 [103].

Table 19.3 Radio access, modulation methods, and frame duration.

System	Radio Access Methods	Modulation Scheme used over Air Interface	Radio Frame Duration, Timeslots
GSM	TDMA, FDMA	GMSK	4.62 ms, 8 slots
UMTS	WCDMA	QPSK	10 ms, 15 slots
LTE	OFDMA, SC-FDMA	QPSK,16 QAM,64 QAM, 256 QAM	10 ms, 20 slots
5G NR	OFDMA	QPSK,16 QAM,64 QAM, 256 QAM	10 ms. Timeslot Varies.

Table 19.4 NR operating frequency ranges and duplexing modes.

Parameters	FR1	FR2
Frequency Range	410–7125 MHz	24 250–52 600 MHz
Duplexing Mode	FDD, TDD	TDD
Channel Bandwidth (MHz)	5, 10, 15, 20, 25, 30, 40, 50,60, 70, 80, 90, and 100	50, 100, 200, and 400

- **Physical Layer Specifications**

NR physical layer is described by several technical specifications, covering the above functions and procedures, as summarized below. Reading should start from TS 38.201 [105].

- TS 38.201 – covers a brief description of the NR physical layer.
- TS 38.202 – describes the services provided by the NR physical layer
- TS 38.211 – describes the physical channels and their modulation details.
- TS 38.212 – describes the multiplexing and channel coding details.
- TS 38.213 – describes the physical layer procedure for various aspects of controlling or signaling.
- TS 38.214 – describes the physical layer procedure for various aspects of data transfer.
- TS 38.215 – describes the various aspects of measurements performed by the physical layer.

19.6.3 OFDM Symbol

Similar to the LTE system, over the NR air interface, the OFDM method is used for transmission and reception between a UE and its NG-RAN in the DL and UL directions. In this section, the OFDM symbol concept will be described first before we describe the other aspects of the NR physical layer.

The higher layer information, either user or signaling, is converted into digital or stream of bits containing 0 and 1. These streams of bits can be transmitted over the air interface in the form of a radio wave using a particular carrier frequency through a process called *modulation*. At the receiving end, the original information is recovered through a reverse process called *demodulation*. Prior to transmission over the air, the streams of bits are modulated in terms of frequency or amplitude or phase. The modulation technique being used, e.g. PSK, QPSK, and QAM, determines the number of bits that can be transmitted over a particular carrier frequency. The number of bits to be transmitted over a particular carrier frequency is called a *data symbol*. In other words, a carrier frequency represents or carries a *data symbol*.

For example, the QPSK modulation technique can be used to map and transmit 2 bits per data symbol per carrier frequency. Similarly, the 16 QAM modulation technique may be used to map and transmit 4 bits ($2^4 = 16$) per data symbol per carrier frequency. The data symbols of a particular modulation technique may be represented through its symbol constellation. Figure 19.20 shows the typical symbol constellation for the 64 QAM modulation technique where each data symbol carries 6 ($2^6 = 64$) bits of information.

Depending on the physical layer channel type being used, the BPSK, QPSK, and QAM modulation techniques are used over the NR air interface. A data symbol is called as the *OFDM symbol*. The term *OFDM symbol* will be used in the subsequent sections below.

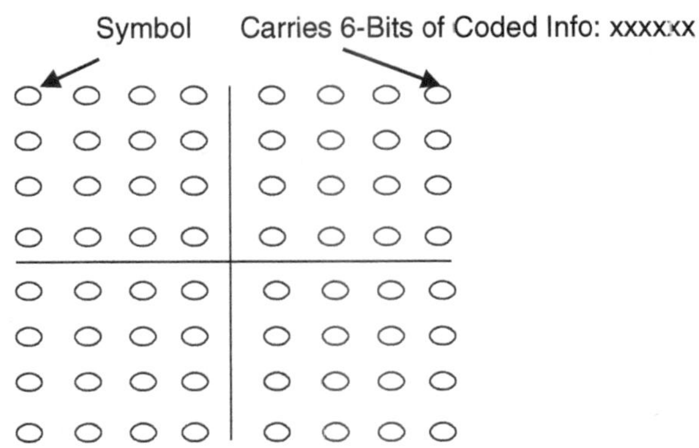

Figure 19.20 Illustration: symbol constellation diagram for 64 QAM modulation.

19.6.4 NR Frame and Slot Format

DL and UL transmissions between UE and the NG-RAN take place in terms of a radio frame over the NR air interface. The NR frame structure is similar to an LTE frame as shown in Figure 19.21.

Each NR frame, Figure 19.21, has a 10 ms duration. Each frame is divided into 10 subframes. Further, each frame is divided into two half frames of equal sizes. In the LTE system, separate frame types are used for FDD and TDD operations. However, in the NR system, a common frame type is used in FDD and TDD modes, as illustrated in Figure 19.21. Each NR subframe has a 1 ms duration which may contain a variable number of timeslots depending on the SCS used, as described below.

19.6.4.1 Subcarrier Spacing (SCS)/Numerologies (μ)

In the LTE system, only one type of SCS, i.e. 15 kHz, is used. However, in the NR system, different types of SCS, in multiples of 15 kHz, are used. The SCS types used in the NR system are 15, 30, 60, 120, and 240 kHz. Each SCS type is called *numerology (μ)*, from 0 to 4, in the NR physical layer specifications. Numerology 0 corresponds to 15 kHz, numerology 1 corresponds to 30 kHz, and so on. A normal cyclic prefix (CP) is used with all the numerologies, except in numerology 2 where normal as well as an extended CP can be used; refer to Table 4.2-1 in TS 38.211 [106]. NR system can use different numerologies within a particular CBW; refer to Table 5.3.2-1 in TS 38.104 [103]. Further, using the numerology 4 (SCS: 240 kHz), UL and DL data cannot be transmitted over the PDSCH and PUSCH. Similarly, using numerology 2 (SCS: 30 kHz), synchronization information cannot be transmitted over the physical layer synchronization signals; refer to Table 5.1-1 in TS 38.300 [111]. SCS information to be used by a UE is notified by different RRC signaling messages by the RRC layer; refer to TS 38.331 [116]. The maximum number of subcarriers in the NR system is 3300, although the maximum CBW is 400 MHz as listed in Table 19.4.

19.6.4.2 Slots per NR Frame and Subframe

In the LTE system, the number of slots (2) per subframe or number of slots (20) per frame is fixed. However, in the NR, the number of slots per frame and subframe is not fixed but varies according to the numerology or SCS used, as listed in Table 19.5.

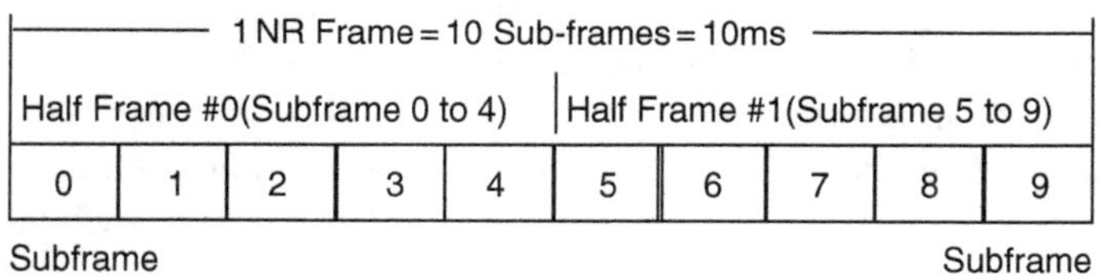

Figure 19.21 Illustration: NR frames, subframes, and durations.

Table 19.5 NR frames and slots for different NR subcarrier spacings.

Numerologies (μ)/ Subcarrier Spacings	Frame Duration (a)	Subframe Duration (b)	Slots Per Frame (c)	Slots Per Subframe (d)	FR1/FR2 (e)
0(15 kHz)	10 ms	1 ms	10	1	FR1
1(30 kHz)	10 ms	1 ms	20	2	FR1
2(60 kHz)	10 ms	1 ms	40	4	FR1, FR2
3(120 kHz)	10 ms	1 ms	80	8	FR2
4(240 kHz)	10 ms	1 ms	160	16	FR2

Table 19.6 NR frames and slots for different NR subcarrier spacings.

Numerologies (μ)/ Subcarrier Spacings	Slot Duration	#of Symbols Per Slot	#of Symbols Per Subframe
0(15 kHz)	1 ms	14	14(14*2^{μ})
1(30 kHz)	500 μs	14	28(14*2^{μ})
2(60 kHz)	250 μs	14	56(14*2^{μ})
3(120 kHz)	125 μs	14	112 (14*2^{μ})
4(240 kHz)	62.5 μs	14	224(14 *2^{μ})

With numerology:0/SCS: 15 kHz, the number of slots per frame is 10, and the slot per subframe is 1. With numerology:1/SCS: 30 kHz, the number of slots per frame is 20, and the slot per subframe is 2. For more information on the slot structures of different numerologies used in the NR system, refer to Table 4.3.2-1 in TS 38.211 [106]. In case of transmission with the normal CP, 14 OFDM symbols per slot are used, and 12 OFDM symbols per slot are used in case of transmission with an extended CP.

It may be noted that the duration of an NR frame and its subframes are the same in all the numerologies. Continuing from Table 19.5, the duration of a slot, symbols per slot, as well as per subframe are listed in Table 19.6 in the entire numerologies/SCS configuration.

An NR frame structure with different SCS/numerologies, number of subframes per NR frame, slots per subframe, and OFDM symbols per slot is illustrated in Figure 19.22.

As illustrated, the number of slots per subframe and frame increases as the SCS increases. For example, there is only 1 slot per subframe/10 slots per frame with SCS: 15 kHz, whereas there are 16 slots per subframe/160 slots per frame, with SCS: 240 kHz. Due to this, the duration of a timeslot decreases or becomes shorter as the SCS increases, i.e. duration from 1 ms with SCS:15 kHz to 62.5 microseconds with SCS: 240 kHz.

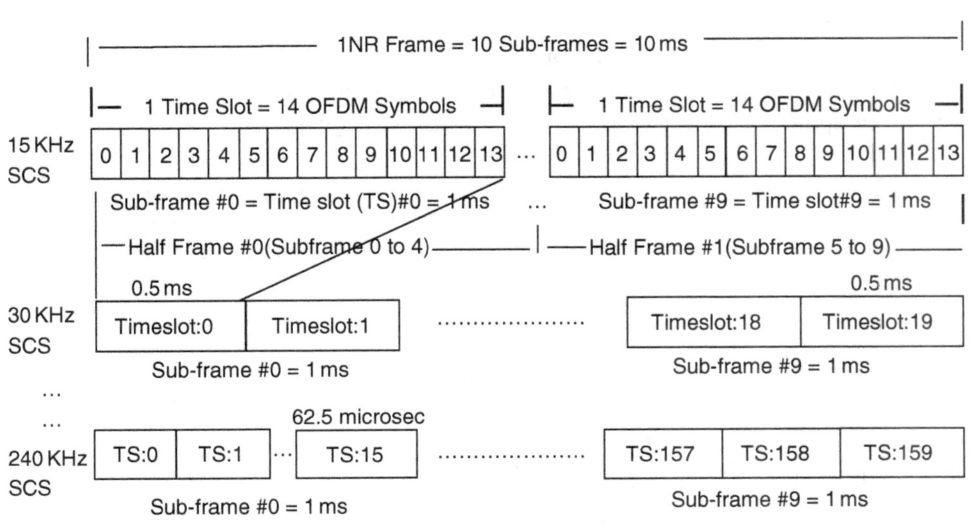

Figure 19.22 Illustration: NR numerologies, frame, subframes, and slots.

19.6.4.3 Slot Formats in TDD Mode

Similar to the LTE system, a timeslot is a basic unit for transmission in the NR system. In the LTE TDD system, a timeslot and all of its OFDM symbols are used to transmit information either in the DL or UL direction only. However, in the NR TDD system, there are more flexibilities of transmitting information in the DL (D), UL (U), or mixed/flexible (F) direction even at an OFDM symbol level also. In the NR TDD mode, all the OFDM symbols of a slot can be used to transmit in UL direction only or DL direction only or mixed, that is, either in UL or in DL direction. In the DL direction, a UE can transmit through OFDM symbols configured as "D" as well as "F". Similarly, in the UL direction, a UE can transmit through the OFDM symbol configured as "U" as well as "F". All such possibilities of transmissions/receptions through OFDM symbols of a slot are defined under a particular format of a slot.

Slot formats differ from each other based on the number of OFDM symbols that are used for transmissions/receptions in UL or DL direction. UE is notified about the slot format to be used, either dynamically or semi-statically. A slot format information contains the number of DL or UL slots along with the number of symbols in each slot assigned.

- Semi-static Configuration of Slot Format

A cell- or UE-specific timeslot formation information can be provided by the RRC layer signaling through the IEs: *tdd-UL-DL-ConfigurationCommon* and *tdd-UL-DL-ConfigurationDedicated*. Refer to TS 38.331 [116] for more information on the contents of these IEs. Cell-specific common slot format information provided over the *tdd-UL-DL-ConfigurationCommon* with the "flexible" only symbols can be overridden using the UE-specific slot format, as per UE requirements, provided through the IE: *tdd-UL-DL-ConfigurationDedicated*. As an illustration, consider a cell-specific TDD timeslot format configuration information provided by the NG-RAN/RRC layer using the IE: *tdd-UL-DL-ConfigurationCommon*. Also, consider the contents of this IE: *tdd-UL-DL-ConfigurationCommon* as follows:

TDD-UL-DL time slots and symbols allocation pattern = pattern 1;
referenceSubcarrierSpacing = 60 kHz, i.e., $\mu = 2$.

Further for pattern 1, consider the following:

- *dl-UL-TransmissionPeriodicity* = P = 2.5, nrofDownlinkSlots =4,
- *nrofUplinkSlots = 4, nrofDownlinkSymbols = 11, nrofUplinkSymbols = 0.*

Based on the description provided in Section 11 in TS 38.213 [108],
Total Number of Time Slot $S = P \times 2^{\mu} = 10$.

Using these timeslots and the above typical time allocation pattern 1 information, the DL and UL timeslots (total 10) and DL symbols (total 11) may be represented, as shown in Figure 19.23, for this particular allocation pattern.

For more details on the procedure used by UE and its expected behavior toward the processing of slot format information and transmission/reception using it, refer to Section 11 in TS 38.213 [108].

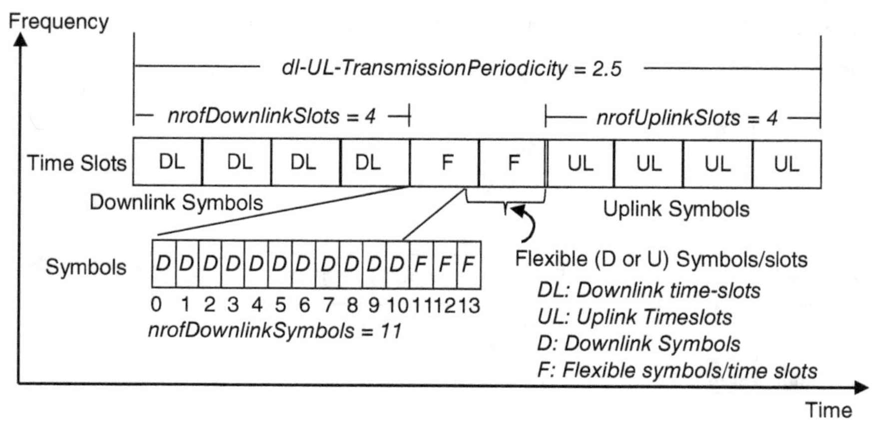

Figure 19.23 Illustration: NR TDD: DL-UL timeslot format allocation pattern 1.

- Dynamic Configuration of Slot Format

Dynamically, a UE or group of UEs is also provided with the slot *format indicator (*SFI*)* information through DCI 2_0 over PDCCH whose CRC is scrambled with *SFI-RNTI*. For this purpose, NG-RAN may configure such dynamic configuration on a per cell basis through the RRC layer IE: *SlotFormatIndicator*, which further contains the IE: *SlotFormatCombinationsPerCell*. The IE: *SlotFormatCombinationsPerCell* contains information from a predefined Table 11.1.1-1 in TS 38.211 [106] for intimating the slot format to be used by a UE or group of UEs over DCI 2_0. It should be noted that Table 11.1.1-1 in TS 38.211 [106] defines 56 predefined formats (DL, UL, flexible, etc.) of timeslots and their symbols. If DCI 2_0 contains 255 as the SFI format, this will indicate to the UE/UEs that the slot format intimated over the *tdd-UL-DL-ConfigurationCommon/ tdd-UL-DL-ConfigurationDedicated* should be used instead.

Figure 19.24 illustrates three-slot formats (0, 10, 20) in TDD mode with normal CP in terms of possible transmission directions for all the OFDM symbols of a slot.

As illustrated in Figure 19.24, slot format 0 may be used for DL heavy data transmission purposes; slot format 10 may be used for UL heavy data transmission purposes; and slot format 20 may be used for mini-slot or small data transmission purposes. For the complete slot formats, refer to table 11.1.1-1 in TS 38.211 [106].

Different formats of timeslot and its symbols, available for TDD transmission and reception, make it possible to receive data as well as provide an ACK within a slot in a faster manner in the 5G NR system. If a UE is not provided with the IE: *tdd-UL-DL-ConfigurationCommon/tdd-UL-DL-ConfigurationDedicated* over the RRC signaling *or* not configured to monitor the DCI 2_0 for slot format information, due to its radio access capability reason, then the UE will receive or transmit as per the timeslot/symbol indicated over the respective UL or DL DCI, i.e. DCI 0_0/0_1 or DCI 0_1/1_1 provided by NG-RAN, thus, achieving a dynamic TDD capability in 5G NR as described next.

19.6.4.4 Dynamic TDD

As described in the previous section, the NR TDD mode provides different timeslot formats that enable it to allocate and switch timeslot based on the dynamic and instantaneous traffic load in the network. That is, a timeslot can be used for UL only or DL, the only transmission based on the instantaneous traffic load in the network. The flexibility of such fast switching between UL and DL directions and vice versa enables NR to support a different kind of services and their traffic dynamically. This is possible through the allocation of dynamic time-domain radio resources over a DCI, through the field: *Time-domain resource assignment*, from NG-RAN to UE. Dynamic TDD differentiates the 5G NR from the LTE TDD mode which uses, unlike NR TDD, a predefined pattern of

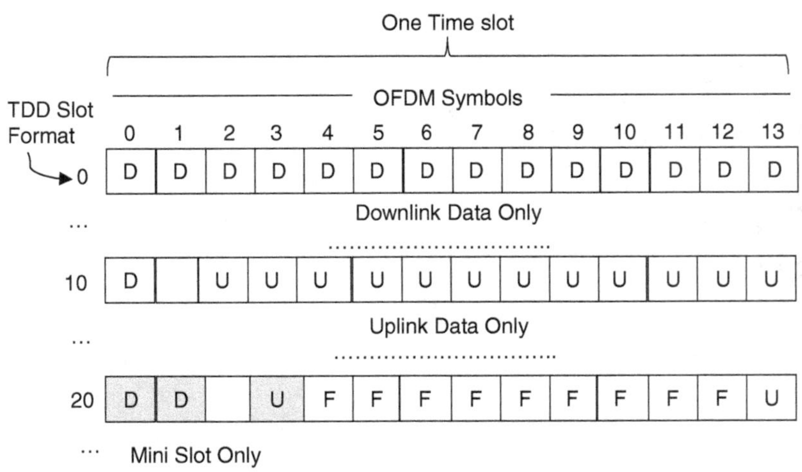

Figure 19.24 Illustration: NR TDD: slot formats of OFDM symbols of a slot.

timeslots (1 timeslot = 1 ms subframe) for UL transmission and DL reception by a UE. Therefore, no time-domain resource is provided dynamically to UE over a DCI in the LTE system in its TDD mode. Also, unlike the LTE system, NR can provide a timeslot format to a UE or group of UEs through a dedicated DCI Format 2_0, as described in the previous section. If a UE is not provided with a timeslot format information through DCI 2_0 or RRC layer signaling, UE will receive in DL/transmit in UL using the timeslot format/resource as indicated through its concerned DCI received from NG-RAN.

19.6.5 Resource Grid and Resource Block

Similar to the LTE system, the NR also organizes the physical resources or subcarriers of particular transmission bandwidth into a resource grid and its RBs. A resource grid contains several RBs, defined for particular numerology, in the frequency domain and OFDM symbols in the time domain. A RE *(k,l)* is the smallest unit or basic granularity of a physical resource grid which is made up of one subcarrier(indexed by k) in the frequency domain and one OFDM symbol (indexed by *l*) in the time domain. However, there is a difference between the LTE and NR RB. Unlike the LTE system, where a resource block is defined as a two-dimensional resource – in the frequency domain (12 subcarriers) as well as time domain (7 OFDM symbols) – an NR resource block is defined as a one-dimensional resource – in frequency domain only consisting of 12 subcarriers. This implies that the minimum duration of an NR resource block is one OFDM symbol in the time domain which provides flexible scheduling in terms of different time duration, whereas an LTE timeslot has a fixed duration with 14 OFDM symbols with a normal CP. Each NR slot also contains 14 OFDM symbols with a normal CP. Figure 19.25 illustrates an NR resource grid, its resource blocks, and OFDM symbols in general with a normal CP.

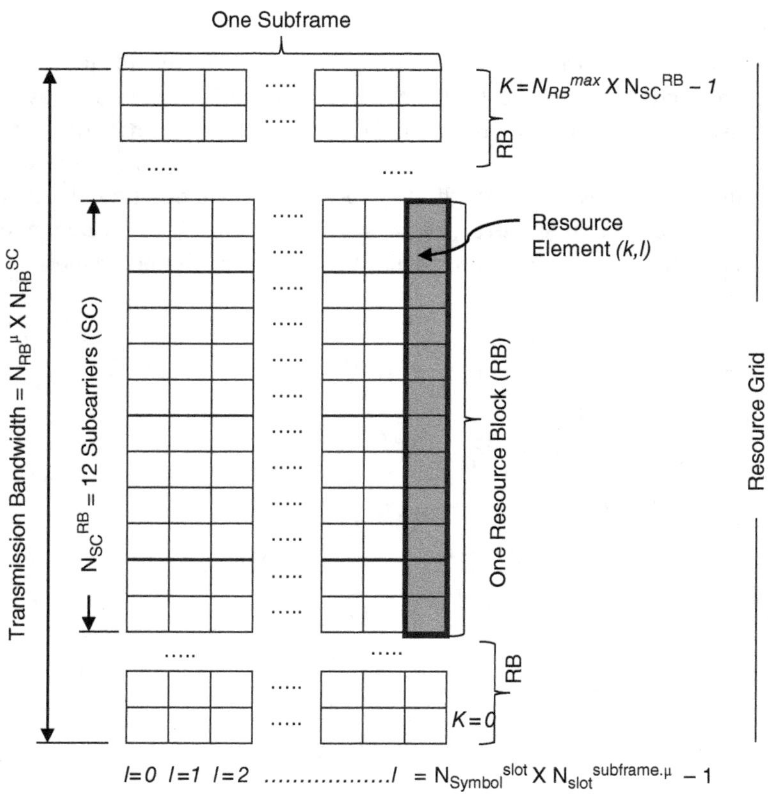

Figure 19.25 Illustration: NR resource grid, resource element, and resource blocks with a normal cyclic prefix.

Unlike the LTE system resource block with an SCS of 15 kHz only, the NR system support transmissions with multiple subcarrier spacings (15, 30, 60, 120, and 240 kHz) or numerologies ($\mu = 0, 1, 2, 3, 4$). Also, in the LTE system, each subframe has two timeslots. However, in the NR system, the number of timeslots per subframe in particular numerology, starting from $\mu = 1$, becomes twice the number of timeslots per subframe from the preceding numerology.

It may be noted that the bandwidth of a resource block in the LTE system is fixed, i.e. 180 kHz, whereas in the NR system, the bandwidth of a resource block can range from 180 to 2880 kHz for different SCS/numerologies, as listed in Table 19.6. For more information on the resource grid, resource blocks, and different numerologies, refer to TS 38.211 [106].

19.6.5.1 Control Resource Set (CORESET)

In the LTE as well as the NR systems, the PDCCH is used to provide UL/DL scheduling and other control information to a UE. In the LTE system, a UE decodes the PDCCH from a *control region* which consists of a collection of *control channel elements* (CCE). Similarly, in the NR system also, a UE decodes the PDCCH from a resource area, in the frequency and time domain, which is known as the *Control Resource Set* (CORESET). A CORESET indicates the search space for a UE of a group of UEs to decode DL control information received over a PDCCH. For example, the MIB message broadcasted by the NG-RAN contains the *CORESET* 0 with *SearchSpaceZero*. Further, based on CORESET 0, UEs in a serving cell read and process the SIB1.

An NR *CORESET* also contains a collection of CCEs, which in turn contains a set of *Resource Element Groups* (REGs). However, there are differences between the LTE control regions and NR *CORESET*. In the case of the LTE system, PDCCH spans across the carrier bandwidth in the frequency domain. On the other hand, an NR *CORESET* does not span the whole carrier bandwidth as a UE may not support the whole carrier bandwidth, which is as large as up to 400 MHz. Thus, in the NR system, frequency domain resources for a CORESET are restricted and allocated over a subset of the carrier bandwidth only along with time-domain resources. A *CORESET* is allocated in frequency as well as time domain through RRC signaling IE *ControlResourceSet*.

- Frequency Domain Resources for CORESET

In the frequency domain, an NR *ControlResourceSet*, among other information, consists of RB, in multiples of 6, as specified by the RRC IE: *frequencyDomainResources*, that is, a CORESET consists of six RBs. An RB is organized into an REG (12 subcarriers) in the frequency domain. Further, six REGs are organized into a CCE in the NR frequency domain.

- Time Domain Resources for CORESET

In the LTE system, a PDCCH can occupy up to three symbols of a slot or four symbols in case of narrowband, i.e. carrier bandwidth less than 10 RBs. However, in the NR time domain, a *CORESET* can be allocated up to three OFDM symbols. In the NR system, up to four CORESETs can be configured per BWP. Similarly, a UE can be configured with several CORESETs. A *ControlResourceSet* also contains *CCE to REG* mapping type, i.e. interleaved or non-interleaved and its size. Refer to TS 38.211 [106] for more information on the contents of a CORESET as configured by the RRC layer.

Figure 19.26 illustrates the allocation of two CORESETs for a typical PDCCH with only one CCE allocation to each. One CCE and its REGs (left side) contain one OFDM symbol only, and another CCE and its REGs (right side) contain two OFDM symbols in the time domain. Further, Figure 19.26 illustrates the hierarchies among the CCE, REG, and RB.

A CORESET also contains the following attributes, as defined by the RRC layer, TS 38.331 [116].

- Control set ID;
- CCE to REG mapping type, i.e. interleave or non-interleaved mapping; and
- Interleave size.

More about the PDCCH search space used by UE is described later in Section 19.6.10.

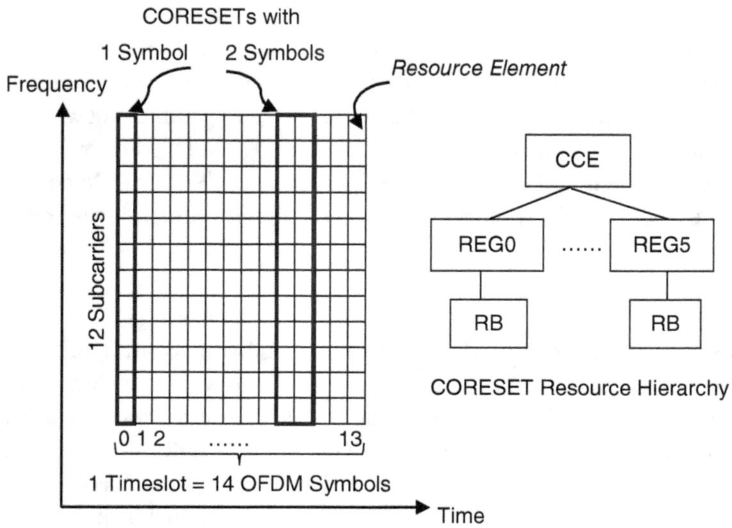

Figure 19.26 Illustration: CORESETs allocation for PDCCH.

19.6.5.2 Common Resource Blocks (CRB)

A common resource block (CRB) is a new concept in the NR system that provides the RBs numbering, starting from CRB:0, in the frequency domain over the entire CBW. For example, there are 273 CRBs over the 100 MHz CBW in FR1, TS 38.104 [103]. Common RBs are common for all the UEs operating with a particular channel or carrier bandwidth in the serving cell. However, UE will be allocated RBs from the allocated BWP only, which will be described later in Section 19.6.7, containing the actual PRBs.

19.6.5.3 Physical Resource Block (PRB)

PRBs are the RBs over which actual transmissions/receptions take place over the NR air interface. Thus, unlike the LTE system where the bandwidth of a PRB is fixed, i.e. 12×15 kHz = 180 Hz, in the NR system, the bandwidth of a PRB varies and depends on its subcarrier spacing being used, which can take place from 15 to 240 kHz. A set of PRBs forms a BWP and is defined within a particular active BWP for a particular numerology or subcarrier spacing only. Physical RB numbers are defined within an active BWP only, with numbering starting from PRB:0 to the size of the BWP ($N_{BWP}{}^{SIZE}$) – 1. A PRB0 of a BWP starts at a particular frequency offset from the CRB0. Such an offset, *offsetToPointA*, is provided by the RRC layer.

19.6.5.4 Virtual Resource Block (VRB)

NR also defines virtual resource blocks (VRBs) within a BWP only which are numbered from the VRB:0 to the size of the BWP ($N_{BWP}{}^{SIZE}$) – 1. DL and UL radio resources for PDSCH or PUSCH are allocated to a UE in terms of VRB only by the NG-RAN, i.e. modulated symbols are mapped to VRBs. Using VRBs, the NG-RAN uses two types (0 and 1) of resource allocation schemes to allocate radio resources to a UE in UL and DL directions.

19.6.5.5 Interleaved and Non-interleaved PRB Allocation

Depending on how a VRB is mapped to an actual PRB, resource allocation made to a UE in terms of a VRB can be either interleaved or non-interleaved, i.e. contiguous. Interleaved or non-interleaved VRB allocation is indicated through a DCI to a UE from the NG-RAN/gNB. Only the non-interleaved RBs allocation scheme is used in case of UL transmission from a UE, whereas interleaved as well as non-interleaved RBs allocation scheme may be used in case of DL reception from the NG-RAN. Non-interleaved resource allocation may be used to achieve a frequency diversity in DL direction in the NR system. VRB to PRB mapping is described next.

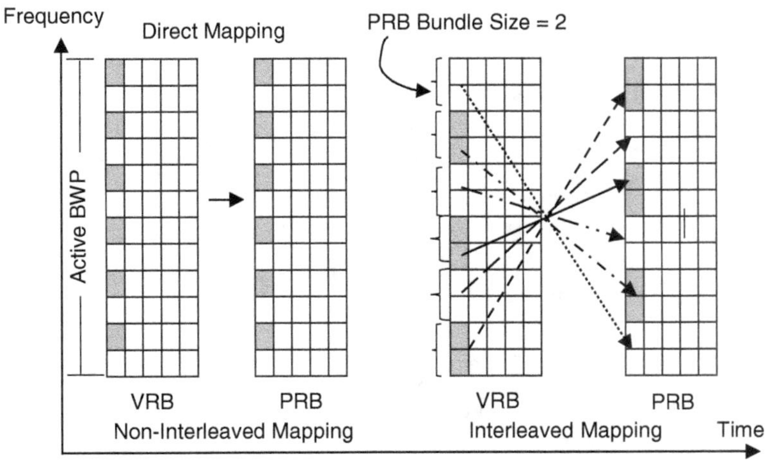

Figure 19.27 Illustration: VRB to PRB mapping in NR.

19.6.5.6 PRB Bundling and VRB to PRB Mapping

In the case of a non-interleaved PRB block allocation, one-to-one and direct VRB to PRB mapping are used. In the case of interleaved PRB allocation for PDSCH reception by a UE from the NG-RAN, VRB to PRB mapping is performed with the help of *PRB bundling* which consists of a set of consecutive resource blocks as configured by the NG-RAN RRC layer. The number of RBs in a PRB bundle is indicated through the RRC layer IE: *vrb-ToPRB-Interleaver*, which is provided to a UE as part of PDSCH configurations by the NG-RAN/gNB. If this IE is found to be absent in PDSCH configuration RRC signaling information to a UE, the non-interleaved resource allocation and direct VRB to PRB mapping are used. A PRB bundle size can be 2 or 4 RBs, see TS 38.331 [116]. Further, depending on the bundle size, a PRB bundling type can be a static or dynamic one as configured by the RRC layer using the IE: *prb-BundlingType*. PRB bundle-related information is provided to a PDSCH scheduled UE over the DCI 1_1. Interleaved and non-interleaved VRBs to PRBs mapping processes are illustrated in Figure 19.27. For the actual mapping process applied by a UE in case of DL for PDSCH transmission, refer to TS 38.211 [106] for the VRB to PRB interleaved resource allocation. In this illustration, the PRB bundle size (L) is assumed to 2.

19.6.5.7 Reference Point "A"

Reference point "A" serves as a common reference point for all the common RB grids of all the NR SCSs. Reference point "A" coincides with the center of the subcarrier 0 of the common RB CRB: 0 of all the SCS. A BWP, described later in Section 19.6.7, containing the actual PRBs on a common RB grid is defined with reference to the point A. The reference point "A" is illustrated in Figure 19.28 for 15, 30, and 60 kHz SCSs in case of the FR1 only. As illustrated, the bandwidth of a CRB with SCS: 30 kHz is twice that of a CRB with SCS: 15 kHz and so on. To determine the position of the reference point "A", the RRC layer at the NG-RAN end provides the frequency domain offset information *offsetToPointA* under the system information IE *FrequencyInfoDL-SIB* to the UEs.

19.6.6 Channel and Transmission Bandwidths

PRB and different SCS/numerologies have been described in the previous sections. Also, CBWs used in the NR were shown earlier in Table 19.4. Further, within a CBW, NR also defines the available transmission bandwidths to be used in FDD as well as TDD modes in terms of the number of RBs (N_{RB}^{μ}) in each numerology. In the case of FR1, the minimum transmission bandwidth is 4.5 MHz, against the CBW 5 MHz, with the number of RBs as 25 for SCS 15 kHz; the maximum transmission bandwidth is up to 98.2 MHz, against the CBW 100 MHz, with a

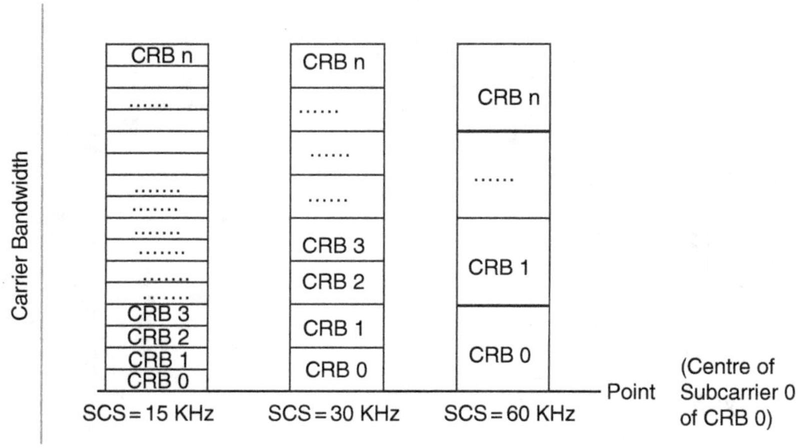

Figure 19.28 Illustration: NR common resource blocks and their reference point "A" in FR1.

maximum number of RBs as 273 for SCS 30 kHz. In the case of FR2, the maximum transmission bandwidth is up to 380.16 MHz, against the CBW 400 MHz, which differentiates distinctly the 5G NR from the LTE air interface. It may be noted that, due to the guard band provisioning requirements at the edges of CBW, the actual transmission bandwidth is a bit less than the CBW. Further, in a particular NR operating band, all the numerologies/SCS as well as the different transmission bandwidths are not supported. Refer to Tables 5.3.2-1/5.3.2-2/5.3.5-1/5.3.5-2 in TS 38.104 [103] for more information on NR channel and transmission bandwidths and their supported RBs.

- Transmission Bandwidth with Carrier Aggregation (CA)

As mentioned above, the maximum carrier or CBW supported in the 5G NR system under the FR1 is 100 MHz and under the FR2 is 400 MHz. But depending on UE capabilities, it may not be able to transmit and receive over such a wider carrier bandwidth. However, for high data rates/throughputs requirements, a UE can also transmit and receive multiple carriers of narrow bands. Such a feature and capability of transmissions and reception over multiple carriers by a UE is called CA, which is supported by the LTE system also from Release 10 onwards. Each carrier in CA is called a *component carrier* (CC). Necessary guard bands are provided between CCs of a CA. CA with intra-band, both contiguous and non-contiguous CC, and inter-band frequency bands are possible in FR1 and FR2. CA is also possible in case of interworking with the LTE system through the dual connectivity feature. Figure 19.29 illustrates a transmission/reception for two UEs - UE1 with wider carrier bandwidth capability and UE2 with intra-band CA capability for a typical carrier bandwidth of 60 MHz.

As illustrated in Figure 19.29, UE2 uses two CCs: CC0 and CC1, as per its CA bandwidth class, to achieve an aggregated carrier bandwidth of which should be less than the upper limit of 50 MHz as per the UE2 CA bandwidth class.

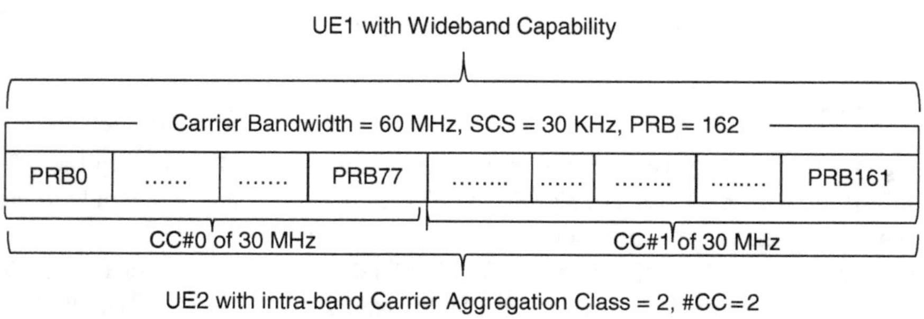

Figure 19.29 Illustration: UEs transmission/reception with wideband as well as an intra-band CA capabilities.

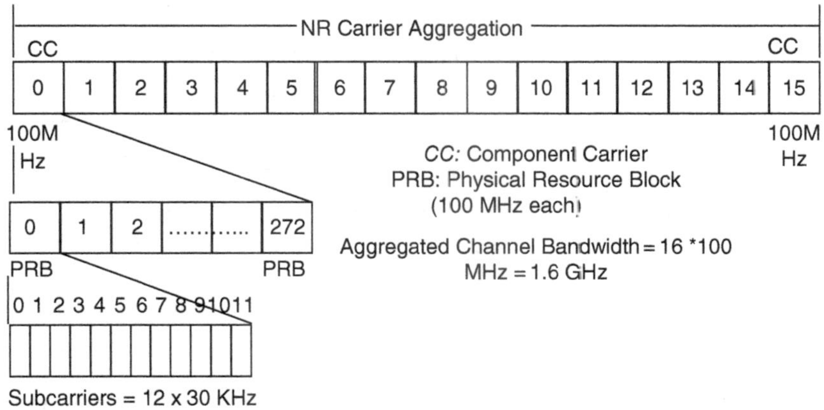

Figure 19.30 Illustration: aggregated channel bandwidth of CA in the NR FR1 with SCS = 30 kHz, CBW = 100 MHz.

In the case of the LTE system, under CA, a maximum of 32 (in Release 13) CCs are supported; in the 5G NR system, however, depending on UE CA bandwidth classes, maximum of 16 CCs may be supported in DL and UL directions. As the 5G NR already supports transmission and reception with wider carrier/channel as well as transmission bandwidth as mentioned above, transmission/reception with CA feature may not be configured in NG-RAN as CA capability involves signaling overhead also.

It may be noted that not all the operating bands of FR1 can be used for CA. Figure 19.30 illustrates the aggregated theoretical CBW, under CA in the NR system, which is 1.6 GHz, with maximum CC bandwidth of 100 MHz in the case of FR1. Note that guard bands are not shown in this illustration for theoretical calculation purposes.

A UE provides CA support information, in terms of CA bandwidth class, as part of its radio access capability *(UE-EUTRA-Capability, UE-NR-Capability, and UE-MRDC-Capability)* information to the NG-RAN. If a UE radio access capability does not support wideband transmissions/receptions, it may use the CA feature for delivering high data rates/throughputs to its user.

19.6.7 Bandwidth Part (BWP)

In the LTE system, a UE is required to support its transmission bandwidth as large as the carrier bandwidth (max 20 MHz) allocated in a cell. For example, in the LTE system, a UE is required to monitor the PDCCH channel over the entire cell bandwidth. Such a wideband operation results in higher power consumption, which is disadvantageous from a UE point of view. Also, all the UEs in the NR system may not support its transmissions/receptions through RF processing with transmission bandwidth as large as the cell or carrier bandwidth, from 5 to 400 MHz, defined for different numerologies. In the NR system, the operating bandwidth of a UE can be adjusted or adapted depending on the requirements of the service, e.g. control channel monitoring or large data transfer. Refer to Tables 5.3.2-1/5.3.2-2 in TS 38.104 [103] for the possible carrier and transmission/reception bandwidths for different numerologies in the NR system.

In the NR system, transmission/reception between UE and the NG-RAN takes place over a subset or smaller portion, which is called BWP, of the total carrier bandwidth defined for a particular SCS. A BWP is a fundamental concept in the NR system, which consists of a set of contiguous PRB within a carrier bandwidth. Different BWP can be defined with different SCSs and is configurable by the RRC layer. Switching between a narrowband, e.g. for control channel monitoring, and wideband, e.g. a large amount of data, BWP transmissions/reception and vice versa, under different SCSs is achieved through a BWP adaptation by a UE. Dynamic bandwidth adaptation is achieved by informing the UE on the active BWP to be used among the configured BWPs. A UE is not allowed to transmit or receive outside the frequencies of the active BWP irrespective of the UE capability, i.e. whether it supports wideband transmissions or not. This, in turn, reduces the power consumption of a UE as a BWP enables it to

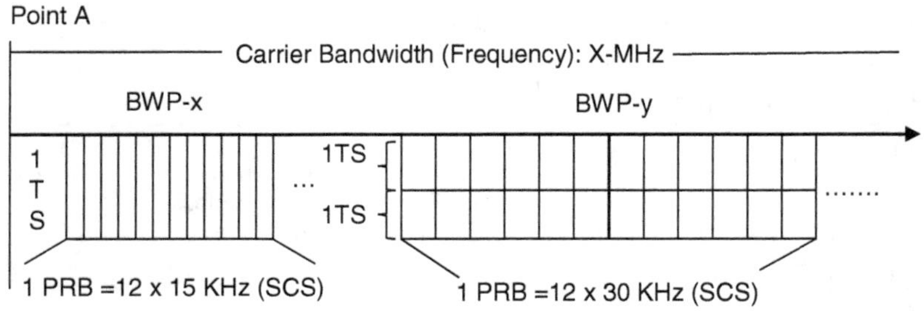

Figure 19.31 Illustration: configuration of BWPs and PRBs with different SCSs.

transmit/receive only over a smaller bandwidth. Figure 19.31 illustrates the configurations of two BWPs, having only 1 PRB each, with two different SCSs of 15 and 30 kHz within a carrier bandwidth, say 5 MHz, allocated to a cell.

19.6.7.1 Types of BWP

The NG-RAN RRC layer configures the following types of BWP to a cell/UE in the UL and DL directions separately for transmission and reception in FDD mode.

- Initial BWP

In each serving cell, the NG-RAN configures an initial BWP, over the PBCH, which is identified by the id:0, i.e. BWP-0, in UL and DL directions. Initial BWP contains the cell-specific parameters which are used by a UE for its initial access, i.e. to perform a random access procedure, to the NG-RAN. Initial BWP parameters are identified by the RRC layer parameters *initialDownlinkBWP and initialUplinkBWP*, which are provided as part of the cell-specific System Information Block (SIB) Type 1 or *RRCReconfiguration* message under the IE *dedicatedSIB1-Delivery* containing the SIB1 message from the NG-RAN to UEs. Initial BWP information contains the frequency domain location and bandwidth of the BWP and SCS for the allocated BWP.

The UE assumes that the *initialDownlinkBWP* is equal to the RBs configured for the *ControlResourceSet-0* (*CORESET-0*). A *CORESET-0* is received as part of the MIB information from the NG-RAN. The RBs allocated to a *CORESET*-0 varies depending on the SCS being used in the serving cell; refer to Tables 13.1–13.10 in TS 38.213 [108]. Using these RBs as the PDCCH search space and from the scheduled PDSCH, a UE acquires the SIB1. The SIB1contains the *initialDownlinkBWP* and *initialUplinkBWP* information also. Using the *initialUplinkBWP*, the UE performs a random access procedure to the NG-RAN to establish an RRC connection and acquire additional and default BWP.

- Additional BWP

In addition to the initial BWP, additional BWP, with the index starting from 1, may be also configured to UEs by the NG-RAN as follows:

a) Cell-specific, which is configured by the RRC parameters: *BWP-DownlinkCommon* and *BWP-UplinkCommon;*
b) UE-specific, which is configured by the RRC parameters: *BWP-DownlinkDedicated* and *BWP-UplinkDedicated.* Such a dedicated BWP is assigned after a UE enters into the RRC CONNECTED state.

- Default DL BWP

A default DL BWP may be also configured, under the RRC IE *defaultDownlinkBWP-Id*, to a UE from anyone of the already configured DL BWPs by the NG-RAN. If no default DL BWP is configured, the initial BWP will be treated as a default BWP. The size of the initial as well as the default BWP may be less than that of an additional BWP.

19.6.7.2 BWP Configuration

A BWP consists of contiguous PRBs and is a subset and smaller than the carrier bandwidth configured in a cell within particular numerology or SCS; refer to Figure 19.31. As mentioned in the previous section, the initial BWP of a cell has the id:0, i.e. BWP-0. A UE can be configured up to four BWPs, with different bandwidths, depending on its capability, with a non-zero index starting from 1, in UL or DL direction per carrier bandwidth per serving cell.

The generic definition of a BWP consists of the following information for UEs; refer to TS 38.331 [116].

- SCS or numerology;
- Frequency domain location and bandwidth size in terms of PRBs; and
- CP, i.e. extended or normal.

In addition to the initial bandwidth BWP-0, the NG-RAN can also configure a maximum of four additional BWPs of additional and default type to a UE dynamically through the RRC layer *RRCReconfiguration* signaling message. Such a BWP configuration contains the following types of parameters:

- Cell-specific common parameters, e.g. *BWP-DownlinkCommon* containing information for the random access purpose, SCS, bandwidth, frequency domain location, CP information, and so on;
- UE-specific dedicated parameters – *BWP-UplinkDedicated* containing information for UL transmission over PUSCH or PUCCH, SRS configuration, beam failure recovery configuration; and *BWP-DownlinkDedicated* containing information for DL reception over the PDSCH or PDCCH, radio link monitoring, and semi-persistent scheduling configurations.

Note that only one BWP can be active at a time in UL or DL direction. Further, a UE is not allowed to transmit over PUSCH and PUCCH or receive over PDSCH and PDCCH, outside the frequencies of the active BWP. However, one exception is that a UE can perform radio resource measurement, for UE mobility purposes, in DL direction outside an active BWP during the measurement period only, described later in Section 19.6.16.

Figure 19.32 illustrates the allocation of one BWP#0 (though possible to allocate up to 4 BWP) for two UEs, UE1 and UE2, with the same subcarrier spacing (SCS) = 30 kHz, $N_{SC}^{RB} = 12$, and CBW of 5 MHz. BWP#0 for UE2 contains 3 PRBs, and for UE1, it is shown to be varied (N). For SCS = 30 kHz, the transmission bandwidth or the number of RBs is $N_{RB} = 11$. The physical and virtual RB mapping of the BWP for UE1 is also illustrated. Inside the BWP#0, physical and virtual RBs are numbered from PRB/VRB:0 to PRB/VRB: N ($<N_{RB}$) as the size of the BWP#0. A physical RB (PRB0) is shown to be located at four common RBs from the Point "A" for UE1; for UE2, PRB0 is shown to be located at five common RBs from the Point "A". Also, the size of the BWP allocated for UE2 is illustrated to be less than that of the UE1.

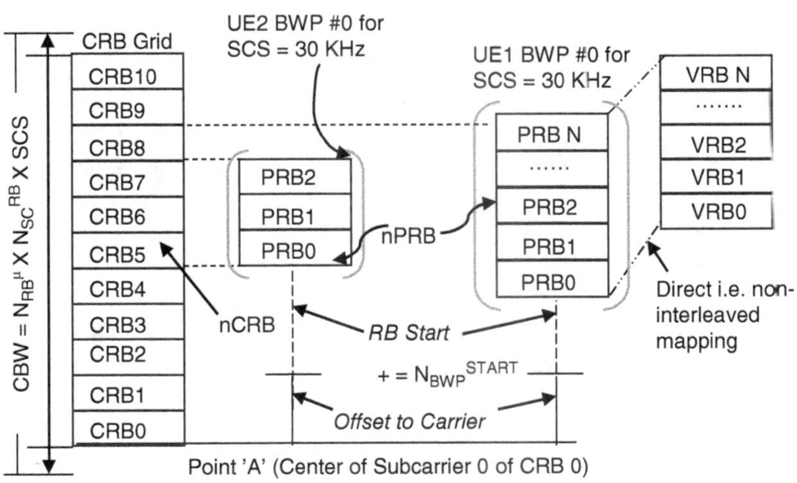

Figure 19.32 Illustration: an NR bandwidth part, CRB, PRB, and VRB mapping.

A common RB(n_{CRB}) can be derived from a starting point of a BWP(N_{BWP}^{START}) and its physical RB(n_{PRB}) using the relationships between them as described in TS 38.211 [106], Section 4.4.4.4. The N_{BWP}^{START} is the common RB where a particular BWP starts in relative to the CRB:0. Further, the common RB N_{BWP}^{START} for each BWP can be derived using the formula as described in Section 12, TS 38.213 [108] which consists of an *offset-To-Carrier* (max:2200) plus starting location of the common RB (RB Start)of a BWP. These two parameters are provided by the RRC layer through *OffsetToCarrier* IE. The RBstart is derived from the *locationAndBandwidth* (max: 37950) information which is provided by the NG-RAN as part of generic BWP information to UEs in a serving cell. Usages of these parameters are illustrated in Figure 19.32.

19.6.7.3 BWP Switching and Associated Delay

Different types of BWP, with common and UE-specific information, as may be configured by the NG-RAN in a serving cell, were described in Section 19.6.7.1. BWP switching refers to the activation of an inactive and deactivation of an active BWP dynamically at a time at UE end, that is, switch the active BWP to default or initial BWP and vice versa. In the case of FDD mode, BWP can switch independently in DL and UL directions because of the usages of a separate frequency band in each direction. In the case of TDD mode, BWP switching takes place simultaneously in DL and UL directions because of usages of a single frequency band.

A BWP switching at a UE end can take place due to the following events:

- Activation/deactivation as part of DL assignment or UL grant through DL control information – DCI 0_1 for UL; DCI 1_1 in the DL, over the PDCCH. DCI 0_1 and DCI 1_1 contain a *bandwidth part indicator* field which is of 2 bits in length to support up to four BWPs; refer to TS 38.212 [107].
- RRC signaling-based switching, i.e. activation through *RRCReconfiguration*.
- Through the MAC CE on the initiation of the random access procedure. The BWP switching, in this case, depends on whether or not PRACH occasions are configured for the active UL BWP. If the PRACH occasion on the active UL BWP is configured, no UL BWP switching is performed and if the serving cell is special, then the DL BWP is switched to a DL BWP whose BWP-id same as the UL BWP. On the other hand, if the PRACH occasion on the active UL BWP is not configured, UL BWP is switched to the *initialUplinkBWP*, and if the serving cell is special, DL BWP is switched to the *initialDownlinlinkBWP*. Refer to TS 38.321 [113].
- Timer-based activation/deactivation of DL BWP.
  This is a recovery mechanism in case a UE fails to decode or does not receive the DL BWP assigned to it by the NG-RAN over the respective DCI. To handle such an erroneous scenario, the NG-RAN may configure a DL BWP inactivity timer called *bwp-inactivityTimer* at UE MAC layer. This timer is started/restarted upon switching to an active DL BWP. At the expiry of this timer, i.e. the UE is not scheduled by the NG-RAN, the active DL BWP at UE is switched to its default or initial DL BWP, in case the default DL BWP is not configured.

It may be noted that to start transmitting (PUSCH on new UL BWP) or receiving (PDSCH over new DL BWP), there is an associated delay for a UE to switch to a new BWP in each of the events mentioned above. Such a delay is called a BWP switching delay, which also depends on a UE capability; TS 38.306 [112]. For the switching delay for a UE to complete a BWP switching in each of the events mentioned above, refer to Table 8.6.2-1 in TS 38.133 [104]. For more information on the BWP operation, refer to TS 38.211 [106], 38.213 [108], and TS 38.331 [116] for BWP parameters as configured by the RRC layer.

An overview of the NR physical layer resources architectures and their organizations into a resource grid, physical and virtual RB, BWP, and timeslot has been described above. The allocation of those resources in the frequency and timer domain in FDD and TDD modes are described in the next sections.

19.6.8 NR Resource Allocations

In this section, NG-RAN resource allocation types used to allocate NR radio resources to UEs in the frequency domain and time domain as well as in UL and DL directions will be described. NR radio resource allocation procedures also differ based on the duplex mode, i.e. FDD and TDD modes, being used by the NG-RAN. The following NR radio resource allocations aspects will be described:

- FDD – Frequency domain radio resource allocation to UEs for PDSCH reception and PUSCH transmission.
- FDD – Time-domain radio resource allocation to UEs for PDSCH reception and PUSCH transmission.
- TDD – Time-domain radio resource allocation to UEs for PDSCH reception and PUSCH transmission.

19.6.8.1 Frequency Domain Resource Allocation for FDD Transmission

At a high level, the physical layer resource allocation types used in the NR system are similar to the LTE system resource allocation types. Though the aspects of the physical layer are different, a resource allocation type indicates the way the LTE or NR scheduler allocates resources to a UE for transmissions and receptions in the UL, over PUSCH, or DL, over PDSCH, direction. However, there is a difference between LTE and NR physical layer resource allocation procedure as far as the carrier bandwidth is concerned. In the LTE system, physical layer resource allocation may span across the carrier bandwidth, for example, the PDCCH and BCCH control regions are allocated throughout the carrier bandwidth in the frequency domain. But in the NR system, only a subset or small portion of the total carrier bandwidth is allocated in terms of a BWP as described in Section 19.6.7. Up to four BWP can be allocated to a UE in DL or UL direction, but only one BWP is active at a time, as described in Section 19.6.7.2 Further, a UE is allocated to UE-specific resources block from the active BWP only and not from the CRB which covers the entire carrier bandwidth.

The NG-RAN/gNB also uses two types of resource allocations types, i.e. Type 0 and Type 1, TS 38.214 [109], in the frequency domain to allocate resources by the scheduler to UEs for transmission in UL and reception in DL directions. Radio resources are allocated in terms of VRB. Radio resource allocation type information is provided as part of a DCI to a UE, as listed in Table 19.7, along with the associated channel, i.e. PUSCH or PDSCH.

In radio resource allocation type, 0 or 1, information may be provided over an RRC layer signaling also through the IE: *resourceAllocation*. If this IE contains the value *dynamicSwitch*, radio resource allocation type will be provided over the DCIs mentioned in Table 19.7.

- **Resource Allocation Type 0 (Based on Bitmap of Resource Block Group)**

Similar to the LTE system, Type 0 resource allocation in the NR UL and DL is based on a bitmap of a resource block group (RBG). An RBG is a set of consecutive virtual RBs. The number of consecutive RBs in an RBG is defined in Tables 5.1.2.2.1-1/6.1.2.2.1-1 in TS 38.214 [109], which depends on the BWP size. There are two variants, configuration 1 and configuration 2, of an RBG size as defined in this table. For example, for BWP size 1–36, an RBG can contain two or four RBs. RRC layer provides such RBG configuration information to be used by a UE through the IE: *PDSCH-Config/PUSCH-Config*. RRC layer indicates the configuration 1 or configuration 2 using the IEL: *PDSCH-Config -> rbg-Size or PUSCH-Config -> rbg-Size.*

Table 19.7 NR resource allocation and VRB-To-PRB mapping types.

Resource Allocation Types	Method of Allocation	DCIs Used to Indicate to UE	Interleaved/Non-Interleaved (VRB-To-PRB)
Type 0	Bitmap Based	0_1(PUSCH),1_1(PDSCH)	Non-Interleaved
Type 1	RIV Based (Resource Indication Value)	0_0,0_1(PUSCH), 1_1, 1_0(PDSCH)	Both

A value 1 in RBG bitmap indicates that the corresponding RBG is allocated to a UE. A bitmap of RBG is used instead of a bitmap of the individual RB. This is because the grouping of RBs into an RBG and allocation of an RBG to UE through a bitmap reduces signaling overhead on the NR air interface as less number of information, or reduced bitmap size, in the number of bits, would be transmitted. Frequency domain resource allocation information through RBG bitmap is provided to a UE over a DCI and its field *frequency domain resource assignment* (FDRA), TS 38.212 [107], whose length in bits varies depending on the size of the UL or DL BWP. To determine the number of RBGs/size of FDRA bitmap, refer to the formula described in Section 51.2.2.1/6.1.2.2.1 in TS 38.214 [109]. The same formula and calculation method are used for PUSCH and PDSCH resource allocations. Examples of FDRA using these formulae are illustrated below, see Examples 19.1 to 19.2.

Example 19.1 Frequency Domain Resource Assignment for PDSCH Reception with Resource Allocation Type 0 and Unequal RBs

Consider the FDRA to a UE for PDSCH reception by it with a BWP of eight RBs. Further, assume that the RB starts at VRB #3. From the BWP size = 8, the RBG configuration #1 (RRC: *PDSCH-Config ->rbg-Size)* will be applicable based on Table 5.1.2.2.1-1 in TS 38.214 [109], i.e. the number of RBs per RBG is P = 2.

Number of RBG/bitmap size (FDRA) = Ceil ((8+ [3 mod 2])/2) = 5 (Section 5.1.2.2.1 in TS 38.214 [109])

Number of resource block in first RBG = 2− (2 mod 2) = 1 (Section 5.1.2.2.1 in TS 38.214 [109])

Number of resource block in last RBG = (3 + 8) mod 2 = 1 [Section 5.1.2.2.1 in TS 38.214 [109]]

Number of resource block in all other RBGs (1−3) = P = 2.

The FDRA to a UE with the above information is illustrated in Figure 19.33. As calculated above, the first and last RBG contain only 1 RB each, and RBGs 1−3 contain equal RBs. In this illustration, RBGs 1−3, and their resources blocks, are shown to be as allocated (shaded) whose RBG bitmap value is 1. RBG 0 and RBG 4 are not allocated.

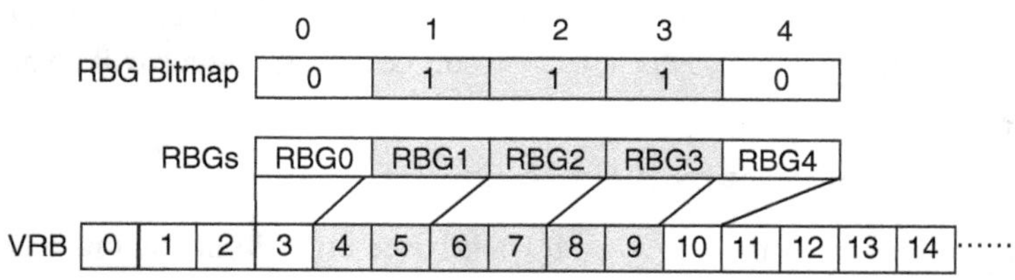

Figure 19.33 Illustration: resource allocation Type 0 in the NR frequency domain.

Example 19.2 Frequency Domain Resource Assignment for PDSCH with Resource Allocation Type 0 and Equal RB

Consider another FDRA to a UE for PDSCH reception by it with a BWP of 16 RBs. Further, assume that the RB starts at VRB #10. From the BWP size = 16, the RBG configuration #1 will be applicable based on Table 5.1.2.2.1-1 in TS 38.214 [109], i.e. number of RBs per RBG is P = 2.

Number of RBG/bitmap size (FDRA) = Ceil ((16+ [10 mod 2])/2) = 8 (Section 51.2.2.1 in TS 38.214 [109])

Number of resource block in first RBG = 2− (10 mod 2) = 2 (Section 51.2.2.1 in TS 38.214 [109]).

Number of resource block in last RBG = 2 since (10 + 16) mod 2 = 0 (Section 51.2.2.1 in TS 38.214 [109]).

Number of resource blocks in all other RBGs = P = 2.

As calculated above, each RBG contains two RBs; thus, the total RB allocated is 8 × 2 = 16. Frequency domain resources assigned to a UE with the above information can be represented similarly as illustrated in Figure 19.33 in the previous example.

- **Resource Allocation Type 1 (Based on Start and Length of Resource Block)**

On the other hand, Type 1 resource allocation is based on the allocation of an active BWP which consists of a set of contiguous PRBs of a BWP. Resource allocation type is specified using the RRC signaling IE *resourceAllocation* as part of *PDSCH or PUSCH-Config* configuration provided by the NG-RAN/gNB toward a UE.

The following information is provided to a UE as part of a BWP allocation; TS 38.331 [116].

i) SCS
ii) CP
iii) BWP identifier
iv) Common and dedicated parameters related to the allocated BWP
v) A common RB and several contiguous RBs as the bandwidth or size of the allocated BWP

The starting RB and the number of RBs for the allocated BWP are jointly encoded into a single parameter called *locationAndBandwidth*, which is provided by the RRC layer as part of the generic BWP information; TS 38.331 [116]. Type 1 resource allocation to a UE in UL or DL direction contains a piece of information called *Resource indication Value (*RIV*)* which is derived from the *locationAndBandwidth* of the BWP information. Since the actual resources are allocated to a UE in terms of a VRB, the RIV contains the starting VRB and length in terms of the number of consecutive RBs with the BWP allocated. The starting RB is denoted by RB_{start}, and the number of consecutive RBs of a BWP is denoted by L_{RB}, as mentioned in TS 38.214 [109]. The calculation of RB_{start}, and L_{RBs} are illustrated below, see Examples 19.3 to 19.4, using the formula described in Section 5.1.2.2.2 in TS 38.214 [109].

Example 19.3 Frequency Domain Resource Assignment for PDSCH Reception with Resource Allocation Type 1 and SCS: 15 kHz

Assume that the *locationAndBandwidthprovided* information provided by the RRC layer as part of the BWP allocation is 1000, i.e. RIV = 1000, for SCS:15 kHz and carrier bandwidth of 20 MHz. The maximum size of a BWP in terms of number PRB (N_{BWP} size) is 275, as defined in TS 38.213 [108].

Maximum number of PRBs for SCS: 15 kHz and carrier bandwidth: 20 MHz is, say, Y = 106 RB (TS 38.104 [103])

Also, say, X = Floor ($N_{BWP}{}^{size}/2$) = 137 PRBs (Section 5.1.2.2.2 in TS 38.214 [109]).

Since Y<X and as per the first equation in Section 5.1.2.2.2 in TS 38.214 [109], calculate L_{RBs} and RBstart as follows: $L_{RBs} = (RIV/N_{BWP}{}^{size}) + 1$ and $RB_{start} = (RIV \% N_{BWP}{}^{size})$ which gives

a) $L_{RBs} = (1000/275) + 1 = 4$ VRBs of the allocated BWP
b) $RB_{start} = (1000\% \ 275) = 175$

The above resource allocation to a PDSCH, in terms of the number of contiguous VRB and its starting VRB, is illustrated in Figure 19.34.

Figure 19.34 Illustration: Type 1 resource allocation to a PDSCH.

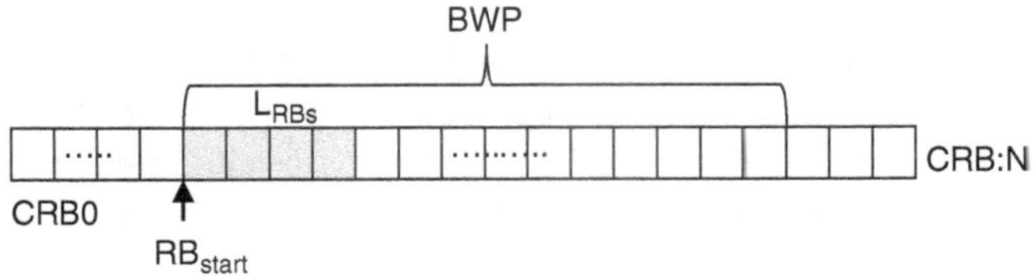

Example 19.4 Frequency Domain Resource Assignment for PDSCH with Resource Allocation Type 1 and SCS: 30 kHz

Similarly, assume that NG-RAN/RRC layer allocated the *locationAndBandwidthprovided* = RIV = 34 924 to a UE for SCS:30 kHz and carrier bandwidth of 60 MHz.

Maximum number of PRBs for SCS = 30 kHz and carrier bandwidth 60 MHz: Y = 162 RB (TS 38.104 [103])

Also, X = Floor ($N_{BWP}^{size}/2$) = 137 PRBs (Section 5.1.2.2.2 in TS 38.214 [109])

Since Y>X and as per the second equation in Section 5.1.2.2.2 in TS 38.214 [109], calculate L_{RBs} and *RBstart* as follows: $L_{RBs} = (N_{BWP}^{size} + 1) - RIV/N_{BWP}^{size}$ and $RB_{start} = N_{BWP}^{size} - 1 - RIV \% N_{BWP}^{size}$, which gives

a) $L_{RBs} = 275 + 1 - 34924/275 = 150$
b) $RB_{start} = 275 - 1 - (34924\% \ 275) = 0$

19.6.8.2 Time-Domain Resources Allocation for FDD Transmission

Similar to frequency domain resources allocation, resource allocation in the time domain, for FDD, can also vary depending on the network configuration such as different SCSs used for PDSCH or PUSCH and PDCCH transmissions. The network requires to provide dynamically scheduled time-domain resources allocation information on the timing aspects for transmission/reception of data by a UE. Frequency domain and time-domain resources for the PDSCH or PUSCH transmission are provided dynamically to a UE through a DCI over the PDCCH. A DCI contains the *time-domain resource assignment* (4 bits length) information field, in UL as well as DL, that carries a *row index* into a predefined lookup table/array defined for resource allocation. Refer to the DL and UL resources allocation lookup tables described in Sections 5.1.2.1.1 and 6.1.2.1.1 in TS 38.214 [109]. The lookup tables have 16 rows (4 bits length) containing different possibilities of resource allocation in DL and UL directions.

Using the predefined lookup tables mentioned above, time-domain resource allocations to UE for the PDSCH and PUSCH transmissions contain combinations of three types of information as mentioned below. Note that time-domain resources, i.e. OFDM symbols, are allocated within a slot boundary only containing 14 OFDM symbols with a normal CP:

- *Slot offset*, which is used between the PDCCH/DCI and the scheduled PDSCH or PUSCH. In the case of PDSCH allocation, slot offset K_0 is used; slot offset K_2 is used in case of PUSCH allocation; refer to TS 38.214 [109]. Using the slot offset K_0, UE determines the slot where the PDSCH reception should be performed. If $K_0 = 0$, this means PDSCH reception should be performed on the same slot, but with different OFDM symbols, where PDCCH/DCI was received. Similarly, using the K_2, UE determines the slot where PUSCH transmission should be performed. Offsets K_0 and K_2 can have different values as described in Sections 5.1.2.1.1 and 6.1.2.1.1 in TS 38.214 [109].
- Start of OFDM symbol "S" (0–13), i.e. first OFDM symbol for the reception of PDSCH or transmission of PUSCH.
- Consecutive OFDM symbols allocation of length "L" (1–14) in the UL or DL direction.
- PDSCH or PUSCH mapping type, which specifies the allowed starting position of the OFDM symbol and its length in a slot.

Depending on the valid combinations of values of the *S and L*, there are two types – Type A and Type B – of resource allocations schemes for PDSCH or PUSCH in the time domain for FDD with normal as well as extended cyclic prefixes. For example, in Type A resource allocation for PDSCH, the allowed value of S is {0, 1, 2, 3} and of L is {3...14}. For normal CP, the maximum value of (S + L) is 14 symbols; for extended CP, it is 12. Refer to Tables 5.1.2.1-1/6.1.2.1-1 in TS 38.214 [109].

Further, the information – *S and L* – are also combined and jointly encoded into a single parameter called the *Start and Length indicator Value* (SLIV). The maximum value range of SLIV is 0–127. Given the SLIV, the formula to calculate the S and L are described in Sections 5.1.2.1 and 6.1.2.1 in TS 38.214 [109]. Allocation of time-domain resources through default as well as RRC signaling methods with the usages of the above offsets and SLIV are described in the subsequent discussions.

- **Time-Domain Resources Allocation (FDD): Default versus RRC Signaling**

NR physical layer provides default as well as RRC layer signaling options to allocate time-domain resources in DL and UL directions in FDD mode for the parameters (K_0, K_2, S, L, and mapping type) mentioned above.

– Default Resource Allocation Method

Default time-domain resources allocation is described and predefined in lookup Tables 5.1.2.1.1-2, 5.1.2.1.1-3, 5.1.2.1.1.-4, 5.1.2.1.1-5 in TS 38.214 [109] for DL direction through DCI 1_0/1_1 and Table 6.1.2.1.1-2 in TS 38.214 [109], for UL direction through DCI 0_0/0_1.

– RRC Signaling-Based Resource Allocation Method

Time-domain resources can be also allocated through dedicated RRC signaling message and its IE: *pdsch-ConfigCommon/ pdsch-Config -> pdsch-TimeDomainAllocationList -> PDSCH-TimeDomainResourceAllocation* for DL direction and pusch-*ConfigCommon/ pusch-Config ->pusch-TimeDomainAllocationList ->PUSCH-TimeDomainResourceAllocation* for UL direction. In this case, the NG-RAN provides the slot offset K_0, mapping type (A/B) of allocation and SLIV to UE through the IE *PDSCH-TimeDomainResourceAllocation or PUSCH-TimeDomainResourceAllocation*. These are array-like IEs that can have N rows/elements as shown below:

```
Entries in PDSCH-TimeDomainResourceAllocation/list or array: refer to TS
38.331 [116]
{
   {K0, mappingType, startSymbolAndLength },/* K0: downlink, K2:Uplink */
    {K0, mappingType, startSymbolAndLength },
    {K0, mappingType, startSymbolAndLength },
      ...............
    {K0, mappingType, startSymbolAndLength }
}; /*N entries */
```

Based on the number of rows/entries (N) in the IE: *PDSCH-TimeDomainResourceAllocation* or *PUSCH-TimeDomainResourceAllocation*, the number of bits required to provide the time-domain resources allocation through a DCI field is calculated as follows:

Bit widths for *Time-domain resource assignment* field in DCI 1_1 or DCI 0_1 = Ceil (log2 (N)) bits [TS 38.212 [107]].

For example, consider that there are eight entries in the *PDSCH-TimeDomainResourceAllocation* IE. This means that three bits will be required to provide the *Time-domain resource assignment* field of DCI 1_1 or DCI 1_0. If a UE receives a *Time-domain resource assignment* as value "000", it indicates the first row/entry of the *PDSCH-TimeDomainResourceAllocation* IE.

Based on several factors such as the RNTI type, PDCCH search space, *Synchronization Signals Block* (SSB) and CORESET multiplexing pattern, a UE will use the appropriate lookup table from a default or RRC signaling-based time-domain resource allocation method as described in Table 5.1.2.1.1-1 (DL)/Table 6.1.2.1.1-1 (UL) in 38.214 [109]. Further, from these tables, there are three (A/B/C) default time-domain resource allocation types in the DL direction, whereas there is only one, i.e. Default A, type of resource allocation in the UL direction. Also, the default time-domain resource allocation method is used in case the allocation through RRC layer signaling based is not configured through *pdsch-Config/*pdsch-*ConfigCommon/*pusch-*ConfigCommon/pusch-Config*. Time-domain resource allocation in FDD using default and RRC signaling are illustrated below, see Examples 19.5 to 19.7.

Example 19.5 Default Resource Allocation in the Time Domain (FDD mode) for PDSCH Reception with Different Subcarrier Spacings for PDCCH and PDSCH

One example is illustrated below for the calculation of a timeslot allocation for PDSCH reception by UE in FDD mode using the default method with normal CP, i.e. 14 OFDM symbols per slot and PDSCH mapping Type B. Also, assume that

- Time-domain resource is allocated using the default allocation of Type B with row index = 6; refer to Table 5.1.2.1.1-4 in TS 38.214 [109]. Thus, S = 2, L = 2, and $K_0 = 1$.
- For PDCCH, numerology = 0 (15 kHz), and for PDSCH, numerology = 1 (SCS 30 kHz).
- DCI is received over the PDCCH subcarrier on slot #1, i.e. n = 1.

Since the numerologies used for PDCCH and PDSCH are different, the timeslot for the PDSCH will be calculated as follows; refer to Section 5.1.2.1 in TS 38.214 [109] for the formula.

PDSCH Timeslot = Floor $(n * [2^{\mu PDSCH}/2^{\mu PDCCH}]) + K_0$ = Floor $(1*[2^1/2^0]) + 1 = 3$ (on PDSCH SCS).

From this calculation, UE PDSCH reception will take place on a timeslot 3 with starting symbol (S): 2 and length (L): 2, as illustrated in Figure 19.35. That is, UE will receive PDSCH over the timeslot 3 only and symbols 2 and 3. Further, using the formula mentioned in Section 5.1.2.1 in TS 38.214 [109], the *SLIV* is calculated to be 16.

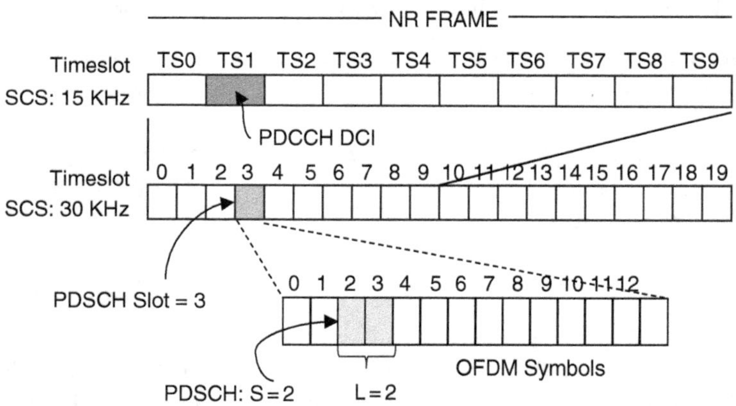

Figure 19.35 Illustration: time-domain resource allocation for PDSCH by RRC signaling.

Example 19.6 Default Resource Allocation in the Time Domain (FDD) for PDSCH Reception with the Same SCS for PDCCH and PDSCH

Consider that PDCCH and PDSCH are allocated over the same SCS of 15 kHz with normal CP, i.e. 14 OFDM symbols per slot. Also, assume that time-domain resource is allocated using the default allocation of Type B with row index = 6; refer to Table 5.1.2.1.1-4 in TS 38.214 [109], giving *S = 2, L = 2, and* $K_0 = 1$.

Further, assume that the DCI is received over PDCCH timeslot #1(n = 1) and its OFDM symbol 0.

Thus, PDSCH Timeslot = Floor $(n * [2^{\mu PDSCH}/2^{\mu PDCCH}]) + K_0$ = Floor $(1*[2^0/2^0]) + 1 = 2$

PDCCH and PDSCH resource allocations described above can be illustrated as shown in Figure 19.36.

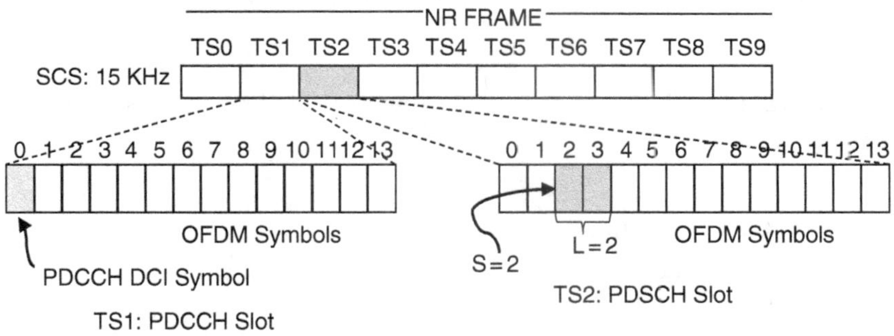

Figure 19.36 Illustration: (FDD) time-domain resource allocation for PDSCH: default allocation.

Example 19.7 RRC Signaling-Based Time-Domain Resource Allocation, for FDD Mode, with Different SCS for PDCCH and PDSCH

As mentioned above, NG-RAN can also provide time-domain resources through the RRC layer signaling message for the PDSCH reception by the UE.

Given the $K_0 = 1$, SLIV = 16 over the RRC IE: *PDSCH-TimeDomainResourceAllocation* and PDSCH timeslot = 3, which was also computed in the previous example as part of a default time-domain resource allocation method, the PDSCH start symbol (S) location and its consecutive symbols length (L) can be calculated by the UE using the formula mentioned in Section 5.1.2.1 in TS 38.214 [109], as follows:

L = (16/14) + 1 = 2 Symbols for PDSCH
S = 26%14 = 2 (PDSCH starting symbol)

Time-domain resource allocation for PDSCH using the above examples was illustrated earlier in Figure 19.35.

Radio resource allocation in frequency and time domain in the NR FDD mode has been described and illustrated above. Radio resource allocation in the time domain in the NR TDD mode will be described next.

19.6.8.3 Time-Domain Resources Allocation for TDD

Configurable timeslot formats used in the NR system for TDD mode transmission/reception were explained earlier in Section 19.6.4.3. A UE will use the TDD resources (slots and symbols) provided over the respective DCI, i.e. DCI 0_0/0_1 and DCI 1_0/1_1, for UL transmission or DL reception under the following conditions:

- If a UE or group of UEs is not configured or UE capability does not support to monitor SFI information provided over the DCI 2_0 with *sfi-RNTI* or
- TDD resources are also not provided over the RRC layer IE: *TDD DL-UL-Config*.

Note that if a UE finds conflict in the TDD resources allocated by the NG-RAN/gNB through the TDD DL-UL configuration and also through the respective DCIs in UL or DL direction, then the UE will not transmit over the PUCSH or receive over the PDSCH using the conflicting resources/timeslot scheduled using the respective DCIs. For example, suppose a UE has been scheduled dynamically using a DCI for DL PDSCH reception over timeslots 1 and 2. Also, the UE was allocated the timeslot 1 through the TDD DL-UL IEs which contains at least one symbol in the UL direction. In this case, the UE will not receive PDSCH over the timeslot 1 allocated through a DCI.

For more details on the procedure used by UE and its expected behavior toward the processing of slot format information and transmission/reception using it, refer to Section 11 in TS 38.213 [108].

19.6.9 Transport Channels and Their Processing Chain

- NR Transport Channels

Similarly, as in the LTE system, 5G NR supports the following transport channels in DL as well as UL direction over its air interface:

- DL-SCH
- BCH in the DL direction
- PCH in the DL direction
- UL-SCH
- RACH in the uplink direction

Using the above transport channels, the NR physical layer provides services to the MAC layer. BCH carries the system information, received through the BCCH, as broadcasted by the NG-RAN in a serving cell. PCH carries the paging message information, received through the PCCH, as broadcasted by the NG-RAN in a serving cell. DL-SCH and UL-SCH carry multiplexed data and control information between UE and NG-RAN and vice versa. Table 19.2 in Section 19.5.10 shows the mapping and multiplexing of different logical channels onto their corresponding transport channels in the NR system.

- NR Transport Channels Processing Chain

Higher layer information, either control or user data, is exchanged in the form of a *transport block* from the MAC layer to the physical layer through the DL-SCH or UL-SCH transport channels. The physical layer transfers the transport blocks over the respective physical channels, e.g. PDSCH and PUSCH. The transport blocks that are transmitted over the air interface between a sender and a receiver are likely to get affected because of various environmental-related reasons, e.g. noisy channels. Thus, the condition of a radio channel between a transmitter and a receiver is not always perfect. To avoid transmission-related errors over an imperfect channel condition, the NR physical layer at the sending end attempts to improve the reliability of information exchanged over the air interface so that the receiver can decode and extract the transport block correctly. In this regard, the NR physical layer at the sending end performs several functions, called *channel coding*, in terms of the preparation of the received transport block from a higher layer as well as its actual transmissions over the NR air interface.

Figure 19.37 illustrates the overall and general functions performed as part of the physical layer processing tasks for transmission of a transport block in UL and DL directions over the NR air interface between UE and NG-RAN.

As part of the preparation stage, the NR physical layer at the sending end performs several subtasks. Those subtasks collectively form the *channel coding* procedure, producing a *code word* from an original transport block. These aspects shall be described in this section. Next, as part of the actual physical channel processing and signal generation stage, the NR physical layer performs several subtasks or procedures. Those subtasks collectively form the *scrambling and modulation* procedure, which shall be described in the next section. The NR physical layer at

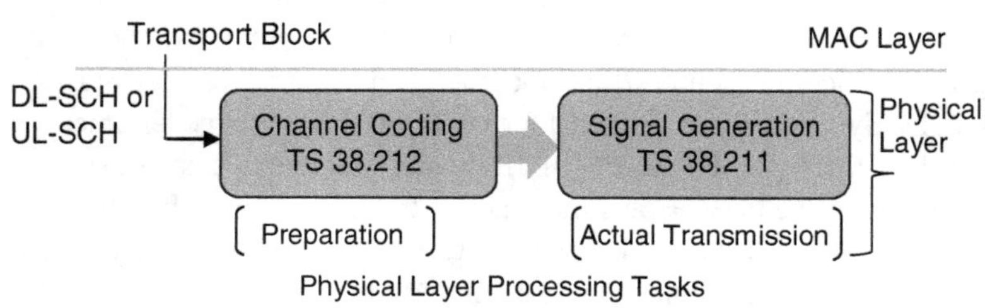

Figure 19.37 Illustration: physical layer processing stages for a transport block.

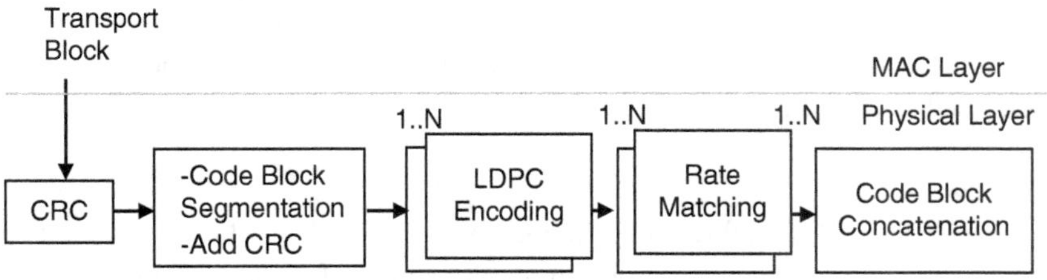

Figure 19.38 Illustration: physical layer processing chain for a transport block.

the receiving end performs the opposite of the above functions, called *decoding*, to extract the transport block and pass it to the MAC layer. Receiving MAC layer further demultiplexes the transport block data and passes it to the RLC layer. Note that depending on the multiple antenna system configurations in use between an NG-RAN and UE (e.g. spatial multiplexing transmissions), two transport blocks may be used for channel coding purposes.

The NR transport channel processing chain performed at the physical layer on a transport block received from the MAC layer contains the following steps, in general, and in sequence in UL and DL directions at the transmitting end; see Figure 19.38 for illustration.

i) Calculation of CRC for error detection purposes (at receiver end) and append it to a transport block;
ii) Code block segmentation into the number of blocks (1...N) and add additional CRC to each segmented block;
iii) Channel encoding for error correction purposes. The LDPC method for encoding of user data and the polar coding method for encoding of control or signaling information is used in the NR;
iv) Rate matching of each encoded block; and
v) Concatenation of the rate matched code blocks (1...N).

The above processing chain resembles with transport channel processing chain used in the LTE system. It may be noted that not all of the above processing steps are applied to process all types of transport channels. An overview of the above transport channel processing steps is described in the next subsections for the UL-SCH transport channel only. Further, to illustrate, the multiplexing of the UL-SCH transport channel carrying user data and UCI from UE to NG-RAN is also described.

19.6.9.1 CRC Calculation and its Attachment to a Transport Block

To detect an error in a transport block carrying MAC layer PDUs, a CRC is attached to it. The length (L) of the CRC parity bits depends on the transport block payload size (A). If the size is more than 3824 bits, CRC parity size (L) of 24 bits is used; otherwise, parity bits size (L) of 16 bits is used and appended to a transport block. Further, as defined by the 3GPP standard in TS 38.212 [107], a base graph is used to generate an LDPC code. Two graphs, base graph 1 and base graph 2, are used to generate an LDPC code depending on the payload size (A) and coding rate. If the transport block payload size (A) is less than or equal to 292 bits or if payload size (A) is less than or equal to 3824 and the coding rate is less than or equal to 0.67 or if the coding rate is less than or equal to 0.25, then the LDPC base graph 2 is used; otherwise, LDPC base graph 1 is used; refer to TS 38.212 [107]. Base graph 1 is used for large payload, i.e. transport block size, whereas the base graph 2 is used for small payload size.

19.6.9.2 Code Block Segmentation

A transport block with CRC attached to it may be segmented again, resulting in several code blocks. A transport block along with its CRC shall be segmented if the combined inputs bits length (B) is greater than the maximum size of a code block. The maximum size of a code block depends on the base graph used to generate an LDPC code. If base graph 1 is used, the maximum size of a code block is 8448 bits, and in the case of base graph 2, the maximum size is 3840. An additional CRC parity bit of 24 bits length is again added into each segmented code block if the transport block, with CRC, was segmented; otherwise, no more additional CRC is added. If the combined size of the transport block and its CRC is less than the maximum size of a code block, i.e. 8448 or 3840 bits, the number of the code block is one. Otherwise, the number of code blocks and the number of bits in each segmented block are determined using the formula described in Section 5.2.2 in TS 38.212 [107].

19.6.9.3 Channel Encoding with LDPC

Each segmented code block is then encoded using the LDPC coding method, which is a key difference from the LTE system as far as the DL and UL SCH processing chain is concerned. LDPC is also a forward error correction method applied over a noisy transmission channel. LDPC code was invented by Robert G. Gallager. The resulting LDPC encoded code block is a linear block code that contains information bits and parity bits, as illustrated in Figure 19.39.

LDPC code works by using a parity check matrix, H, such that for a given LDPC code word C, the relationship $HxC^{T}=0$. Here, C^{T} is the transposed matrix version of the code word C. Parity matrix H is a sparse matrix containing mostly 0s and a small number of 1s.

- Base Graphs (BGs) for LDPC Encoding

LDPC uses two BGs and matrices for channel encoding purposes of varying transport block sizes. BG matrix 1, BG1, has the dimension of 46 rows (Nr)×68(Nc) columns. BG matrix 2, BG2, has the dimension of 42 rows (Nr)×52(Nc) columns; refer to Tables 5.3.2-2 and 5.3.2-3 in TS 38.212 [107]. BG1 and BG2 are illustrated in Figure 19.40.

As illustrated in Figure 19.40, the base graph 1 – BG1 – has the information bits size of Kb = 22 columns and base graph 2 – BG2 – has the information bits size of Kb = 10 columns. It may be noted that the first two columns of the BG1 and BG2 are not transmitted, but they are used for puncturing purposes. Thus, the number of columns available for LDPC encoding using BG1 is 66, and for BG2, it is 50. The dotted rectangles illustrate how varying code rates (CR), CR1, CR2, and so on, such as 1/3 (=22/66), 2/3 (=22/33) with BG1, or 1/5 (=10/50) with BG2, can be achieved with the LDPC encoding method using a BG and its matrix.

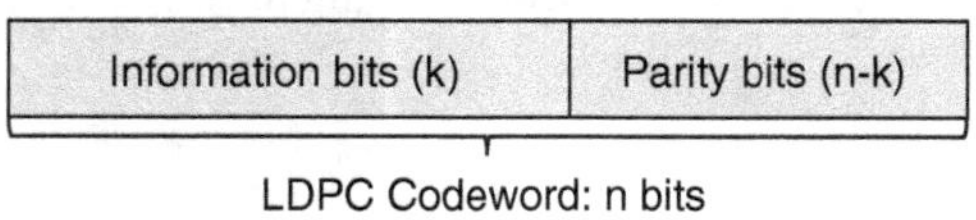

Figure 19.39 Illustration: components of an LDPC codeword.

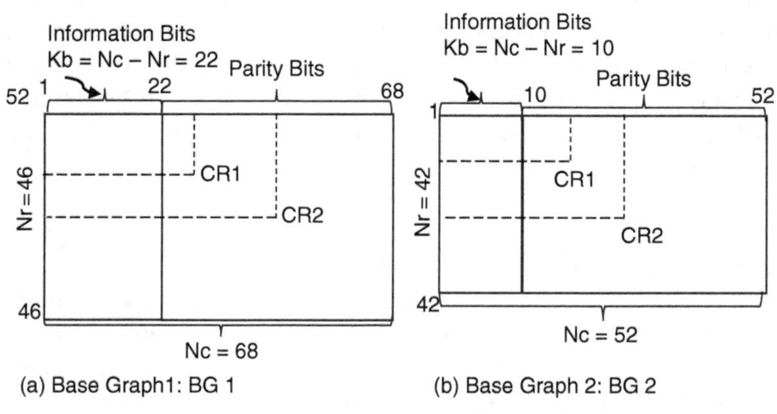

Figure 19.40 Illustration: LDPC base graphs: BG1 and BG2.

- Permutation Matrix for LDPC Encoding

Further, a permutation matrix P, which is a circularly right-shifted identity matrix of size Z×Z, with the dimension of expansion/*lifting size* (Z×Z) is also used. As defined in Table 5.3.2-1 in TS 38.212 [107], there are a total of 51 different lifting or expansion sizes (Z) (right side) which can be used for different LDCP encoding purposes for different payloads or transport block sizes. They are organized into eight sets (i_{LS}) (left side). Using a selected lifting size (Z), an element of a BG matrix: BG1 or BG2 can be further expanded by the size Z×Z. From a given set of lifting sizes provided (right side) in Table 5.3.2-1 in TS 38.212 [107], select the minimum lifting size (Z) in such a way that it is greater than or equal to the information bits divided by the number of base graph columns (22 for BG1 and 10 for BG2) allocated for the information bits. Refer to this condition in Section 5.2.2 in TS 38.212 [107].

If an entry in the base matrix BG1 or BG2 has a value 0, it is replaced by a *zero matrix* of size Z×Z. If an entry in the base matrix has a non-zero value, it is replaced by an *identity matrix* of size Z×Z with circular shifted to the right. The amount of circular shifting to the right to be performed on the *identity matrix* is determined through a modulo operation of the concerned value *(Vij)* and the selected lifting size *Z*. Value for Vij is defined for each selected column of the concerned BG1 or BG2. Using Table 5.3.2-2 (BG1) or Table 5.3.2-3 (BG2) in TS 38.212 [107], the values of *Vij* are selected for a particular row of BG1 or BG2 and a particular index set (i_{LS}).

- Construction of Parity Check Matrix

A parity check matrix H for LDPC encoding purpose can be constructed from a particular BG matrix, BG1 or BG2, and permutation matrix P. It should be noted that Section 5.3.2 in TS 38.212 [107] describes the LDPC encoding procedure as well as the generation of a parity check matrix H using a base matrix BG1 or BG2 along with a permutation matrix P of expansion/lifting size Z.

19.6.9.4 Rate Matching and Concatenation

Finally, the LDPC encoded set of bits is rate matched as per the allocated radio resources and concatenated sequentially to produce a concatenated block as described in TS 38.212 [107]. The processing chain applicable to the different types of NR transport channels is summarized in Table 19.8.

The entire processing steps of a UL-SCH transport channel described in Table 19.8 are illustrated in Figure 19.41. Further, multiplexing of UCI, if required, along with higher data transmitted over the UL-SCH is also illustrated. No such multiplexing of DL control information with higher layer data exists in case of a DL transmission from NG-RAN to UE.

Table 19.8 NR transport channels and their processing chain and coding method.

Transport Channel	UL/DL	Transport Channel Processing Chain	Coding Scheme
UL-SCH	UL	CRC generation and attachment, code block segmentation and CRC attachment, channel coding, rate matching, code block concatenation, multiplexing of control, e.g. HARQ ACK/NACK, and user data	LDPC coding
BCH	DL	CRC generation and attachment, channel coding, and rate matching	Polar coding
DL-SCH	DL	CRC generation and attachment, code block segmentation and CRC attachment, channel coding, rate matching, and code block concatenation	LDPC coding
DCI	DL	CRC generation and attachment, channel coding, and rate matching	Polar coding
UCI	UL	Code block segmentation and CRC attachment, channel coding, rate matching, and code block concatenation	Polar coding

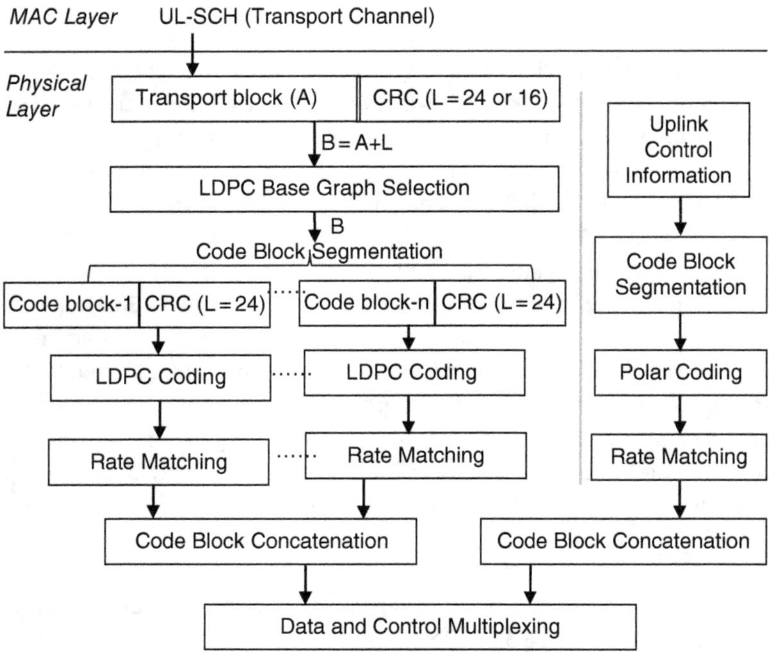

Figure 19.41 Illustration: NR UL-SCH and UCI processing chains and their multiplexing.

Figure 19.42 Illustration: output of transport block coding process at physical layer.

The output of the entire channel coding process performed by the physical layer on a transport block received from the MAC layer is termed as the *code word*. This is illustrated in Figure 19.42. The code word has error detection and protection information, in terms of parity bits, in it because of which its size is more than the original MAC layer transport block. An error protected *code word* is used as the input to the next stage of the physical layer channel coding process, which shall be described in the next section, for its transmission through the respective physical channels, e.g. PDSCH, PUSCH, and so on, over the NR air interface.

As illustrated in Figure 19.38, only one transport block is being used for transport channel coding purposes in the UL direction, producing one code word. However, two transport blocks, Figure 19.42, may be used for transport channel coding purposes in the DL direction in case a spatial multiplexing transmission scheme through multiple antenna systems is used. In this case, two code words, one for each transport block, are produced.

19.6.9.5 Multiplexing of UL-SCH Data and Uplink Control Information

Unlike the PDSCH, a PUSCH can also transfer physical layer UCI such as the HARQ ACK and CSI from a UE to NG-RAN in a multiplexed way with normal user data. Multiplexing of upper-layer information as well as physical layer UL control information (right side) is illustrated in Figure 19.41. As illustrated in Figure 19.41, physical layer UL control information also goes through the typical transport channel coding chain. Refer to Section 6.2.7 in TS 38.212 [107] for details on the multiplexed transmission of data and UL control information over PUSCH.

19.6.9.6 LDPC Encoding Examples

Based on the description provided in Section 19.6.9.3, the construction of a Parity Check Matrix for LDPC encoding is illustrated through the following examples. A graph representation of the parity check matrix is also illustrated. See Examples 19.8 to 19.9.

Example 19.8 Construction of Parity Check Matrix

Based on the steps described in TS 38.212 [107] as well as using a BG matrix mentioned in Section 19.6.9.3, the generation of a parity check matrix is described below.

Consider that the size (B) of the transport block received from the MAC layer is 10 bits. Thus, the number of the code block (C) is one (1), and also the BG BG2 selected (Step #2 above) in this case as the input bit string length (10) is less than the maximum code block size 3840. Thus, as per Section 5.2.2 in TS 38.212 [107], the value of Kb = 6 for base graph BG2. Next, find the value of BG lifting (Z) or expansion size. From the condition – $Kb \times Z \geq B$, the minimum value of Z = 2. Against Z = 2, the value of the set index (iLS) = 0 from Table 5.3.2-1 in TS 38.212 [107]. Now, using the iLS = 0 and row index = 0 and column indices: 01, 2, 3, 6, 9, 10, 11 from Table 5.3.2-3 in TS 38.212 [107], construct the base matrix as follows. This example illustrates only the first row of the BG2.

```
Row-Col: 0-0 P(i=0,j=0) = mod(Vij, Z) [ Vij : Table 5.3.2-3, TS 38.212 [107]]
=mod(9,2) =1
Row-Col: 0-1 P(i,j)= mod(Vij, Z) =mod(117,2) =1
Row-Col: 0-2 P(i,j)= mod(Vij, Z) =mod(204,2) =0
Row-Col: 0-3 P(i,j)= mod(Vij, Z) =mod(26,2) =0
Row-Col: 0-6 P(i,j)= mod(Vij, Z) =mod(189,2) =1
Row-Col: 0-9 P(i,j)= mod(Vij, Z) =mod(205,2) =1
Row-Col: 0-10 P(i,j)= mod(Vij, Z) =mod(0,2) =0
Row-Col: 0-11 P(i,j)= mod(Vij, Z) =mod(0,2) =0
Columns(Vij): BG2
              0   1   2   3   4   5   6   7   8   9  10  11
 Row(i)-0     1   1   0   0           1           1   0   0
...........   ...........................................
Row: 51       ...........................................
```

In the above base matrix BG2, a row/column entry with value 1 will be replaced by a 2 × 2 identity matrix with circularly shifted to the right by 1 time, i.e. [{0,1},{1,0}]. Similarly, if a row/column entry has any other non-zero value, the identity matrix will be circularly right-shifted by those times. A row/column entry with value 0 will be replaced by a zero matrix of 2 × 2 size, i.e. [{0,0},{0,0}].

Example 19.9 Graph Representation of Parity Check Matric of LDPC Code

A parity check matrix H (left side) for an LDPC code and its bipartite graph (right side) representation is illustrated in Figure 19.43. The bipartite graphical representation consists of two nodes – variable nodes and checks nodes. Variable nodes (N = 8) are equal to the number of columns of the parity check matrix H or total number of bits being transmitted. Each variable node represents one of the bits (labeled as C0–C7) LDPC encoded code word. Similarly, the number of parity bits/columns (P): C5, C6, and C7 are equal to the number of rows of the parity check matrix H. A row represents the parity check constraint or equation. In this illustration, the coding rate will be R = (Number of message bits: k excluding the parity bits)/Number of bits transmitted (n), i.e. 5/8. As the number of parity or protection bits increases, the code rate decreases.

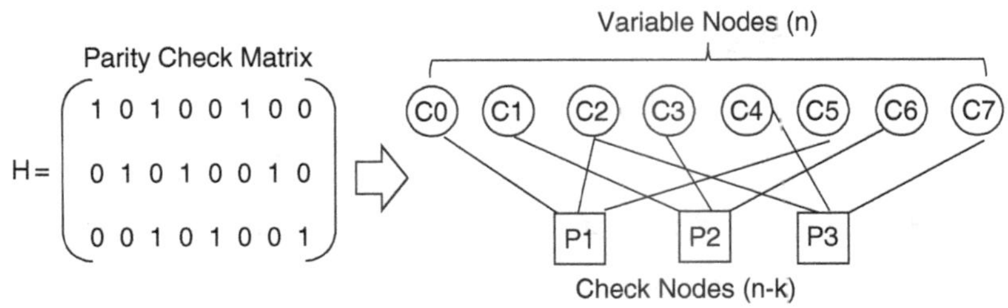

Figure 19.43 Illustration: an LDPC parity check matrix and its graph representation.

The parity check constraints can be expressed as follows:

C0	C1	C2	C3	C4	C5	C6	C7	
1⊕	0⊕	1⊕	0⊕	0⊕	1⊕	0⊕	0	= C0 + C2 + C5 (Exp1)
0⊕	1⊕	0⊕	1⊕	0⊕	0⊕	1⊕	0	= C1 + C3 + C6 (Exp2)
0⊕	0⊕	1⊕	0⊕	1⊕	0⊕	0⊕	1	= C2 + C4 + C7 (Exp3)

Consider that the information bit string to be transmitted is 11001, which is labeled from C0 to C4. The parity bits C5, C6, and C7 will be calculated as follows using the above expressions: Exp1, Exp2, and Exp3. Note that the bits are added using the modulo-2 sum (⊕)operator.

From the given input string: 11001

C0	C1	C2	C3	C4
1	1	0	0	1

With these input bit strings and using the Exp1, Exp2, and Exp3 shown above, the parity bits C5, C6, and C7 can be calculated as follows:

1⊕0⊕1 = 0 (C0 + C2 + C5 -> C5 must be 1 to satisfy the condition)
1⊕0⊕1 = 0 (C1 + C3 + C6 -> C6 must be 1 to satisfy the condition)
0⊕1⊕1 = 0 (C2 + C4 + C7 -> C7 must be 1 to satisfy the condition)

Thus, from the above modulo-2 sum, the parity bits C5, C6, and C7 are 1,1,1. The complete LDPC encoded bit strings to be transmitted with these parity bits will be 11001111, which is further represented graphically in Figure 19.44.

At the receiving end, the LDPC decoder will find that the modulo −2 (⊕) sum of all 1's (11001111) is equal to 0, which indicates that the message bits, as well as the parity bits, are received correctly. The original message bits (11001) are retrieved and passed to the higher layer.

19.6.10 Physical Channels and Their Processing Chain

19.6.10.1 Physical Channels

The NR physical layer uses the following physical channels, in UL and DL directions to transfer the contents, i.e. a *code word*, of the corresponding transport channel after processing it. Each of these physical channels is assigned a set of REs to transmit higher layer information.

- PDSCH

The PDSCH transfers user data as well as signaling information such as paging information to UE in a multiplexed manner.

- PDCCH

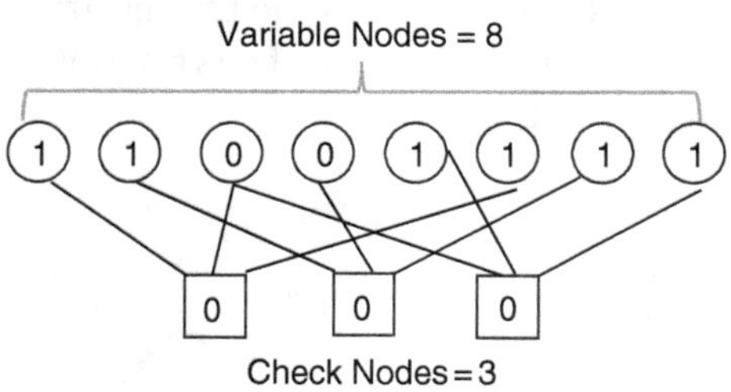

Figure 19.44 Graphical representation of LDPC encoded bit string: 11001111.

The PDCCH transfers scheduling assignment for DL PDSCH reception/UL PUSCH transmission and other control information from Ng-RAN to UEs in a serving cell. PDCCH and DCI are described later in Section 19.6.10.5.

- PBCH

The PBCH transfers minimum and key system information (MIB) which is read by UEs in a serving cell to access an NG-RAN and register with the 5GC. Using the information provided in the MIB, a UE further reads the remaining minimum system information (RMSI) transmitted by the NG-RAN/gNB in every 160 ms.

- Physical Random Access Channel (PRACH)

As part of a RACH procedure, the PRACH transfers random access request preamble from a UE to NG-RAN due to several events such as UE initial network registration, RRC connection re-establishment requirements, and on-demand system information request (new in the NR) by UE. Similar to the LTE system, there are a total of 64 PRACH preambles. Zadoff–Chu sequences are used to generate a RACH preamble sequence. However, unlike LTE, and also due to the different SCSs in the NR system, there are two types of preamble formats – *long and short* – with different preamble sequence lengths (839 and 139); refer to Tables 6.3.3.1-1 and 6.3.3.1-2 in TS 38.211 [106]. Preamble sequence length 139 supports nine formats and is used with SCSs 1.25 and 5 kHz. Preamble sequence length 839 supports four formats and is used with SCSs 15, 30, 60, and 120 kHz.

- PUSCH

The PUSCH transfers user data as well as piggybacked UL control information, such as HARQ ACK and CSI, from a UE to NG-RAN. A UE performs PUSCH transmission upon detection of DCI 0_0 and DCI 0_1 over PDCCH from NG-RAN. Further, using PUSCH, a UE can perform codebook-based (over multiple antenna ports or layers) and non-codebook (using single antenna port)-based transmission in the UL direction to the NG-RAN.

- Physical Uplink Control Channel (PUCCH)

As part of reporting of control or signaling information from UE to NG-RAN, the PUCCH transfers UL control information such as the HARQ, SR, and CSI from UE to NG-RAN. Similar to LTE PUCCH, there are several formats of PUCCH, Format 0–4.

It may be noted that in the LTE physical layer, there are additional DL control channels, namely *Physical Control Format Indicator Channel* (PCFICH) and *Physical* Hybrid ARQ *Indicator Channel* (PHICH), which are not available in the NR physical layer. A PCFICH provides the PDCCH OFDM symbol information to UEs from the LTE control region in the time domain. In the NR system, such a channel is not required as the OFDM symbol allocation information for the PDDCH is provided as part of a *CORESET*, which is described later in Section 19.6.10.5. A PHICH provides the status of the transport blocks received by the E-UTRAN from UE.

19.6.10.2 Channel Mappings

A transport channel may carry the multiplexed data from different logical channels containing higher layer data. A transport channel is mapped to its corresponding physical channel for transmission over the NR air interface in the DL or UL direction as listed in Table 19.9. Transmission of DL and UL control information through their respective physical channel is also listed. Such control information is generated at the physical layer itself and transmitted through the respective physical channel over the NR air interface.

Table 19.9 NR transport channel to physical channel mappings.

Transport Channel	Control Information	Physical Channel	Direction
DL-SCH, PCH	–	PDSCH	Downlink
BCH	–	PBCH	
–	DCI	PDCCH	
UL-SCH	–	PUSCH	Uplink
RACH	–	PRACH	
–	UCI	PUCCH, PUSCH	

The spatial multiplexing through multiple antenna transmissions and receptions and its related aspects are described first in the next section as they are essential to understand the tasks associated with a physical channel processing chain.

19.6.10.3 Multiple Physical Antenna Transmissions

In a conventional communications system, a single antenna system is used at the transmitter and receiver end to transmit and receive information between them. However, to offer higher data throughput to users or to provide a robust communication link in varying radio conditions, the NR system uses the multiple physical antenna systems at the NG-RAN/gNB and UE ends, which is one of the main features introduced over its physical layer radio air interface access path. Multiple antenna configurations used in NR NG-RAN are classified into the following types:

- Single Input Single Output (SISO);
- Multiple Input Single Output (MISO);
- Single Input Multiple Output (SIMO); and
- Multiple Input Multiple Output (MIMO).

The aforementioned multiple antenna configurations used over the NR air interface are illustrated in Figure 19.45.

In a SISO configuration, only one physical antenna is used by each transmitter and a receiver. SISO is a kind of default antenna configuration used between a transmitter and receiver in any communications system. However, the reliability of signal strength and its quality received at the receiver end using a SISO antenna system may be low.

- Purposes of Multiple Physical Antenna Systems

Transmissions and receptions using multiple antenna (MISO, SIMO, and MIMO) systems over the physical layer air interface may be used for the following purposes:

- Diversity gain, where the same data is transmitted from multiple antennae (MISO) or received at multiple antennae (SIMO) to increase the reliability, e.g. for different services under the URLLC use case, of data reception at the receiver end. Diversity gain may be used to recover the original signal from fading losses as a result of its radio wave propagations through different surrounding environmental conditions.
- Spatial multiplexing, for achieving a high data rate over the NR air interface.

- Spatial multiplexing.

In a conventional communications system, different physical antenna systems transmit in different frequencies. However, like in the LTE system, in the NR system also, data transmissions/receptions between UE and NG-RAN/gNB can be performed through the usages of the MIMO physical antenna configurations over the same frequency and time resource at the same time, i.e. a resource grid/block. Such an arrangement of transmissions/receptions over multiple

Figure 19.45 Illustration: NR multiple antenna configurations and transmissions.

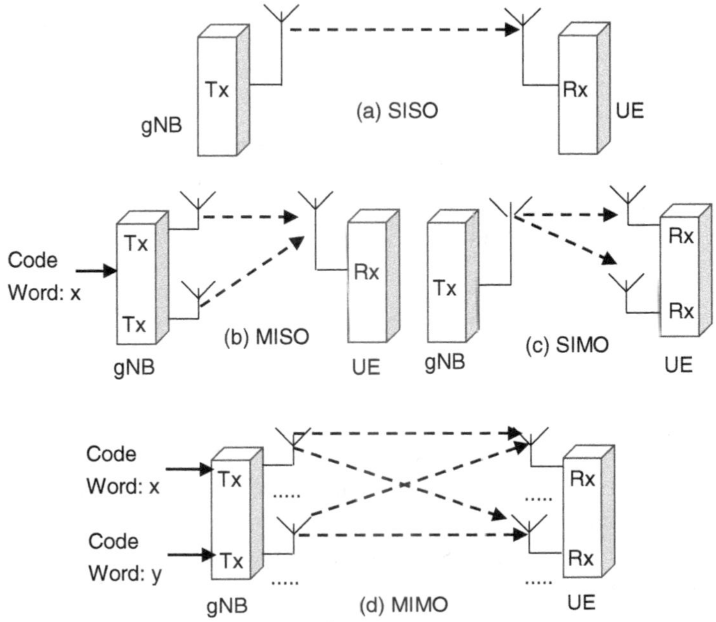

antenna systems is called *a Spatial Multiplexing* and may be used as another method to achieve high data rates between a transmitter and its receiver. Such high data rates with spatial multiplexing are achievable through transmissions of multiple independent parallel data streams, resulting from either a single code word or two code words, between a transmitter and a receiver. Transmissions with spatial multiplexing targeted to a single UE are called *Single user MIMO (SU-MIMO)*, whereas transmissions to multiple UEs are called *Multiuser MIMO (*MU-MIMO). For a MIMO configuration to work, at least two antennas are required at the transmitter and receiver ends, as illustrated in Figure 19.45d.

Unlike the MISO and SIMO configurations where a *single code word* is transmitted, see Figure 19.45b, in the case of MIMO configuration, up to two code words of a user data can be transmitted (Tx) and received (Rx) through multiple antennas. Refer to the illustration shown in Figure 19.45d. Thus, the user data throughput gained in the case of MIMO configuration is higher than the MISO and SIMO configurations. A MIMO configuration is specified as T×R where T is the number of transmit antennas and R is the receiving antennas. In Figure 19.45d, the MIMO configuration is referred to as the 2(at gNB)×2(at UE) matrix dimension. There can be an 8×8, 4×4, 3×3, and so on, MIMO configurations in the NR system, which is derived based on Demodulation Reference Signal (DMRS; as described in a later Section 19.6.12) port as specified at various tables in TS 38.212 [107]. The different paths taken by data transmissions between a transmitter and receiver using a spatial multiplexing method are mathematically represented through a matrix called *precoding matrix*. Refer to TS 38.211 [106] for such a precoding matrix used for spatial multiplexing transmissions purposes.

- Logical Antenna Ports

In the preceding paragraphs, NR physical layer transmission through a spatial multiplexing method using multiple physical antenna systems was illustrated and described briefly. NR physical layer also defines the logical antenna ports through which the individual physical channels and signals are transmitted. For example, antenna ports starting with 1000 are used for PDSCH transmission, antenna ports starting with 4000 for SS/PBCH block transmission, and so on; refer to TS 38.211 [106]. On top of the starting port number, the actual antenna port numbers to be used are derived by adding the respective port numbers used for the particular DMRS. Refer to the different antenna ports and DMRS ports used in DL and UL directions as specified in TS 38.212 [107] for a single as well as two code words transmissions.

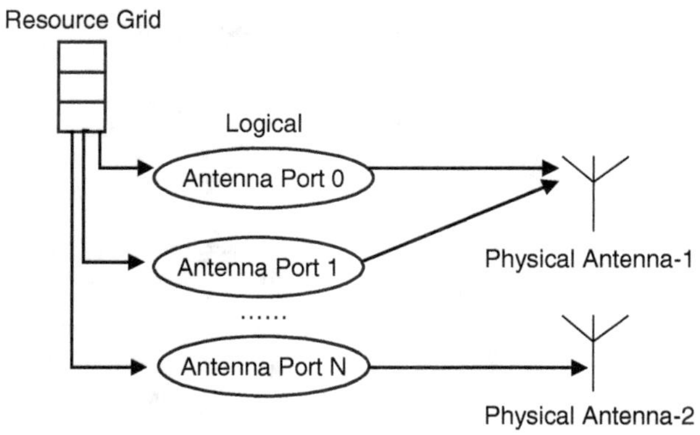

Figure 19.46 Illustration: mapping of logical antenna port to a physical antenna for a 2 × 2 MIMO.

Information through a particular physical channel and signal is transmitted through its defined antenna port only. There can be a one-to-one mapping between a logical antenna port and a physical antenna, or multiple logical antenna ports to a single physical antenna. For example, in the case of SISO configuration, a single code word is transmitted through a single logical antenna port and physical antenna. Figure 19.46 illustrates the mapping of logical antenna ports to a physical antenna for a typical 2 × 2 by MIMO transmission.

As illustrated, antenna ports of a particular physical channel or signal are associated with a resource grid having the same resources in the frequency and time domain. An analogy for an antenna port can be drawn from the network interface (NIC) card of a desktop computer using which multiple processes can send and receive information through its TCP or UDP port. Similarly, a server computer can have more than one NIC to provide higher throughput or failover protection. Air interface transmissions and receptions through multiple antenna systems and ports in NR NG-RAN may be considered similarly.

- Transmissions Layers and Mapping

Similar to the LTE system, another function defined by the NR physical layer as part of the physical channel MIMO processing is the *transmission layer* which is an independent data stream. A MIMO transmission layer has a one-to-one relationship with a logical antenna port. A scrambled and modulated (described in the next section) *code word* is divided and mapped into multiple layers for transmission in UL or DL direction with a spatial multiplexing method. NR supports the transmission of up to eight layers in the DL direction and up to four layers in the UL direction; refer to Table 7.3.1.3-1 in TS 38.211 [106]. This means up to eight, in the DL, or four, in the UL, parallel independent streams of data from two code words or one code word may be transmitted.

The number of transmission layers is less than equal to the logical antenna ports available at the NR physical layer. Transmission layers mapping may be also used to achieve transmit diversity. Figure 19.47 illustrates a single, and same, code word to multiple layers, either two or four, along with its logical antenna ports mapping in case of a DL transmission using transmit diversity, which is used for reliability decoding purposes at the receiver end.

On the other hand, the MIMO configuration shown in Figure 19.48 has two antenna ports or two layers for one code word and five antenna ports or layers for two code words: 1 and 2 mapping. As illustrated, the code word is divided accordingly between the layers for parallel transmissions through different antenna ports. Such transmissions and receptions of code words of a transport block result in a higher throughput for a UE.

As illustrated, in case of two code words transmissions, the first code word is divided into two layers, whereas the second code word is divided into three layers. Refer to Table 7.3.1.3-1: TS 38.211 [106] for the code word to different layers mapping for spatial multiplexing purposes. The preferred number of layers to be used by the gNB in the DL direction may be also reported and recommended by the UE to the gNB. Such recommendation is based on the *Rank Indicator* K *(*RI*)* which is the part of the CSI from UE to gNB. A UE-reported RI may be greater or smaller than the number of layers in use in the current DL transmission by the gNB. For example, if the current number of layers in

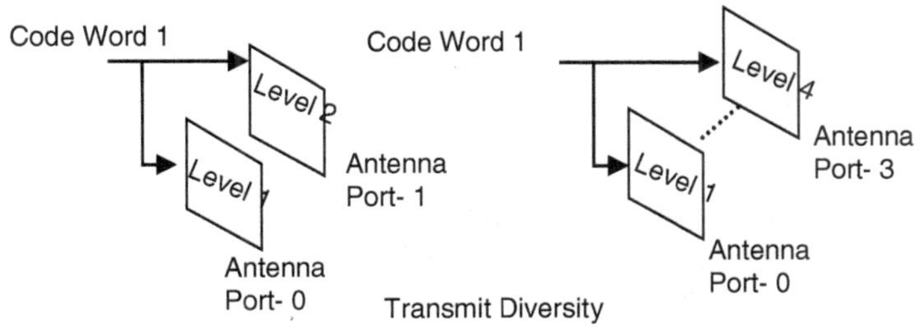

Figure 19.47 Illustration: transmit diversity: code word to layer mappings.

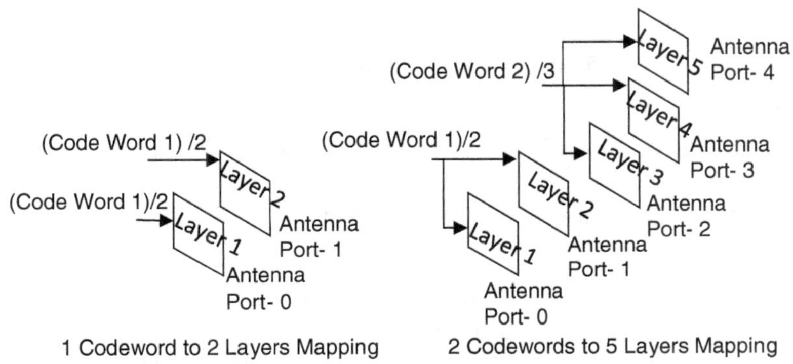

Figure 19.48 Illustration: spatial multiplexing: code words to layers mappings.

use in the DL direction is 2 but the UE reports the RI = 1, i.e. one layer, this would mean a poor DL channel state as perceived by the UE. For more supplementary information, the reader may further refer to R1-1702188 [149].

19.6.10.4 Physical Channel Processing Chain

Similarly, as in the case of the LTE system, the NR physical channel processing chain also contains the following steps in general, and in sequence, at the transmitting end. However, the exact tasks performed and the inputs used in each processing stage vary depending on the physical channel type as well as its direction, UL/DL, being used. In this section, an overview of UL PUSCH processing-related tasks performed at the physical layer is described below:

- PUSCH Processing Chain

i) Scrambling of LDPC encoded and rate matched code word, which is received from a transport channel processing chains, as shown in Figure 19.41, on a transport block.

A code word from the transport channel coding steps is provided to the scrambling process whose purpose is to reject the inter-cell interference from different sources in wireless communications systems. A block of scrambled bits is generated using a pseudo-random sequence generator; refer to Section 5.2.1 in TS 38.211 [106]. The same pseudo-random sequence generator is used by the different physical channels, i.e. PDSCH, PUSCH, PDCCH, and PUCCH, of the NR system but with a channel-specific initializer. An LDPC encoded code word, scrambling identity, and the RNTI of a UE are used as inputs to a scrambling process and generate a scrambled code word. The pseudo-random generator is initialized with a scrambling identity whose value can range from 0 to 1023 as configured by the RRC layer through IR *dataScramblingIdentityPUSCH*. If it is not configured, then the NR PCI with a value range from 0 to 1007 is used as the scrambling identity. Refer to TS 38.211 [106], Section 6.3.1.1 for more information on the NR physical layer scrambling process.

ii) Modulation of a scrambled code word

Using a particular modulation scheme such as QPSK, 16/64/256 QAM, Table 6.3.1.2-1 in TS 38.211 [106], the scrambled code word/binary sequence is mapped using the corresponding expression as defined in TS 38.211 [106] to produce complex-valued modulation symbols.

iii) Layer mapping of the modulated code word for different transmission layers

A code word may be mapped to more than one antenna ports or layers, in case of a MIMO transmission with spatial multiplexing over the PDSCH or PUSCH. In the UL direction, only one code word can be transmitted and is mapped up to a maximum of four layers (1...4), or parallel data streams, through different antenna ports as part of PUSCH transmission using the same time and frequency resources. On the other hand, using DL PDSCH, one or two code words can be transmitted. In the DL direction, one code word can be mapped up to four layers, and two code words can be mapped up to eight layers or parallel data streams, according to rules defined in Table 7.3.1.3-1 in TS 38.211 [106] for PDSCH transmission using the same time and frequency resources. A similar code word to layer mapping, for diversity and spatial multiplexing, procedure is used in the case of the LTE system physical layer also.

iv) Generation of transform precoded symbols for modulation symbols in case of DFT-s-OFDM transmission in the UL direction. It is an optional stage as configured by the RRC layer through the IE *transformPrecoder*. If it is enabled, PUSCH transmission is performed using SC-OFDM/DFT-S-OFDM; otherwise, normal OFDM transmission method is used. If transform precoding is enabled, an additional modulation scheme π/2-BPSK of order 1 is used.

v) Precoding

As part of the precoding stage, the data received from the previous stage, i.e. layer mapping, is mapped to different antenna ports. The data from different layers are multiplied by a precoding matrix. The precoding matrix to be multiplied depends on the transmission schemes being used for PUSCH, i.e. codebook based and non-codebook based, as configured by the RRC layer under the IE *PUSCH-Config -> txConFig*. For non-codebook-based transmission, the precoding matrix to be used is an identity matrix, and a UE derives precoding matric to be used from the *CSI-Reference Signal (RS)* measurement. However, for codebook-based PUSCH transmission, the precoding matrix is provided by the gNB in predefined Tables 6.3.1.5-1–6.3.1.5-7 in TS 38.211 [106] for different transmission layers. In these tables, the number of columns is equal to the number of layers used in the UL transmission, and Tables 6.3.1.5-2–6.3.1.5-7, TS 38.211 [106], are used if transform precoding is disabled by the RRC layer under the IE: *transformPrecoder*. The exact table to be applied depends on factors such as the number of layers and antenna ports used for UL transmission. Thus, the predefined table to be applied for precoding multiplication purpose is provided by the NG-RAN/gNB to a UE over the DCI 1_0 using the *Precoding information and number of layers* field.

vi) Mapping of the precoded symbols into the virtual RB

Complex valued symbols are mapped to the resource allocated in the frequency domain as well as time-domain over the DCI-0_0/0_1 from the NG-RAN/gNB to a UE.

vii) Mapping of the virtual RB to the physical virtual RB.

The opposite of the above steps is performed for a typical PUSCH at the receiving physical layer end and passes the information to the higher layer. However, note that not all of the above processing steps are applied to process all the types of physical channels. For example, in the case of PDSCH, transform precoding and precoding steps are not performed. Table 19.10 summarizes the different processing steps associated with the different types of physical channels, i.e. DL and UL control channels and shared channels. Refer to TS 38.211 [106] for the inputs used in each processing stage such as the initializer used to generate a block of scrambling of bits, and modulation scheme for each type of channel. In the case of PUSCH, non-interleaved VRB to PRB mapping type is used. In the case of PDSCH, both the interleaved and non-interleaved VRB to PRB mapping types are used.

viii) Summary of physical layer channel processing steps

From the descriptions of the previous paragraphs, the entire physical layer channel processing steps and procedures can be summed up at a high level, as shown in Figure 19.49. Note that the antenna port layer mapping stage is applicable for the PDSCH and PUSCH only.

Table 19.10 Transport channels and their coding steps and schemes.

Physical Channels	UL/DL	Physical Channel Processing Steps
PUSCH	UL	Scrambling, modulation, layer mapping, transform-precoding, precoding, mapping to VRB, and mapping VRB to non-interleaved PRB.
PDSCH	DL	Scrambling, modulation, layer mapping, mapping to VRB, and mapping VRB to interleaved/non-interleaved PRB.
PDCCH/PBCH/PUCCH	DL	Scrambling, modulation, and mapping to PRB.

Figure 19.49 Illustration: physical layer channel processing steps.

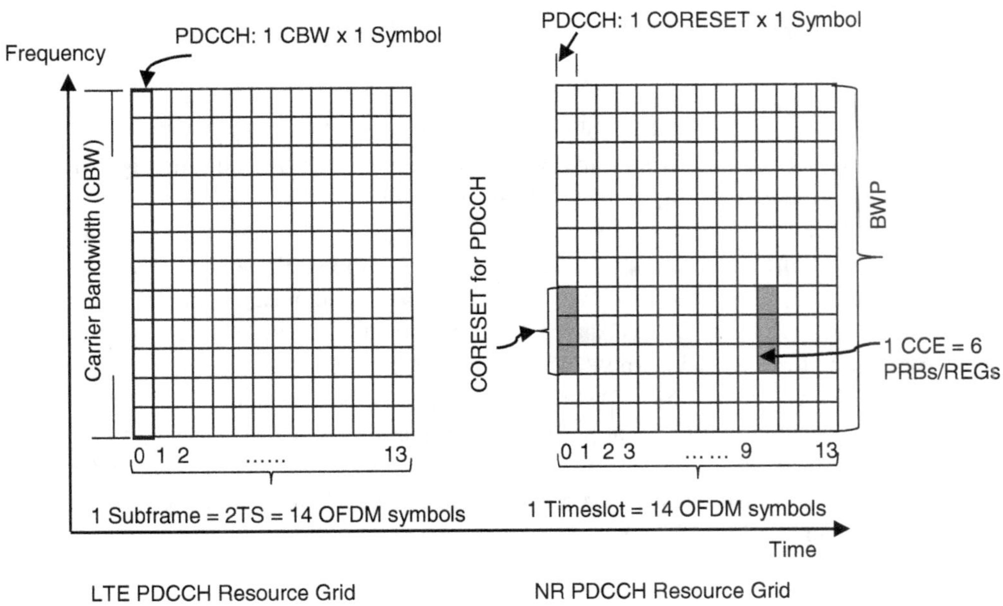

Figure 19.50 Illustration: LTE vs. NR PDCCH resource allocation.

19.6.10.5 Physical Downlink Control Channel (PDCCH)

An overall description of the NR physical layer DL control channel is provided in this section. PDCCH is the only DL control channel in the NR that provides vital information to UEs. A PDCCH transfers information on the allocation of resources, and other control information, for transmission and reception of data in UL and DL directions.

From a comparative study point of view, Figure 19.50 illustrates the appearance of a PDCCH on a resource grid in frequency and time domain in the LTE and NR systems.

In the case of the LTE system, a PDCCH is mapped to the RBs/elements of the whole carrier bandwidth (CBW) in the frequency domain along with the allocation of up to three or four OFDM symbols in the time domain. On the other hand, in the NR system, a PDCCH is restricted and mapped to a certain RB only, called *CORESET*, of a BWP in the frequency domain. In the time domain, a PDCCH may be allocated up to three OFDM symbols. An NR BWP, as described earlier, is a small subset of entire carrier bandwidth with a contiguous set of RBs. NR *CORESET* was described earlier in Section 19.6.5.

It may be noted that unlike the PDCCH in the LTE system, a DMRS is also transmitted along with an NR PDCCH within the same REG of a CCE. A description of various radio resources associated with a PDCCH is described below.

- Frequency domain resources for PDCCH

In the NR, RBs as well as the OFDM symbols of a PDCCH *CORESET* are configured by the RRC layer under the IE: *PDCCH-Config* (UE-specific)/*PDCCH-ConfigCommon* (cell-specific); refer to TS 38.331 [116]. Further, RBs, from the allocated BWP, are assigned to a *CORESET* in terms of a bitmap, *called frequencyDomainResources*. Each bit of *frequencyDomainResources* indicates six RBs or REG, which are organized into one CCE as described earlier in Section 19.6.5. A bit value of one indicates that the CCE, or six RBs/REGs, is allocated to a *CORESET*. Thus, a UE or group of UEs requires to monitor candidates PDCCH over a small part only of the entire carrier bandwidth. Figure 19.50 illustrates (on the right side) the configuration of two *CORESET*s for a UE as its candidates PDCCHs by the NR RRC layer within an allocated BWP.

- Time-domain resources for PDCCH

A PDCCH can occupy up to three consecutive OFDM symbols of a timeslot in the time domain.

- PDCCH aggregation level

A PDCCH can transfer a varying amount of DL control information from NG-RAN to a UE. Due to this, the frequency domain resource requirements may also vary in terms of the number of CCE requirements to transfer the contents of a PDCCH. This is specified through a PDCCH *Aggregation level*, where each level contains the specified consecutive number of CCEs. Similar to the LTE system, the NR system also defines PDCCH aggregation level up to 1, 2 4, 8, or 16 CCEs. Such CCEs are provided through a 45-bit long bitmap called *frequencyDomainResources*, from NG-RAN to a UE, as mentioned above. More aggregation level means more CCE resources are allocated to a PDCCH. Table 19.11 lists the PDCCH Aggregation Level and Candidates PDCCH and the number of RE available after discounting the RE used by the PDDCH DMRS. A DMRS is also transmitted along with a PDCCH which occupies three REs within an REG. A DMRS for PDCCH is described later in Section 19.6.12.1.

Within an REG, PDCCH DMRS occupies three REs at positions #1, 5, 9 as per the PDCCH DMRS to RE mapping definition provided in TS 38.211 [106]. Thus, within an REG, only 12–3 = 9 RE (REs) are available for PDCCH transmission. Table 19.11 also illustrates the number of bits that can be transmitted over a PDCCH in each aggregation level with QPSK modulation (2 bits per OFDM symbol).

Table 19.11 PDCCH aggregation level and candidates PDCCH.

PDCCH Aggregation Level/#of CCE	#of REG	Total #RE	#RE for PDCCH = *Total RE - #RE for PDCCH DMRS*	#Bits with QPSK (2 Bits) Modulation	#of Candidates PDCCH
1	6*1 = 6	72	72 − 6*3 = 54	2*54 = 108	0,1,2,3,4,5,6,8
2	6*2 = 12	144	144 − 12*3 = 108	2*108 = 216	0,1,2,3,4,5,6,8
4	6*4 = 24	288	288 − 24*3 = 216	2*216 = 432	0,1,2,3,4,5,6,8
8	6*8 = 48	576	576 − 48*3 = 432	2*432 = 864	0,1,2,3,4,5,6,8
16	6*16 = 96	1152	1152 − 96*3 = 864	2*864 = 1728	0,1,2,3,4,5,6,8

– PDCCH REG bundle size

The RE groups (REG0 to REG5) of a CCE are bundled together with a particular bundle size as configured by the RRC layer. Bundle size can be 2, 3, or 6 REGs.

– CCE to REG mapping

Further, CCEs of a particular aggregation level are mapped to different REG bundles in an interleaved or non-interleaved way. For a description of the hash function which is used for interleaved mapping of CCE to REG, refer to TS 38.211 [106]. CCE to REG mapping type, i.e. interleave or non-interleaved, used for a PDCCH is configured by the RRC layer IE: *cce-REG-MappingType*. Interleaved mapping of CCE to REG may be used to achieve frequency diversity within a BWP.

– UE PDCCH monitoring: search space

A PDCCH transmits vital DCI which is monitored by a UE or group of UEs in a DL direction. As the NR support larger carrier bandwidth and to avoid monitoring of a PDCCH over the entire carrier bandwidth, which is performed in the LTE system, PDCCH monitoring in the NR is restricted to a smaller part, which is called as a *search space*, of an active BWP. Usages of the NR BWP and its types were described earlier in Section 19.6.7.

A PDCCH search space is a region in the DL resource grid, and up to three symbols long in the time domain, as illustrated in Figure 19.50, which is configured by the RRC layer. A search space provides all the possible locations of a PDCCH in the frequency domain. Further, each possible location of a PDCCH is called a candidate PDCCH. For example, in Figure 19.50, there are two possible locations (symbol #0 and 9) of an NR PDCCH, and in each location, there is only candidate PDCCH. Using a search space, a UE or group of UEs comes to know about the candidate PDCCH to be monitored and decodes the one being addressed to a UE or group of UEs.

A search space configuration from NG-RAN to UE provides information about the contiguous CCEs of a particular aggregation level being allocated for each candidate PDCCH. CCE aggregation level for a PDCCH can be 1, 2 4, 8, or 16 CCEs as defined in the NR. A PDCCH search space also contains the periodicity of its monitoring as configured by the RRC layer. Using the candidates PDCCH from an RRC configured search space, a UE or group of UEs uses the blind decoding method to decode a candidate PDCCH intended for the UE or group of UEs using the corresponding RNTI as listed in Table 19.12.

Two types of search spaces for PDCCH monitoring purposes are defined in the NR – *UE specific (*USS*) and common search space (*CSS*);* refer to Table 19.12. NG-RAN can configure up to 10 search spaces, including USS and

Table 19.12 PDCCH search space, purposes, and their RNTIs.

PDCCH Search Space	Type	Purpose of Search Space	RNTI Used to Scramble CRC of DCI
Type 0	CSS	SIB Type 1	SI-RNTI
Type 0A	CSS	Other System Information	SI-RNTI
Type 1	CSS	Random Access Response	RA-RNTI or a TC-RNTI
Type 2	CSS	Paging	P-RNTI
Type 3	CSS	Common Signaling	INT-RNTI, SFI-RNTI, TPC-PUSCH-RNTI, TPC-PUCCH-RNTI, or TPC-SRS-RNTI
UE-specific	USS	UE-specific	C-RNTI MCS-C-RNTI, SP-CSI-RNTI, or CS-RNTI

CS, to be supported by a UE per BWP. UE-specific PDCCH search is carried out to decode UE-specific information such as DL data reception. Common PDCCH search is carried out by a group of UEs to decode common signaling data such as system information, paging message, and random access response sent by the NG-RAN. A candidate PDDCH search is carried out based on a CRC, of the transmitted DCI, which is scrambled with the corresponding RNTI as part of a DCI processing chain at the physical layer.

Note that depending on the payload size of a DL control information transferred over by a PDCCH, the CORSET of different aggregation levels may be assigned to USS or CSS.

- Maximum number of candidates PDCCH per SCS

Based on the SCS being used in a cell, NR also defines the maximum number of candidate PDCCH to be monitored by a UE. As defined in Table 10.1-2 in TS 38.213 [108], the maximum number of candidate PDCCH to be monitored by a UE is 44, 36, 22, and 20 for the SCS of 15, 30, 60, and 120 kHz, respectively.

For a detailed UE PDCCH monitoring procedure, refer to TS 38.213 [108]. Example 19.10 illustrates the decoding of PDCCH by a UE using its search space.

Example 19.10 UE Decoding of PDCCH from a UE-Specific Search Space

In 5G NR, the number of CORESETs, common as well as UE-specific, supported per BWP is three. Thus, the maximum number of control sets to be supported by a UE and its serving cell is 3* number of maximum BWP = 3*4 = 12. Further, the number of PDCCH search spaces allowed per BWP is 10. Thus, the maximum number of search spaces to be supported by a UE in the NR system is 4 BWPs * 10 = 40 search spaces.

With the above background, decoding of a PDCCH over the allocated BWP to a UE with one CORESET and one allocated search space, containing two candidates PDCCH, is illustrated in Figure 19.51. The following typical configuration is being used in this illustration.

- Frequency domain and time-domain resource allocation for CORESET 1 is as follows:
  CORESET ID = 1, with non-contiguous frequency domain resources between candidates PDCCH.
  Frequency Domain Resources = (CCE0 and CCE1) and (CCE7 and CCE8)
  Allocated CCEs bitmap (RRC IE: *frequencyDomainResources*) in frequency domain = 11000001100...(1: CCE Allocated, 0: CCE not allocated)
  Allocated duration = 1 (ODFM symbol)
  Allocated slot = 0
  Allocated symbol = 0, i.e. 0th OFDM symbol of monitored timeslot starting with slot #0 and allocated slot periodicity (#4) for PDCCH search space monitoring
  Allocated periodicity = 4, i.e. search space monitoring repeats at every fifth slot (sl5)
  CCE to REG Mapping Type = Non-interleaved
  REG bundle size = 6 resource blocks
  Number of REG bundles = 2
- Frequency domain and time-domain resource allocation for PDCCH search space, SS1, is as follows:
  Allocated Search Space (SS) ID = SS1
  Monitoring slot periodicity = 4 (RRC IE: *monitoringSlotPeriodicityAndOffset*)
  Monitoring symbol = 0 (RRC IE: *monitoringSymbolsWithinSlot*), i.e. 0th OFDM symbol of monitored timeslot starting with slot #0 for search space monitoring
  No. of candidates PDCCH = 2
  CCE aggregation level = 2, i.e. number of CCEs for each candidate PDCCH in CORESET 1

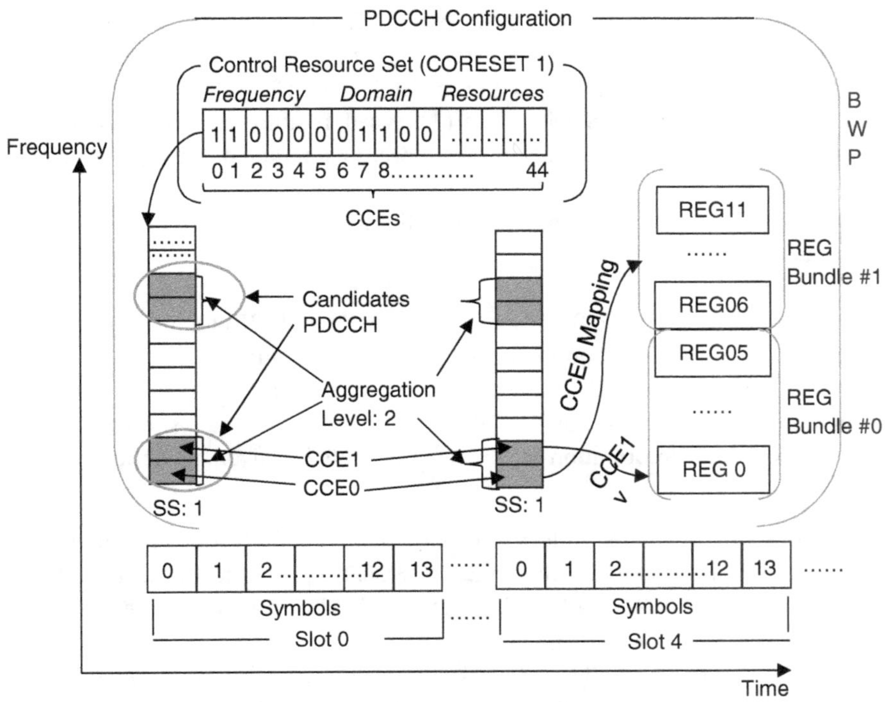

Figure 19.51 Illustration: PDCCH monitoring using search space with aggregation level 2 and interleave mapping for CCE to REG.

Thus, with the above typical CORESET and search space as configured by the RRC layer within the assigned BWP, the allocation of frequency domain and time-domain resources for a PDCCH monitoring and its decoding by a UE will be as follows:

Allocated CORESET ID = 1
Allocated search space = SS1

As illustrated in Figure 19.51, CCE0 is mapped to REG bundle 1, and CCE1 is mapped to REG bundle 0 in an interleaved way. An actual interleaved mapping between a CCE to its corresponding REG bundle takes place using the formula described in TS 38.211 [106]. For more information on the other information carried by a PDCCH, CORESET, Search Space, and BWP, refer to TS 38.331 [116].

Transmission of DL control information, within a CORESET, dynamically from NG-RAN to a UE or group of UEs over a PDCCH in a serving cell is described next.

- DCI

Similarly, as in the LTE system, the 5G NR system also allocates radio resources and scheduling information and other physical layer signaling information in the form of a DL control information from the NG-RAN to a UE over the PDDCH described above. The NG-RAN allocates UL/DL resources, either Type 0 or Type 1, for PDSCH or PUSCH and also provides other signaling information for different purposes such as power control commands, DL, and UL HARQ-related information, paging, slot format indication, and so on, to a UE. Using a DCI, NG-RAN can also allocate semi-persistently scheduled resources to a UE. Different types of information carried by a DCI, as may be configured by the RRC layer, are summarized at a high level below:

i) DCI identifier (0: UL DCI, 1: DL DCI);
ii) BWP and UL/DL physical resources allocation in frequency and time domain for PDSCH reception and PUSCH transmissions by UE. Unlike a DCI used in the LTE system, a DCI in the NR system contains the

time-domain resource allocation information also from NG-RAN to a UE. Depending on the user traffic requirements, a timeslot can be switched and used for DL only or UL only transmission. Thus, an NR DCI contains the dynamic TDD information also unlike a DCI used in the LTE system;

iii) Transport block-related information such as the MCS being used/to be used, redundancy version, new data indicator to indicate whether a transport block is a new transmission or retransmission of the previous block;
iv) PUCCH and PUSCH resource-related information;
v) UE PDSCH reception to UL HARQ-ACK transmission timing information;
vi) UL power control-related information;
vii) Timeslot format-related information;
viii) Virtual to PRB mapping type (interleaved/non-interleaved) used for DL transmission and PRB bundling information;
ix) Multiple antenna, or MIMO, transmissions configuration information such as the antenna port and number of layers used in UL and DL transmissions between UE and NG-RAN and vice versa. Note that in the case of the LTE system, such information is provided by the E-UTRAN over an RRC signaling message to a UE. However, in NR, MIMO transmission information is derived by a UE based on predefined Tables 7.3.1.2.2x/7.3.1.2.2x in TS 38.212 [107].

Refer to TS 38.212 [107] for more details on the different information carried by different DCIs, and differences among them, under the above categories of information. Further, refer to TS 38.213 [108] and TS 38.214 [109] on the usages, in terms of resource allocations by NG-RAN/gNB and its interpretation by a UE, of the different information carried by a DCI.

A 24-bit CRC is attached to the information to be transmitted over a DCI during its physical layer processing chain. Further, the parity bit of the CRC is scrambled with the corresponding RNTI. An RNTI can be UE-specific, allocated as a result of a successful RRC connection establishment with NG-RAN, or common RNTI, addressing a group of UEs. An RNTI indicates the intended mobile device and kind of information being transmitted to a UE or UEs. Thus, a UE or group of UEs monitors the PDCCH and decodes the contents of a DCI addressed to it using its RNTI. Table 19.13 lists the different types of DCIs used in the NR system, their purposes, and the corresponding RNTI used to scramble its CRC at the NG-RAN end and descramble it at UE end. The SI-RNTI and P-RNTI have fixed or static value, i.e.

Table 19.13 NR DCIs formats, purposes, and their RNTIs.

DCI formats	Purpose	RNTIs for Scrambling
0_0	Uplink scheduling information for PUSCH with basic NR features	C-RNTI or CS-RNTI or MCS-C-RNTI
0_1	Uplink scheduling information for PUSCH with more NR features	C-RNTI or CS-RNTI or SP-CSI-RNTI or MCS-C-RNTI:
1_0	Downlink scheduling information for PUSCH with basic NR features	C-RNTI or CS-RNTI or MCS-C-RNTI, P-RNTI, SI-RNTI, RA-RNTI
1_1	Downlink scheduling information for PUSCH with more NR features	C-RNTI or CS-RNTI or MCS-C-RNTI
2_0	Used for timeslot format indication to a UE in TDD mode	SFI-RNTI
2_1	Used for preemption indication to ongoing PDSCH transmission and intimate a UE that the PRBs/OFDM symbols allocated are not carrying any useful information for it	INT-RNTI
2_2	Used for transmission of closed-loop power control commands to UE for PUCCH and PUSCH	TPC-PUSCH-RNTI or TPC-PUCCH-RNTI
2_3	Used for transmission of closed-loop power control commands for SRS transmissions by one or more UEs	TPC-SRS-RNTI

0xFFFF and 0xFFFE, while the other RNTIs have value in the range from 0x0001 to 0xFFEF, refer to TS 38.321 [113].

The amount of information provided to a UE or UEs over a DCI, and hence its size, varies depending on the RNTI type used to scramble the CRC parity bits. Some of the fields, for example, MCS, of a DCI have fixed length, while other fields, for example, FDRA, *time-domain resource assignment*, and so on, have variable length, as well as maximum, allowed length. The length of such fields depends on the amount of configuration information provided by the RRC layer in the form of lists and number of entries (rows, columns) in it, through the corresponding IEs such as the *pdsch-TimeDomainAllocationList* as illustrated earlier in Section 19.6.8. The maximum number of time-domain resource allocation for PDSCH in *pdsch-TimeDomainAllocationList* is 16, i.e. it is a16×3 array. Thus, the sizes of the payload of different DCI formats may vary depending on the contents carried by them. Due to this, a DCI payload size alignment is performed by appending "0" or truncation of few bits of the FDRA field. The minimum size of the DCI payload is 12 bits.

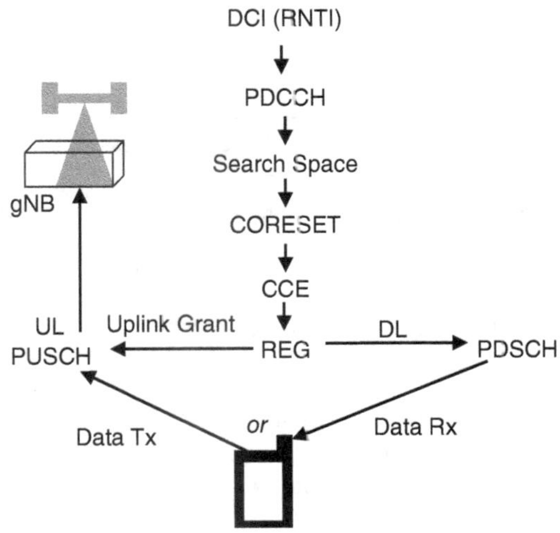

Figure 19.52 Illustration: transmission flow of a DCI.

Figure 19.52 illustrates, in general, the transmission flow and usages of a DCI from NG-RAN to UE. If a DCI carries a DL assignment, UE receives the DL data from the NG-RAN over the PDSCH and its corresponding radio resources indicated in the DCI. If a DCI carries the UL assignment for PUSCH transmission, UE transmits UL data over the PUSCH using the corresponding radio resources granted over the DCI.

DCI 0_0 and DCI 1_0 contain the basic information such as frequency and time domain resources allocated by the NG-RAN to a UE, HARQ-process number, and so on. DCI 0_1 and DCI 1_1 contain, in addition to the basic, other information of NR transmissions and receptions as may be configured by the RRC layer such as CBG transmission, the number of antenna ports and a number of layers to be used, phase tracking RS information, and so on.

– DCI and PDCCH Processing Chain

As listed in Table 19.13, a DCI for a UE or group of UEs contains different types of information and also undergoes a channel coding process similar to the coding of transport channels that carry higher layers of information. The typical DCI processing chain at the gNB end contains the CRC attachment (24 bits length), polar encoding, and a rate matching. Rate matching is essential to ensure that the DCI payload +CRC after its polar encoding matches the number of REs available after discounting the resource element consumed by PDCCH DMRS as mentioned in the previous paragraphs. All these steps produce a code word. A CRC is added to a DCI to detect an error in it by a UE. Polar encoding of a DCI also differentiates it from a DCI transmitted in the LTE system. Note that the CRC parity bits are scrambled with, through modulo-2 operation, the corresponding RNTI of a UE or group of UEs for common data so that only the intended UE or group of UEs processes the DCI. Tasks associated with a DCI processing chain is illustrated in Figure 19.53.

Figure 19.53 also illustrates the processing chain of a PDCCH, taking a DCI code word as the input bit string. The typical processing chain for the transmission of DCI code word over the PDCCH contains the task of scrambling, QPSK modulation, and its mapping to the allocated REs. A PDCCH is mapped to REs of an REG which are not used for its DMRS. DMRS for a PDCCH is described later in Section 19.6.12.1. Scrambling sequence initialization depends on the type of search space being used. If USS is used, the scrambling sequence is initialized with the RRC IE *pdcch-DMRS-ScramblingID* and UE C-RNTI. If the IE *pdcch-DMRS-ScramblingID* is not provided by the RRC layer, the scrambling sequence for USS is initialized with the corresponding physical layer cell ID of the serving cell, as used in the case of CSS also.

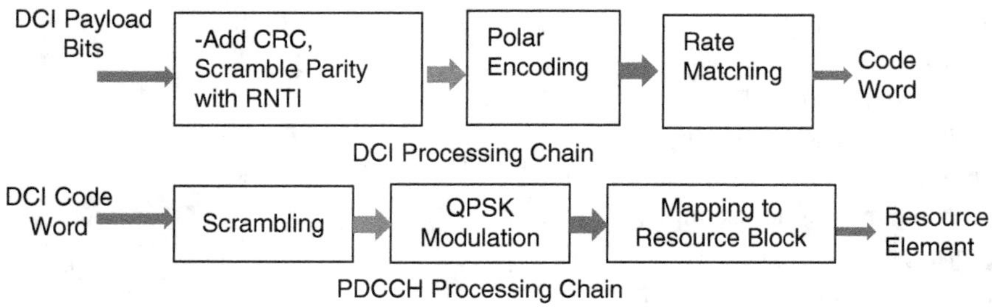

Figure 19.53 Illustration: DCI processing chain.

19.6.10.6 Physical Uplink Control Channel (PUCCH) and Information (UCI)

In the NR, a UE reports certain information to the NG-RAN/gNB in the form of UCI, which is similar to the LTE system. A UCI transmitted is from a UE to the NG-RAN over the PUCCH and may contain one or more or a combination of the following information in a multiplexed way:

i) HARQ ACK/NACK only as an acknowledgment for the previously received transport block from the NG-RAN. NR physical layer HARQ ACK/NACK procedure is described later in Section 19.6.11;
ii) HARQ ACK/NACK and SR to the NG-RAN;
iii) CSI only with respect to a PDSCH reception. CSI shall be described later in Section 19.6.17; and
iv) HARQ ACK/NACK, SR, and CSI.

As mentioned in the previous section, a DCI is transmitted over the PDCCH only from NG-RAN/gNB to a UE with the variable number of information. Compared to a DCI, less number of control information is carried by a UCI over the PUCCH in the UL direction. A UCI can be transmitted over the PUCCH as well as PUSCH (in a multiplexed way with upper-layer data over PUSCH; Section 19.6.9.5) from a UE to NG-RAN. A UCI transmitted from a UE to the NG-RAN over the PUSCH can contain HARQ ACK/NACK only as well as CSI. Refer to TS 38.212 [107] for bit sequences used by the various UCI. Depending on the contents/payloads and their length, several PUCCH formats are used to send UCI from a UE to NG-RAN as described in the next paragraph.

- PUCCH Formats and Their Lengths

Depending on the length of the payload, i.e. UCI bits and type of control information being transmitted from a UE to NG-RAN, different types of PUCCH Formats 0–4 are used to transfer a UCI mentioned above. PUCCH Formats 0 and 2 are called *short PUCCH format* and are allocated one or two OFDM symbols. PUCCH Formats 1, 3, and 4 are called *long PUCCH format* and are allocated 4–14 OFDM symbols.

Using the PUCCH Format 0 and 1 UCI length at most 2 bits can be transmitted, whereas using the other PUCCH formats (2/3/4), UCI length of more than 2 bits can be transmitted. For example, an SR or HARQ-ACK/NACK from a UE to NG-RAN can be sent/reported using PUCCH Format 0. Similarly, PUCCH Format 2 can be used to report a CSI from a UE to NG-RAN. For more information on PUCCH formats and its time-domain resource allocation, i.e. OFDM symbols, refer to TS 38.211 [106], and for UE reporting of single or multiple control information to NG-RAN over the PUCCH, refer to TS 38.214 [109].

- PUCCH Resource Configurations

Several parameters such as the starting PRB, number of PRBs, starting symbol, number of symbols, and so on are used by the different PUCCH formats to report a UL control information, i.e. HARQ-ACK, SR, and CSI, from UE to NG-RAN. Two types, *dedicated*, i.e. UE-specific, and *common*, i.e. cell-specific, of methods are used to allocate and configure resources to these parameters from the NG-RAN to a UE for transmissions of UCI over the different PUCCH formats. Further, there are two ways – a predefined table in TS 38.213 [108] as well as dynamically over RRC signaling – to allocate frequency and time domain, i.e. OFDM symbols, resources to different PUCCH formats. A set of dedicated resources, which belongs to a particular BWP, can be provided dynamically to a UE over RRC layer signaling IE *PUCCH-Config -> PUCCH resource set* to the PUCCH parameters. On the other hand, the allocation of common or cell-specific PUCCH resources is indicated to UEs through an index (0–15) *PUCCH-ConfigCommon -> pucch-ResourceCommon* into a predefined Table 9.2.1-1 in TS 38.213 [108], which is used by UEs during the initial network access made by it. These common PUCCH resources are used till dedicated PUCCH resources are not provided to a UE by the NG-RAN.

HARQ-ACK-related PUCCH resources are configured by the RRC layer through the IE PUCCH-Config -> *dl-DataToUL*-ACK in association with the *PUCCH resource indicator* and *PDSCH-to-HARQ_feedback timing indicator field* provided over DCI 1_0 and 1_1. Similarly, SR-related PUCCH resources are configured by the RRC layer through the IE *PUCCH-Config -> schedulingRequestResourceToAddModList*. Using these resources, a UE reports the respective UL control information, as described in TS 38.213 [108].

It may be noted that not all the PUCCH formats use the same value for a particular parameter. Each PUCCH format can have a format-specific parameter and its resource/value as assigned by the RRC layer. For example, the *number of PRB* parameters is applicable for the PUCCH Formats 2 and 3 only, whereas the *number of OFDM symbol allocation* parameters is common, with different values, and applicable for all types of PUCCH formats. For more information on these, refer to *PUCCH-Config* IE in TS 38.331 [116].

- UCI and PUCCH Processing Chain
  - UCI

Multiplexing transmission of UCI along with higher layer data over the PUSCH was illustrated earlier in Figure 19.41. The different tasks associated with the processing chain of a UCI depend on its payload lengths. As illustrated in Figure 19.41, the typical processing chain for a UCI contains the code block segmentation with CRC attachment if the UCI payload size is greater than 11 bits, polar encoding, rate matching, and code block concatenation to produce a code word. If the payload size is less than 11 bits, no CRC is attached. If the UCI payload size is less than 11 bits, it is encoded with repetition coding, Simplex coding, or Reed Muller coding method.

  - PUCCH

The processing steps used for PUCCH vary depending on its format, as mentioned above. PUCCH Format 0 contains the scrambling sequence generation and its mapping to RBs tasks. PUCCH Format 1 contains the modulation, sequence generation, and its mapping to RBs tasks. PUCCH Format 2 contains the coding, scrambling, modulation, and its mapping to RBs tasks. PUCCH Formats 3 and 4 contain the coding, scrambling, modulation, spreading, transform precoding, and its mapping to RB tasks. The modulation technique used may be either QPSK or BPSK depending on the PUCCH format, e.g. short or long, being used. Also, different scrambling sequences are used in different PUCCH formats. Refer to TS 38.211 [106] for more information on the scrambling sequences and initializers used by different PUCCH formats.

19.6.11 Code Block Group-Based Transmissions and Receptions

- **Downlink HARQ-ACK/NACK for Code Block Transmissions/Receptions**

CBG-based transmission in the UL (PUSCH) and DL (PDSCH) directions is a new feature in the NR physical layer which does not exist in the LTE system physical layer. As part of a transport channel processing chain at the NR or LTE physical layer, a large transport block (beyond up to 8448 bits in case of NR) is segmented into several code blocks. Each code block goes through the entire channel coding processing chain which consists of encoding, rate matching, and so on. At the end of the channel coding processing chain, the transport block is transmitted, after scrambling and modulating it, to the receiving physical layer. If a segmented code block is detected as an erroneous one by the receiver, the entire transport block is retransmitted by the sender, in general. A transport block is retransmitted (NG-RAN to UE) based on the HARQ-NACK feedback from the receiver (UE) to the sender (NG-RAN), which is similar as in the case of the LTE system also. Only one HARQ-ACK/NACK bit is provided at a transport block level in such a retransmission scenario. The sender retransmits the entire transport block even though some of the segmented code blocks may have been received correctly. This is fine for a small-size transport block, but for a large transport block, such retransmission may result in wastages in the air interface resources/ less spectral or bandwidth efficiency, as is the case with the LTE system.

To optimize air interface resources usages resulting in high spectral efficiency, the NR physical layer may perform CBG-based transmissions and retransmissions for larger transport blocks, which may be used to provide enhanced mobile broadband services to users. In CBG-based transmission and reception procedure, segmented code blocks may be organized into several CBGs. Further, the receiver provides HARQ-ACK/NACK status bit to the sender at CBG instead of the transport block-level as mentioned in the previous paragraph. If some of the CBGs are received erroneously by the receiver, HARQ-NACK will be provided to the sender for those erroneous CBGs. The sender will retransmit the erroneous CBGs only, thus avoiding the retransmission requirements of the entire transport block of large size.

It may be noted that as compared to the providing of a single HARQ-ACK/NACK bit for an entire transport block, multiple HARQ-ACK/NAK status bits are provided in case of CBG-based transmission in the NR system. This results in a HARQ-ACK/NACK signaling overhead over the NR air interface. To reduce such overhead, the CBGs-based transmission, reception, and retransmission requirements may be scheduled dynamically by the NR-RAN/gNB over the RRC layer signaling. To indicate/enable such transmissions/ receptions/retransmissions, the RRC layer configures a parameter called *codeBlockGroupTransmission* which is provided to a UE as part of its PDSCH and PUSCH configurations using the IEs *PDSCH or PUSCH-ServingCellConfig*. The absence of the *codeBlockGroupTransmission* parameter as part of a PDSCH or PUSCH allocation from NG-RAN to UE disables the CBG-based transmission/reception. Following this, only one HARQ-ACK/NACK bit is provided, as usual in the LTE system, per transport block. A transport bock is segmented into several code blocks; each is of the maximum allowed size; Section 5.2.2 in TS 38.212 [107]. Code blocks are further organized into several CBGs. The maximum number of granularity (i.e. 2, 4, 6, and 8) of CBGs per transport block is provided to a UE through the RRC layer IE -> *PDSCH or PUSCH-CodeBlockGroupTransmission* -> *maxCodeBlockGroupsPerTransportBlock*.

Initially, all the code blocks of a transport block are transmitted by the sender to the receiver. A receiver provides a HARQ ACK/NACK status for each of the CBG of a transport block received by it from the sender. If a code block of a CBG has been received as erroneous, the receiver reports a HARQ-NACK to the sender. The affected CBG is retransmitted by the sender instead of the whole transport block, which leads to efficient usage of the air interface resource in case of larger size transport block transmission. Retransmitted CBGs are indicated by the sender through the presence of bit "1" in the CBGTI *(Code block group transmission information)* bitmap field of DCI_1_1 (for PUSCH) or DCI 0_1 (PUSCH) for each CBG (TS 38.212 [107]). Each bit (0 or 1) of the CBGTI bitmap field corresponds to a CBG of the associated transport block is transmitted. Another associated parameter that is provided as part of the DCI 1_1 for PDSCH transmission only is the CBG flush indicator (CBGFI). This is a 1-bit length parameter which is used to indicate whether the CBG indicated by CBGTI that is already available in the soft combining buffer should be flushed out (value 0), say corrupted or pre-empted (INT-RNTI), or the CBG should be used (value: 1) for soft combining with the existing CBGs.

The grouping of transport code blocks (CBs) into CBGs for PDSCH transmission/retransmission and multiple HARQ-ACK/NACK bits is illustrated in Figure 19.54. As illustrated, consider that a transport block is segmented into eight (=C) code blocks. Also, the *maxCodeBlockGroupsPerTransportBlock* as provided by RRC Layer is 3 (=N). The actual process used for the calculation of the number of CBGs (CBG 0...CBG-n) and the number of code blocks (CB0...CBn) per CBG for transmission of a transport block is described in Section 5.1.7.1 in TS 38.214 [109]. Using this procedure, the number of CBGs and CBs in each CBG is calculated below with C = 8, N = 3.

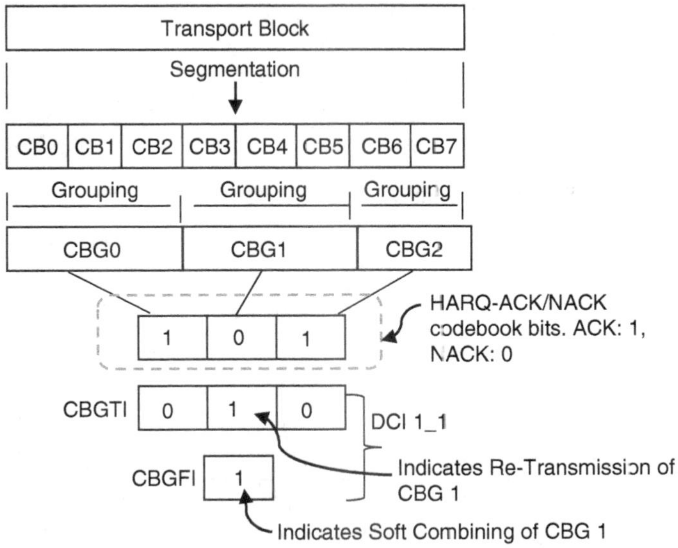

Figure 19.54 Illustration: code block group-based downlink transmission and its HARQ ACK/NACK.

M = min (N, C) = 3; M1 = mod (C, M) = 2; K1 = floor(C/M) = 3; K2 = Ceil (C/M) = 2
Since M1 > 0,
CBG 0 has 3 code blocks with index = 0,1,2 (as per Section 5.1.7.1 in TS 38.214 [109])
CBG 1 has 3 code blocks with index = 3,4,5 (as per Section 5.1.7.1 in TS 38.214 [109])

Further, CBG 2 has two code blocks with index = 6,7 (as per Section 5.1.7.1 in TS 38.214 [109]).

Segmentation of the above transport block into CBs 0–7 and their grouping into three CBGs 0–2 is illustrated in Figure 19.54.

Figure 19.54 also illustrates the HARQ-ACK (1)/NACK (0) codebook method of providing the status, from UE to NG-RAN, of the decoded CBGs. A HARQ-ACK/NACK codebook contains multiple bits that carry the CBGs decoding status from UE to NG-RAN. Further, a HARQ-NACK is reported to NG-RAN for the CBG 1 due to the reception of at least one code block as erroneous and its corresponding retransmission by NG-RAN as indicated through the CBGTI field of DCI 1_1.

- **Semi-Static and Dynamic HARQ-ARQ Codebook**

The number of bits used for HARQ-ACK (1)/NACK (0) status information, from UE to NG-RAN, in case of normal PDSCH transmission or CBG-based PDSCH transmission/reception, was described above. There are also other types of codebook determination procedures – Type 1 and Type 2 – which determine the number of bits to be used for the HARQ-ACK/NACK codebook reporting purposes that also vary depending on a UE configuration by the RRC layer. Type 1 is called the *semi-static*, and Type 2, which is used for reporting DL HARQ-ACK/NACK information from UE to NG-RAN, is called the *dynamic* codebook. The type of HARQ-ACK codebook to be used is indicated by the RRC layer through the IE: *pdsch-HARQ-ACK-Codebook* with value either semi-static or dynamic. Type 1 HARQ-ARQ codebook is reported over the PUCCH or PUSCH, whereas Type 2 HARQ-ARQ codebook is reported over the PUCCH only to the NG-RAN. For more information on these procedures to determine the DL HARQ-ACK codebook to be used by a UE, refer to Section 9 in TS 38.213 [108].

- **Timing for Downlink HARQ-ACK (UE to NG-RAN)**

  - FDD mode

As far as the timing for HARQ ACK/NACK, from UE to NG-RAN, for DL data reception at UE end is concerned, there is a difference between LTE HARQ and NR HARQ processes. In the case of the LTE HARQ process, the timing

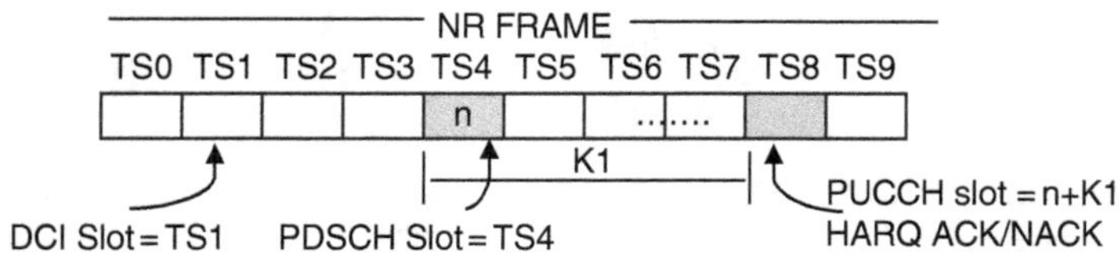

Figure 19.55 Illustration: FDD: PDSCH to its HARQ-ACK timing.

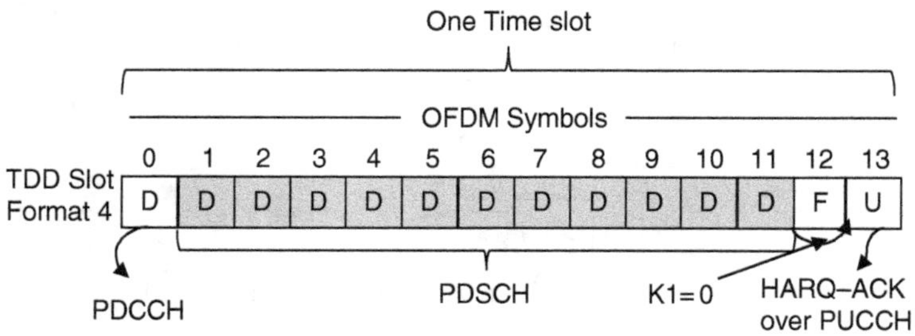

Figure 19.56 Illustration: TDD: PDSCH to its HARQ-ACK timing with K1 = 0.

for HARQ ACK/NACK is fixed, i.e. 4 ms in case of FDD mode; and for TDD mode, timing is provided in a predefined table; refer to Table 9.1.2-1 in TS 36.213 [91]. That is, in the case of LTE FDD mode, a transport block will be retransmitted, if required, after 8 ms of round trip time only. However, in the case of the NR DL HARQ process, the timing to report its ACK/NACK from UE to NG-RAN is configurable and controlled by the RRC layer, thus providing a flexible timing between a DL data/PDSCH reception at UE end and its corresponding ACK transmission over the PUCCH in the UL direction. The RRC layer provides a 3-bit length IE: *dl-DataToUL-ACK* as part of a PUCCH resource allocation to a UE. The IE: *dl-DataToUL-ACK* is used as an index into an RRC-configured Table 9.2.3-1 in TS 38.213 [108], which provides timing (K1) of transmitting a HARQ-ACK/NACK from UE to NG-RAN against the DL data/PDSCH reception by UE from the NG-RAN. Note that UE transmits the HARQ-ACK/NACK to NG-RAN with relative ACK timing: *K1 slot offset* to the DL data reception over PDSCH slot (n), as illustrated in Figure 19.55.

Slot offset or timing for transmission of DL HARQ-ACK/NACK from UE to NG-RAN may be also provided over the DCI_1_0/1_1 and its *PDSCH-to-HARQ_feedback timing indicator* field. If the DCIs do not contain this field, UE will provide a HARQ ACK/NACK to NG-RAN according to the information: K1 provided over RRC layer IE: *dl-DataToUL-ACK*. Thus, the round-trip time for PDSCH to HARQ-ACK/NACK from UE to NG-RAN in 5G NR is not fixed, unlike it is in the case of the LTE system.

- TDD mode

Figure 19.56 shows the transmission of DL data and its ACK in the same timeslot or subframe in TDD mode with the necessary guard period between them. Using the TDD slot format 4, this figure illustrates the flexible timing between a DL data/PDSCH reception by a UE and its corresponding ACK transmission over the PUCCH in the UL direction. As illustrated, the following information is being multiplexed over the different symbols of a TDD slot with the format 4.

(i) DCI (PDCCH) over symbol 0,
(ii) DL data reception over PDSCH, symbols 1–11, and
(iii) Corresponding HARQ-ACK, with slot offset K1 = 0, over a short PUCCH format #0 using only one OFDM symbol 13.

Further, as illustrated above, the timing between a DL data/PDSCH and its corresponding ACK transmission over a PUCCH is not fixed; rather, it is configurable, which is based on the slot/symbol offset: K1. Due to this, the DL/UL switching in TDD is fast because the PDSCH reception and its ACK transmission take place within the same slot boundary only, Figure 19.55, which results in a lower latency in the NR than the LTE system. For more information on the UE procedure for reporting HARQ-ACK to NG-RAN using the timing offset *K1*, refer to TS 38.213 [108].

- **NR Uplink Retransmission Requirement Indication Through NDI**

Unlike in the LTE system, where a separate DL channel called PHICH is used to send a UL HARQ-ACK status of transport block received over the PUSCH, no such specific DL control channel exists in the NR system to report UL HARQ-ARQ requirements from NG-RAN to UE. To indicate a UL retransmission requirement of an erroneously decoded transport block or CBG from UE to NG-RAN, the NG-RAN schedules the UE again with a *new data bit indicator* flag which is provided as part of a UL resource grant over a DL control information: DCI 0_0/0_1 to the UE. Further, UL retransmission from UE to NG-RAN can be CBG based, provided that the RRC layer configures a CBG-based transmission/reception for the UE. That is, a CBGTI field (DCI 0_1) is used to indicate a UL CBG-based retransmission, which is the same as the retransmission in case of the DL HARQ procedure described above. However, unlike the DL HARQ-ACK procedure, no CBGFI field is required as part of a UL HARQ-ACK procedure. Similar to the DL HARQ-ACK procedure, the UL HARQ-procedure is also asynchronous.

19.6.12 Physical Signals

Similar to the LTE system, the NR physical layer defines several types of physical signals in terms of RSs and synchronization signals in UL and DL directions which are used for different purposes by the UE as well as the NG-RAN.

- A DMRS for every physical channel, i.e. PUSCH, PUCCH, PDSCH, PDCCH, and PBCH;
- CSI reference signal (CSI-RS) in the DL direction;
- SRS in the UL direction;
- Phase-tracking reference signals (PT-RS) for PUSCH and PDSCH; and
- Secondary synchronization signal (SSS) and primary synchronization signal (PSS) in the DL direction.

A physical signal mentioned above is transmitted along with its corresponding physical channel. Physical signals are assigned a set of predefined REs to transmit information generated by the physical layer. But each physical signal has different densities, i.e. occurrences, within the frequency and time-domain resources allocated for the respective physical channel.

It may be noted that the physical channels and signals used over the 5G NR and LTE air interfaces have the same names. However, unlike in the LTE system, there is no "always on" cell-specific reference signal (CRS) in the NR system. Thus, the absence of a CRS makes the NR and NG-RAN an energy-efficient air interface. A UE uses the CSI-RS and DM-RS for channel estimation and measurement purposes only. A DMRS is used, instead, as an RS by the NR physical channels for its decoding by a UE. A DMRS is transmitted only when user data or signaling information is required to be transmitted. Similarly, the "always on" primary and secondary DL synchronization signals are used by a UE for synchronization and physical layer cell search procedures with the NG-RAN, obtaining the cell identity, and frame timing. In the NR, the PSS, SSS, and PBCH are collectively called *SS block*, which is allocated on specific REs of a DL resource grid.

Similar to the LTE system, predefined resources are assigned to each NR physical channel, except PUSCH AND PDSCH, and signals in time and frequency domains for transmission of transport channel carrying control information over the NR air interface. The individual transport channel is mapped to a particular physical channel, as listed earlier in Table 19.9.

19.6.12.1 Reference Signals Transmitted as Part of Physical Channels

Reference signals used by different physical channels in the NR system are described in this subsection.

- **Demodulation Reference Signals (DMRS)**

Unlike the LTE system, there is no DL cell-specific RS in the NR system. To help a UE in channel condition estimation and demodulating the associated UL and DL transmissions over different physical channels correctly, a UE-specific DMRS is used instead. A DMRS is used for the acquisition and processing of PDSCH, PUSCH, PUSCH, PUCCH Format 2, PDCCH, and PBCH. This means a DMRS is available only when user data or signaling information transmission is available. Depending on the physical channel being used, its corresponding DMRS may occupy either one or two OFDM symbols in the time domain. The DMRSs are generated from the same binary sequences, as defined in TS 38.211 [106]. However, the PUCCH Format 1 and PUCCH Formats 3 and 4 use different binary sequences. DMRSs are initialized with different parameters as configured by the RRC layer. DMRSs for PDSCH and PBCH are described below.

- PDSCH-DMRS

A PDSCH DMRS is a UE-specific RS which is transmitted in the same RB allocated for the PDSCH transmissions. For faster demodulation, DMRS for PDSCH is front-loaded, i.e. time-domain resource for a DMRS is allocated at the beginning of the RB allocated to its associated PDSCH. RRC layer configures the frequency domain and time-domain resources for the DMRS, like its type, symbol positions, length, and so on, through the IE: *DMRS-DownlinkConfig*. DMRS resources configuration by the RRC layer consists of the following information for a PDSCH scheduled UE:

(a) DMRS mapping type

This indicates the starting OFDM symbol of a slot allocated in the time domain for a PDSCH DMRS. Two mapping types are used – Type A and Type B – for a DMRS time-domain resource allocation. In DMRS mapping Type A, the position of the DMRS OFDM symbol is either 2 or 3 (indicated by the *dmrs-TypeA-Position* in MIB) depending on the length of the DL PDCCH. For PDCCH OFDM symbol length = 2, which uses the OFDM symbols 0 and 1, the OFDM symbol 2 will be used for DMRS. For PDCCH OFDM symbol length = 3, which uses the OFDM symbols 0, 1 and 2, the OFDM symbol 3 will be used for DMRS. Refer to Table 5.1.2.1.1-2–5 in TS 38.214 [109]. For mapping Type B, DMRS always occupies the first (i.e. Symbol 0th) OFDM symbol of a slot. Refer to TS 38.211 [106].

(b) PDSCH DMRS configuration type (*DMRS-DownlinkConfig -> dmrs-Type)*

This indicates the number or density of REs allocated in the frequency domain for a DMRS in the associated PDSCH RB. Two configuration types – Type 1 and Type 2 – are used to allocate REs to a PDSCH DMRS in the frequency domain; refer to Tables 7.4.1.1.2-1/2 in TS 38.211 [106].

(c) PDSCH DMRS length (*DMRS-DownlinkConfig -> maxLength)* in the time domain of one OFDM symbol or two consecutive OFDM symbols.
(d) Additional OFDM symbol positions (0, 1, and 3) for additional DMRS for channel condition estimation in case of high-speed mobility scenario. Refer to Tables 7.4.1.1.2–3/4 in TS 38.211 [106].

Figure 19.57 illustrates the mapping of a DMRS (represented by a filled rectangle) to the frequency and time domain RE (k, l) of its associated PDSCH RB. In this illustration, DMRS resource configurations used are as follows:

i) PDSCH DMRS mapping type A and OFDM position l = 2 in the time domain;
ii) PDSCH DMRS configuration Type 1 (left side with OFDM *maxLength* =1) and Type 2 (right side with OFDM *maxLength* =2) in the frequency domain; and
iii) PDSCH DMRS *maxLength* =1 and 2 OFDM symbols in the time domain.

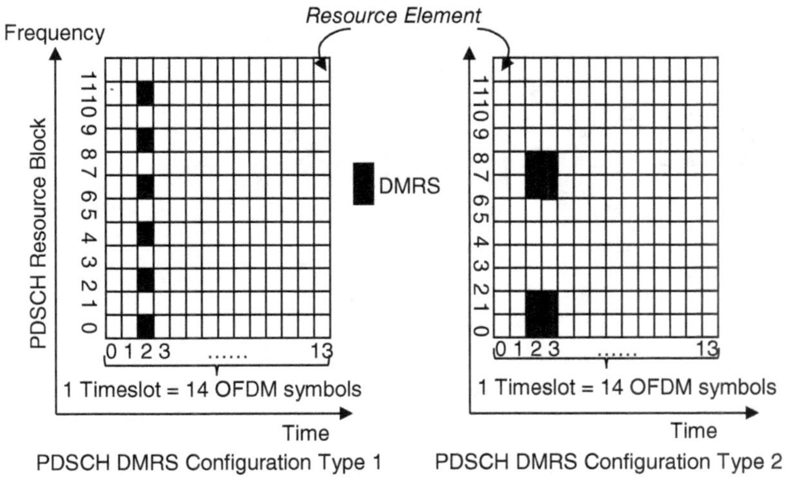

Figure 19.57 Illustration: mapping of DMRS to resource element and OFDM symbols.

In the frequency domain, the position (k) of the RE (represented by a filled rectangle) for the occurrences of DMRS is calculated using the expression described in Section 7.4.1.1.2 in TS 38.211 [106] in association with Table 7.4.1.1.2-1/2 in TS 38.211 [106]. Based on these expressions and as illustrated in Figure 19.57, a DMRS symbol occupies every alternate RE, starting from RE:0, in configuration Type 1. In configuration Type 2, a DMRS symbol occupies two consecutive REs, starting from RE:0, and is placed six REs apart. Further, it is observed that the density of DMRS in configuration Type 1 is more than the configuration Type 2.

Based on the DMRS configuration type and length (*DMRS-DownlinkConfig -> dmrs-Type and maxLength*) mentioned above, a UE finds the MIMO configuration being used in the DL or UL direction. For this purpose, NG-RAN provides the antenna port and layer information through the DCI 0–1, for PUSCH transmission or 1_1, for PDSCH reception, to a UE. Using this information and in combination with the predefined Tables 7.3.1.1.2x/7.3.1.2.2x in TS 38.212 [107], a UE finds the MIMO antenna and layer configuration being used in UL and DL direction transmissions.

- PDCCH DMRS

Transmission of a DCI over a PDCCH was described earlier in Section 19.6.10.5. A DMRS is also transmitted along with a PDCCH transmission. PDCCH DMRS is transmitted within the same REGs where PDCCH is transmitted. A PDCCH DMRS occupies three REs within an REG. Thus, the overhead due to a DMRS while transmitting a PDCCH within an REG is 25%. Due to this, not all the REs of REGs are available for transmission of a PDCCH. The positions of a DMRS or its mapping to REs within the REGs are fixed as described in TS 38.211 [106]. The positions start at Resource Element #1, 5, 9, and so on, as illustrated in Figure 19.58.

A DMRS scrambling ID *pdcch-DMRS-ScramblingID* is used to initialize a DMRS by the NG-RAN, which is configured by the RRC layer and is provided to a UE as part of a CORESET of a PDCCH. If *pdcch-DMRS-ScramblingID* is not available at UE end, the physical layer cell ID is used instead to scramble/descramble the PDCCCH DMRS.

- PBCH – DMRS

A DMRS is also transmitted as part of PBCH transmission, for its demodulation, in the same OFDM symbols: 1, 2, and 3 but with different RE/subcarriers in the frequency domain. Thus, each PBCH has a different DMRS. Mapping of PBCH-DMRS to the frequency and time-domain resources is described later in Section 19.6.13.

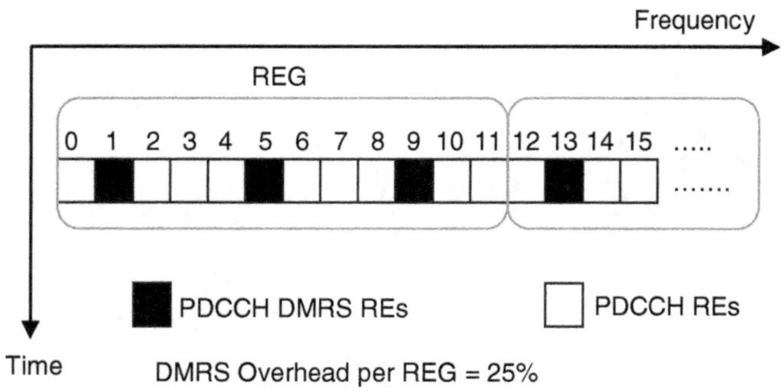

Figure 19.58 Illustration: mapping of PDCCH and its DMRS to REs and an OFDM symbol.

- **Phase Tracking Reference Signal (PTRS)**

5G NR system envisages providing high-speed mobile data services through transmission and reception of data at extremely high or mmWave length frequency. However, one disadvantage of operating at mmWave frequency is that transmitter and receiver are subject to suffer from unwanted phase noise which is caused by the mismatch in transmitter and receiver frequency oscillators. 5GR NR system introduces the PTRS in UL and DL directions to track the phase of the frequency oscillator and mitigate the performance loss due to phase noise.

Similar to the DMRS, a PTRS is associated with a PDSCH or PUSCH and is transmitted in the same RBs allocated to them. To a UE, the NG-RAN indicates the transmission/reception requirements of PTRS with PUSCH/PDSCH as part of their configurations by the RRC layer through the IEs *DMRS-Downlink or Uplink Config -> phase-TrackingRS*. DL PTRS is generated from the DMRS associated with a PDSCH, which is further scaled by a factor as described in TS 38.211 [106]. For more information on the process of allocation of frequency and time-domain resources as well as UE requirements of transmission/reception along with the DMRS of PDSCH or PUSCH, refer to TS 38.214 [109] in association with UL and DL PTRS configurations through the RRC layer IEs *PTRS-DownlinkConfig and PTRS-UplinkConfig*.

19.6.12.2 Sounding Reference Signals

Similar to other wireless communications system, in a mobile communication system, the different characteristics of a channel over which information is transmitted are subjected to undergo various radio environmental conditions such as the path loss, and multipath effects. Such factors further impact the overall performance of a communication system. It is important to measure the channel characteristics by the receiver and report it to the sender so that the different transmission parameters can be adjusted in subsequent transmissions with respect to the present channel states. The result of channel estimation in one direction can be also used for the decision to allocate resources in other directions, for example, in the case of the TDD system, where a single frequency is used for transmission/reception in the UL/DL direction. But in the case of the FDD system, separate frequencies are used in DL and UL directions between a UE and its base station. The UE reports back the status of the DL channel state as measured by it to its base station. Thus, a channel sounding is the process of measurement and estimation, by the receiver, i.e. UE, and acquisition by the transmitted, i.e. gNB, on the characteristics or knowledge of a channel based on a known sequence of an RS.

Channel condition estimation using DMRS, which is available only when scheduled data transmission is available, by a UE within the allocated bandwidth resource for PDSCH/PUSCH was described above. However, a channel sounding process does not involve the transmission of a physical channel, i.e. PDSCH or PUSCH. Due to this, a channel sounding method can be used to estimate channel state outside the allocated, and small, bandwidth resource as well as in the absence of scheduled data transmissions in UL or DL direction. Two channel sounding methods used in the NR system through DL and UL RSs are described below.

– Downlink Channel Status Information Reference Signal (CSI-RS)

In the 5G NR system, the CSI-RS, over a BWP, is used as the channel SRS in the DL direction. A UE uses the CSI-RS to measure the DL channel state and reports it to the gNB. Such a CSI is used by the gNB to acquire channel state for an efficient and channel-dependent scheduling of UE in the DL direction. CSI-RS is also used for the Layer 1/layer beam management as well as for mobility, through handover procedure, management purposes at Layer 3, i.e. RRC layer. NR system specifies zero-power (ZP), i.e. nothing is transmitted in mapped REs in time and frequency domain, and non-zero-power (NZP) CSI-RSs. Like in the LTE system, in NR too a UE measures the DL channel qualities using the CSI-RS and reports the result to the NG-RAN/gNB. Please note that Table 7.4.1.5.3-1 in TS 38.211 [106] defines the possible locations of an NR CSI-RS, in terms of a resource element in the frequency domain and an OFDM symbol in the time domain, which can be derived by a UE. This table contains the CSI-RS resource information such as the number of antenna port(s); densities of 0.5, 1, and 3 in the frequency domain; code domain sharing (CDM) type; *frequencyDomainAllocation;* and *firstOFDMSymbolInTimeDomain*. From Table 7.4.1.5.3-1 in TS 38.211 [106], the location of a CSI-RS in time and frequency domain is indicated to a UE by the NG-RAN through the RRC layer IE *CSI-RS-ResourceMapping*.

The density information contains the occurrences, in the frequency domain, of the CSI-RS in a given RB of a BWP. CSI-RS density of 1 indicates its appearance in every RB, whereas a density of 0.5 indicates its appearance in every other RB. In the time domain, a CSI-RS can be periodic (between 4 and 640 slots, refer to *periodicityAndOffsetperiodicityAndOffset* TS 38.331 [116]), aperiodic (indicated through DCI 0_1), and semi-persistent (indicated through DCI 0_1 and MAC CE) in type. Using these resources and Table 7.4.1.5.3-1 in TS 38.211 [106], a UE determines the CSI-RS currently the NG-RAN is transmitting in the DL direction. Refer to TS 38.211 [106] for more information on the processes of a CSI-RS sequence generation from a pseudo-random sequence and their mapping to REs based this table. It may be noted that the transmission of a CSI-RS is limited to a maximum of 32 different antenna ports (from 3000 to 3032), which is indicated by the RRC IE: *nrofPorts*.

– UL SRS

The SRS is used as the reference sounding signal in the UL direction because using a UL DMRS, only the associated physical channel state, i.e. PUSCH and PUCCH, can be estimated. An SRS is not transmitted with any other UL physical channel. Using an SRS in the UL, which is equivalent to the CSI-RS in the DL, channel state can be estimated at a different frequency outside UE allocated resource also. SRS estimation result is used by the NG-RAN to determine channel quality in the UL direction, and based on it, suitable radio resources with a good channel state are allocated to a UE for transmission in the UL direction.

Similarly, as in the case of the LTE system, an SRS in the NR system can be a periodic as well as aperiodic type. An SRS can be transmitted in the frequency hopping as well as non- frequency hopping mode. RRC layer configures the UE SRS resource through the IE: *SRS-Config*. Bandwidth allocation for the SRS can be between 4 and 272 RBs as defined in Table 6.4.1.4.3-1 in TS 38.211 [106]. This means minimum bandwidth allocation for SRS is 4 RBs and thereafter in multiples of 4. A row from this table is selected using the information provided as part of *SRS-Config* information. In the time domain (l), an SRS can occupy a maximum of four OFDM symbols *(SRC-config ->...nrofSymbols)* within the last six symbols of a slot as configured by the RRC layer *SRC-config -> SRS-Resource -> resourceMapping -> startPosition*. The starting position of the last symbol is 0; the starting position of the second last symbol is 1; and so on.

Further, in the frequency domain, the SRS appears to be a comb-like structure, i.e. SRS is transmitted at a particular subcarrier in comb style. The comb values (*SRS-Config -> transmissionComb)* as configured by the RRC layer are 2 and 4. That is, SRS is transmitted at either every second or fourth subcarrier. Figure 19.59 illustrates an SRS transmission with comb value = 4, i.e. every fourth subcarrier, along with the time-domain resources of four OFDM symbols allocated from the *resourceMapping*, starting from 1 to 4.

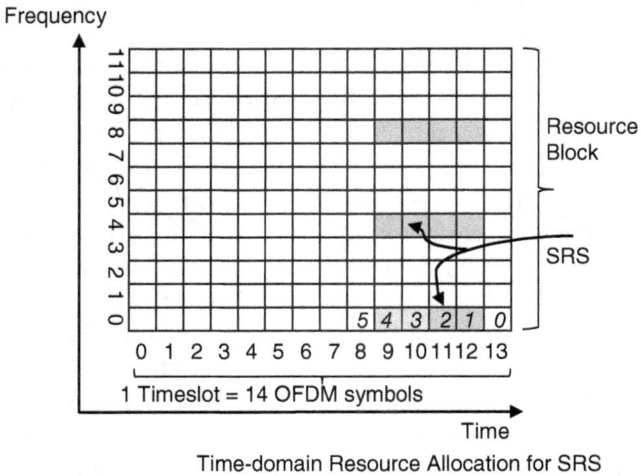

Figure 19.59 Illustration: time-domain resource allocation for SRS.

Refer to TS 38.211 [106], 38.331 [116] for more information on the frequency domain resource allocation and mapping process applied to UL SRS. SRS generations from a base sequence and their mapping to REs are described in TS 38.211 [106]. It may be noted that the transmission of an SRS is limited to only four antenna ports.

19.6.13 Downlink Synchronization

Similarly, as in the case of the LTE system, a UE upon its power on requires acquiring time (i.e. symbol and timeslot) and frequency synchronizations with an NR network to select a suitable cell and provide 5G services to users. This task is performed by a UE as part of its physical layer cell search and selection procedure, which is further followed by the acquisition of system information from the NG-RAN. To help a UE to synchronize with the NG-RAN, the NR physical layer transmits the synchronization signals in DL direction within a cell with a predefined signal and sequence. This procedure is similar to the LTE system. Further, to help a UE get synchronized with an NR network, the synchronization signals are put into predefined RE and OFDM symbols in the frequency and time domain. The overall process of UE synchronization within an NR system is described below.

- **Physical Layer Cell identity (PCI)**

Similar to the LTE system, a UE derives the PCI from the DL PSS and SSS. Also, in the NR system, the number of PCI groups is 336 (0–335) as compared to 168 in the LTE system. The PCI group (0 to 335) is obtained from the SSS. Each group contains three unique cell identities (0, 1, and 2) which indicates three PSS sequences within the PCI group. A cell identity can be obtained from the PSS. Refer to Section 7.4.2 in TS 38.211 [106] for the formula used to calculate the NR PCI.

- **PSS, SSS, and Synchronization Signals Block (SSB) for Cell Search Procedure**

The PSS is the first signal to be detected through correlation by a UE, which is followed by the detection of the SSS. In the LTE system, the PSS is generated from a Zadoff–Chu sequence, whereas in the NR system, the PSS is generated from an m-sequence, as described in Section 7.4 in TS 38.211 [106]. Further, in the LTE system, the PSS is mapped to 72 subcarriers, whereas the NR PSS is mapped to 127 subcarriers. Using the combination of PSS and SSS, the UE derives the PCI of the associated cell. Once a UE is synchronized with NR-RAN, it searches for the PBCH DMRS, which is used to demodulate the PBCH in association with the derived PCI since the information transmitted over PBCH is initialized with PCI by the NG-RAN. UE decodes the PBCH bits and reads the minimum system information transmitted by the NR-RAN over the PBCH, containing the MIB for UEs in the selected cell.

Unlike the LTE system, in the NR system, the PSS, SSS, and PBCH are bundled and transmitted together as a single block which is called the *Synchronization Signals and PBCH block (SS Block – SSB)*. An *SS block*, which starts with the PSS, is transmitted periodically through a set of resources in the frequency and time domain of RB over the Antenna Port #4000. A UE uses an SS block for the following purposes:

- To synchronize with the NG-RAN;
- To read the system information broadcasted by the NG-RAN;
- To perform beam management-related procedures, which are described in the next section; and
- To find a cell as part of its initial cell search or during its idle or mobility state within the NR coverage area.

For an overall structure of the SS block in frequency and time domain, refer to figure 5.2.4-1 in TS 38.300 [111].

An SS block has a fixed size in the frequency and time domain of a carrier frequency. A complete SS block occupies 240 subcarriers (from 0 to 239) or 20 RBs in the frequency domain and four consecutive OFDM symbols, starting from 0 to 3 *within the SS block*, in the time domain. Refer to Section 7.4.3/Table 7.4.3.1-1 in TS 38.211 [106]. From this table, the time and frequency domain resource which are mapped to the PSS, SSS, and PBCH/DMRS are as follows:

- PSS is transmitted in the first OFDM symbol location index: 0 within *the SS block* and is mapped to center 127, out of 240, subcarriers/REs starting from 56 to 182 in the frequency domain. Each PSS symbol generated using the sequence generator mentioned in Section 7.4.2 in TS 38.211 [106] is mapped to one of these subcarriers.
- SSS is transmitted in the third OFDM symbol location index: 2 *within the SS block* and is mapped to center 127, out of 240, subcarriers/REs starting from 56 to 182 in the frequency domain. Each SSS symbol generated using the sequence generator mentioned in Section 7.4.2 in TS 38.211 [106] is mapped to one of these subcarriers.
- PBCH and its associated DMRS are transmitted with three OFDM symbol locations: 1, 2, and 3 *within the SS block*. When PBCH is transmitted in OFDM symbols 1 and 3, it occupies the entire 240 subcarriers, i.e. 0–239. However, the DMRS is also frequency multiplexed and transmitted in the same OFDM symbols 1, 2, and 3, which start at RE #0.
- Some of the subcarriers are empty such as 0–55 in OFDM symbol position #0.

It may be noted that each PSS, SSS, and PBCH and its DMRS is scaled by its corresponding factor, as described in Section 7.4.3 in TS 38.211 [106].

- **SCS and Bandwidth of SS Block**

SCS numerology 0 (15 kHz), 1 (30 kHz) for operating band <6 GHz and 3(120 kHz), 4(240 kHz) for operating band >6 GHz with the normal CP are used for transmission of an NR SS block. Accordingly, the bandwidth of an SS block for these SCSs is 3.6 (=240 * 15 KHz) MHz, 7.2, 28.8, and 57.6 MHz, respectively. SCS for an SSB block is indicated through RRC layer IE: *ServingCellConfigCommon -> ssbSubcarrierSpacing*. The duration of the SS block also depends on the SCS being used, as illustrated below.

- **SS Block Periodicity**

It is configured by the RRC layer through the IE: *ServingCellConfigCommon/ServingCellConfigCommonSIB -> periodicityServingCell*. If the RRC layer does not configure the periodicity, the default value of the periodicity of the transmission of the SS block is assumed to be 5 ms. In the case of initial cell selection, the periodicity of transmission of SS block may be assumed as 20 ms.

- **Location of PBCH DMRS**

From Table 7.4.3.1-1 in TS 38.211 [106], it is observed that PBCH and its DMRS is transmitted in the same OFDM symbols 1, 2, and 3 in the time domain in a multiplexed way. However, depending on the *PCI* (last row, last column from Table 7.4.3.1-1 in TS 38.211[106]) of a cell, the position of the DMRS varies in frequency domain across the different cells. The starting position of the PBCH DMRS can be RE #0, 1, 2, or 3 in the frequency domain. This is calculated using the PCI and formula (*PCI mode 4)* as mentioned in Section 7.4.3.1 in TS 38.211 [106]. Using the PCI, which has values from 0 to 1008, the starting position of the PBCH DMRS in different cells repeats as 0 1 2 3, 0 1 2 3, ... From this calculation, it is observed that in the frequency domain, the position of the DMRS across the different cells with different PCIs can be at RE #0, RE #1, RE #2, RE #3, RE #0, RE #1, and so on.

As an example, and based on Table 7.4.3.1-1 in TS 38.211 [106], a visual illustration on the allocation and length of the PSS, SSS, and PBCH/DMRS in frequency and time domain is illustrated in Figure 19.60 for physical cells satisfying the condition: *mod (PCIs, 4) = 0*. In this illustration, the starting position of the DMRS is shown to be at RE #0 with OFDM symbols 1, 2, 3 for typical cell IDs 0, 4,8,12, and so on.

In the above illustration, the time-domain positions of PSS, SSS, and PBCH/DMRS are concerning the SS block only. Depending on the SCS being used, the actual stating location of the first OFDM symbol of each SS block varies in the time domain, which will be described later at the end of this section. The frequency domain position of PSS and SSS is fixed, while it varies for PBCH and DMRS across different cells based on its PCI. For example, for physical cells satisfying the condition: mod *(PCIs, 4) = 1*, the starting position of the PBCH DMRS is RE #1, for a *mod (PCIs, 4) = 2*, the starting position of the PBCH DMRS is RE #2, and lastly, for *a mod (PCIs, 4) = 3*, the starting position of the PBCH DMRS is RE #3.

- **SS Burst and SSB Index**

In the NR system, PSS, SSS, PBCH, and its DMRS are transmitted in the form of a burst. A burst contains several SS blocks with a total duration of 5 ms. Recall the structure of an NR frame which is of 10 ms duration containing 10 subframes with 10 slots. An NR frame is divided into two half frames, and each has a 5 ms duration. An SS burst has a duration of a half-frame, i.e. five subframes, of an NR frame. Each slot has 14 OFDM symbols in the case of

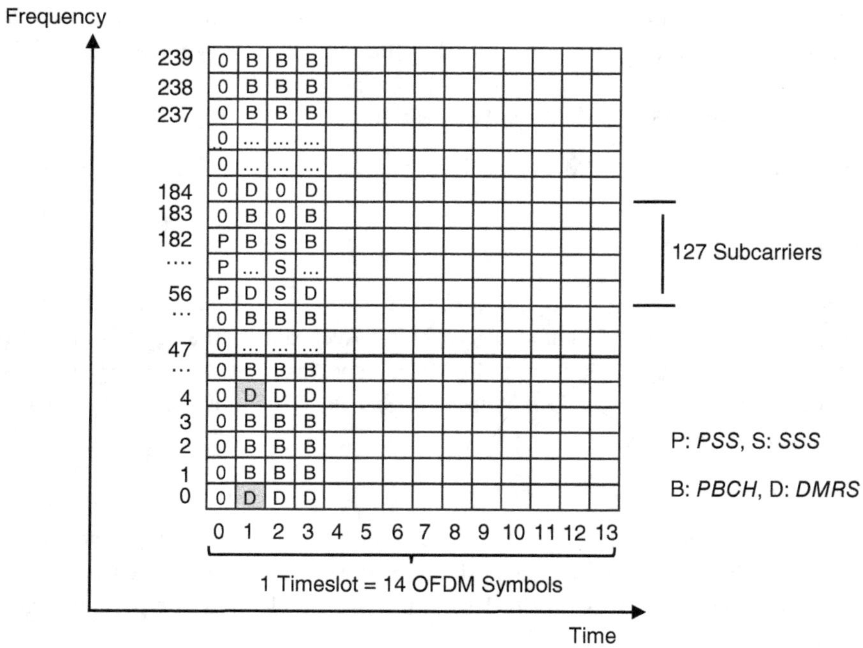

Figure 19.60 Illustration: position of PSS, SSS, PBCH-DMRS within an SS block for cells with a mod (PCIs, 4) = 0.

a normal CP. However, as listed in Table 19.5, the number of slots per subframe depends on the SCS or numerology being used. Therefore, the number of symbols in an SS burst also varies depending on the SCS being used. For example, the number of slots per subframe is 8 in case the SCS 120 kHz is used; for SCS 240 kHz, the number of slots is 16.

Consider that the SCS 120 kHz/numerology 3 is used in a cell. The total number of OFDM symbols to be transmitted for an SS burst can be calculated as follows:

Number of OFDM symbols in SS burst with 120 kHz SCS = OFDM symbols per slot or frame * number of slots per subframe * SS burst duration = 14 * 8 *5 = 560.

Thus, an array of 240 × 560 dimensions can be used to represent the above SS burst.

SS blocks of an SS burst per half-frame are identified through a bitmap which is provided as part of a cell configuration by the NG-RAN. Depending on the SCS/operating carrier frequency range being used in a cell – short, medium, and long formats of a bitmap are used to identify different SS blocks of an SS burst. A particular bitmap indicates the maximum possible SS blocks configured and transmitted by the NG-RAN in a particular cell. Such a bitmap indicating the SS blocks of an SS burst is provided to UEs in a cell through the RRC layer IEs: *ServingCellConfigCommon -> ssb-PositionsInBurst and ServingCellConfigCommonSIB -> ssb-PositionsInBurst.*

The IE *ServingCellConfigCommon -> ssb-PositionsInBurst* provides the bitmap length (Lmax) of 4 bits, 8 bits, and 64 bits, i.e. the maximum number of possible SS blocks in each bitmap is 4, 8, and 64 per SS burst per half-frame or 5 ms window. On the other hand, the IE *ServingCellConfigCommonSIB -> ssb-PositionsInBurst* provides SS blocks bitmap length of (*Lmax*) 8 bits, but each bit further represents a group of 8 SS blocks, giving a total 64 SS blocks per half frame. This bitmap may be used in case millimeter-wave radio frequency is used in a cell for transmission and reception between UE and NG-RAN through the beamforming method. Beamforming shall be described later in Section 19.6.14.

SS blocks bitmap length of 4 bits may be used for operating carrier frequency range up to 3GHz, 8 bits may be used for operating carrier frequency ranging from 3 to 6GHz, and 64 bits may be used for operating carrier frequency ranging from 6 to 52.6 GHz. Figure 19.61 illustrates the time-domain allocation of the maximum number of SS blocks for the above bitmap lengths: 4/8/64 configured under the IE: *ServingCellConfigCommon -> ssb-PositionsInBurst.*

Figure 19.62 illustrates the time-domain allocation of SS blocks for the above bitmap lengths: 8 bits configured under the IE: *ServingCellConfigCommonSIB -> ssb-PositionsInBurst.*

- **SS Block Pattern in the Time Domain for Different SCS**

It has been mentioned above that an SS block, comprising PSS, SSS, and PBCH/DMRS, occupies four consecutive OFDM symbols in the time domain, i.e. within an SS block the position of PSS: 0th index, SSS: 2nd index, and PBCH/DMRS: 1st, 2nd, and 3rd indices. The number of candidate SS blocks transmitted per SS burst may also vary, which is indicated by the corresponding bitmap (4/8/64 bits) mentioned above. To decode a candidate SS block, a UE requires to know the patterns, i.e. the occurrences and positions, of the SS blocks within an SS burst that spans across five subframes. Further, the first symbol, i.e. PSS, of each candidate SS block within the 5 ms window depends on the SCS frequency being used in a cell. For example, if the SCS 15 kHz is used in a cell, the

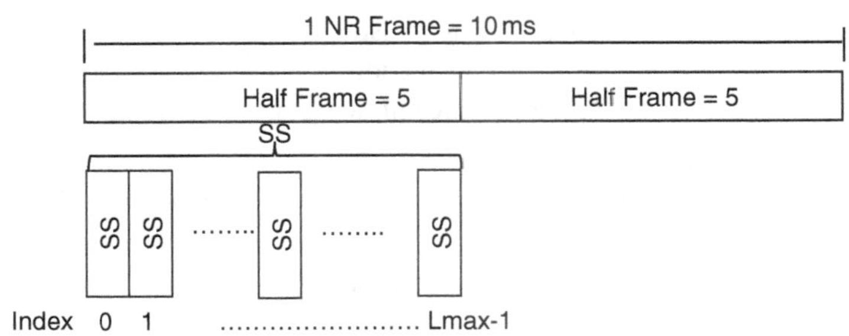

Figure 19.61 Illustration: organization of SS blocks/burst in the time domain for bitmap length:4/8/64 bits.

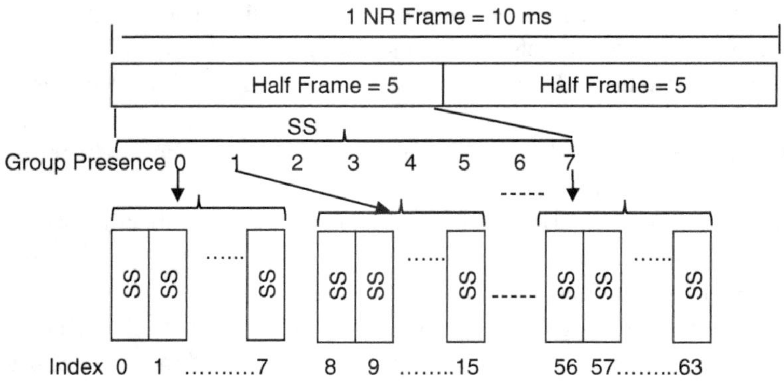

Figure 19.62 Illustration: organization of SS blocks/burst in the time domain for bitmap group length: 8 bits.

total number of OFDM symbols is 70 (5 TS, i.e. half frame for SS block × 14 symbols per TS). Within these 70 OFDM symbols for 15 kHz SCS, the PSS of each candidate SS block may appear at different OFDM locations (i.e. 2, 8,16, 22) in the time domain within the 5 ms window. This is true for other SCSs also where the number of starting positions of SS blocks is different and increases with an increase in SCS with a maximum of 64 for SCS 240 kHz.

As described in Section 4 in TS 38.213 [108], there are five cases, A–E, defined for FR1 and FR2 bandwidths, that provide the patterns of SS blocks in the time domain using which a UE can detect and decode an SS block of an SS burst transmitted by the NG-RAN. Case A SS block pattern is defined for SCS 15 kHz; Case B and Case C for 30 kHz; Case D for 120 kHz; and Case E for 240 kHz. Such an SS block pattern is used by a UE as part of its synchronization and cell search procedure with NG-RAN.

Consider Case A with 15 kHz SCS and 4 bits long-short bitmap to be used for SS blocks of SS burst. Also, consider that the NG-RAN transmits every SS block of the SS burst, i.e. all the bits of the short bitmap are 1. The index (zero-based) for the first symbol, i.e. for PSS, for each SS block, can be derived with the following inputs:

- Case – A, L = 4 blocks per burst, i.e. *ssb-PositionsInBurst -> shortBitmap* = 1 1 1 1
- SCS = 15 kHz
- n = 0,1

For n = 0, first OFDM symbol index for each SS block = [2, 8] + 14 × n = [2 + 14 × 0,8 + 14x0] = [2,8]
For n = 1, first OFDM symbol index for each SS block = [2, 8] + 14 × n = [2 + 14 × 1, 8 + 14.1] = [16,22]
Thus, SS blocks starting OFDM symbol indexes are = 2,8,16 22

Positions of each SS block (indexed from 0 to 3 with shaded boxes) with the above indices in the time domain are illustrated in Figure 19.63. Similarly, for the positions of SS blocks with SCS 30, 120, and 240 kHz for the cases B–E, refer to Section 4 in TS 38.211 [106].

At this point, a UE synchronized with a physical layer cell and acquired the system information from the NG-RAN. The next step for a UE is to register itself with the 5G core network through a random access procedure with NG-RAN, which is described in Section 19.6.20.

In summary, a UE performs the following tasks in sequence as part of an NR physical layer cell search procedure:

- PSS and SSS search
- PBCH DMRS search
- PBCH demodulation
- Decoding of BCH transport channel
- Read MIB

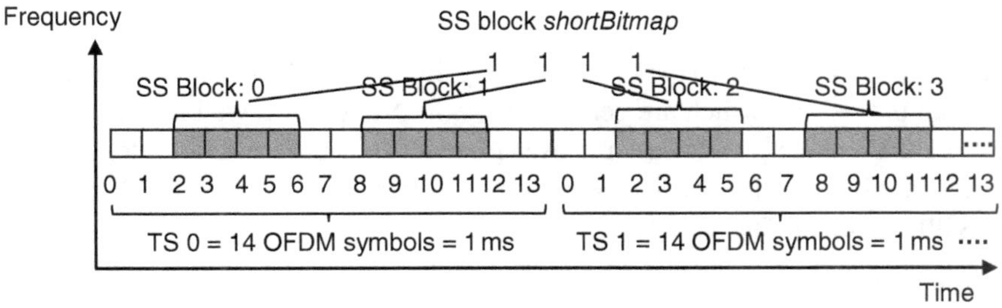

Figure 19.63 Illustration: positions of SS blocks in the time domain for Case A and SCS: 15 kHz with a short bitmap.

19.6.14 Millimeter Wave Transmission, Beamforming, and Its Management

One of the fundamental differences between the LTE and 5G NR systems is the provision for transmissions using a very high or mmWave carrier frequency, which is above 24 GHz, in the case of the NR system. Transmission with mmWave carrier frequency, whose wavelength is in millimeter, envisages providing high data rates communication services to users in the NR system in urban areas. At the same time, transmission with mmWave carrier frequency suffers from path losses along with other environmental conditions or factors such as obstacles and rain due to which the mmWave signal strength gets attenuated/reduces at the receiver end, resulting in poor data communication services. To help and overcome such path losses and reduced signal strength issues, a directional or *beamforming* transmission technique is used when the mmWave frequencies are used in the NR system. A *beamforming* technique is achieved by increasing or decreasing the gain of antennas in a specific direction through the usages of a large number of antennas, called *massive MIMO*, at the base station. Massive MIMO antennas generate a composite signal/beam which is transmitted in a particular direction in a cell. The number of antennas in massive MIMO can be up to 256, TR 38.913 [27]. The size of an antenna depends on the wavelength of a radio frequency being transmitted. Being a short wavelength of a mmWave frequency (recall frequency × wavelength = speed of light), the size of the antenna used to transmit it is also small, which makes it feasible to put a large number of antennas within a given antenna panel.

Transmissions using low and mid carrier frequency range, like in the LTE system, take place through a single omnidirectional antenna system. A single omnidirectional antenna covers an entire cell coverage area. On the other hand, to achieve coverage and capacity, transmissions using a very high frequency/mmWave take place through the directional antenna system using the beamforming technique. This is because mmWave frequency transmission suffers from large path losses along its propagation paths which further reduces the signal strength at the receiver end.

The direction or shape of a transmission pattern depends on the type of antenna being used. For a directional antenna, the transmission pattern is a *lobe shape*, whereas for an omnidirectional antenna, the transmission patterns are in a doughnut-like shape. In the case of an omnidirectional antenna system, the transmission energy is radiated in all the directions of a cell, i.e. covers the entire coverage of a cell. On the other hand, in the case of a directional antenna system, the transmission energy is radiated in a particular direction only of a cell, and hence its lobe shape pattern. A directional transmission pattern increases the coverage distances of a cell, but it decreases the beam-width or coverage area. Thus, a directional transmission does not cover the entire coverage area of a cell or all the UEs located at different locations within a cell. To overcome this situation and provide network coverage across the entire cell area, the *beamforming technique* is applied in the NR system in case of transmission with mmWave carrier frequency. Transmissions/receptions with the beamforming method deployed in an NR cell have several aspects as described below.

- **Beam Management**

Beams are transmitted by the NG-RAN/gNB, called a Transmit/Receive point (TRxP). Beam management is a set of physical and MAC functions and procedures which are performed by the NR MAC as well as the physical layer for transmissions/receptions in DL/UL directions in case mmWave frequency is used in a cell. Beam management also deals with the recovery of a beam failure which is detected between a UE and the NG-RAN. The scope of beam management tasks is illustrated in Figure 19.64.

- **Reference Signals for Beam Management**

DL and UL RSs are used to carry out beamforming and its management-related tasks between a UE and the NG-RAN in UE RRC idle as well as the connected state. In UE idle state, beam management is performed using an SS block, transmitted by the NG-RAN in the DL, over the initial DL BWP only. On the other hand, in UE RRC connected state, UE beam management is performed using the CSI-RS transmitted by the NG-RAN in the DL direction over the active BWP. The SRS is used for beam management purposes in the UL direction (RRC IE: *SRS-Config -> usages*). Beam management using these RSs is performed over the respective and allocated BWP only, other than the initial BWP. Beam management functions and procedures are described below.

- **Beam Sweeping: NR Initial Beam Selection in UE Idle State**

DL SS block transmission was described earlier in Section 19.6.13 as part of the initial NR physical layer cell selection procedure. In the NR system, several SS blocks are transmitted, in the form of an SS burst, in the DL direction at different times within a 5 ms window or half-frame duration. The number of DL SS blocks of an SS burst depends on the operating carrier frequency range being used in a cell. For the sub-6-GHz frequency range, more than one SS block may be used. In case mmWave frequency range is used, several SS blocks per SS burst are used, which are provided and configured by the NG-RAN through SS block bitmap as described earlier in Section 19.6.13. Each SSB represents one beam in a particular direction within a cell and is identified through an *SS block index*. Thus, to cover UEs located at different locations of a cell, multiple SS blocks with index: 1...Lmax-1 are configured and transmitted by the NG-RAN in DL direction in different directions at different and predetermined times within the 5 ms or half-frame window time. This process of transmitting multiple SS blocks, each covering a part of a cell, in different directions at different times in DL direction is called *Beam Sweeping* in the NR system, which is illustrated in Figure 19.65.

- **Beam Determination**

Among the available SS blocks within a cell, a UE selects an SS block or beam as part of its initial cell search procedure to communicate with the NG-RAN. For a beam determination and selection purpose, a UE uses the DL *SS-RSRP* (secondary synchronization signal RS received power). SS-RSRP is determined from the DMRS which is transmitted along with the PBCH of an SS block. Note that the CSI-RS may be also used to determine the

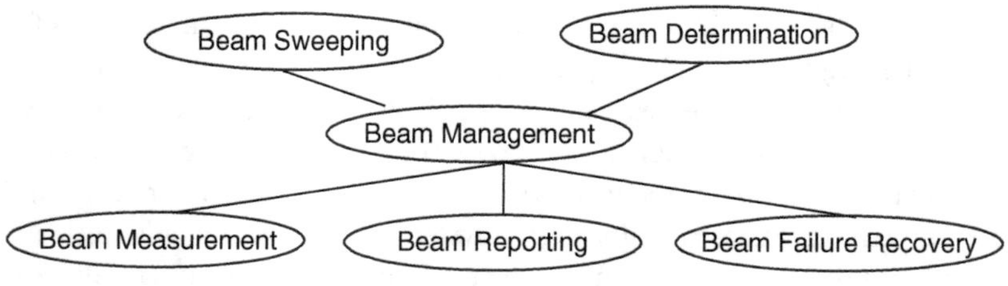

Figure 19.64 Illustration: NR beam management procedures.

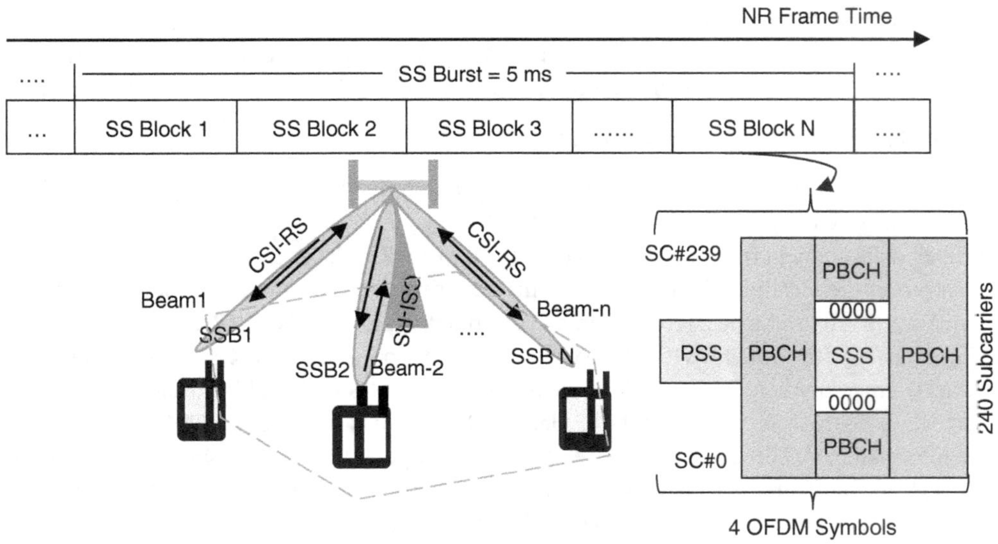

Figure 19.65 Illustration: NR beam sweeping with multiple beams.

SS-RSRP. Further, an RSRP threshold, called *rsrp-ThresholdSSB*, for a beam selection is also configured by the RRC layer. UE selects an SS block or a beam, as illustrated above, whose SS-RSRP is greater than the threshold *rsrp-ThresholdSSB* as configured by the RRC layer. The process of selection of an SS block or beam by a UE is called as *P-1 procedure* in the 5G NR system; refer to TR 38.912 [26], 3GPP R4-1814656 [140]. Further, due to user movement or UE rotation or other environmental factors, a UE requires to select a better TRxP beam from time to time to maintain desired beam quality between the gNB/TRxP and UE. This procedure performed by a UE is called *Beam Refinement (P-2)* procedure, TR 38.912 [26], 3GPP R4-1814656 [140]. Thus, the procedures P-1 and P-2 deal with beam sweeping at the gNB end. There is another procedure which is called as *P-3*, TR 38.912 [26], 3GPP R4-1814656 [140], where the gNB transmit the same beam, but the UE uses different received beams for UE beam sweeping and Rx beam refinement purposes to select the best received (Rx) beam by it. Thus, the purposes of the P-1, P-2, and P-3 procedures are to find the best Transmit (Tx) beam at the gNB end and Receive (Rx) beam at the UE end. For more supplementary information, the reader may refer to R1-1610239 [141].

- Beam Correspondence

Among the various capabilities of a UE, one mandatory and key capability is the *Beam Correspondence*, TS 38.101 [102], and RP-182580 [138] for FR2. A *Beam Correspondence* is the capability of a UE to select a beam for its UL transmission based on the DL signal measurements. As a result of a beam correspondence, the gNB and a UE use the same beam for their transmissions as well as receptions and vice versa. A beam correspondence is illustrated in Figure 19.65 where a beam is used for transmission and receptions, shown by arrows, between the gNB and a UE. A UE indicates the support of a beam correspondence capability to the NG-RAN, either through the selection of a beam autonomously or through a UL beam sweeping/refinement method. A UE provides the *beamCorrespondenceWithoutULBeamSweeping* capability information to the NG-RAN. If a UE supports a beam correspondence with NG-RAN, UE UL beam refinement is not required. For more information on the UE beam management-related capabilities indication to the NG-RAN, refer to TS 38.306 [112].

Using the selected beam through a beam correspondence, a UE transmits the random access preamble to the NG-RAN as part of the MAC random access procedure described earlier in Section 19.5.3. The NG-RAN allocates necessary radio resources and provides them to a UE over the same beam on which the RACH procedure was received from the UE.

Usages of a beam correspondence between the gNB and a UE result in the improved network performances in terms of reduction of signaling overheads and time to select a right beam during the initial network access by a UE. For more information on the benefits of usages of a beam correspondence in NR, refer to 3GPP RP-182452 [139].

- **Beam Level (Downlink) Measurement and Reporting**

Similar to a serving cell-level DL signal quality measurement configuration, the RRC layer may configure UEs to perform beam-level measurement (refer to IE: *MeasurementReport -> MeasResultNR* in TS 38.331 [116]) and report the results to the NG-RAN. Such measurement results may be used for the Layer 3/RRC layer, mobility management purposes through a handover procedure. Beam level DL measurements from several beams are used to derive the corresponding cell-level DL quality. Beam-level measurement can be performed while a UE is in RRC idle or connected state. In UE idle state, SS blocks, over the initial BWP, are used, whereas in the connected state, CSI-RS, over the active BWP, is used for beam level DL measurement purposes. Since each SS block or CSI-RS is identified through an index, a measurement report from UE to NG-RAN is provided at the SS block or CSI-RS index level. Refer to TS 38.215 [110] and TS 38.133 [104] for more information on SSB and CSI-RS-based NR physical layer intra- and inter-frequency measurements; refer to TS 38.331 [116] for configuration and reporting processes of various measurements as indicated by RRC layer to UEs. NR measurements for Layer 3 mobility purposes shall be described further in Section 19.6.16.

- **Link Recovery: Beam Failure Detection (BFD)**

Due to varying environmental and RF channel state, and other obstacles, the signal strength of the current/serving beam in use by a UE may become degraded and is no longer possible to continue using to communicate with the NG-RAN. In such a scenario, the UE MAC layer requests and triggers the beam recovery procedure by transmitting a random access preamble toward the NG-RAN. Before initiating a beam recovery procedure by the UE MAC layer, the UE physical layer detects a beam failure instance (BFI) and indicates it to the MAC layer. An instance of a beam failure is indicated from a UE physical layer to its MAC layer which is based on the RS resources – CSI-RS or SS block configured under the RRC layer IE: *failureDetectionResources*. The *Synchronization Signal-based Reference Signal Received Power (*SS-RSRP*)* metric is used for SSB measurement, and the CSI-RSRP metric is used for CSI-RS measurement purposes; refer to TS 38.215 [110].

For beam failure detection purposes, DL radio link quality is determined by comparing the monitored results against a link recovery threshold, called *Qin and Qout*; refer to TS 38.133 [104]. Further, the thresholds are linked to a target *block-level error rate (BLER = #of erroneous transport block/Total transport block transmitted)* of a hypothetical, not actual, PDCCH transmission in a DL direction. By comparing the DL radio link quality, which is based on the measured SSBs and/or CSI-RSs RSs, to the link recovery thresholds, a UE detects a beam failure or a recovery instance. The default BLER-level criteria for consideration of a BFI is 10% (i.e. worse quality), as mentioned in Tables 8.5.2.1-1/8.5.3.1-1 in TS 38.133 [104]. At this BLER level, a DL radio link cannot be received reliably by a UE, thus leading to a BFI between UE and the NG-RAN.

A UE performs the DL radio link quality measurement, based on periodic CSI-RS, within the active DL BWP only to detect a BFI. If the UE MAC layer finds that the number of such BFIs, considering the qualities of all the CSI-RSs, from the physical layer has exceeded the maximum value as configured under the RRC layer IE: *beamFailureInstanceMaxCount*, the UE MAC layer considers it as beam failure. UE MAC layer beam failure detection-related parameters are configured by the RRC layer through the IE: *RadioLinkMonitoringConfig*.

- **Link Recovery: Beam Failure Recovery (BFR)**

As described above, whenever the DL radio link quality of the DL RSs, i.e. CSI-RS or SSB, transmitted by the NG-RAN exceeds the configured BLER-level threshold of PDCCH, the UE will consider it as a BFI. The UE physical layer will attempt to recover a radio link by sending a BFI indication to the UE MAC layer. The UE will select a new candidate beam whose SS-RSRP or CSI-RSRP is greater than its respective threshold, i.e. *rsrp-ThresholdSSB* and *rsrp-ThresholdCSI-RS*.

As part of a beam recovery procedure, a UE MAC layer selects a candidate beam from the *BeamFailureRecoveryConfig -> candidateBeamRSList* provided by the RRC layer and transmits a contention free RACH preamble to the NG-RAN. The RACH preamble, its index, and the PRACH occasion correspond to the selected beam from the candidate list. If no such information is provided from the NG-RAN/gNB, the UE will perform a contention-based random access procedure. Further, the contention free RACH procedure is controlled by the timer *beamFailureRecoveryTimer*. On the expiry of this timer, a contention-based RACH procedure is performed by the UE. In both the cases, i.e. contention-free access and contention-based random access, if the UE does not receive a random access response from the NG-RAN/gNB even after the maximum retries with transmission power being ramped up, the UE will declare a radio link failure (RLF) and re-establishes an RRC connection with NG-RAN/gNB in the serving cell. The RRC layer, at gNB, configures UE MAC layer beam failure recovery-related parameters through IE: *BeamFailureRecoveryConfig*. Detection of a beam failure and its recovery procedures together constitutes the link recovery procedure in the NR system.

Thus, the NR beam failure detection and recovery processes, which are performed at UE Physical (L1) and MAC layer (L2), as described above, can be summarized and consist of the following steps with the illustration shown in Figure 19.66:

- Detection of a beam failure
- Selection of a new beam from the given candidate beams (*candidateBeamRSList)*
- Beam failure recovery attempt by a UE through a random access procedure, which can be contention free (using the RACH information provided in RRC IE: *BeamFailureRecoveryConfig*) or contention based, to the NG-RAN/gNB
- Beam failure recovery response through a random access response from the NG-RAN/gNB to UE. If a BFR fails, an indication is sent to the RRC layer.

Figure 19.66 Illustration: NR beam failure recovery procedure.

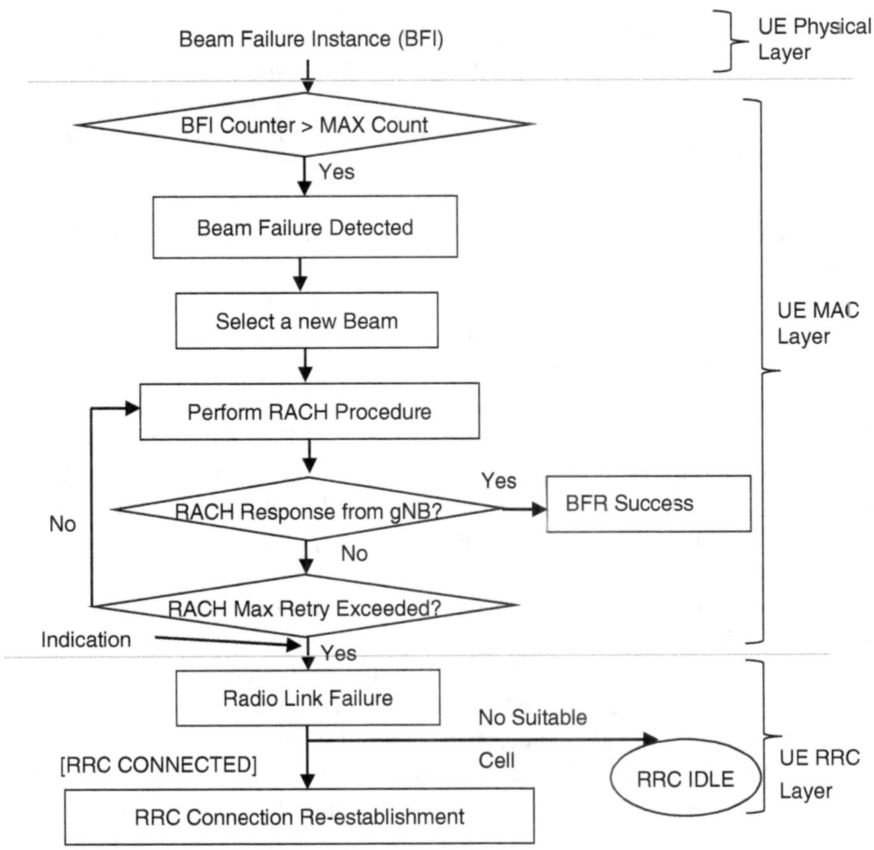

For more information on the NR BFD ad BFR procedure performed by its physical and MAC layer, refer to TS 38.213 [108]/TS 38.133 [104] and TS 38.321 [113].

The next section describes another similar procedure that is performed at a cell level by the NR UE physical layer and RRC layer and is called a Radio Link Monitoring procedure.

19.6.15 Cell-Level Radio Link Monitoring (RLM)

DL radio link quality measurements, which are based on the SS-RSRP or CSI-RSRP or both, for beam failure detection and its recovery purposes, have been described in the previous section. A radio link recovery procedure, which consists of beam failure detection and its recovery tasks, is performed at the NR Layer 1 (Physical) and Layer 2 (MAC). However, a DL radio link monitoring may be also performed at NR Layer 3, while a UE is in the RRC CONNECTED state. This is due to the varying and dynamic nature of an RF channel state where a UE in its RRC CONNECTED state requires to monitor the DL radio link quality over the active BWP of the cell.

A UE performs radio link monitoring so that services are not interrupted to its user. UE uses the DL *RLM reference signals (*RLM-RS*)*, i.e. CSI-RS based or SSB based or both, to monitor the DL radio link quality in a cell. In both of these methods, DL radio quality is determined by comparing the monitored results against two thresholds, *Qin* and *Qout*, as described in TS 38.133 [104], which is similar to the threshold used for a beam failure detection and recovery purposes as described in the previous section. Further, the thresholds are linked to a target *block-level error rate (BLER = #of erroneous transport block/Total transport block transmitted)* of a hypothetical, not actual, PDCCH transmission in a DL direction. The thresholds are used to detect whether a DL radio link with quality as perceived by a UE can be considered as a cell-level *out of synchronization (*OOS*)*, i.e. problem at the physical layer and hence an RLF, or *in synchronization (*IS*)*, i.e. physical layer problem recovered and no more RLF.

OOS for a UE in RRC CONNECTED state indicates a problem at NR physical layer, and hence a radio link cannot be received reliably by a UE. *In-synchronization* indicates a normal condition, as well as no more problem at the physical layer, and hence the radio link can be received more reliably by a UE. The default target BLER level for out-of-synchronization consideration criteria is 10% (i.e. worse quality), and for in-synchronization or normal consideration criteria, the target BLER level is 2% (i.e. better quality), as mentioned in Table 8.1.1-1 in TS 38.133 [104]. The UE RRC layer is intimated accordingly for such physical layer events, i.e. out-of-sync, in-sync, based on which either an RLF is declared or recovered or enter into RRC idle state by the UE as specified in TS 38.331 [116].

BLER level for an out-of-synchronization condition is detected based on typical PDCCH transmission parameters as mentioned in Table 8.1.2.1-1 (for SSB-based link monitoring) and Table 8.1.3.1-1 (for CSI-RS-based link monitoring) in TS 38.133 [104]. Similarly, to consider the DL radio link as in-synchronization *or normal*, the target BLER-level requirement is 2% (i.e. better quality), as mentioned in Table 8.1.2.1-2 (SSB based) and Table 8.1.3.1-2 (CSI-RS based), in TS 38.133 [104]. Note that the BLER level for out-of-synchronization and in-synchronization threshold may be provided by the RRC layer also through the IE: *rlmInSyncOutOfSyncThreshold*. Further, depending on the operating frequency range, i.e. FR1 or FR2, being used, the RRC layer configures the reference signals (CSI-RS or SSB) to be used for radio link monitoring purposes through the IE: *RadioLinkMonitoringConfig -> failureDetectionResources;* refer to Table 5-1 in TS 38.213 [108] and Table 8.1.1-2 in TS 38.133 [104]. The periodicity to measure and indicate a radio link as out-of- synchronization or in-synchronization from UE physical layer to the RRC layer is defined in Table 8.1.2.2-1/2 (FR1/FR2) TS 38.133 [104], for SSB based, and in Table 8.1.3.2-1/2 (FR1/FR2) TS 38.133 [104], for CSI-RS-based DL radio link monitoring.

The result of DL RLM is reported to the UE RRC layer with status either as *out of sync* or *in sync* based on thresholds as mentioned above. Upon detection of an RLF, the UE, in RRC CONNECTED state, *re-establishes the RRC connection* by sending the *RRCReestablishmentRequest* message to the NR-RAN in the serving cell or may enter into the RRC idle state if a suitable cell is not found. RRC layer configures RLM along with the beam failure recovery-related parameters through the IE: *RadioLinkMonitoringConfig*. The purpose of configuration, i.e. beam failure or RLF and the detection resource, i.e. SS block or CSI-RS, is specified using the IE: *RadioLinkMonitoringRS*. Refer to TS 38.331 [116] for more information on these parameters and UE actions taken upon the detection of an RLF.

Figure 19.67 illustrates the RLF detection process at UE end, in RRC CONNECTED state, using RLF-related timers and counters. An RLF declaration by a UE is based on the expiry of the RRC layer T310 timer at UE end. Timer T310 at the UE RRC layer expires as a result of consecutive indications of the UE physical layer problem, i.e. out-of-synchronization, to its RRC layer. Also, counters such as the N310/311 and timers T310/311 are used at the UE RRC layer for detection of an RLF issue in an NR cell, and they are configured by the RRC layer at NG-RAN through the IE: *RLF-TimersAndConstants*, TS 38.331 [116].

From Figures 19.66 and 19.67, we can note one difference between beam failure detection and its recovery and the RLF monitoring processes that in the case of the former one, a beam failure indication from the physical layer is counted by the MAC layer. On the other hand, in the case of an RLF monitoring process, an out-of-sync or in-sync indication from the physical layer is counted by the RRC layer. However, both the processes use the same RS as well as the BLER limit of a hypothetical PDCCH. It may be noted that a UE may also declare the occurrence of an RLF if the maximum retransmission count of an SDU by the RLC layer in its AM has reached the threshold.

In summary, in the NR system also, an RLF at UE end may occur due to the failure of a random access procedure during a BFR or a physical layer issue, i.e. out-of-sync, or exceeding of maximum retransmission limits of an RLC layer SDU. The occurrences and clearance of an RLF problem may be notified to the operation and maintenance (O&M) operator through a suitable alarm, as described earlier in Chapter 15.

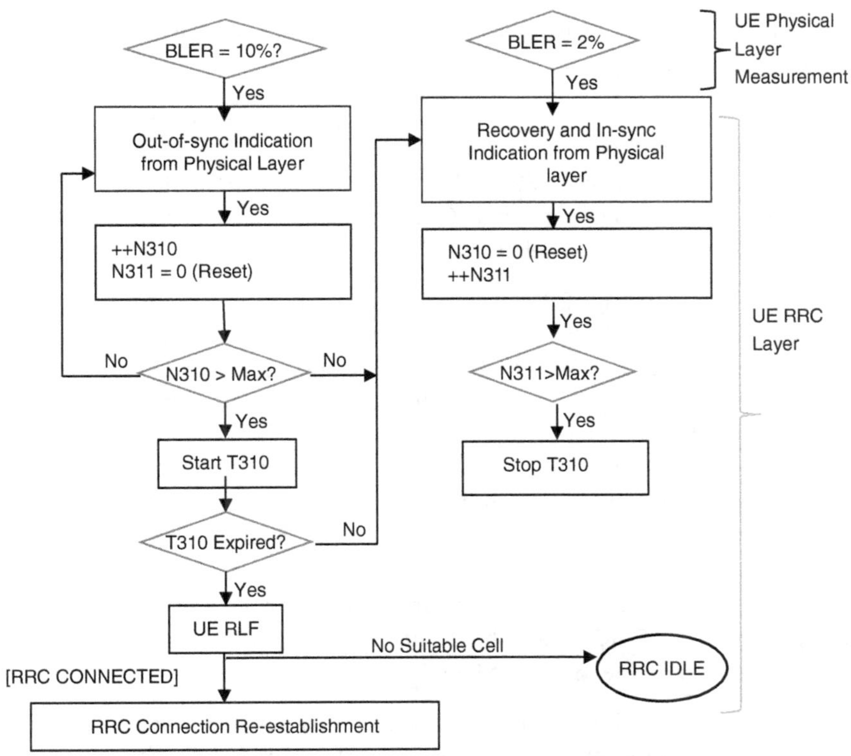

Figure 19.67 Illustration: UE RLF due to out-of-synchronization indication/BLER.

19.6.16 RRM Measurements for UE Mobility

UE beam failure detection and its recovery as well as radio link monitoring have been described in the previous Sections 19.6.14 and 19.6.15 which are based on the DL radio link quality measurements using the SS-RSRP and CSI-RSRP RS. Such a failure detection and its recovery at an NR beam level (intra cell) are performed at the UE Layer 1, i.e. physical layer, and Layer 2, i.e. MAC layer, for an idle and stationary UE. In a beam level (intra cell) failure detection and recovery process, no signaling connection exists at the RRC (Layer 3) level between a UE and the NG-RAN.

Similar to the previous generation mobile communication systems, one basic functionality of the NR system for the continuation of ongoing communication services of a UE is to support the UE and its user mobility at a cell level as the UE/user moves from one cell to another cell within a coverage area. A UE/user mobility at a beam level, within a cell, is also required to be supported through the beam management procedures, as described in Section 19.6.14. A cell-level mobility shall be required to be supported for an idle as well as connected UE and its user. For a UE in RRC CONNECTED state, its cell-level mobility is supported by transferring the RRC signaling connection to the target cell through a handover procedure as described earlier in Section 18.5.5. For a UE in RRC idle state, its cell-level mobility is supported through a cell selection and reselection procedure which is performed by the UE. Such a UE mobility requirement is accomplished by measuring the DL radio link quality of the serving as well as neighbor cells by the UE. Further, for a UE in RRC CONNECTED state having an RRC signaling connection with the NG-RAN, the UE reports the measurement results to the NG-RAN. Requirements for reporting of measurement results are configured by the NG-RAN to a UE. Based on the measurement results and reports, the NG-RAN may decide to handover a call of a UE in its RRC CONNECTED state. Such measurement result inputs used by the NG-RAN to manage the mobility of a UE through a handover procedure is called as *Radio Resource Management (*RRM*) measurements*, which is also performed in the previous generation mobile communication systems. Thus, the basic principles of overall RF measurements in the NR system are similar to the procedure used in the LTE system. However, there are differences also between the LTE and NR measurement procedures because of multibeam-based operations and the absence of cell-specific RSs in the NR system. Different aspects, i.e. basis of measurements, framework, and overall working procedure, of an RRM procedure used in the NR system are described next.

19.6.16.1 RRM Measurement Signals and Their Quantities

In the NR system, an RRM measurement may be based on the SS block(s) or CSI-RS signal. The NG-RAN configures a UE to perform an SS block or CSI-RS-based measurement in RRC idle, inactive, and connected states using their respective measurement metrics/quantities mentioned below.

- *Reference Signal Received Power* (RSRP): The metric SS-RSRP is used for SS-based measurement and CSI-RSRP is used for CSI-RS-based measurement. SS-RSRP provides the linear average over the power contributions of the REs that carry *Secondary Synchronization Signals. C*SI-RSRP provides the linear average over the power contributions of the REs of the antenna port(s), e.g. 3000, 3001, and so on, which carry CSI reference signals configured for CSI-RS-based RSRP measurements.
- *Reference Signal Received Quality* (RSRQ): The metric SS-RSRQ is used for SS-based measurement and CSI-RSRQ is used for CSI-RS-based measurement. RSRQ is the ratio of respective RSRP to the RSSI (received signal strength indication) over several resource blocks within the measurement bandwidth.
- *Signal-to-Noise and Interference Ratio* (SINR): The metric SS-SINR is used for SS-based measurement and CSI-SINR is used for CSI-RS-based measurement. SINR is derived from the linear average over the power contribution of the REs that carry the SSSs or CSI-RS divided by the linear average of the noise and interference power contribution over the REs that carry the SSSs or CSI-RS.

For the definition of the above metrics/quantities of SS block or CSI-RS-based measurements, refer to TS 38.215 [110]. In the LTE system also, similar metrics for RRM purposes are used. Measurements of the above metrics/quantities depend on the state of the RRC layer at UE end. The SINR is measured in RRC CONNECTED state only. Similarly, the CSI-RSRP and CSI-RSRQ are measured in UE RRC CONNECTED state only. Using these metrics, the NG-RAN may configure a UE to perform the following types of DL radio link measurements.

- NR intra-frequency, where the center frequency as well as the SCS used in the serving cell as well as neighbor cells are the same;
- NR inter-frequency measurements; and
- Inter-RAT, LTE/E-UTRAN only, measurements.

19.6.16.2 RRM Measurements Framework

Similar to the LTE system, a measurement framework between a UE and the gNB is used in the NR system also to perform an RRM measurement by the UE, and its reporting to the gNB, using the metrics described above. An RRM measurement framework in NR also consists of the following components to perform intra/inter-frequency and inter-RAT measurements.

- Measurement objects for LTE/E-UTRAN and 5G NR, i.e. carrier frequency, subcarrier spacing (for NR), SS-block, CSI-RS, and so on, to be measured by a UE. Refer to IE *MeasObjectNR*, TS 38.331 [116];
- Reporting configurations, i.e. periodical or event-based reporting, RS to be measured (i.e. SSB or CSI-RS);
- Measurement identities;
- Quantity configurations, i.e. the SS block or CSI-RS to be measured as well as the Layer 3 filtering coefficient to be applied for a particular measurement quantity;
- Measurement gaps; and
- SS Block Measurement Time Configuration (SMTC) window.

Figure 19.68 illustrates the possible associations among the above components of the NR RRM measurement framework.

Figure 19.68 illustrates a list (1...N) of measurement objects and reporting configurations. A measurement object and a reporting configuration are assigned with a measurement identity. Further, a measurement object may be associated with more than one reporting configurations with unique measurement identities. All such logical objects of a measurement framework are configured by the NG-RAN to a UE through the IE: *measConfig* in the *RRCReconfiguration* message; TS 38.331 [116].

- RRM Measurements Events for Reporting

Based on the configuration information provided by the NG-RAN under the above measurement framework, a UE performs the DL radio link measurements in a cell. A UE reports the measurement results to the NG-RAN based on certain events, which are similar to the LTE system. Such events may be classified into two categories as mentioned below.

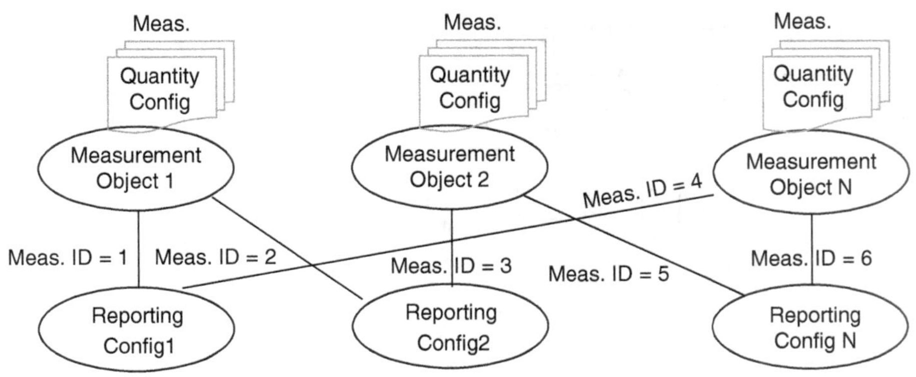

Figure 19.68 Illustration: NR RRM measurements framework.

- Events (A1–A6) used for reporting intra and inter NR frequency measurement results to the NG-RAN;
- Events (B1–B3) used for reporting inter-RAT frequency measurement result in the NG-RAN.

Each event mentioned above has its exit and entry criteria; refer to TS 38.331 [116]. An event is triggered on fulfilling its criteria, and the measurement result is reported from a UE to the NG-RAN. Prior to the evaluation of the criteria of an event, a Layer 3/RRC layer filtering is used.

- Measurement Gap

While a UE is engaged in transmission/reception within its active BWP in its serving cell, it may not be able to perform DL radio link measurements for the intra- and inter-frequency as well as inter-RAT signals and their quantities provided in a measurement object to the UE. For such UEs, a measurement gap is provided by the NG-RAN during which the UE does not transmit/receive in its serving cell but measures the DL radio link quality of signals and their quantities, outside the active BWP, provided in the measurement object to it. The NG-RAN provides measurement gap information in the form of a measurement gap pattern; refer to Table 9.1.2-1 in TS 38.133 [104]. A measurement gap pattern consists of *Measurement Gap Length (*MGL) and *Measurement Gap Repetition Period* (MGRP), both are specified in milliseconds. As shown in Figure 19.69, a measurement gap includes RF switching time also at the beginning and end of a measurement process.

- SMTC window duration

The SMTC window is a new concept that has been introduced as part of the NR RRM measurement process. An SMTC window is defined, with a periodicity, offset, and duration, within a measurement gap. Within an SMTC window duration only, a UE measures the DL radio link signal strength and quality for the serving cell as well as its neighbor cells as configured by the NG-RAN to the measuring UE. Only the SSB blocks that fall within the SMTC window are measured. Figure 19.69 illustrates the usages of a measurement gap and its SMTC window (indicated by the rounded dotted rectangle), which are used during an NR RRM measurement process, for a typical serving cell and one neighbor cell.

As illustrated in Figure 19.69, within an SMTC window, the UE measures four SS blocks in its serving cell, which can be specified through the IE: *MeasObjectNR -> SSB-ToMeasure*, TS 38.331 [116].

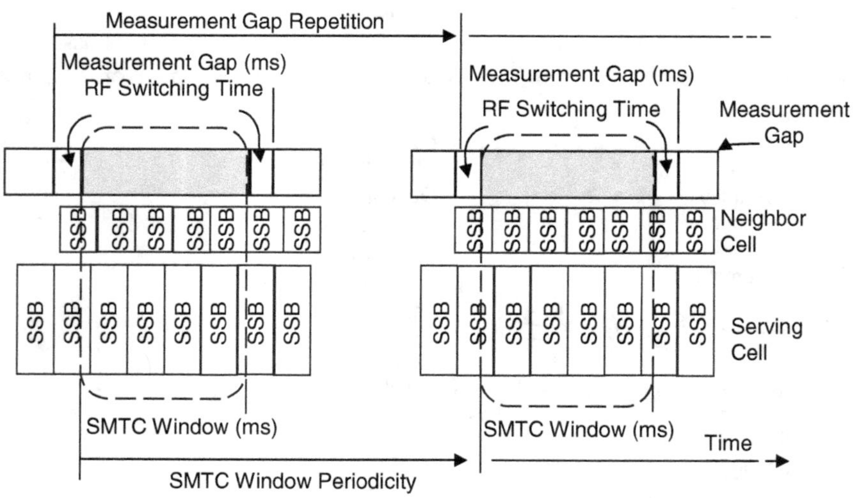

Figure 19.69 Illustration: NR RRM measurement gap and SMTC window.

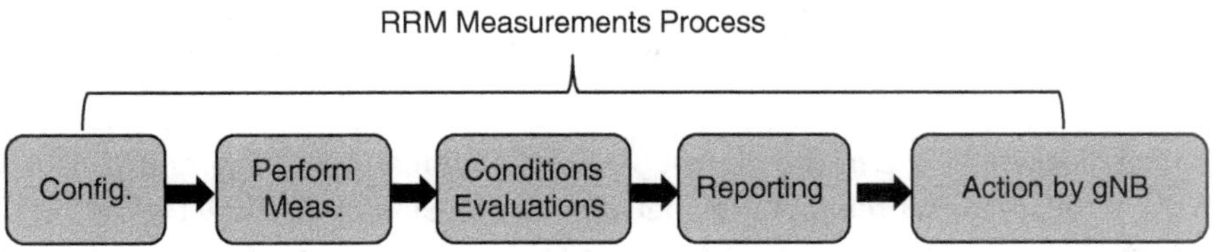

Figure 19.70 Illustration: NR measurements process.

19.6.16.3 Overall RRM Process

In the previous Sections 19.6.16.1 and 19.6.16.2, NR measurement metrics and their framework have been described. Using the measurement matrices and its framework, the overall NR measurement process, which involves both the UE and gNB, contains several stages, as illustrated in Figure 19.70.

The gNB configures the measurement requirements and its reporting criteria toward a UE using the RRC layer signaling message described below.

- Measurements Configurations

In this step, using the *RRCReconfiguration* message, the gNB configures the measurement-related information to a UE such as the criteria, i.e. periodical or event-driven, for measurements reporting to gNB; the measurement gap to be used by a UE; and the type of cells, i.e. serving cell, listed cell, or detected cell. The eNodeB provides the measurement-related information as well as other information such as thresholds and hysteresis to a UE through the RRC layer signaling message IE: *MeasConfig*. Each measurement configuration is identified by a corresponding measurement identifier that is used during the reporting step from the UE to the gNB.

- Perform Measurements

In this step, a UE performs the actual measurement of various matrices described earlier.

- Conditions Evaluations

In this step, a UE evaluates the reporting conditions of measurement results to be reported to the gNB. The conditions are based on the measured value of the received signal strength and quality as well as the other information provided during the configuration of a particular measurement. A UE measurement and reporting is termed as an *event (A1–A6, B1–B3)*, as described earlier.

- Measurement Reporting

In this step, a UE reports the measurement result performed by it to the gNB using the RRC layer message *MeasurementReport*. The measurement result contains the measured value of the RSRP and RSRQ for the serving cell and its neighbor cells.

- Measurement Action

In this step, for a UE in its RRC connection state, the gNB may decide to initiate a handover of the ongoing call to another cell or network based on the measurement report provided by the UE. For a UE in its RRC idle state, it may select the best cell among the measured ones and perform a cell reselection to make it as the serving cell.

Example 19.11 illustrates a practical scenario where the DL signal level is measured and displayed by a UE from its live network.

Example 19.11 Intra-/Inter-Frequency NR and Inter-RAT DL Signal-Level Measurements by Mobile Device with Multiple SIMs

Use a mobile app for network signal measurement purposes. Insert two SIMs of different RATs, e.g. one 4G SIM and one 5G SIM. In case if both the SIMs are 5G, set the preferred network for one SIM as 4G. Using the mobile APP, one can observe the intra-/inter-frequency NR and inter-RAT DL signal-level measurements as follows.

- RSRP – DL signal strength (in dB) for LTE E-UTRAN and 5G NG-RAN.
- RSSI - DL signal strength (in dB), for LTE E-UTRAN and 5G NG-RAN.

The mobile device will measure and display the DL signal levels and quality (RSRQ) received from the E-UTRAN and 5G base stations.

19.6.17 Channel State Information (CSI)

In a wireless communications system, the RF channel state does not remain the same but varies from time to time due to various environmental factors. Also, as a mobile user moves from one area to another area, the mobile device experiences different signal quality and strength. It also depends on the current position of the user. Under such conditions, the transmitter or the base station needs to know the current state of the DL channel for effective transmissions with multiple antennas and good signal quality. The base station adapts the subsequent data transmissions with the current channel states/properties to provide reliable and high data rates services, especially with multiantenna transmissions with the MIMO method, to a UE/user. Similar to the LTE system, a DL channel state is measured at a receiver, i.e. UE, end and is reported back to the transmitter, i.e. gNB. To manage such dynamic environmental conditions for transmissions with multiple antennas and MIMO method, a UE measures the DL non-zero powered (NZP) *CSI Reference Signal* (NZP CSI-RS*) and CSI-Interference Management* (CS-IM) signals received, for a given number of antenna ports, from the gNB over a particular BWP. The CSI-RS was described earlier in Section 19.6.12. Using these reference signals, a UE measures and reports the current channel state as experienced to be best by it to the gNB over the PUCCH or PUSCH. Based on the CSI and its recommendations, the gNB may adjust the current resources allocated to a UE dynamically in terms of the number of transmission layers for DL single-user or multiuser MIMO transmissions, allocation of a more robust coding scheme, and so on so that error-free information is delivered to the UE.

- Framework of CSI Measurements and Reporting

The NR framework of CSI reporting from a UE to the gNB consists of two components – CSI Reporting Configurations and CSI Resource Configurations. Both are configured by the RRC layer using its IEs: *CSI-ReportConfig* and *CSI-ResourceConfig*, containing the frequency domain and time-domain information to be measured and reported by a UE to gNB.

- CSI Report Configuration

CSI report configuration includes the parameters, and their periodicity, to be reported by a UE to the gNB. The names of CSI parameters (IE: *reportQuantity*) are Channel Quality Indicator (CQI), Precoding Matrix Indicator (PMI), SS/PBCH Block Resource indicator (SSBRI), CSI-RS resource indicator (CRI), RI, Physical Layer RSRP, and Layer indicator (LI). The time-domain information of CSI reporting configuration contains the periodicity of reporting which can be periodic, aperiodic, and semi-persistent. A periodic CSI report can be sent over the PUCCH only; a semi-persistent CSI report can be sent over the PUCCH or PUSCH; and an aperiodic CSI report can be sent over the PUSCH only, which is activated by the gNB over a DCI. The frequency domain information of CSI reporting settings contains the flexible granularity of the frequency band to be measured and reported, which can be a wideband or sub-band, i.e. small portion. Similar frequency band granularity is used in the LTE system also. A semi-persistent or aperiodic CSI report based on a wideband and sub-band measurement can be transmitted over the PUSCH. An example of a wideband measurement includes the combinations of the CSI parameters such as the CRI, RI, PMI, CQI, or CRI, LI, PMI, and CQI. Further, a CSI report configuration includes codebook configuration also, which can be of Type I or Type II.

- CSI Resource Configuration

The CSI resource configuration includes the RS, i.e. CSI-RS (NZP), or Synchronization Signal Blocks (SSB) and CSI-IM along with the BWP to be used for CSI measurement at UE end. A CSI resource configuration contains more than one CSI resource set (RRC IE: *NZP-CSI-RS-ResourceSet*). A UE can be configured with a CSI resource set consisting of a single (with multiple ports, up to Max: 32 ports) or multiple CSI-RS resources, or SS block resource(s). SSB resources are used for beam measurements and reporting purposes. Based on these resources sets, the current DL channel state is measured by a UE and is reported to the gNB. The periodicity of a CSI resource configuration can be periodic, semi-persistent, or aperiodic in type.

- CSI Parameters Reported by UE (RRC IE: *reportQuantity*)

PMI/RI – A PMI and RI indicate the appropriate precoding matrix for a particular rank or layer that may be used for the next DL MIMO transmission, over the PDSCH, by the gNB to the reporting UE. The purpose of a precoding matrix was mentioned earlier in Section 19.6.10.4. A precoding matrix, containing the complex value (j), used for MIMO transmission, is defined in terms of a *codebook*. A codebook is defined for each layer of a MIMO transmission. The number of columns in a precoding matrix is equal to the number of layers used for a MIMO transmission.

Layer Indicator – The LI parameter indicates the particular layer, or column in a codebook, that the UE measures to be the best layer among the layers reported as part of a PMI.

CQI – A CQI from a UE indicates its desired MCS to be used by the gNB for DL transmission over the PDSCH. CQI indices are defined for the different MCSs, i.e. QPSK–256 QAM. Refer to Table 5.2.2.1-2 to 4 in TS 38.214 [109] for CQIs for different MCS. A UE reported CQI is used as part of an NR DL link adaption procedure, which shall be described later in Section 19.6.19.

CRI – The CRI parameter indicates the preferred beam as measured and considered by the UE in case multiple beams transmission is used by the gNB.

SSBRI – The SSBRI parameter indicates the preferred SS block as measured and considered by the UE in case multiple beams transmission is used by the gNB.

A typical reporting of the above CSI parameters from a UE to the gNB is illustrated in Figure 19.71 for four CSI-RSs, with two ports each, and four beams using a single panel antenna structure. As illustrated in Figure 19.71,

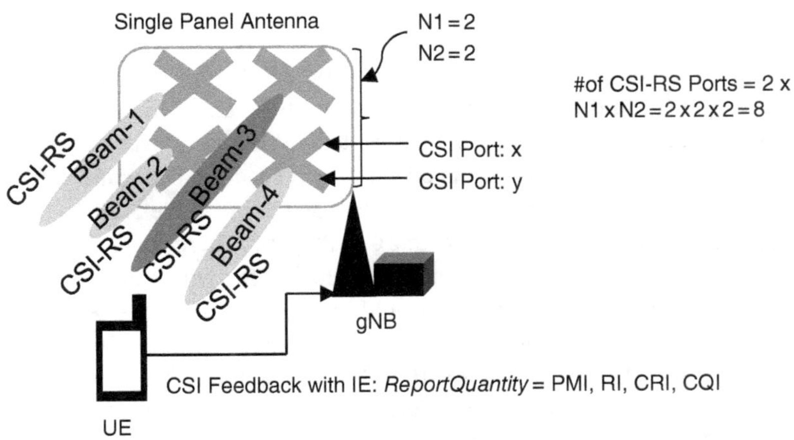

Figure 19.71 Illustration: Type I CSI reporting with a single-panel antenna structure.

the best or strongest beam as considered by the UE is shown to be the dark one, i.e. Beam #3. Thus, the UE will report the CRI along with other CSI such as the PMI and RI of the strongest beam to the gNB.

Types of CSI reporting and antenna ports layout dimensions, N1 and N2, are described below.

- Types of CSI Reporting

Based on the usages for different types of MIMO transmissions, a CSI reporting from a UE to gNB is classified as Type I, for a SU-MIMO transmission, and Type II, for MU-MIMO transmission, as identified by the RRC layer IE: *codebookType*. Further, based on the cross-polarized antenna panel structures being used for MIMO transmission, Type I CSI is divided into a single panel and multipanel CSI. A MIMO antenna panel contains the arrangement of physical antennas having an array of N1 rows and N2 columns. The number of antenna panels (Ng) used, along with N1 and N2, in case of multipanel MIMO transmissions is specified by the RRC layer as part of CSI codebook configuration toward a UE. For a single panel transmission, the N1 and N2 parameters are specified by the RRC layer through its IE: *CodebookConfig ->...n1-n2*. For multipanel transmissions, the N1, N2, and Ng parameters are specified by the RRC layer through its IE: *CodebookConfig ->...ng-n1-n2*.

For a given number of CSI antenna ports (4, 8, 12, 16, 24, and 32 ports) and using the N1 and N2 dimensions, there can be more than one way of arranging the physical antennas on an antenna panel, both single and multiple panels. For example, from Table 5.2.2.2.1-2 in TS 38.214 [109], the physical antennas used for a MIMO transmission can be arranged in (4,4), (8,2), or (16,1) ways on a single panel with 32 antenna ports being used for the CSI-RSs. Refer to Table 5.2.2.2.1-2, for a single panel, and Table 5.2.2.2.2-1, for multipanel, TS 38.214 [109], for the supported MIMO antenna configurations for given CSI-RS antenna ports. In the case of the single panel MIMO transmission, the number of antenna ports used for the CSI-RSs is equal to $2 \times N1 \times N2$. In the case of multipanel MIMO transmission, the number of antenna ports used for the CSI-RSs is equal to $2 \times Ng \times$ `$N1 \times N2$. Figure 19.71 illustrates a Type I CSI reporting with four beams using a single panel antenna structure and eight CSI-RS ports, with two ports for each CSI-RS. Refer to Table 5.2.2.2.2-1, TS 38.214 [109] for the possible layouts of eight CSI-RS antenna ports with $N1 = N2 = 2$.

- Number of Rank Indication in CSI

The maximum rank indication (RI) or transmission layers that can be reported by a UE as part of a codebook-based CSI to the gNB depend on the type of the CSI. For the Type I single panel codebook-based CSI reporting, a UE can report up to eight layers or rank. For the Type I multipanel codebook-based CSI reporting, a UE can report up to four layers or rank. For the Type II codebook-based CSI reporting for MU-MIMO transmissions, a UE can report up to two layers or rank only, i.e. the gNB can transmit over two layers only per UE.

- Codebooks for CSI Reporting

A *codebook* contains a precoding matrix (W), containing the complex value (j), for MIMO transmission. A precoding matrix is used to map the data from single or multiple layers for transmissions through multiple antennae with appropriate weights or scaling factors such as ½ or $1/\sqrt{2}$. A codebook is defined for each layer of MIMO transmission and contains a precoding matrix in a predefined table. A codebook or precoding matrix index corresponds to an index into a predefined table. For example, consider the codebook defined in Table 5.2.2.2.1-1 in TS 38.214 [109], which is used for reporting of a Type I Single panel CSI with 2 CSI-RS ports. There are four indices (0–3) for a one-layer codebook and two indices (0 and 1) for a two-layer codebook. But as the number of layers and the CSI-RS ports increases, the size of a codebook also increases, which would increase the payload size for UL control information. To avoid this, a UE reports certain indices, as part of a Type I and Type II CSI reporting, based on which the gNB computes the desired codebook or PMI to be used for DL MIMO transmission. For such different indices reported by a UE to gNB for deriving a Type I or Type II codebook, refer to TS 38.214 [109]. For example, in the case of Type II CSI reporting, a UE reports indices for sub-band amplitude value set mapping, oversampling factors (O1, O2), combinatorial coefficient, number of beams configured (L), and so on. Refer to TS

38.214 [109] for more information on the entire CSI reporting framework used by the NR and TS 38.331 [115] for CSI-related configurations performed by the RRC layer. For more supplementary information, the reader may refer to the 3GPP TDocs [142–150] listed in the references section.

19.6.18 Modulation and Coding Schemes (MCSs)

As with other communications systems, modulation is another fundamental function performed by the NR physical layer as part of the physical layer channel processing chain, in both the DL and UL directions, described in Section 19.6.10. A modulation process converts the encoded information into complex-valued symbols for further mapping into the REs of a PRB for transmission over an antenna port. Each subcarrier of a PRB can be modulated using different modulation schemes such as the QPSK, 16 QAM, 64 QAM, and 256 QAM to transmit a different number of bits, which is the *modulation order*, of a particular modulation scheme. For example, on an SC, 8 bits of information can be transmitted over an OFDM symbol using the 256 QAM. Similarly, 7 bits of information can be transmitted over an OFDM symbol using the 64 QAM. Thus, as the modulation order increases, the data throughput as well as the spectral efficiencies also increases accordingly over a particular carrier frequency.

Table 19.14 lists the different MCSs used over the NR air interface for different types of physical channels in the UL and DL directions.

From Table 19.14, it appears that the control channels (PDCCH, PBCH, and PUCCH) have fixed MCS at a lower rate, whereas the physical channels, i.e. PUSCH and PDSCH, carrying user traffic are allocated MCS with varying data rates based on radio conditions. Generally signaling information is sent using the lower coding scheme that includes more redundant error recovery information. For example, the QPSK modulation scheme is used to transmit information over the PBCH; the 16 QAM or 64 QAM or 256 QAM modulation scheme may be used to transmit information over the PDSCH from the NG-RAN/gNB to a UE depending on prevailing radio conditions.

- UE Determination of Transport Block Size

The higher the coding scheme, or modulation order, the greater the throughput of data services to a UE and its subscriber as less error recovery but more user traffic/information is transmitted per unit of scheduling. Similar to the LTE system, in the NR system also, information sent by a UE is scheduled in a given TTI. In the LTE system, the TTI is fixed i.e. 1ms. However, in the NR system, TTI varies from 1ms to a fraction of it depending on the subcarrier spacing being used. Over a TTI, up to two transport blocks may be transmitted. But the amount of information to be carried in a transport block is determined from the corresponding MCS index and order which is assigned to a UE through the DCI from the NG-RAN/gNB. The NR physical layer defines 31 such MCS indexes/orders along with their code rates. Refer to Tables 5.1.3.1-1/3 and 6.1.4.1-1/2 in TS 38.214 [109] for the list of MCS indexes defined for the NR DL and UL shared channel. From a given MCS index and the number of RBs allocated to a UE, the corresponding transport block size can be derived in association with Table 5.1.3.2-1 in TS 38.214 [109].

Table 19.14 NR physical channel types and their MCSs.

Channels/Signals	MCSs	Multiplexing
PUSCH	QPSK, 16/64 QAM	Spatial Multiplexing
PUCCH	π/2-BPSK, QPSK	–
PDSCH	QPSK, 16/64/256 QAM	Spatial Multiplexing
PBCH, PDCCH	QPSK	–
Reference Signal	No	No
PSS, SSS	No	No

19.6.19 Link Adaptation Procedure

One of the most troublesome areas of a mobile communications network is the air interface between MS/UE and its radio access network (RAN). Air interface issues arise from poor radio conditions, which are dynamic in nature because of various environmental reasons. This would result in erroneous data at the receiving end and further lead to poor throughput as experienced by mobile users because of retransmission requirements of information between MS/UE and the RAN. The RRM procedure at the RAN end should be able to adapt dynamically to the changing radio conditions by adjusting the current transmission parameters being applied to the existing radio resources which were allocated to a UE. Similar to the previous generation mobile communication systems, the procedure to adapt to the prevailing radio conditions by adjusting the current transmission parameters, between a UE and NG-RAN and vice versa, over the air interface is called *Link Adaptation (LA) procedure*. An LA is performed in DL as well as the UL directions of data transmission between UE and RAN and vice versa. An LA procedure works closely with the scheduling procedure of an NG-RAN.

An LA for the ongoing user/UE data session is performed in terms of the modifications of the existing transmission parameters of an MCS that was allocated by the NG-RAN to a UE. Note that an LA is performed on the channels carrying user traffic only whose MCS can be modified to adapt to the changing radio conditions. However, no LA is performed on the channels carrying signaling information as they are allocated lower order MCS, Table 19.14, which remains fixed with sufficient error correction information so that signaling information is delivered reliably. As a result of the LA procedure, a new MCS with a higher or lower modulation order may be allocated to UE. The modulation order is the number of bits transmitted per modulated symbol or resource element of a PRB. The NG-RAN informs the allocation of a new MCS to a UE as a DCI over the physical layer signaling PDCCH. Each MCS is associated with a transport block size at the MAC layer. Higher the MCS index, the higher the coding rate, transport block size, which would result in a higher throughput over a transmission time interval (TTI). MCS index and its transport block size determination have been described in Section 19.6.18.

- Downlink LA

DL LA procedure at the NG-RAN/gNB works based on the DL channel quality information (CQI) as measured and reported by an MS/UE over the UL PUCCH or PUSCH as part of a CSI to the gNB. A UE measures the quality of the DL RS and reports the CQI index to gNB, indicating the MCS whose transport BLER does not exceed a certain threshold. The gNB may use the UE reported CQI/MCS in the next DL transmission. NG-RAN CQI indexes along with their modulation schemes, and coding rate are defined in Tables 5.2.2.1-2, 5.2.2.1-3, and 5.2.2.1-4 in TS 38.214 [109]. The code rate shown in these tables represents the number of user data or information bits transmitted per 1024 bits. The code rate increases as the modulation scheme increases. This means that the higher the modulation scheme, the lesser the redundancy information (leaving more room for user data) added to the user data by the physical layer during the channel coding process. The efficiency of a particular modulation scheme can be obtained by multiplying the code rate by the modulation order.

The relationship between CQIs and their throughputs is illustrated in Figure 19.72 from QPSK to 64 QAM. The coding rate decreases whenever there is a change in the modulation order from QPSK to 16 QAM and 16 QAM to 64 QAM. Following this, the coding rate increases again as the MCS increases; see Figure 19.72. The coding rate is defined as the ratio of the (MAC layer transport block size + Length of CRC) to the number of RE available for user data transmissions.

- UL LA

UL LA procedure at the NG-RAN/gNB works based on the *UL* SRS sent by a UE to the gNB. Using the SRS measurements, channel quality over a section of bandwidth is measured over the prevailing various environmental conditions, and an appropriate MCS is selected for transmission in the UL direction. UL SRS was described earlier in Section 19.6.12.2.

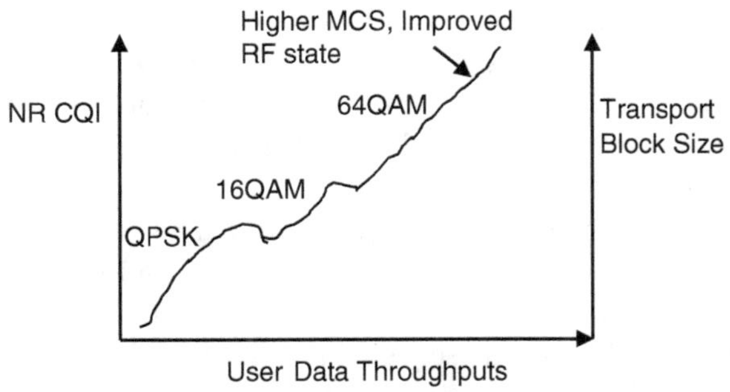

Figure 19.72 Illustration: NR CQI (MCS) vs. data throughput.

19.6.20 Random Access (RACH) Procedure

As mentioned earlier in Section 19.5.3, UE MAC and physical layer perform their tasks associated with an NR random access procedure. Also, the UE MAC layer controls the random access procedure-related tasks performed by its physical layer. Various aspects of the random access procedure concerning the UE physical layer are described below.

- **Purpose of a RACH Procedure and its Resources**

Following the successful selection of a physical layer cell as described in Section 19.6.13, a UE is further required to register itself with the core network to provide communication services to its user. To register with the core network, a UE performs the random access procedure toward the NG-RAN which is similar to the LTE system. Before the RACH procedure, a UE acquires the contents of the System Information Block Type 1 (SIB1) which is sent by the NG-RAN, in the DL direction. SIB1 contains the RACH procedure-related information *(RRC layer: SIB 1 -> ServingCellConfigCommonSIB)* for UEs of the selected cell using which a UE requests the NG-RAN to allocate UL resources to it.

- **Difference Between an NR and LTE Random Access Procedure**

As described earlier in Section 19.5.3, the overall message (Msg1–Msg4) flow of a random access procedure is the same in both LTE and NR systems. However, as far as the mmWave transmissions/receptions are concerned, there is a difference between an NR and LTE random access procedure. In the case of mmWave transmissions/receptions in the NR system, a UE requires to detect and select the best beam, as described earlier in Section 19.6.14. An NR UE initiates a random access procedure over the selected beam. However, no such mmWave transmission/reception and function for detection and selection of a beam exists in the case of the LTE system. Another difference is the possibility of usages of multiple RACH occasions, unlike one RACH occasion in the LTE system, in the NR system within a particular subframe which will be described in the subsequent section.

- **RACH Preamble**

Similar to the LTE system, an NR UE RACH procedure begins by transmitting a preamble in the UL direction toward the NG-RAN. A preamble is generated from cyclic shifts of root Zadoff–Chu sequence which is also used to generate a preamble sequence in the LTE system. The length of a preamble sequence can be 839 or 139, which is also the preamble sequence length used in the LTE system.

There are 64 preambles (0...63) defined for an NR cell. All of the 64 preambles may not be available for use during a contention-based as well as contention-free random access procedure by UE. Some of the preambles may be used for other purposes also. The total number of preambles available for contention-based and contention-free random access procedure is configured by the RRC layer through IE *RACH-ConfigCommon -> totalNumberOfRA-Preambles*, are further grouped into two groups – Group A and Group B – similarly as in the case of the LTE system. Further, the total number of RACH preambles available under Group A is configured by the RRC layer IE: *RACH-ConfigCommon -> numberOfRA-PreamblesGroupA*. Thus, the total number of RACH preambles available under Group B will be *totalNumberOfRA-Preambles – numberOfRA-PreamblesGroupA*. All these parameters are used by the MAC layer as part of its random access procedure at UE end.

A preamble is transmitted over the PRACH using the antenna port 4000. Depending on the SCS used, a PRACH preamble can have two formats: long preamble (sequence length = 839) and short preamble (sequence length = 139), as described in Tables 6.3.3.1-1 and 6.3.3.1-2 in TS 38.211 [106]. Preamble format with sequence length 839 and FR1 SCS 1.25 and 5 kHz has four formats. Preamble format with sequence length 139 and FR1 SCS 15 and 30 kHz and FR2 SCS 60 and 120 kHz has nine formats. Further, preamble formats can be of restricted as well as unrestricted types. Preamble formats 0–3 with sequence length = 839 are in restricted types – Type A and Type B, and preamble formats A1-A3, B1-B4, C0, and C2, with sequence length = 130, are unrestricted type as listed in Tables 6.3.3.1-1 and 6.3.3.1-2 in TS 38.211 [106]. The type of preamble to be used in a cell is specified by the RRC layer through the IE: *RACH-ConfigCommon → restrictedSetConfig*.

- **Preamble Sequence and CP Length**

Similar to the LTE preamble, the NR preamble also consists of a CP and the Zadoff–Chu sequence. The length of the CP and the preamble sequence for each preamble format vary across the SCSs as well as different preamble formats. Further, in a particular preamble format, the preamble sequence is repeated, as described in tables 6.3.3.1-1 and 6.3.3.1-2 in TS 38.211 [106].

Figure 19.73 illustrates the structure of preamble formats, in the case of long sequence (839), with CP (Ncp) and repetition of a preamble sequence (Nu) for SCS 1.5 and 5 kHz. For the actual duration of the CP and preamble sequence, refer to Table 6.3.3.1-1 in TS 38.211 [106].

Example 19.12 illustrates the calculation of the duration of the cyclic prefix and the preamble sequence of a particular preamble format.

- **Time-Domain Resource for PRACH**

Similar to the mapping of other physical channels into time and frequency resources, PRACH used for random access procedure is also mapped to the NR resource grid in the time domain. Time-domain resource information such as the preamble format, subframe, timeslots, start symbol, the number of PRACH symbols within a slot, and so on is provided through RRC layer IE: *RACH-ConfigGeneric -> prach-ConfigurationIndex(0...255)*. A PRACH configuration index points to a particular row of Tables 6.3.3.2-2, 6.3.3.2-3, and 6.3.3.2-4 in TS 38.211 [106] giving the time-domain resources for a RACH procedure under FR1, FR2 as well paired (FDD)and unpaired (TDD) spectrum.

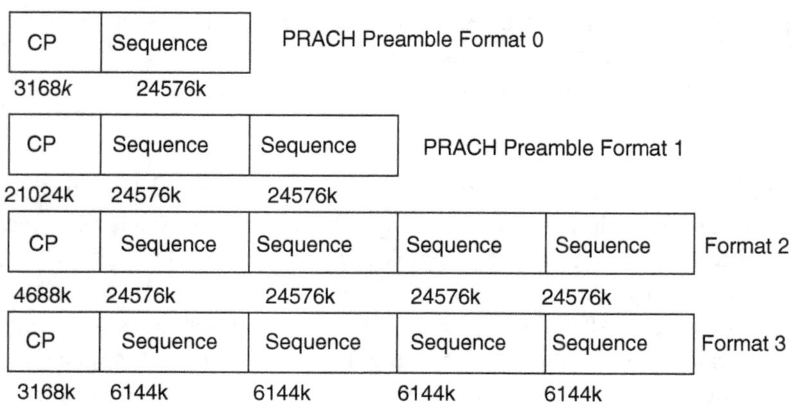

Figure 19.73 Illustration: NR RACH: CP and preamble sequences lengths for SCS: 15 and 5 kHz.

Example 19.12 Consider the Calculation of the Duration of CP and Preamble Sequence for the Preamble Format 3 with SCS 5 kHz.

Time Unit = Tc = 1/(480000* 4096) seconds (TS 38.211 [106], Section 4.1)
From Table 6.3.3.1-1 in TS 38.211 [106]:
Duration of Cyclic Prefix = 3168k = 3168 * 64 Samples = 3168*64*Tc Seconds = (3168 *64)/(480000*4096) = 0.103125 millisecond
Duration of Preamble Sequence = 4 * 6144 *k = 4*6144*64 samples =4*6144*64*Tc Seconds = 0.8 millisecond.

- **Frequency Domain Resource for PRACH**

Frequency domain resources for the RACH procedure are provided to UE over the RR layer IEs: *RACH-ConfigGeneric -> msg1-FDM and msg1-FrequencyStart*. IE *msg1-FDM* provides the number of PRACH occasions which is frequency division multiplexed (FDMed) in a one-time instance. IE *msg1-FrequencyStart provides* the offset of lowest PRACH transmission occasion, within the active UL BWP, in the frequency domain concerning PRB 0.

- **RACH Occasion**

Another difference between LTE and NR system RACH procedure is the usages of multiple RACH occasions in the NR system. A RACH occasion indicates the frequency and time-domain resources or location during which a UE transmits a RACH preamble to the NG-RAN. In the LTE system, there is only RACH occasion, occupying 6 RBs in the frequency domain and 1 subframe in the time domain. On the other hand, in the NR system, multiple RACH occasions are possible in the frequency domain as well as in the time domain. This can be observed by comparing Tables 6.3.3.2-2–6.3.3.2-4 in TS 38.321 [113] and Tables 5.7.1-2 and 5.7.1-3 in TS 36.321 [93].

- Number of RACH occasions in the frequency domain in a one-time instance

The maximum number of RACH occasions, in a one-time instance only, in the *frequency domain* in the NR system is 8, which is specified by the RRC layer using the *RACH-ConfigGeneric -> msg1-FDM*. Refer to TS 38.331 [116].

- Number of RACH occasions in different time instances in the time domain

The RRC layer provides the RACH resources to be used by the UEs through the IE: *prach-ConfigurationIndex*, which is an index into a row of Tables 6.3.3.2-2–6.3.3.2-4 in TS 38.321 [113] for FR1 as well as FR2 and paired (FDD) as well as unpaired (TDD) spectrum. From these tables, a *prach-ConfigurationIndex* provides the following RACH-related resources information to a UE:

a) Preamble format, for short (A1...A3, B1...B4, C0, C2) and long format (0–3);
b) Odd or even frame, which is derived based on the system frame number;
c) Subframe number in case of FR1 with 15 kHz SCS and timeslot within a subframe in case of FR2 with 60 kHz SCS (which may contain a varying number of timeslots, based on an SCS or numerology; refer Figure 19.22;
d) Starting symbol within a slot;
e) Number of PRACH slots within a subframe;
f) Number of time-domain PRACH occasions within a PRACH slot; and
g) PRACH duration.

In the time domain, the number of RACH occasions within a slot ((e) above) is specified under the column *number of time-domain PRACH occasions within a PRACH slot ((f) above)*. A slot may contain one RACH occasion.

As seen from Table 6.3.3.2-4 in TS 38.211 [106], the maximum number of PRACH occasions within a slot is 7. Multiplication of (e) and (f) shall provide the total number of RACH occasions in the time domain.

Example 19.13 illustrates the calculation of the actual starting location of the OFDM symbol for transmission of a RACH preamble for a given preamble format and RACH occasion.

- Number of SSB per RACH occasion

As described earlier in Section 19.6.14 and illustrated in Figure 19.65, an SSB is associated with a particular beam, and a UE requires selecting the best beam as part of its beam selection procedure. A UE transmits the RACH preamble to the NG-RAN over the selected beam within a particular RACH occasion. Thus, the NG-RAN learns about the beam being selected by the UE based on the RACH occasion used to transmit the preamble with the associated RA-RNTI. For this purpose, the NG-RAN RRC layer configures the number of SSB per RACH occasion and preambles per SSB to the UE through the RRC layer IE: *RACH-ConfigCommon -> ssb-perRACH-OccasionAndCB-PreamblesPerSSB*, which can start from OneEight(1/8) SSBs to sixteen (16) SSBs per RACH occasion.

Example 19.14 describes and illustrates the association of SS blocks and distribution of the entire 64 RACH preambles of NR to different RACH occasions.

Example 19.13 Calculation of OFDM Symbol Locations for RACH Preamble Occasions Within a PRACH Subframe and its Timeslots

Let us calculate the OFDM symbol position (t_{start} l) for each PRACH occasion within a subframe and its timeslot(s) where a RACH preamble is being transmitted using FR2 in FDD for a given *prach-ConfigurationIndex* = 41, Table 6.3.3.2-4 in TS 38.211 [106]. The OFDM symbol positions (t_{start} l) can be calculated using the formula described in Section 5.3.2 in TS 38.211 [106].

From Table 6.3.3.2-4 in TS 38.211 [106] and for *prach-ConfigurationIndex = 41*

a) Preamble format = A2
b) Frame: Even (as y = 0)
c) TSLs = 19, 39
d) Starting symbol within a slot(l0) = 5
e) Number of PRACH slot within a subframe n_{slot}^{RA}=1
f) Number of time-domain PRACH occasions within a PRACH slot n_t^{RA}=2, i.e. Two (0,1) RACH occasions in a slot
g) PRACH format A2 duration N_{dur}^{RA} = 4

From Section 5.3.2 in TS 38.211 [106], the actual starting OFDM symbol position for each RACH occasion (total two RACH occasions as per (f) above) with the slot, as per (e) above, is given by the following formula.

$$l = l_0 + n_t^{RA} N_{dur}^{RA} + 14 n_{slot}^{RA}$$

Using the above equation,

Actual starting OFDM symbol location for transmitting preamble with format A2 for PRACH occasion #1: $l = 5 + (0 \times 4) + 14 \times 1 = 19$;

Actual starting OFDM symbol location for transmitting preamble with format A2 for PRACH Occasion #2: $l = 5 + (1 \times 4) + 14 \times 1 = 23$.

The above RACH occasions are illustrated in Figure 19.74.

Figure 19.74 Illustration: NR RACH occasions for FR2 and FDD modes.

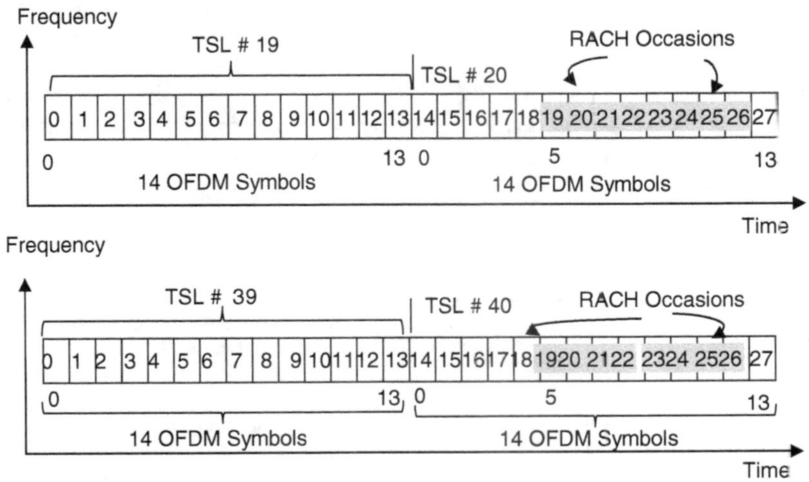

Depending on the actual OFDM symbol location of a PRACH occasion as calculated above, a preamble may be transmitted in the next immediate TSL, e.g. TSL 20 and TSL 40 above, following the slot provided under the *prach-ConfigurationIndex* ((c) above). From the tables mentioned above, different possibilities of RACH occasions in the NR system can be represented graphically as illustrated above under FR1 and FR2 with paired and unpaired spectrum used in different cells.

- **Processing RAR**

After transmitting a RACH preamble, a UE is expected to receive a RACH response from the NG-RAN within the RACH response window *(ra-ResponseWindow)* time. For this purpose, the UE physical layer monitors the PDCCH from the NG-RAN for a DL control information (DCI 1_0) whose CRC is scrambled with the corresponding RA-RNTI which was transmitted by UE along with preamble. If the UE physical layer receives a transport block, correctly sent over PDSCH, it will pass the same to the MAC layer for its further processing. On the other hand, if the UE physical layer does not receive a RAR within the RACH response window time, the UE physical layer will retransmit the PRACH to the NG-RAN. Thus, a separate acknowledgment, i.e. HARQ-ACK from UE to NG-RAN, is not required for a successfully processed RAR decoded from the PDSCH transmitted by the NG-RAN. A PRACH may be also retransmitted if the transport block provided over the PDSCH as part of RAR is not correct. For more information on the task performed by the physical layer as part of the processing of a RAR, refer to TS 38.213 [108].

- **Acknowledging on Successful Decoding of CRI**

On successfully decoding its own CRI from the DL PDSCH as described earlier in Section 19.5.3 and Figure 19.10, the physical layer of a UE responds with a UL control information, i.e. HARQ-ACK over PUCCH to the NG-RAN, as part of a successful contention-based random access procedure. For more information on the task performed by the physical layer as part of a CRI, refer to TS 38.213 [108].

19.6.21 NR Radio Resources Management (RRM) Procedure

One of the important functions performed by the base station of a mobile communication network, in general, in uplink and downlink directions is the Radio Resources Allocations and Managements (RRM) procedure. An RRM

Example 19.14 Association of NR SS blocks to different RACH occasions.

Consider the case of 4 RACH occasions #0–3 in the frequency domain, i.e. *RACH-ConfigGeneric ->msg1-FDM = 4* and SSB per RACH occasion as well as preambles per SSB as 4, i.e. *RACH-ConfigCommon–>ssb-perRACH-OccasionAndCB-PreamblesPerSSB = 4*. The associations of SSBs to different RACH occasions in frequency (*msg1-FDM*) and time domain for the entire 64 preambles available in the NR system can be represented as illustrated in Figure 19.75 by following the rules mentioned in Section 8.1 in TS 38.213 [108].

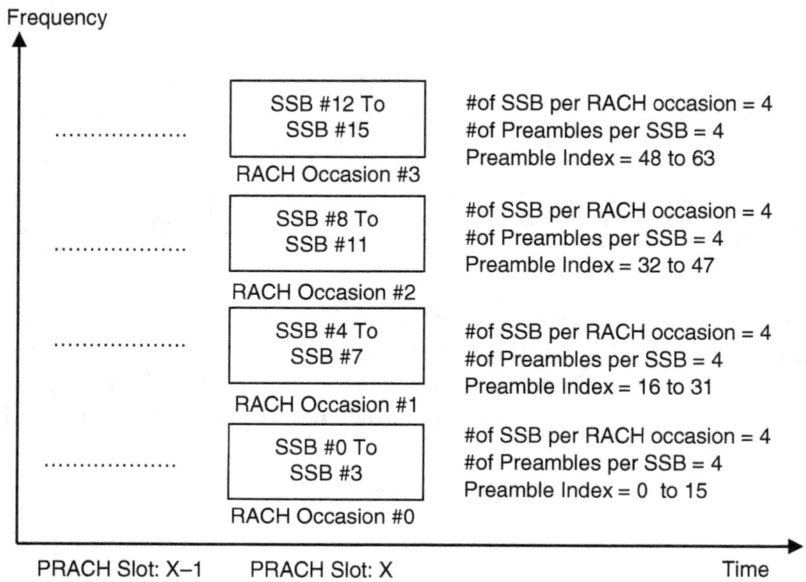

Figure 19.75 Illustration: NR RACH occasions in frequency domain.

procedure deals with the allocation, re-allocation, and release management of the precious and shared radio resources or spectrum among the users/UE of a mobile communication network in uplink and downlink directions. An RRM procedure is an implementation-dependent and system-specific (i.e. GSM, GPRS, UMTS, E-UTRA, and NG-RAN) procedure that is performed by a base station of a mobile communication network in association with an MS/UE in its idle, inactive (5GS only), and connected state. An RRM procedure is required to be an efficient one for optimum utilization of the system as well as the radio spectrum resources of an operator. This is especially important for the 5G NR system, considering its different network slices using which different types of communication services under its different use cases are provided to users. Thus, an RRM procedure plays a critical role as it may directly impact the performance of its base station as well as the quality of services to users.

The previous generation of mobile communication systems and networks were a kind of one-size-fits-all network, providing primarily voice and data services, requiring a simpler RRM procedure. However, the 5G system provides different types of services, through different network slices, with specific QoS characteristics and requirements under a particular use case – i.e. eMBB, or URLLC, or mIoT, etc. Further, each use case of the 5G system serves different industry verticals and deployment scenarios (indoor, private network, etc) with different types of traffics generated by heterogeneous devices.

The entire RRM procedure of the 5G system, which consists of several collective sub-tasks or areas that are responsible for different areas/aspects of an RRM procedure, is illustrated in Figure 19.76. Considering the different use cases, industry verticals, and deployment scenarios along with the usages of multiple subcarrier spacings architecture and mmWave transmissions and receptions over NR, the network slice-aware RRM procedure for

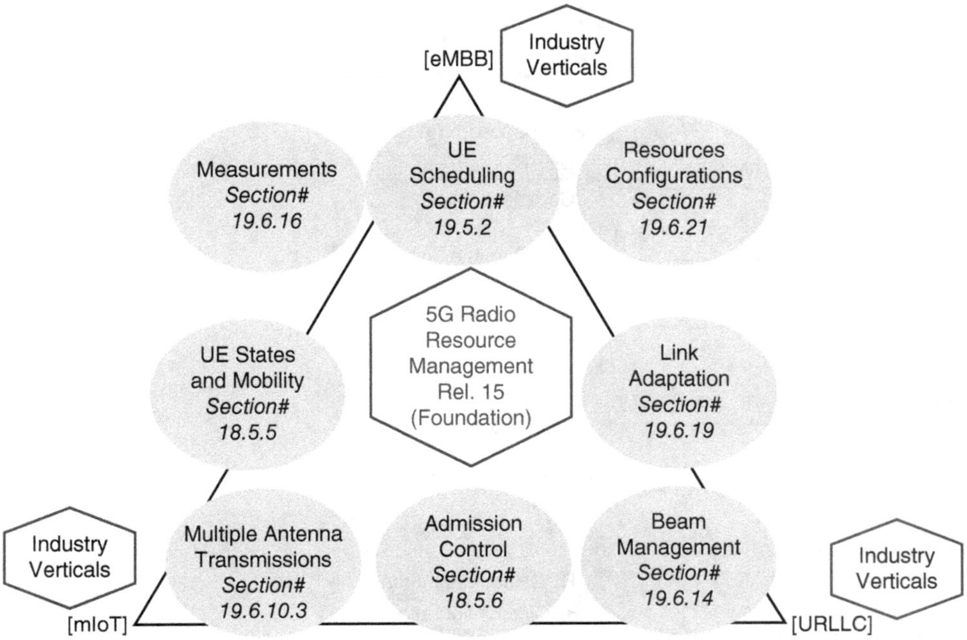

Figure 19.76 Illustration: 5G NR RRM procedure and its sub-tasks for different use cases.

the 5G NR system is much more complex than the previous generation's mobile communication system RRM procedure. Except for the 5G system–specific RRM procedure and its sub-tasks or areas as illustrated in Figure 19.76, the remaining RRM procedure and its sub-tasks or areas are found, in general, across the different systems, i.e. GSM/GPRS, UMTS, LTE, etc. However, the typical RRM sub-tasks or areas shown in Figure 19.76 are system implementation–specific across the GSM to the 5G system. The NR beam management, as described earlier in Section 19.6.14, is new to the 5G system and its RRM procedure.

Different tasks associated with an NR RRM procedure, except the resources configuration shown in Figure 19.76, have been already described earlier under different sections with respect to the 5G system. The corresponding section of an RRM sub-task is also indicated in Figure 19.76, against each sub-task. From these descriptions, it is clear that the entire RRM procedure spans across several layers of the NR, i.e. RRC, MAC, and physical layer, and their technical specifications.

The typical resources configurations task (see Figure 19.76) associated with an NR RRM procedure used in a 5G system is described in the following text.

- Radio Resources Configurations and Allocations

The radio resources for different cell sites are planned and configured by the network planning and optimization (RNPO) group of a network operator. Using these radio resources, the RRM procedure of a base station grants admissions and allocates resources to UEs dynamically. Further, radio resource modifications, i.e upgrade or downgrade, are also performed dynamically by the RNPO group. Apart from the RRC, MAC, and physical layer and their processes, the entire RRM procedure may consist of several other processes, each one handling a particular area, such as an uplink scheduler, downlink scheduler, Link Adaptation, intra-RAT or inter-RAT RF measurements, etc., of the RRM procedure. Figure 19.77 illustrates a central O&M RRM process at the NG-RAN end, which configures the network slice–specific or slice/service type (SST) radio resources, in terms of BWP, to the other NR RRM processes at the gNB end. A BWP can be an initial, default, or an additional/dedicated type, as described earlier in Section 19.6.7.

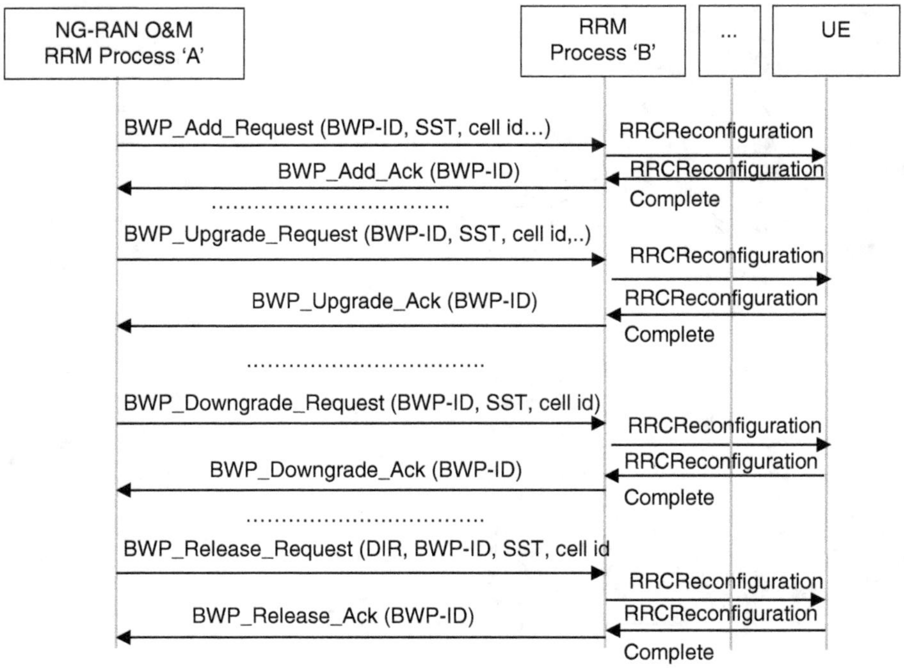

Figure 19.77 Illustration: Configuration of RRM Messages Flow in NG-RAN/gNB

Figure 19.77 also illustrates the conceptual (as well as implementation-dependent message) flow between two NR RRM processes for allocation and modifications (upgrade/downgrade) of an additional BWP to a UE as part of the RRM procedure. Releasing of a BWP is also illustrated in Figure 19.77.

The receiving RRM process at the gNB end will pass-on slice-specific radio resources information to other RRM processes, say the RRC layer, of the gNB. The RRC layer, using its signaling messages, configures and provides radio resources information to a UE or UEs, through system information, in a cell. Such a signaling message includes common, i.e. cell-specific initial BWP, or dedicated, i.e. UE specific additional BWP, radio resources information as part of the NR RRM procedure. The RRC layer signaling message and its IE: *downlinkBWP-ToAddModList or uplinkBWP-ToAddModList* are used to allocate an additional BWP to a UE in the downlink or uplink direction. An existing BWP may be also modified – i.e. increase or decrease the bandwidth of the bandwidth part using the same RRC layer signaling message. On the other hand, a BWP may be released using the RRC signaling message and its IE: *downlinkBWP-ToReleaseList or uplinkBWP-ToReleaseList*, and may be re-assigned to other traffic purposes. To ensure resources (i.e. BWP), isolations, and protections among the different traffic classes, a corresponding network slice/service type (SST) information (e.g. eMBB, URLLC, mIOT) may be added as part of a BWP information.

- UE-specific RRM procedure

The NR RRM procedure may also apply UE-specific configuration information while allocating radio resources to it. One such configuration information is the *Index to RAT/Frequency Selection Priority*, called an *RFSP index* (refer to TS 23.501[40]). An RFSP index is provided by the AMF to NG-RAN, which is used to prioritize the allocation of resources by the NR RRM procedure. The RFSP index is used while UE is in an idle state (to help in cell reselection), or connected state (to help in a handover to different RAT).

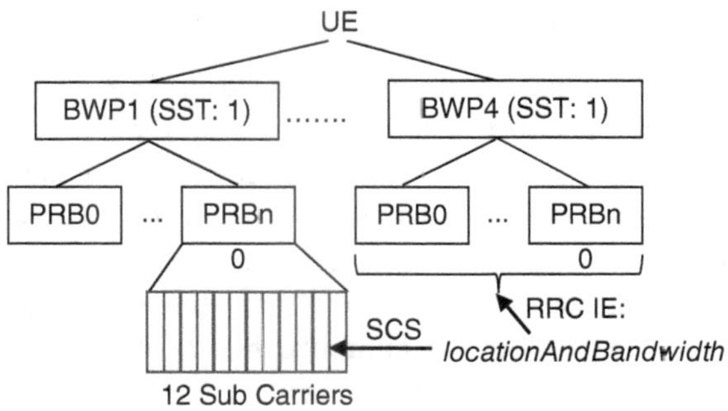

Figure 19.78 Illustration: RRM Allocation of Four BWPs to a UE for eMBB Use Case

- Radio Resources Allocation to UE

As mentioned earlier, the NR RRM procedure deals with the allocation, re-allocation, and release management procedures of radio resources dynamically at the cell as well as UE level, which may spread across the RRC, MAC, and Physical layer. Figure 19.78 illustrates the allocation of four BWPs to a UE for the eMBB use case purpose only (Slice/Service Type: SST=1).

Radio resource isolation in NR is important to ensure that the services of one use case and its network slice do not impact the services of another user case and its network slice. Thus, the 5G NR RRM procedure must be a network slice–aware procedure. The bandwidth or the number of PRBs in a BWP and its subcarrier spacing to be used is provided through the RRC Layer IE: *locationAndBandwidth* from the gNB to a UE or UEs in a cell.

- Evolutions of the 5G RRM procedure

The NR RRM procedure and its sub-tasks described earlier are the foundations for the 5G system as far as its Release 15 is concerned. The 5G system is further evolving into subsequent 3GPP releases, i.e. Release 16, Release 17, and beyond, with enhancements of existing features as well as introductions of new features, as described in Chapter 22, on top of Release 15. In parallel to this evolution, the NR RRM procedure and its associated sub-tasks illustrated in the preceding text will also evolve accordingly, creating new requirements on top of the NR Release 15 RRM procedure to support new industry verticals, deployment scenarios, etc. The NR RRM of Release 15 shall act as the foundation RRM procedure for the subsequent 3GPP releases of the 5G system. Due to this, the NR Release 15 RRM procedure should be flexible/re-usable enough to customize and incorporate release-specific enhancements or new features, as described in Chapter 22, being introduced in the subsequent releases of the 5G system. The evolution of the NR RRM procedure for subsequent releases of the 5G system is illustrated in Figure 19.79.

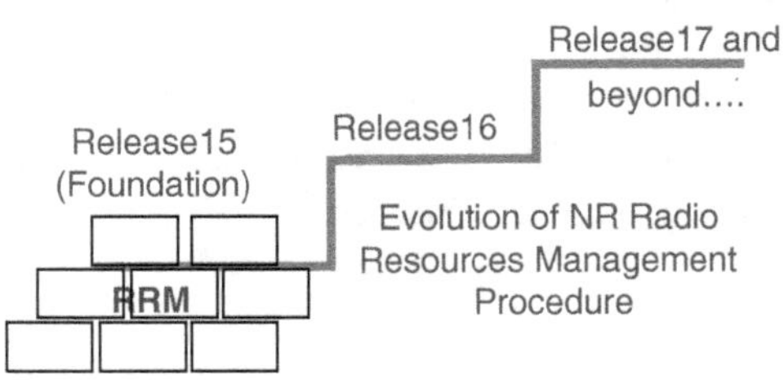

Figure 19.79 Illustration: Evolution of the 5G NR RRM Procedure

19.6.22 UE Transmit Power Control

One of the important functions performed by the base station in a mobile communications system is the UE transmit power control (TPC) procedure that is required to maintain a proper radio link between an MS/UE and a base station. Mobile devices at a distance location require to transmit with a higher power in the UL than those closer to the base station. It is the responsibility of the base station power control procedure to ensure that all the mobile devices, especially those closer to the base station, do not transmit with the same power which would otherwise affect telecommunication services in the serving as well as interfere with the neighbor cells. The goal of a power control procedure is to minimize the transmitting, TX, the power required by the MS/UE or the base station while, at the same time, maintaining a good radio link between an MS/UE and base station/RAN element so that the information exchanged between them can be decoded successfully. Apart from this, a proper power control procedure between a UE/MS and the RAN is applied to reduce the impacts of the following in the serving network:

- Interference to other mobile users and neighboring cells;
- Spectrum inefficiencies;
- Triggering of unnecessary handover operation from one cell to another; and
- The battery power consumption of a UE.

UE TPC procedure is applied across the different mobile communication systems. However, the procedure differs from GSM to the NR system. For example, though UE transmits power control procedure is similar in the LTE and NR systems, there is a fundamental difference between them. Refer and observe the formula used for the calculation of transmit power by a UE in LTE and NR systems in TS 36.213 [91] and TS 38.213 [108]. This is due to the beamforming transmission method being used in the case of the NR system. In the NR system also, a UE estimates a DL path loss as part of its transmission power adjustment. However, such a path loss calculation uses the beam-based RS also which can be based on either SS/PBCH block or CSI-RS. A set of RSs (RRC IE: *PUSCH-PathlossReferenceRS/PUCCH-PathlossReferenceRS*) to be used by a UE in path loss calculation is provided by the NG-RAN as part of its power control configurations/parameters to a UE. UE TPC parameters, RRC IE: *PUSCH/PUCCH-PowerControl*, are intimated along with the UL radio resources allocation messages sent to an MS/UE from the NG-RAN. An exclusive command for power control procedure may be also sent from the base station/RAN element to an MS/UE.

19.6.22.1 Types of Power Control Procedure in NR

In the NR system, UE TPC is applied for transmission over the physical channels – PUSCH PUCCH and PRACH – as well its physical signal – SRS. Based on the availability of feedback information from the NG-RAN to a UE, two types of UE TPC procedures are applied in an NR network.

- Open Loop Power control

As the name implies, this procedure has no feedback information from an NG-RAN to a UE. It is used during the initial network access through the random access procedure made by an MS/UE with initial allowed power. Similar to the LTE system, in 5G also, a UE sends the initial RACH preamble to make a random access request with the initial power specified by the *preambleReceivedTargetPower* parameter. A UE may retransmit the RACH preamble with ramped-up power, as specified by the parameter *powerRampingStep*. These parameters are communicated by the NG-RAN, i.e. RRC layer, as part of its system information to the UEs in a serving cell. Similarly, UE also adjusts its SRS transmission power in the UL direction.

- Closed-Loop Power control

In a closed-loop UE power control method, the NG-RAN/gNB communicates and controls the transmit power of a UE or group of UEs in the UL direction through a 2-bit long TPC command that is included as part of a DL control information to a UE; refer to TS 38.212 [107]. A UE adjusts its transmission power to be used for the PUSCH or PUCCH, over the active BWP, as per the feedback received in a TPC command from the NG-RAN/gNB. Refer to Tables 7.1.1-1 (PUSCH) and Table 7.2.1-1 (PUCCH) in TS 38.213 [108] for more information on the mapping of TPC command to power adjustment to be applied by a UE for its PUSCH and PUCCH transmissions.

19.6.22.2 UE Transmit Power Determination Procedure in NR

In a serving cell and for a particular transmission occasion over the physical channels – PUSCH, PUCCH, and PRACH – and physical signal – SRS, a UE uses the minimum of the following power value as its transmit power:

- Maximum transmit power configured and allowed by the NG-RAN for a carrier frequency: f in a serving cell: c, in association with TS 38.101 [102] and
- Transmit power calculation and adjustments.

The following parameters are taken into account, in general, to calculate the UE transmit power calculation with adjustments. For the formulae and their actual parameters to be used for the determination of a UE transmit power over the PUSCH, PUCCH, RACH, and SRS in its serving cell, refer to Section 7 in TS 38.213 [108].

1) Target power to be received by NG-RAN/gNB, which is derived from several power calculation components as configured through RRC layer signaling IE: *PUSCH/PUCCH-PowerControl*;
2) The amount of UL resource allocation in terms of RBs;
3) DL path loss calculation, taking into account beam-based RS, i.e. SS/PBCH block or CSI-RS, and its index represented by q_d in the power calculation formula;
4) Transport format, i.e. MCS to be used;
5) Power adjustment, taking into account the active UL BWP, its carrier frequency, and the TPC command received through a DL control information from the NG-RAN.

If the calculated transmit power is smaller than the maximum transmits power configured by the NG-RAN, then a UE will use the calculated transmit power for UL transmission; otherwise, the UE will use the maximum transmit power configured by the NG-RAN.

From the comparisons of a UE TPC procedure and their formulae used in LTE and NR systems, it can be observed that in the case of the LTE system, the UE power control procedure is applied over a subframe only in a serving cell. On the other hand, in the NR system, the UE power control procedure is applied in smaller granularity, i.e. at a UL BWP level with different SCSs (μ). Further, in the case of NR, a UE transmit power calculation also depends on the type of the UE UL scheduling (indicated by j in the formulae), which can be a dynamic grant or CS grant (*ConfiguredGrantConfig*), being used for granting UL resources to the UE.

19.6.23 Effect of Physical Layer on Data Throughputs

The throughput of user data transmitted over the NR air interface may depend on the several typical factors of the physical layer, as highlighted below:

- Amount of radio resources allocated to the physical DL and UL shared channels;
- MCS assigned to a UE to modulate a resource element of an RB;
- Amount of redundant information added by the physical layer for error recovery, at the receiver end, even within a particular MCS;
- Usages of spatial multiplexing, e.g. 8×8, 4×4, 3×3, and so on, through multiple antenna transmissions; and
- Amount of radio resources consumed by the various physical signals such as the DMRSs.

As described and illustrated earlier in Section 19.6.12, a PRB and its RE also carry the information of various physical signals and channels over the NR air interface. The number of RE occupied by the physical signals and channels is different, which further reduces the RE available for transmission of actual user traffic. Thus, the resource elements consumed by the individual physical signals and channels are considered as the overhead, which collectively reduces the overall bandwidth or throughput offered to users within a particular system bandwidth.

Chapter Summary

This chapter presented the NR air interface user plane protocol layers of the 5G system. The reader has learned that the NR Air interface Layer 2 consists of four sub-layers, namely the SDAP, PDCP, RLC, and MAC layer, whose functions and procedures are configured by the NR RRC at Layer 3. 5G NR air interface protocol layers perform functions and procedures that are similar to the LTE air interface protocol layers in several aspects. However, the NR air interface has also introduced several new functionalities based on which it envisages providing different services or uses cases as described in Chapter 16. These new functions and procedures greatly differentiate the 5G NR air interface even from the LTE system air interface.

As described in this chapter and to provide low-latency communication services by the NR system, some of the functions and procedures performed by RLC and MAC layers have been enhanced or reorganized in comparison to the same protocol layers found in the LTE system. Formats of PDU for different protocol layers were described in this chapter and indicated how such reorganization aims to achieve low-latency communication services by the NR system. QoS model supported in the 5G system and differences in comparison to the LTE system was described.

There are startling differences between the 5G NR and LTE systems at the physical layer level. Multiple SCSs or numerologies with different transmission bandwidths and the structure of RBs, transmission, and reception in terms of active bandwidth used in the 5G NR system were described. Transport code block group-based retransmission, which is a new functionality in the 5G NR system, was described. Another major difference from the LTE system is the usages of mmWave transmissions/reception through the beamforming method, as described in this chapter. Various aspects of beam management and its impacts on the NR RACH procedure were also described.

20

5G Core Network Architecture

Introduction

The 5G core (5GC) network architecture also differentiates the overall 5G system from the core networks of the previous generation mobile communication systems. First of all, in 5GC, the control plane as well as user plane functions performed by a core network element are separated into software-based entities that are based on the 3rd Generation Partnership Project (3GPP) Release 13 feature called Control Plane and User Plane Separation (CUPS). Further, unlike the previous generation's core network systems, the 5GC network entities or functions work in a service-oriented model where each network entity or function produces/consumes different network services. The 5GC network functions (NFs) communicate over HTTP, which is a new introduction into the core network of a mobile communication system by 3GPP. Also, 5GC enables the provisioning of different services with different Quality of Service (QoS), as described in Chapter 18.

This chapter covers some of the key concepts of the 5GC and begins with the Release 13 feature called CUPS as the basis for the 5GC. Then, the service-based architecture (SBA) of 5GC is described as the framework of operation of different NFs. The chapter ends with the description of network function virtualization (NFV) that enables a 5G network as the means for scalability, maintenance, and network slicing to provision different services requirements.

20.1 Control Plane and User Plane Separation – CUPS

Till the 3GPP Release 13, the Long-Term Evolution (LTE)/Evolved Packet Core (EPC) network entities, i.e. Serving Gateway (SGW) and Packet Data Network Gateway (PGW), performed control plane as well as user plane functions such as the allocation of an IP address to a User Equipment (UE), bearer management, GPRS Tunneling Protocol (GTP) path managements, and forwarding of user plane data of a UE/subscriber. Performing control plane as well as user plane functions by a single EPC network element has some form of limitations in terms of non-flexibility in deployment as well as non-scalability to increase the existing network capacity/sizing requirements that may arise from time to time. To alleviate such limitations, 3GPP introduced the CUPS feature, which is an architectural enhancement of LTE/EPC network elements, as part of its Release 14 feature. The control plane as well as user plane tasks performed by an LTE/EPC network entity have been separated/decoupled into two software-based entities/nodes of the same network element. For example, the control plane functions of SGW are performed by the SGW-C entity. Similarly, the user plane functions of SGW are performed by the SGW-U entity. The control plane functions of PGW are performed by the PGW-C entity. Similarly, the user plane functions

Mobile Communications Systems Development: A Practical Introduction to System Understanding, Implementation, and Deployment, First Edition. Rajib Taid.

of PGW are performed by the PGW-U entity which is connected to an external IP network. Refer to TS 23.214 [34] for the tasks performed by the LTE/EPS SGW and PGW before and after the CUPS architecture.

CUPS architecture enables a *Software-Defined Networking (*SDN*)* capability for a 5GC network, which results in a flexible core network in terms of enabling an operator to provide services dynamically to meet the businesses' and consumer's needs. From an OEM/system vendor and operator points of view, some of the benefits provided by the CUPS architecture/feature of a core network are as follows:

- Evolution and enhancement of the control plane functions and procedures independently of user plane functions and vice versa;
- Expansion or scaling of network capacity through the deployment of additional user plane nodes only to meet increasing or on-demand data traffic;
- Deployment of user plane nodes in centralized or distributed topology, that is, nearer to a Next-Generation Radio Access Network (NG-RAN) and UE, which would reduce time, in terms of latency, to transfer user application data; and
- An OEM/system vendor may provide a user plane node as a separately licensed solution to an operator.

CUPS feature becomes the baseline/reference architecture for the 5G core network also where the control plane and user plane are separated into distinct entities or NFs. A single control plane entity of a network element can be deployed to cater to low signaling activities. On the other hand, more than one user plane entity/node can be deployed to meet heavy user traffic demands or to meet the scalability requirement of a 5G network. Due to multiple deployment scenarios of a user plane entity, its control plane entity uses a selection method to select a user plane entity. Thus, a control plane entity uses the deployment scenario of a user plane entity as one of its selection criteria. Also, the control plane entity provides information or rules which control the tasks, such as packet detection and forwarding functions and policy and charging control (PCC) functions performed by a user plane entity. A control plane entity, in CUPS, can control multiple user plan entities.

20.1.1 Impacts of CUPS Feature

CUPS feature introduces new reference points and protocol between its control plane and user plane nodes, as described below.

- New Reference Points for CUPS

  As part of CUPS feature and architecture, 3GPP defines the following new logical reference points, on top of the User Datagram Protocol (UDP)/IP network, for communication between the separated control plane and user plane functions of a network element.

  - Sxa reference point between the LTE/EPC SGW-control plane and SGW-user plane;
  - Sxb reference point between the LTE/EPC Packet Data Network (PDN) gateway control plane and PDN gateway–user plane; and
  - N4 reference point between 5GC Session Management Function (SMF) and User Plane Function (UPF) NFs.
  - New Protocol – Packet Forwarding Control Protocol (PFCP)

Over the Sxa, Sxb, and N4 reference points, the PFCP is used to exchange user plane controlling information between the separated control plane and user plane network elements/nodes of LTE/EPC or NFs of 5GC. User plane controlling information from the control plane function could be packet forwarding rule, lawful interception, and charging control to be used by the user plane node or network function. To exchange such information, certain procedures using the PFCP are performed between the separated control plane and a user plane of 5GC network function or LTE/EPC network elements/nodes. Such PFCP procedures are of two types – node management procedures and session management procedures.

PFCP Node management procedures are used to manage LTE/EPC network nodes or 5GC NFs/entities. Typical PFCP node management–related procedures are: PFCP node association setup, update, release, heartbeat information, load control, and overload control information. PFCP session management procedures between the control plane and user plane function are used to manage sessions, such as its creation, update, or removal, between the separated control plane and user plane NFs of 5GC (i.e. SMF and UPF) or LTE/EPC nodes. A PFCP association, i.e. a node management procedure, between the control plane and user plane function is required to be setup prior to establishing a PFCP session on the user plane function. The control plane function provides the necessary information, as part of a PFCP session creation, to the user plane function on how to handle user plane data/traffic for a PFCP session. A PFCP session is uniquely identified by a Session Endpoint Identifier (SEID). Refer to TS 29.244 [69] for more information on these PFCP procedures.

- Protocol Stack of PFCP

Protocol stack-wise, PFCP control plane messages, i.e. node management and session management, are transported on top of UDP/IP (port 8805). On the other hand, user plane data between the control plane and user plane NFs or network nodes are forwarded using GTP-U on top of UDP. Refer to TS 29.244 [69] for more information on the PFCP protocol stack and message header contents and their formats.

Typical messages exchanged between the control plane and the user plane entities, as part of the CUPS feature, of LTE/EPC and 5GS network elements using the PFCP are illustrated in the next two sections.

20.1.2 CUPS in the LTE/EPC Network

Figures 20.1 and 20.2 illustrate the differences in the typical signaling message flow for a UE-initiated network ATTACH request procedure, along with PDN connectivity request, sent to an LTE/EPS network. Figure 20.1 shows the partial signaling message flow for normal and without the CUPS feature. Figure 20.2 shows the partial signaling message flow with the CUPS feature with PFCP signaling messages, showing the interactions between the control plane and user plane nodes.

In the case of the CUPS feature (Figure 20.2), more than one user plane node may associate with a control plane node. Due to this, a control plane selects a particular user plane node. SGW control plane node selects an SGW user plane node. Similarly, the PGW control plane node selects an SGW user plane node. The respective control plane node creates a PFCP session, e.g. *Sx Session Establishment Request*, toward the respective user plane node, as illustrated in Figure 20.2. In response, the user plane node creates the session and replies *Sx Session Establishment*

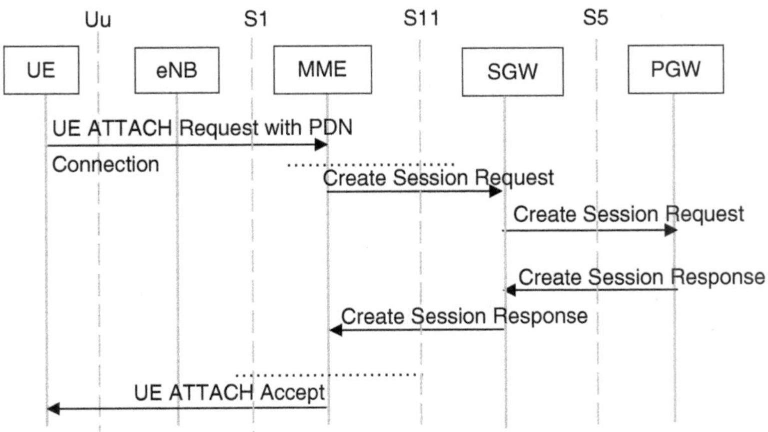

Figure 20.1 Illustration: LTE/EPS UE ATTACH procedure without CUPS.

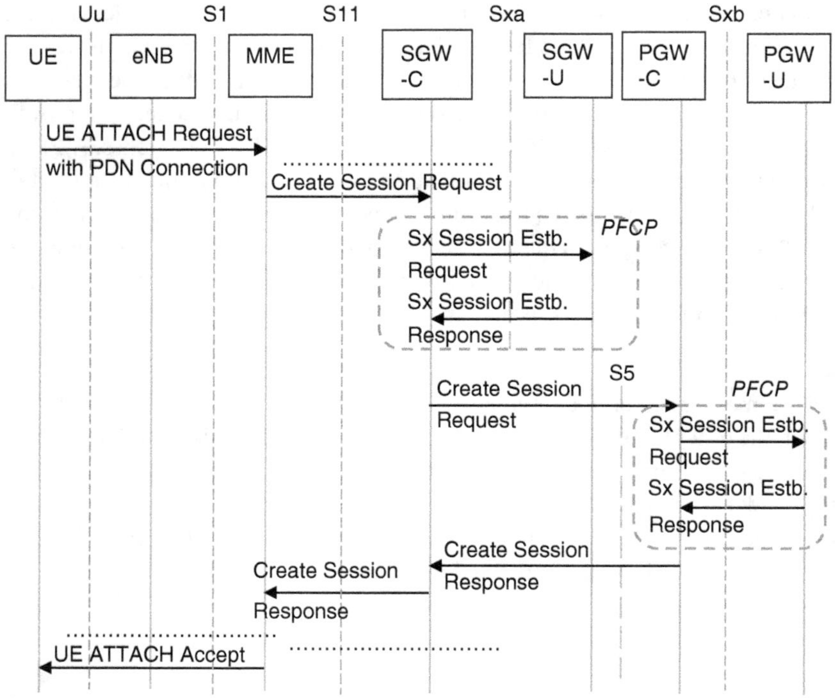

Figure 20.2 Illustration: LTE/EPS UE ATTACH procedure with CUPS feature.

Response to the control plane node. Refer to TS 23.214 [34] for information on the user plane selection procedure and PFCP session procedures over the Sx reference points.

A control plane node provides the following rules, as part of the packet forwarding model over the N4 or Sx reference points, to the user plane node through the PFCP session creation message, as illustrated in Figure 20.2. These rules are used by the user plane node to process and transfer user data/packets through different QoS flows.

- Packet Detection Rule (PDR): This rule specifies how to detect incoming packets/PDUs and how to classify the data. A PDR contains the packet detection and classification information in terms of filters, e.g. IP filters. Separate PDRs are used in uplink and downlink directions.
- Usage Reporting Rule (URR): This rule contains information on the measurement mechanism to be applied by a user plane node on the usage of network resources, e.g. data volume, and its reporting to the control plane node.
- Forwarding Action Rule (FAR): This rule contains information on how to forward user data packets in the downlink (to NG-RAN) or uplink (to DNN) direction, dropping, or buffering user data packets by the user plane node.
- QoS Enforcement Rule (QER): This rule contains information on QoS enforcement to be applied to user traffic.

Each of the preceding rules contains information that controls the user data/packets processing in the user plane function. Refer to TS 23.214 [34] for more information on these packet forwarding rules applied on a PFCP session used by the user plane node.

20.1.3 CUPS Feature in 5G Core Network

Based on 3GPP Release 14 CUPS feature/architecture, the 5GS also separates its core NFs into the control plane and the user plane functions, which will be described in this section. A comparative study of the LTE/EPC and 5GC network elements is described below with respect to the CUPS architecture.

Recall that the LTE/EPC Mobility Management Entity (MME) performs UE control plane functions, i.e. mobility as well as the session management functions (SMFs). But the 5GC *Access and Mobility Management Function* (AMF) performs the UE mobility management functions only. The 5GC *Session Management Function* (SMF) performs UE session management tasks. Note that the SMF also performs similar functions to the SGW-C of LTE/EPC. The LTE/EPC SGW-U and PGW-U perform the UE/user plane functions. But in 5GS, UE user plane tasks are performed by the User Plane network Function (UPF) only.

In the LTE/EPS, UE mobility and session management-related Non-access Stratum (NAS) signaling messages terminates at the MME end. In 5GS, UE mobility management-related NAS signaling messages terminate at the AMF end, whereas the session management-related NAS signaling messages terminate at the SMF end. The SMF communicates session management-related information to a UE via the AMF only. The control plane NFs (e.g. AMF and SMF) of 5GC communicate with each other through service-based interfaces using the HTTP protocol, which is a new introduction into the 5GC by the 3GPP. Unlike the previous-generation core network systems, no GTP control plane protocol is used in the 5GC. However, the communication between the control plane, i.e. SMF, and user plane, i.e. UPF, NFs takes place using the PFCP to exchange control plane messages between them.

Figure 20.3 illustrates the interaction between the SMF (control plane) and UPF (user plane) NFs over the N4 reference point as part of the CUPS feature. This is an illustration of a UE-initiated typical protocol data unit

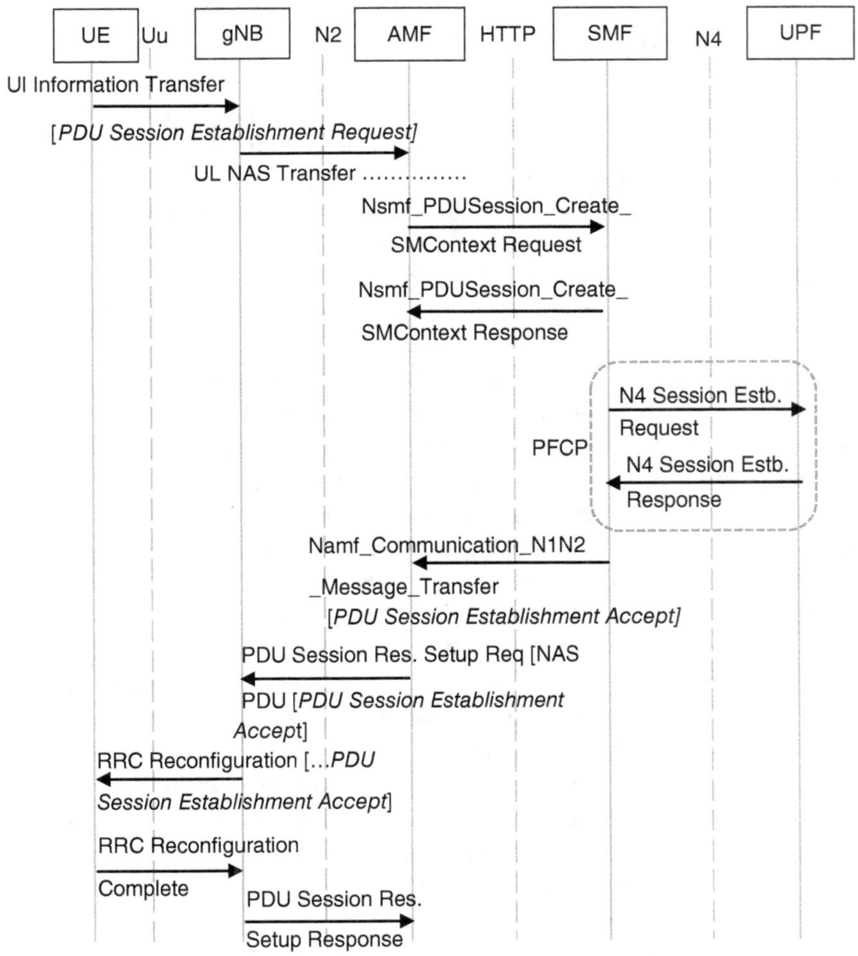

Figure 20.3 Illustration: 5GC CUPS: interaction between control plane and user plane network functions during UE-initiated PDU Session Establishment Request.

(PDU) session establishment request made to the 5GC. This figure also illustrates the exchange of PDU session-related messages between the AMF and SMF through service-based interfaces, for example, *Nsmf_ PDUSession_ Create_SMContextRequest*. Service-based 5GC architecture is described in the next section.

As illustrated in Figure 20.3, the UE sends the UL NAS TRANSFER with Payload container = *PDU Session Establishment Request* NAS layer message, *Payload container type = N1 SM Information, Request Type = Initial Request*, to the 5G Base Station (gNB), which is further forwarded to the AMF. Since it is an initial request from the UE, the AMF forwards the *PDU Session Establishment Request* using the service interface, *Nsmf_ PDUSession_Create_SMContext Request,* of SMF. The SMF sends the *N4 Session Establishment Request*, containing the PDR, FAR, QER, and URR, to the UPF using the PFCP protocol. UPF responds with *N4 Session Establishment Response* to the SMF. Finally, SMF responds with the *PDU Session Establishment Accept* message to the UE to confirm the acceptance of the session establishment request sent by the UE by following this signaling message path – *SMF*(Namf_Communication_N1N2_Message_Transfer[..])→*AMF([*PDU Session Res. Setup Req[..])→*gNB(*DL Information Transfer[..])→*UE*. Replace the square bracket with the *PDU Session Establishment Accept* message. The 5GC/AMF sends the *PDU Session Resource Setup Request* message to the NG-RAN/gNB. The NG-RAN/gNB allocates necessary resources to the UE initiated PDU session being established.

20.2 Service-Based Architecture (SBA)

The different core network entities and the services provided by them in the previous generations' mobile communication networks are based on OEM/proprietary and dedicated hardware with specific software, performing the functions and procedures of the core network protocols. However, in a 5GC network, OEM/proprietary hardware-based network entities, along with specific software, used in modern mobile communication networks, e.g. GSM, 3G, or 4G, are replaced by software-based network elements due to the CUPS feature, as described in the previous section, that is, 5GC NFs and procedures are implemented in software. Such a network element implemented in software performs the specific functions and procedures of an OEM/proprietary hardware-based network element.

A network element implemented in software, instead of dedicated hardware, is the basis for the service-based or service-oriented architecture 5G core network. A software-based network element, or service producer, provides services to other, or services consumers, software-based network elements through a standard communication protocol, e.g. HTTP, over an existing transport network, e.g. IP. A software-based network element, performing a particular 5GC protocol layer's functions and procedures, can be added/removed dynamically to scale out/scale in independently and expand the overall network capacity in the 5G system. Dynamic services provisioning through a software-based network element in 5GC is faster than an OEM/proprietary hardware-based network element. Such capabilities are possible only and realized due to the CUPS architecture of the 5GC.

The 5GC network architecture is also based on the 3GPP Release 14 CUPS feature as described in the previous section where the core network functional areas, such as mobility management, session management, and user data transfer, are separated into a different control plane and user plane functional blocks. Such functional blocks are called as *NFs* with well-defined behaviors in terms of performing protocol layer functions and procedures. Separation of control and user plane tasks as separate NFs along with their service-oriented architectures results in a customizable, scalable, and modular 5GC network. NFs provide ease of scalabilities, i.e. add or remove, enabling a service provider to deploy more than one instance of the same network function in the same Public Land Mobile Network (PLMN). NFs can be deployed in a centralized or distributed manner.

The 5GC NFs communicate through service-based interfaces using which a producer provides or exposes its services to a consumer of various network services. Due to this, the 5GC network architecture is also called a *service-based architecture*, which is one of the major differences between the 5GC and its previous generation's core network architecture. All the software-based components, i.e. producers/consumers, can reside and provide communication services running on a single host with virtualized infrastructures.

The subsequent sections describe the service-oriented architecture as well as various other aspects of the NFs of the 5GC network.

20.2.1 Network Functions and Its Instances

As part of an SBA, the 5GC contains software-based components which are called *NFs* as mentioned in the previous section. An NF may perform the functions and procedures of a specific protocol layer. An NF may also perform other functions and procedures such as performing the functions of central registering and authenticating entity for other NFs, user data management, and logical network provisioning of the 5GC. A network function is a "stateless" entity, i.e. functions and procedures performed by a network function are independent of its state and can provide services to other NFs as well as a consumer of services produced by other NFs. The control plane and user plane NFs used as part of the SBA of the 5GC network are described below.

- Control Plane NFs
  - *AMF,* which performs similar functions and procedures of the LTE/EPC MME. 5G NAS layer signaling over the NR air interface (N1) terminates at the AMF end. It also interfaces with NG-RAN over N2. It performs the tasks of 5G MM such as UE registration, connection management, and security management, such as UE authentication/authorization. An AMF can serve multiple network slices or multiple AMFs may be deployed based on the services offered by a 5G network.
  - Session Management Function (SMF), which performs 5G NAS layer session management functions, similar to LTE/EPC MME. But SMF performs the UE session management task at the PDU session level only. 5G NAS layer session management signaling messages over the NR air interface (N1) terminate at the SMF end. Session management signaling messages include a PDU session establishment, modification, release, and so on. SMF performs the tasks of IP address allocation, selection of a UPF, creation of GTP tunnel between gNB and UPF, user traffic configuration at UPF end, and so on. Multiple SMFs may be deployed based on the network slices and services offered by a 5G network.
  - Authentication Server Function (AUSF), which performs an authentication task during a UE registration either from a 3GPP or non-3GPP access network.
  - *Network Exposure Function (*NEF*),* which enables to expose the services and capabilities provided by 5GC NFs in a secured manner to other applications within the 5GC network or external applications like the Policy Control Function (PCF).
  - *Network Repository Function (*NRF*),* which registers and maintains the profiles of the different 5GC NFs and their services instances. NRF provides the service discovery facility to other NFs. Any network function can query the NRF to discover a network function and the services offered by it. NRF provides security-related services also by providing access tokens to network functions during their registration process. Multiple NRFs may be deployed based on the services offered by a 5G network.
  - *Network Slice Selection Function (*NSSF*),* which select a network slice instance from a set of logical networks or provide network slices for a UE to be served, based on its subscription information. A network slice is a key enabler feature for the 5G system, as described earlier in Chapter 16. NSSF provides a target or list of candidate AMF and NRF to the current or requesting AMF as part of the selection of a network slice instance during a UE initial registration procedure or a PDU session establishment procedure or UE configuration update procedure or an inter-PLMN mobility procedure.
  - *PCF, which* is similar to the Policy Charging and Restriction Function (PCRF) of LTE/EPC. PCF defines and maintains the rules or policies to be used by a UE or to be applied on UE by the 5G network control plane functions while providing communication services to a UE. AMF uses the access and mobility management-related policies from PCF. SMF uses the session management-related policies from PCF. UPF uses the service data flow-related policies, which are provided via the SMF, to transfer data to a UE.

- *Application Function (*AF), which provides *PCC*, per network slice or Data Network Name (DNN) or both, rules to PCF for services delivered through visited PLMN, i.e. local breakout, in case of a roaming subscriber.
- *Unified Data Management (*UDM*),* which is similar to the Home Subscriber Server (HSS) of LTE/EPC and holds subscriber and subscription database. It performs user identification management, generation of UE authentication credentials, and subscription management tasks, e.g. subscribed network slices information, in association with the subscriptions data stored in Unified Data Repository (UDR). UDM provides security-related services also by generating and providing an authentication vector to the AUSF network function during a UE authentication procedure.
- *UDR*: As the name implies, UDR stores the subscriptions data, policy data, application data, and so on, which are retrieved and used by other NFs such as the UDM, PCF, and NEF.

- User Plane Network Function

5GC contains the following user plane component or network function, which only consumes network services and does not produce any network service for other NFs.

- *User Plane Function (UPF).* A UPF performs tasks that are similar to the tasks performed by the LTE/EPC MME SGW-U and PGW-U planes. UPF performs user data packets routing/forwarding between gNB and data network, packet inspection, and QoS handling. Such functionalities are controlled by the SMF based on the packet handling information provided by the SMF to the UPF. The UPF acts as the interface to an external data network and also acts as an anchor point to support intra- as well as inter-Radio Access Technology (RAT) UE mobility. Multiple UPFs may be deployed based on the network slices and services offered by a 5G network.

For the detailed description of the tasks performed and services provided by the above 5GC NFs, refer to TS 23.501 [40] and 23.502 [41].

Various tasks performed and services to be provided by a 5GC network function to other NFs can be derived from its corresponding 5G system procedures as defined by TS 23.502 [41]. The services to be provided by a network function are also derived from the other related technical specifications. Thus, a particular 5G system procedure will involve the interactions among different NFs through service-based interfaces.

It may be noted due to the 5G network deployment requirements that there can be multiple instances of a particular network function in the control plane, e.g. AMF, as well as the user plane, e.g. UPF. The CUPS feature described earlier in Section 20.1 enables such a deployment of a 5GC network. Each instance provides its designated services to other NF instances. As an analogy, a network function can be compared with a *class,* and an instance can be compared with an *object* of the C++ programming language.

20.2.2 Network Functions (NFs) and Their Services Interfaces

As part of the SBA, each 5GC control plane network function has its well-defined service interface through which a producer network function exposes its various services capabilities for other NFs. A consumer NF can consume the services of a producer NF through its service interface. Only one service interface is defined per network function. Table 20.1 lists the service interfaces, along with allowed operations, of 5G control plane network functions as part of its SBA.

For example, the AMF NFs have the *Namf_Communication, Namf_EventExposure,Namf_MT, Namf_Location* service interfaces using which they expose various services to other consumer NFs of 5GC. A service interface of a network function is realized through a RESTful API, which shall be described later in Section 20.2.6. For the description of the services provided by a network function over a particular service interface, refer to TS 23.501 [40].

- Representations of 5G System Architecture

There are two ways to represent the 5G system architecture, as described in TS 23.501 [40].

Table 20.1 5G network functions and their service interface name.

5G Control Plane Network Function Names	Interface Names	Operations
Access and Mobility Management Function (AMF)	Namf	Request
Session Management Function (SMF)	Nsmf	Response
Unified Data Management (UDM)	Nudm	
Network Repository Function (NRF)	Nnrf	Subscribe
Network Slice Selection Function (NSSF)	Nnssf	Notify
Authentication Server Function (AUSF)	Nausf	
Network Exposure Function (NEF)	Nnef	
User Plane Function (UPF)	Nudr	
Policy Control Function (PCF)	Npcf	

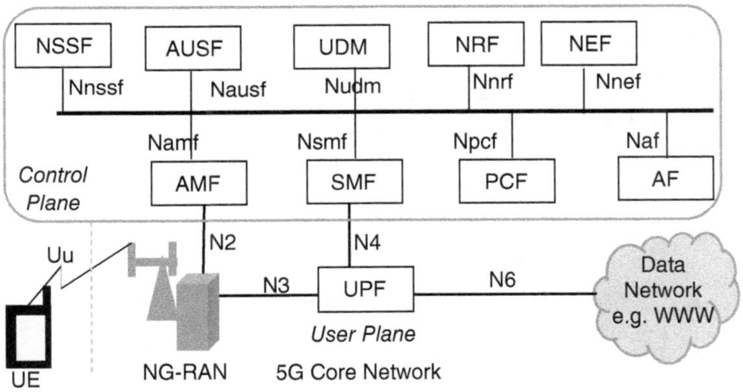

Figure 20.4 Illustration: service interfaces-based architecture of 5G system.

- Service Interfaces-Based Representation

NFs interact and are interconnected through predefined service interfaces, as in Table 20.1 above. Note that service interfaces are used among the control plane NFs only. The service interfaces-based representation of a 5G system architecture for a nonroaming scenario is shown in Figure 20.4. For a 3GPP access network and reference points-based representation of a 5G architecture for a nonroaming scenario, refer to Figure 4.2.3-2 in TS 23.501 [40].

- Reference Points-based Representation

In the previous generation's core networks, a reference point was used to interconnect two network elements. However, in a 5GC network, a reference point is used to interconnect two NFs as well as a network function and a network element; see Figure 20.5. For example, the N1 reference point is used between a UE and an AMF network function.

Both the 3GPP access and non-3GPP access network 5G system architectures, described in Chapter 16, use reference points naming conventions between two NFs. In the case of the 3GPP access network, reference points are named from N1 to N52; see Figure 20.5. For example, the N4 reference point connects the SMF (control plane) and UPF (user plane) NFs. Similarly, the N6 reference point connects the UPF and external data networks. Reference points – N1 (UE-AMF), N2 (NG-RAN – AMF), N3 (NG-RAN – UPF), N4 (SMF – UPF), N6 (UPF - DN),

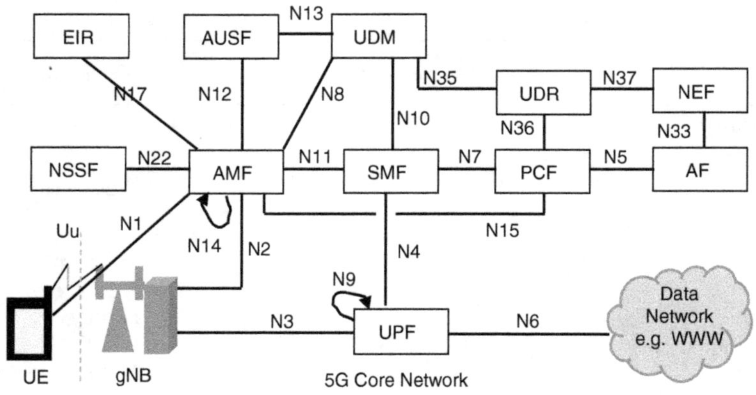

Figure 20.5 Illustration: reference points-based architecture of 5G system.

Table 20.2 5GS reference points and their protocol and transport networks.

5GS Reference Point	Protocol and Transport Network
N1	NAS over Air interface: Uu
N2	NG-AP over SCTP/IP
N3	NG-U/GTP-U over UDP/IP
N4	PFCP/GTP-U over UDP/IP
N6	IP
Others, i.e. N5, N7, N8. . .	Services produce/consumed through the respective service-based interface over TCP/IP with TLS

and N9 (UPF – UPF) – are realized using their corresponding logical interface, protocol stack, and transport network. The remaining reference points are realized using their respective service interface mentioned in Table 20.1. Note that a *Reference Point* representation is used for the control plane as well as user plane NFs, but service-based interface representation is used among the control plane NFs only. Table 20.2 summarizes the different reference points and their protocol and transport networks.

In the case of the non-3GPP access network also, reference points are named as Y1, Y2, Y4, Y5, NWu, NWt, Ta, and Tn; refer to TS 23.501 [40]. The Y4, Y5, NWt, Ta, and Tn reference points were introduced as part of the 3GPP Release 16.

20.2.3 5G System Architecture with NF

- Services Architecture/Interface-Based Nonroaming Architecture of 5G System

Similar to the LTE/EPS system architecture, roaming and nonroaming architectures are also possible within the 5G system. The nonroaming architecture of the entire 5G system comprising 5GC *NFs* along with the services interfaces among them was shown in the previous section; see Figure 20.4.

- Services Architecture/Interface-Based Roaming Architecture of 5G System

In the case of a roaming architecture, refer to Figure 4.2.4-1 in TS 23.501 [40], there are two options for routing of user traffic between two 5G PLMNs.

- Local breakout, where the roaming subscriber traffic is routed from the visited network or PLMN using its UPF and SMF NFs and data network. Home PLMN provides subscriber information, authentication, and policy-related information to the visited PLMN.
- Home routed, where the roaming subscriber traffic is routed back from the visited PLMN to the home network or PLMN. In this case, the home and visited PLMN's UPF and SMF NFs are used. UPF of the visited and home PLMNs is interconnected through the *N9* interface for user traffic routing purposes. For roaming architecture and interconnections of 5G systems, refer to Section 4.2.4 in TS 23.501 [40].

20.2.4 Network Functions and Their Services and Operations

Each 5GC producer network function provides or exposes a set of services through its service interface as mentioned above. Further, there are service operations, e.g. register, update, notify, and so on, which are associated with each service provided by a network function. Using a service interface, a producer network function exposes such services and their associated operations to consumer NFs. Network function services are designed to be self-contained and reusable.

Figure 20.6 illustrates the relationship (one to many – 1 : M) between a network function interface and different services provided under it. It also shows the association of a service and its different operations.

For example, the NRF has its service interface, which is called: *Nnrf*; it has three services called *Nnrf_NFManagement, Nnrf_NFDiscovery,* and *Nnrf_AccessToken under the Nnrf interface.* Each service has one or several operations. For example, the *Nnrf_NFManagement* services have *NFRegister*, *NFUpdate*, *NFDeregister*, and so on service operations.

Service operations provided by a 5GC network function can be compared with a 3GPP protocol layer procedure which consists of exchanging of signaling messages between two network elements of a mobile communication network. For the successful completion of a 3GPP protocol layer procedure, several request/response signaling messages are exchanged among the network elements. Similarly, a particular service operation between a consumer and producer network function can be accomplished through the following messages:

- *Request*, from a consumer to a producer;
- *Response*, from a producer to a consumer;
- *Subscribe*, from a consumer to a producer. A consumer network function may subscribe to a service, say the occurrence of an interesting event, provided by a procedure network function. For example, a consumer network function instance may be interested to get notified by the NRF on the failure of a producer network function service instance; and
- *Notify*, from a producer to a consumer. A producer network function will notify the availability of an interesting service to a consumer network function.

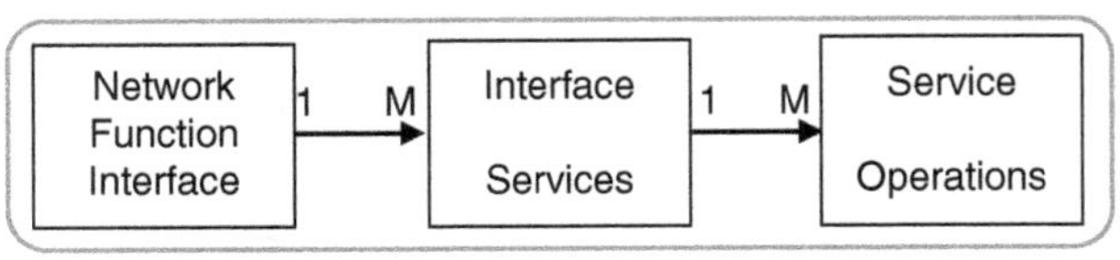

Figure 20.6 Illustration: network function interface, services, and its operations.

Figure 20.7 illustrates the above service operation flow method applicable between two NFs of 5GC.

For more description of the 5GC NFs along with their service interfaces, different services offered by them under each interface, their respective operations, design guidelines, and guidelines used for the nomenclature of a service and its operations, refer to TS 23.501 [40] and 23.502 [41].

20.2.5 Network Functions Services Framework

The 5GC NFs produce/consume various network services and interact with each other through a common framework that consists of the following procedures:

- Register
- Authorization
- Discovery
- Deregister

The 5GC NFs services framework is illustrated in Figure 20.8.

Within the above service framework, the *NRF* acts as the registration and access authentication server for the rest of the 5GC NFs. The NRF provides management, *discovery*, and authorization services to the other NFs. NRF uses the token-based OAuth 2.0 service authorization framework to authorize a consumer network function. A consumer network function registers with the NRF using the *Nnrf_NFManagement_NFRegister* service operation. A consumer network function provides its profile information to the NRF in terms of network function type, instance id, and services that it exposes for other NFs.

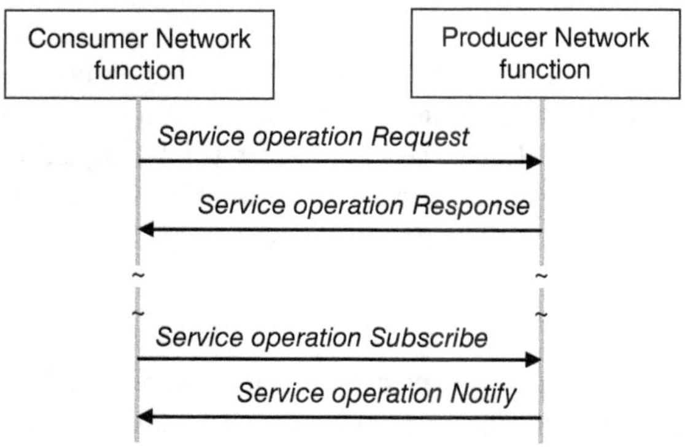

Figure 20.7 Illustration: 5GC network functions service operation flow.

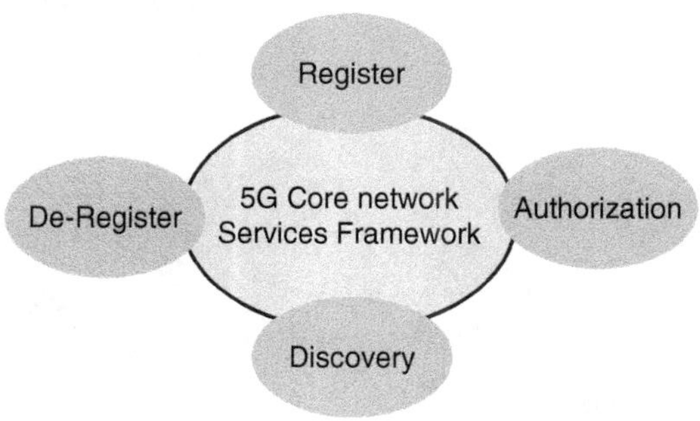

Figure 20.8 Illustration: 5GC network functions services framework.

A consumer network function must be registered with and authenticated by the NRF before the consumption of other network services. Once registered with the NRF, a consumer network function executes the *Discovery* procedure toward the NRF to discover a particular network function instance and its exposed services. A consumer network function may discover instances of one network function and a particular service, or it may discover the instances of all the NFs and their services. Further, before consuming network services, a consumer network function must be *authenticated* by sending the *Nnrf_AccessToken_Get_* service request to the NRF. The NRF uses the Oauth2 authentication protocol and provides an access token to the consumer network function for its subsequent services requests to producer NFs.

For the description of each of the above processes, shown in Figure 20.8, of the 5GC NFs services framework, refer to TS 23.501 [40] and 23.502 [41]. Figure 20.9 further illustrates the above service framework through the *registration, discovery, authorizations, and deregistration* processes for some of the network functions such as AMF, SMF, AUSF Service Communication Proxy (SCP), and NRF. In Release 15 5G core network, NFs can make direct communication with each other over HTTP. The SCP, which was introduced as part of the 3GPP Release 16, however, provides a delegated, or an indirect, network function discovery mechanism for other network functions. The arrow shows the corresponding request/response between a network function and the NRF.

As part of the registration process, a consumer network function, say AMF, uses the *Nnrf_NFManagement_NFRegister* service to register with the NRF. The service registration request contains the profile of the consumer network function, which consists of mandatory as well as optional input parameters. Some of the mandatory parameters supplied as part of the registration request message are network function type, instance id, PMLN id, and list of supported services. As part of the response to the registration request, the NRF provides a list of NFs and their profiles to the AMF. As illustrated, a profile of a network function consists of its ID, type, list of services offered by it, and so on. For more details on the input parameters to be provided by each of the NFs to NRF as well as the response/output parameters from the NRF to a network function as part of the 5GC service framework and its operations mentioned above, refer to TS 29.510 [76].

- Status of Network Function (NF) Service Instance

Multiple instances of a particular network function can exist, which also enables to support NFV capability in 5GC. To inform the current operational status, each network function service instance contacts and exchanges

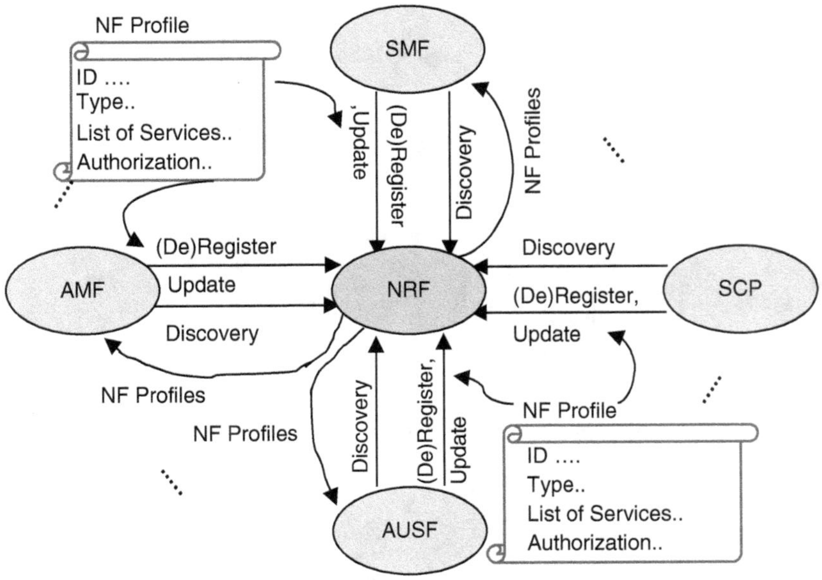

Figure 20.9 Illustration: registration, authorization, and discovery of network functions.

heartbeat information periodically with the NRF. The periodicity of heartbeat information from a registered network function instance to the NRF is operator configurable. The periodicity of heartbeat information is defined as part of the network function profile parameter which is called the *heartBeatTimer* and defined in seconds. Figure 20.10 illustrates the statuses maintained by the NRF for each of the network function instances. As illustrated, a network function instance can be in one of the following states, as maintained by the NRF:

- Registered

In this state, a network function service instance is registered in the NRF and can be discovered by other NFs.

- Suspended

In this state, a network function service instance is registered in the NRF but not in an operational state. NRF marks a registered network function instance in a SUSPENDED state if there is no heartbeat information periodically from it. Also, in SUSPENDED status, a network function instance cannot be discovered by other NFs.

- Undiscoverable

In this state, a network function service instance is registered in the NRF, which is also in an operational state. But it cannot be discovered by other NFs. NRF may mark a registered network function instance in undiscoverable states if it is shut down because of Operation and Maintenance (O&M) operator actions.

Figure 20.11 further illustrates the usages of the heartbeat timer at the NRF end. At the expiry of the heartbeat timer, the NRF marks the status of the particular network function instance as suspended.

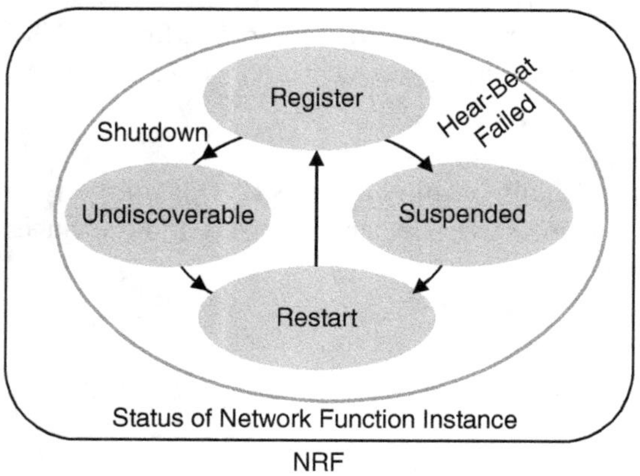

Figure 20.10 Illustration: status of network function instance at NRF.

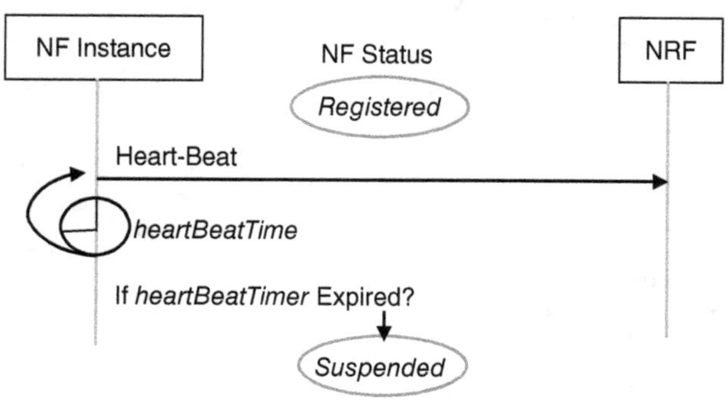

Figure 20.11 Illustration: usages for heart beat timer for NF status.

O&M operator action, such as a restart, may be required to bring a network function service instance into a registered and operational state again. Refer to TS 23.527 [43] for more details on the service interface-based network function instances restoration processes.

The subsequent sections describe the protocol layers and methods used to realize the 5GC service framework architecture with the various NFs described in Section 20.2.1..

- **Protocol Stack (Control Plane) for Service Interface-Based Communications**

Earlier, we have described the various physical as well as logical interfaces used by the previous generation's mobile communications networks. For example, between the LTE eNodeB and its EPC, user plane (S1-U) data are transported through GTP/UDP and control plane (S1-AP) data are transported through the Stream Control Transmission Protocol (SCTP)[17] over the existing IP transport network. We have also described in Table 20.1. on the various service interfaces provided by the different 5GC NFs in the control plane. Using the service interfaces (Figure 20.4), the 5GC NFs exchange their application protocol information among them in terms of request/response, subscribe/notify communications as described above; refer to Figure 20.7.

Service interface-based communication among the 5GC control plane NFs also uses the existing IP transport network. In addition to this, the HTTP/2 protocol is used to pass network function-specific application protocol information among the different 5GC control plane NFs. Further, the HTTP/2 protocol uses the Transmission Control Protocol (TCP) as the transport layer along with the transport layer security (TLS) layer to provide an encrypted communication among the control plane NFs. Figure 20.12 illustrates the usages of the HTTP/2 protocol stack to achieve service-interfaces-based communications among the 5GC control plane NFs.

The service interfaces supported by the 5GC control plane NFs were described earlier in Section 20.2.2. Each control plane network function performs its application protocol specific tasks. Communication for any request or response for a desired resource or information is passed as a payload over the HTTP/2. Note that the error reporting mechanism of HTTP/2 protocol is used to report any error between a consumer and producer network function which will be described and illustrated in later sections. HTTP/2 is also used by the World Wide Web (WWW) for communications between client and server systems over the internet. HTTP/2 standard methods or verbs such as the PUT, POST, GET, and DELETE are used to make service operation requests and responses between a consumer and producer network function. These HTTP/2 methods are used by the WWW also for communication between a client and server system over the internet. The method used to communicate between a consumer and producer network function within the 5GC services-based framework is described below.

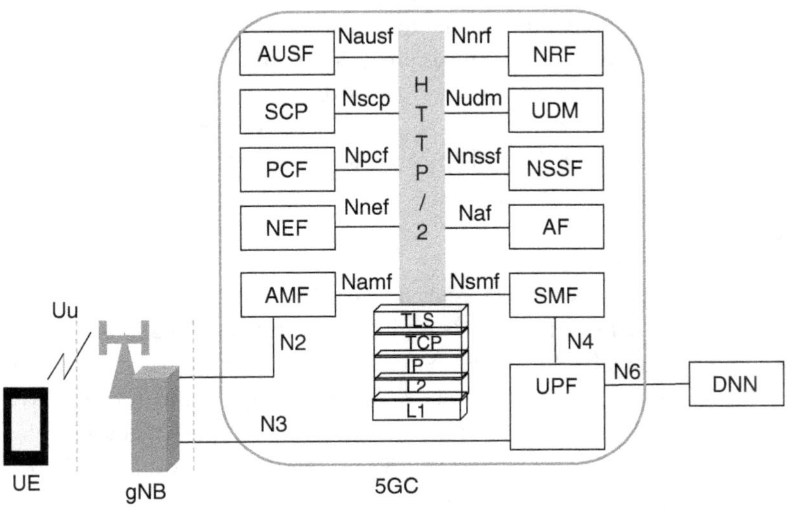

Figure 20.12 Illustration: communications among 5GC control plane network functions using HTTP/2 protocol.

20.2.6 Services API for Network Functions

- **Service Application Programming Interfaces** (API) **for Communications Among Network Functions**

In a 5GC service-based framework, the consumer as well as producer network function use the *API* method of communication among them. In day-to-day life, mobile users use mobile applications for different types of services such as weather services and social media services. Such mobile applications request the desired information or resources from the concerned web server on behalf of the mobile user. The client or the mobile application presents the information received from the web server in a human-readable format to the mobile user. Such a request/response operation between a web client and its server is accomplished through the usages of APIs using the HTTP over the internet.

As far as the 5GC is concerned, an API is defined for each of the services supported by a particular network function. A service API operates on a resource which can be a network function instance or collection of all the NFs instances and so on. A resource can be a logical object upon which a service API may perform its desired action/operation. Using a service API, a consumer network function consumes a service produced by a producer network function. To accomplish this, a consumer network function invokes the desired service API over the HTTP toward the producer network function.

A service API enables a consumer network function to specify the desired action to be performed by the producer network function on a particular resource through one of the standard methods or verbs of HTTP such as the PUT, POST, GET, and DELETE. Such operations applied on a network function resource can be also described using the acronym *CRUD* where C stands for Create, i.e. POST, a resource; R for read, i.e. GET, a resource; U for an update, i.e. PUT, a resource; and D for delete, a resource. Each service API has a predefined set of resources on which the desired operation can be performed through one of the HTTP methods mentioned above. Refer to the corresponding network function services' description of technical specification, e.g. TS 29.510 [76] for NRF services, TS 29.518 [77] for AMF services, and so on, for more details on such resources used by each service API.

- **Definition, Descriptions, Format, and Data Types for Service API**

As part of the 5GC SBA framework, NFs communicate with each other through different service APIs. Such APIs are defined in Open API 3.0.0 specification. Further, the YAML Ain't Markup Language (YAML– a recursive acronym) format is used to describe a service API. The information elements, information elements identifier, and types used in the air interface layer 3 messages were described earlier in Chapter 4 of this book. Similarly, through a service API and its corresponding operation, different types of information are exchanged between a 5GC consumer and a producer network function. Due to this, different data types are used for different types of information. 5GC SBA framework defines a simple data type, enumeration type, and structured types that are used by such service APIs and its operations of different NFs. Such data types to be used by the service operations of different APIs are described in the 3GPP TS 29.501 [74], TS 29.571 [78] and are also described further by the corresponding network function service description of technical specification.

- **Design Style of 5GC Service API: RESTFul**

One important aspect of the 5GC NFs and their service APIs is their design style used over the HTTP. Services APIs are designed in *RESTful* style, which stands for *Representational State Transfer. RESTful* style APIs are used by the 5GC network functions for communications over the different service based interfaces. Representation specifies the format to be used to represent or describe a resource as part of a request/response API. In the case of a 5GC service API, a *Java Script Object Notation* (JSON) format of structuring data, having Key (on the left side) – Value (on the right side) pair, is used to represent a core network resource. The JSON representation format is simpler than the XML format. A profile of a typical network function can be represented as follows using the JSON format:

```
{
"nfInstanceId": "xxxxxxxx-xxxx-xxxx-xxxx-xxxxxxxxxxxx,"
        "nfType": "NRF,"
        "nfStatus": "REGISTERED"
        ..........................................
}
```


A consumer network function specifies the representation style being used under the "Content-Type" field of HTTP header in its API request made to a producer network function. In an Open API specification of a service API, resource representation in JSON format is specified under content-type: "application/json" for a successful scenario, and for an erroneous scenario, the content-type "application/problem + json" is used. For more details on HTTP headers to be supported by a consumer and producer network function, refer to TS 29.500 [73].

There are several characteristics based on which an API is classified to be a RESTful API. Some of the characteristics are Caching, Stateless, Client–Server, and so on. Depending on the purposes, the APIs supported by a particular 5GC network function may or may not exhibit all the characteristics of a *RESTful* API. A particular response/information from a producer network function may no longer be valid at the consumer end after a certain interval. This is specified through "Caching" or "max-age" as part of the HTTP header of the corresponding response message from the producer network function. The "Caching" period is applicable for the consumer network function till it considers the information received from the producer network function as valid. For more information on the RESTful characteristics of 5GC APIs, refer to the corresponding 3GPP TS for network function service description. The NRF discovery service and UDR data repository service APIs are the examples of APIs that exhibit the RESTful caching characteristics. For more information on the characteristics of *RESTFul* API, refer to TR 28.891.

RESTFul representation of API was introduced by Dr. Roy Fielding in his 2000 Doctorate Dissertation "Architectural Styles and the Design of Network-based Software Architectures.

- **Unique Resource Identifier** (URI) **for Service API**

As described above, a consumer NFs may request the information of a particular resource or perform other actions/operations on the existing resource available at the producer network function end. Each resource is identified through a *URI* using which a consumer network function performed the desired operation through one of the HTTP methods/verbs described above. As an analogy, consider the path of a directory and its files. Using a unique path, a new file can be added to the directory or delete an existing particular file.

A URI contains the concatenation of the following components separated by "/":

- API Root, e.g. https://localhost:12345, with a fully qualified domain name (FQDN);
- API Name, e.g. "nnrf-disc" API name, is used for NRF discovery service. Similarly, the "nnrf-nfm" API name is used for the NRF network management service;
- API Version, with, major, minor, and patch versions. Only v1 as a major version is used; and
- API-Specific URI part, e.g. ../nf-instances which represent a collection of network function instances. Similarly, .../nf-instances/{nfInstanceId}(Profile) which represent a particular network function instance.

For more information on the API name of various services offered by the different NFs, refer to the corresponding network function service description of technical specification, e.g. TS 29.510 [76] for NRF services, TS 29.502 [75] for sessions management services, and TS 29.518 [77] for access and mobility management services. Each service API has a predefined set of URI and resources on which the desired operation can be performed through one of the HTTP methods mentioned above. Refer to the corresponding network function service description of technical specification for more details on such resources, e.g. TS 29.510 [76] for NRF services, TS 29.502 [75] for sessions management services, and TS 29.518 [77] for access and mobility management services. Usages of URIs and HTTP methods in the 5GC service framework are illustrated through Examples 20.1 to 20.5.

Example 20.1 NF Registration with NRF Using HTTP PUT Method

Figure 20.13 illustrates the usages of the HTTP PUT method and its URI for registration of a particular consumer network function instance, e.g. SMF, with the registration and authorization server, i.e. NRF. An instance of a network function must be registered with the NRF so that other NFs can discover the requested network function and its services. Figure 20.13 also illustrates the usages of the FQDN as part of the URI along with the API name, i.e. nnrf-nfm; version, i.e. v1; and service resource-specific URI, i.e.. */nf-instance/{nfInstanceID}*. Usages of a particular port, e.g. 12 345, and the status code, either success code: 201 OK or error code: 4XX/5XX, returned from the NRF to the SMF are also illustrated.

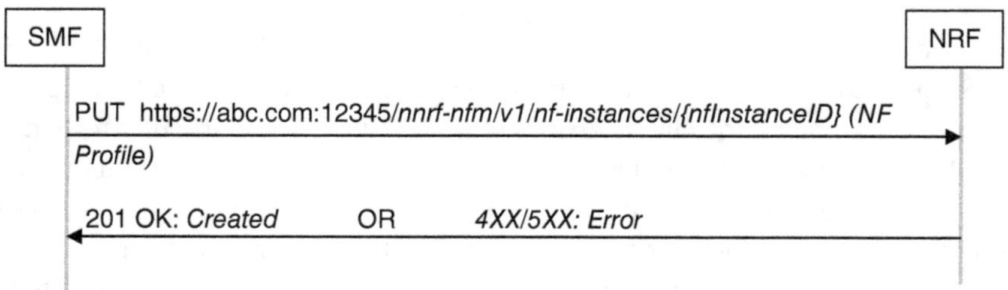

Figure 20.13 Illustration: usages of a service API and HTTP PUT method for SMF registration with NRF.

nfInstanceID refers to a *Universally Unique Identifier (*UUID*)* which is assigned to the NF instance, i.e. SMF, to be registered with NRF. A *UUID*, RFC 4122 [15], of 128 bits length is allocated to each instance of a network function. Note that a UUID is also allocated as a service instance identifier to an NF service. An UUID takes the format of: xxxxxxxx-xxxx-xxxx-xxxx-xxxxxxxxxxxx where "x" can be 0–9 or hexadecimal: a–f. The profile, e.g. *NFProfile*, of the requesting network function, e.g. SMF, is also included as part of the HTTP PUT request.

A profile of a network function instance contains mandatory, conditional, as well as optional parameters. Some of the parameters contained in a profile of a network function instance are listed below:

- Identifier, i.e. UUID, for the network function instance;
- Type of the network function instance;
- Status of the network function instance;
- Name of the network function instance;
- Heartbeat timer;
- IP address and so on.

The typical definition of a network function profile through the sample program is illustrated later in Chapter 21. Data types and their contents, used by each service API and its operations, are described by their corresponding technical specification. For NRF, refer to TS 29.510 [76] that describes all such data types, i.e. *NFProfile*, *nfInstanceID*, as mentioned above.

If the network function instance was created and registered successfully with the NRF, it will return the HTTP status code: 201; otherwise, appropriated HTTP error code: 4xx/5xx will be returned to the requesting network function instance, e.g. SMF. Refer to TS 29.500 [73] for the supported HTTP error codes used in 5GC service-based interfaces among the different NFs.

For more information on the services provided by the NRF and their operations, refer to TS 29.510 [76].

Example 20.2 Retrieval of a Profile of an NF Instance from NRF Using HTTP GET Method

Figure 20.14 illustrates the retrieval of the profile of a network function instance, given its instance identifier *nfInstanceID,* from the NRF using the HTTP PUT method. The NRF returns the desired NF profile instance in case of a successful retrieval with status code: OK or an error code: 4XX/5XX to the consumer network function if the provided *nfInstanceID* is an invalid one.

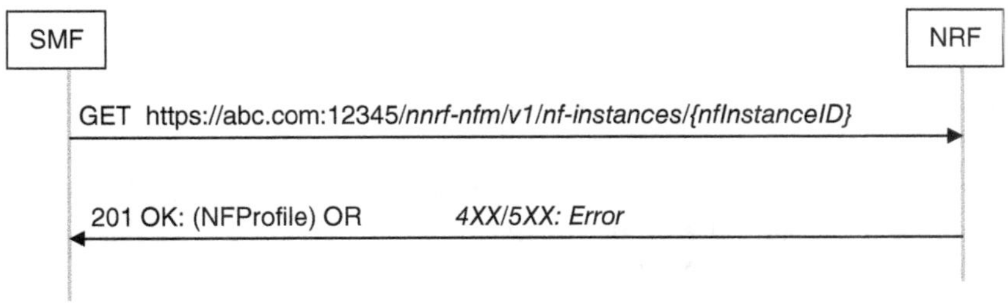

Figure 20.14 Illustration: usages of a service API and HTTP GET method for retrieval of NF profile.

Example 20.3 Deregistration of an NF Instance from NRF Using HTTP DELETE Method

Figure 20.15 illustrates the deregistration of a network function instance, given its instance identifier *{nfInstanceID}*, from the NRF using the HTTP DELETE method. The NRF returns the HTTP code: 204 in case of a success scenario with an empty response body to the consumer or an error code to the consumer network function if the provided *nfInstanceID* is an invalid one.

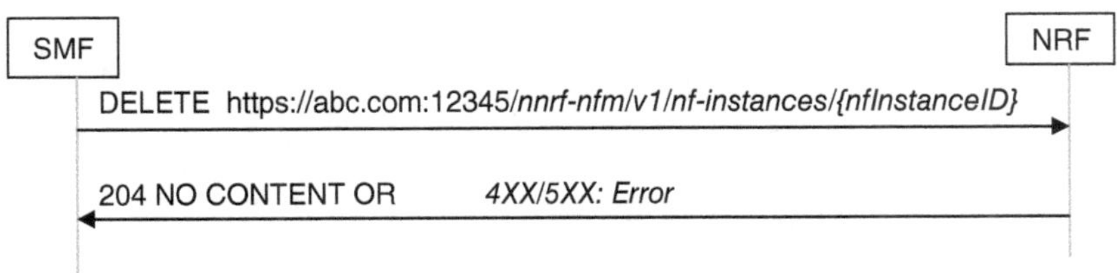

Figure 20.15 Illustration: usages of a service API and HTTP DELETE method to deregister an NF profile.

Example 20.4 Discovery of a Network Function Instance from NRF Using HTTP GET Method

Figure 20.16 illustrates the discovery of a network function instance and its services by another network function. Consider that SMF wants to discover the availability of the AMF network function and its services. To discover a network function instance, the SMF uses the HTTP GET method with its API URI, as illustrated in Figure 20.16. Note that several query parameters can be specified as part of the URI of the discovery procedure of a network function and its services. In this typical illustration, the SMF attempts to discover the *namf-comm* service of the AMF which is specified as part of the query parameters of the URI. The NRF returns the HTTP code: 200 to the SMF, or consumer network function, in case of successful discovery of the target or producer network function instances. The response also contains the validity of the target or procedure NFs instances so that the requester or consumer network function can store and cache it.

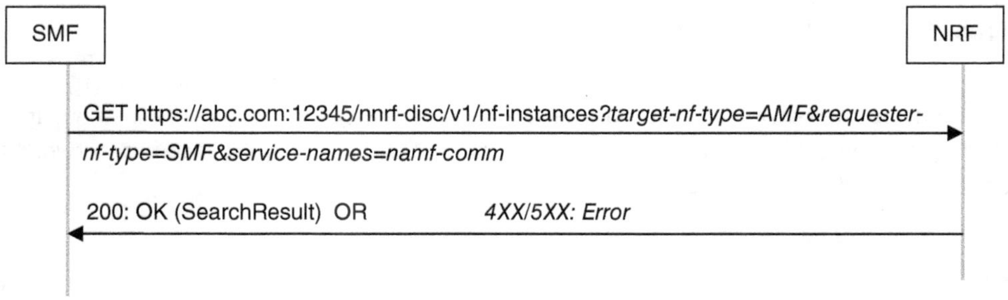

Figure 20.16 Illustration: discovery of a network function instance and service using HTTP GET.

A typical response from the NRF for the discovery procedure illustrated in Figure 20.16 is shown in JSON format below. The status of the AMF network function instance and its offered service, i.e. *namf-comm*, is being shown as the REGISTERED state. The unique instance identifiers for the AMF NF instance and its service *namf-comm* are also shown.


```
 {
"validityPeriod": 1234
  "nfInstances": [{
      "nfInstanceId": "xxxxxxxx-xxxx-xxxx-xxxx-xxxxxxxxxxxx,"
      "nfType": "AMF",
      "nfStatus": "REGISTERED",
     .........................
      "nfServices": [{
          "serviceInstanceId": "yyyyyyyy-yyyy-yyyy-yyyy-yyyyyyyyyyyy",
          "serviceName": "namf-comm",
          "versions": [{
              "apiVersionInUri": "v1",
             ....................
            }],
          "scheme": "https",
          "nfServiceStatus": "REGISTERED",
           ..............................
        }]}]}
```


For more information on the NF service discovery procedure using API and query as well as its response parameters offered by the NRF, refer to TS 29.510 [76].

Example 20.5 UE Context Transfer Request from Source to Target AMF

As a mobile user moves from one location to another location, a UE may be required to register with an AMF in its new location. The UE performs a registration request with its GUTI to the new AMF. Based on the GUTI, the new, i.e. target, AMF requests the old, i.e. source, AMF to transfer the UE context. This is accomplished through the *Namf* communication service and its transfer operation from the target AMF to source AMF, as illustrated in Figure 20.17.

As illustrated in Figure 20.17, a UE context information transfer request is made through the HTTP POST method which uses the structure *UeContextTransferReqData* as the content of its body. The context of the UE is provided from the old/source AMF to the new/target AMF through the *UeContextTransferRspData* structure as the body of

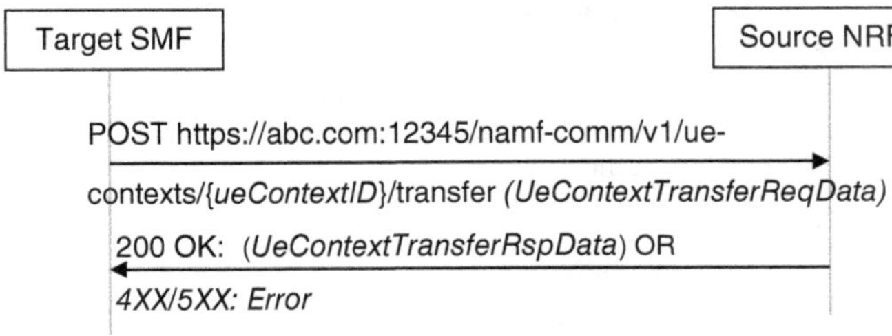

Figure 20.17 Illustration: usages of a service API and HTTP POST method to transfer UE context information.

the 200 : OK response. The {*ueContextID*} represents the context ID of the UE whose context information is requested from old AMF. For more information on the *Namf* communication service and transfer operation provided by NRF, refer to TS 29.518 [77].

- Security for Service-Based Interface: OAuth 2.0 Framework

The security mechanisms applied over some of the logical interfaces of the previous generation mobile communication systems and the network were described in Chapter 9 of this book. Further, the 5G security system and its services were described earlier in Section 16.11. Similarly, security measures are also applied over the service-based interfaces of the 5GC NFs services framework as described in the previous section. As part of the overall 5G security architecture, TS 33.501 [87], the security measures applied in the services-oriented architecture of 5GC is based on the OAuth 2.0 authorization framework. OAuth 2.0 is defined in RFC 6749 [20]. In this framework, the authorization server, e.g. 5GC NRF, issues an access token to a registered client, i.e. consumer network function. A consumer network function obtains an access token before accessing different NFs and their services.

As far as the 5GC is concerned, a consumer network function is required to be authorized and then obtain an access token from an authorization server, i.e. NRF. A consumer NF can obtain an access token for several producer NFs as specified under the scope of the access token request communication made to the NRF. The access token provided by the authorization server, i.e. NRF, is a JSON web token (JWT) type, as described in RFC 7519 [23]. A JWT is secured through a digital signature or a message authentication code (MAC). Table 20.3 shows the various authorization roles defined under the OAuth2.0 authentication framework along with the 5GC NFs performing the corresponding role. For more information on the OAuth 2.0 authentication framework, refer to the RFC 6749 [20].

In an OAuth 2.0 framework, the authorization server issues an access token to a client as part of an authorization grant. Under the OAuth 2.0 framework, there are four types of authorization grants between a client or consumer network function and a resource server. Refer to RFC 6749 [20] for more details on different authorization grant types. However, the 3GPP specifies the client credential authorization type to be used between a consumer and producer network function. In the 5GC network, the authorization server, i.e. NRF, issues an access token based on the credentials of a client, i.e. consumer network function. The allocated access token is included through the HTTP standard header field called authorization in a service operation request from a consumer to a producer network function. For more information on the authorization and access token allocation process, contents of access token request, and response, refer to TS 29.501 [74]. For illustrations on the network function authorization process and the allocation of an access token over 5GC service-based interfaces, refer to TS 33.501 [87].

- **Error Handling and HTTP Responses in Service API communication**

The protocol layer error handling mechanism was described earlier in Chapter 15 of this book. For example, to report an unexpected protocol error detected over the LTE air interface, GPRS Gb interface, and so on, a STATUS PDU is used by the UE, the E-UTRAN, Serving GPRS Support Node (SGSN) or Base station controller (BSC).

Table 20.3 Authorizations roles of 5GC network functions with respect to OAuth 2.0.

Role of Network Function as Per 5GC Service-based Framework	Role as Per OAuth 2.0 Framework
Network Resource Function (NRF)	Authorization Server
Consumer	Client
Producer	Resource Server
Network Function Instance ID	Client ID

Similarly, an error handling mechanism is also provisioned in case of a 5GC service framework and interface-based communications over the HTTP to report the status of the executed service operation. For these purposes, any application protocol-specific error detected by a producer network function is reported to a consumer network function by mapping to a standard error code of HTTP. For example, during a service operation request, if an invalid service API name or version error is detected by a producer network function, the error will be mapped to the HTTP standard error code: "400 Bad Request" and reported to the consumer network function. Further, the producer network function will provide additional information under the "ProblemDetails" information element on the error being detected to the consumer network function. This is similar to the inclusion of the whole erroneous PDU in STATUS PDU in case a protocol error is detected over the LTE air interface or GPRS Gb interface.

HTTP status code: 2xx implies successful scenarios, whereas the status code:4xx/5xx implies failure scenarios. Refer to the illustration shown in Figure 20.17. A successful scenario provides the requested information about a resource from producer to a consumer network function whose content-type in the HTTP response header will be set to *application/json*. However, in case of a failure scenario, such as due to an invalid instance id, the content-type in the HTTP response header will be set to *application/problem+json* with additional information put under the "ProblemDetails" information element. *ProblemDetails* contain several helpful information, such as the type, title, status, detail, and cause, as described in TS 29.571 [78]. 5GC error handling will be defined and illustrated in the next chapter through a sample software program.

20.2.7 Network Function Selection

Core network element selection function, i.e. selection of a mobile switching center (MSC), SGSN, or MME from a pool, used in the previous generation mobile communication networks, was described earlier in Section 7.1.2. Such a selection function is called the NAS Node Selection function (NNSF). In Section 20.2.1, different NFs of 5GC along with the functions and procedures performed by each network function were described briefly. In a 5GC network, due to the virtualization of NFs, multiple instances of a particular network function may be configured and run to activate or deactivate a service dynamically depending on the business or on-demand events requirements. In case of a configuration and deployment of multiple instances of a network function, a network function instance selection function shall be also required, which is similar to the NNSF, to serve a particular service or a network slice. Similar to the NNSF which is used to select a core network element, several parameters or factors are considered to select a network function instance also.

Figure 16.14 illustrated earlier shows the NFs instance selection for a typical network slice. The arrow lines (1, 2, 3, 4, 5, 6) show the selection of a particular network function instance based on several parameters or factors. For example, based on the requested Network Slice Selection Assistance Information (NSSAI), a gNB selects an AMF to forward UE NAS layer messages. An AMF selects an AUSF which performs authentication between the UE and 5GC. AMF also selects SMF and UDM. Further, an SMF selects a UPF to transfer user data corresponding to a UE PDU session. As illustrated in Figure 16.14, several instances of a network function can be deployed for redundancy purposes and are grouped into an NF set. If an instance is not available either in the control plane or user plane, another network function instance may be selected.

In summary, the following NFs instance selection functions are required to be made available in the 5GC. Some of the typical parameters to be used for the selection of a particular network function instance are also mentioned. Operator-specific 5GC configuration information may be also used while selecting a particular network function instance.

- AMF Selection

Input parameters for selection: Operator configuration, e.g. dynamic load, 5G-S-TMSI or Globally Unique AMF ID (GUAMI), requested NSSAI, AMF region ID, and AMF set ID

- SMF Selection

Input parameters for selection: Data network name, Single Network Slice Selection Assistance Information (S-NSSAI), Network Slice Instance Identifier (NSI-ID), subscription information, and UE location

- UPF Selection

Input parameters for selection: Operator configuration, e.g. dynamic load, deployment location, i.e. centralized or distributed, S-NSSAI, SSC mode, and subscription information.

- UDM Selection

Input parameters for selection: Operator configuration, UDM group ID, Subscription Permanent Identifier (SUPI), and PLMN

- AUSF Selection

Input parameters for selection: Operator configuration, AUSF group ID, SUPI, and PLMN

- PCF Selection

Input parameters for selection: Operator configuration, PCF group ID, SUPI, PLMN, S-NSSAI, and DNN

- UDR Selection

Input parameters for selection: Operator configuration, UDR group ID, and SUPI

20.3 Network Function Virtualization (NFV)

In a typical enterprise IT system architecture, a general-purpose hardware server may be dedicated to run only one instance of an operating system along with an application such as a database. On the other hand, a dedicated general-purpose hardware server can be also deployed to run multiple instances of an operating system or different operating systems as well as run different applications on top of them. This is accomplished through the server virtualization technology where virtual computing resources in terms of CPU cores, memory, NICs, and storage can be allocated differently as per the requirements of different operating systems and applications to be run on top of it. Another example is the emergence of a next-generation firewall, which provides, in addition to protecting a network, other security services like antivirus, intrusion prevention or load balancing, and so on, all on a single box. Earlier, each of these security services was provided through separate hardware appliances from different Original Equipment Manufacturer (OEM)/vendors.

Similar to the virtualized IT system architecture of an enterprise, the 5G system core network architecture would also enable operators to provide communication services through virtualizations of different core NFs on a general-purpose hardware server. Earlier, the individual core network element, such as the MSC, SGSN, and Gateway GPRS Support Node (GGSN), was deployed on a dedicated and tightly coupled proprietary hardware with specific tasks, e.g. switching, routing, etc., performed by them. However, as far as the SBA of the 5GC is concerned, a protocol layer or a functional area can be deployed as an individual software-based entity or functional block performing its designated NFs and procedures, either control plane or data plane, as explained earlier in Section 20.2. Such software-based entities or functional block designed for 5GC NFs can be decoupled from its underlying hardware and deployed on top of virtualized infrastructures provided by a general-purpose hardware server. Virtualized network functions (VNFs) provide ease of independent (of underlying hardware) scalability, on-demand network capacity expansion, customize virtualized as well as physical resources for different network slices, has low operation and maintenance cost due to software-based/software defined NFs, and so on.

Figure 20.18 illustrates the NFs of virtualizations through the deployment of 5GC NFs on a virtualized hardware platform.

As illustrated, two SMFs and four UPFs are shown to be used in this (Figure 20.18) typical NFVs deployment scenario. The SMFs can be used to deploy two network slices, assigning one SMF to each slice. Similarly, the UPFs can be used to connect to four data networks, assigning one UPF to each data network. Thus, a UE can be served by four data networks through two PDU sessions and their network slices. Note that a UE will be served only AMF and NRF for all of its PDU sessions and their network slices.

Further, as illustrated, NFVs consists of two layers:

- Virtualized infrastructure layer and
- VNFs layer.

The virtual infrastructure layer consists of physical computing hardware resources and a special-purpose software layer that provides virtual computing resources in terms of multiple logical CPU cores, memory, NICs, and storages to the virtual NFs. The special-purpose software component is called as *Hypervisor*, which forms the *virtualization layer* of the virtualized infrastructure. A hypervisor interacts directly with the physical computing resources, i.e. CPU, memory, NICs, and storages, and provides a hardware abstraction to the VNFs of 5GC.

As the CUPS feature, described earlier in Section 20.1, separates/decouples the control plane and user plane of a network element, similarly the virtualization layer, i.e. hypervisor, decouples/separates the 5GC NFs from the actual physical computing resources and provides the virtual computing resources to them, as illustrated in Figure 20.18.

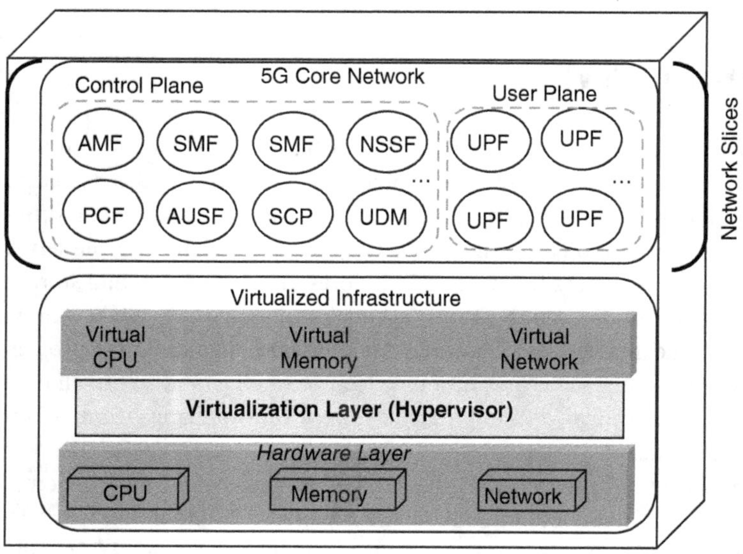

Figure 20.18 Illustration: 5G core network functions virtualizations.

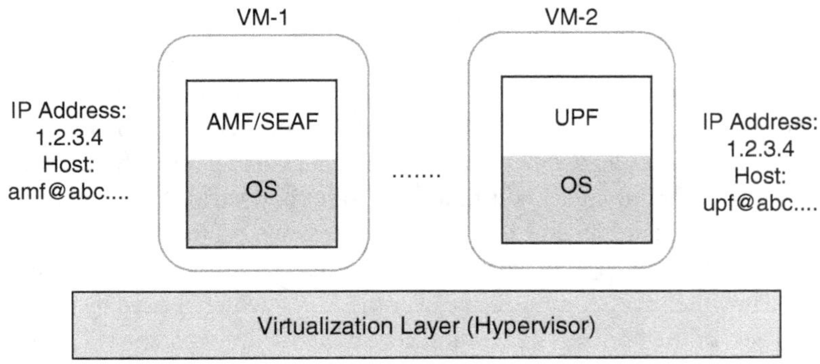

Figure 20.19 Illustration: virtualization with virtual machines.

Virtual NFs consist of the different NFs of 5GC on top of the virtual infrastructure layer. As an example, the UDM, similar to Home Location Register (HLR), may be provisioned with more storage capacity than the other NFs. AMF and SMF may be provisioned with more CPU cores to handle a large number of signaling messages. Similarly, multiple instances of the UPF can be deployed to handle user plane traffic for different use cases or network slices and multiple data networks.

- Virtual Machine (VM)

A physical desktop or server contains physical computing hardware resources. As mentioned above, the hypervisor creates a virtualized computing facility by providing virtualized or logical resources, i.e. processor, memory, network, and so on, to the different NFs. Such a virtualized computing facility is called a VM that behaves like a physical machine. Each VM can be assigned an IP address, a name with an FQDN. Depending on the requirements, different applications or NFs can be run on different VMs. A VM can also run multiple applications or NFs, as illustrated in Figure 20.19.

As the AMF and Security Anchor Functionality (SEAF) functions are co-located in 5GC, they run on the same VM-1. UPF, which handles user plane data transfer, runs on a separate VM-2.

- Management and Orchestration (MANO) of NFV

MANO of network slices, of a 5G network, for managing its lifecycle was described earlier in Chapter 16. Similarly, the management of NFV is also required to manage the lifecycle of its various components or layers, i.e. virtualized infrastructures, VNFs, such as its instantiations, scale-out/in, termination, and so on, by system administrators. On the other hand, an orchestration of the entire NFV ecosystem with other support systems such as operation support systems and business support systems as deployed by an operator is also required. For interoperability, open interfaces/reference points are used between the NGV MANO and other support systems. Such open interfaces/reference points are standardized by European Telecommunications Standards Institute (ETSI).

As standardized by the ETSI [7, 8], an NFV framework consists of the following domains:

- VNF;
- NFV infrastructure (NFVI); and
- NFV MANO.

VNF, NFVI, and MANO of an NFV framework have been described briefly in the preceding paragraphs.

For more information on NFV reference architecture and its building blocks, requirements, and so on, as standardized by ETSI, refer to [7, 8].

Chapter Summary

This chapter presented some of the fundamental aspects of the 5GC network system architecture, mainly the separation of control and user plane functionalities of a core network element into separate software-based entities; SBA, for ease of scalability; and NFVs, for creation of logical networks through network slicing. These architectures shall become the backbone for the 5GC network along with its new radio to support various use cases and services envisaged to provide through a 5G system. The architecture, logical interfaces, and protocols used by the 5GC system are different from the previous generation's core networks. 5GC introduces the web programming and communication paradigm among its NFs in which they communicate over the HTTP which is used by different hosts also to communicate over the internet. In the previous generation core network systems, the GTP control plane (GTP-C) or other protocol, e.g. Diameter, were used for signalling communications among the network elements. However, in the 5GC, the network functions uses the newly introduced HTTP only for signaling communications among them,which is a key difference from previous generation core networks. A physical node may have been used for a network element in the previous generation's core network, but in 5GC, several logical instances for a particular network function may be used and deployed either through a distributed or localized configuration. This further enables the 5GC to support NFVs, as well as network slicing on top of it.

21

5G System

Low-level Design

Introduction

In Chapter 14, we presented several core aspects of the design and development of protocol layers as far as the 3rd Generation Partnership Project (3GPP) technical specifications are concerned. We covered protocol layer message formats and their data structures using which a layer-to-layer or peer-to-peer layer communication takes place among the network elements of mobile communications networks. Apart from these, all the relevant information and configuration data as required by a protocol layer are described by its concerned 3GPP TS.

This chapter further presents the low-level design (LLD) requirement phase of a typical Software Development Life Cycle (SDLC) in terms of identification, the definition of logical objects, and parameters that are derived from the companion technical specifications of a protocol layer. The LLD document may be used as the basis to formally start the protocol layer software development process.

This chapter begins with the 5G Core Network (5GC) service interface description and presents how a network function service interface can be designed and realized using the C++ programing language features. Then, how the network functions of the 5GC service framework can be modeled and visualized using the Unified Modelling Language (UML) in association with the C++ class diagram is presented. The usages of the C++ standard template library (STL) classes are also presented, which can be used as the data structures of the 5GC network functions and their instances. Data types of the information elements used by 5GC network function services and their application programming interfaces (APIs) are also covered in this chapter. Further, the conceptual software architecture of the Next-Generation Radio Access Network (NG-RAN) and 5GC will be also described. The definition of a profile of a network function, its services, as well as the design of the 5GC service interface through sample software code shall also be described from the LLD point of view.

21.1 Design of 5GC Service Interface and Its Operations

As described in Chapter 20, in 5GC service-based architecture framework, a network function exposes its services and operations to other network functions through different service interfaces. Each network function provides several services and their interfaces through which a consumer network function consumes different services. For example, the Network Repository Function (NRF) provides services – *nnrf-nfm* and *nnrf-disc* – for network function management and discovery services to other consumer network functions. Further, a network function service interface provides service operations, in terms of request and response procedure, under it.

Mobile Communications Systems Development: A Practical Introduction to System Understanding, Implementation, and Deployment, First Edition. Rajib Taid.

As far as the design, development, and implementation of a network function and its service interfaces and its operations are concerned, C++ programming language features such as the abstract class and inheritance can be used, as described in the next section.

21.2 Design of 5GC NF Service Interface Using UML and C++ Class Diagram

The 5GC network function (NF) service interfaces and their operations can be modeled and visualized using the general-purpose *UML*. A UML model consists of several diagrams using which the structure and behavior of a system can be prepared. One such diagram is the *class diagram* which consists of different classes, representing the different aspects or entities or functional building blocks of a particular system along with the appropriate relationship and interactions among the classes. Such diagrams are drawn with proper nomenclatures which are part of a UML modeling technique. Thus, using a UML class diagram, a simplified software system design blueprint for the entire 5GC network functions, their instances, services, operations, and interactions can be modeled. Such a representational model would help in visualizing the entire 5GC and its expected behavior and also provides enough relevant information to a developer. Correct modeling of a system using the UML method is important, as it would lead to its proper implementation as well as achieve the expected system behavior at the end.

Different network functions are the building blocks for the 5GC service-based network architecture. Such a building block or a network function can be represented through a C++ class. Also, an instance of a network function can be represented through an instance of the corresponding network function class. Further, the service interfaces and their operations of a network function can be designed and realized by using the central features of the C++ programming language. One such feature is the usages of abstract class along with pure virtual functions in it. Such abstract classes can be modeled and defined to realize each service interface, and its operations, exposed by a particular network function. A consumer network function can consume such interfaces and their services by deriving/inheriting itself from the interested abstract classes which provide the desired services.

For example, consider the NRF network function which acts as the registration and authorization server in 5GC service-based architecture framework. Every network function is required to register with as well as authorized by the NRF. The interfaces for the network function management and discovery services provided by NRF can be designed using the C++ abstract classes – *Nnrf_NFManagement* and *Nnrf_NFDiscovery* – with their service operations such as the register, update, deregister, and discover, which can be declared as pure virtual functions in it. Other network functions and their classes can derive themselves from these abstract classes of NRF. The derived/inherited classes of the other network functions shall implement the virtual functions of the abstract base classes of NRF along with network function-specific functions/information elements. Using the UML class diagram nomenclature, the design and usages of C++ abstract classes to implement service interfaces of NRF are illustrated in Figure 21.1.

A UML class diagram contains the class name, attributes, and methods, e.g. service operations. Further, other derived classes of network functions, such as Access and Mobility Management Function (AMF), and Session Management Function (SMF) can also define and implement network function-specific tasks such as initialization and startup procedure, as illustrated in Figure 21.1. Note that the inheritance or "is a" relationship that exists between the base class and its derived classes are also shown in Figure 21.1.

Each network function of the 5GC service-based framework provides certain services to other network functions. Figure 21.2 illustrates the "Has-a" relationship, with cardinality 1 and 1..*, between network function and its service class using the *composition* diagram of UML. A UML composition and "Has-a" relationship are represented through a filled "diamond"; Figure 21.2.

C++ classes with appropriate relationships, such as *is-a*, and *has-a*, among them are required to be modeled and defined through UML to represent various resources of the 5GC service-based architecture framework. A

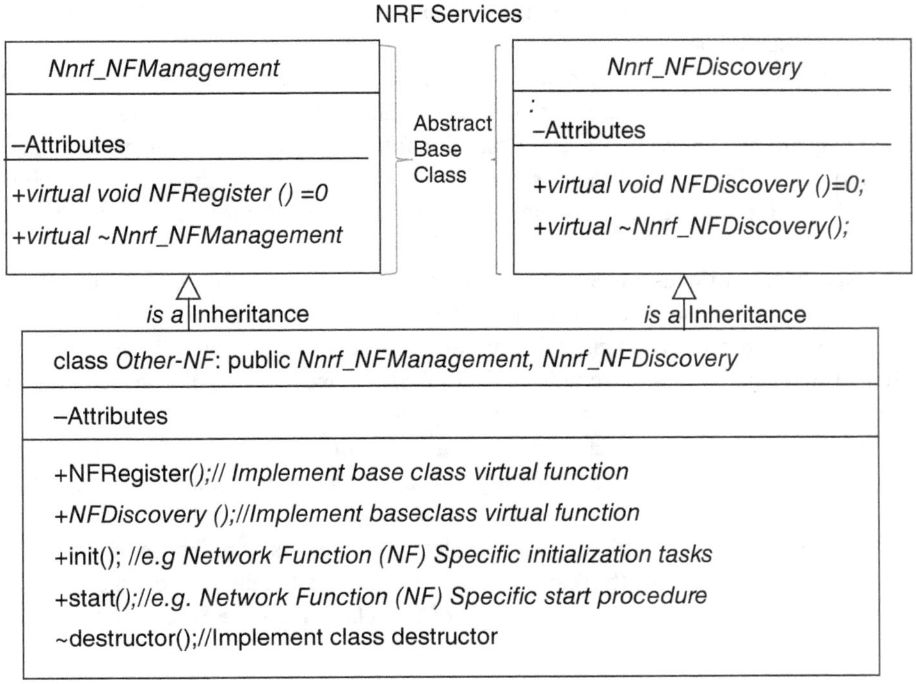

Figure 21.1 Illustration: design of NF service interface using UML/C++ class diagram.

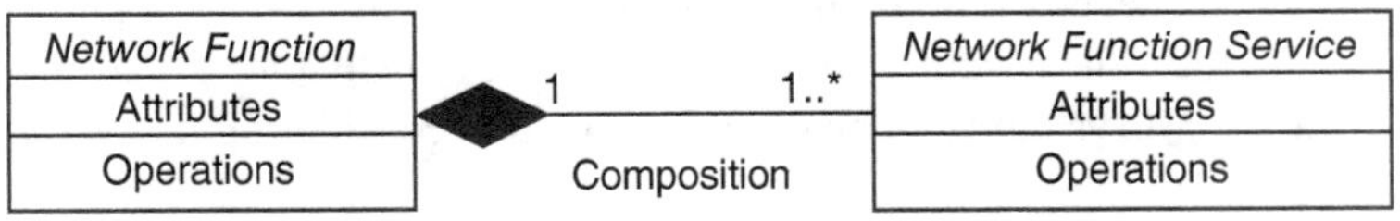

Figure 21.2 Illustration: UML composition with "Has-a" relationship class diagram relationship.

proper representational relationship between the participating classes is important for the expected behavior of the concerned network functions at runtime. Also, appropriate data structures shall be required to be selected to store the objects of different classes as modeled and designed using the UML and C++ class diagram illustrated above. The selection of an appropriate data structure is important from the performance point of view also for each of the 5GC network functions. One such data structure which is called the *Standard Template Library* (STL) is described below.

21.3 Usages of C++ Standard Template Library (STL)

For faster LLD and implementation of the different network functions of the 5GC service framework architecture, the STL of the C++ programming language can be used. STL is a set of C++ template classes that comprise containers, algorithms, and iterators. An STL container is nothing but a C++ object that stores information of a logical or physical object. STL template classes are generic and provide common and powerful data structures and operations for ease of storage, retrieval, i.e. search, and update of information of user-defined data types such as C- structures or C++ classes. STL template classes can be also used to store information on standard data types such as integer and string. Thus, by using C++ STL, a considerable portion of the overall time and effort required to develop the 5GC network functions can be saved.

STL provides different kinds of containers, in terms of data structures, such as the list, vector, set, map, and multimap, using which data of a user-defined or standard type can be stored. Each one of these data structures stores information in a different way and also provides different functionalities/operations to manage, navigate, and retrieve the stored information. Depending on how data is stored in memory, an STL container can be sequential, e.g. array, or associative, i.e. key-value, in nature. An appropriate container shall be required to be chosen accordingly.

Usages of STL containers are especially helpful as far as the runtime growing memory requirements for storing of information are concerned. For example, consider an array. The size of the array is fixed, which is determined at compile time, and there is no provision for its size to grow during the runtime of a program. STL, on the other hand, takes care of such runtime growing memory requirements and adjusts/manages the size of a particular data structure automatically, which is used to store information of logical or physical objects.

STL algorithms can be used for different purposes, such as the sorting, searching, count, find, maximum, and minimum, on the information stored in container classes. Such algorithms are becoming handy and helpful at different times of operation of a network function.

An STL iterator is like a pointer using which navigation through a collection of objects stored in a container can be performed. Further, using an iterator, an STL algorithm can update or manipulate the data stored in container classes and its object which is being pointed by the iterator. The usage of STL classes to store 5GC network functions is illustrated later in Section 21.5.

21.4 Software Architecture for 5G System

Software architecture, organization, execution model, i.e. pipelining, run-to-completion, and choice of a platform for design and developments of a mobile communication system were described earlier in Chapter 18. In this section, a brief on the software architecture for the 5G system shall be described further, considering its overall system architecture.

21.4.1 NG-RAN Logical Nodes Software Architecture

As described earlier in Section 16.4, the NG-RAN architecture contains logical nodes in terms of one central unit (CU) node, i.e. gNB-CU-CP, and multiple gNB-CU-UP nodes and multiple distribution units (DU) nodes. Multiple instances of a gNB-DU as well as gNB-CU-UP nodes, can be run to serve one or multiple cells by a gNB.

The CU node performs control plane-related functions and procedures, whereas the DU nodes perform the user plane functions and procedures such as User Equipment (UE) scheduling, resource allocation, and data transfer.

Based on the above logical nodes of an NG-RAN, a prototype of its software system may be developed and simulated to test the performance of such architecture. The NG-RAN/gNB software system architecture will comprise the following groups:

- Centralized control plane group consisting of the Radio Resource Control (RRC) and Packet Data Convergence Protocol (PDCP) layer;
- Centralized data plane group consisting of the Service Data Application Protocol (SDAP) and PDCP layer; and
- DU group consisting of the Radio Link Control (RLC), Medium Access Control (MAC), and physical layer.

Between two logical groups, the communication shall take place based on the defined interfaces, i.e. F1 or E1, between them, as described earlier in Section 16.4. Between two protocol layers, the communication shall take place through Service Access Point (SAP). Several aspects of the software architecture of NG-RAN and its logical nodes are described below:

- CPU Core Assignments

Deployment hardware platform-wise, an embedded system with a multicore processor will be more suitable for design development and deployment of an NG-RAN. On an embedded system with a multicore processor, an NG-RAN system designer can take full control of the functionalities provided by its software development kit (SDK) in terms of allocation and management of various hardware resources, i.e. cores, memory, registers, interrupts, network interfaces special hardware units, coprocessor, and so on.

On a multicore processor board, CPU core assignments can be made based on the following factors at a high level:

- Control plane path
- Data plane path
- NG-RAN slice specific

To support different network slices, the available DU cores may be further partitioned into NG-RAN slice-specific cores. Figure 21.3 illustrates a conceptual software system architecture for the NG-RAN and its 3GPP standardized slices of a 5GS for its realization on an embedded system with a multicore processor. Only a few cores are illustrated. There are commercial multicore processors with 16, 32, or 64 cores. An NG-RAN system can be designed and developed accordingly.

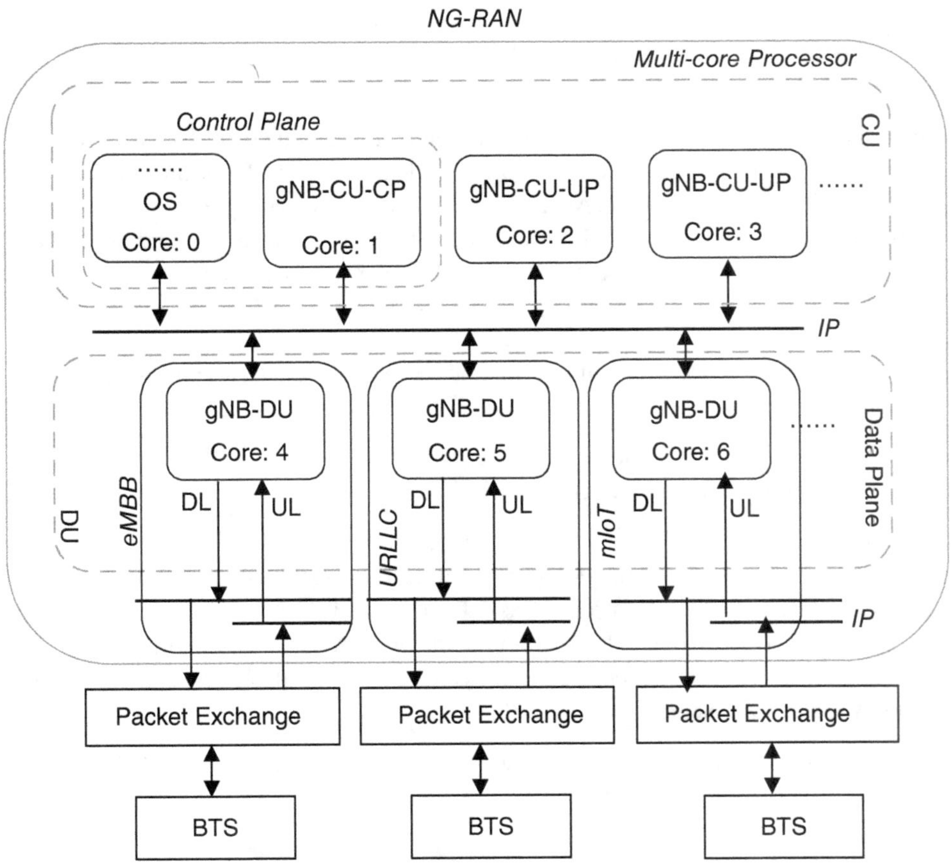

Figure 21.3 Illustration: software prototype architecture and cores assignment to logical units of NG-RAN slices of 5GS.

Capacity sizing/dimensioning of Radio Network (RNW) resources of a base station has a close relationship with the available core counts of a multicore processor. RNW resources may be physical objects such as Physical Resource Block (PRB), a base transceiver station (BTS), a trans-receiver (TRX), a timeslot, or a logical object such as a Virtual Resource Block (VRB). RNW resources are also identified from the related technical specifications of NR air interface protocol layers, e.g. RRC layer TS 38.331 [116] and physical layer 38.21x series. An RNW resource may be identified through a suitable parameter name. Typical capacity sizing of all the RNW parameters may be considered while designing the NR air interface control plane and data plane paths, and the number of cores is allocated accordingly for transmission and reception of packets between UE and NG-RAN and vice versa. Example 21.1 illustrates the typical sizing and allocation of maximum capacity of radio network resources to the per core of a CPU.

Example 21.1 Radio Network (RNW) Resources Sizing and Cores Allocation

As an example, the maximum number of RNW resources to be handled by a core to process air interface user packets in uplink and downlink directions may be dimensioned as illustrated below:

a) Maximum number of physical resource blocks in the NR: NPRB = 275 (table 5.3.2-1, TS 38.104 [103]);
b) Maximum number of subcarriers: $N_{SC} = 3300$ (N_{PRB} x12);
c) Maximum number of gNB-DU cores: $N_{CORE} = N$ (as per availability of cores on processor);
d) Maximum number of PRB per gNB-DU core: $N_{PRB}^{CORE} = N_{PRB}/N_{CORE}$;
e) Maximum number of subcarriers per gNB-DU core: $N_{SC}^{CORE} = N_{SC}/N_{CORE}$;
f) Maximum number of TSLs per gNB-DU core: $N_{TSL}^{CORE} = 10 * 32 = 320$ for SCS: 480 KHz.

Above typical sizing, parameters are to be used during the software design and implementation phase to create necessary logical objects for each physical (e.g. a TRX representing a subcarrier) or logical (e.g. VRB) resource so that a gNB supports the maximum system capacity being dimensioned. Allocating more RNW resources, to process by a core, than the typical maximum sizing/dimensioned limits as illustrated above would be reported as an error. Accordingly, processor cores will be required to be allocated based on an actual RNW resource being provisioned in a particular gNB.

- Performances

Performances of the prototype design of an NG-RAN and its logical nodes can be simulated and measured against expected benchmarks. Typical benchmarks to be evaluated can be as follows:

- End-to-end latency in the control plane, data plane, resources assignments
- Maximum theoretical data throughputs, and so on, per network slice

Further information on the minimum technical performance requirements in the case of the 5G system as defined by the International Telecommunication Union—Radio communications sector (ITU-R) can be found in their report ITU-R M.2410. For example, the IMT-2020 vision [9], as defined by the ITU-R, has the requirements for the 5G New Radio (NR) system to support a peak data rate of 20 Gbps in downlink and 10 Gbps in uplink for a UE.

The theoretical maximum data throughput to be supported by the 5G NR can be simulated and computed based on the combinations of the following factors, as described in TS 38.306 [112]:

1) A maximum number of component carriers (#16) in case carrier aggregation is used
2) Scaling factor
3) Number of layers
4) Modulation and coding scheme
5) Different subcarrier spacings

6) Different operating frequency bands for FR1 and FR2
7) Transmissions bandwidth from 5 to 400 MHz
8) Overheads due to various reference signals

- Runtime Choices

An SDK of a multicore embedded system may provide different runtime choices also to run applications differently. Each runtime choice may have different overheads in terms of processor time, memory requirements, and so on. Accordingly, an appropriate runtime choice may be selected for a particular network function depending on its nature and role. For example, an instance of gNB-DU can be just run as a standalone instance on a particular core. An instance of gNB-CU-CP can be run along with an operating system on a particular core. Alternatively, an application per core may be also run.

A standalone application will run faster than the application being run with an operating system as there will be overhead associated with scheduling by the operating system. Accordingly, a system designer can partition and assign the available cores to different logical units of NG-RAN as mentioned above.

- Event-Driven Vs. Interrupt-Driven Packet Processing

Further, information packet processing by a core may be based on either an event-driven or polling-based or interrupt-driven method. An event-driven or polling-based method of packet processing is more efficient than the interrupt-driven method. In the case of the polling method, a core executes and completes a task when it is available; when there is no work to do, a core keeps looking or looping for a work. In the case of the interrupt-driven method of packet processing, there may be a time delay between the arrival of a packet and its actual time of notification to a core. Accordingly, an appropriate runtime packet processing method may be used for a particular core.

21.4.2 5GC Software Architecture

As described earlier in Section 20.2, a 5GC is built on the service-based architecture which consists of multiple network functions as well as multiple instances of a network function that can be run to meet the requirements of different use cases and network slices of 5GS. For example, separate SMF may be run for each network slice. Similarly, separate UPF may be run to transfer user traffic between a UE and its data network.

Figure 20.18 shown earlier illustrates the software system architecture of the 5GC for its realization on a virtualized hardware platform with a multicore processor. Here, more cores are assigned to handle user plane data traffic at the 5GC end. The UPF network function, i.e. application software, is being run on 4 cores, resulting in a parallel packet processing to transfer user plane data or to provide redundant transmissions paths for user plane data, i.e. duplicating the user plane data of a PDU session of a service that belongs to a use case such as an Ultra Reliable and Low-Latency Communications (URLLC) service.

Good software architecture and design should provide more scalability, represented by dots in Figure 20.18, in terms of provisioning of adding more cores, to meet growing demands of high data throughput requirements or provision of new services over time. If a prototype simulation result does not meet the expected performances of the NG-RAN or 5GC, the overall software architecture along with their core assignments must be revisited as part of an iterative system design round or process on a given multicore platform.

21.5 Data Types Used in 5GC SBI Communications

In this section, typical data types used in 5GC services-based interface (SBI) communications shall be covered. As part of 5GC service-based architecture, network functions communicate with each other through their service interfaces and their APIs over the HTTP. Network functions and their application protocol-specific information are passed as a payload over the HTTP. Such application protocol-specific communications which are made through network function service APIs contain information elements of different types.

Data types of information elements and their contents, used by each service API and its operations, are described pretty clearly by the corresponding technical specification. In general, the type of information element or data passed using an API for a particular service operation is classified into the following categories:

- Simple, primitive, or basic data types such as the integer, Boolean, and string type;
- Enumeration type: Usages of enumeration type over preprocessor using #define are preferred as the compiler can perform a data type check over an enumerated data;
- Structured type, which can contain simple, enumerated type also. Protocol information, such as UE context, mobility and session context and network function instance profile, is packed into a structured type and passed as a payload through HTTP from a source to a destination network function.

Common data types for information elements used in the 5GC service framework API-based communications are defined in the 3GPP TS 29.571 [78]. Such simple or primitive data types used in the 5GS conform to the open API standard specification. In addition to this, the data types for other information elements that are specific to a particular network function service/API are defined by the corresponding technical specification, i.e. 3GPP TS 29.5xx [78] series. For data types that are readily available and defined by such technical specifications can be converted into an LLD for their implementation through a computer program. Definitions of such data types are illustrated below, starting from the simple or primitive type, in the form of sample software code. To illustrate the 5GC NF service interface design using the C++ class diagram shown in Figure 21.1, the definition of abstract classes and their usages are also provided through sample software code.

- Type definition of some of the simple data type for generic usages

```
typedef string Mnc, Mcc;
typedef string Uri;
typedef unsigned integer Uint32;
typedef string Ipv4Addr, Ipv6Addr;
typedef string Bytes , Binary.
typedef String Date, DateTime.
typedef Unsigned integer PduSessionId
```

- Definition of common enumeration data type

```
/* Define the enumeration type for the status of a network function and its
services */
enum NFStatus_NFServiceStatus{REGISTERED,SUSPENDED,UNDISCOVERABLE};
/*Define the enumeration type for UE PDU session creation request type */
enum RequestType {"INITIAL_REQUEST","EXISTING_PDU_SESSION","INITIAL_EMERGENCY_
REQUEST","EXISTING_EMERGENCY_PDU_SESSION"};
/*Define the enumeration type for UE PDU session update, i.e. modification,
release and so on, request type */
enum RequestIndication {"UE_REQ_PDU_SES_MOD","UE_REQ_PDU_SES_REL","PDU_SES_
MOB","NW_REQ_PDU_SES_AUTH","NW_REQ_PDU_SES_MOD","NW_REQ_PDU_SES_REL","EBI_
ASSIGNMENT_REQ"};
/*Define the enumeration for different types, i.e. 21, of network functions */
enum NFType{NRF,UDM,AMF,SMF,AUSF,NEF,PCF,SMSF,NSSF,UDR,LMF,GMLC,EIR_5G,SEPP,UP
F,N3IWF,AF,UDSF,BSF,CHF,NWDAF};
/*Define the enumeration type for different services, around a total of 32, of
each network functions. Refer to table 6.1.6.3.11-1, TS 29.510 [76] */
```

```
enum NFService{NNRF_NFM, NNRF_DISC,................};
/*Define the enumeration type for URI scheme */
enum scheme {http, https);
```

- Definition of a public land mobile network

```
/*Public Land Mobile Network: used in GSM, GPRS, UMTS,
LTE and 5G system */
typedef struct /* Definition of a PLMN structure */
{
Mnc mobile_country_code;
Mcc mobile_network_code;
} plmn_type;
```

- Definition of structure – *problem* details – that can be used as a data type to report HTTP error cause code and its reason

```
typedef struct {//TS 29.571 [78]
string param;
string reason;
}InvalidParameter;//Structure for Invalid Parameter

typedef struct {//TS 29.571 [78]
        string type;
        string title;
        Uint32 status;
        string detail;
        string instance;
        string cause;
        InvalidParameter  InvalidParam;

} ProblemDetails; //Structure for ProblemDetails report
```

- Definition of a *NFProfile* structure, containing some of the Information Elements (IEs) of a network function profile, which is used as part of network function management service; refer to TS 29.510 [76].

```
typedef struct
{//Beginning of Definition of Network Function Profile Structure
        string nfInstanceID;
        NFType nfType;
        NFStatus nfStatus;
        string nfInstanceName;
        UINT32 heartBeatTimer;
        PlmnId plmnList[];
        Fqdn fqdn; // fqdn is a string type
        Fqdn interPlmnFqdn;
        Ipv4Addr ipv4Addresses;
        Ipv6Addr ipv6Addresses;
        NFService nfServices[];
        .......
//Define the common helper member functions to set/read the IEs of a Network
function Profile
```

```
void SetnfInstanceID(string nfinstanceid)
{//For Register/Deregister/Update  a NF instance
nfInstanceID = nfinstanceid;
.............
}/////

void SetNFType(NFType nftype)
{
nfType = nftype;
.............
}////////
void SetNFStatus(NFStatus  nfstatus)
{
nfStatus = nfstatus;
.............
}///////
void SetnfInstanceName(string nfinstnacename)
{
nfInstanceName = nfinstnacename;
.............
}/////
void SetHeartBeatTimer(string nfHeartbeattimer)
{
heartBeatTimer = nfHeartbeattimer;
.............
}/////

typedef struct
{//Beginning of Definition of Network Function Profile Structure
       string nfInstanceID;
       NFType nfType;
       NFStatus nfStatus;
       string nfInstanceName;
       UINT32 heartBeatTimer;
       PlmnId plmnList[];
       Fqdn fqdn; // fqdn is a string type
       Fqdn interPlmnFqdn;
       Ipv4Addr ipv4Addresses;
       Ipv6Addr ipv6Addresses;
       NFService nfServices[];
       .......
//Define the common helper member functions to set/read the IEs of a Network
function Profile

void SetnfInstanceID(string nfinstanceid)
{//For Register/Deregister/Update  a NF instance
nfInstanceID = nfinstanceid;
.............
}/////
```

```
void SetNFType(NFType nftype)
{
nfType = nftype;
.............
}////////
void SetNFStatus(NFStatus  nfstatus)
{
nfStatus = nfstatus;
.............
}///////
void SetnfInstanceName(string nfinstnacename)
{
nfInstanceName = nfinstnacename;
.............
}/////
void SetHeartBeatTimer(string nfHeartbeattimer)
{
heartBeatTimer = nfHeartbeattimer;
.............
}/////

//..continued from previous page..
void SetServicesName(NFService Services[])
{//list of services prvided by  a network function
nfServices = Services;
.............
}/////
}NFProfile;//End of Definition of Network Profile Structure
```

- The definition of an *NFService* structure containing some of the IEs of a network function service instance, which is used as part of the network function profile defined above, is defined below. Like the NF profile, each NF service instance is assigned a unique instance identifier, which is also a *Universally Unique Identifier* (UUID). Refer to TS 29.510 [76] for more details on IEs of the *NFService* structure.

```
typedef struct
{//Beginning of Definition of Service Instance Structure
       string serviceInstanceId;
       ServiceName serviceName;
       NFServiceVersion versions;
       UriScheme scheme;
       NFServiceStatus nfServiceStatus;
       .......
//Define the common helper member functions to set/read the IEs of a Network
function Profile

void SetnfServiceInstanceID(string nfServinstanceid)
{//For Register/Deregister/Update  a NF instance
serviceInstanceId = nfServinstanceid;
.............
} NFService; //End of Definition of NF Instance Structure
```

```
typedef struct
{//Beginning of Definition of Service Instance Structure
        string serviceInstanceId;
        ServiceName serviceName;
        NFServiceVersion versions;
        UriScheme scheme;
        NFServiceStatus nfServiceStatus;
        .......
//Define the common helper member functions to set/read the IEs of a Network
function Profile

void SetnfServiceInstanceID(string nfServinstanceid)
{//For Register/Deregister/Update  a NF instance
serviceInstanceId = nfServinstanceid;
.............
} NFService; //End of Definition of NF Instance Structure

typedef struct
{//Beginning of Definition of Service Instance Structure
        string serviceInstanceId;
        ServiceName serviceName;
        NFServiceVersion versions;
        UriScheme scheme;
        NFServiceStatus nfServiceStatus;
        .......
//Define the common helper member functions to set/read the IEs of a Network
function Profile

void SetnfServiceInstanceID(string nfServinstanceid)
{//For Register/Deregister/Update  a NF instance
serviceInstanceId = nfServinstanceid;
.............
} NFService; //End of Definition of NF Instance Structure
```

- Definition of a *SubscriptionData* and *NotificationData* structure, containing the IEs of an event subscribed by a network function, which is used as part of network function management service; refer to TS 29.510 [76]

```
typedef struct
{
Uri nfStatusNotificationUri; //Uri is a string type

.......

}SubscriptionData;//End of Definition of Network Profile Structure
//Define a NotificationData structure containing the status of Network
function. Refer to TS 29.510 [76]
typedef struct
{
NotificationEventType event; //NotificationEventType is enumeration type
}NotificationData;//End of Definition of Network Profile Structure
```

- Definition of abstract class – *Nnrf_NFManagement* – as the pure interface for the network function management service of NRF; refer to TS 29.510 [76].

  The abstract class *Nnrf_NFManagement* declares pure virtual functions such that they can be used as an interface to access the network function management service of NRF by other network functions. The virtual functions shall be implemented by the derived classes of the respective network functions. Note that NF instance-specific parameters, e.g. *UDRInfo*, *AUSFInfo*, and *UDMInfo*, are to be defined at the derived class of a particular network function. Some typical helper member functions, e.g. get and set, to read/write general/mandatory parameters of NF Instances are also illustrated through sample code below:

```
class Nnrf_NFManagement //Abstract Class for NF Management Service of      NRF
{
//Pure virtual functions for Request operation only.
public:
virtual void NFRegister(NFProfile nfProfile)=0;
virtual void NFUpdate(NFProfile nfProfile)=0;
virtual void NFDeRegister(string nfInstanceid)=0;

virtual void NFProfileRetrieval(string nfInstanceid)=0;
virtual void NFStatusSubscribe(SubscriptionData subscriptionData)=0;

virtual void NFStatusNotify(NotificationData notificationData)=0;
virtual void NFStatusUnSubscribe(string subscriptionid)=0;
};//End of Nnrf_NFManagement
```

- Definition of abstract class – *Nnrf_NFDiscovery* – as the pure interface for the network function discovery service of NRF; refer to TS 29.510 [76]. NF instance-specific parameters are to be defined at the derived class of the specific network function.

```
class Nnrf_NFDiscovery
{
//Define the IEs used by the Network Function Discovery Service of NRF. Refer
to TS 29.510 [76].
protected:
NFType  targetNFtype;
NFType  requesterNFtype;
NFProfile nfProfile[];
//Pure virtual functions: Request/Response
public:
virtual void DiscoveryRequest(NFType  targetnftype, NFType  reqNFtype)=0;
virtual void DiscoveryResponse(NFProfile nfprofile[])=0;
.............
void SetTargetNFType(NFType nftype)
{//For discovery of a NF instance
targetNFtype=nftype;
.............
}////
void SetRequestingNFType(NFType nftype)
{//For discovery of a NF instance
requesterNFtype=nftype;
```

```
.............
}///
};//End of Nnrf_NFDiscovery
```

- Definition of abstract class – *Nnrf_AccessToken* – for access token allocation service of NRF to other network function; refer to TS 29.510 [76].

```
class Nnrf_AccessToken
{
//Defining the Mandatory and Conditional IEs of Access Token //Request/Response
used by the Network Function Access token Service
protected:
AccessTokenReq accessTokenReq;
AccessTokenRsp accessToken;//contains JSON Web Token
//Pure virtual functions
public:
virtual void AccessTokenRequest(AccessTokenReq accessTokenRequest)=0;
virtual void AccessTokenResponseAccessTokenRsp accesstoken )=0;
void SetAccessTokenRequesting(AccessTokenReq token)
{//For authorization of a  NF instance
accessTokenReq= token;

.............
}//
};//End of Nnrf_AccessToken
```

The NRF abstract classes – *Nnrf_NFManagement, Nnrf_NFDiscovery,* and *Nnrf_AccessToken* – defined above can act as the service interfaces for the other network functions of 5GC. Consider that the Access and Mobility Management (AMF) network function desires to use the different services of NRF and their operations through these interfaces. To accomplish this, an AMF class can be derived from these abstract base classes, as illustrated below:

```
class AMF: public Nnrf_NFManagement, Nnrf_NFDiscovery, Nnrf_AccessToken
{
private:
//Define AMF specific member variables/IEs
string nfProfile.nfInstanceID;
public:

//Implement the virtual functions outside the class
void NFRegister(NFProfile nfProfile);
void NFUpdate(NFProfile nfProfile);
void NFDeRegister(string nfInstanceid);
void NFProfileRetrieval(string nfInstanceid);
void NFStatusSubscribe(SubscriptionData subscriptionData);
void NFStatusNotify(NotificationData notificationData);
void NFStatusUnSubscribe(string subscriptionid);
void DiscoveryRequest(NFType  targetnftype, NFType  reqNFtype);
void DiscoveryResponse(NFProfile nfprofile[]);
void AccessTokenRequest(AccessTokenReq accessTokenRequest);
void AccessTokenResponseAccessTokenRsp accesstoken);
```

```
.............
Virtual ~AMF();
void initialization (…);//Initialize AMF NF
void startAMF(…);//start the NF
//Declare AMF specific member functions
};
```

The AMF class implements the above virtual functions, i.e. NRF services operations, in actual, as illustrated below:

```
//NRF: NF Management Service and its operations- Only Request Part
void AMF::NFRegister(NFProfile nfProfile)
{
nfProfile.nfInstanceID=GenerateUUID();//External function to generate UUID
and assigned to NF instance ID
nfAMFInstanceID = nfProfile.nfInstanceID;
.............
//CALL HTTP PUT passing the nfProfile
}
//Update an AMF instance to NRF
void AMF::NFUpdate(NFProfile nfProfile)
{
.............
//CALL HTTP PUT passing the nfProfile
}

//DeRegister  an AMF instance from NRF
void AMF::NFDeRegister(string nfInstanceid)
{
.............
//CALL HTTP DELETE passing the nfInstanceid of Network function instance
}
//Retrieve an AMF instance from NRF
void AMF::NFProfileRetrieval(string nfInstanceid)
{
.............
//CALL HTTP GET passing the nfInstanceid of Network function instance
}
//Subscribe events to be notified from NRF
void AMF::NFStatusSubscribe(SubscriptionData subscriptionData)
{
.............
//CALL HTTP POST passing the subscriptionData containing the event
}
//Process Subscribed events notified from NRF
void AMF::NFStatusNotify(NotificationData notificationData)
{
.............
//Process Subscribed events notified  from NRF using the HTTP POST
}
```

```
//Cancel subscription of an event subscribed earlier to NRF
void AMF::NFStatusUnSubscribe(string subscriptionid)
{
.............
//CALL HTTP DELETE passing the subscriptionid of subscribed event
}
/////NRF NF Discovery Service and its operations
void AMF::DiscoveryRequest(NFType  targetnftype, NFType  reqNFtype)
{
.............
//CALL HTTP GET to   discover  the Network functions instances and their ser-
vices offerings
}

void AMF::DiscoveryResponse(NFProfile nfprofile[])
{
.............
//Process the network functions instances and their services offerings returned
by  the NRF
}
/////NRF NF Access Token Allocation Service and its operations
void AMF::AccessTokenRequest(AccessTokenReq accessTokenRequest)
{
.............
//CALL HTTP POST to  make Access Token Request to the NRF
}
void AMF::AccessTokenResponse(AccessTokenRsp accesstoken)
{
.............
//Process the Access token returned by  the NRF
}
```

- Usages of STL container to store network function information

As mentioned briefly earlier in Section 21.2, STL provides different types of data structures to store, update, and retrieve information efficiently with runtime memory management of its own. Consider the STL associative container *map* for storing of network function profile and network function instance. The following structure/C++ classes were defined earlier.

- *NFProfile*, a C-structure containing information of a network function profile
- *NFService*, a C-structure containing information of a network function service
- AMF, a C++ class representing the AMF network function

The network function information mentioned above can be stored through a *map* data structure, which is an associative container provided by the STL, as illustrated below:

```
void storeNF(string nfInstanceID,.....)//Store NF instances in  map
{
map< nfInstanceID, NFProfile> nfProfile;//declare map container
map<nfInstanceID, NFProfile>::iterator itr; //declare map iterator
NFProfile abc;
```

```
abc.nfInstanceID=nfInstanceID;//UUID
.............
nfProfile[abc.nfInstanceID]=abc; //store NF profile: abc in STL map
for (itr= nfProfile.begin(); itr!= nfProfile.end(); ++itr) {
 cout << "\nNF Instance Id: "<< itr->first;
}
map< nfAMFInstanceID, AMF>nfAMF;//declare map container: nfAMF
AMF abc;//instantiate AMF NF instance
Abc.nfAMFInstanceID ="12345......abcd"; //UUID
nfAMF[Abc. nfAMFInstanceID]=abc; //store AMF NF instance: abc in map
```

In the typical illustration shown above, the network function instance identifier, i.e. *nfInstanceID* or *nfAMFInstanceID,* is being used as the key for the value of user-defined type *NFProfile* or *AMF*. These user-defined types are stored in the respective STL *map* container, i.e. *nfProfile* or *nfAMF*. The usages of an iterator of the map container are also illustrated to navigate through the collection of network function instances and print the instance identifiers.

Similarly, each service provided by a network function has its service instance identifier. The service instance identifier *serviceInstanceId* is being used as the key for the value of user-defined type *NFService* being stored in its STL *map* container *nfService*, as illustrated below:

```
map< serviceInstanceId, NFService> nfService;//declare map container
NFService abc;
Abc.serviceInstanceId="12345......abcd"; //UUID
nfService[Abc.serviceInstanceId]=abc; //store NF Service: abc in map
```

A standard/primitive data type, e.g. int., string, or user-defined value/data type, e.g. structure or class, is stored against a corresponding key in an STL map container as illustrated above. However, if an object of a class is stored in a map container, then certain overloaded operators, such as the comparison operator, equal operator, copy, and default constructors, are to be provided as part of the class definition. Providing such customized overloaded operator functions ensures the expected behavior of a network function at runtime.

- Definition of 5GC Quality of Service (QoS)

A QoS rule consists of, among other things, a QoS flow identifier (QFI), the number of packet filters to be used, and a list of packet filters and their purposes, i.e. create, delete, or modify a QoS rule. A QFI identifies a QoS flow description that contains the QoS-related parameters, as illustrated below:

```
/* Refer to TS 24.501 [47] */
#define 5GS_MAX_SESSIONS 0XFF

/*----Define QoS rule identifier--- */
#define QRI                    0x0 /* No QoS identifier Assigned */
#define QRI1                   0x1

....................
#define QRI255                 0xFF

/*---Define QFI----------------------*/
#define QFI                    0x0 /* No QFI identifier Assigned */
#define QRF1                   0x1

.............
#define QRF63                  0x3F
```

```
/*---Define Packet filters----*/
#define PFI0                0x0

.............
#define QRF15              0xF

/* --------Define 5GS PDU Session Types-- */

#define IPV4_SESSION_TYPE    0x1
#define IPV6_SESSION_TYPE    0x2
#define IPV4V6_SESSION_TYPE  0x3
#define EHERNET_SESSION_TYPE 0x5

Typdef struct{/* QoS flow Description */
       uchar qfi;/ * QFI: 6 bits */
       uchar opcode;/ * operation code: 3 bits */
       uchar num_paramet;/ * number of parameters*/
      qfi_param param[];
}qfi_flow_description;

Typdef struct{/*define QFI parameter */
uchar param_id;
uchar param_value;
}qfi_param;

/* define param_id */
#define  5QI 0x1
#define  GFBR_UPLINK 0x2

.............
#define  EPS_BEARER_ID 0x7
```

- Network slice-related definitions

```
/*Define  Network Slice Type and Value: TS 23.501 [40]*/
#define NS_SST_eMBB 1
#define  NS_SST_URLLC 2
#define NS_SST_mIoT 3
#define NS_SST_V2X 4
/* Refer to TS 24.501 [47] */

Typedef struct{
              char  iei_type:4;
           char  spare:1
           char  spare:1
              char  dcni:1; / * Default configured NSSAI indication */;

char  NSSCI:1; / * Network slicing subscription change indication*/
}NSI_t; /* Network slicing indication*/
#define MAX_ALLOWED_NSSAI 8
#define MAX_CONFIGURED_NSSAI 16
```

```
#define SD_RESEREV 0xFFFFFF
typedef struct{
            string amfid;/*Associated AMF identifier */
            string tac; ;/*Associated TAC */
            PlmnId plmnid; ;/*Associated PLMN identifier */
            char sst_type;
            char sd[3];/* optional */
            char mapped_hplmn_sst;/* optional */
            char mapped_hplmn_sd[3];/* optional */
} s_nssai_t;
typedef struct{
              s_nssai_t s-nssai[MAX_ALLOWED_NSSAI];
              } NSSAI_allowed;
typedef struct{
              s_nssai_t s-nssai[MAX_CONFIGURED_NSSAI];
              } NSSAI_configured;
typedef  strcut {
            char rejected_s-nssai:4;
            char  reject_cause:4;
              NSSAI_allowed rejected_NSSAI;
              }rejected_s-nssai_t;
Typedef struct{
            char  iei_type4;
            char  spare:1
            char  spare:1
            char  nim:2;/ * NSSAI inclusion mode;
            }NSSAI inclusion mode;
```

- 5G mobility management and session management-related definitions

```
/*Define  5G MM and SM NAS Layer messages*/

/* 5GS mobility management messages: TS 24.501 [47]*/
#define REGISTRATION_REQUEST 0X41
#define REGISTRATION_ACCEPT  0X42
#define REGISTRATION_COMPLETE 0X43
/* 5GS session management messages*/
#define PDU_SESSION_ESTABLISHMENT_REQUEST 0XC1
#define PDU_SESSION_ESTABLISHMENT_ACCEPT  0XC2
#define PDU_SESSION_AUTHENTICATION_COMMAND 0XC5
#define PDU_SESSION_AUTHENTICATION_COMPLETE 0XC6
```

Chapter Summary

This chapter presented some of the important aspects of the LLD requirements of 5GC functions. As described earlier in Chapter 18, the development platform, language, and tools have to be chosen depending on the requirements of a particular system and its applications. As far as the 5GC architecture and its network functions are concerned, the C++ programming language is found to be the appropriate development language for their implementation and realization along with a suitable modeling tool such as the UML. Using a modeling language, the expected behaviors or functionalities and interactions by a network function can be visualized through different building blocks, e.g. a C+ class diagram, object diagram, and so on with appropriate relationships among the building blocks. Services produced/exposed and consumed by a particular network function may be modeled as the operations or methods supported by the C++ class. A network function service interface may be represented through an interface provided by a C++ class. Further, a C++ class also allows code reusability, which is appropriate as far as the 5GC functions and their instances are concerned. This chapter also presented the usages of the STL, a powerful development tool/library, which reduces development time as well as provides ease of maintenance of 5GC network functions over time.

A brief description of the multicore processor-based software architecture and system capacity and sizing aspects of NG-RAN and 5GC was also covered. This chapter also presented the definition of the typical profile of a network function through a sample C-structure derived from the concerned 3GPP technical specification. Design and implementation of network functions, their services, and interfaces through the abstract and derived class concept of C++ programming language were also presented.

22

3GPP Release 16 and Beyond

Introduction

The 5G system (5GS) has been described in Chapters 16–21 based on its first base Release 15 as defined by the 3rd Generation Partnership Project (3GPP). These chapters have attempted to provide overall concepts to the reader on the various key aspects of the 5G system, its use cases, and features such as the network slicing, network functions virtualizations, and so on, along with the architecture of its radio access network as well as the core network. Further, the protocol stacks and their different layers of the 5G New Radio (NR) air interface and core network have been described. Some of the important functions and procedures of the different protocol layers of 5G NR and their differences from the legacy Long-Term Evolution (LTE) system have also been described.

In this chapter, a brief highlight of the various enhancements of existing features and the addition of new services or capabilities requirements, in terms of reliability, mobility, latency, and so on, added as part of the 3GPP Release 16 and Release 17 of the 5G system shall be described. Such enhancements and the addition of new features/capabilities spread across the different use cases as well as the radio access network and the core network of the 5G system.

22.1 5GS Enhancements as Part of Release 16

Some of the enhancements made, as part of 3GPP Release 16, in the existing features of Release 15 features of the 5G system are described below:

- Enhancement of the 5G core network to support enhanced Ultra Reliable and Low Latency Communications (URLLC) through redundant transmissions for high reliability, Quality of Service (QoS) monitoring, enhancement of session service continuity (SSC), and so on for new application areas or verticals. User plane redundant transmission paths may be realized in several ways, for example, by creating a redundant tunnel (N3) between the NG-RAN/gNB and a User Plane Function (UPF); or end-to-end redundant transmissions paths are created from a UE to a UPF based on UE dual connectivity; or by inserting an intermediate UPF (I-UPF) and creating redundant tunnels (N3 and N9) between NG-RAN/gNB and an I-UPF and between I-UPF and UPF. SSC was described earlier in Chapter 18.
- Enhancement of the physical layer to support more stringent reliability requirements for some services, such as Vehicle to Everything (V2X), under the URLLC use case. Enhancement has been introduced to the uplink and downlink control channels, physical uplink shared channel (PUSCH), and so on.
- Enhancement of multiple-input multiple-output (MIMO) system from increasing system capacity and multiple beam management points of view. Enhancements such as the enhanced Type-II codebook containing precoding matrix for multiuser MIMO, multiple transmission/reception point (TRP), i.e. 5G Base Station (gNBs),

Mobile Communications Systems Development: A Practical Introduction to System Understanding, Implementation, and Deployment, First Edition. Rajib Taid.

transmissions, physical layer signal-to-noise and interference ratio (L1-SINR) measurement and reporting, and so on. In the NR Release 15, a User Equipment (UE) can report up to two layers as part of its Type-II Channel State Information (CSI) report to gNB, whereas the UE can report up to four layers in Release 16. As part of MIMO enhancement, a UE can expect to monitor a maximum of two Physical Downlink Control Channels (PDCCHs) (RRC IE: *coresetPoolIndex*) from two TRPs, i.e. gNB, for data reception over multiple PDSCHs. Similarly, UE performs beam failure recovery for a secondary cell (SCell) also.

- UE battery power saving by adapting dynamically, in its Radio Resource Control (RRC) CONNECTED, IDLE, and INACTIVE states, to different power saving features such as bandwidth part, usages of maximum MIMO layers, reducing of radio resources measurements, and so on.
  - Enhancement of service-based architecture of the 5G core network. The enhancement allows an indirection communication mechanism, e.g. a message router, between two network functions of the 5G Core Network (5GC) through a *Service Communication Proxy (*SCP*)* network function. A 5GC consumer network function may pass on the task of the discovery of a producer network function to the SCP, which is called *delegated discovery* procedure within the 5GC in Release 16. In Release 15 5G core network, network functions can make direct communication with each other over HTTP. Further, the concepts of Network Function Set (NF Set) and Network Service Set, which consists of network function instances or network service instances of the same type, have been introduced as part of the 3GPP Release 16. Network function or service instances within an NF set or network service set have access to various context information, e.g. UE context information. Usage of AMF set (in Release 15) was illustrated earlier in Chapter 7/Section 7.1 and Chapter 16/ Section 16.9.

    5GC was also enhanced to support better mobility of a UE whenever it moves out of its initial location. To handle such mobility requirements of UEs in a location where no user plane connectivity (i.e. N3 tunnel) exists between the serving NG-RAN/gNB and current User Plane Function (UPF), an intermediate Session Management Function (SMF) (I-SMF) and intermediate UPF (I-UPF) was added as part of the 5GC architecture in 3GPP Release 16. The I-SMF is inserted to ensure session continuity of the UE whenever the current SMF cannot control the I-UPF where the N3 tunnel terminates from the serving NG-RAN/gNB.
- Architectural enhancement to the 5G system to support V2X services for vehicular communications, which is already part of the legacy LTE/Evolved Packet System (EPS) system. Interworking between the 5G and LTE/ EPS for the V2X services will be also supported. Around 25 use cases of V2X vertical are being identified, such as vehicle platooning, to form dynamically a group of vehicles; vehicles coordinated communications using extended sensors for avoiding vehicle collision and traffic safety; and so on. A new network slice/service type (SST) = 4 was added as part of 3GPP Release 16 to support the V2X enhancement.
- Enhancement of network slicing for interworking between LTE/EPS and 5G system by enabling the 5G core Access and Mobility Management Function (AMF) network function to support all the data sessions of a UE from the source to the target system. Further, unlike Release 15, 3GPP Release 16 also allows AMF and SMF reallocation when a UE moves from LTE/EPS to the 5G system. In addition to this, an enhancement to perform per network slice-specific authentication and authorization requirement has been added.
- Provision for voice call continuity, through the Single Radio Voice Call Continuity (SRVCC), from the 5G Next-Generation Radio Access Network (NG-RAN) to the 3G UMTS terrestrial radio access network (UTRAN) and not vice versa will be supported. Support of the legacy SRVCC was not available in Release 15 of the 5G system. Voice call continuity from the LTE/EPS to the 3G system was described earlier in Chapter 6.

22.2 5GS New Features as Part of Release 16

Some of the new features for different application areas or verticals such as the industrial internet of things and electrical power distribution factory automatons that are being envisaged as part of the 3GPP Release 16 of the 5G system are described below:

- Support for LAN-type services

5G system supports LAN-type services through a 5G network. A 3GPP Release 16-compliant 5G system would enable its operator to provide voice, data, multimedia services, beyond its Public Land Mobile Network (PLMN), to UE and its user over a local area network type of services deployed at different locations such as a resident, office, and factory. 5G LAN-type services may be deployed over the licensed as well as unlicensed spectrums.

- Support of the Ethernet Transport Services

5G system supports the Ethernet Transport Services for an Ethernet PDU session type. This feature is required to be supported by a UE to enable it to exchange Ethernet frames, over 5G LAN-type services mentioned above, with a device connected to an Ethernet data network.

- Support of industrial Internet of Things (IoT)

5G system supports industrial automation through the *industrial IoT* and its different use cases such as factory automation, transport, and electrical power distributions. The 5G system will support integrations and communications with such industrial time-sensitive networks (TSNs), requiring accurate time synchronization, through highly deterministic time-sensitive data packets communications, for example, between a TSN controller and TSN end device. These features impact the NR Layer 2 and Layer 3 protocol layers to support low-latency and high-reliability communications.

- Support of a Non-Public Network (NPN)

5G system supports NPN for private network usage purposes by an entity or organization with its dedicated resources. Such an NPN may be operated independently without relying on a PLMN. However, an NPN may be integrated with a PLMN also. A brief on the NG-RAN sharing with an NPN was described earlier in Chapter 7.

- Two-step Random Access Channel (RACH) procedure

5G system supports a two-step RACH procedure to reduce signaling overhead and latency in comparison to the four-step RACH procedure in the Release 15 of 5G NR. In a two-step RACH procedure, the number of signaling messages between a UE and gNB is reduced during connection setup and resume procedure initiated by UEs/devices across the different use cases of the 5G system.

- Enhancement of UE mobility

5G system supports enhanced UE mobility through the *Dual Active Protocol Stack (*DAPS*)* handover procedure where a UE maintains an RRC connection with the source gNB until the UE is successfully handed over to the target gNB. DAPS is a kind of make-before-break kind of handover procedure. UE mobility enhancement in Release 16 aims to improve the performance of a handover procedure in the 5G system, especially in the high-frequency range (FR2) of NR. To improve the robustness of a handover procedure, a *Conditional Handover* procedure is also specified as part of the NR Release 16.

- NR operations in unlicensed spectrum

5G system supports its services over a shared/unlicensed spectrum also to increase capacity, by allowing the NR to operate in 5 and 6 GHz frequency bands. Currently, such unlicensed frequency bands are also used by other technologies such as the Wi-Fi communications which use the 5 GHz frequency band. Using an unlicensed spectrum, the NR can operate as a standalone system only. On the other hand, using an unlicensed spectrum, the NR can also operate in paired/anchored with a carrier on a licensed frequency band.

- Support of an Integrated Access and Backhaul (IAB)

5G system supports an *IAB* solution for the NR to provide both wireless access and backhauling connectivity between base stations. Thus, instead of using fixed fiber optics-based backhaul connectivity currently in use, the

NR IAB can be used as an alternative and cost-efficient wireless-based backhaul solution to increase the 5G service coverages, especially through base stations using the mmWave frequencies. An NR IAB node/base station can serve normal UEs access as well as backhaul data traffic services. The NR IAB feature is realized through the split architecture of an NG-RAN, consisting of a centralized unit and distributed unit, as described earlier in Chapter 16. The IAB feature also introduces a new Layer 2 *Backhaul Adaptation Protocol* (BAP) layer on top of the Radio Link Control (RLC) layer.

- UE positioning services

5G system support for UE positioning services in NR has been added for various use cases such as regulatory, commercial, factory automation, vehicular positioning, and so on, requirements for both the outdoors and indoor deployment areas. For UE positioning purposes, a new reference signal called *Positioning Reference Signal (*PRS*)* has been added at NR physical layer in a downlink direction. In the uplink direction, the existing *Sounding Reference Signal,* with additional capabilities, in uplink direction is used for UE positioning purposes.

- Evolution of IP Multimedia Subsystem (IMS) for voice over NR

To provide voice services over the NR, the IMS has also evolved as part of the 3GPP Release 16. The IMS also supports service-based architecture(SBA) in the 3GPP Release 16 where the Proxy - Call Session Control Function (P-CSCF) and Home Subscriber Server (HSS) are discoverable dynamically through the NRF network function of the 5G core.

22.3 3GPP Release 17

The 3GPP Release 17 also enhances the existing features as well as introduces new features to address new applications/verticals, use cases, and so on on top of the Release 16 described above. Release 16 existing candidate features that have been identified for enhancement as part of the Release 17 are the IAB, V2X, Non-Public Network, IIoT, URLCC, UE positioning, and so on.

3GPP Release 17 introduces further evolution to the 5G system new radio (NR) by allowing the NR to operate at high frequencies (52.6–71 GHz). Release 17 also supports NR operation over Non-Terrestrial Network (NTN) and multi-SIM UEs. All such enhancements and the addition of new features are spread across the different use cases of the 5G system, i.e. eMBB, URLLC, and mIoT. For more information, refer to the Release 17 description available on the 3GPP site.

Chapter Summary

Several enhancements made on top of the 5G system baseline Release 15 as well as the new features added as part of the 3GPP Release 16 have been highlighted above. Further, some of the enhancements and new features being introduced as part of the Release 17 have been highlighted above. For more information on the complete list of enhancements and features added as part of the 3GPP Release 16 and Release 17, reading should start with the TR 21.916/917 [25] for Release 16 and Release 17 descriptions. The reader can also refer to TS 38.300 [111], Release 16 or Release 17 version, for an overview of each enhancement and feature added on top of the 3GPP Release 15 of the 5G system. Such enhancements and new features have led to the added impact to the various protocol layers as well as the system architecture of the 5G system. The changes introduced in a particular protocol layer may be studied from the corresponding change request number, for example, refer to change request # RP-200349 [151] for the affected protocol layers and their changes due to the IAB feature. TR 21.916/917 [25] mentions the technical specifications of the affected protocol layers, existing or new, or other areas of the 5G system due to the 3GPP Release 16 and Release 17. The "Change History" section of each of the affected technical specification contains the technical document (TDoc) number of the corresponding feature or enhancement added into a particular 3GPP release. Using the TDoc number, the reader can obtain further supplementary information for a particular enchantment or feature.

Test Yourself!

Introductions

This book has covered the various aspects of the design and development of mobile communications systems and networks drawn from the GSM, GPRS, UMTS, LTE, and 5G systems. As part of a self-assessment on the topics covered in this book, several relevant exercises are provided below to broaden the knowledge of the reader on the various aspects of the 5G mobile communication system and network.

The exercises are divided into the following groups:

- 5G Mobile Communications and Systems Concepts
- Software Program Development Exercises
  - Generic Utility and Reuseable Software
  - 5G System Protocol Layer Development

A.1 5G Mobile Communications and Systems Concepts

1) What is the purpose of Protocol discriminator (PD) and Message Discriminator in the air interface Layer 3 protocol stack? Why SMS and Supplementary services have the same PD as Call Control messages?
2) The NR Physical Downlink Shared Channel (PDSCH) multiplexes the DL-SCH and PCH transport channels. How does a UE differentiate the type of data it receives through the PDSCH?
3) How does the air interface Layer 3 protocols perform routing functions among the protocol layers?
4) Why is the SCTP used, in place of TCP, for call controlling signaling purposes over IP?
5) What is the SCTP port id that is used for transporting Xn-AP messages over the Xn logical interface?
6) How does a packet encapsulation method help in routing the user data packets of a mobile user in a mobile communications network?
7) Compare FDD and TDD modes of transmission with respect to the following factors.
   - Spectrum efficiencies
   - Complexity in terms of their implementations at MAC and Physical Layer
   - Latency due to switching time requirements in TDD
8) Why are interleaved resource blocks used and allocated to a UE by NG-RAN?
9) Why is there no CBGFI flag in case of uplink HARQ-ARQ?
10) Make a comparative study of the overhead caused by different physical signals with respect to their frequency and time-domain density.

Mobile Communications Systems Development: A Practical Introduction to System Understanding, Implementation, and Deployment, First Edition. Rajib Taid.

11) How does a UE know whether beamforming is used in a cell?
12) List the network slice-related information stored at UE and 5G core networks.
13) Explain how a RAN slicing supports a network slicing.
14) Identify the frequency bands and carrier frequencies suitable for different use cases of the 5G system.
15) Identify the various deployment scenarios and KPIs for different use cases i.e. eMBB, mMTC, and URLLC. Refer to TR 38.913 [27].
16) Make a comparative study of the LTE and NR air interfaces with respect to their physical layer.
17) Make a prototype design of NG-RAN for different use cases and network slices of 5GS.
18) Compare the SOAP, RPC, and REST styles of API design and representation.
19) List the HTTP methods and API URI used for services operations of different network functions.
20) Make a comparative study of the functions and procedures performed by the 5GC AMF and LTE/EPS MME.
21) Whys are DMRSs used in NR? Calculate the overhead associated with typical DMRSs of various physical channels in NR.
22) Why is there no Length Indicator field in the RLC layer header?
23) Why cannot the DMRS be used for channel sounding purposes?
24) Compare the similarities and differences between radio link failure/radio link monitoring and beam failure and its recovery procedures.
25) Compare the differences of RACH procedure performed by a UE during an initial cell access and during a beam failure recovery.
26) How does the RRC_INACTIVE state reduce signaling overhead and meet low latency requirements over the NR air interface?
27) How does the RRC layer contribute to an energy-efficient in the 5G NG-RAN base station?
28) What is the NG-RAN paging and 5G core paging in 5G? How does a UE mobility tracking take place in RRC INACTIVE state?

A.2 Software Program Development Exercises

This section contains several coding exercises for the development of typical utility and reuseable software that can be used across the protocol stack implementation. Protocol Layer-specific coding exercises are also provided.

A.2.1 Generic Utility and Re-Useable Software

1) Given an Information Element (IE), develop a function or a macro to extract the
   - Type(T) of the IE
   - Length(L) of the IE
   - Value(V) of the IE
2) Given an IE, develop a function or a macro to set its length field.
3) Given air interface Layer 3 message, develop a function or a macro to extract its
   - Message type or PDU type
   - Protocol Discriminator and Extended Protocol Discriminator
   - Skip Indicator
   - Transaction Identifier and Its Flag
4) Develop a function or macro to determine the length, i.e. one octet or two-octet, of an IE.
5) Develop functions or macros to extract
   - N-bits from a frame of bits
   - N-Octets from a frame of octets
6) Develop a function to generate a CRC polynomial of N-degree.
7) Develop a function or macro to extract the 4-bits or nibble from an octet.

8) Develop a function or macro to combine two 4-bits or nibbles to form an octet. For example, combine the protocol discriminator (4-bits) and the skip indicator (4-bits) to form an octet of the Layer 3 header.
9) Develop a function or macro to add and extract the N- padded bits into an octet or word.
10) Develop a function or macro to combine two 3-bits information and also add two padded bits to form an octet. For example, combine the Network Colour Code (NCC) and Base Station Colour Code (BCC), each is of 3-bits in length, used as the Base Station Identity Code (BSIC) in the GSM system.
11) Develop and define C-structures using bit-field for the following network identities:
    - A PLMN Id, consisting of {MCC (3 bits), MNC (2 bits)}.
    - IMSI of a subscriber consisting of MCC (3-bits), MNC (2-bits), and MSIN (10-bits).
    - A GUTI (Globally Unique Temporary UE Identity) of UE is allocated in the case of the LTE/EPS system. A GUTI consists of GUMMEI (MCC, MNC, MME Identifier {MME Group ID (16bits), MME Code (8-bits)}), and M-TMSI (32 bits).
    - An RAI is allocated in the case of GPRS/UMTS. An RAI consists of {MCC, MNC, LAC, and RAC}.
    - An LAI is allocated in the case of the GSM system. An LAI consists of {MCC, MNC, LAC (2 octets)}.
    - IMEI of MS. An IMEI consists of {TAC (8-bits), Serial Number (SNR) (6-bits), CD_SD (1-bit)}.
    - UMTS Core Network CS domain id (CN_CS_ID) consisting of {PLMN ID, LAC}.
    - An RNC identifier, consisting of {PLMN ID, RNC ID}.
    - An LTE/EPS Tracking Area Code, consisting of {MCC, MNC, TAC (2 octets)}.
12) Develop generic C-functions of the following categories; these utility functions can be used across the protocol layers:
    - Timer
      - Start a Timer with a time reference object/structure containing the timer id, timer duration, timer call-back function, and application or protocol layer id.
      - Stop a Timer with the supplied timer id
      - Expiry of a timer with a particular timer id
    - Counter
      - Initialize and update a particular counter identifying a call processing event.
      - Reset a particular counter identifying a call processing event.
    - Alarm
      - Raise a particular alarm, with the inputs: alarm id, any supplementary information
      - Clear a particular alarm, with the inputs: alarm id.
    - Inter-process communications between network elements or application processes
      - Initialize Windows or Unix domain TCP and UDP sockets
      - Write to a TCP socket
      - Write to a UDP socket
      - Read from a TCP socket
      - Read from a UDP socket
      - Write and Read a message through Message Queue
      - Write and Read a message through Shared Memory
13) Develop a generic C-Function to prepare a message or a buffer containing the various parameters of messages of different protocol layers, e.g. air interface Layer 3. Input is the message type and its related IEs.
14) One common identity of an MS across the cellular systems is the IMSI. Develop a C-function to decode or encode the IMSI in a protocol message.
15) In the user plane, the application layer's data goes through several pre-processing like compression, segmentation, and so on before it's transmitted over the air interface. Develop a utility function that processes the application's data and
    - Segments it, at the sender end, into N-octets before passing it down to the next layer.
    - Reassembles it before passing it to the concerned application at the receiving end.

16) Implement a Hash Table for storing, searching, and retrieving structures containing MS/UE information.
17) Implement a linked list for storing, searching, and retrieving structures containing radio network element configuration information.
18) Air interface Layer 2 protocols header/data block/PDU contains bit-level information. Develop a C-macro or function for the following.
    - Set a particular bit
    - Clear a particular bit
    - N-bit left shift
    - N-bit right shift
19) An integer random value is used during an initial access attempt made by a mobile device, for transmitting either signaling or user traffic. For e.g., in NR, an integer random value in the range $0–2^{39}$ -1 is used in the *RRCSetupRequest* message. Develop a C-function that generates a random value in the range from 0 to $2^{(\text{Number of bits})}-1$.
20) Develop functions to (a) read the first octet of a message and (b) write into the last octet of a message.
21) Develop generic functions/macro for encoding and decoding the individual bits for a 32-bits identifier, i.e. XX_ENCODE_1_BIT, XX_ENCODE_2_BIT, XX_ENCODE_3_BIT, and so on.
22) Compare the relative merits and demerits of TLV, CSN.1, and ASN.1 encoding and decoding methods of protocol layer messages.

A.2.2 5G System Protocol Layer Development

This section contains several coding exercises for the development of the actual logics for 3GPP-complaint 5G system Protocol layers, its functions, procedures, and algorithms.

1) Develop a C-macro to calculate the RA-RNTI associated with the Random-Access Response for a given UE in the NR system using the formula mentioned in TS 38.321 [113].
2) Develop C-structures for the definition of the following
    - NR RLC Unacknowledged mode PDU with x-bits and y-bits sequence number.
    - NR RLC Acknowledged mode PDU with x-bits and y-bits sequence number.
    - NR MAC layer, consisting of the MAC Header, MAC control elements, and MAC Service data units.
3) The Downlink Control Information (DCI) format of the NR physical layer contains information to be used by a UE for data transmission/receptions. Refer to TS 38.212 [107] and derive the length of the individual information from its description and develop C-structures and code for the following:
    - C-structure definitions, at bit-level, to pack the various scheduling and other information under each of the DCI formats.
    - C-functions to encode/decode DCI formats.
4) The RLC layers contain several state variables for the Acknowledged Mode, Unacknowledged Mode, and Transparent mode data transmission and reception. Define and establish the relationship among the state variables in the code.
5) Develop C-function/macro to Conceal and De-Conceal a UE SUPI to SUCI.
6) Develop an IMSI hash algorithm to be used for a shared RAN under the MOCN feature; refer to TS 23.251 [37].
7) Develop C-functions for the following NR UE MAC layer procedures:
    - Scheduling Request
    - Buffer Status Report
    - Semi-persistent Scheduling
    - Hybrid Automatic Repeat Request (HARQ)

8) Develop C-functions for the following NR UE PDCP layer procedures:
   - User data transfer over the DRB
   - Control plane data transfer over the SRB
9) Develop C-functions for the following NR UE RLC layer procedures:
   - Transparent Mode data transfer
   - Acknowledged Mode data transfer
   - Unacknowledged Mode data transfer
   - Hybrid Automatic Repeat Request (HARQ)
10) Prepare a high-level design document for the following layers:
    - MAC layer and its entity
    - RLC layer and its entity
    - Service Access Points (SAP) between RLC and MAC layers
11) Develop timer-based service routines for scheduling and transmitting the following NR air interface downlink information from the gNB to UEs in a cell.
    - Master Information Block, periodicity – every 'X' milliseconds
    - System Information Block – periodicity – every 'X' milliseconds
12) Assume that the transmission timer interval (TTI) used over the NR air interface is 1 millisecond, which is equal to 1 sub-frame in FDD/TDD mode. Develop suitable functions/procedures to process, i.e. read/write, packets every TTI from the Ethernet frame received from the underlying IP stack.
13) Consider the PDSCH channel and its transmission of an NR transport block using the MCS 'X' and 64QAM modulation scheme, normal cyclic prefix over a single resource block with no control channels on it. Calculate the effective channel coding rate of this transport block transmission. Remember to add a 24-bits CRC to the transport block coding.
14) List the different data types used by network function services and their operations and develop suitable data structures to represent them.
15) Develop and model the 5GC network functions using the UML class diagrams with correct relationships among them.
16) Identify the sequential and associative STL containers. Select the appropriate container which can be used as the data structures to store 5GC network functions information.
17) Develop an algorithm to generate a random number-based UUID for NF instances.
18) Develop functions for discovery and selection of different network functions.
19) Develop C-functions for generations of modulation symbols and scrambling sequences for the following physical channels:
    - PBCH
    - PDCCH
    - PDSCH
20) Develop C-functions for the encoding and decoding of DCIs.
21) Develop C-structure for BWP consisting of
    - Subcarrier Spacing
    - Number of Resource Block
    - Cyclic Prefix
22) Develop C-structure for a CORESET consisting of
    - Allocated PRB i.e. frequency domain resources
    - Duration
    - CCE to REG mapping
    - REG bundle size
    - Interleaved size
    - Shift index

23) Develop a C-structure for a PDCCH search space consisting of
    - Monitoring slot periodicity
    - Monitoring symbol
    - Number of candidates PDCCH = 2
    - CCE aggregation level
24) Develop a C-structure to store a UE-specific PDCCH configuration.
25) Design an appropriate data structure to store a 240-by-4 matrix representing the SS/PBCH block.
26) Given the LDPC lifting size of say, 8, generate the BG2.
27) Develop suitable C-data structures to store the DMRS information, such as type, position, and length, for the following channels:
    - PDSCH, PUSCH, and PBCH
28) Design prototype NR scheduling algorithms for different use cases of 5GS.
29) Compare the pros and cons of the LDPC encoding method used for different natures of, i.e. large and small, traffic generated from the 5G use cases eMBB, mMTC, and URLLC.
30) Develop a C-function for the following:
    - Store the NSSAI and S-NSSAI, and their associated DNN list.
    - Network Slice Selection Function (NSSF).
31) Compare the different types of STL C++ container classes from their performance point of view.
32) Develop a suitable data structure to store UE PDU Session information with the following details.
    - PDU Session Identity
    - PDU Session Type
    - Network Slice Identifier (S-NSSAI)
    - Data Network Name
    - Session and Service Continuity (SSC) mode
33) Design and develop appropriate data structures to store network function profiles, at NRF, of various network functions of the 5G core.
34) Design and develop RESTful APIs for the following services provided by the 5GC NRF network function (NF) to its consumer network functions.
    - NF register and deregister
    - NF update
    - NF status subscribe
    - NF status Notify and un-subscribe
    - NF list retrieval
    - NF profile retrieval
    - Any user-defined services
35) For each of the network function of the 5G core, identify the following typical
    - Set of features supported, e.g. voice over IMS support at SMF, by a network function.
    - Mandatory, e.g SSC mode at SMF, and optional/custom features of a network function.
    - Interworking requirements, e.g. fall-back to LTE/EPS to support voice over IMS.
    - Resources management, e.g. allocate an IP address to UE.
    - Policy and Charging requirements.
36) Refer to Chapters: 10 and 11 on the alarm, fault, and performance management methods/processes used in a mobile communication network. Using these methods, design and develop appropriate functions and procedures for the following typical use cases of a Self-Organizing Network (SON), refer to TS 28.861.
    - Detection and automated recovery of cell outages, or a sleeping cell, as part of the self-healing function of a SON.
    - Traffic distribution among neighbor cells, in terms of usages of radio resources, as part of the load balancing optimizations function of a SON.

References

All the 3GPP Technical Specifications being referred and listed below for the 5G system are based on its 1st Release 15. Other 3GPP Technical Specifications being referred and listed below for the legacy systems (GSM/GPRS/UMTS/LTE) are based on the Release 10.

1 www.3gpp.org
2 www.3gpp.org/specifications/specification-numbering
3 www.3gpp.org/ftp/Information/WORK_PLAN/Description_Releases/
4 www.3gpp.org/specifications-groups
5 www.3gpp.org/specifications/specification-numbering/81-version-numbering-scheme.
6 www.3gpp.org/ftp/Specs/latest
7 ETSI GS NFV 002 Network Functions Virtualisation (NFV); Architectural Framework
8 ETSI GS NFV 004 Network Functions Virtualisation (NFV); Virtualization Requirements
9 https://www.itu.int/dms_pubrec/itu-r/rec/m/R-REC-M.2083-0-201509-I!!PDF-E.pdf
10 CSN.1 [www.csn1.info]; ASN, [visit: www.itu.int/ITU-T/studygroups/com17/languages/X.690-0207.pdf.]
11 ITU-T [X.691] ASN.1 Packet Encoding Rule
12 ITU-T Recommendation G.703 for E1 interface
13 RFC 3095 RObust Header Compression (ROHC)
14 RFC 3748 Extensible Authentication Protocol
15 RFC 4122 A Universally Unique IDentifier (UUID)
16 RFC 4815 RObust Header Compression (ROHC)
17 RFC 4960 Stream Control Transmission Protocol
18 RFC 5246 The Transport Layer Security (TLS)
19 RFC 5448 Improved Extensible Authentication Protocol Method
20 RFC 6749 The OAuth 2.0 Authorization Framework
21 RFC 7515 JSON Web Signature (JWS)
22 RFC 7516 JSON Web Encryption (JWE)
23 RFC 7519 JSON Web Token (JWT)
24 TR 21.905 Vocabulary for 3GPP Specifications
25 TR 21.916/917 Release description; Release 16/ Release description; Release 17
26 TR 38.912 Study on New Radio (NR) access technology
27 TR 38.913 Study on Scenarios and Requirements for Next Generation Access Technologies
28 TS 22.261 Service requirements for the 5G system
29 TS 23.002 Network architecture
30 TS 23.003 Numbering, addressing and identification

Mobile Communications Systems Development: A Practical Introduction to System Understanding, Implementation, and Deployment, First Edition. Rajib Taid.

31 TS 23.060 General Packet Radio Service (GPRS); Service description; Stage 2
32 TS 23.122 Non-Access-Stratum (NAS) functions related to Mobile Station (MS) in idle mode
33 TS 23.203 Policy and charging control architecture
34 TS 23.214 Architecture enhancements for control and user plane separation of EPC nodes
35 TS 23.216 Single Radio Voice Call Continuity (SRVCC); Stage 2
36 TS 23.228 IP Multimedia Subsystem (IMS); Stage 2
37 TS 23.251 Network Sharing Architecture and Functional Description
38 TS 23.272 Circuit Switched (CS) fall back in Evolved Packet System (EPS); Stage 2
39 TS 23.401 General Packet Radio Service (GPRS) enhancements for Evolved Universal Terrestrial Radio Access Network (E-UTRAN) access
40 TS 23.501 System Architecture for the 5G System
41 TS 23.502 Procedures for the 5G System
42 TS 23.503 Policy and charging control framework for the 5G System (5GS); Stage 2
43 TS 23.527 5G System; Restoration procedures
44 TS 24.007 Mobile radio interface signaling layer 3; General aspects
45 TS 24.008 Mobile radio interface Layer 3 specification; Core network protocols; Stage 3
46 TS 24.301 Non-Access-Stratum (NAS) protocol for Evolved Packet System (EPS); Stage 3
47 TS 24.501 Non-Access-Stratum (NAS) protocol for 5G System (5GS); Stage 3
48 TS 24.502 Access to the 3GPP 5G Core Network (5GCN) via non-3GPP access networks
49 TS 25.301 Radio Interface Protocol Architecture
50 TS 25.331 (UMTS) Radio Resource Control (RRC); Protocol specification
51 TS 25.401 UTRAN overall description
52 TS 25.410 UTRAN Iu Interface: general aspects and principles
53 TS 25.412 UTRAN Iu: Signalling Transport
54 TS 25.413 UTRAN Iu interface Radio Access Network Application Part (RANAP) signaling
55 TS 28.310 Management and orchestration; Energy efficiency of 5G
56 TS 28.530 Management and orchestration; Concepts, use cases and requirements
57 TS 28.531 Management and orchestration; Provisioning
58 TS 28.533 Management and orchestration; Architecture framework
59 TS 28.540 Management and orchestration; 5G Network Resource Model (NRM); Stage 1
60 TS 28.541 Management and orchestration; 5G Network Resource Model (NRM); Stage 2 and stage 3
61 TS 28.546 Management and orchestration of networks and network slicing; Fault Supervision (FS); Stage 2 and stage 3
62 TS 28.550 Management and orchestration; Performance assurance
63 TS 28.552 Management and orchestration; 5G performance measurements
64 TS 28.555 Management and orchestration; Network policy management for 5G mobile networks
65 TS 28.556 Management and orchestration; Network policy management for 5G mobile networks; Stage 2 and stage 3
66 TS 29.018 Serving GPRS Support Node (SGSN) -Visitors Location Register (VLR); Gs interface network service specification
67 TS 29.060 GPRS Tunnelling Protocol (GTP) across the Gn and Gp interface
68 TS 29.118 Mobility Management Entity (MME) - Visitor Location Register (VLR) SGs interface specification
69 TS 29.244 Interface between the Control Plane and the User Plane nodes
70 TS 29.274 Evolved General Packet Radio Service (GPRS) Tunnelling Protocol for Control plane (GTPv2-C)
71 TS 29.280 Evolved Packet System (EPS); 3GPP Sv interface (MME to MSC, and SGSN to MSC) for SRVCC
72 TS 29.281 General Packet Radio System (GPRS) Tunneling Protocol User Plane (GTPv1-U)
73 TS 29.500 Technical Realization of Service Based Architecture

74 TS 29.501 Principles and Guidelines for Services Definition
75 TS 29.502 5G System; Session Management Services; Stage 3
76 TS 29.510 Network Function Repository Services
77 TS 29.518 Access and Mobility Management Services
78 TS 29.571 Common Data Types for Service Based Interfaces
79 TS 32.101 Telecommunication management; Principles and high-level requirements
80 TS 32.250 CS Domain Charging
81 TS 32.410 Key Performance Indicators (KPI) for UMTS and GSM
82 TS 32.450 Key Performance Indicators (KPI) for Evolved Universal Terrestrial Radio Access Network
83 TS 32.451 Key Performance Indicators (KPI) E-UTRAN: Requirements
84 TS 32.455 Key Performance Indicators (KPI) EPC
85 TS 33.102 3G Security; Security architecture
86 TS 33.401 3GPP System Architecture Evolution (SAE); Security architecture
87 TS 33.501 Security architecture and procedures for 5G system
88 TS 34.12* UE Conformance Specifications
89 TS 36.201 LTE physical layer; General description
90 TS 36.211 Evolved Universal Terrestrial Radio Access (E-UTRA); Physical Channels and Modulation
91 TS 36.213 Evolved Universal Terrestrial Radio Access (E-UTRA); Physical layer procedures
92 TS 36.300 (E-UTRAN); Overall description
93 TS 36.321 Evolved Universal Terrestrial Radio Access (E-UTRA); Medium Access Control (MAC) protocol specification
94 TS 36.331 Evolved Universal Terrestrial Radio Access (E-UTRA); Radio Resource Control (RRC); Protocol specification
95 TS 36.401 (E-UTRAN); Architecture description
96 TS 36.410 S1 general aspects and principles
97 TS 36.413 Evolved Universal Terrestrial Radio Access Network (E-UTRAN); S1 Application Protocol (S1AP)
98 TS 36.423 Evolved Universal Terrestrial Radio Access Network (E-UTRAN); X2.application protocol (X2AP)
99 TS 36.52* Series UE Conformance Specifications
100 TS 37.324 Evolved Universal Terrestrial Radio Access (E-UTRA) and NR; Service Data Adaptation Protocol (SDAP) specification
101 TS 37.340 NR; Multi-connectivity; Overall description; Stage-2
102 TS 38.101 NR; User Equipment (UE) radio transmission and reception
103 TS 38.104 NR; Base Station (BS) radio transmission and reception
104 TS 38.133 NR; Requirements for support of radio resource management
105 TS 38.201 NR; Physical layer; General description
106 TS 38.211 NR; Physical channels and modulation
107 TS 38.212 NR; Multiplexing and channel coding
108 TS 38.213 NR; Physical layer procedures for control
109 TS 38.214 NR; Physical layer procedures for data
110 TS 38.215 NR; Physical layer measurements
111 TS 38.300 NR; Overall description; Stage-2
112 TS 38.306 User Equipment (UE) radio access capabilities
113 TS 38.321 NR; Medium Access Control (MAC) protocol specification
114 TS 38.322 NR; Radio Link Control (RLC) protocol specification
115 TS 38.323 NR; Packet Data Convergence Protocol (PDCP) specification
116 TS 38.331 NR; Radio Resource Control (RRC); Protocol specification
117 TS 38.401 NG-RAN; Architecture description

118 TS 38.410 NG-RAN; NG general aspects and principles
119 TS 38.413 NG-RAN; NG Application Protocol (NGAP)
120 TS 38.415 NG-RAN; PDU Session User Plane Protocol
121 TS 38.420 NG-RAN; Xn general aspects and principles
122 TS 38.423 NG-RAN; Xn Application Protocol (XnAP)
123 TS 38.425 NG-RAN; NR user plane protocol
124 TS 38.50/52/53 Series UE Conformance Specifications
125 TS 38.460 NG-RAN; E1 general aspects and principles
126 TS 38.470 NG-RAN; F1 general aspects and principles
127 TS 38.473: NG-RAN; F1 application protocol (F1AP)
128 TS 42.009 Security aspects
129 TS 44.008 Mobile radio interface layer 3 specification
130 TS 44.018 Mobile radio interface layer 3 specification; Radio Resource Control (RRC) protocol
131 TS 44.060 General Packet Radio Service (GPRS); Mobile Station (MS) - Base Station System (BSS) interface; Radio Link Control/Medium Access Control (RLC/MAC) protocol
132 TS 44.065 Sub-Network Dependent Convergence Protocol (SNDCP).
133 TS 48.006 Signalling transport mechanism Specification for the Base Station System - Mobile Services Switching Centre (BSS - MSC) interface
134 TS 48.008 Mobile Switching Centre - Base Station System (MSC-BSS) interface; Layer 3 specification.
135 TS 48.016 Base Station System (BSS) - Serving GPRS Support Node (SGSN) interface; Network Service
136 TS 48.018 Base Station System (BSS) - Serving GPRS Support Node (SGSN) interface; BSS GPRS Protocol (BSSGP)
137 TS 51 Series MS Conformance Specifications
138 3GPP TSG RAN Meeting #82 RP-182580 UE support for NR Beam Correspondence
139 3GPP TSG RAN Meeting #82 RP-182452 Evaluation of beam correspondence as mandatory UE feature.
140 3GPP TSG-RAN WG4 Meeting #89 R4-1814656 Beam Correspondence for Rel-15
141 3GPP TSG-RAN WG1 #86 Bis R1-1610239 On beam management in NR – procedures
142 3GPP TSG RAN WG1 Meeting #87 R1-1611664 Multi-panel-based DL MIMO transmission
143 3GPP TSG RAN WG1 Meeting #87 R1-1611665 Multi-panel-based UL MIMO transmission
144 3GPP TSG RAN WG1 Meeting #87 R1-1611666 Codebook design for multi-panel structured MIMO in NR
145 3GPP TSG RAN WG1 Meeting #88 R1-1701691 DL Codebook design for multi-panel structured MIMO in NR
146 3GPP TSG RAN WG1 Meeting #88 R1-1701694 CSI-RS design for beam management
147 3GPP TSG RAN WG1 Meeting #88 R1-1701695 CSI-RS design for CSI acquisition
148 3GPP TSG RAN WG1 Meeting #88 R1-1701807 On Type I codebook design for NR MIMO
149 3GPP TSG RAN WG1 Meeting #88 R1-1702188 On codeword to MIMO layer mapping for NR
150 3GPP TSG RAN WG1 Meeting #88 R1-1702266 Aperiodic CSI Reporting for NR MIMO
151 RP-200349 CR to 38.300 on Integrated Access and Backhaul for NR

Further Readings

About 3GPP Meeting Discussions/Study Items/Work Items

In addition to the list of references mentioned above, the reader is further recommended to go through the different study items, work items, and so on, as mentioned in [138–150] in the references section above. These study items and work items are available in the form of meeting discussions and their technical documents (TDocs) within a particular 3GPP specification group. Given a TDoc number or its title, the link to search and download it is https://portal.3gpp.org/ngppapp/TdocList.aspx. A TDoc can be also searched using the "search" option provided on the 3GPP website. A study item contains valuable information about a particular area of discussion and can be used to supplement existing or to acquire additional knowledge by a reader.

Index

Mobile Communications Systems Development: A Practical Introduction to System Understanding, Implementation, and Deployment, First Edition. Rajib Taid.

g

www.ingramcontent.com/pod-product-compliance
Lightning Source LLC
LaVergne TN
LVHW080056220926
848484LV00007BA/1771
9781119778684